Lecture Notes in Computer Science 14077

The series Lecture Notes in Computer Science (LNCS), including its subseries Lecture Notes in Artificial Intelligence (LNAI) and Lecture Notes in Bioinformatics (LNBI), has established itself as a medium for the publication of new developments in computer science and information technology research, teaching, and education.

LNCS enjoys close cooperation with the computer science R & D community, the series counts many renowned academics among its volume editors and paper authors, and collaborates with prestigious societies. Its mission is to serve this international community by providing an invaluable service, mainly focused on the publication of conference and workshop proceedings and postproceedings. LNCS commenced publication in 1973.

Jiří Mikyška · Clélia de Mulatier ·
Maciej Paszynski · Valeria V. Krzhizhanovskaya ·
Jack J. Dongarra · Peter M. A. Sloot
Editors

Computational Science – ICCS 2023

23rd International Conference Prague, Czech Republic, July 3–5, 2023 Proceedings, Part V

Editors
Jiří Mikyška
Czech Technical University in Prague
Prague, Czech Republic

Maciej Paszynski
AGH University of Science and Technology
Krakow, Poland

Jack J. Dongarra
University of Tennessee at Knoxville
Knoxville, TN, USA

Clélia de Mulatier
University of Amsterdam
Amsterdam, The Netherlands

Valeria V. Krzhizhanovskaya
University of Amsterdam
Amsterdam, The Netherlands

Peter M. A. Sloot
University of Amsterdam
Amsterdam, The Netherlands

ISSN 0302-9743 ISSN 1611-3349 (electronic)
Lecture Notes in Computer Science
ISBN 978-3-031-36029-9 ISBN 978-3-031-36030-5 (eBook)
https://doi.org/10.1007/978-3-031-36030-5

This Springer imprint is published by the registered company Springer Nature Switzerland AG
The registered company address is: Gewerbestrasse 11, 6330 Cham, Switzerland

Preface

Welcome to the 23rd annual International Conference on Computational Science (ICCS - https://www.iccs-meeting.org/iccs2023/), held on July 3–5, 2023 at the Czech Technical University in Prague, Czechia.

In keeping with the new normal of our times, ICCS featured both in-person and online sessions. Although the challenges of such a hybrid format are manifold, we have always tried our best to keep the ICCS community as dynamic, creative, and productive as possible. We are proud to present the proceedings you are reading as a result.

ICCS 2023 was jointly organized by the Czech Technical University in Prague, the University of Amsterdam, NTU Singapore, and the University of Tennessee.

Standing on the Vltava River, Prague is central Europe's political, cultural, and economic hub.

The Czech Technical University in Prague (CTU) is one of Europe's largest and oldest technical universities and the highest-rated in the group of Czech technical universities. CTU offers 350 accredited study programs, 100 of which are taught in a foreign language. Close to 19,000 students are studying at CTU in 2022/2023. The Faculty of Nuclear Sciences and Physical Engineering (FNSPE), located along the river bank in Prague's beautiful Old Town (Staré Mesto) and host to ICCS 2023, is the only one in Czechia to offer studies in a broad range of fields related to Nuclear Physics and Engineering. The Faculty operates both fission (VR-1) and fusion (GOLEM Tokamak) reactors and hosts several cutting-edge research projects, collaborating with a number of international research centers (CERN, ITER, BNL-STAR, ELI).

The International Conference on Computational Science is an annual conference that brings together researchers and scientists from mathematics and computer science as basic computing disciplines, as well as researchers from various application areas who are pioneering computational methods in sciences such as physics, chemistry, life sciences, engineering, arts, and humanitarian fields, to discuss problems and solutions in the area, identify new issues, and shape future directions for research.

Since its inception in 2001, ICCS has attracted increasingly higher-quality attendees and papers, and this year is not an exception, with over 300 participants. The proceedings series have become a primary intellectual resource for computational science researchers, defining and advancing the state of the art in this field.

The theme for 2023, "**Computation at the Cutting Edge of Science**", highlights the role of Computational Science in assisting multidisciplinary research. This conference was a unique event focusing on recent developments in scalable scientific algorithms; advanced software tools; computational grids; advanced numerical methods; and novel application areas. These innovative novel models, algorithms, and tools drive new science through efficient application in physical systems, computational and systems biology, environmental systems, finance, and others.

ICCS is well known for its excellent lineup of keynote speakers. The keynotes for 2023 were:

- **Helen Brooks**, United Kingdom Atomic Energy Authority (UKAEA), UK
- **Jack Dongarra**, University of Tennessee, USA
- **Derek Groen**, Brunel University London, UK
- **Anders Dam Jensen**, European High Performance Computing Joint Undertaking (EuroHPC JU), Luxembourg
- **Jakub Šístek**, Institute of Mathematics of the Czech Academy of Sciences & Czech Technical University in Prague, Czechia

This year we had 531 submissions (176 to the main track and 355 to the thematic tracks). In the main track, 54 full papers were accepted (30.7%); in the thematic tracks, 134 full papers (37.7%). A higher acceptance rate in the thematic tracks is explained by the nature of these, where track organizers personally invite many experts in a particular field to participate in their sessions. Each submission received at least 2 single-blind reviews (2.9 reviews per paper on average).

ICCS relies strongly on our thematic track organizers' vital contributions to attract high-quality papers in many subject areas. We would like to thank all committee members from the main and thematic tracks for their contribution to ensuring a high standard for the accepted papers. We would also like to thank *Springer, Elsevier,* and *Intellegibilis* for their support. Finally, we appreciate all the local organizing committee members for their hard work in preparing for this conference.

We are proud to note that ICCS is an A-rank conference in the CORE classification.

We hope you enjoyed the conference, whether virtually or in person.

July 2023

Jiří Mikyška
Clélia de Mulatier
Maciej Paszynski
Valeria V. Krzhizhanovskaya
Jack J. Dongarra
Peter M. A. Sloot

Organization

The Conference Chairs

General Chair

Valeria Krzhizhanovskaya	University of Amsterdam, The Netherlands

Main Track Chair

Clélia de Mulatier	University of Amsterdam, The Netherlands

Thematic Tracks Chair

Maciej Paszynski	AGH University of Science and Technology, Poland

Scientific Chairs

Peter M. A. Sloot	University of Amsterdam, The Netherlands | Complexity Institute NTU, Singapore
Jack Dongarra	University of Tennessee, USA

Local Organizing Committee

LOC Chair

Jiří Mikyška	Czech Technical University in Prague, Czechia

LOC Members

Pavel Eichler	Czech Technical University in Prague, Czechia
Radek Fučík	Czech Technical University in Prague, Czechia
Jakub Klinkovský	Czech Technical University in Prague, Czechia
Tomáš Oberhuber	Czech Technical University in Prague, Czechia
Pavel Strachota	Czech Technical University in Prague, Czechia

Thematic Tracks and Organizers

Advances in High-Performance Computational Earth Sciences: Applications and Frameworks – IHPCES

Takashi Shimokawabe, Kohei Fujita, Dominik Bartuschat

Artificial Intelligence and High-Performance Computing for Advanced Simulations – AIHPC4AS

Maciej Paszynski, Robert Schaefer, Victor Calo, David Pardo, Quanling Deng

Biomedical and Bioinformatics Challenges for Computer Science – BBC

Mario Cannataro, Giuseppe Agapito, Mauro Castelli, Riccardo Dondi, Rodrigo Weber dos Santos, Italo Zoppis

Computational Collective Intelligence – CCI

Marcin Maleszka, Ngoc Thanh Nguyen

Computational Diplomacy and Policy – CoDiP

Michael Lees, Brian Castellani, Bastien Chopard

Computational Health – CompHealth

Sergey Kovalchuk, Georgiy Bobashev, Anastasia Angelopoulou, Jude Hemanth

Computational Modelling of Cellular Mechanics – CMCM

Gabor Zavodszky, Igor Pivkin

Computational Optimization, Modelling, and Simulation – COMS

Xin-She Yang, Slawomir Koziel, Leifur Leifsson

Computational Social Complexity – CSCx

Vítor V. Vasconcelos. Debraj Roy, Elisabeth Krüger, Flávio Pinheiro, Alexander J. Stewart, Victoria Garibay, Andreia Sofia Teixeira, Yan Leng, Gabor Zavodszky

Computer Graphics, Image Processing, and Artificial Intelligence – CGIPAI

Andres Iglesias, Lihua You, Akemi Galvez-Tomida

Machine Learning and Data Assimilation for Dynamical Systems – MLDADS

Rossella Arcucci, Cesar Quilodran-Casas

MeshFree Methods and Radial Basis Functions in Computational Sciences – MESHFREE

Vaclav Skala, Samsul Ariffin Abdul Karim

Multiscale Modelling and Simulation – MMS

Derek Groen, Diana Suleimenova

Network Models and Analysis: From Foundations to Complex Systems – NMA

Marianna Milano, Pietro Cinaglia, Giuseppe Agapito

Quantum Computing – QCW

Katarzyna Rycerz, Marian Bubak

Simulations of Flow and Transport: Modeling, Algorithms, and Computation – SOFTMAC

Shuyu Sun, Jingfa Li, James Liu

Smart Systems: Bringing Together Computer Vision, Sensor Networks and Machine Learning – SmartSys

Pedro Cardoso, Roberto Lam, Jânio Monteiro, João Rodrigues

Solving Problems with Uncertainties – SPU

Vassil Alexandrov, Aneta Karaivanova

Teaching Computational Science – WTCS

Angela Shiflet, Nia Alexandrov

Reviewers

Zeeshan Abbas
Samsul Ariffin Abdul Karim
Tesfamariam Mulugeta Abuhay
Giuseppe Agapito
Elisabete Alberdi
Vassil Alexandrov
Nia Alexandrov
Alexander Alexeev
Nuno Alpalhão
Julen Alvarez-Aramberri
Domingos Alves
Sergey Alyaev
Anastasia Anagnostou
Anastasia Angelopoulou
Fabio Anselmi
Hideo Aochi
Rossella Arcucci
Konstantinos Asteriou
Emanouil Atanassov
Costin Badica
Daniel Balouek-Thomert
Krzysztof Banaś
Dariusz Barbucha
Luca Barillaro
João Barroso
Dominik Bartuschat
Pouria Behnodfaur
Jörn Behrens
Adrian Bekasiewicz
Gebrail Bekdas
Mehmet Belen
Stefano Beretta
Benjamin Berkels
Daniel Berrar
Piotr Biskupski
Georgiy Bobashev
Tomasz Boiński
Alessandra Bonfanti
Carlos Bordons
Bartosz Bosak
Lorella Bottino
Roland Bouffanais
Lars Braubach
Marian Bubak
Jérémy Buisson
Aleksander Byrski
Cristiano Cabrita
Xing Cai
Barbara Calabrese
Nurullah Çalık
Victor Calo
Jesús Cámara
Almudena Campuzano
Cristian Candia
Mario Cannataro
Pedro Cardoso
Eddy Caron
Alberto Carrassi
Alfonso Carriazo
Stefano Casarin
Manuel Castañón-Puga
Brian Castellani
Mauro Castelli
Nicholas Chancellor
Ehtzaz Chaudhry
Théophile Chaumont-Frelet
Thierry Chaussalet
Sibo Cheng
Siew Ann Cheong
Lock-Yue Chew
Su-Fong Chien
Marta Chinnici
Bastien Chopard
Svetlana Chuprina
Ivan Cimrak
Pietro Cinaglia

Noélia Correia
Adriano Cortes
Ana Cortes
Anna Cortes
Enrique Costa-Montenegro
David Coster
Carlos Cotta
Peter Coveney
Daan Crommelin
Attila Csikasz-Nagy
Javier Cuenca
António Cunha
Luigi D'Alfonso
Alberto d'Onofrio
Lisandro Dalcin
Ming Dao
Bhaskar Dasgupta
Clélia de Mulatier
Pasquale Deluca
Yusuf Demiroglu
Quanling Deng
Eric Dignum
Abhijnan Dikshit
Tiziana Di Matteo
Jacek Długopolski
Anh Khoa Doan
Sagar Dolas
Riccardo Dondi
Rafal Drezewski
Hans du Buf
Vitor Duarte
Rob E. Loke
Amir Ebrahimi Fard
Wouter Edeling
Nadaniela Egidi
Kareem Elsafty
Nahid Emad
Christian Engelmann
August Ernstsson
Roberto R. Expósito
Fangxin Fang
Giuseppe Fedele
Antonino Fiannaca
Christos Filelis-Papadopoulos
Piotr Frąckiewicz
Alberto Freitas
Ruy Freitas Reis
Zhuojia Fu
Kohei Fujita
Takeshi Fukaya
Wlodzimierz Funika
Takashi Furumura
Ernst Fusch
Marco Gallieri
Teresa Galvão Dias
Akemi Galvez-Tomida
Luis Garcia-Castillo
Bartłomiej Gardas
Victoria Garibay
Frédéric Gava
Piotr Gawron
Bernhard Geiger
Alex Gerbessiotis
Josephin Giacomini
Konstantinos Giannoutakis
Alfonso Gijón
Nigel Gilbert
Adam Glos
Alexandrino Gonçalves
Jorge González-Domínguez
Yuriy Gorbachev
Pawel Gorecki
Markus Götz
Michael Gowanlock
George Gravvanis
Derek Groen
Lutz Gross
Tobias Guggemos
Serge Guillas
Xiaohu Guo
Manish Gupta
Piotr Gurgul
Zulfiqar Habib
Yue Hao
Habibollah Haron
Mohammad Khatim Hasan
Ali Hashemian
Claire Heaney
Alexander Heinecke
Jude Hemanth

Marcin Hernes
Bogumila Hnatkowska
Maximilian Höb
Rolf Hoffmann
Tzung-Pei Hong
Muhammad Hussain
Dosam Hwang
Mauro Iacono
Andres Iglesias
Mirjana Ivanovic
Alireza Jahani
Peter Janků
Jiri Jaros
Agnieszka Jastrzebska
Piotr Jedrzejowicz
Gordan Jezic
Zhong Jin
Cedric John
David Johnson
Eleda Johnson
Guido Juckeland
Gokberk Kabacaoglu
Piotr Kalita
Aneta Karaivanova
Takahiro Katagiri
Mari Kawakatsu
Christoph Kessler
Faheem Khan
Camilo Khatchikian
Petr Knobloch
Harald Koestler
Ivana Kolingerova
Georgy Kopanitsa
Pavankumar Koratikere
Sotiris Kotsiantis
Sergey Kovalchuk
Slawomir Koziel
Dariusz Król
Elisabeth Krüger
Valeria Krzhizhanovskaya
Sebastian Kuckuk
Eileen Kuehn
Michael Kuhn
Tomasz Kulpa
Julian Martin Kunkel
Krzysztof Kurowski
Marcin Kuta
Roberto Lam
Rubin Landau
Johannes Langguth
Marco Lapegna
Ilaria Lazzaro
Paola Lecca
Michael Lees
Leifur Leifsson
Kenneth Leiter
Yan Leng
Florin Leon
Vasiliy Leonenko
Jean-Hugues Lestang
Xuejin Li
Qian Li
Siyi Li
Jingfa Li
Che Liu
Zhao Liu
James Liu
Marcellino Livia
Marcelo Lobosco
Doina Logafatu
Chu Kiong Loo
Marcin Łoś
Carlos Loucera
Stephane Louise
Frederic Loulergue
Thomas Ludwig
George Lykotrafitis
Lukasz Madej
Luca Magri
Peyman Mahouti
Marcin Maleszka
Alexander Malyshev
Tomas Margalef
Osni Marques
Stefano Marrone
Maria Chiara Martinis
Jaime A. Martins
Paula Martins
Pawel Matuszyk
Valerie Maxville

Pedro Medeiros
Wen Mei
Wagner Meira Jr.
Roderick Melnik
Pedro Mendes Guerreiro
Yan Meng
Isaak Mengesha
Ivan Merelli
Tomasz Michalak
Lyudmila Mihaylova
Marianna Milano
Jaroslaw Miszczak
Dhruv Mittal
Miguel Molina-Solana
Fernando Monteiro
Jânio Monteiro
Andrew Moore
Anabela Moreira Bernardino
Eugénia Moreira Bernardino
Peter Mueller
Khan Muhammad
Daichi Mukunoki
Judit Munoz-Matute
Hiromichi Nagao
Kengo Nakajima
Grzegorz J. Nalepa
I. Michael Navon
Vittorio Nespeca
Philipp Neumann
James Nevin
Ngoc-Thanh Nguyen
Nancy Nichols
Marcin Niemiec
Sinan Melih Nigdeli
Hitoshi Nishizawa
Algirdas Noreika
Manuel Núñez
Joe O'Connor
Frederike Oetker
Lidia Ogiela
Ángel Javier Omella
Kenji Ono
Eneko Osaba
Rongjiang Pan
Nikela Papadopoulou
Marcin Paprzycki
David Pardo
Anna Paszynska
Maciej Paszynski
Łukasz Pawela
Giulia Pederzani
Ebo Peerbooms
Alberto Pérez de Alba Ortíz
Sara Perez-Carabaza
Dana Petcu
Serge Petiton
Beata Petrovski
Toby Phillips
Frank Phillipson
Eugenio Piasini
Juan C. Pichel
Anna Pietrenko-Dabrowska
Gustavo Pilatti
Flávio Pinheiro
Armando Pinho
Catalina Pino Muñoz
Pietro Pinoli
Yuri Pirola
Igor Pivkin
Robert Platt
Dirk Pleiter
Marcin Płodzień
Cristina Portales
Simon Portegies Zwart
Roland Potthast
Małgorzata Przybyła-Kasperek
Ela Pustulka-Hunt
Vladimir Puzyrev
Ubaid Qadri
Rick Quax
Cesar Quilodran-Casas
Issam Rais
Andrianirina Rakotoharisoa
Célia Ramos
Vishwas H. V. S. Rao
Robin Richardson
Heike Riel
Sophie Robert
João Rodrigues
Daniel Rodriguez

Marcin Rogowski
Sergio Rojas
Diego Romano
Albert Romkes
Debraj Roy
Adam Rycerz
Katarzyna Rycerz
Mahdi Saeedipour
Arindam Saha
Ozlem Salehi
Alberto Sanchez
Ayşin Sancı
Gabriele Santin
Vinicius Santos Silva
Allah Bux Sargano
Azali Saudi
Ileana Scarpino
Robert Schaefer
Ulf D. Schiller
Bertil Schmidt
Martin Schreiber
Gabriela Schütz
Jan Šembera
Paulina Sepúlveda-Salas
Ovidiu Serban
Franciszek Seredynski
Marzia Settino
Mostafa Shahriari
Vivek Sheraton
Angela Shiflet
Takashi Shimokawabe
Alexander Shukhman
Marcin Sieniek
Joaquim Silva
Mateusz Sitko
Haozhen Situ
Leszek Siwik
Vaclav Skala
Renata Słota
Oskar Slowik
Grażyna Ślusarczyk
Sucha Smanchat
Alexander Smirnovsky
Maciej Smołka
Thiago Sobral
Isabel Sofia
Piotr Sowiński
Christian Spieker
Michał Staniszewski
Robert Staszewski
Alexander J. Stewart
Magdalena Stobinska
Tomasz Stopa
Achim Streit
Barbara Strug
Dante Suarez
Patricia Suarez
Diana Suleimenova
Shuyu Sun
Martin Swain
Edward Szczerbicki
Tadeusz Szuba
Ryszard Tadeusiewicz
Daisuke Takahashi
Osamu Tatebe
Carlos Tavares Calafate
Andrey Tchernykh
Andreia Sofia Teixeira
Kasim Terzic
Jannis Teunissen
Sue Thorne
Ed Threlfall
Alfredo Tirado-Ramos
Pawel Topa
Paolo Trunfio
Hassan Ugail
Carlos Uriarte
Rosarina Vallelunga
Eirik Valseth
Tom van den Bosch
Ana Varbanescu
Vítor V. Vasconcelos
Alexandra Vatyan
Patrick Vega
Francesc Verdugo
Gytis Vilutis
Jackel Chew Vui Lung
Shuangbu Wang
Jianwu Wang
Peng Wang

Katarzyna Wasielewska
Jarosław Wątróbski
Rodrigo Weber dos Santos
Marie Weiel
Didier Wernli
Lars Wienbrandt
Iza Wierzbowska
Maciej Woźniak
Dunhui Xiao
Huilin Xing
Yani Xue
Abuzer Yakaryilmaz
Alexey Yakovlev
Xin-She Yang
Dongwei Ye
Vehpi Yildirim
Lihua You
Drago Žagar
Sebastian Zając
Constantin-Bala Zamfirescu
Gabor Zavodszky
Justyna Zawalska
Pavel Zemcik
Wenbin Zhang
Yao Zhang
Helen Zhang
Jian-Jun Zhang
Jinghui Zhong
Sotirios Ziavras
Zoltan Zimboras
Italo Zoppis
Chiara Zucco
Pavel Zun
Karol Życzkowski

Contents – Part V

Simulations of Flow and Transport: Modeling, Algorithms and Computation

Smart Systems: Bringing Together Computer Vision, Sensor Networks and Artificial Intelligence

Quantum Computing

Searching B-Smooth Numbers Using Quantum Annealing: Applications to Factorization and Discrete Logarithm Problem

Olgierd Żołnierczyk(✉) and Michał Wroński

Military University of Technology, Kaliskiego Str. 2, Warsaw, Poland
{olgierd.zolnierczyk,michal.wronski}@wat.edu.pl

Abstract. Integer factorization and discrete logarithm problem, two problems of classical public-key cryptography, are vulnerable to quantum attacks, especially polynomial-time Shor's algorithm, which has to be run on the general-purpose quantum computer. On the other hand, one can make quantum computations using quantum annealing, where every problem has to be transformed into an optimization problem, for example, the QUBO problem. Currently, the biggest available quantum annealer, D-Wave advantage, has almost 6,000 physical qubits, and therefore it can solve bigger problems than using general-purpose quantum computers. Even though it is impossible to run Shor's algorithm on a quantum annealer, several methods allow one to transform factorization or discrete logarithm problems into the QUBO problem. Using a D-Wave quantum annealer, the biggest factored integer had 20 bits, and the biggest field, on which it was possible to compute a discrete logarithm problem using any quantum method, had 6 bits. This paper shows how to transform searching for B-smooth numbers, an important part of the quadratic sieve method for factorization and index calculus for solving discrete logarithm problems, to the QUBO problem and then solve it using D-Wave Advantage quantum solver. The linear algebra step for integer factorization and index calculus methods has been solved using classical computations. Using our method, we factorized the 26-bit integer and computed the discrete logarithm problem over the 18-bit prime field. Therefore we broke the current records in factorization using quantum annealing by 6 bits and in discrete logarithm problem, using any quantum method, by 12 bits.

Keywords: Integer factorization · discrete logarithm problem · D-Wave · quantum annealing · cryptanalysis

This work was supported by the Military University of Technology's University Research Grant No. 858/2021: "Mathematical methods and models in computer science and physics".

J. Mikyška et al. (Eds.): ICCS 2023, LNCS 14077, pp. 3–17, 2023.
https://doi.org/10.1007/978-3-031-36030-5_1

1 Introduction

The integer factorization problem (IFP) and discrete logarithm problem (DLP) over finite fields are some of the most widespread problems in modern classical public-key cryptography. Let us recall these notions, denoting the set of prime numbers as $\mathbb{P}$.

Definition 1 (Integer factorisation problem). *The task of finding* $p_1, p_2, ..., p_k \in \mathbb{P}$*, such that* $\prod_{i=1}^{k} p_i = n$*, being given* n *only; is called the integer factorisation problem.*

Definition 2 (Discrete logarithm problem over finite field). *Let* $q = p^e, p \in \mathbb{P}, e \in \mathbb{N}_{>0}$ *and* $\langle g \rangle = G \leq \mathbb{F}_q^*$*. The task of finding* $y \in \{0, 1, \ldots, ord(g) - 1\}$*, being given an element from multiplicative subgroup* $h \in G$*, such that* $g^y = h$*; is called discrete logarithm problem over a finite field.*

In general, the most efficient, non-quantum method to solve the integer factorization problem is the general number field sieve (GNFS) method (but in this work, we use a simpler version - the quadratic sieve method). The improved method – number field sieve for discrete logarithm is used to solve DLP.

Since 1994 it has been known, according to Shor [13], that there exists a quantum algorithm that can solve integer factorization and discrete logarithm problem over finite fields in polynomial time. Moreover, in 2003, Proos and Zalka [12] presented how to apply polynomial-time Shor's algorithm to solve the discrete logarithm problem on elliptic curves. Since then, there have been many efforts to implement Shor's algorithm practically on existing quantum computers. Unfortunately, till now, the record of the integer factorization problem, computed using Shor's algorithm, is number 21 [9]. Applications of Shor's algorithm for DLP over finite fields and elliptic curves have not been reported.

On the other hand, many more methods solve these problems. One such method is transforming the given problem into the Ising or QUBO problem. Such problems may be solved using adiabatic quantum computers. Using such a method, it was possible to make factorization of 40-bit length number $N = 1,099,551,473,989$ using superconducting quantum computer [3], and 20-bit integer $N = 1,028,171$ using quantum annealing [14]. Using quantum annealing, it was also possible to compute DLP over a 6-bits prime field $\mathbb{F}_{59}$. Secondly, the latest research shows [16] it is possible to solve the integer factorization problem by disposing of approximately $\frac{(\log n)^2}{4}$ logical qubits and DLP over $\mathbb{F}_p$, disposing of $2(\log p)^2$ logical qubits. These methods, however, do not outperform Shor's algorithm. In the case of ECDLP, using a hybrid quantum solver, searching for relations in the index calculus method also has been computed quantumly for 8-bit prime $p = 251$ [15]. The second step of this method, the linear algebra step, has been computed using a classical computer.

It should be noted that a proposal for solving the IFP using the Schnorr method and QAOA has also emerged [17]. However, determining its effectiveness (similarly to the aforementioned examples) requires a deeper analysis.

Let us define the problem whose solution we can obtain using the quantum annealing machine. The problem can be presented as a multivariate binary polynomial $f(x_i, \ldots x_n)$ of degree less or equal to 2 with real coefficients. The task is to find the values of variables for which the polynomial takes the minimal value (in particular, one of such vectors).

Having constrained the degree of the polynomial, we could denote coefficients as follows: a_i – terms of degree one, $b_{i,j}$ – terms of degree two. Such a problem is called the QUBO problem – Quadratic Unconstrained Binary Optimization and can be expressed as follows:

$$f(x_1, \ldots, x_n) = \sum_i a_i x_i + \sum_{i<j} b_{i,j} x_i x_j. \tag{1}$$

In this paper, we use the modified factorization procedure, presented by [5], as a subroutine to check if an integer, randomly chosen in a quadratic sieve or index calculus method, has all factors less than or equal to the given bound B.

It is worth noting that quadratic sieve and index calculus methods have limited applications. The complexities of these methods in the classical case are subexponential. In the quantum case, it is also clear that the quadratic sieve and index calculus method for finite fields will not outperform Shor's algorithm. Nevertheless, the methods presented below may show some promise compared to classical methods, depending on a more thorough investigation of the complexity of quantum annealing.

Even if the presented by us method has some limits in its applications, we factorized the 26-bits integer and computed the DLP over the 18-bits prime field using this method. Therefore we broke the current records in IFP using quantum annealing by 6 bits and in DLP, using any quantum method, by 12 bits. As reported later, a hybrid classical-quantum application of a general number field sieve for both IFP and DLP should allow finding solutions in about 50-bit cases without increasing the number of logic variables in given QUBO problems. These researches require, however, much more effort.

2 Classical Methods for Integer Factorization and Discrete Logarithm

2.1 Quadratic Sieve Method

One of the most crucial integer factorization method is the quadratic sieve method, proposed by Carl Pomerance in 1981. Up to an approximately 340-bit length of factored number, the QS is the most efficient method since, for larger inputs, the general number field sieve method is more efficient.

An observation underlying this method is the following fact. For a given factorization problem (as defined by Definition 1), having a pair of numbers for which holds:

$$a^2 \equiv b^2 \pmod{n} \tag{2}$$

and

$$a \not\equiv \pm b \pmod{n}, \tag{3}$$

implies the possibility of computing $d \,|\, n$, (it is because $a^2 - b^2 \equiv 0 \pmod{n} \vdash n \,|\, (a+b)(a-b) \vdash gcd(a \pm b, n)$ – non-trivial divisor of n, where $\vdash$ denotes a conditional assertion, so $\mathcal{P} \vdash \mathcal{Q}$ means that from $\mathcal{P}$, we know that $\mathcal{Q}$.

The naive way of searching the above relation $(a,\, b)$ is replaced by splitting this task into more steps. Namely, we explore many weakened relations of the form $a_i^2 \equiv b_i$, where all prime factors of b_i are less or equal a bound B – it means b_i is B-smooth. Finally, we multiply the congruences corresponding to the selected pair (a, b), creating primary relation (Eq. 2), which we need. We choose the base of prime factors for this purpose:

$$\mathcal{B} = \{ -1, p_2, p_3, \ldots, p_k \mid \mathbb{P} \ni p_i < B \}, \tag{4}$$

fixing a well adjusted bound B and next, denoting $\#\mathcal{B} = k$.

Using the following polynomial:

$$T(c) = (m+c)^2 - n \equiv (m+c)^2 \pmod{n}, \tag{5}$$

where $m = \lfloor \sqrt{n} \rfloor$, we can generate the set of weakened relations: $T(c) = b_c \equiv a_c^2$. We select only these, where b_c is B-smooth, and we vary $-M < c < M$, to obtain suitable cardinality of set $\mathcal{A} \ni b_i$. If we establish each prime factorization $\mathcal{A} \ni b_i = \prod_{j=1}^{k} p_j^{e_j}$, we are able to obtain a non-trivial solution of the following system of equations:

$$\begin{cases} e_{11}x_{11} + e_{21}x_{21} + \cdots + e_{k1}x_{k1} & = 0, \\ e_{12}x_{12} + e_{22}x_{22} + \cdots + e_{k2}x_{k2} & = 0, \\ \vdots & \\ e_{1A}x_{1A} + e_{2A}x_{2A} + \cdots + e_{kA}x_{kA} & = 0, \end{cases} \tag{6}$$

over a prime field $\mathbb{F}_2$. The simplest way for this linear algebra step is Gauss elimination, based on transforming the matrix to row echelon form by adding multiplication (in the general case over $\mathbb{F}_p$) of a row to picked rows. Although the cost of the Gauss algorithm is too high ($O(B^3)$), in practice, we exploit the fact that we work with a sparse matrix in the quadratic sieve method. As a result, the application of an efficient general procedure as Lanczos's and Wiedemann's algorithms [4], achieve complexity $O\left(B^{2+o(1)}\right)$, equaling the time bound for sieving.

Thus, the solution will indicate the set of indices S, such that $\prod_{i \in S} b_i = b^2$. Additionally, due to the form of the polynomial, we have $\prod_{i \in S} a_i = a^2$ and $a^2 \equiv b^2 \pmod{n}$. Eventually, we obtain primary relation (Eq. 2). It is worth noting, because of the preceding, that we neglect the condition from Eq. 3; however, the possibility of gaining a non-trivial divisor is relatively high (>0.5).

The quadratic sieve method has a few vital improvements: one big prime (two big primes), a multi-polynomial variant, and self-initialization. The complexity of the method, according to [2], is $O\left(\exp\left(\sqrt{\ln n \ln \ln n}\right)\right)$.

2.2 Index Calculus Method

An analogous method for DLP, corresponding to QS, is the index calculus method, proposed by Leonard Adleman in 1979. It has similar proprieties, the same main idea based on searching B-smooth number, and the same complexity. This is the simplest variant of the method, presented in two stages.

First Stage. Let $\langle g \rangle = \mathbb{F}_p^*$ for some prime p and our goal is to find $y : g^y \equiv h \pmod{p}$. The factor base

$$\mathcal{B} = \{p_1, p_2, \ldots, p_k \mid \mathbb{P} \ni p_i < B\} \tag{7}$$

consists of k small primes. For random $x \in \{1, 2, \ldots, p-2\}$ we try to obtain relations of the form:

$$\mathbb{F}_p^* \ni g^x = p_1^{e_1} p_2^{e_2} \ldots p_k^{e_k}, \tag{8}$$

by finding g^x smooth over $\mathcal{B}$. Each relation obtained implies, that for certain indexes $i_{p_j} : \mathbb{F}_p \ni g^{i_{p_j}} = p_j$, holds

$$e_1 i_{p_1} + e_2 i_{p_2} \ldots e_k i_{p_k} \equiv x \pmod{p-1}. \tag{9}$$

Collecting enough relations (coefficient matrix to be of full column rank), we will be able to set and solve the system of equations modulo $p - 1$:

$$\begin{cases} e_{11} i_{p_1} + e_{12} i_{p_2} + \cdots + e_{1k} i_{p_k} & \equiv x_1 \pmod{p-1} \\ e_{21} i_{p_1} + e_{22} i_{p_2} + \cdots + e_{2k} i_{p_k} & \equiv x_2 \pmod{p-1} \\ \vdots & \\ e_{a1} i_{p_1} + e_{a2} i_{p_2} + \cdots + e_{ak} i_{p_k} & \equiv x_a \pmod{p-1}. \end{cases} \tag{10}$$

If so, we uniquely get the indices i_{p_j}. Detailed computation involves factoring $p-1$, solving the system modulo each prime power separately, and combining all solutions using the Chinese Remainder Theory to obtain the unique one. We use the same techniques announced in Subsect. 2.1. This fulfills the first stage.

Second Stage. The second stage is possible due to collected values $i_{p_1}, i_{p_2}, \ldots, i_{p_k}$. Namely, we keep picking $z \in \{1, 2, \ldots, p-1\}$ randomly to find one such that:

$$\mathbb{F}_p^* \ni hg^z = p_1^{a_1} p_2^{a_2} \ldots p_k^{a_k}, \tag{11}$$

so until hg^z will be B-smooth. As a result, we can compute searched y, indeed

$$y \equiv a_1 i_{p_1} + a_2 i_{p_2} + \cdots + a_k i_{p_k} - z \pmod{p-1}. \tag{12}$$

Therefore the goal has been achieved.

3 Hybrid Methods

It is well known that the relation-collection phase determines the running time of the methods above and all algorithms based on searching B-smooth numbers. Given the quantum factorization technique with limited input, one can apply it to check smoothness in one of these methods, making them a hybrid with the quantum smoothness searching stage.

3.1 Known Results and Previous Work

In 2017 Daniel Bernstein et al. proposed a low-resource quantum factoring algorithm [1]. This algorithm uses improved Shor's [13] procedure for integer factorization as a search criterion function in Grover's searching algorithm [6] to create a quantum circuit for detecting B-smooth numbers. Eventually, this circuit is a subroutine in the general number field sieve. Due to some improvements, Shor's algorithm can take as an input superposition of many sieved numbers. Thus, Bernstein's method is a trade-off between quantum and classical resources. As a result, this very efficient approach has classical complexity $O(L^{\sqrt[3]{\frac{3}{8}}+o(1)})$, where $L = \exp\left((\ln n)^{\frac{1}{3}}(\ln\ln n)^{\frac{2}{3}}\right)$ (n is a number to be factored) and uses $O\left((\log n)^{\frac{2}{3}}\right)$ qubits. In terms of quantum computing development, this method can be used with fewer quantum resources than Shors's algorithm to factorize numbers greater than using the general number field sieve method.

Similar hybrid ideas exist for quantum annealing. Michele Mosca et al. presented a significant conceptual result in 2019 [10]. Mosca's method also uses GNFS and assumes searching of smoothness quantumly. However, in this method, the quantum stage consists of the transformation elliptic curve method for factorization (ECM) to the QUBO problem (in the original paper, more general to the SAT problem). ECM complexity depends on the least divisor of factorized number. Thus, the algorithm is useful for splitting probably B-smooth number. The QUBO problem, equivalent to the smoothness problem, is made from a sequence of ECM blocks connected input-output. This way is realized by extracting all smooth divisors. Mosca's method achieves better complexity than classical GNFS if only quantum annealing reaches non-trivial speedup compared to classical solvers. Under maximal optimistic conditions, this method has computational complexity asymptotically equal to Bernstein's method.

We present a similar concept below, using QS and index calculus method.

3.2 Our Result – Factorization by Quantum Annealing as a Subroutine

We present two hybrid methods based on the quadratic sieve and the index calculus algorithm. We use the modified quantum annealing factorization technique in both to detect B-smooth numbers. The complexity of these methods cannot be better than subexponential (due to the number of quantum subroutines required), so they do not significantly outperform known results. However, the proposed methods allow for examining the efficiency of one of the quantum-classical solutions in practice and break factorization and discrete logarithm records with quantum annealing. Furthermore, their more detailed comparison is currently unknown due to the lack of a deeper analysis of the complexity of quantum annealing.

The difference between the direct and presented quantum annealing factorization method is the following. Based on the idea of [11], we try to find a lot of small factors instead of finding the factorization of semiprime. Therefore, we introduce a different definition of the input problem. The second biggest difference is the nested way we apply the factorization using quantum annealing. So presented quantum annealing factorization technique below is a subroutine of both hybrid methods. As a result, even though the direct method has better complexity, applying the sieve methods allows solving bigger instances of IFP and DLP than direct methods (with quantum annealing power available today). Our idea is also some intermediate step between classical sieve methods and the propositions of Bernstein et al. [1] and Moska et al. [10].

Checking smoothness in presented hybrid methods is performed by splitting the candidate to be smooth until all the divisors are less or equal to B, or factorization will return a false divisor (not smooth). In the factorization subroutine, we try to find a divisor of the factorized number less or equal to some bound B. We achieve this by setting suitable bit lengths of both searched divisors, so the smaller one is $\leq B$. However, if factorization fails, we treat obtained outputs as primes greater than B. The essential idea of the quantum annealing factorization technique is to present the factorization problem as a QUBO problem. It will be done in the following main steps.

Establishing Multiplication Table. Let n be the number to be factored (the candidate for being smooth). First, the procedure requires constant bit lengths of factors b, d, where $d < b$, denoted as l_d, l_b (we will generally indicate the bit length in this way). The expected maximal length depends on the least and the largest primes in the base: $l_b = l_{\lfloor n/p_1 \rfloor}, l_d = l_{p_t}$ (assuming p_1 is the smallest prime in the base and p_t is the biggest one). In practice, the base used in quantum annealing can also be reduced by applying initial naive division on a classic machine.

Then we follow the steps discussed in [11], so division into column blocks should be established in the multiplication table. The width of each column block, denoted by the W_i for an i-th block, can be 2 or 3 (except the first block width, which always equals 1). We consider bits d_i, b_i, n_i of, respectively: d, b, n and carry bits c_i. Preceding the quantum annealing phase by splitting powers of 2 from the input number, we get the following variables and column blocks, represented on Table 1.

Counting Carry Bits. We aim to express the factorization problem as a cost function. We split the problem into smaller parts by treating each block separately. This technique reduces the values of bias coefficients and the number of

Table 1. Multiplication table.

				d_{l_d}	·	·	·	d_3	d_2	1
				b_{l_b}	·	·	·	b_3	b_2	1
			c_1	d_{ld}	·	·	·	d_3	d_2	1
		c_2	$d_{l_d}b_1$	·	·	·	d_3b_2	d_2b_2	b_2	
	·	$d_{l_d}b_3$	·	·	·	d_3b_3	d_2b_3	b_3		
	· ·						·			
	· ·					·				
c_k	·				·					
$d_{l_d}b_{l_b}$	· · ·	$d_3b_{l_b}$	$d_2b_{l_b}$	b_{l_b}						
n_{l_n}		·		·		·		n_3	n_2	1
s-th block								2. block		1. block

logical qubits needed to solve this problem. Thus, we sum terms with reduced column weights: from 2^0 to 2^{W_i-1}. The notation is the following:

- N_i – the number read from bits of n narrowed to an i-th block with shifted weights 2^0 to 2^{W_i-1} (called target value),
- $F_i(d_2, \ldots d_{l_d}, b_2 \ldots, b_{l_b})$ – multiplication result polynomial for an i-th block,
- $C_i(c_1, \ldots c_k)$ – carry variables polynomial for an i-th block, the sum of variables c from i-th block with shifted weights 2^0 to 2^{W_i-1}.

Eventually, we formulate the following expressions from the multiplication table:

$$\begin{aligned} f_i &= F_i(d_2, \ldots d_{l_d}, b_2 \ldots, b_{l_b}) + C_i(c_1, \ldots c_k) - 2^{W_i} C_{i+1}(c_1, \ldots c_k) - N_i, \\ f &= f_2^2 + f_3^2 \cdots f_s^2 \end{aligned} \tag{13}$$

(we assume $C_{s+1} = 0$), receiving the formula $f(d_2, \ldots, d_{l_d}, b_2, \ldots, b_{l_b}, c_1, \ldots c_k) \rightarrow 0$ as an equivalent to IFP.

While numbers l_d, l_b, l_n are known, we need to determine the number of carry bits k, especially the number of carry bits in i-th block K_i (surely, there is no gap between c_i within one block). Let us express the maximal value of the i-th block $F_i(d_2, \ldots d_{l_d}, b_2 \ldots, b_{l_b})$ as Max_i. Obviously, a congruence $F_i(d_2, \ldots d_{l_d}, b_2 \ldots, b_{l_b}) \equiv N_i \pmod{2^{W_i}}$ must hold. So, we define Max_i as:

$$Max_i = max\{F_i(d_2, \ldots d_{l_d}, b_2 \ldots, b_{l_b}) | F_i(d_2, \ldots d_{l_d}, b_2 \ldots, b_{l_b}) \equiv N_i \; (mod\; 2^{W_i})\}. \tag{14}$$

Therefore we want to know if $F_i(d_2, \ldots d_{l_d}, b_2 \ldots, b_{l_b})$ takes any value between minimum and maximum. The blocks with the exact position of columns signed by d_{l_d}, b_{l_b}, d_2, within this block, and exact width, turn out to have the same properties - similarly to common properties shown in [11]. For example, all blocks in type $\{left, 1, right\}$ with width 3 have the same properties. Thus, the set of taken values for F_i could be precomputed for all types of blocks in practice, and the exact value of K_i can be computed taking into consideration C_i, which can take any values between minimum and maximum.

Linearisation. We must reduce the polynomial to quadratic form to transform the cost function to the QUBO. Therefore, if in the polynomial f_i there are monomials of degree ≥ 2, then we need to linearise them so that they will be quadratic after the polynomial is squared.

Let us consider the monomial of binary variables $\gamma x_1 x_2 x_3$ with real coefficient γ. We aim to reduce the degree of this monomial by one (finally, to degree equals 1). We create an auxiliary variable $a \in \{0, 1\}$ to replace $x_1 x_2$. Adding a new expression, the so-called penalty

$$P_a = 2(x_1 x_2 - 2a(x_1 + x_2) + 3a), \tag{15}$$

to monomial, we get the equivalence of the following transformation

$$\gamma x_1 x_2 x_3 \longrightarrow \gamma a x_3 + P_a. \tag{16}$$

To be more precise, the minimal value of both expressions is the same, and the set of point for which the left expression take minimal value is equal to the set of point for which the right expression takes minimal value.

We can prove this by considering all possible values of these expressions.

Table 2. Table of all values taken by considered expression $\gamma a x_2 + P_a$. The minimal values are taken only when $x_1 x_2 = a$, so $P_a = 0$.

x_1	x_2	x_3	$\gamma x_1 x_2 x_3$	a	$\gamma a x_3 + P_a$	
0	0	0	0	0	0	$x_1 x_2 = a$
0	0	1	0	0	0	
0	1	0	0	0	0	
0	1	1	0	0	0	
1	0	0	0	0	0	
1	0	1	0	0	0	
1	1	0	0	1	0	
1	1	1	γ	1	γ	
0	0	0	0	1	6	$x_1 x_2 \neq a$
0	0	1	0	1	$\gamma + 6$	
0	1	0	0	1	2	
0	1	1	0	1	$\gamma + 2$	
1	0	0	0	1	2	
1	0	1	0	1	$\gamma + 2$	
1	1	0	0	0	2	
1	1	1	γ	0	2	

It is worth noting that our expression takes minimal value only when $P_a = 0$. However, since we are interested in minimal values of a squared polynomial, we should consider the absolute value of the sum of such monomials. We observe that the absolute value of the sum of many of these monomials has a minimum only when $P_a = 0$ also. Moreover, excluding P_a from the absolute value brackets does not change these properties. In other words, we obtain equivalence in the following transformation $f^2 \rightarrow f_{lin}^2 + P$, where f_{lin} is the linear form of the polynomial f, obtained from many substitutions $x_i x_j \rightarrow a_l$ performed on f.

The conclusions for the linearisation procedure follow. We can transform a cost function f_i to the QUBO form by keeping on substitution $x_i x_j \rightarrow a_l$ and summing up all penalties separately, as long as f has a degree greater than 1. Finally, we square f_{lin}, add the sum of penalties, and $deg(f_{lin}^2 + P) = 2$.

Formulating QUBO Problem. To summarise, formulating QUBO to find divisors of n consists of the following steps, preceded by precomputation of lists of values taken by the function F_i from each type of block. One can also assume that n is nondivisible by a few small primes, for example, $\{2, 3, 5, 7\}$.

1. Establish the multiplication table (see Table 1) by initializing an array of subsequent indices of the first columns of each block.
2. Initialize $f = 0, P = 0$.
3. For second block fix $C_2 = 0$.
4. For each non-empty, subsequent block, do:
   - Fix the carry bits number K_{i+1}, exiting i-th block, knowing target value N_i and list of values taken by function F_i (see Sect. 3.2). Initialise corresponding polynomial C_{i+1} (knowing K_{i+1}).
   - Substitute d_2 by $n_2 \oplus b_2$ in F_i.
   - Linearise polynomial F_i, saving all triples (x_1, x_2, a) and adding penalties to P (see Sect. 3.2).
   - Fix the polynomial $f_i^2 = \left(F_i + C_i - N_i - 2^{W_i} C_{i+1}\right)^2$ for this block.
   - Add $f \leftarrow f + f_i^2$.
5. Add $f \leftarrow f + P$.

For simplicity, we do not stand out the last step, which one can reduce to the substitution of d_1, linearisation, fixing f_i^2 and adding $f + f^2$.

Requesting Quantum Computer. The Quantum annealing sampler can be used now to find the solution to the QUBO problem, determining divisors d, b due to returned low-energy states of the objective function.

3.3 Quantum Annealing Stage – Summary

With suitably adjusted l_d and B, we no longer have to split d to check smoothness, but it is generally sufficient to compare d to B. The nested applications of the factorization subroutine above lead us to examine if all prime factors of n belong to $\mathcal{B}$ or not. Potentially failure is simply detected by verifying if $d \mid n$ is considered as a negative response in checking smoothness. Thus we obtain smoothness checking by the quantum annealing stage.

4 Experiments

The experiments investigated the maximal size of solvable factorization and discrete logarithm problems by applying the above method on the most potent quantum annealing computer available today, D-Wave. We have established current size records of solved discrete logarithm problem: 18-bits, using any quantum method, integer factorization problem: 26-bits, using quantum annealing.

All quantum computations have been made involving direct QPU solver *Advantage_system4.1* from D-Wave. The quantum computer was requested using Ocean SDK and *D-Wave.system* library. Sampling has been carried out with the number of reads equaling 10,000, and requested problems were formulated as QUBO problems, as described above. Coefficients in QUBO were autoscaled by solver API to hardware ranges to meet the requirements of the QPU. The used quantum machine has the following main properties: the number of working qubits: 5627, weight (qubits coefficients) range: $[-4, 4]$, strengths (couplers coefficients) range: $[-1, 1]$, annealing time range: $[0.5\,\mu s, 2\,\text{ms}]$, default annealing time: 20 μs.

4.1 Results for Integer Factorization

We apply our hybrid method for integer factorization based on the variant of the quadratic sieve method described in Subsect. 2.1. As shown above, the hybrid method involved modified factoring by annealing procedure used as a subroutine in the sieving stage from the quadratic sieve.

We have adopted the following methodology to select sample input data for each input size. Firstly, we used a parameter β in bounding arguments for generating polynomial from Subsect. 2.2. Secondly, trying different β, B, n values, each complete performance of the method was initially realized on the classical computer as many times as needed to find an instance with sieved numbers size according to asymptotic constraint. Then, for fixed β, the candidates for being smooth were generated via this polynomial with random arguments from range $[0, \beta]$ as far as a full rank matrix of relations has been completed. In some cases, the procedure was repeated several times. We have excluded the possibility of simultaneously finding the whole relation of two squares. The best approach is to use the official RSA challenge numbers, but it is impossible due to the current state of development of quantum annealing computers available today.

The final results are presented in two tables. In Table 3 are listed specifications of problems, so the quantities from the classical part of the method, while in Table 4 are listed values describing the usage of quantum machine for every problem.

Table 3. Results of experiments with solving integer factorization problem by hybrid quadratic sieve: classical part.

n	problem size [bit]	β	sieved numbers size [bit]	matrix dimensions	base
$874219 = 1013 \times 863$	20	4	4–14	2×5	$[-1, 2, 3, 5, 11]$
$3812491 = 2029 \times 1879$	22	4	5–15	2×6	$[-1, 2, 3, 5, 7, 11]$
$8732021 = 2953 \times 2957$	24	5	8–15	2×6	$[-1, 2, 3, 5, 7, 11]$
$42273409 = 6709 \times 6301$	26	4	7–16	3×6	$[-1, 2, 3, 5, 13, 17]$

Columns of Table 3 contain the following values, respectively, from left: 1. input number n, to be factored, 2. the bit length of input number n, 3. parameter

Table 4. Results of experiments with solving integer factorization problem by hybrid quadratic sieve: quantum part.

problem size [bit]	number of logical qubits	number of physical qubits	QPU total time [ms]
20	9–45	14–261	182
22	9–49	14–326	796
24	29–41	60–367	678
26	22–77	67–772	8705

β fixed during selecting examples used for bound random range for candidates to being smooths, 4. ranges of bit length of numbers, from every nested factorization subroutine step, sieved in smoothness checking stage, 5. dimension of the matrix of coefficients from the system of congruences, 6. base $\mathcal{B}$ used to sieving.

The above values show that with currently available quantum annealing resources, we can factor up to 26-bit semiprime by the procedure of quantumly splitting many 16-bit numbers, applied in a nested way in the quadratic sieve.

Columns of Table 4 contain the following values, respectively, from left: 1. the bit length of input number n denoting the same problem from Table 3, 2. number of variables from the QUBO problem, from every nested factorization subroutine step, 3. number of qubits used by QPU to make embedding of the requested problem from every nested factorization subroutine step, 4. total working time of QPU in milliseconds. Table 4 shows that for 26-bit semiprime, 772 qubits of quantum resources have been used. Despite the total working qubits numbers being much greater, attempts to solve bigger problems have failed because of chain lengths (the number of physical qubits required to represent one logical qubit).

4.2 Results for Discrete Logarithm Problem over Prime Field

The second application of quantum annealing smoothness checking was the discrete logarithm problem over a prime field. In this case, we have used the basic index calculus method (see Subsect. 2.2). As in the case of IFP, modified factoring by annealing procedure was used as a subroutine in the sieving stage from the index calculus method.

The whole method was performed on the classical computer many times, and, unlike in the case of the integer factorization, all randomly picked candidates to be B-smooth were saved from each trial. We established the maximum number of trials: 150, and we have chosen the shortest one from these tries and have realized them on the quantum computer, in some cases, several times to succeed.

Each problem has been solved in the whole multiplication subgroup of the prime field (it means $\langle g \rangle = \mathbb{F}_p^*$), and each of p is cryptographic prime ($p - 1 = 2q, q \in \mathbb{P}$).

The final results were divided into two tables also. In Table 5, the problems are characterized, while in Table 6 are listed similar to previous problem values, presenting performing of quantum computation for every problem.

Table 5. Problems chosen for experiments with solving discrete logarithm problem over a prime field by hybrid index calculus.

p	problem size [bit]	$p-1$	$g^x = y$	sieved numbers size [bit]	base
8543	14	2×4271	$186^x = 7986$	4–14	$[2, 3, 5, 7, 11]$
23399	15	2×11699	$17856^x = 2525$	4–15	$[2, 3, 5, 7, 11]$
33623	16	2×16811	$25065^x = 25932$	4–15	$[2, 3, 5, 7]$
79943	17	2×39971	$16657^x = 9503$	4–17	$[2, 3, 5, 7, 11, 13]$
147047	18	2×73523	$8962^x = 38492$	3–18	$[2, 3, 5, 7, 11, 13]$

Table 6. Results of experiments with solving discrete logarithm problem by hybrid quadratic sieve: quantum part.

problem size [bit]	number of logical qubits	number of physical qubits	QPU total time [ms]
14	10–54	12–203	903
15	10–58	14–229	1427
16	6–48	8–172	2292
17	10–68	12–273	6297
18	9–71	11–292	17331

Columns of table Table 5 contain the following values, respectively, from left: 1. characteristic p of prime field $\mathbb{F}_p$, 2. the bit length of characteristic p, 3. factorization of multiplicative subgroup order $p - 1$, 4. the value of discrete logarithm, 5. ranges of bit length of numbers, from every nested factorization subroutine step, sieved in smoothness checking stage, 5. the base used in sieving.

These values mean that with currently available quantum annealing resources, we can solve DLP up to an 18-bit prime field by the procedure of quantumly splitting many 18-bit numbers, applied in a nested way in the index calculus method.

Columns of Table 6 contain the following values, respectively, from left: 1. the bit length of input number n denoting the same problem from Table 5, 2. number of variables from the QUBO problem, from every nested factorization subroutine step, 3. number of qubits used by QPU to make embedding of the requested problem from every nested factorization subroutine step, 4. total working time of QPU in milliseconds.

Table 6 shows that for 18-bit DLP, 292 qubits of quantum resources were used. It is much less than in the case of IFP because of the size of the sieving numbers. Similarly, the length of the chains did not allow to solve a bigger problem.

5 Summary

This paper aimed to show how one can apply classical integer factorization methods and discrete logarithm problem computation using quantum annealing. Both for quadratic sieve and index calculus methods, searching for B-smooth numbers has been performed using quantum annealing and modified factorization method presented by [5,8], applied as a subroutine. The second part of both algorithms, the linear algebra step, has been computed using classical methods. It is worth noting that the linear algebra step could also be implemented using quantum annealing, especially in the case of quadratic sieve, where we operate on the matrix defined over $\mathbb{F}_2$.

Using our method, we factorized the 26-bits integer and computed the discrete logarithm problem over the 18-bits prime field. Therefore we broke the current records in factorization using quantum annealing by 6 bits and in discrete logarithm problem, using any quantum method, by 12 bits. One can easily estimate the number of qubits needed to run presented methods for larger-scale consideration. Our modified factorization subroutine uses the same number ($\frac{\log^2 n}{4}$) of qubits, as in [8]. Thus, sieving numbers $\sim \sqrt{n}$ in IFP and $\sim p$ in DLP, we need, respectively, $\frac{\log^2 n}{16}$ and $\frac{\log^2 p}{4}$ logical qubits.

Further works should include the general number field sieve algorithm applications for integer factorization and discrete logarithm problem computation. Our estimations show that it should be possible to factorize a 50-bit integer or compute a discrete logarithm problem over a 50-bit prime field in such a case. For this achievement, we do not need more resources. The needed memory is about 700-800 physical qubits. Moreover, the number of logical qubits needed to solve IFP and DLP by hybrid methods with GNFS is about $\frac{\sqrt[3]{\log^5 n \log\log n}}{4}$, this is also because of the size of sieved numbers: $\sim n^{\frac{1}{d}}$, where $d \sim \sqrt[3]{\frac{3\log n}{\log\log n}}$ (see [7]). For example, a 400-bit problem (both IFP and DLP) requires 1.500 logical qubits versus 40.000 logical qubits in the case of the direct method.

It is an open question if applying quadratic sieve, index calculus method, or general number field sieve may give better performance on quantum computers than on the classical one using for B-smooth numbers searching different algorithm than Shor's.

References

1. Bernstein, D.J., Biasse, J.F., Mosca, M.: A low-resource quantum factoring algorithm. Cryptology ePrint Archive, Paper 2017/352 (2017). https://eprint.iacr.org/2017/352
2. Crandall, R., Pomerance, C.: Prime Numbers. A Computational Perspective. Springer, New York (2005). https://doi.org/10.1007/0-387-28979-8
3. Crane, L.: Quantum computer sets new record for finding prime number factors. New Scientist (2020)
4. Das, A.: Computational Number Theory. CRC Press Taylor and Francis Group (2013)
5. Dattani, N.S., Bryans, N.: Quantum factorization of 56153 with only 4 qubits (2014). https://doi.org/10.48550/ARXIV.1411.6758
6. Grover, L.K.: A fast quantum mechanical algorithm for database search (1996). arXiv: https://arxiv.org/abs/quant-ph/9605043

7. Jarvis, F.: Algebraic Number Theory. Springer, Switzerland (2014). https://doi.org/10.1007/978-3-319-07545-7
8. Jiang, S., Britt, K.A., McCaskey, A.J., Humble, T.S., Kais, S.: Quantum annealing for prime factorization. Sci. Rep. **8**(1), 1–9 (2018)
9. Martin-Lopez, E., Laing, A., Lawson, T., Alvarez, R., Zhou, X.Q., O'brien, J.L.: Experimental realization of Shor's quantum factoring algorithm using qubit recycling. Nat. Photonics **6**(11), 773–776 (2012)
10. Mosca, M., Vensi Basso, J.M., Verschoor, S.R.: On speeding up factoring with quantum sat solvers (2019). arXiv: https://doi.org/10.48550/arXiv.1910.09592
11. Peng, W.C., et al.: Factoring larger integers with fewer qubits via quantum annealing with optimized parameters. Sci. China Phys. Mech. Astron. **62**(6), 1–8 (2019). https://doi.org/10.1007/s11433-018-9307-1
12. Proos, J., Zalka, C.: Shor's discrete logarithm quantum algorithm for elliptic curves. arXiv preprint: https://doi.org/10.48550/arXiv.quant-ph/0301141 (2003)
13. Shor, P.W.: Algorithms for quantum computation: discrete logarithms and factoring. In: Proceedings 35th Annual Symposium on Foundations of Computer Science, pp. 124–134. IEEE (1994)
14. Wang, B., Hu, F., Yao, H., Wang, C.: Prime factorization algorithm based on parameter optimization of Ising model. Sci. Rep. **10**(1), 1–10 (2020)
15. Wroński, M.: Index calculus method for solving elliptic curve discrete logarithm problem using quantum annealing. In: Paszynski, M., Kranzlmüller, D., Krzhizhanovskaya, V.V., Dongarra, J.J., Sloot, P.M.A. (eds.) ICCS 2021. LNCS, vol. 12747, pp. 149–155. Springer, Cham (2021). https://doi.org/10.1007/978-3-030-77980-1_12
16. Wroński, M.: Practical solving of discrete logarithm problem over prime fields using quantum annealing. In: Groen, D., de Mulatier, C., Paszynski, M., Krzhizhanovskaya, V.V., Dongarra, J.J., Sloot, P.M.A. (eds.) Computational Science - ICCS 2022, pp. 93–106. Springer, Cham (2022). https://doi.org/10.1007/978-3-031-08760-8_8
17. Yan, B., et al.: Factoring integers with sublinear resources on a superconducting quantum processor (2022)

Classification of Hybrid Quantum-Classical Computing

Frank Phillipson[1,2(✉)], Niels Neumann[1], and Robert Wezeman[1]

[1] Department Applied Cryptography and Quantum Algorithms, TNO, The Hague, The Netherlands
{frank.phillipson,niels.neumann,robert.wezemen}@tno.nl
[2] School of Business and Economics, Maastricht University, Maastricht, The Netherlands

Abstract. As quantum computers mature, the applicability in practice becomes more important. Quantum computers will often be used in a hybrid setting, where classical computers still play an important role in operating and using the quantum computer. However the term hybrid is diffuse and multi-interpretable. In this work we define two classes of hybrid quantum-classical computing: vertical and horizontal hybrid quantum-classical computing. The first is application-agnostic and concerns using and operating quantum computers. The second is application-specific and concerns running an algorithm. For both, we give a further subdivision in different types of hybrid quantum-classical computing and we introduce terms for them.

Keywords: hybrid quantum computing · classification · hybrid quantum algorithm · workflow

1 Introduction

Quantum computing is the technique of using quantum mechanical phenomena such as superposition, entanglement and interference for doing computational operations. The type of devices which are capable of doing such quantum operations are still being actively developed and named quantum computers. We distinguish between two paradigms of quantum computing devices: gate-based quantum computers and quantum annealers. A gate-based quantum computing system uses basic quantum circuit operations on qubits, similar to the classical operations on regular bits, that can be put together in any sequence to form algorithms. A quantum annealer brings a collection of qubits into an equal superposition and then applies a problem specific magnetic field. The qubits will interact under this magnetic field and move towards the state with the lowest energy, which encodes the solution of an optimisation problem.

Theory predicts that quantum computers will solve specific problems much faster than classical computers. Where classical computers have been under development for decades and are therefore quite mature, quantum computers

J. Mikyška et al. (Eds.): ICCS 2023, LNCS 14077, pp. 18–33, 2023.
https://doi.org/10.1007/978-3-031-36030-5_2

are still in the early stages of development. They are not yet capable of solving real world problems, due to the low number of qubits and their unstable nature causing noise, errors and loss of information. This current state of quantum computers is called the noisy intermediate-scale quantum (NISQ) era introduced by Preskill [42], where quantum computers have 50–200 qubits and their noise places serious limitations on their capabilities. Only recently IBM passed the 100-qubit barrier for the first time with its 127-qubit Eagle processor and they plan a 1,000-qubit chip, the Condor, in 2023 [26].

Other aspects than number of qubits affect the capabilities of quantum computers in practice, such as parallelism of operations and the topology layout of the qubits. Researchers are investigating innovative ways to solve valuable problems using already available NISQ systems and to achieve quantum advantage by demonstrating a significant performance advantage over today's classical computers. The latter was only quite recently claimed for the first time for an artificially created problem [2]. As quantum computers mature in the coming years, their computational power will increase and they can be applied in more settings and actually provide help in specific practical areas.

For the first practical applications in the near future, quantum computers will only execute a small part of a larger total workflow where a classical computer executes the other steps. See for example the workflow of retrieving data, training a classification algorithm and evaluating the obtained classifier [55], as shown in Fig. 1. In this example only a small part of the workflow is performed on a quantum computer. Already in 2005 such a combination was described in literature and was named hybrid quantum computing that "combine both classical and quantum computing architectures in order to leverage the benefits of both" [31].

Murray Thom, the Vice President of Software and Cloud Services at D-Wave, compared this with jets and "normal" vehicles: "Consider that while jet airplanes transformed the way we travel long distances, we still need vehicles that take us to our front door." Thus, "quantum applications will always and only be hybrid" [37].

This term, hybrid quantum-classical computing, was and is used in various contexts, each time used for (slightly) different settings. This is confusing for many researchers and practitioners in quantum computing, both on the application side as on the hardware side. The interest in quantum computing causes an enormous amount of research output[1] and, resulting, survey papers. One example of this kind is [14], where a survey of NISQ era hybrid quantum-classical machine learning research on hybrid quantum-classical systems is given, without explaining what is meant by hybrid quantum-classical exactly. This is totally left to the reader. Our goal is to provide a classification framework where authors can refer to, to make the scope of their contribution more clear to the reader. For this, we connect quantum research with classical computer science and workflow terminology. So, many of the proposed terminology in this paper exists already in computer science, think of compilation between computer languages, and work-

[1] Google scholar already gives 2090 results for the search on 'hybrid quantum-classical computing' for the period January–October 2022.

flow research, think of decomposition and activities. Our contribution is bringing them together in a clear framework for hybrid quantum-classical computing.

In this contribution we aim to describe clearly the various contexts hybrid quantum-classical computing can have in literature and to name these different approaches clearly and appropriately. In Sect. 2 we give an overview of various forms of hybrid quantum-classical computing that can be found in literature. Next, in Sect. 3, we distinguish a number of different types of hybrid quantum-classical computing from this overview and provide examples for each type. We end this paper with some conclusions and ideas for further research.

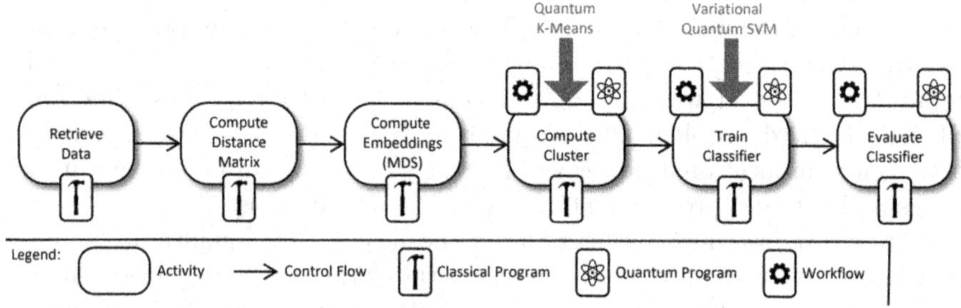

Fig. 1. An example workflow of a hybrid quantum-classical application where quantum computers perform only a small part of the computations [55]

2 Literature

We consider a global situation where we have a collection of computational tasks in which both the quantum computer and the classical computer are used. As such, hybrid forms of computing that allow for both discrete and continuous variables [34] and hybrid quantum-classical models of molecules in chemical and biological studies [22] are out of scope. We do not try to give an exhaustive overview of all research done on this topic. Our goal here is to give an overview, based on some examples, of the various meanings of the term hybrid quantum-classical computing in literature. This overview will be the basis of the proposed classification later on in this work.

Lanzagorta and Uhlmann presented one of the first hybrid algorithms that used both classical and quantum computers [31]. Later, research appeared on computing schemes and architectures to optimise the interactions between the different type of computers when executing hybrid quantum algorithms. A first example presents a candidate framework to analyse hybrid computations by fully integrating the quantum and classical resources and processes used for measurement-based quantum computing, where the feed-forward of classical measurement results is an integral part of the quantum design [25,27]. A second example proposes a quantum co-processor to accelerate a specific subroutine of a larger task. This is most often seen as the main reason for hybrid algorithms, for example in [5,33,36,41]. The work by Li et al. [33] results in a system-level software infrastructure for hybrid quantum-classical computing. Endo [18,19] indicates that for early quantum applications, a large portion of the computational burden is performed on a classical computer and hence fully coherent deep quantum circuits are not required. As the quantum computer takes on more computational load, noise of the quantum computer will result in more errors, which will have more impact on the total calculation. This in turn requires qubits of higher quality and error mitigation routines to suppress noise.

An important type of hybrid computing appeared with the introduction of Variational Quantum Algorithms (VQA). VQAs use a classical optimiser to train a parameterised quantum circuit and provide a framework to tackle a wide array of tasks, as shown in the extensive overview by Cerezo et al. [9]. Examples include finding ground states and excited states of molecules, optimisation, solving linear systems of equations and machine learning. The first VQA, the variational quantum eigensolver (VQE) algorithm of [39], appeared in 2014. Some papers see this group of algorithms as "a novel class of hybrid quantum-classical algorithms", without explicitly diving in other groups of algorithms [7]. Also, [11] sees this class of algorithms as a specific example of hybrid quantum-classical computing in a noisy environment. In [54] the main reason given for hybrid quantum-classical computing is the size of the problems in combination with the available hardware. They distinguishes two types of hybrid computing: 1) Pre- or post-processing of a quantum computation on a classical computer. Examples are algorithms by Shor [50] and Simon [51] that use classical post-processing. 2) Algorithms that perform multiple iterations of quantum and classical computations. Thereby, the output of the quantum computation is improved in each iteration until the result reaches the required accuracy. An example they give using this approach is the quantum approximate optimisation algorithm (QAOA) [20].

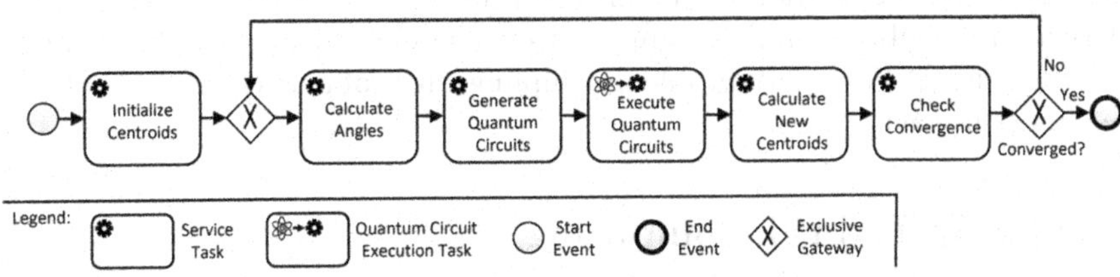

Fig. 2. The sub-workflow of the "Compute Cluster activity" shown in Fig. 1 [55]

In the works [7,8], the term 'quassical' computing is coined and motivated by "Classical computing and quantum computing have obvious complementary strengths, so instead of opposing them it might be better to combine them into a new type of computing." They give two reasons for the combination: First, most quantum computing algorithms "require some preliminary classical pre-processing to shape the problem into one the quantum computer can recognise and then to receive the data returned by the [quantum computer] and shape it into the answer the engineer needs." Second, "all the quantum computers we have heard of are designed as cyber-physical systems, quantum mechanical systems controlled by digital controllers", meaning they are quassical in a trivial sense. In this light, you can also think of the classical steps needed to transform a quantum circuit to an execution as performed for example by an openQL framework [29]. They expect that this combination will stay, also when the quantum computer is in full maturity. The first remark is also mentioned by [16], who indicate that "while hybrid algorithms and platforms may just be the best first step, it is reasonable to assume that quantum applications will always be hybrid", for example, by the need of a pre-processing step which prepares data for a quantum algorithm or a post-processing step which handles data coming from a quantum algorithm.

Another view on hybrid algorithms is given by the idea that a problem or circuit that is too big to be executed on (noisy) quantum processors of intermediate size, is partitioned automatically into smaller parts that are evaluated separately. Suchara et al. [53] suggest such an approach for gate-based quantum computers: "We advocate using a hybrid quantum-classical architecture where larger quantum circuits are broken into smaller sub-circuits that are evaluated separately, either using a quantum processor or a quantum simulator running on a classical supercomputer." The D-Wave hybrid solvers should also be seen in this light. They offer the functionality to partition the problem into smaller pieces that fit the current chip size and are solved sequentially. The outcomes of the subroutines form the resulting (probably sub-optimal) solution. They also provide hybrid approaches, where multiple branches of a process solve the problem, some of them with a classical solver, others using the QPU, and then return the first or best solution [3]. Here, the quantum and classical computers compete in parallel to find a solution to the problem. In [10] the D-Wave Kerberos solution is used, a hybrid built-in sampler, that combines Tabu search, simulated annealing, and D-Wave sub-problem sampling on problem variables that have high-energy impact.

3 Types of Hybrid Computing

To make a clear distinction between all the different views on hybrid quantum-classical computing, we use the workflow approach presented in [55]. They propose workflows to specify the (partial) order of a collection of activities needed to execute a hybrid quantum-classical application and combine this with topologies to reveal the overall structure of hybrid quantum applications. An activity in such a workflow can be further expanded in sub-workflows. Typically, the activities are represented as nodes in a directed graph with the control flow dependencies the directed edges of the graph. Figure 1 shows an example, where the workflow of a hybrid quantum-classical application in quantum machine learning is shown. The presented quantum application performs clustering on a set of input data, and based on the clustering results, trains a classifier for the classification of future data. The "gear" icon indicates sub-workflows. An example of the "Compute Cluster" sub-workflow is shown in Fig. 2. They consider two dimensions of hybrid computing, the vertical provisioning engine and the horizontal workflow-engine, from a software architecture point of view. This insight and the workflow technology are the basis of our proposed classification framework. However, we elaborate further on this and expand it into sub-categories.

We distinguish two main categories of hybrid computing:

1. Vertical hybrid quantum computing: All controlling activities required to control and operate a quantum circuit on a quantum computer, as was the case in classical computing providing compilation and controlling in the stack. An example of a quantum stack can be found in [43]. These steps are application-agnostic.
2. Horizontal hybrid quantum computing: All operational activities required to use a quantum computer and a classical computer to perform an algorithm. These steps are application-dependent. Here we use the classical workflow terminology as proposed in [17].

We can subdivide both categories further, as shown in Table 1, and explained further in the next sections. Note that some use-cases might show signs of more than one type of hybrid quantum-classical computing. It is also important to stress that these main categories are not mutually exclusive. The vertical category is mainly about computing and computing stack, the horizontal category is mainly about algorithms.

Table 1. Overview of the main categories and their sub-categories

Vertical hybrid	Horizontal hybrid
Decomposition hybrid	Processing hybrid
Implementation hybrid	Micro hybrid split
Controlling hybrid	Macro hybrid split
	Parallel hybrid
	Breakdown hybrid

3.1 Vertical Hybrid Quantum Computing

The types of vertical hybrid quantum computing contain the classical steps that have to be taken to let the quantum computer run the quantum routine. A general overview, starting from a single activity in a sub-workflow is depicted in Fig. 3. The specific steps are described in the next sub-sections. These steps are in some way similar to the layers in the OpenQL framework [29], however, we distinguish more between steps that are topology and technology-agnostic and steps that are not.

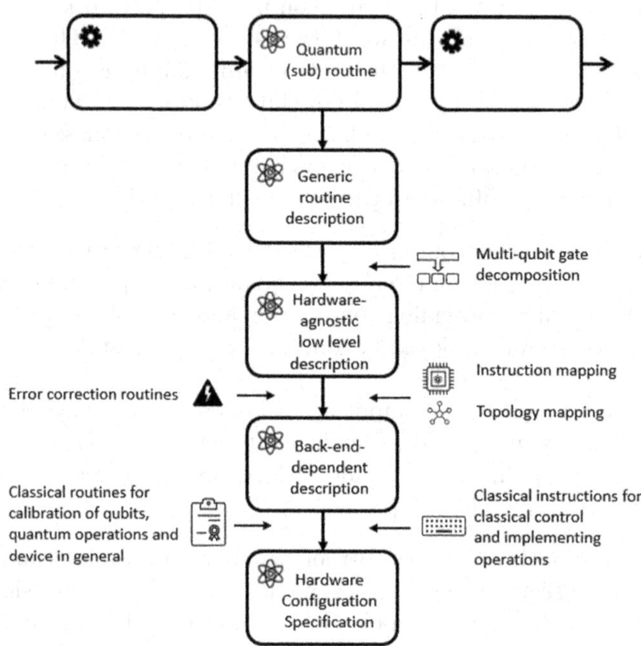

Fig. 3. Schematic view on relation between horizontal and vertical dependencies

Decomposition Hybrid. The workflows of decomposition hybrid consider a higher level algorithm description, which is then decomposed in classical instructions and quantum instructions, see Fig. 4. A high-level quantum routine is broken down into low-level hardware-agnostic quantum instructions. The higher level algorithm description can be any commonly used classical language, enhanced with classical routines, or a dedicated quantum routine. For gate-based devices, this hardware-agnostic instruction set can for instance include single qubit rotations and some two-qubit gates such as the CNOT-gate and controlled-phase-gates. Libraries can help decomposing quantum instructions to low level hardware-agnostic instructions [6].

Examples of decomposition hybrid include decomposing algorithmic instruction to classical instructions and quantum instructions and mapping both to low-level instructions. This includes decomposing high-level instructions not suited for the hardware to lower-level instructions with a more direct mapping to hardware. Classical routines can help with this decomposition [30]. The vertical hybrid can be separated into the following parts, which are depicted in Fig. 4.

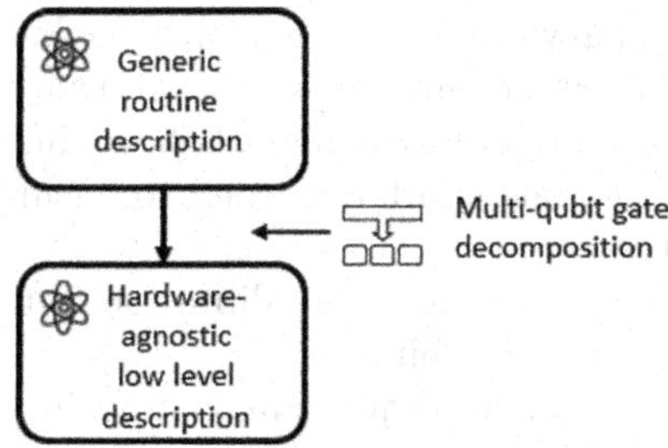

Fig. 4. Schematic view on *decomposition hybrid*: a high-level quantum routine is broken down into low-level hardware-agnostic quantum instructions

Implementation Hybrid. The workflows of implementation hybrid consider all steps to map and implement the operations on a quantum computer. This workflow specifically aims to map the hardware-agnostic low-level instructions to hardware-specific instructions, see Fig. 5. This includes both the classical and quantum instructions.

Examples of implementation hybrid include mapping instructions to hardware. This includes assigning operations to qubits and taking into account the topology of the hardware backend. This workflow also outputs a time-schedule for which hardware instructions should be applied to which qubits. If necessary, this workflow also adds error correcting routines together with the classical feedback loop of these routines.

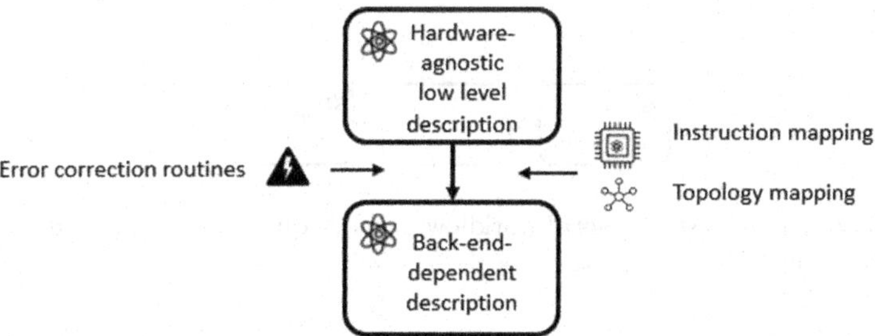

Fig. 5. Schematic view on *implementation hybrid*: low-level quantum instructions are mapped to a specific hardware-backend. This includes creating a time-schedule, assigning operations to qubits and, if necessary, apply error correcting routines

Controlling Hybrid. The workflows of controlling hybrid consider all steps to operate and control a quantum computer, see Fig. 6. Due to the intricate nature of quantum computers, their operations might behave differently over time than intended. Therefore, continuous effort is needed to ensure that quantum computers behave as expected.

Examples of controlling hybrid include calibration routines of the qubits and of elementary operations on the qubits.

All of the steps in vertical hybrid quantum computing can include optimisation steps. Sometimes, optimised approximation methods yield better performing implementation then exact full implementations, for instance see [47].

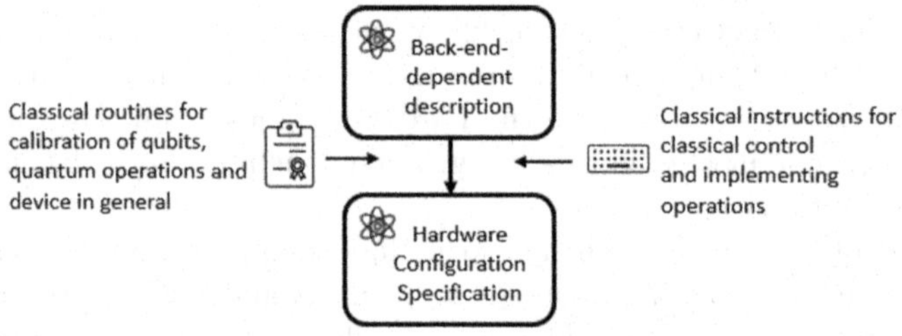

Fig. 6. Schematic view on *controlling hybrid*: this considers all steps to operate and control a quantum computer, including the actual mapping of hardware instructions to qubits and calibrating the quantum computer

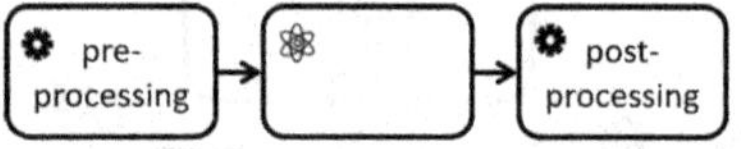

Fig. 7. Schematic *processing hybrid* workflow: a single quantum routine with additional classical pre- and post-processing

3.2 Horizontal Hybrid Quantum Computing

The types of horizontal hybrid quantum computing distinguish between the variety of orderings quantum and conventional computing steps within an algorithm.

Processing Hybrid. The workflows of processing hybrid have a single quantum block, combined with classical pre-processing and classical post-processing [7,16]. A schematic representation is shown in Fig. 7.

Examples of processing hybrid algorithms are the algorithms by Shor and Simon [53]. Another example is the distance based classifier [49,56], where data standardisation and normalisation are the pre-processing steps and translating the measurements to the desired kernel classifier the post-processing step.

Micro Hybrid Split. The workflows of a micro hybrid split consider a single activity of a larger workflow. The workflow shown in Fig. 2 is a micro hybrid split of Fig. 1. Within the single activity, some operations are quantum and others are classical, possibly in an iterative fashion. A schematic representation is shown in Fig. 8.

Examples of micro hybrid splits are variational algorithms [7]. Measurement-based quantum computing can be seen as a special member of this class, as future measurements and operations depend on previous measurements and classical operations.

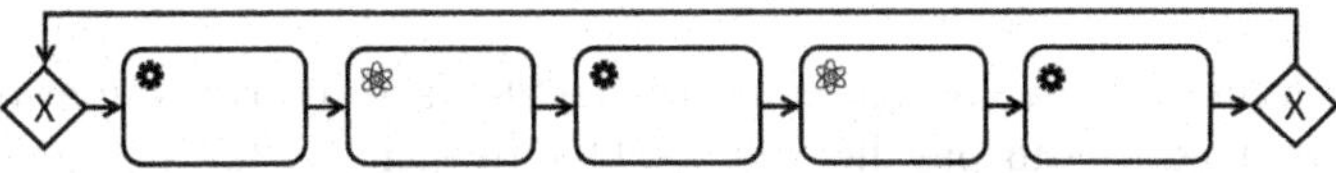

Fig. 8. Schematic *micro hybrid split* workflow: a single activity has both classical and quantum operations

Macro Hybrid Split. The workflows of a macro hybrid split consider different tasks that belong to different activities within a larger algorithm. The difference with micro hybrid split workflows is small and depends mainly on the granularity with which the workflow is observed: A micro hybrid split in one workflow can be a small part of a larger macro hybrid split workflow. The activities in a macro hybrid split can also iterate and each task can be hybrid in itself. A schematic representation is shown in Fig. 9. A possible relation between macro and micro hybrid splits is shown in Fig. 10.

Examples of macro hybrid splits are the hybrid quantum machine learning approach in the domain of humanities [55] and the workflow shown in Fig. 1.

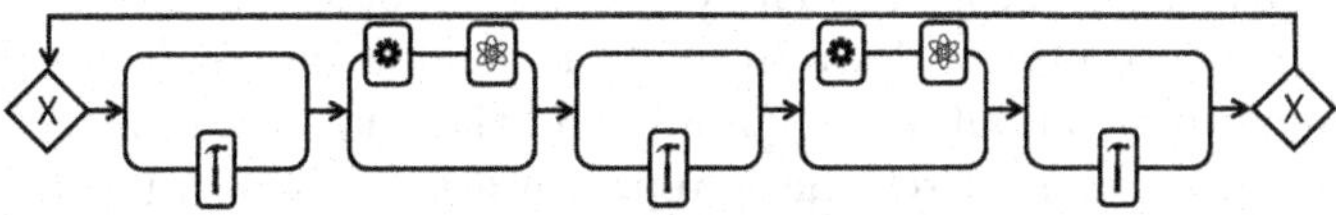

Fig. 9. Schematic *macro hybrid split* workflow: each block is a specific task that can have both quantum and classical operations. A single block can be further subdivided in a micro hybrid split workflow

Parallel Hybrid. The workflows of parallel hybrid have multiple independent branches to solve a specific problem. Each branch tries to solve the problem independently and the first (or best) solution found is returned. Each branch can use different solvers. A schematic representation is shown in Fig. 11.

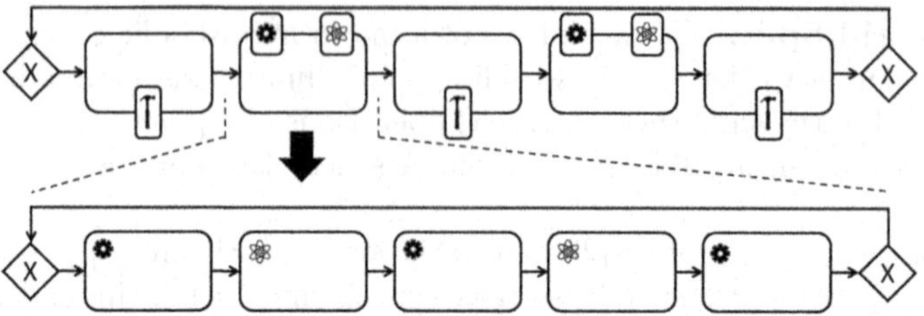

Fig. 10. Schematic relationship between *macro hybrid split* and *micro hybrid split* workflow: a block in the *macro hybrid split* can be specified as a *micro hybrid split*

Examples include the configuration of the D-Wave-hybrid framework, where samples are parsed to four parallel solvers. One branch can for instance be a classical tabu search that either returns with certainty an answer, or is interrupted by another finished branch [3,10].

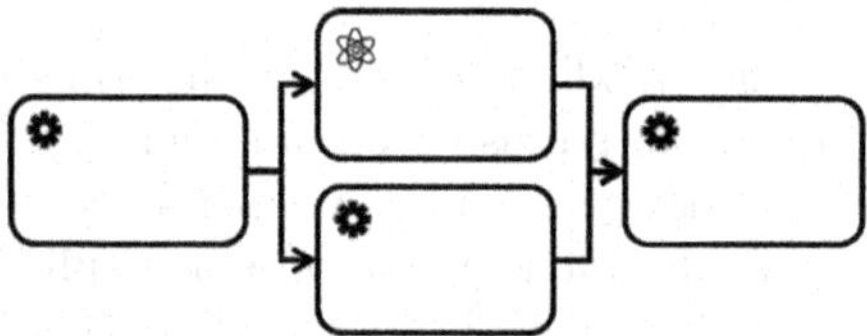

Fig. 11. Schematic *parallel hybrid* workflow: a task is processed by multiple independent branches. The answer is returned based on some criteria, for instance, coming from the branch that finishes first

Breakdown Hybrid. The workflows of breakdown hybrid consider multiple small parts of a larger problem. The considered problem is too large to solve directly and is hence broken down in multiple smaller parts. Each smaller part is run on a quantum computer sequentially and the final answer is reconstructed from the partial answers. A schematic representation is shown in Fig. 12.

Examples are the gate-based approach in [53], where they advocate using a hybrid quantum-classical architecture where larger quantum circuits are broken into smaller sub-circuits that are evaluated separately and specific options within D-Wave's hybrid solvers [3], where the problem divided into several parts that are solved using classical or quantum annealing approaches. Note that these classes are not disjoint. Here, one of the breakdown parts might be run on a classical computer in parallel with one or more quantum tasks. This would make it a combination of breakdown hybrid and parallel hybrid.

Fig. 12. Schematic *breakdown hybrid* workflow: a large problem is decomposed in smaller problems, each of which is run on a quantum computer, the final answer is reconstructed from the partial answers

4 Application

As indicated in the introduction, the number of papers that are categorised under hybrid quantum-classical computing is enormous. We will not give an exhaustive

Table 2. Example classification of recent papers

Title	Classification
NetQASM-a low-level instruction set architecture for hybrid quantum-classical programs in a quantum internet [13]	Decomposition
QuantumPath: A quantum software development platform [23]	Decomposition
An LLVM-based C++ compiler toolchain for variational hybrid quantum-classical algorithms and quantum accelerator [28]	Implementation
Diversity of hybrid quantum system [24]	Controlling
Quantify-scheduler: An open-source hybrid compiler for operating quantum computers in the NISQ era [12]	Controlling
The optimization landscape of hybrid quantum-classical algorithms: from quantum control to NISQ applications [21]	Micro split
A hybrid quantum-classical CFD methodology with benchmark HHL solutions [32]	Micro split
Considerations for evaluating thermodynamic properties with hybrid quantum-classical computing work flow [52]	Micro split
Hyperparameter optimization of hybrid quantum neural networks for car classification [45]	Macro split
Hybrid quantum-classical algorithms for approximate graph coloring [4]	Macro split
A hybrid quantum-classical neural network architecture for binary classification [1]	Macro split
Hybrid quantum-classical algorithm for computing imaginary-time correlation functions [46]	Macro split
Hybrid quantum-classical algorithm for hydrodynamics [58]	Macro split
Hybrid quantum-classical search algorithms [44]	Macro split
A quantum algorithm of k-means toward practical use [38]	Macro split
A hybrid quantum-classical algorithm for robust fitting. [15]	Macro split
Graph-$\|Q><C\|$, a graph-based quantum/classical algorithm for efficient electronic structure on hybrid quantum/classical hardware systems: Improved quantum circuit depth performance [57]	Parallel
Optimization of robot trajectory planning with nature-inspired and hybrid quantum algorithms [48]	Parallel
A quantum approach for tactical capacity management of distributed electricity generation [40]	Parallel
Hybrid quantum-classical unit commitment [35]	Breakdown

overview of all papers and their classification. As an example, we selected a few papers from 2022 that have this terms in their title or key words to illustrate the classification.

In Table 2, this classification is shown. In this table, but also in reality, the majority of papers are within the micro and macro hybrid split classes. We could not find any papers within the processing hybrid class that use the terminology hybrid quantum-classical. Papers in this class mostly use the term quantum algorithm.

5 Conclusions

It is expected that quantum computing will always need some form of classical computing to enable the calculations and the execution on the hardware platforms. This is often named hybrid quantum-classical computing. The term hybrid is however diffuse and multi-interpretable. We showed in this paper that in literature this term covers many concepts. Based on this literature and concepts from workflow approach and classical computer science, we distinguished between horizontal and vertical hybrid quantum computing and defined and named various specific types within these classes. This can help researchers and practitioners in quantum computing to make clear what they mean when using the general term 'hybrid quantum-classical computing' and can help in developing more concise tools within the quantum computing stack. We do not assume to be complete in our overview and categorisation. We encourage scientists and practitioners to complement this framework as part of future research on this topic.

References

1. Arthur, D., et al.: A hybrid quantum-classical neural network architecture for binary classification. arXiv:2201.01820 (2022)
2. Arute, F., Arya, K., et al.: Quantum supremacy using a programmable superconducting processor. Nature **574**(7779), 505–510 (2019)
3. Booth, M., Reinhardt, S.P., Roy, A.: Partitioning optimization problems for hybrid classical. quantum execution. Technical report, pp. 01–09 (2017)
4. Bravyi, S., Kliesch, A., Koenig, R., Tang, E.: Hybrid quantum-classical algorithms for approximate graph coloring. Quantum **6**, 678 (2022)
5. Bravyi, S., Smith, G., Smolin, J.A.: Trading classical and quantum computational resources. Phys. Rev. X **6**(2), 021043 (2016)
6. Van den Brink, R., Phillipson, F., Neumann, N.M.: Vision on next level quantum software tooling. In: Computation Tools (2019)
7. Calude, C.S., Calude, E., Dinneen, M.J.: Guest column: adiabatic quantum computing challenges. ACM SIGACT News **46**(1), 40–61 (2015)
8. Calude, C.S., et al.: Quassical computing. Int. J. Unconv. Comput. **14**(1), 43–57 (2018)
9. Cerezo, M., Arrasmith, A., et al.: Variational quantum algorithms. Nat. Rev. Phys. **3**(9), 625–644 (2021)

10. Chiscop, I., Nauta, J., Veerman, B., Phillipson, F.: A hybrid solution method for the multi-service location set covering problem. In: Krzhizhanovskaya, V.V., et al. (eds.) ICCS 2020. LNCS, vol. 12142, pp. 531–545. Springer, Cham (2020). https://doi.org/10.1007/978-3-030-50433-5_41
11. Córcoles, A.D., Kandala, A., et al.: Challenges and opportunities of near-term quantum computing systems. arXiv:1910.02894 (2019)
12. Crielaard, D., De Jong, D., et al.: Quantify-scheduler: an open-source hybrid compiler for operating quantum computers in the NISQ era. Bull. Am. Phys. Soc. **4**, 1–29 (2022)
13. Dahlberg, A., van der Vecht, B., et al.: NetQASM-a low-level instruction set architecture for hybrid quantum-classical programs in a quantum internet. Quantum Sci. Technol. **7**, 035023 (2022)
14. De Luca, G.: A survey of NISQ era hybrid quantum-classical machine learning research. J. Artif. Intell. Technol. **2**(1), 9–15 (2022)
15. Doan, A.D., Sasdelli, M., et al.: A hybrid quantum-classical algorithm for robust fitting. In: Computer Vision and Pattern Recognition, pp. 417–427 (2022)
16. Edwards, M.: Towards Practical Hybrid Quantum/Classical Computing. Master's thesis, University of Waterloo (2020)
17. Ellis, C.A.: Workflow technology. In: Computer Supported Cooperative Work. Trends in Software Series, vol. 7, pp. 29–54 (1999)
18. Endo, S.: Hybrid quantum-classical algorithms and error mitigation. Ph.D. thesis, University of Oxford (2019)
19. Endo, S., Cai, Z., et al.: Hybrid quantum-classical algorithms and quantum error mitigation. J. Phys. Soc. Jpn. **90**(3), 032001 (2021)
20. Farhi, E., Goldstone, J., Gutmann, S.: A quantum approximate optimization algorithm. arXiv:1411.4028 (2014)
21. Ge, X., Wu, R.B., Rabitz, H.: The optimization landscape of hybrid quantum-classical algorithms: from quantum control to NISQ applications. arXiv:2201.07448 (2022)
22. Henelius, P., Fishman, R.S.: Hybrid quantum-classical Monte Carlo study of a molecule-based magnet. Phys. Rev. B **78**(21), 214405 (2008)
23. Hevia, J.L., Peterssen, G., Piattini, M.: QuantumPath: a quantum software development platform. Softw. Pract. Exp. **52**(6), 1517–1530 (2022)
24. Hirayama, Y.: Diversity of hybrid quantum systems. In: Hirayama, Y., Hirakawa, K., Yamaguchi, H. (eds.) Quantum Hybrid Electronics and Materials. Quantum Science and Technology, pp. 1–14. Springer, Singapore (2022). https://doi.org/10.1007/978-981-19-1201-6_1
25. Horsman, C., Munro, W.J.: Hybrid hypercomputing: towards a unification of quantum and classical computation. arXiv:0908.2181 (2009)
26. IBM: IBM's roadmap for scaling quantum technology (2020). https://research.ibm.com/blog/ibm-quantum-roadmap
27. Jozsa, R.: An introduction to measurement based quantum computation. arXiv:quant-ph/0508124 (2005)
28. Khalate, P., Wu, X.C., et al.: An LLVM-based C++ compiler toolchain for variational hybrid quantum-classical algorithms and quantum accelerators. arXiv:2202.11142 (2022)
29. Khammassi, N., Ashraf, I., et al.: OpenQL: a portable quantum programming framework for quantum accelerators. ACM J. Emerg. Technol. Comput. Syst. (JETC) **18**(1), 1–24 (2021)
30. Kitaev, A.Y.: Quantum computations: algorithms and error correction. Russ. Math. Surv. **52**(6), 1191–1249 (1997)

31. Lanzagorta, M., Uhlmann, J.K.: Hybrid quantum-classical computing with applications to computer graphics. In: ACM SIGGRAPH 2005 Courses, p. 2-es. ACM (2005)
32. Lapworth, L.: A hybrid quantum-classical CFD methodology with benchmark HHL solutions. arXiv:2206.00419 (2022)
33. Li, J., Yang, X., Peng, X., Sun, C.P.: Hybrid quantum-classical approach to quantum optimal control. Phys. Rev. Lett. **118**(15), 150503 (2017)
34. Lloyd, S.: Hybrid Quantum Computing. In: Braunstein, S.L., Pati, A.K. (eds.) Quantum Information with Continuous Variables, pp. 37–45. Springer, Dordrecht (2003). https://doi.org/10.1007/978-94-015-1258-9_5
35. Mahroo, R., Kargarian, A.: Hybrid quantum-classical unit commitment. In: Texas Power and Energy Conference (TPEC), pp. 1–5. IEEE (2022)
36. McCaskey, A.J., Lyakh, D.I., Dumitrescu, E.F., Powers, S.S., Humble, T.S.: XACC: a system-level software infrastructure for heterogeneous quantum-classical computing. Quantum Sci. Technol. **5**(2), 024002 (2020)
37. Murray, T.: Three truths and the advent of hybrid quantum computing, June 2019. https://medium.com/d-wave/three-truths-and-the-advent-of-hybrid-quantum-computing-1941ba46ff8c
38. Ohno, H.: A quantum algorithm of k-means toward practical use. Quantum Inf. Process. **21**(4), 1–24 (2022)
39. Peruzzo, A., McClean, J., et al.: A variational eigenvalue solver on a photonic quantum processor. Nat. Commun. **5**(1), 1–7 (2014)
40. Phillipson, F., Chiscop, I.: A quantum approach for tactical capacity management of distributed electricity generation. In: Phillipson, F., Eichler, G., Erfurth, C., Fahrnberger, G. (eds.) Innovations for Community Services, I4CS 2022. Communications in Computer and Information Science, vol. 1585. Springer, Cham (2022). https://doi.org/10.1007/978-3-031-06668-9_23
41. Possignolo, R.T., Margi, C.B.: A quantum-classical hybrid architecture for security algorithms acceleration. In: Trust, Security and Privacy in Computing and Communications, pp. 1032–1037. IEEE (2012)
42. Preskill, J.: Quantum computing in the NISQ era and beyond. Quantum **2**, 79 (2018)
43. Riesebos, L., Fu, X., et al.: Quantum accelerated computer architectures. In: Circuits and Systems (ISCAS), pp. 1–4. IEEE (2019)
44. Rosmanis, A.: Hybrid quantum-classical search algorithms. arXiv:2202.11443 (2022)
45. Sagingalieva, A., Kurkin, A., et al.: Hyperparameter optimization of hybrid quantum neural networks for car classification. arXiv:2205.04878 (2022)
46. Sakurai, R., et al.: Hybrid quantum-classical algorithm for computing imaginary-time correlation functions. Phys. Rev. Res. **4**(2), 023219 (2022)
47. Schalkers, M.A., Möller, M.: Learning based hardware-centric quantum circuit generation. In: Phillipson, F., Eichler, G., Erfurth, C., Fahrnberger, G. (eds.) Innovations for Community Services, I4CS 2022. Communications in Computer and Information Science, vol. 1585. Springer, Cham (2022). https://doi.org/10.1007/978-3-031-06668-9_22
48. Schuetz, M.J., Brubaker, J.K., et al.: Optimization of robot trajectory planning with nature-inspired and hybrid quantum algorithms. arXiv:2206.03651 (2022)
49. Schuld, M., Fingerhuth, M., Petruccione, F.: Implementing a distance-based classifier with a quantum interference circuit. EPL (Europhys. Lett.) **119**(6), 60002 (2017)

50. Shor, P.W.: Polynomial-time algorithms for prime factorization and discrete logarithms on a quantum computer. SIAM J. Comput. **26**(5), 1484–1509 (1997)
51. Simon, D.R.: On the power of quantum computation. SIAM J. Comput. **26**(5), 1474–1483 (1997)
52. Stober, S.T., Harwood, S.M., et al.: Considerations for evaluating thermodynamic properties with hybrid quantum-classical computing work flows. Phys. Rev. A **105**(1), 012425 (2022)
53. Suchara, M., Alexeev, Y., et al.: Hybrid quantum-classical computing architectures. In: Post-Moore Era Supercomputing (2018)
54. Weder, B., Barzen, J., Leymann, F., Salm, M., Vietz, D.: The quantum software lifecycle. In: Architectures and Paradigms for Engineering Quantum Software, pp. 2–9 (2020)
55. Weder, B., Barzen, J., Leymann, F., Zimmermann, M.: Hybrid quantum applications need two orchestrations in superposition: a software architecture perspective. In: Web Services, pp. 1–13. IEEE (2021)
56. Wezeman, R., Neumann, N., Phillipson, F.: Distance-based classifier on the quantum inspire. Digitale Welt **4**(1), 85–91 (2020)
57. Zhang, J.H., Iyengar, S.S.: Graph-$|Q><C|$, a graph-based quantum/classical algorithm for efficient electronic structure on hybrid quantum/classical hardware systems: improved quantum circuit depth performance. J. Chem. Theor. Comput. **18**(5), 2885–2899 (2022)
58. Zylberman, J., Di Molfetta, G., et al.: Hybrid quantum-classical algorithm for hydrodynamics. arXiv:2202.00918 (2022)

Solving (Max) 3-SAT via Quadratic Unconstrained Binary Optimization

Jonas Nüßlein[1(✉)], Sebastian Zielinski[1], Thomas Gabor[1], Claudia Linnhoff-Popien[1], and Sebastian Feld[2]

[1] Institute for Informatics, LMU Munich, Munich, Germany
jonas.nuesslein@ifi.lmu.de

[2] Faculty of Electrical Engineering, Mathematics and Computer Science, TU Delft, Delft, The Netherlands

Abstract. We introduce a novel approach to translate arbitrary 3-SAT instances to Quadratic Unconstrained Binary Optimization (QUBO) as they are used by quantum annealing (QA) or the quantum approximate optimization algorithm (QAOA). Our approach requires fewer couplings and fewer physical qubits than the current state-of-the-art, which results in higher solution quality. We verified the practical applicability of the approach by testing it on a D-Wave quantum annealer.

Keywords: QUBO · quantum annealing · satisfiability · 3-SAT

1 Introduction

In recent years, many well-known optimization and decision problems have been translated to the model of quadratic unconstrained binary optimization (QUBO) [13,18]. The main motivation behind this is that QUBO models can be used as a problem specification for various early quantum algorithms, most notably the quantum approximate optimization algorithm (QAOA) [9,24] and quantum annealing (QA) [16,17]. Current quantum computers are noisy and limited in size; thus it is important to encode problems as efficiently as possible. However, quantum hardware is conjectured to further grow in capability and a first demo application recently suggested that it might already have a substantial advantage over classical hardware for specific tasks [1].

The most promising problems to be solved using quantum algorithms certainly include problems of the complexity class NP-hard, which are hard to solve for classical computers (unless P = NP) [7,11]. Many NP-hard problems like scheduling [23], quadratic assignment [19], or travelling salesman [10] are of immense practical importance and practical instances often challenge current computing hardware. Thus, the eventual benefit of making these kinds of problems faster to solve may be especially appealing.

The canonical problem for the class NP-complete is 3-satisfiability (3-SAT), which we focus on in this paper [8]. A 3-SAT instance is a formula in Boolean algebra and its solution is the binary answer to whether the formula is satisfiable.

J. Mikyška et al. (Eds.): ICCS 2023, LNCS 14077, pp. 34–47, 2023.
https://doi.org/10.1007/978-3-031-36030-5_3

Our contributions in this paper are:

- We present two novel 3-SAT-to-QUBO translations: NÜSSLEIN^{2n+m} and NÜSSLEIN^{n+m}
- We empirically show that NÜSSLEIN^{2n+m} performs slightly better than CHANCELLOR^{n+m} despite the bigger QUBO matrix
- We show that NÜSSLEIN^{n+m} requires fewer couplings and fewer physical qubits than the current state-of-the-art approach CHANCELLOR^{n+m}
- We empirically show that NÜSSLEIN^{n+m} performs best, compared to three other 3-SAT-to-QUBO translations

2 Foundations

In this section, we introduce the mathematical foundations of the problems involved in the translation algorithms: 3-SAT and QUBO.

2.1 Satisfiability Problems

The satisfiability problem (SAT) of propositional logic is informally defined as follows: Given a Boolean formula, is there any assignment of the involved variables so that the formula is reduced to "true"? The problem occurs in every application involving complex constraints or reasoning, like (software) product lines, the tracing of software dependencies, or formal methods [12].

All SAT problem instances can be reduced with only polynomial overhead to a specific type of SAT problem called 3-SAT, in which the input propositional logic formula has to be in conjunctive normal form with all of the disjunctions containing exactly three literals.

Definition 1 3-SAT. *A* 3-SAT *instance with* n *variables and* m *clauses is given as (i) a list of variables* $(v_j)_{0 \leq j \leq n-1}$, *from which a list of literals* $(l_i)_{0 \leq i \leq 3m-1}$ *can be built of the form*

$$l_i \in \bigcup_{0 \leq j \leq n-1} \{v_j, \neg v_j\},$$

and (ii) a list of clauses $(c_k)_{0 \leq k \leq m-1}$ *of the form*

$$c_k = (l_{3k} \vee l_{3k+1} \vee l_{3k+2}).$$

A given 3-SAT *instance is* satisfiable *iff there exists a variable assignment given by the structure* $(v_j \mapsto b_j)_{0 \leq j \leq n-1}$ *with* $b_j \in \{\top, \bot\}$ *so that*

$$\bigwedge_{0 \leq k \leq m-1} c_k$$

reduces to $\top$ *when interpreting all logical operators as is common. The problem of deciding whether a given* 3-SAT *instance is satisfiable is called* 3-SAT.

For example, we may write a 3-SAT instance as Boolean formula $\mathcal{F} = (a \vee b \vee c) \wedge (a \vee \neg c \vee \neg d)$ consisting of $m = 2$ clauses and featuring the $n = 4$ distinct variables $\{a, b, c, d\}$. Obviously, $\mathcal{F}$ is satisfiable, for example via the variable assignment $(a \mapsto \bot, b \mapsto \top, c \mapsto \top, d \mapsto \bot)$.

3-SAT was the first problem to be shown to be NP-complete, which means that all problems in NP can be reduced to 3-SAT [8]. In fact, as many proofs for NP-completeness for other problems build upon their reduction to 3-SAT, 3-SAT solvers can be used as tools to solve many different decision problems.

As 3-SAT is central to many proofs of NP-completeness, it is somewhat surprising that when we generate random 3-SAT instances with random amounts of variables n and clauses m, most of these instances will be really easy to solve for standard SAT solvers. It is only as the ratio of clauses per variable approaches $\frac{m}{n} \approx 4.2$ that we can see the problems take exponential computing time. Knowing that many 3-SAT instances are relatively easy to solve even for classical computers, we focus our attention regarding new methods (like quantum-based ones) on the critical 3-SAT instances with $\frac{m}{n} \approx 4.2$.

MAX-3-SAT is an optimization problem that corresponds to the decision problem 3-SAT. Instead of checking whether an assignment exists that fulfils the whole formula, i.e., reduces all clauses individually to $\top$, we try to find the assignment that fulfils as many clauses as possible. Note that MAX-3-SAT is a generalization of 3-SAT as MAX-3-SAT's optimal result is an assignment that fulfils all clauses and thus proves the satisfiability of the whole formula.

Definition 2 MAX-3-SAT. *A MAX-3-SAT instance is given the same way as a 3-SAT instance (cf. Definition 1). The objective of a MAX-3-SAT instance is to find a variable assignment of the structure $(v_j \mapsto b_j)_{0 \leq j \leq n-1}$ with $b_j \in \{\top, \bot\}$ so to*

$$\textit{maximize} \sum_{k=0}^{m-1} \begin{cases} 1 & \textit{if } c_k \textit{ reduces to } \top, \\ 0 & \textit{otherwise.} \end{cases}$$

2.2 Quadratic Unconstrained Binary Optimization

In quadratic unconstrained binary optimization (QUBO) we are looking for a binary vector $\mathbf{x} = \langle x_i \rangle_{0 \leq i \leq k-1}$ of length k that minimizes the value of a formula that at most contains quadratic terms in x.

Definition 3 QUBO. *A QUBO instance with k variables is given as a $k \times k$ matrix $Q \in \mathbb{R}^{k \times k}$. The objective of a QUBO instance is to find a binary vector $\mathbf{x} \in \mathbb{B}^k$ so to*

$$\textit{minimize} \quad H(\mathbf{x}) = \sum_i Q_{ii} x_i + \sum_{i<j} Q_{ij} x_i x_j.$$

$H(\mathbf{x})$ is also called the *energy* of a QUBO solution $\mathbf{x}$. Note that the lower triangle of the matrix Q is always empty (since its values do not occur in the formula for the energy H). Finding the ideal solution vector $\mathbf{x}$ of a given QUBO Q is NP-hard.

When solving QUBO instances using a quantum annealer, the solution vector $\mathbf{x}$ is mapped to a set of qubits. These qubits have connections whose strength can be manipulated to emulate the values in the QUBO matrix. As the limiting factor in current hardware is the size of problems that can be solved, we seek translations to QUBO that require as few qubits (i.e., minimal size of the QUBO matrix) and as few connections between them (i.e., minimal density within the QUBO matrix) as possible.

3 Related Work

There are currently two main approaches for translating 3-SAT to QUBO, which we refer to as CHANCELLOR^{n+m} [5] and CHOI^{3m} [6]. We will review them in more detail in Subsects. 3.1 and 3.2 respectively.

In [12], the authors examined the critical region of the problem domain for 3-SAT, i.e., instances with $\frac{m}{n} \approx 4.2$. They observed that the clause-to-variable ratio has a great impact on the solution quality even on the quantum annealers.

Quantum annealing has previously been regarded as a solution to satisfiability problems: [2] focuses on embedding an originally SAT-related QUBO into the architecture of the most common quantum annealing chip. [20] shows a method to derive formulation for the optimization energy and proves mathematical bounds for the mapping of general k-SAT problems. Similarly, [14] shows an approach justifying feasibility but provides no empirical data. In [15], Grover's search algorithm was used to solve k-SAT. In [21], a QUBO formulation for k-SAT is proposed which only scales logarithmically in k compared to the linear scaling in [6] and [5]. In [22], a method is proposed to not hard-code a QUBO to SAT translation but to learn it using gradient-based methods.

3.1 Chancellor^{n+m}

Let $a_i^{(l)}$ be the i-th literal of clause $a^{(l)}$. The idea in [5] is to present a QUBO formulation for an arbitrary clause that assigns the energy g to the *one* variable assignment which does not fulfil the clause and the energy 0 to all other possible variable assignments. The energy spectrum is therefore given by:

$$Spec(\{a^{(l)}\}) = \begin{cases} g & a_i^{(l)} = 0 \ \forall i, \\ 0 & otherwise. \end{cases}$$

Thus we can create the QUBO formulation for the whole 3-SAT formula by superimposing all clause-formulations: $H = \sum_l Spec(\{a^{(l)}\})$. For $g > 0$ the minimum energy bit-string will always be the one which satisfies the most clauses.

To move from logical values to spin variables, one can map each logical variable $a_i = 0$ to a spin variable with value $\sigma_i^z = -1$ and each logical variable $a_i = 1$ to a spin variable with value $\sigma_i^z = +1$. Negation of the logical variable is then implemented through gauges on the spin variables. More precisely, a_i is mapped to $c(i)\sigma_i^z$ with $c(i) = 1$ and $\neg a_i$ to $c(i)\sigma_i^z$ with $c(i) = -1$.

The authors subsequently present the clause-formulation in the following way: The energy spectrum of the clause $(a_1 \vee a_2 \vee a_3)$ can be rewritten as $a_1 + a_2 + a_3 - a_1a_2 - a_1a_3 - a_2a_3 + a_1a_2a_3$ (with $a_i \in \{-1, +1\}$). The terms $a_1, a_2, a_3, a_1a_2, a_1a_3$ and a_2a_3 can be directly inserted into the Ising Hamiltonian. For the triple term $a_1a_2a_3$, however, an ancilla qubit is necessary. The authors then present an Ising Hamiltonian for the triple term:

$$H = h\sum_{i=1}^{3} c(i)\sigma_i^z + J^a \sum_{i=1}^{3} c(i)\sigma_i^z\sigma_a^z + h^a\sigma_a^z$$

in which up to the gauge choice $c(i) \in \{-1, 1\}$ the 3 variables σ_i^z are coupled with equal strength J^a to the same ancilla spin variable σ_a^z.

There are some constraints for the choice of the hyperparameters h, J^a, h^a, and J. We chose $h = g = 1, h^a = 2h = 2, J = 5$ and thus $J^a = 2J = 10$ as values for the variables in the "specials cases" section of Chancellor et al. [5]. It is important to note that the choice of these values has no influence on the number of couplings needed. Any clause-translation will produce a fully-connected Ising/QUBO matrix (Note that Ising and QUBO are isomorphic).

For each clause exactly one ancilla qubit C_i is needed. Thus the whole QUBO matrix will have size $n+m$. The 3-SAT formula $(\neg a \vee \neg b \vee \neg c) \wedge (a \vee b \vee c)$ would, for example, be represented by the QUBO matrix in Table 1.

Table 1. QUBO matrix using CHANCELLOR^{n+m} for the 3-SAT formula $(\neg a \vee \neg b \vee \neg c) \wedge (a \vee b \vee c)$.

	a	b	c	C_1	C_2
a	−88	48	48	40	40
b		−88	48	40	40
c			−88	40	40
C_1				−56	0
C_2					−64

3.2 Choi^{3m}

Choi [6] provides a translation of 3-SAT to QUBO that takes up $3m$ qubits, i.e., three qubits per clause in the original 3-SAT formula (or one qubit per literal). It is inspired by the maximum independent set problem (to which 3-SAT is first reduced, then to QUBO). Given a 3-SAT instance with m clauses and n variables, CHOI^{3m} reserves a qubit $x_{k,i}, 0 \leq k < m, 0 \leq i \leq 2$, for every literal. Thus CHOI^{3m} needs $3m$ qubits in total. One can interpret a solution candidate x for this QUBO formulation in the following way:

- If $x_{k,i} = 1$ and the corresponding literal $l_{3k+i} = v$ for some variable v, then we add the assignment $(v \mapsto \top)$ to the solution candidate for 3-SAT.
- If $x_{k,i} = 1$ and the corresponding literal $l_{3k+i} = \neg v$ for some variable v, then we add the assignment $(v \mapsto \bot)$ to the solution candidate for 3-SAT.
- If $x_{k,i} = 0$, then we do nothing.

Note that a solution candidate for the QUBO may thus be illegitimate from the 3-SAT perspective when it assigns different truth values to the same variable. Further note that CHOI^{3m}, even when returning the perfectly optimal solution, does not necessarily assign a truth value to every variable that occurs in the original formula.

For the detailed algorithm, we refer to [6] and will instead provide a small example. Note that the incentive and penalty values X, Y, Z can be chosen rather freely as long as $Y > 2|X|$ and $Z > 2|X|$. Given the example 3-SAT instance $(a \vee b \vee c) \wedge (a \vee b \vee \neg c)$, we can then write a QUBO matrix as follows:

$(a \vee b \vee c) \wedge (a \vee b \vee \neg c)$

Q	$x_{0,0}$	$x_{0,1}$	$x_{0,2}$	$x_{1,0}$	$x_{1,1}$	$x_{1,2}$
$x_{0,0}$	$-X$	Y	Y			
$x_{0,1}$		$-X$	Y			
$x_{0,2}$			$-X$			Z
$x_{1,0}$				$-X$	Y	Y
$x_{1,1}$					$-X$	Y
$x_{1,2}$						$-X$

Intuitively, we need to penalize setting a pair of qubits from the same clause (Y) and penalize setting a pair of qubits which correspond to contradicting literals of the same variable (Z). Since so far we only assigned penalties, we need to set negative energy values on the diagonal ($-X$) in order to incentivize setting any qubits at all (and avoid the trivial solution $\mathbf{x} = \mathbf{0}$).

4 Approaches

We now describe two new approaches for translating a given 3-SAT instance to QUBO. We introduce a new approach NÜSSLEIN^{2n+m} in Sect. 4.1, which uses $2n+m$ logical qubits, where n is the number of variables and m is the number of clauses. In Sect. 4.2 we then propose another formulation NÜSSLEIN^{n+m}, which requires $n+m$ qubits. This is on par with the state-of-the-art CHANCELLOR^{n+m}; however, NÜSSLEIN^{n+m} uses fewer couplings, which leads to a reduction of physical qubits.

4.1 A $2n + m$ Approach

We now introduce a novel approach for the translation of 3-SAT to QUBO: NÜSSLEIN^{2n+m}. Like CHOI^{3m} and CHANCELLOR^{n+m}, NÜSSLEIN^{2n+m} actually solves MAX-3-SAT by trying to accumulate as many solvable clauses as possible. We build on the idea of [21] to use an algorithm to describe the QUBO translation instead of an arithmetic notation.

We use the qubits in the following way:

- For each variable $v_j, 0 \leq j \leq n-1$, occurring in the 3-SAT instance, we use two qubits to encode if the variable is to be assigned $\top$ or if the variable is to be assigned $\bot$. Thus, $(v_j \mapsto \top)$ occurs in the variable assignment if $x_{2j} = 1$. Likewise, $(v_j \mapsto \bot)$ occurs in the variable assignment if $x_{2j+1} = 1$. Note that assigning both $v_{2j} = v_{2j+1}$ the same value makes for an illegitimate 3-SAT solution candidate.
- Beyond those qubits, we further use one qubit for every clause in the 3-SAT instance.

Effectively, the approach then uses $2n + m$ qubits for a 3-SAT instance with n variables and m clauses. This may be less or more than the $3m$ qubits used in CHOI^{3m}; however, consider that difficult 3-SAT instances are categorized by $\frac{m}{n} \approx 4.2$ (cf. Sect. 2). Thus, for the 3-SAT instances which actually require extensive computations on classical computers, NÜSSLEIN^{2n+m} manages to generate substantially smaller matrices. For the detailed instructions of Algorithm 1, we first need to introduce the following definitions:

- We write $L = (v_0, \neg v_0, ..., v_{n-1}, \neg v_{n-1})$ for the list containing all possible literals given variables $(v_j)_{0 \leq j \leq n-1}$. Note that $|L| = 2n$.
- We write $v_j \in c_k$ when clause c_k contains a literal of the form v_j. Likewise, we write $\neg v_j \in c_k$ when c_k contains a literal of the form $\neg v_j$. We subsequently write $L_i \in c_k$ when c_k contains the literal L_i.
- We define
$$R(L_i) = \sum_{k=0}^{m-1} \begin{cases} 1 & \text{if } L_i \in c_k, \\ 0 & \text{otherwise.} \end{cases}$$
Thus $R(L_i)$ is counting how often the literal L_i occurs in the formula.
- We define
$$R(L_i, L_{i'}) = \sum_{k=0}^{m-1} \begin{cases} 1 & \text{if } L_i \in c_k \text{ and } L_{i'} \in c_k, \\ 0 & \text{otherwise.} \end{cases}$$
Thus $R(L_i, L_{i'})$ is the number of occurrences of the literals L_i and $L_{i'}$ together in the same clause.

Intuitively, NÜSSLEIN^{2n+m} (cf. Algorithm 1) encodes how many clauses are fulfilled by the solution. For example, if the minimal energy H^* of a given NÜSSLEIN^{2n+m}-QUBO is -20 this means that 20 clauses are fulfilled. If the formula, however, has more than 20 clauses this means that the formula is not

Algorithm 1. NÜSSLEIN^{2n+m}

procedure NÜSSLEIN^{2n+m}
  $Q = \mathbf{0} \in \mathbb{R}^{2n+m \times 2n+m}$
  for $i := 0$ **to** $2n + m$ **do**
    for $j := i$ **to** $2n + m$ **do**
      if $i = j$ **and** $j < 2n$ **then**
        $Q_{ij} := -R(L_i)$
      else if $i = j$ **and** $j \geq 2n$ **then**
        $Q_{ij} := 2$
      else if $j < 2n$ **and** $j - i = 1$ **and** $i \mod 2 = 0$ **then**
        $Q_{ij} := m + 1$
      else if $i < 2n$ **and** $j < 2n$ **then**
        $Q_{ij} := R(L_i, L_j)$
      else if $j \geq 2n$ **and** $i < 2n$ **and** l_i **in** c_{j-2n} **then**
        $Q_{ij} = -1$
      end if
    end for
  end for
  return Q
end procedure

satisfiable. We can consider the example formula $(a \vee b \vee \neg c) \wedge (a \vee \neg b \vee \neg c)$ and its translation to QUBO using NÜSSLEIN^{2n+m}:

$$(a \vee b \vee \neg c) \wedge (a \vee \neg b \vee \neg c)$$

Q	a	$\neg a$	b	$\neg b$	c	$\neg c$	$(a \vee b \vee \neg c)$	$(a \vee \neg b \vee \neg c)$
a	−2	3	1	1	0	2	−1	−1
$\neg a$		0	0	0	0	0	0	0
b			−1	3	0	1	−1	0
$\neg b$				−1	0	1	0	−1
c					0	3	0	0
$\neg c$						−2	−1	−1
$(a \vee b \vee \neg c)$							2	0
$(a \vee \neg b \vee \neg c)$								2

4.2 An $n + m$ Approach

In this section we present NÜSSLEIN^{n+m}, which is a 3-SAT (again actually MAX-3-SAT) to QUBO translation, which only requires $n + m$ logical qubits. This is on par with CHANCELLOR^{n+m}. However, we will show that our approach requires fewer couplings, which leads to a reduction of needed physical qubits in the hardware embedding.

We use the qubits in the following way:

- For each variable $v_j, 0 \leq j \leq n-1$, occurring in the 3-SAT instance, we use one qubit to encode the value it is assigned. Thus, $(v_j \mapsto \top)$ occurs in the variable assignment iff $x_j = 1$. This implies that $(v_j \mapsto \bot)$ occurs in the variable assignment iff $x_j = 0$.
- Beyond those qubits, we again use one qubit for every clause in the 3-SAT instance.

For the algorithm, we start with an empty QUBO matrix as a canvas and then add specific patterns of values for each clause. As these pattern stack, we acquire the final value of Q_{ij} as a sum of all stacked values. The algorithm thus needs to iterate over all clauses and repeatedly update the QUBO matrix while doing so. As we need to look at each clause individually, we can assume without loss of generality that all clauses are sorted, i.e., all negated literals appear as far towards the back of the clause as possible. This leaves us with only four possible patterns for clauses:

$$(a \vee b \vee c), (a \vee b \vee \neg c), (a \vee \neg b \neg c), (\neg a \vee \neg b \vee \neg c)$$

We now want to arrange the energy levels for each of the four cases such that a satisfied clause (no matter in which way it was satisfied, i.e., with one literal, with two, or with three) has the energy H^* and the *one* state which does not satisfy the clause has the energy $H^+ = H^* + 1$. See Table 2 for all pattern matrices that might occur. The final QUBO matrix is then constructed by adding the pattern matrices' values to the cells in the QUBO matrix that correspond to the involved variables. For a 3-SAT formula with p clauses where there are no negated literals and q clauses where there are only negated literals, a variable assignment that satisfies the entire formula has the energy $H^* = -p - q$.

We can now consider the example formula $(a \vee b \vee c) \wedge (a \vee \neg b \vee \neg c)$ and its translation to QUBO using NÜSSLEIN^{n+m}:

$$(a \vee b \vee c) \wedge (a \vee \neg b \vee \neg c)$$

Q	a	b	c	$(a \vee b)$	$(a \vee \neg b)$
a	$0+2$	$2-2$	$0+0$	-2	-2
b		$0+0$	$0+0$	-2	2
c			$-1+1$	1	-1
$(a \vee b)$				1	0
$(a \vee \neg b)$					0

Table 2. Pattern matrices for the four different types of clauses.

$(a \vee b \vee c)$, $H^* = -1$

	a	b	c	$(a \vee b)$
a		2		-2
b				-2
c			-1	1
$(a \vee b)$				1

(b) $(a \vee b \vee \neg c)$, $H^* = 0$

	a	b	c	$(a \vee b)$
a		2		-2
b				-2
c			1	-1
$(a \vee b)$				2

(c) $(a \vee \neg b \vee \neg c)$, $H^* = 0$

	a	b	c	$(a \vee \neg b)$
a	2	-2		-2
b				2
c			1	-1
$(a \vee \neg b)$				

(d) $(\neg a \vee \neg b \vee \neg c)$, $H^* = 1$

	a	b	c	$(\neg a \wedge \neg b \wedge \neg c)$
a	-1	1	1	1
b		-1	1	1
c			-1	1
$(\neg a \wedge \neg b \wedge \neg c)$				-1

A possible optimal solution to this QUBO would be $\mathbf{x} = \langle 1, 0, 0, 1, 1 \rangle$, which corresponds to the variable assignment: $(a \mapsto \top, b \mapsto \bot, c \mapsto \bot)$ with the energy $H(\mathbf{x}) = -1$. Note that this QUBO matrix uses 5 logical qubits and 6 couplings (non-zero weights in the QUBO matrix). We can compare that to the CHANCELLOR^{n+m} formulation, which requires 5 logical qubits as well but 9 couplings:

$(a \vee b \vee c) \wedge (a \vee \neg b \vee \neg c)$

Q	a	b	c	C_1	C_2
a	-88	40	40	40	40
b		-88	48	40	40
c			-88	40	40
C_1				-64	0
C_2					-64

Another notable feature of NÜSSLEIN^{n+m} is the possibility to use the same clause-qubit for more than one clause. For example in the formula $(a \vee b \vee c) \wedge (a \vee b \vee \neg c)$ the logical sub-formula $(a \vee b)$ appears in both clauses thus we just need one clause-qubit instead of two. In total, we thus would need 4 logical qubits instead of 5.

5 Empirical Evaluation

To empirically verify that NÜSSLEIN^{n+m} requires fewer couplings than CHANCELLOR^{n+m} we created random 3-SAT formulas, applied both approaches, and counted the number of non-zero elements in the corresponding QUBO matrices. The results are shown in Fig. 1. The x-axis describes the number of variables V in the 3-SAT formula. We then created random formulas with $\lceil 4.2V \rceil$ clauses. As can be seen in the charts, for both approaches the number of non-zero couplings in the QUBO matrices scales linearly in the number of variables V of the 3-SAT formula. However, NÜSSLEIN^{n+m} only requires roughly 0.7 of the couplings that CHANCELLOR^{n+m} needs.

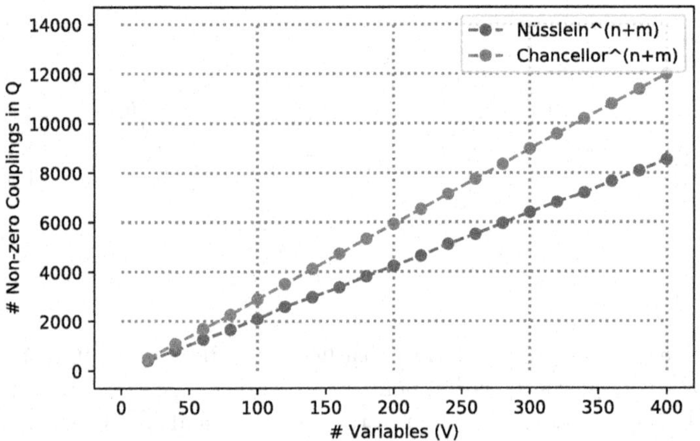

Fig. 1. Relation of the number of variables in the 3-SAT formula to the number of non-zero couplings in the QUBO matrix for the approaches CHANCELLOR^{n+m} and NÜSSLEIN^{n+m}.

In the next experiment, we evaluated how this reduction of couplings translates to a reduction of physical qubits. Note that both approaches NÜSSLEIN^{n+m} and CHANCELLOR^{n+m} require $n+m$ *logical* qubits. However, to run a QUBO on a quantum annealer the QUBO has to be embedded into the hardware graph, which currently follows the Pegasus graph design [3]. We again created random 3-SAT formulas for different V, applied both approaches to create the corresponding QUBO matrices and then ran the minorminer to find an embedding [4]. Finally, we counted how many *physical* qubits were needed. The results (Fig. 2) show that for both approaches the number of physical qubits scales linearly with V but the chart of CHANCELLOR^{n+m} has again a bigger gradient than the chart of NÜSSLEIN^{n+m}. The line represents the median of 20 formulas and the shaded areas enclose the 0.25 and 0.75 quantiles.

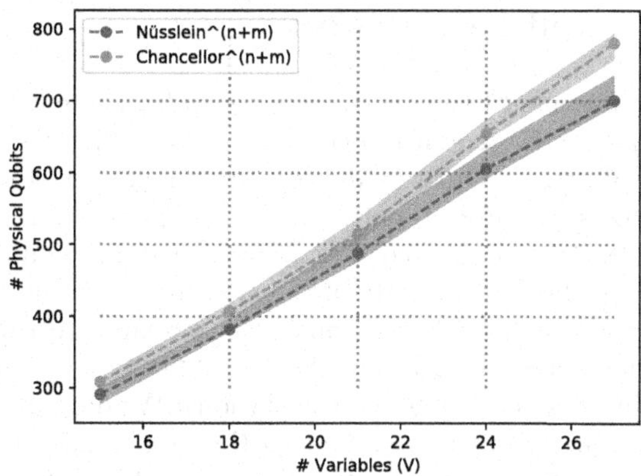

Fig. 2. Relation of the number of variables in the 3-SAT formula to the number of needed physical qubits for the approaches CHANCELLOR^{n+m} and NÜSSLEIN^{n+m}. The shaded areas enclose the 0.25 and 0.75 quantiles.

In a final experiment, we created random 3-SAT formulas and solved them with all four methods on the D-Wave Quantum Annealer. We tested three sizes for the 3-SAT formula and for each we created 20 random formulas. Table 3 shows the mean number of fulfilled clauses with the best-found variable assignment. For example for the size $(V = 5, C = 21)$ we created a random formula and solved it using NÜSSLEIN^{2n+m} on the D-Wave. For the best answer of the D-Wave, we calculated the variable assignment and how many clauses are fulfilled with this assignment. We repeated this procedure for 20 3-SAT formulas. As can be seen, NÜSSLEIN^{n+m} was the best approach for every size of the formula. Another very interesting result is that CHOI^{3m} was mostly better than NÜSSLEIN^{2n+m} and NÜSSLEIN^{2n+m} was mostly better than CHANCELLOR^{n+m} which indicates that the size of the QUBO matrix is not an optimal predictor for performance. The code for all four approaches can be found here: https://github.com/JonasNuesslein/3SAT-with-QUBO.

Table 3. Performance of four 3-SAT to QUBO translations on random formulas. The values represent the mean number of fulfilled clauses of the best-found solution vector.

	$(V = 5, C = 21)$	$(V = 10, C = 42)$	$(V = 12, C = 50)$
NÜSSLEIN^{2n+m}	20.4	39.0	45.0
NÜSSLEIN^{n+m}	**20.6**	**41.2**	**49.0**
CHANCELLOR^{n+m}	18.6	37.8	47.0
CHOI^{3m}	20.0	39.8	47.2

6 Conclusion and Future Work

In this paper, we presented two new approaches to translate 3-SAT instances to QUBO. Despite the smaller size of the QUBO, the first approach NÜSSLEIN^{2n+m} showed worse results than CHOI^{3m} in the experiments, which indicates that the size of a QUBO is not an optimal predictor for performance. For the other approach NÜSSLEIN^{n+m}, we showed that it requires fewer couplings and fewer physical qubits than the current state-of-the-art CHANCELLOR^{n+m}. We empirically verified that NÜSSLEIN^{n+m} performs best compared to three other 3-SAT to QUBO translations. The structure of the NÜSSLEIN^{n+m} approach also shows a new paradigm in constructing QUBO translations: We did not derive a formulation from the original problem by adapting the mathematical framework; the QUBO matrix of NÜSSLEIN^{n+m} was instead constructed from the ground up with the sole goal of mirroring 3-SAT's global optimum. We hope that NÜSSLEIN^{n+m} can thus also inspire more new QUBO translations in the future.

Regarding 3-SAT, it needs to be further investigated whether in general or for special cases even more favorable QUBO formulations for 3-SAT exist. This investigation could also be formulated as an optimization problem (more precisely as Integer Linear Program), where all solutions of the 3-SAT together with the energetically most favorable choice of auxiliary qubits must have the energy H^* and all non-solutions together with the energetically most favourable choice of auxiliary qubits must have an energy $H^+ > H^*$. However, the choice of the "most energetically favorable auxiliary qubits can be formulated by a series of linear inequalities (all other choices of auxiliary qubits with the same variable assignment must have a greater or equal energy). Following this approach, we may even be able to automatically generate new and efficient QUBO translations for practically relevant problems.

References

1. Arute, F., et al.: Quantum supremacy using a programmable superconducting processor. Nature **574**(7779), 505–510 (2019)
2. Bian, Z., Chudak, F., Macready, W., Roy, A., Sebastiani, R., Varotti, S.: Solving SAT and MaxSAT with a quantum annealer: foundations and a preliminary report. In: Dixon, C., Finger, M. (eds.) FroCoS 2017. LNCS (LNAI), vol. 10483, pp. 153–171. Springer, Cham (2017). https://doi.org/10.1007/978-3-319-66167-4_9
3. Boothby, K., Bunyk, P., Raymond, J., Roy, A.: Next-generation topology of D-Wave quantum processors. arXiv preprint arXiv:2003.00133 (2020)
4. Cai, J., Macready, W.G., Roy, A.: A practical heuristic for finding graph minors. arXiv preprint arXiv:1406.2741 (2014)
5. Chancellor, N., Zohren, S., Warburton, P.A., Benjamin, S.C., Roberts, S.: A direct mapping of max k-sat and high order parity checks to a chimera graph. Sci. Rep. **6**(1), 1–9 (2016)
6. Choi, V.: Adiabatic quantum algorithms for the NP-complete maximum-weight independent set, exact cover and 3SAT problems. arXiv preprint arXiv:1004.2226 (2010)

7. Cook, S.: The P versus NP problem. In: The Millennium Prize Problems, pp. 87–104 (2006)
8. Cook, S.A.: The complexity of theorem-proving procedures. In: Proceedings of the 3rd Annual ACM Symposium on Theory of Computing. ACM (1971)
9. Farhi, E., Goldstone, J., Gutmann, S.: A quantum approximate optimization algorithm. arXiv preprint arXiv:1411.4028 (2014)
10. Feld, S., et al.: A hybrid solution method for the capacitated vehicle routing problem using a quantum annealer. arXiv preprint arXiv:1811.07403 (2018)
11. Fortnow, L.: The status of the P versus NP problem. Commun. ACM **52**(9), 78–86 (2009)
12. Gabor, T., et al.: Assessing solution quality of 3SAT on a quantum annealing platform. In: Feld, S., Linnhoff-Popien, C. (eds.) QTOP 2019. LNCS, vol. 11413, pp. 23–35. Springer, Cham (2019). https://doi.org/10.1007/978-3-030-14082-3_3
13. Glover, F., Kochenberger, G., Du, Y.: A tutorial on formulating and using QUBO models. arXiv preprint arXiv:1811.11538 (2018)
14. Hen, I., Spedalieri, F.M.: Quantum annealing for constrained optimization. Phys. Rev. Appl. **5**(3), 034007 (2016)
15. Hogg, T.: Adiabatic quantum computing for random satisfiability problems. Phys. Rev. A **67**(2), 022314 (2003)
16. Johnson, M.W., et al.: Quantum annealing with manufactured spins. Nature **473**(7346), 194–198 (2011)
17. Kadowaki, T., Nishimori, H.: Quantum annealing in the transverse Ising model. Phys. Rev. E **58**(5), 5355 (1998)
18. Lucas, A.: Ising formulations of many np problems. Front. Phys. **2**, 5 (2014)
19. McGeoch, C.C., Wang, C.: Experimental evaluation of an adiabiatic quantum system for combinatorial optimization. In: Proceedings of the ACM International Conference on Computing Frontiers, pp. 1–11 (2013)
20. Mooney, G.J., Tonetto, S.U., Hill, C.D., Hollenberg, L.C.: Mapping NP-hard problems to restricted adiabatic quantum architectures. arXiv preprint arXiv:1911.00249 (2019)
21. Nüßlein, J., Gabor, T., Linnhoff-Popien, C., Feld, S.: Algorithmic QUBO formulations for K-SAT and Hamiltonian cycles. arXiv preprint arXiv:2204.13539 (2022)
22. Nüßlein, J., Roch, C., Gabor, T., Linnhoff-Popien, C., Feld, S.: Black box optimization using QUBO and the cross entropy method. arXiv preprint arXiv:2206.12510 (2022)
23. Venturelli, D., Marchand, D.J., Rojo, G.: Quantum annealing implementation of job-shop scheduling. arXiv preprint arXiv:1506.08479 (2015)
24. Zahedinejad, E., Zaribafiyan, A.: Combinatorial optimization on gate model quantum computers: a survey. arXiv preprint arXiv:1708.05294 (2017)

Black Box Optimization Using QUBO and the Cross Entropy Method

Jonas Nüßlein[1(✉)], Christoph Roch[1], Thomas Gabor[1], Jonas Stein[1], Claudia Linnhoff-Popien[1], and Sebastian Feld[2]

[1] Institute of Computer Science, LMU Munich, Munich, Germany
jonas.nuesslein@ifi.lmu.de

[2] Faculty of Electrical Engineering, Mathematics and Computer Science, TU Delft, Delft, Netherlands

Abstract. Black-box optimization (BBO) can be used to optimize functions whose analytic form is unknown. A common approach to realising BBO is to learn a surrogate model which approximates the target black-box function which can then be solved via white-box optimization methods. In this paper, we present our approach *BOX-QUBO*, where the surrogate model is a QUBO matrix. However, unlike in previous state-of-the-art approaches, this matrix is not trained entirely by regression, but mostly by classification between "good" and "bad" solutions. This better accounts for the low capacity of the QUBO matrix, resulting in significantly better solutions overall. We tested our approach against the state-of-the-art on four domains and in all of them *BOX-QUBO* showed better results. A second contribution of this paper is the idea to also solve white-box problems, i.e. problems which could be directly formulated as QUBO, by means of black-box optimization in order to reduce the size of the QUBOs to the information-theoretic minimum. Experiments show that this significantly improves the results for MAX-k-SAT.

Keywords: QUBO · Black-Box · Quantum Annealing · SAT · Ising

1 Introduction

The goal of black-box optimization is to minimize a function $E(x)$, where this function is not known analytically. Like an oracle, this function can only be queried for a given x: $y = E(x)$. A commonly used solution for this problem is to create a surrogate model $E'(x)$. This surrogate model is trained to provide the same outputs as $E(x)$. Since, unlike $E(x)$, $E'(x)$ is a white-box model, it is easier to search for the best solution $x^* = argmin_x\ E'(x)$. Black-box optimization then iterates between searching the best solution x^* for the surrogate model $E'(x)$, asking the oracle for the actual value $y = E(x^*)$ and re-training the surrogate model E' to output y for the input x^* using the mean squared error as the loss function: $loss = (E'(x^*) - y)^2$. We use the terms "oracle" and "black-box function" interchangeably in this paper.

J. Mikyška et al. (Eds.): ICCS 2023, LNCS 14077, pp. 48–55, 2023.
https://doi.org/10.1007/978-3-031-36030-5_4

Following the tradition of machine learning, we use the term 'capacity' to refer to the amount of information a model can memorize. In black-box optimization, there is a trade-off between the capacity of the surrogate model (the higher the capacity of surrogate model $E'(x)$, the better it can approximate the actual values y) and the difficulty of the optimization (the higher the capacity of surrogate model $E'(x)$, the more difficult the optimization $x^* = argmin_x\ E'(x)$). A frequent choice for the model $E'(x)$ recently fell [2,13] on the Quadratic Unconstrained Binary Optimization (QUBO) matrix Q [4], which seems like a good trade-off for the capacity.

2 Background

2.1 MAX-SAT

Satisfiability (SAT) is a canonical NP-complete problem [11]. Let $V = \{v_1, ..., v_n\}$ be a set of Boolean variables and let f be a Boolean formula represented in conjunctive normal form: $f = \bigwedge_{i=1}^{|f|} \bigvee_{l \in C_i} l$; with l being a literal and C_i being the i-th clause. The task is to find an assignment $\underline{V}$ for V such that $f(\underline{V}) = 1$. MAX-SAT is a variant of SAT, where not all clauses have to be satisfied, but only as many as possible. In k-SAT, each clause consists of exactly k literals.

2.2 Feedback Vertex Set (FVS)

Feedback Vertex Set is an NP-complete problem [11]. Let $G = (V, E)$ be a directed graph with vertex set V and edge set E which contains cycles. A cycle is a path (following the edges of E) in the graph starting from any vertex v_S such that one ends up back at vertex v_S. The path can contain any number of edges. The task in Feedback Vertex Set is to select the smallest subset $V' \subset V$ such that graph $G = (V \backslash V', E')$ is cycle-free, where $E' = \{(v_i, v_j) \in E : v_i \notin V' \wedge v_j \notin V'\}$.

2.3 MaxClique

Maximum Clique is another NP-complete problem [11]. Let $G = (V, E)$ be an undirected graph with vertex set V and edge set E. A clique in graph G is a subset $V' \subset V$ of the vertices, so all pairs of vertices from V' are connected with an edge. That is, $\forall (v_i, v_j) \in V' \times V' : (v_i, v_j) \in E$. The task in MaxClique is to find the largest clique of the graph. The size of a clique is the cardinality of the set V': $|V'|$.

2.4 Quadratic Unconstrained Binary Optimization (QUBO)

Let Q be an upper-triangular $(\mathrm{n} \times \mathrm{n})$-matrix Q and x be a binary vector of length n. The task of Quadratic Unconstrained Binary Optimization [16] is to solve the following optimization problem: $x^* = argmin_x(x^T Q x)$. QUBO is NP-hard [3] and has been of particular interest recently as special machines, such as the Quantum Annealer [4,8] or the Digital Annealer [1], have been developed to solve these problems. Therefore, in order to solve other problems with these special machines, they must be formulated as QUBO. Numerous problems such as MAX-k-SAT [5,6,19,20], MaxClique [17,18], and FVS [18] have already been formulated as QUBO.

2.5 Cross-Entropy Method

The cross-entropy method [7] is a general iterative optimization method to optimize an oracle $g(x)$ using a parameterized probability distribution $f(x; v)$. The method iterates between sampling p solutions: $X \sim f(x; v)$, sorting them by their values $g(X)$ in increasing order and adjusting the parameters v such that better solutions x get a higher probability $f(x; v)$. This makes it more likely to sample better solutions in the next iteration, starting again with sampling from the probability distribution: $X \sim f(x; v)$, with the new values v.

3 Related Work

The main Related Work to our approach is Factorization Machine Quantum Annealing (*FMQA*) [10,13]. *FMQA*, like our approach, is based on an iteration of three phases: 1) sample from the surrogate model (a QUBO matrix); 2) retrieve the value from the oracle (either via experiments or simulation); 3) update the QUBO matrix. This is in fact a cross-entropy approach with f being the surrogate model (the QUBO matrix). The main difference to our approach is that *FMQA* solely uses regression to train the model, while we use simultaneous regression and classification. The authors subsequently use *FMQA* in their paper to design metamaterials with special properties. A similar approach to *FMQA* is Bayesian Optimization of Combinatorial Structures (BOCS) [2] which also uses a quadratic model as a surrogate model. BOCS additionally uses a sparse prior to facilitating optimization. However, BOCS also only uses regression to optimize the model. In the original paper [2], BOCS was solved using only non-quantum methods (including Simulated Annealing [12]), Koshikawa et al. [14], however, also tested BOCS for optimization with a quantum annealer. In [15] Koshikawa et al. used BOCS for a vehicle design problem and found that it performed slightly better than a random search. In our experiments, we use the white box QUBO formulations for MAX-k-SAT [6], FVS [18] and MaxClique [18] as baselines. In [9], surrogate QUBO solvers were trained to simplify the optimization of hyperparameters arising in the relaxation of constrained optimization problems. Roch et al.

4 Black Box Optimization with Cross Entropy and QUBO (BOX-QUBO)

Main idea: The QUBO matrix is a surrogate model with relatively low capacity since it has only $n(n+1)/2$ trainable parameters for a matrix of size $(n \times$

n). Previous approaches to black-box optimization with QUBO always attempt full regression on all training data. However, the loss becomes larger the more training data is available for the same size of the QUBO matrix due to the limited number of trainable parameters. The main idea behind our approach is to perform a regression only on a small part of the training data and classification on the remaining data since classification requires less capacity of the model than regression. For this purpose, the training vectors x are sorted according to their solution quality (also called energy) $E(x)$ and divided into two sets: the set of vectors with higher energies H and the set of vectors with lower energies L. There is now a threshold τ such that:

$$\forall x \in H : E(x) \geq \tau$$
$$\forall x \in L : E(x) < \tau$$

Note that the goal is to find vector x^* with minimum energy: $\forall x : E(x) \geq E(x^*)$. The goal of the classification is to classify whether a vector x is in set H or L. The regression is performed exclusively on L. The size of L is determined by a hyperparameter k. For example, using $k = 0.03$, L contains 3% of all data: $|L| = 0.03 \cdot |D|$, thus a more precise regression is possible. *BOX-QUBO* requires as input the oracle (black-box function) $E(x)$ and an initial training set D. The output will be the vector x^* with minimum energy $y^* = E(x^*)$. First, the QUBO matrix Q is randomly initialized. After that, the following three phases alternate:

1. Search the p best solution vectors X^* for the current QUBO matrix Q (e.g. using simulated or quantum annealing)
2. Query the oracle for actual values $E(X^*)$
3. Append $(X^*, E(X^*))$ to the training set D and retrain Q on D

The training of Q using the training set D proceeds as follows. First, D is sorted by the energies $E(x)$ and then subdivided into the two non-overlapping sets H and L. After splitting, Q is trained for *nCycles* iterations. In each cycle, the *temporary* training set T is created dynamically: all vectors $x \in L$ are always part of T, vectors $x \in H$, however, are only part of T if their current prediction $y_{predict} = x^T Q x$ is smaller than τ and thus would violate the classification. The loss function on this temporary training set is now the mean squared error between $y_{predict} = x^T Q x$ and y_{target}. y_{target}, for a vector x, is equal to $E(x)$ if $x \in L$ and it is equal to τ if $x \in H$:

$$Q^* = \underset{Q}{\operatorname{argmin}} \left(\sum_{x \in T} (y_{predict} - y_{target}) \right)^2$$

We optimize Q using gradient descent with respect to this loss function. The complete algorithm is listed in Algorithm 1.

Algorithm 1: BOX-QUBO

```
Data: Set of initial training data D = {x, E(x)}; Oracle E(.)
Result: Best solution (x*, y*) of the oracle

Q ← init QUBO matrix
Q ← train(Q, D)

for trainingLength do
    X ← sample best k solutions from Q (e.g. using quantum annealing)
    Y = E(X)                                   // query oracle ∀x ∈ X
    D.add(X, Y)
    Q ← train(Q, D)
end

return (x*, y*) ∈ D          // Return best y and corresponding x

Function train(Q, D):
    H, L, τ ← sort and split D      // see chapter Splitting H and L
    for nCycles do
        T, Y_target = L, E(L)              // ensure regression ∀x ∈ L
        for x ∈ H do
            y_predict = x^T Qx          // ensure classification ∀x ∈ H
            if y_predict < τ then
                T ← x
                Y_target ← τ
            end
        end
        L(x, y_target) = (x^T Qx − y_target)^2          // the loss function
        optimize Q via gradient descent w.r.t. ∇_Q L(T, Y_target)
    end
    return Q
```

Splitting H and L: We divide the training data D into the 'good' solutions L and the 'bad' solutions H based on their energies $E(x)$. For this, we use a hyperparameter k (a percentage): *BOX-QUBO(k%)*. $k\%$ of the training data are inserted into the set L and the remaining data are inserted into H. In addition, we use the notation *BOX-QUBO(k%\invalids)* to state that all invalid solutions are inserted into the set H. To solve a problem (e.g., MaxClique) with QUBO, the problem must be formulated as QUBO and then the QUBO solution x must be translated back into a problem solution (for example, the set of vertices forming the clique). We call a QUBO solution x 'invalid' if it does not correspond to a valid problem solution (for example, if the QUBO solution x corresponds to a set of vertices V' which is not a clique).

5 Experiments

In the experiments, we tested whether the approaches *BOX-QUBO* and *FMQA* are able to solve the domains MAX-k-SAT, FVS and MaxClique when these domains are solely accessible as black-box functions. That is, the concrete Boolean formula (for MAX-k-SAT) or the concrete graph (for FVS and MaxClique) are unknown. The oracles return an indirectly proportional value to the solution quality and the value 1 if the QUBO solution x does not translate into a valid problem solution. The goal of *BOX-QUBO* and *FMQA* was then to minimize the black-box functions (oracles). We have chosen as oracles:

$$SAT_{oracle}(x) = -(\#\ satisfied\ clauses\ with\ variable\ assignment\ x)$$
$$FVS_{oracle}(x) = if\ G\backslash x\ still\ has\ cycles\ then\ 1\ else\ (\Sigma_i x_i - |V|)$$
$$MaxClique_{oracle}(x) = if\ x\ is\ a\ valid\ clique\ then\ (-\Sigma_i x_i)\ else\ 1$$

where G is a graph. In MAX-k-SAT, V is the number of boolean variables and C is the number of clauses. In FVS and MaxClique, V is the number of vertices of the graph and E is the number of edges. $(\Sigma_i x_i)$ calculates the number of 1s in the vector x, which represents in MaxClique the size of the clique and in FVS the number of deleted vertices. For each of the four domains, 15 random instances (with different seeds) were created. That is, for Max-4-SAT 15 random formulas were created and for FVS and MaxClique 15 different graphs were generated each. Then, the *BOX-QUBO* and *FMQA* approaches were applied to each of the 15 instances (i.e. on the corresponding oracle functions). Both algorithms used the same initial training set. The best solution from this initial set served as the baseline *base*. In addition, specialized white-box methods were used to determine the true optimal solution to each instance, which we refer to as *optimum*. To better compare performance between domains, we normalized the best solution found for each of the *BOX-QUBO* and *FMQA* approaches, using the formula: $f(y^*) = (y^* - base)/(optimum - base)$. This gives $f(y^*) = 1$ if $y^* = optimum$ and $f(y^*) = 0$ if $y^* = base$, where y^* is the value of the best solution (x^*) found so far. The results are shown in Fig. 1.

The first result is that black-box optimization (both *BOX-QUBO* and *FMQA*) found significantly better solutions than Choi's white-box solution for Max-4-*SAT*. Using the black-box optimization, we have reduced the size of the QUBO matrix to the information-theoretic minimum. For example, in the first experiment, we considered formulas with $V = 30$ variables and $C = 400$ clauses. Here, Choi's QUBO matrix had size $n = 4 \cdot 400 = 1600$, while the QUBO matrices for *BOX-QUBO* and *FMQA* corresponded to the information-theoretic minimum ($n = V = 30$). However, solving with black-box methods is not always better than solving with white-box methods, as can be seen in the results for *FVS*. The second result is that *BOX-QUBO* was always more successful than *FMQA*. In *FVS*, for example, *BOX-QUBO(10%\invalids)* reached a mean performance of roughly 0.87 after 15 iterations while *FMQA* only reached roughly 0.03. A similar picture emerged for MaxClique, with all *BOX-QUBO* variants

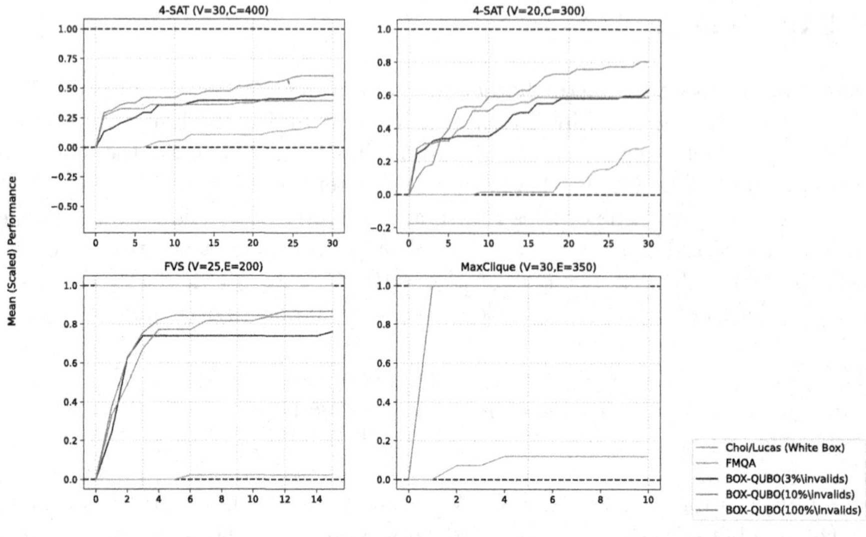

Fig. 1. Normalized performance of *BOX-QUBO*, *FMQA* [13] and the white-box solutions of Choi [6] and Lucas [18]. The x-axes describe the iteration number and the y-axes describe the (normalized) mean performance over 15 instances. In MaxClique, the graphs of *BOX-QUBO(100%\invalids)*, *BOX-QUBO(10%\invalids)* and *BOX-QUBO(3%\invalids)* overlap.

reaching the optimum already after the first iteration. The code for *BOX-QUBO* is available on GitHub [https://github.com/JonasNuesslein/BOX-QUBO].

6 Conclusion and Future Work

In this paper, we studied black-box optimization using QUBOs as surrogate models. We presented the *BOX-QUBO* approach, which is characterized by the idea to use a simultaneous classification and regression, instead of regression alone. We have tested our approach on the MAX-k-SAT, FVS, and $MaxClique$ domains, and the experiments showed that *BOX-QUBO* consistently outperformed *FMQA*. Besides introducing *BOX-QUBO*, we also presented the idea of solving white-box problems using black-box optimization to reduce the size of the QUBO matrices to the information-theoretic minimum. The experiments on MAX-4-SAT showed that the solutions found using black-box optimization were significantly better than those found using the white-box QUBO formulation.

Acknowledgements. This publication was created as part of the Q-Grid project (13N16179) under the "quantum technologies - from basic research to market" funding program, supported by the German Federal Ministry of Education and Research.

References

1. Aramon, M., Rosenberg, G., Valiante, E., Miyazawa, T., Tamura, H., Katzgraber, H.G.: Physics-inspired optimization for quadratic unconstrained problems using a digital annealer. Front. Phys. **7**, 48 (2019)
2. Baptista, R., Poloczek, M.: Bayesian optimization of combinatorial structures. In: International Conference on Machine Learning, pp. 462–471. PMLR (2018)
3. Barahona, F.: On the computational complexity of Ising spin glass models. J. Phys. A: Math. Gen. **15**(10), 3241 (1982)
4. Boothby, K., Bunyk, P., Raymond, J., Roy, A.: Technical report: next-generation topology of d-wave quantum processors (2019)
5. Chancellor, N., Zohren, S., Warburton, P.A., Benjamin, S.C., Roberts, S.: A direct mapping of max k-SAT and high order parity checks to a chimera graph. Sci. Rep. **6**, 37107 (2016). https://doi.org/10.1038/srep37107
6. Choi, V.: Adiabatic quantum algorithms for the NP-complete maximum-weight independent set, exact cover and 3SAT problems. arXiv preprint arXiv:1004.2226 (2010)
7. De Boer, P.T., Kroese, D.P., Mannor, S., Rubinstein, R.Y.: A tutorial on the cross-entropy method. Ann. Oper. Res. **134**(1), 19–67 (2005)
8. Denchev, V.S., et al.: What is the computational value of finite-range tunneling? Phys. Rev. X **6**(3), 031015 (2016)
9. Huang, T., Goh, S.T., Gopalakrishnan, S., Luo, T., Li, Q., Lau, H.C.: QROSS: QUBO relaxation parameter optimisation via learning solver surrogates. In: 2021 IEEE 41st International Conference on Distributed Computing Systems Workshops (ICDCSW), pp. 35–40. IEEE (2021)
10. Kadowaki, T., Nishimori, H.: Quantum annealing in the transverse Ising model. Phys. Rev. E **58**, 5355 (1998)
11. Karp, R.M.: Reducibility among combinatorial problems. In: Miller, R.E., Thatcher, J.W., Bohlinger, J.D. (eds.) Complexity of Computer Computations. The IBM Research Symposia Series. Springer, Boston (1972). https://doi.org/10.1007/978-1-4684-2001-2_9
12. Kirkpatrick, S., Gelatt, C., Vecchi, M.: Optimization by simulated annealing. Science **220**, 671–680 (2000)
13. Kitai, K., et al.: Designing metamaterials with quantum annealing and factorization machines. Phys. Rev. Res. **2**(1), 013319 (2020)
14. Koshikawa, A.S., Ohzeki, M., Kadowaki, T., Tanaka, K.: Benchmark test of black-box optimization using d-wave quantum annealer. J. Phys. Soc. Jpn. **90**(6), 064001 (2021)
15. Koshikawa, A.S., et al.: Combinatorial black-box optimization for vehicle design problem. arXiv preprint arXiv:2110.00226 (2021)
16. Lewis, M., Glover, F.: Quadratic unconstrained binary optimization problem preprocessing: theory and empirical analysis. Networks **70**(2), 79–97 (2017)
17. Lodewijks, B.: Mapping NP-hard and NP-complete optimisation problems to quadratic unconstrained binary optimisation problems. arXiv preprint arXiv:1911.08043 (2019)
18. Lucas, A.: Ising formulations of many NP problems. Front. Phys. **2**(5), 1–15 (2014)
19. Nüßlein, J., Gabor, T., Linnhoff-Popien, C., Feld, S.: Algorithmic QUBO formulations for k-SAT and Hamiltonian cycles. arXiv preprint arXiv:2204.13539 (2022)
20. Nüßlein, J., Roch, C., Gabor, T., Linnhoff-Popien, C., Feld, S.: Black box optimization using QUBO and the cross entropy method. arXiv preprint arXiv:2206.12510 (2022)

Sub-exponential ML Algorithm for Predicting Ground State Properties

Lauren Preston[1(✉)] and Shivashankar[2]

[1] Delta Blue, Oxford, USA
lpreston@deltabluelabs.com
[2] Delta Blue, Bengaluru, India

Abstract. Analysing properties of ground state of a quantum systems, is an important problem with applications in various domains. Recently, Huang et al. [2021] demonstrate how machine learning algorithms can be used to efficiently solve this problem with formal guarantees. However this method requires an exponential amount of data to train. In this work we show a method with improved efficiency for a wide class of energy operator. In particular, we show an ML-based method for predicting ground state properties for structured Hamiltonian with sub-exponential scaling in training data. The method relies on efficiently learning low-degree approximation of the energy operator.

1 Introduction

The challenge of predicting the ground state of a quantum system is an important one that has applications in a variety of fields, such as quantum machine learning [Arunachalam and de Wolf 2017, Biamonte et al. 2017, Schuld and Killoran 2019], variational quantum algorithms [Cerezo et al. 2021, Gibbs et al. 2022], experimental quantum physics [Carleo and Troyer 2017, Sharir et al. 2020] and quantum benchmarking [Scott 2008, Levy et al. 2021]. Huang et al. [2021] show that machine learning can be applied with a classical-shadow [Huang et al. 2020] based representation of quantum states, to effectively handle this problem. Nevertheless, this approach has a sample complexity that is exponential in the amount of training data. Specifically, their proposed algorithm has a sample complexity of $\mathcal{O}(n^{\frac{c}{\epsilon}})$ for a prediction error ϵ. As such, when the prediction error ϵ is small, a significant amount of training data is required to achieve that error. In this work, we propose a method that is more effective than previous approaches for a diverse range of energy operators. The method relies upon low-degree approximation to the energy. Our method uses a representation known as classical shadow, which is a condensed classical description of a many-body quantum state [Flammia and Preskill 2022]. This description can be created in quantum experiments and can be used to predict many of the attributes of the state. Our method improves over existing ML based algorithms for predicting ground state features in terms of sample complexity. Specifically, for structured Hamiltonians with sub-exponential scaling of training data.

Classical ML could be used to generalize from training data that are obtained from either quantum experiments or classical simulations; the same rigorous performance guarantees apply in either case. Even if the training data are generated classically, it

J. Mikyška et al. (Eds.): ICCS 2023, LNCS 14077, pp. 56–63, 2023.
https://doi.org/10.1007/978-3-031-36030-5_5

could be more efficient and more accurate to use ML to predict properties for new values of the input x, rather than doing new simulations which could be computationally very demanding and of unverified reliability. Promising insights into quantum many-body physics are already being obtained using classical ML based on classical simulation data [Deng et al. 2017, Nomura et al. 2017, Zhang et al. 2017, Vargas-Hernández et al. 2018, Schütt et al. 2019, Zhang et al. 2020, Kawai and Nakagawa, 2020].

2 Preliminaries and Related Work

2.1 Formulation

Consider an m-dimensional vector $x \in [-1, 1]^m$ that parameterizes an n-qubit gapped geometrically local Hamiltonian given as

$$H(x) = \sum_j h_j(\boldsymbol{x}_j), \tag{1}$$

$\boldsymbol{x}_1, \ldots, \boldsymbol{x}_L$ are $\mathcal{O}(1)$ sized vectors parameterizing the few-body interaction $h_j(\boldsymbol{x}_j)$. For example, x_j might be the coupling coefficients between a node and its neighbours in an Ising model. Let $\rho(x)$ be the ground state of $H(x)$ and O be a sum of geometrically local observables with $\|O\|_\infty \leq 1$.

The goal is to learn a function $h^*(x)$ that approximates the ground state property $\mathrm{tr}(O\rho(x))$,

$$(x_\ell, y_\ell), \quad \forall \ell = 1, \ldots, N, \tag{2}$$

where $y_\ell \approx \mathrm{tr}(O\rho(x_\ell))$ records the ground state property for $x_\ell \in [-1, 1]^m$ sampled from an arbitrary unknown distribution $\mathcal{D}$.

The setting considered in this work is very similar to that in Huang et al. [2021], but we assume the geometry of the n-qubit system to be known.

Adiabatic quantum computation [Farhi et al. 2000, Aharonov et al. 2008, Wan and Kim 2020], focuses on finding ground states of special Hamiltonian to perform computation. However, unlike these works we would not use any quantum memory, or explicit description of the operator O or any information about an adiabatic path to hamiltonian H.

We prove that given $\epsilon = \Theta(1)$, the improved ML algorithm can use a dataset size of

$$N = \mathcal{O}\left(\log\left(n\right)\right), \tag{3}$$

to learn a function $h^*(x)$ with an average prediction error of at most ϵ,

$$\mathop{\mathbb{E}}_{x \sim \mathcal{D}} \left|h^*(x) - \mathrm{tr}(O\rho(x))\right|^2 \leq \epsilon, \tag{4}$$

The correctness of the method relies upon results on optimizing k-local Hamiltonians Dinur et al. [2006], Barak et al. [2015], Harrow and Montanaro [2017], Anshu et al. [2021], Flammia and Preskill [2022]. The efficiency of the method is based upon recent results in quantum Bohnenblust-Hille inequalities Rouzé et al. [2022], Bohnenblust and Hille [1931].

2.2 Classical Shadows

Classical shadows are an efficient classical representations of quantum systems. The fundamental idea of this representations is similar to random projections. We use the terminology borrowed from earlier works van Enk and Beenakker [2012], Ohliger et al. [2013], Paini and Kalev [2019], Huang et al. [2020]. An n-qubit quantum state ρ can be approximated by performing randomized single-qubit Pauli measurements on T copies of ρ. If we measure every qubit of the state ρ in a random Pauli basis X, Y or Z, and collect the observations, and repeat the procedure T times, we are left with a set of measurement

$$S_T(\rho) = \{|s_i^{(t)}\rangle : i \in \{1, \ldots, n\},\ t \in \{1, \ldots, T\}\}$$

where each $|s_i^{(t)}\rangle \in \{|0\rangle, |1\rangle, |+\rangle, |-\rangle, |\mathrm{i}+\rangle, |\mathrm{i}-\rangle\}$ corresponds to an eigenstate of the corresponding Pauli operator.

Each element is a highly structured single-qubit pure state, and there are nT of them in total. So, $3nT$ bits suffice to store the entire collection in classical memory. The randomized measurements can be performed in actual physical experiments or through classical simulations. Resulting data can then be used to approximate the underlying n-qubit state ρ:

$$\rho \approx \sigma_T(\rho) = \frac{1}{T}\sum_{t=1}^{T} \sigma_1^{(t)} \otimes \cdots \otimes \sigma_n^{(t)} \quad \text{where} \quad \sigma_i^{(t)} = 3|s_i^{(t)}\rangle\langle s_i^{(t)}| - \mathbb{I}, \tag{5}$$

and $\mathbb{I}$ denotes the 2×2 identity matrix. This *classical shadow* representation Huang et al. [2020] asymptotically reproduces the global density matrix. By the Hoeffding-Chernoff bound, one can also show that with $\bar{T} = \mathcal{O}(\log(n)/\epsilon^2)$ one can get an ϵ-accurate approximation of the density matrix as well. This, implies that with $T > \bar{T}$ experiments, we can use $\sigma_T(\rho)$ to predict local functions (like expectation values).

As detailed next, this classical shadow representation is utilized by Huang et al. [2021] to build an ML algorithm for estimation of local properties of ground states ρ.

2.3 Predicting Ground States of Quantum Many-Body Systems

Huang et al. [2021] consider the task of predicting ground state properties for finite many-body systems. For this purpose they propose training an ML algorithm on a dataset collected from quantum experiments over a parametric family of Hamiltonians $H(x)$. Before the training of the ML algorithm, many Hamiltonians $H(x)$ are sampled, the classical shadow of the corresponding ground state $\rho(x)$ of $H(x)$ is obtained. The full training data of size N is given by $\{x_\ell \to \sigma_T(\rho(x_\ell))\}_{\ell=1}^{N}$, where T is the number of measurements in the construction of the classical shadows at each value of x_ℓ.

The ML models is trained on this size-N training data, such that when given the input x_ℓ, the ML predicts a vector representation $\hat{\sigma}(x)$ that approximates $\sigma_T(\rho(x_\ell))$.

In particular, they use a Nadaraya-Watson estimator [Nadaraya 1964, Watson 1964] (a version of nearest-neighbour regression) with a kernel function [Bierens 1988].

$$\hat{\sigma}(x) = \frac{1}{N}\sum_{\ell=1}^{N} \kappa(x, x_\ell)\sigma_T(\rho(x_\ell)). \tag{6}$$

where $\kappa(x, x_\ell)$ is a kernel function [Bierens 1988]. The ground state properties are then estimated using these predicted classical representations $\hat{\sigma}(x)$. Specifically, $f_O(x) = \mathrm{tr}\,(O\rho(x))$ can be predicted efficiently whenever O is a sum of few-body operators. They provide guarantees by using a truncated Fourier (also known as Dirichlet kernel) $\kappa(x, x_\ell) = \sum_{k\in\mathbb{Z}^m, \|k\|_2\leq\Lambda} \cos(\pi k \cdot (x - x_\ell))$ with cutoff Λ. Their method guarantees that $\mathbb{E}_x \mid \mathrm{tr}(O\hat{\sigma}(x)) - f_O(x)|^2 \leq \epsilon$ for $N = m^{\mathcal{O}(1/\epsilon)}$.

3 Proposed Method

3.1 Idea

Suppose that O is an arbitrary and unknown n-qubit observable, and a distribution $\mathcal{D}$ of n-qubit quantum states. O will correspond to a local ground state property as described earlier. The distribution $\mathcal{D}$, in our case, would correspond to a shadow representation of the ground state of Hamiltonians $H(x)$. More specifically for each $x \in [-1, 1]^m$ we have $\sigma_T(\rho(x))$ in $\mathcal{D}$. The probability over $\mathcal{D}$ is the one naturally induced by this transformation on the distribution over x. Our goal is to find a function $h(\rho)$ which predicts the expectation value $\mathrm{tr}(O\rho)$ of the observable O on the state ρ with a small mean squared error:

$$\mathop{\mathbb{E}}_{\rho\sim\mathcal{D}} |h(\rho) - \mathrm{tr}(O\rho)|^2 \leq \epsilon.$$

We will assume that we can access training data of the form

$$\{\rho_\ell, \mathrm{tr}\,(O\rho_\ell)\}_{\ell=1}^{N}, \tag{7}$$

where ρ_ℓ is sampled from the distribution $\mathcal{D}$. In practice, though, we cannot directly access the exact value of the expectation value $\mathrm{tr}\,(O\rho_\ell)$; instead, we might measure O multiple times in the state ρ_ℓ to obtain an accurate estimate of the expectation value.

A critical aspect of the is that the distribution $\mathcal{D}$, has a specific structure which allows us to learn an efficient approximation to the entire process.

Specifically, the distribution of classical shadow representations $\mathcal{D}$ is *locally flat*. This means that the distribution is unmodified (i.e., the distribution appears flat) when we locally rotate any one of the qubits by a Clifford gate. To see this, recall that the Clifford group normalizes Pauli operators. Hence the composition of a Clifford gate C with a Pauli gate P is equivalent to composition with a different Clifford gate C' and Pauli gate P'. Since the classical shadow $\sigma_T(\rho)$ is obtained by applying Pauli operator P to the state ρ, applying a Clifford gate C to $\sigma_T(\rho)$, is equivalent to applying a different Clifford gate C' with a different Pauli operator P'. However since the classical shadow representation is obtained by applying a randomly chosen Pauli operator to the quantum state, the distribution of classical shadow representations is invariant to applying any Clifford gate, as applying a Clifford gate to the quantum state simply corresponds to applying a different Clifford gate and a different Pauli operator to the classical shadow representation.

An arbitrary observable O can be expanded in terms of the Pauli operator basis:

$$O = \sum_{P\in\{I,X,Y,Z\}^{\otimes n}} \alpha_P P. \tag{8}$$

Though there are 4^n Pauli operators, if the distribution $\mathcal{D}$ is locally flat and O has a constant spectral norm, we can approximate the sum over P by a truncated sum

$$O^{(k)} = \sum_{P \in \{I,X,Y,Z\}^{\otimes n} : |P| \leq k} \alpha_P P. \tag{9}$$

including only the Pauli operators P with weight $|P|$ up to k, those acting nontrivially on no more than k qubits. The mean squared error incurred by this truncation decays exponentially with k. Therefore, to learn O with mean squared error ϵ it suffices to learn this truncated approximation to O, where $k = \mathcal{O}(\log(1/\epsilon))$.

Furthermore, using recent work on Bohnenblust-Hille Inequalities [Slote et al. 2023, Volberg and Zhang 2022] one can show that for O with low norms, only a few large coefficients α_P are relevant. Hence, instead of the complexity of $n^{\mathcal{O}(k)}$ in order to learn O one can get an efficient approximation with $\mathcal{O}(\log n)$ samples.

To make predictions about the expectation value of an operator O for a new quantum state ρ drawn from a distribution $\mathcal{D}$, we need to have some information about ρ. Both ρ and O can be represented using Pauli operators. When we only consider the truncated part of O, the prediction is based only on the corresponding part of ρ. If the reduced density matrices of the states in $\mathcal{D}$ are known, the prediction can be calculated classically. If the states in $\mathcal{D}$ are unknown, the reduced density matrices can be learned efficiently (for small k) using classical shadow tomography, after which the classical calculation can be performed to make a prediction about the expectation value of $\mathrm{tr}(O\rho)$.

3.2 Algorithm Details

Suppose we have obtained a classical dataset by performing N randomized experiments. Recall that a randomized Pauli measurement measures each qubit of a state in a random Pauli basis (X, Y or Z) and produces a measurement outcome of $|\psi^{(\mathrm{out})}\rangle = \bigotimes_{i=1}^{n} |s_i^{(\mathrm{out})}\rangle$, where $|s_i^{(\mathrm{out})}\rangle \in \mathrm{stab}_1 \triangleq \{|0\rangle, |1\rangle, |+\rangle, |-\rangle, |y+\rangle, |y-\rangle\}$. We denote the classical dataset of size N to be

$$S_N \triangleq \left\{ \left|\psi_\ell^{(\mathrm{in})}\right\rangle = \bigotimes_{i=1}^{n} \left|s_{\ell,i}^{(\mathrm{in})}\right\rangle, \left|\psi_\ell^{(\mathrm{out})}\right\rangle = \bigotimes_{i=1}^{n} \left|s_{\ell,i}^{(\mathrm{out})}\right\rangle \right\}_{\ell=1}^{N}, \tag{10}$$

where $|s_{\ell,i}^{(\mathrm{in})}\rangle, |s_{\ell,i}^{(\mathrm{out})}\rangle \in \mathrm{stab}_1$. Each product state is represented classically with $\mathcal{O}(n)$ bits. Hence, the classical dataset S_N is of size $\mathcal{O}(nN)$ bits.

Let O be an observable with $\|O\| \leq 1$ that is written as a sum of few-body observables, where each qubit is acted by $\mathcal{O}(1)$ of the few-body observables. We denote the Pauli representation of O as $\sum_{Q \in \{I,X,Y,Z\}^{\otimes n}} a_Q Q$. By definition of O, there are $\mathcal{O}(n)$ nonzero Pauli coefficients a_Q. We consider a hyperparameter $\tilde{\epsilon} > 0$; roughly speaking $\tilde{\epsilon}$ will scale inverse polynomially in the dataset size N. For every Pauli observable $P \in \{I, X, Y, Z\}^{\otimes n}$ with $|P| \leq k = \Theta(\log(1/\epsilon))$, the algorithm computes an empirical estimate for the corresponding Pauli coefficient α_P via

$$\hat{x}_P(O) = \frac{1}{N}\sum_{\ell=1}^{N} \operatorname{tr}\left(P\bigotimes_{i=1}^{n}|s_{\ell,i}^{(\text{in})}\rangle\langle s_{\ell,i}^{(\text{in})}|\right)\operatorname{tr}\left(O\bigotimes_{i=1}^{n}\left(3|s_{\ell,i}^{(\text{out})}\rangle\langle s_{\ell,i}^{(\text{out})}|-I\right)\right), \tag{11}$$

$$\hat{\alpha}_P(O) = \begin{cases} 3^{|P|}\hat{x}_P(O), & \left(\frac{1}{3}\right)^{|P|} > 2\tilde{\epsilon} \text{ and } |\hat{x}_P(O)| > 2\cdot 3^{|P|/2}\sqrt{\tilde{\epsilon}}\sum_{Q:a_Q\neq 0}|a_Q|, \\ 0, & \text{otherwise.} \end{cases} \tag{12}$$

The computation of $\hat{x}_P(O)$ and $\hat{\alpha}_P(O)$ can both be done classically. The basic idea of $\hat{\alpha}_P(O)$ is to set the coefficient $3^{|P|}\hat{x}_P(O)$ to zero when the influence of Pauli observable P is negligible. Given an n-qubit state ρ, the algorithm outputs

$$h(\rho, O) = \frac{1}{N}\sum_{\ell=1}^{N}\kappa(x, x_\ell)\sum_{P:|P|\leq k}\hat{\alpha}_P(O)\operatorname{tr}(P\rho(x_\ell)). \tag{13}$$

Note that, to make predictions, the ML algorithm only needs the k-body reduced density matrices (k-RDMs) of ρ. The k-RDMs of ρ can be efficiently obtained by performing randomized Pauli measurement on ρ and using the classical shadow formalism Huang et al. [2020].

Theorem 1 (Learning an unknown observable). *Given* $\epsilon, \epsilon', \delta > 0$, $\|O\| < 1$ *from a training data* $\{\rho_\ell, \operatorname{tr}(O\rho_\ell)\}_{\ell=1}^{N}$ *of size*

$$N = \log(n/\delta)2^{\mathcal{O}(\log(\frac{1}{\epsilon})\log(n))}, \tag{14}$$

where ρ_ℓ *is sampled from* $\mathcal{D}$, *we can learn a function* $h(\rho)$ *such that*

$$\mathop{\mathbb{E}}_{\rho\sim\mathcal{D}}|h(\rho) - \operatorname{tr}(O\rho)|^2 \leq (\epsilon + 2\epsilon') \tag{15}$$

with probability at least $1-\delta$.

The proof the theorem and the detailed description of the ML algorithm are given in the longer version of the paper. To measure the prediction error of the ML model, we consider the average-case prediction performance under an arbitrary n-qubit state distribution $\mathcal{D}$ invariant under single-qubit Clifford gates, which means that the probability distribution $f_{\mathcal{D}}(\rho)$ of sampling a state ρ is equal to $f_{\mathcal{D}}(U\rho U^\dagger)$ of sampling $U\rho U^\dagger$ for any single-qubit Clifford gate U.

4 Conclusion

Traditional machine learning (ML) offers a strategy that has the potential to be extremely effective in resolving difficult quantum many-body issues in the fields of physics and chemistry. On the other hand, as these algorithms use an exponential amount of training data[Huang et al. 2021], it is not clear whether there is any guaranteed advantage of using ML based methods. In this study, we show that for certain restricted families ML algorithms do not need exponential amount of data. Our proposed method cuts down the required sample complexity, at the cost of considering only local Hamiltonians. We have focused purely on a theoretical analysis, and experiments need to be conducted to assess the efficiency of our approach against existing methods.

References

Aharonov, D., Van Dam, W., Kempe, J., Landau, Z., Lloyd, S., Regev, O.: Adiabatic quantum computation is equivalent to standard quantum computation. SIAM Rev. **50**(4), 755–787 (2008)

Anshu, A., Gosset, D., Korol, K.J.M., Soleimanifar, M.: Improved approximation algorithms for bounded-degree local Hamiltonians. Phys. Rev. Lett. **127**(25), 250502 (2021)

Arunachalam, S., de Wolf, R.: Guest column: a survey of quantum learning theory. ACM SIGACT News **48**(2), 41–67 (2017)

Barak, B., et al.: Beating the random assignment on constraint satisfaction problems of bounded degree. arXiv preprint arXiv:1505.03424 (2015)

Biamonte, J., Wittek, P., Pancotti, N., Rebentrost, P., Wiebe, N., Lloyd, S.: Quantum machine learning. Nature **549**(7671), 195–202 (2017)

Bierens, H.J.: The Nadaraya-Watson kernel regression function estimator (1988)

Bohnenblust, H.F., Hille, E.: On the absolute convergence of Dirichlet series. Ann. Math. **32**, 600–622 (1931)

Carleo, G., Troyer, M.: Solving the quantum many-body problem with artificial neural networks. Science **355**(6325), 602–606 (2017)

Cerezo, M., et al.: Variational quantum algorithms. Nat. Rev. Phys. **3**(9), 625–644 (2021)

Deng, D.-L., Li, X., Sarma, S.D.: Machine learning topological states. Phys. Rev. B **96**(19), 195145 (2017)

Dinur, I., Friedgut, E., Kindler, G., O'Donnell, R.: On the Fourier tails of bounded functions over the discrete cube. In: Proceedings of the 38th Annual ACM Symposium on Theory of Computing, pp. 437–446 (2006)

Farhi, E., Goldstone, J., Gutmann, S., Sipser, M.: Quantum computation by adiabatic evolution. arXiv preprint arXiv:quant-ph/0001106 (2000)

Flammia, S.T., Preskill, J.: Learning noisy quantum experiments. arXiv preprint (2022)

Gibbs, J., Caro, M., Sornborger, A., Coles, P.: Quantum simulation with machine learning. arXiv preprint (2022)

Harrow, A.W., Montanaro, A.: Extremal eigenvalues of local Hamiltonians. Quantum **1**, 6 (2017)

Huang, H.-Y., Kueng, R., Preskill, J.: Predicting many properties of a quantum system from very few measurements. Nat. Phys. **16**(10), 1050–1057 (2020)

Huang, H.-Y., Kueng, R., Torlai, G., Albert, V.V., Preskill, J.: Provably efficient machine learning for quantum many-body problems (2021)

Kawai, H., Nakagawa, Y.O.: Predicting excited states from ground state wavefunction by supervised quantum machine learning. Mach. Learn. Sci. Technol. **1**(4), 045027 (2020)

Levy, R., Luo, D., Clark, B.K.: Classical shadows for quantum process tomography on near-term quantum computers. arXiv preprint arXiv:2110.02965 (2021)

Nadaraya, E.A.: On estimating regression. Theor. Probab. Appl. **9**(1), 141–142 (1964)

Nomura, Y., Darmawan, A.S., Yamaji, Y., Imada, M.: Restricted Boltzmann machine learning for solving strongly correlated quantum systems. Phys. Rev. B **96**(20), 205152 (2017)

Ohliger, M., Nesme, V., Eisert, J.: Efficient and feasible state tomography of quantum many-body systems. New J. Phys. **15**(1), 015024 (2013)

Paini, M., Kalev, A.: An approximate description of quantum states. arXiv preprint arXiv:1910.10543 (2019)

Rouzé, C., Wirth, M., Zhang, H.: Quantum Talagrand, KKL and Friedgut's theorems and the learnability of quantum Boolean functions. arXiv preprint arXiv:2209.07279 (2022)

Schuld, M., Killoran, N.: Quantum machine learning in feature Hilbert spaces. Phys. Rev. Lett. **122**(4), 040504 (2019)

Schütt, K.T., Gastegger, M., Tkatchenko, A., Müller, K.-R., Maurer, R.J.: Unifying machine learning and quantum chemistry with a deep neural network for molecular wavefunctions. Nat. Commun. **10**(1), 5024 (2019)

Scott, A.J.: Optimizing quantum process tomography with unitary 2-designs. J. Phys. A Math. Theor. **41**(5), 055308 (2008)

Sharir, O., Levine, Y., Wies, N., Carleo, G., Shashua, A.: Deep autoregressive models for the efficient variational simulation of many-body quantum systems. Phys. Rev. Lett. **124**(2), 020503 (2020)

Slote, J., Volberg, A., Zhang, H.: Noncommutative Bohnenblust-Hille inequality in the Heisenberg-Weyl and Gell-Mann bases with applications to fast learning. arXiv preprint arXiv:2301.01438 (2023)

van Enk, S.J., Beenakker, C.W.: Measuring tr ρ n on single copies of ρ using random measurements. Phys. Rev. Lett. **108**(11), 110503 (2012)

Vargas-Hernández, R.A., Sous, J., Berciu, M., Krems, R.V.: Extrapolating quantum observables with machine learning: inferring multiple phase transitions from properties of a single phase. Phys. Rev. Lett. **121**(25), 255702 (2018)

Volberg, A., Zhang, H.: Noncommutative Bohnenblust-Hille inequalities. arXiv preprint arXiv:2210.14468 (2022)

Wan, K., Kim, I.H.: Fast digital methods for adiabatic state preparation. arXiv preprint arXiv:2004.04164 (2020)

Watson, G.S.: Smooth regression analysis. Sankhyā: Indian J. Stat. Ser. A **26**, 359–372 (1964)

Zhang, Y., Melko, R.G., Kim, E.-A.: Machine learning z2 quantum spin liquids with quasiparticle statistics. Phys. Rev. B **96**(24), 245119 (2017)

Zhang, Y., Ginsparg, P., Kim, E.-A.: Interpreting machine learning of topological quantum phase transitions. Phys. Rev. Res. **2**(2), 023283 (2020)

Quantum Factory Method: A Software Engineering Approach to Deal with Incompatibilities in Quantum Libraries

Samuel Magaz-Romero(✉), Eduardo Mosqueira-Rey, Diego Alvarez-Estevez, and Vicente Moret-Bonillo

Universidade da Coruña, CITIC, Campus de Elviña, 15071 A Coruña, Spain
{s.magazr,eduardo,diego.alvareze,vicente.moret}@udc.es

Abstract. The current context of Quantum Computing and its available technologies present an extensive variety of tools and lack of methodologies, leading to incompatibilities across platforms, which end up as inconsistencies in the developed solutions. We propose a design called Quantum Factory Method, based on software engineering and design patterns, to solve these issues by integrating different quantum platforms in the same development. We provide example implementations whose results prove the suitability of the design in different cases, and conclude on how this approach can be expanded for future work.

Keywords: Quantum Computing · Software Engineering · Design Patterns · Rule-Based Systems · Uncertainty

1 Introduction

The exponential growth that has been experienced in recent years of the interest towards the field of Quantum Computing (QC) is undeniable [4]. Despite its theoretical basis being around since the 80s [8], the manufacturing of quantum computers and their availability has put this field in the eyes of many.

Some of the most relevant groups that are leading this new field are IBM [12] and Google [5], both of which offer different commercial solutions for QC. Other important competitors are Amazon [1] or Atos [2]. These are some names in the industry, but more are involved, showing the size of the QC environment so far.

From a Software Engineering (SE) perspective, this varying environment can end up being detrimental. The lack of standardisation makes each group approach their solution differently, and while programming libraries in Python are the most popular solution, these present differences on some core concepts.

Libraries are designed to work on the platform (programming tools and execution stack) of their company. There are some common factors among them, but they still present different philosophies and forms for users to interact with them. This complicates the introduction into the QC world, as well as the work for developers, dealing with the differences between several platforms.

J. Mikyška et al. (Eds.): ICCS 2023, LNCS 14077, pp. 64–71, 2023.
https://doi.org/10.1007/978-3-031-36030-5_6

All these factors make the comparison between options more troublesome; for example, when selecting the appropriate platform for a project. Each platform presents advantages and inconveniences regarding development, maintainability and usage, so the decision should be made taking these into account.

Therefore, the objectives that our approach must accomplish are: (i) to allow for a single problem definition to be used in different platforms, without addressing their philosophies for each case, (ii) to obtain standard outputs across platforms, facilitating their comparison, and (iii) to have the possibility to include new platforms in the future as they are made available to the general public.

2 State of the Art

We present a brief excerpt of QC's state of the art, specifically on the current situation of Quantum Software Engineering and the OpenQASM standard.

2.1 Quantum Software Engineering

While there is a lack of Quantum Software Engineering methodologies, they exist, but there is not a consensus like on classical methodologies. It seems logical due to Quantum Computing being on its early stages, yet it could benefit from tools and procedures in order to keep expanding its horizons [10,15,17].

There is interest in the subject, but no proposal for a methodology of Quantum Software Engineering has firmly established itself as the proper approach, due to: (1) Quantum Computing is not as software-engineer oriented as other branches of computing, drawing along users not familiarised with software engineering methodologies, and (2) a general lack of interest in methodology innovation, leading to not focusing on methodologies as much it would be needed.

In summary, there is no procedure or methodology (yet) to rely on when developing software using Quantum Computing.

2.2 OpenQASM: A Not-so-Standard Standard

Open Quantum Assembly Language (OpenQASM) is an imperative programming language designed for near-term QC algorithms and applications. Programs are described using the measurement-based quantum circuit model with support for classical feed-forward flow control based on measurement outcomes [7].

It was proposed by the IBM Quantum Computing group as an imperative programming language for quantum circuits. Since OpenQASM 2 was introduced, it has become a *de facto* standard in the field.

Nowadays, OpenQASM 3 includes better support for the next phase of quantum system development, as well as to incorporate some of the best ideas that have arisen in other circuit description languages [6].

However, OpenQASM is not a standard *per se*, since each quantum programming library implements it differently. Table 1 illustrates this situation [1–3,5,12]. OpenQASM is a suitable option for cases where only basic features are used, but it is not enough for the variety of problems we want to address here.

Table 1. OpenQASM's features supported in each platform S = Supported P = Partially supp. N = Not supp. X = Unavailable in version 2.0

Platform		Qiskit	Cirq	Braket	Pennylane	myQLM
Version	2.0	S	S	S	S	S
	3.0	S	N	S	S	N
Main types	qubit	P	S	S	P	S
	bit	S	S	S	S	S
	bool	S	X	S	S	X
Secondary types	uint	N	X	S	N	X
	float	N	P	S	N	S
	complex	N	X	S	N	X
Gates	Basis	S	S	S	S	S
	Custom	N	S	S	N	S
	Control (on non-basis gates)	N	X	S	N	X
	Barrier	S	N	S	S	S
Flow	if	S	N	S	S	S
	else	S	X	S	S	X
	else if	N	X	S	N	X
	for	P	X	S	P	X
Subroutines		N	X	S	N	X

3 Proposal

Our proposal is based on the use of design patterns, which eases the elaboration of an object-oriented design and its understanding for other developers.

3.1 Design Patterns

In SE, a design pattern [16] is a general repeatable solution to a commonly occurring problem in software design. It is a description or template for how to solve a problem that can be used in many different situations.

Design patterns provide a common vocabulary that eases the sharing and understanding of concepts, preventing subtle issues that can cause major problems, improving code readability for coders and architects familiar with them. The design patterns commonly known in the general literature are the ones presented on [9]; the ones our solution needs are Factory Method and Adapter (Fig. 1).

Factory Method is a creational pattern to deal with the problem of creating objects without having to specify the exact class. This is done by creating objects by calling a method that encapsulates a constructor. The `Creator` class defines a method to build `Product` objects, but lets the subclasses like `ConcreteCreator` decide which class to instantiate, in this case `ConcreteProduct`.

Adapter is a structural pattern that converts the interface of a class into another, creating an intermediary abstraction that translates the old component to the new system. The `Client` interacts with the `Target` class, from whom expects the interface `doThat()`. The `Adaptee` class offers the interface `doThis()`, which is converted by the `Adapter` class into the expected `doThat()`.

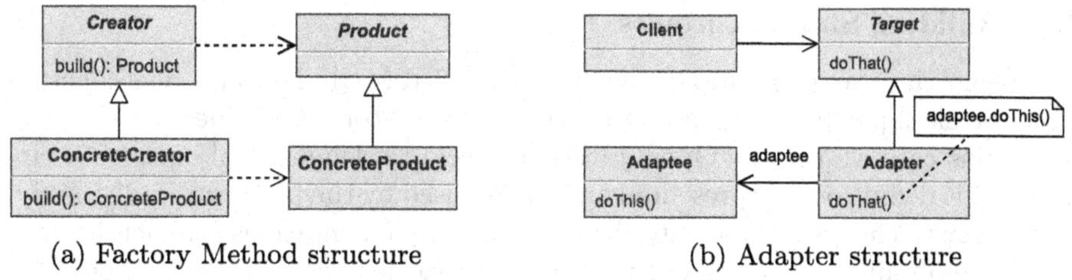

(a) Factory Method structure

(b) Adapter structure

Fig. 1. Design patterns' structures

3.2 Application

We can now introduce the final design of our proposal, illustrated in Fig. 2.

The `Platform` class (representing a quantum platform) acts as the `Creator` from the Factory Method design pattern, and the `Circuit` class (representing a quantum circuit) acts as the `Product`, including a subclass of each per quantum platform. Regarding the Adapter design pattern, the `Circuit` class acts as the `Target`, and each subclass is the `Adapter` of their quantum library.

The `build` method can receive an input to be translated into the quantum circuit. Therefore, with a single definition, the same circuit can be built in different platforms and run on different quantum computers.

The `execute` method can return an object with the results, easing the comparison of platforms, for example for benchmarking studios.

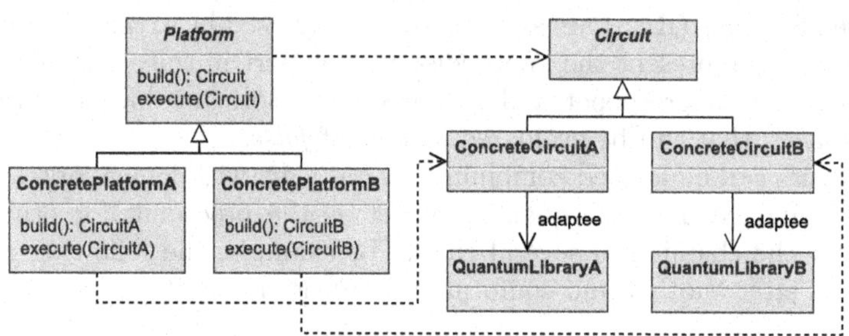

Fig. 2. Quantum Factory Method structure

4 Examples

Two examples were developed, one using OpenQASM to build simple circuits and other building Quantum Rule-Based Systems (QRBS) [14] with inferential circuits. We use Qiskit and Cirq as platforms; code for the implementation is in [13].

4.1 Building Simple Circuits

We define the classes required: one concrete `Platform` class and one concrete `Circuit` class per platform, and a `Result` class to store the values.

In this case we use the OpenQASM standard to define simple quantum circuits. This definition is represented on a string used by the platforms indistinctly. The `Circuit` classes encapsulate the quantum implementations for each library, which are obtained in the `build` method and used in the `execute` method. The `Result` class stores the values from the measurements of each qubit after the execution of the quantum circuit, as real numbers in the range $[0, 1]$.

With these elements defined, we obtain the design illustrated in Fig. 3.

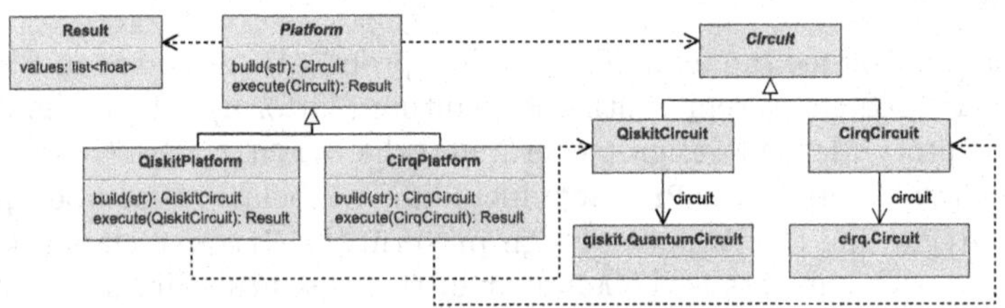

Fig. 3. OpenQASM example's design

4.2 Building Quantum Rule-Based Systems

Rule-Based Systems (RBS) are systems commonly used in Artificial Intelligence that encode and represent the knowledge of an expert in rules [11]. These rules are composed by a precedent and a consequent. Both are logical statements, known as facts, that can be evaluated as *true* of *false*.

Precedents can be formed combining several facts with logical operators like AND, OR, and NOT. Consequents can act as the precedent for other rules, allowing for the chaining of several rules. For example, the following rules are chained one after another and conform an inferential circuit:

$$A \text{ AND } B \Rightarrow C \qquad C \text{ OR } D \Rightarrow E \qquad E \text{ AND } (F \text{ OR } G) \Rightarrow I$$

However, facts can be between *true* or *false* states. This idea is called uncertainty, and can be implemented in RBS. Using QC to represent uncertainty gives birth to QRBS. Overall, the elements that conform a QRBS are:

- **Facts**: single qubits, with state $|0\rangle$ being false and state $|1\rangle$ true.
- **Operators**: quantum operators, as illustrated in Fig. 4.
- **Uncertainty**: a parameterized operator (Eq. 1) that puts a qubit in superposition, mapping its uncertainty degree in $[0, 1]$ to an angle in $[0, \pi/2]$.

$$M(\delta) = \begin{bmatrix} cos(\delta) & sin(\delta) \\ sin(\delta) & -cos(\delta) \end{bmatrix} \tag{1}$$

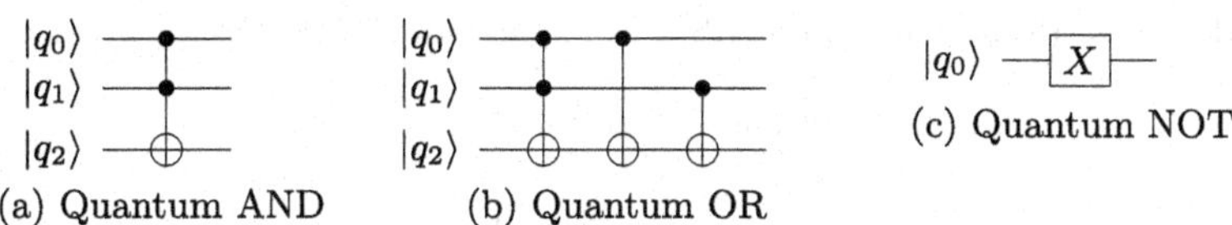

Fig. 4. Quantum logical operators

We can now apply the proposed design to build QRBS: we need to define one concrete `Platform` and `Circuit` per platform we want to implement, and a `Result` class to store the values obtained after measuring. In this case, the `Platform` class provides a `build` method that receives an inferential circuit as the input, and an `execute` method that receives the `Circuit` object and returns a `Result` object with the measured values. To represent classical inferential circuits we use the Composite design pattern, used to model tree structures that represent part-whole hierarchies. For the building process we incorporate the Visitor design pattern, where quantum platforms visit the elements of an inferential circuit to build its quantum circuit. The `Result` class stores the values obtained after measuring in a key-value dictionary, where the key is the tag of the element and the value is the measurement obtained. With these elements defined, we obtain the design shown in Fig. 5.

4.3 Experiments and Results

At this point, we can experiment with the examples as follows: (1) define a proper problem (an OpenQASM string or an inferential circuit), (2) call the `build` methods to obtain the `Circuit` objects, (3) call the `execute` methods to obtain the `Result` objects, and (4) analyse the obtained values to conclude.

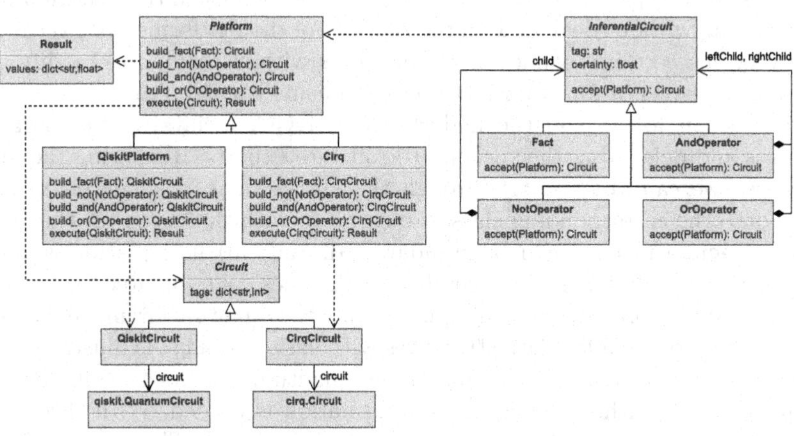

Fig. 5. QRBS example's design

Table 2. Results for OpenQASM example with the Bell State circuit

Platform	State probability	
	$\|00\rangle$	$\|11\rangle$
Qiskit	0.511	0.489
Cirq	0.491	0.509

Table 3. Results for QRBS example with the proposed circuit

Platform	Element								
	A	B	C	D	E	F	G	H	I
Qiskit	1.000	0.331	0.286	0.373	0.416	0.214	0.950	0.838	0.344
Cirq	1.000	0.342	0.295	0.381	0.423	0.232	0.944	0.823	0.341

We define a proper problem for each example (a Bell state quantum circuit for OpenQASM and an inferential circuit with the rules of Sect. 4.2 for QRBS), and execute them with the default parameters of each platform (local simulators, 1024 shots), obtaining the results shown in Tables 2 and 3. The values are similar but not identical, in part due to QC's probabilistic nature, yet they show the relevance of comparing the results obtained by different quantum platforms.

5 Discussion and Conclusions

The design obtained, named Quantum Factory Method, is a robust and solid product, as a result of using design patterns. Its simplicity resides on covering the needs and modelling through the Factory Method and Adapter patterns.

The QRBS example illustrates the potential of our approach, as a single definition of an inferential circuit is built in each platform, without having to rebuild for each experiment. The abstraction allows to design the quantum inferential circuits without dealing with the specifics of the platforms. In this case, we could vary the certainty values of the elements without having to modify their quantum implementation, speeding up experiments carried out.

This design focuses on the higher levels of QC, facilitating the work for end users and delegating the specifics to software experts. Reviewing the initial objectives, we can observe: (i) through the abstractions, a single definition is passed onto the corresponding classes to carry on with the tasks, (ii) the output for the `execute` method can be standardised, and (iii) new platforms can be included into the design without modifying the ones contemplated.

The flexibility provided by the input to build the quantum circuits enables the representation of complex data structures with ease. Its range is illustrated with the examples of Sect. 4, going from simple quantum circuits to complex cases like quantum algorithms, where more information is needed. With this design, one definition is enough for several quantum platforms. This workflow eases development and maintenance, which is key to develop large software products.

Regarding future work, we intend to keep looking for this kind of synergies. We believe that Software Engineering will be interesting for better develop

Quantum Computing algorithms. We hope the ideas presented in this paper conform a step towards that direction.

Acknowledgements. This work has been supported by the European Union's Horizon 2020 under project NEASQC (grant agreement No 951821) and by the Xunta de Galicia (grant ED431C 2022/44) with the European Union ERDF funds and Centro de Investigación de Galicia "CITIC", funded by Xunta de Galicia and the European Union (European Regional Development Fund-Galicia 2014–2020 Program, grant ED431G 2019/01). DAE received funding from the project ED431H 2020/10 of Xunta de Galicia.

References

1. Amazon Web Services: Amazon Braket. https://aws.amazon.com/braket/. Accessed 23 Nov 2022
2. Atos: myQLM. https://myqlm.github.io/. Accessed 23 Nov 2022
3. Bergholm, V., et al.: PennyLane: automatic differentiation of hybrid quantum-classical computations (2018). https://doi.org/10.48550/arxiv.1811.04968. Accessed 23 Nov 2022
4. Biondi, M., et al.: Quantum Computing: An Emerging Ecosystem and Industry Use Cases. McKinsey & Company (2021)
5. Cirq Developers: Cirq, April 2022. https://doi.org/10.5281/zenodo.6599601
6. Cross, A., Javadi-Abhari, A., Alexander, T., Beaudrap, N.D.: OpenQASM 3: a broader and deeper quantum assembly language. ACM Trans. Quant. Comput. **3**(3), 1–50 (2022). https://doi.org/10.1145/3505636
7. Cross, A.W., Bishop, L.S., Smolin, J.A., Gambetta, J.M.: Open quantum assembly language (2017). https://doi.org/10.48550/arxiv.1707.03429
8. Feynman, R.P.: Quantum mechanical computers. Found. Phys. **16**(6), 507–531 (1986). https://doi.org/10.1007/bf01886518
9. Gamma, E., Helm, R., Johnson, R., Vlissides, J.M.: Design Patterns: Elements of Reusable Object-Oriented Software. Addison-Wesley Professional (1994)
10. Gemeinhardt, F., Garmendia, A., Wimmer, M.: Towards model-driven quantum software engineering. In: 2021 IEEE/ACM 2nd International Workshop on Quantum Software Engineering (Q-SE), pp. 13–15. IEEE (2021). https://doi.org/10.1109/Q-SE52541.2021.00010
11. Grosan, C., Abraham, A.: Intelligent Systems: A Modern Approach. Springer, Heidelberg (2011). https://doi.org/10.1007/978-3-642-21004-4
12. IBM: IBM Quantum. https://quantum-computing.ibm.com/. Accessed 26 Oct 2022
13. Magaz: samu-magaz/quantum-factory-method: Quantum Factory Method v1.0.0, January 2023. https://doi.org/10.5281/zenodo.7544539
14. Moret-Bonillo, V., Magaz-Romero, S., Mosqueira-Rey, E.: Quantum computing for dealing with inaccurate knowledge related to the certainty factors model. Mathematics **10**(2) (2022). https://doi.org/10.3390/math10020189
15. Piattini, M., Serrano, M., Perez-Castillo, R., Petersen, G., Hevia, J.L.: Toward a quantum software engineering. IT Prof. **23**(1), 62–66 (2021). https://doi.org/10.1109/MITP.2020.3019522
16. Shvets, A.: Dive into Design Patterns. Refactoring, Guru (2021)
17. Zhao, J.: Quantum software engineering: landscapes and horizons. arXiv preprint arXiv:2007.07047 (2020). https://doi.org/10.48550/arxiv.2007.07047

A Polynomial Size Model with Implicit SWAP Gate Counting for Exact Qubit Reordering

J. Mulderij[1,2], K.I. Aardal[2], I. Chiscop[1], and F. Phillipson[1,3(✉)]

[1] TNO, The Hague, The Netherlands
frank.phillipson@tno.nl
[2] Delft University of Technology, Delft, The Netherlands
[3] Maastricht University, Maastricht, The Netherlands

Abstract. Due to the physics behind quantum computing, quantum circuit designers must adhere to the constraints posed by the limited interaction distance of qubits. Existing circuits need therefore to be modified via the insertion of SWAP gates, which alter the qubit order by interchanging the location of two qubits' quantum states. We consider the Nearest Neighbor Compliance problem on a linear array, where the number of required SWAP gates is to be minimized. We introduce an Integer Linear Programming model of the problem of which the size scales polynomially in the number of qubits and gates. Furthermore, we solve 131 benchmark instances to optimality using the commercial solver CPLEX. The benchmark instances are substantially larger in comparison to those evaluated with exact methods before. The largest circuits contain up to 18 qubits or over 100 quantum gates. This formulation also seems to be suitable for developing heuristic methods since (near) optimal solutions are discovered quickly in the search process.

Keywords: Integer Linear Programming · Nearest Neighbor Architectures · NNC · Optimization · Quantum Circuits · SWAP gate

1 Introduction

The rules that govern physical interactions in a quantum setting allow quantum computing to provide algorithms with a better complexity scaling than their classical counterparts for many naturally arising problems. Exploiting the properties of phenomena such as superposition and entanglement, one can search in a database [18], factor integers [44] or estimate a phase [37] more efficiently than previously possible.

The many advantages of quantum computing come at the price of physical limitations in circuit design. First, relevant coherence times (what is relevant depends on the technology) indicate that information on qubits is perturbed or even lost after some time due to a qubit's interaction with its environment [11]. It is therefore, for a fixed number of qubits, desirable to do calculations with as few

J. Mikyška et al. (Eds.): ICCS 2023, LNCS 14077, pp. 72–89, 2023.
https://doi.org/10.1007/978-3-031-36030-5_7

gates as possible. A second limitation is induced by nearest neighbor constraints, where 2-qubit quantum gates can only be used when the qubits are physically adjacent. The nearest neighbor constraints have been considered in proposals for a range of potential technological realizations of quantum computers such as ion traps [4,30,36], nitrogen-vacancy centers in diamonds [36,53], quantum dots emitting linear cluster states linked by linear optics [9,20], laser manipulated quantum dots in a cavity [24] and superconducting qubits [12,32,38]. They are also considered in realizations of specific types of circuits and architectures, such as surface codes [47], Shor's algorithm [14], the Quantum Fourier Transform (QFT) [46], circuits for modular multiplication and exponentiation [33], quantum adders on the 2D NTC architecture [8], factoring [40], fault-tolerant circuits [31], error correction [15], and more recently, IBM QX architectures [13,48,54,55].

Up to now, the design of quantum circuits consists of manual work in elementary cases and for specific circuits. As the complexity of the algorithms increases, however, manual synthesis will no longer be feasible. When constructing a circuit from scratch, using only the set of elementary gates, even without considering nearest neighbor constraints, one is solving specific instances of the PSPACE-complete Minimum Generator Sequence problem [23], where the group consists of all unitary matrices and the elementary gate operations form the set of generators. Here one tries to find the shortest sequence of generators to map an input to a given output. A lot of work was done in this area using boolean satisfiability [17], template matching [34,41] and methods for reversible circuits [3,51] as all quantum gates perform unitary operations [37]. Other methods consider already designed circuits that do not comply with nearest neighbor constraints. In these approaches, SWAP gates, which swap the information of two adjacent qubits, are inserted into the circuit. The goal herein is to minimize the number of required SWAP gates to make the whole circuit compliant. Within this branch of research there are two approaches to the topic, global and local reordering. Global reordering determines the initial layout of the qubits such that there are as few SWAP gates as possible required in the remainder of the circuit. In order to elude the micromanagement that local reordering is concerned with, the global reordering problem is generally approximated with the NP-complete [16] problem of Optimal Linear Arrangement (OLA) on the interaction graph of the circuit with edge weights taking the Nearest Neighbor Cost [29]. Here the gate sequence is either disregarded [43] or encoded in the weights [28].

The local reordering problem allows for any change in the qubit order before each gate, resulting in a vast feasible region, even for small instances. The more general problem of SWAP minimization where qubits are placed on a coupling graph (two qubits can share a gate if their corresponding nodes share an edge) is shown to be NP-Complete [45] via a reduction from the NP-complete token swapping problem [6,25]. The problem we consider, where the graph is a simple path, is widely believed to be NP-complete (as conjectured in [21]) but to the best of the authors' knowledge, no formal proof is given yet. Many heuristics have been developed including receding horizon [21,27,42,50], greedy [1,21], harmony search [1] and OLA on parts of the circuit [39]. Only a few works have dared to approach the problem with exact methods, all of which embody an explicit factorial scaling in the amount of variables or processed nodes, either through the use of the adjacent transposition graph [35], exhaustive searches [10,21] or explicit cost enumeration for each permutation [52]. The exact approaches have delivered small benchmark instances to compare the heuristics' results to. The size of these benchmark instances typically does not exceed circuits of about 5 qubits and 16 gates due to the vast scaling of the number of variables in the optimization model.

In this work we will provide an exact Integer Linear Programming (ILP) formulation of the Nearest Neighborhor Compliance (NNC) problem that does not entail a factorial scaling in the number of qubits, by implicitly counting the number of required SWAP gates at each reordering step. The power of the commercial solver CPLEX is used to optimally solve the problem for 123 instances from the *RevLib* library [49] and 8 QFT circuits. The considered benchmark instances include the largest circuits to be exactly solved up to this point. They include the QFT for 10 qubits and even a circuit with 18 qubits. The evaluation of the bigger benchmark instances finally allows for heuristics to be compared to exact solutions on larger circuits.

The remainder of this paper is structured as follows. In Sect. 2 we introduce basic concepts of quantum computing. In Sect. 3 the problem of NNC is formulated. Next, in Sect. 4, the proposed mathematical model is introduced. The results are presented and discussed in Sect. 5. Finally, conclusions are drawn in Sect. 6.

2 Background

In this section we will first introduce some basic concepts of quantum computing. A more detailed explanation can be found in [37]. Then, a description of decomposing multi-qubit gates is given.

2.1 Building Blocks of QC

The quantum version of the classical basic unit of computation, the bit, is the quantum bit (qubit). The qubit has the special property that it does not have to take value 0 or 1, but it can be in a superposition of the computational basis states $|0\rangle \equiv [1,0]^T$ and $|1\rangle \equiv [0,1]^T$. The state of a qubit $|\phi\rangle$ is denoted by a vector in $\mathbb{C}^2$ where in general we write

$$|\phi\rangle = \alpha\,|0\rangle + \beta\,|1\rangle\,, \tag{1}$$

where $\alpha, \beta \in \mathbb{C}$. When information about the state's value is extracted by the means of measurement, the state collapses to a single value. If, for example, the measurement is done in the standard basis, one would obtain $|0\rangle$ with probability $|\alpha|^2$ and $|1\rangle$ with probability $|\beta|^2$. Necessarily, $|\alpha|^2 + |\beta|^2 = 1$. In n-qubit systems, the combined state is the tensor product of individual states, which is an element of $\mathbb{C}^{2^n}$. Calculations are done by executing quantum circuits, which consist of a set of qubits and a list of quantum gates. The initial qubit states are the input of the calculation. The gates operate, in order, on specified qubits. Afterwards, a measurement is performed on one or more of the qubits to determine the probabilistic outcome of the calculation. Quantum gates are inherently reversible and are denoted by linear operators in the form of invertible matrices. Their action on the combined qubit state is simply the matrix vector product.

Below we will introduce some of the most common quantum gates, starting with the controlled NOT (CNOT) gate, see Fig. 1.

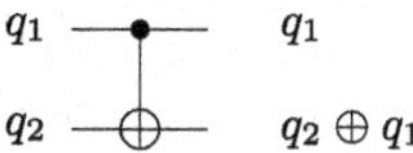

Fig. 1. A CNOT gate. Qubit q_1 is the control qubit and q_2 is the target qubit.

The controlled CNOT gate is one of the most commonly used gates. It is also used to construct the SWAP gate by placing three CNOT gates consecutively such as in Fig. 2.

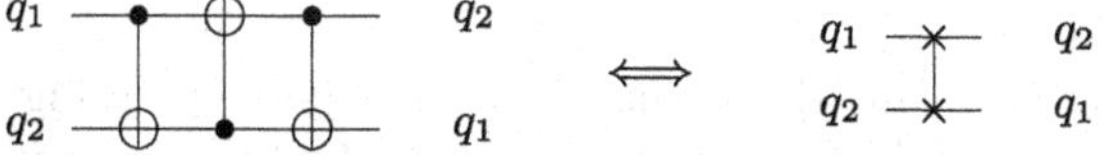

Fig. 2. A decomposed and composite SWAP gate. The operations are equivalent, the gates interchange the states of two qubits.

The SWAP gate will be the main tool used to overcome the physical constraints that limit quantum circuit design. We will, for the remainder of this text, not make a distinction between interchanging two qubits and interchanging the quantum states of two qubits.

2.2 Decomposing Multi-qubit Gates

Many circuits make use of composite gates that resemble entire circuits themselves. They are often performed on more than two qubits at once, take for example the quantum Fourier Transform (QFT), which can act on any number of qubits. In order to describe what it means for a gate to act on adjacent qubits, it only makes sense to consider 2-qubit gates. To achieve this without losing the meaning of the circuit, we have to do a modification in the following two cases:

1. Gates that only act on a single qubit are ignored for the rest of this research. These gates are of no interest in this context.
2. Gates that act on more than two qubits are decomposed into 2-qubit gates. The fact that this is always possible can be found in [37].

The second point can be implemented in a great variety of ways and doing this "optimally" is outside the scope of this work. We therefore make two straightforward design choices: 1) We only consider circuits using multiple-control Toffoli gates, Peres gates and multiple-control Fredkin gates up to a certain size; 2) We always decompose a given circuit in the same way. There is clearly room for improvement here, but the search space we consider is large enough as it is. We ignore all single-qubit gates during the modification to a nearest neighbor compliant circuit. The normal Toffoli gate's decomposition, with two control qubits can be found in [5] in the section "Three-Bit Networks". In the same work, the decomposition of a 3-control Toffoli gate is shown in the section "n-Bit Networks". The decomposition of the 4-control Toffoli is the direct extension of the previous decompositions. The Peres gate is decomposed as in the circuit "peres_8.real" from RevLib [49]. The Fredkin gate is decomposed as in the circuit "fredkin_5.real", also from RevLib. The two-qubit controlled Fredkin gate is decomposed into a controlled-NOT gate, a Toffoli gate and another controlled-NOT gate as shown by [3] in Fig. 2.4c. Larger composed gates do not make an appearance in the circuits that are considered in this work (Fig. 3).

3 Problem Definition

In this section, some basic definitions will be introduced in order to formalize the NNC problem.

For the NNC problem, the actual operation corresponding to a gate that is being used has no influence on the problem. Only the qubits on which the gate acts matter. Some definitions are introduced below. Denote the set Q of n qubits as the set of integers $Q = \{1, \ldots, n\}$. Since all the qubits have one physical location in a one dimensional array, the locations are numbered as $L = (1, \ldots, n)$ and are in a fixed order. To keep track of the location of each qubit before every gate, the notion of a qubit order will be introduced.

Definition 1. *Let $\mathcal{S}_n$ be the permutation group and $[n]$ the vector $(1, \ldots, n)$. Then a qubit order is a permutation denoted by the vector $\tau([n])$ with $\tau \in \mathcal{S}_n$, which maps the qubits to locations. We call τ^t the qubit order before gate t.*

Now that the qubit orders are defined, one needs a way of altering such an order. This is done via the previously mentioned SWAP gates.

Definition 2. *A SWAP gate is an adjacent transposition $\tau \in \mathcal{S}_n$ that permutes a qubit order, $\tau \circ (q_1, \ldots, q_i, q_{i+1}, \ldots, q_n) = (q_1, \ldots, q_{i+1}, q_i, \ldots, q_n)$, by interchanging the positions of two adjacent qubits.*

The number of SWAP gates that one minimally requires to "move" from one qubit order to another is inherently equal to the Kendall tau distance between the corresponding permutations.

Definition 3. *Given two permutations $\tau_1, \tau_2 \in \mathcal{S}_n$ for some fixed n, the Kendall tau distance between τ_1 and τ_2 is defined as*

$$I(\tau_1, \tau_2) \equiv |\{(i, j) \mid 1 \leq i, j \leq n, \tau_1(i) < \tau_1(j), \tau_2(i) > \tau_2(j)\}|. \tag{2}$$

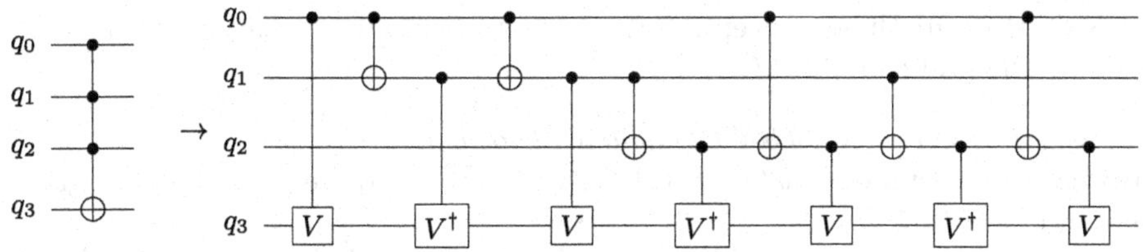

Fig. 3. The decomposition of a Toffoli gate with 3 control qubits and one target qubit into only 2-qubit gates. Here we have $V^4 = X$, where X is the usual Pauli-X gate.

The nearest neighbor interaction constraints can only be formulated once the concept of quantum gates has been properly introduced in this setting.

Definition 4. *Let $q_i, q_j \in Q$ be two qubits such that $i \neq j$. Let g_{ij} be an unordered pair $g = \{q_i, q_j\}$. Then we say that g_{ij} is a quantum gate, or simply a gate, that acts on qubits q_i and q_j. When the specific qubits do not matter in the context, the subscripts may be omitted. When multiple gates are present and their order is important, this will be reflected with a superscript as g^t.*

Please note that this definition only allows for quantum gates that act on pairs of qubits. If a gate (in the more general sense) acts on more qubits, we assume it to be decomposed, whilst if it only works on one qubit, the gate can be ignored.

To describe an entire quantum circuit, multiple gates are needed and their order is important. To this end, a gate sequence is introduced.

Definition 5. *Let $g^1, \ldots, g^m$ be m gates. Let G be the finite sequence of gates $G = (g^1, \ldots, g^m)$, then we say G is a gate sequence of size m.*

We also assume the gate sequence to be given and fixed. Allowing changes in the gate order when some commutative rules are satisfied, as was done in [19,22,35], is beyond the scope of this work.

Now we can introduce the concept of a quantum circuit more formally.

Definition 6. *Let Q be the set of qubits and G be a gate sequence. Let QC be a tuple of the set of qubits and the gate sequence $QC = (Q, G)$. Then we say that QC is a quantum circuit.*

At the core of the problem are the nearest neighbor (NN) constraints. Formalizing these requires a number of the above definitions. These constraints are what make the problem difficult.

Definition 7. *Given are a gate g_{ij}^t and a qubit order τ^t before that gate. We say that the gate complies with the NN constraints if $|\tau(i) - \tau(j)| = 1$, i.e., if the qubits on which the gate acts are adjacent in the qubit order. If, given a qubit order for each gate, all the gates in a quantum circuit's gate sequence comply with the NN constraints, we say that the quantum circuit complies with the NN constraints.*

Now that all these concepts have been formalized, we can continue with defining the problem of NNC.

Problem 1 (Nearest Neighbor Compliance Problem).
Input: A quantum circuit $QC = (Q, G)$ with $|Q| = n$ qubits and $|G| = m$ gates and an integer $k \in \mathbb{Z}_{\geq 0}$.
Question: Do there exist qubit orders $\tau^t, t \in [m]$, one before each gate of QC, such that the sum of the Kendall tau distances between consecutive qubit orders satisfies $\sum_{t=1}^{m-1} I(\tau^t, \tau^{t+1}) \leq k$ and such that the quantum circuit complies with the NN constraints?

In the minimization version of the problem, which we model in the next section, we seek to find the smallest integer k such that Problem 1 is still answered affirmatively. Considering the problem in this way, we do not require the qubits to end up in the same qubit order as they started out in. We also do not allow for changes in the gate order and do not optimize over different ways of decomposing multi-qubit quantum gates. The objective function in the minimization problem simply counts the number of required SWAP gates.

Note that calculating the Kendall tau distance between two permutations can be naively done in $\mathcal{O}(n^2)$ time, following the steps of the bubble sort algorithm [26]. A faster computation of the distance, in $\mathcal{O}(n\sqrt{\log n})$ time, can be found in [7].

We will however not be concerned with explicitly listing the Kendall tau distances for all $n!$ permutations. In order to avoid the listing, the metric should be implicitly calculated in the model. The objective function, variables and constraints that allow us to do so, will be introduced in the next section.

4 Mathematical Model

In this section the proposed ILP formulation of the NNC minimization problem will be discussed in detail. First, the variables and constraints are presented and explained. Finally, the complete model is given, along with a linearization of the constraints.

Given a quantum circuit $QC = (Q, G)$, we introduce integer variables $x_i^t \in L = \{1, \ldots, n\}$ for the location of each qubit $q_i \in Q$ before each gate $g^t \in G$. Since the goal is to avoid the explicit $n!$ scaling in the number of variables and constraints, we make use of the Kendall tau metric to count the number of required SWAP gates when going from one qubit order τ^t to the next τ^{t+1}. To accomplish this, keeping track of the pairwise order of the qubits is essential. We introduce binary variables to do precisely this,

$$y_{ij}^t = \begin{cases} 1 & \text{if location } x_i^t \text{ is before location } x_j^t \text{ in qubit order } \tau^t \\ 0 & \text{else.} \end{cases} \tag{3}$$

Keeping track of changes in the y variables when moving from one qubit order to the next allows us to count the amount of SWAP gates needed. The x and y variables are related through the following big-M type constraints,

$$x_i^t - x_j^t \leq My_{ij}^t - 1 \qquad \forall i,j \in Q, i < j, t \in [m] \tag{4}$$
$$x_j^t - x_i^t \leq M(1 - y_{ij}^t) - 1 \qquad \forall i,j \in Q, i < j, t \in [m] \tag{5}$$

where M is a big enough constant, $M = (n + 1)$ being sufficient in this case. Note that these constraints also enforce two important features:

1. No two qubits can be at the same location at the same time.
2. The definition of the y variables is enforced by the constraints.

For fixed i, j and t, one of the two constraints is always trivially satisfied due to the large value of M. The -1 term in the right-hand side even ensures that the location indices differ by at least one from each other. This allows us, later on, to relax the x variables to be continuous without losing the property that feasible solutions have integer x variables.

To make sure that the result also complies with the NN constraints, the following constraints need to be added:

$$x_i^t - x_j^t \leq 1 \qquad \forall g_{ij}^t \in G \tag{6}$$
$$x_i^t - x_j^t \geq -1 \qquad \forall g_{ij}^t \in G. \tag{7}$$

For each gate that acts on qubits q_i and q_j, the qubit order that is assumed just before the gate, it is required to have the qubits in adjacent locations.

The objective is to minimize the total amount of absolute changes in the y variables,

$$\min \sum_{\substack{i,j \in Q \\ i<j}} \sum_{t \in [m-1]} |y_{ij}^t - y_{ij}^{t+1}|. \tag{8}$$

Note that the objective function exactly computes the Kendall tau distance between every two consecutive qubit orders. Currently, the objective function is not linear, so extra binary variables k_{ij}^t are added to complete the model. These substitute $|y_{ij}^t - y_{ij}^{t+1}|$ in the objective function and are constrained in the following way

$$y_{ij}^t - y_{ij}^{t+1} \leq k_{ij}^t \qquad \forall i,j \in Q, i < j, t \in [m-1] \tag{9}$$
$$y_{ij}^t - y_{ij}^{t+1} \geq -k_{ij}^t \qquad \forall i,j \in Q, i < j, t \in [m-1]. \tag{10}$$

Now the k variables can be substituted into Expression (8), which, together with the constraints, result in the ILP model:

$$
\begin{aligned}
\min \quad & \sum_{\substack{i,j \in Q \\ i<j}} \sum_{t \in [m-1]} k_{ij}^t \\
\text{subject to} \quad & (4), (5), (6), (7), (9), (10) \\
& x_i^t \in \{1, \ldots, n\} && \forall i \in Q, t \in [m] \\
& y_{ij}^t \in \{0, 1\} && \forall i, j \in Q, i < j, t \in [m] \\
& k_{ij}^t \in \{0, 1\} && \forall i, j \in Q, i < j, t \in [m-1]
\end{aligned} \tag{11}
$$

The number of variables in this formulation is equal to

$$\# \text{ variables} = n^2 m - \frac{n^2 - n}{2}, \tag{12}$$

and the number of constraints in the ILP is equal to

$$\# \text{ constraints} = 2(n^2 - n)m - n^2 + n + 2m, \tag{13}$$

which is polynomial in the number of qubits and gates. In order to improve running times in practice, it helps to relax variables to take continuous values. We state the following about this relaxation:

Proposition 1. *Allowing the x- and k-variables to take continuous values does not change the optimal value.*

Proof. The x-variables must take values that are pairwise separated from each other by at least 1 due to constraints (4), (5). There are n variables that all have to take a value in a connected interval of length n, all spaced at least 1 from each other. This can only be done if the x's are all integer and all integer values are taken. The k-variables are constrained by (9), (10). Since the y-variables are binary, their difference is also binary (or -1, in which case $k = 0$ is allowed). Since we are minimizing over the k-variables, their value will always assume the smallest possible allowed value by the constraints, which is integer.

Even though relaxing these variables does not impact the objective value of optimal solutions, it reduces the number of integer-restricted variables which improves the running time in practice.

5 Experimental Results

In this section, the exact solutions provided by the proposed method are compared to the previous best exact approaches in terms of instances they can solve, as well as state-of-the-art heuristic approaches in terms of attained objective value.

5.1 Experimental Setup

The mathematical model as described in the previous section has been implemented in Python and solved with the commercial solver CPLEX 12.7 through the Python API. All but the quantum fourier transform instances, which were constructed following the circuit of [37], were obtained form the RevLib [49] website. The evaluations were conducted using up to 16 threads of 2.4 GHz each, working with 16 GB of RAM. All instances were solved to optimality.

The benchmark instances are subdivided over three tables, according to the amount of qubits addressed. In the first column of each table, the name of the circuit is provided, and in the second column, n denotes the number of qubits in the circuit. In the third column, $|G|$ denotes the number of 2-qubit gates present in the circuit after gate decomposition and the removal of single-qubit gates. The optimal value of the local reordering problem, i.e., the minimum number of needed SWAP gates to make the circuit nearest neighbor compliant, is provided in the fourth column. The column "Time" denotes the run time in seconds. The column entitled "Time E" denotes the running time of other exact methods, also in seconds. Exact running times with subscript a are from [52], subscript b from [35]. Heuristic solution's objective values are presented in the last column, denoted by "# SWAPS H". Here the subscript c indicates the results are from [28], subscript d from [42], subscript e from [2], subscript f from [27] and subscript g from [50]. An asterisk as superscript indicates that for the other exact solution methods, either the objective value differs, or the number of gates differs or they both differ. For the heuristic results, the asterisk indicates that the number of gates differs or the objective value of the heuristic is lower than that of the proposed exact method. These anomalies are believed to find their roots in differing gate decomposition methods, resulting in slightly different instances.

5.2 Results

The running time required to solve the instance is heavily dependent on three factors:

1. The number of qubits in the quantum circuit,
2. The number of gates in the quantum circuit,
3. The minimal number of required SWAP gates.

The number of qubits and gates is expected to heavily influence the running time. The number of qubits is the term that influences the run time the most. This is due to the fact that the number of feasible solutions scales factorially in the number of qubits. Surprisingly, the run time also scales quite badly with the number of required SWAP gates. During the Branch & Bound tree search, the upper bound determined by CPLEX, which is the best feasible solution found up to that point, converges to the optimal value (or close to it) rather quickly. The best known lower bound, however, takes a long time to improve. When the number of required SWAP gates increases, the time needed to improve the lower bound all the way to the optimal value increases as well. This phenomenon is analyzed for two of the benchmark instances that require a lot of SWAP gates:

1. **mod8-10_177.** The search method found a feasible solution with an objective value within 10% of the optimal value in $2.4 \cdot 10^6$ iterations, found an optimal solution in $4.0 \cdot 10^7$ iterations, and proved optimality by a matching lower bound after $1.7 \cdot 10^8$ iterations.

Table 1. Benchmark instances with three or four qubits

Benchmark	n	$\|G\|$	$\|S\|$	Time	Time E	# SWAPS H
QFT_QFT3	3	3	1	0.02	-	-
peres_10	3	4	1	0.14	0.1_a	-
peres_8	3	4	1	0.06	0.1_a	-
toffoli_2	3	5	1	0.12	0.2_a	-
toffoli_1	3	5	1	0.1	0.1_a	-
peres_9	3	6	1	0.02	2463_a	-
fredkin_7	3	7	1	0.16	-	-
ex-1_166	3	7	2	0.08	0.1_a	-
fredkin_5	3	7	1	0.15	$0.1_a, 0.1^*_b$	-
ham3_103	3	8	2	0.04	-	-
miller_12	3	8	2	0.14	$745.6_a, 0.1_b$	-
ham3_102	3	9	1	0.05	0.1^*_a	-
3_17_15	3	9	2	0.04	$630.2_a, 0.1^*_b$	-
3_17_13	3	13	3	0.12	0.1^*_a	$4^*_c, 4_d, 3_e, 6_g$
3_17_14	3	13	3	0.15	0.1^*_a	-
fredkin_6	3	15	3	0.06	4.6_a	-
miller_11	3	17	4	0.15	0.1^*_a	-
QFT_QFT4	4	6	3	0.17	-	-
toffoli_double_3	4	7	1	0.11	$0.9_a, 0.1^*_b$	-
rd32-v1_69	4	8	2	0.16	0.1_a	-
decod24-v1_42	4	8	2	0.12	$7.7_a, 0.1^*_b$	-
rd32-v0_67	4	8	2	0.07	1.6_a	$2_c, 2_d$
decod24-v2_44	4	8	3	0.07	0.1^*_b	-
decod24-v0_40	4	8	3	0.06	0.1^*_b	-
decod24-v3_46	4	9	3	0.09	$0.1_a, 0.1^*_b$	$3_c, 3_d$
toffoli_double_4	4	10	2	0.07	200^2_a	-
rd32-v1_68	4	12	3	0.24	0.4^*_a	-
rd32-v0_66	4	12	0	0.09	0.4^*_a	-
decod24-v0_39	4	15	5	0.53	0.5_a	-
decod24-v2_43	4	16	5	0.23	0.1^*_a	-
decod24-v0_38	4	17	4	0.57	19.2_a	-
decod24-v1_41	4	21	7	0.5	-	-
hwb4_52	4	23	8	0.97	-	$9_c, 10_d, 9_e, 9_f$
aj-e11_168	4	29	12	5.36	-	-
4_49_17	4	30	12	6.1	-	$12^*_c, 12_d, 16_e$
decod24-v3_45	4	32	13	6.25	-	-
mod10_176	4	42	15	7.94	-	-
aj-e11_165	4	44	18	9.36	-	$36_d, 33^*_g$
mod10_171	4	57	24	27.18	-	-
4_49_16	4	59	22	24.23	-	-
mini-alu_167	4	62	27	23.7	-	-
hwb4_50	4	63	23	17.61	-	-
hwb4_49	4	65	23	21.64	-	-
hwb4_51	4	75	28	75.09	-	-

2. **decod24-enable_126.** The search method found a feasible solution with an objective value within 10% of the optimal value in $5.3 \cdot 10^6$ iterations, found an optimal solution in $1.0 \cdot 10^7$ iterations, and proved optimality by a matching lower bound after $5.8 \cdot 10^7$ iterations.

If a 10% optimality gap would suffice, only less than 2% of the total number of iterations would be needed in the first case, and 10% in the second case. This observation indicates that running an incomplete Branch and Bound algorithm might be an interesting and easy-to-implement heuristic algorithm.

The 131 evaluated benchmark instances are listed in Tables 1, 2 and 3. The improvement in computation time with respect to previous exact methods is significant. The results show exact solutions that are obtained for much larger circuits than previously held possible. The largest instance with respect to the number of qubits has as much as 18 qubits. Furthermore, for the first time, NNC has been solved to optimality for circuits with more than 100 quantum gates.

Table 2. Benchmark instances with six or more qubits. No times of other exact methods are known.

Benchmark	n	$\|G\|$	$\|S\|$	Time	# SWAPS H
graycode6_47	6	5	0	0.02	-
graycode6_48	6	5	0	0.02	-
QFT_QFT6	6	15	11	7.43	$11_c, 12_d$
decod24-enable_124	6	21	5	1.86	-
decod24-enable_125	6	21	5	1.83	-
decod24-bdd_294	6	24	7	9.37	-
mod5adder_129	6	71	34	534.38	-
mod5adder_128	6	77	36	1103.51	$45_c^*, 51_d, 46_g^*$
decod24-enable_126	6	86	37	1954.28	-
xor5_254	7	5	3	0.61	-
ex1_226	7	5	3	0.25	-
QFT_QFT7	7	21	16	28.26	$28_c, 26_d, 18_g$
4mod5-bdd_287	7	23	7	4.3	-
ham7_106	7	49	28	495.43	-
ham7_105	7	65	34	1613.33	-
ham7_104	7	83	42	3238.82	56_c^*
QFT_QFT8	8	28	23	334.6	$32_c, 33_d, 31_g$
rd53_139	8	36	11	76.29	-
rd53_138	8	44	11	100.86	-
rd53_137	8	66	35	6271.11	-
QFT_QFT9	9	36	30	1482.53	$52_c, 54_d, 49_g$
QFT_QFT10	10	45	39	39594.99	64_g
mini_alu_305	10	57	23	1711.75	-
sys6-v0_144	10	62	19	887.71	-
rd73_141	10	64	21	845.05	-

Table 3. Benchmark instances with five qubits

Benchmark	n	$\|G\|$	$\|S\|$	Time	Time E	# SWAPS H
4mod5-v1_25	5	7	1	0.26	11705.3_a	-
4gt11_84	5	7	1	0.06	16.6_a	$1_c, 1_d, 1_e$
4gt11-v1_85	5	7	1	0.09	-	-
4mod5-v0_20	5	8	2	0.08	45.5_a	-
4mod5-v1_22	5	9	1	0.08	548.8^*_a	-
QFT_QFT5	5	10	6	0.41	1.6_a	$7_c, 6_d$
mod5d1_63	5	11	2	0.12	-	-
4mod5-v0_19	5	12	3	0.84	55.3^*_a	-
4gt11_83	5	12	3	0.15	9^*_a	-
4mod5-v1_24	5	12	3	0.28	-	-
mod5mils_65	5	12	4	0.26	-	-
mod5mils_71	5	12	2	0.15	-	-
alu-v2_33	5	13	4	0.45	-	-
alu-v1_29	5	13	4	0.61	-	-
alu-v0_27	5	13	4	0.48	-	-
mod5d2_70	5	14	5	0.43	-	-
alu-v3_35	5	14	5	0.38	-	-
alu-v4_37	5	14	5	0.37	-	-
alu-v1_28	5	14	4	0.26	-	-
4gt13-v1_93	5	15	5	0.69	489.3^*_a	$7^*_c, 6_d, 4^*_e$
4gt13_92	5	15	6	0.53	-	-
4gt11_82	5	16	6	0.89	-	-
4mod5-v0_21	5	17	8	2.84	-	-
rd32_272	5	18	7	0.94	-	-
alu-v3_34	5	18	4	0.4	-	-
mod5d2_64	5	19	6	1.81	-	-
alu-v0_26	5	21	8	3.56	-	-
4gt5_75	5	21	6	1.1	-	$9^*_c, 12_d$
4mod5-v0_18	5	23	8	3.35	-	-
4mod5-v1_23	5	24	9	5.06	-	$9_c, 9_d, 15_e$
one-two-three-v2_100	5	24	7	5.37	-	-
one-two-three-v3_101	5	24	7	2.96	-	-
rd32_271	5	26	11	7.37	-	-
4gt5_77	5	28	10	6.2	-	-
4gt5_76	5	29	10	5.45	-	-
alu-v4_36	5	30	9	6.34	-	$15^*_c, 18_d, 17_e$

(*continued*)

Table 3. (*continued*)

Benchmark	n	$\|G\|$	$\|S\|$	Time	Time E	# SWAPS H
4gt13_91	5	30	8	4.46	-	-
4gt13_90	5	34	12	6.77	-	-
4gt10-v1_81	5	34	13	12.38	-	$18^*_c, 20_d, 16_e, 24^*_g$
one-two-three-v1_99	5	36	15	17.27	-	-
4gt4-v0_80	5	36	19	43.45	-	$34_d, 33_f$
4mod7-v0_94	5	38	12	12.83	-	-
alu-v2_32	5	38	16	22.05	-	-
4mod7-v0_95	5	38	14	14.59	-	$19^*_c, 21_d, 22_e$
4mod7-v0_95	5	38	14	14.59	-	$19^*_c, 21_d, 22_e$
4mod7-v1_96	5	38	14	13.49	-	-
one-two-three-v0_98	5	40	15	15.67	-	-
4gt12-v0_88	5	41	20	34.01	-	-
4gt12-v1_89	5	44	22	52.36	-	$35_d, 26_e, 32_f$
sf_275	5	46	18	21.42	-	-
4gt4-v0_79	5	49	22	80.16	-	-
4gt4-v0_78	5	53	26	167.03	-	-
4gt4-v0_72	5	53	24	49.7	-	-
4gt12-v0_87	5	54	22	45.88	-	-
4gt4-v1_74	5	57	29	84.87	-	-
4gt12-v0_86	5	58	26	108.35	-	-
mod8-10_178	5	68	37	389.47	-	-
one-two-three-v0_97	5	71	32	76.8	-	-
4gt4-v0_73	5	89	40	699.65	-	-
mod8-10_177	5	93	48	3650.26	-	72_d
alu-v2_31	5	100	49	2906.35	-	-
hwb5_55	5	101	48	2264.0	-	$59_c, 63_d, 60_e, 66_g$
rd32_273	5	104	50	4631.7	-	-
alu-v2_30	5	112	55	13558.87	-	-

6 Conclusion

In this paper we consider the local reordering scheme for nearest neighbor architectures of quantum circuits. We propose a new mathematical model that counts the number of required SWAP gates implicitly, by using specific properties of the constraints. The implicit counting improves upon previous exact approaches in which costs were explicitly determined for each permutation, leading to a factorial scaling of the model size, and therefore, a high running time. The presented innovations result in a great improvement in the model size, such that the resulting ILP only contains $\mathcal{O}(n^2m)$ variables and constraints.

The benchmark instances with available exact solutions known in the literature were no larger than circuits with five qubits and no more than twenty gates, due to the excessive running times. The proposed method can handle quantum circuits with five qubits and 112 gates or up to eighteen qubits and sixteen gates. In total 131 benchmark instances are evaluated, most of which have not been solved to optimality.

Because the implicit counting is based on counting inversions in permutations, the formulation is not easily translated to the popular higher dimensional cases where qubits are placed on a 2D or 3D grid. To the authors' best knowledge there is no known polynomial time algorithm that, in 2- or 3-dimensional grids, solves the subproblem of calculating the minimum number of required SWAP gates when transforming one qubit order into another. Such a method could have great impact on exact solution methods in the higher-dimensional setting.

Practical experience with the Branch & Bound tree search indicates that finding a (near) optimal feasible solution does not consume the most computation time. This means that solving the ILP heuristically, with a restriction in running time or iteration count for example, could make for a good heuristic solution method.

For future research we propose to look at algorithms dedicated to specific hardware topologies and benchmarks on larger number of qubits and gates, potentially with approximate, nearly-optimal approach and comparison with existing transpilers.

References

1. AlFailakawi, M.G., et al.: Harmony-search algorithm for 2d nearest neighbor quantum circuits realization. Exp. Syst. with Appl. **61**, 16–27 (2016)
2. AlFailakawi, M., AlTerkawi, L., Ahmad, I., Hamdan, S.: Line ordering of reversible circuits for linear nearest neighbor realization. Quantum Inf. Process. **12**(10), 3319–3339 (2013)
3. Alhagi, N.: Synthesis of Reversible Functions Using Various Gate Libraries and Design Specifications. Portland State University, Technical report (2000)
4. Amini, J.M., et al.: Toward scalable ion traps for quantum information processing. New J. Phys. **12**(3), 033031 (2010)
5. Barenco, A., et al.: Elementary gates for quantum computation. Phys. Rev. A **52**(5), 3457–3467 (1995)
6. Bonnet, E., et al.: Complexity of token swapping and its variants. Algorithmica **80**(9), 2656–2682 (2018)
7. Chan, T.M., Pǎtraşcu, M.: Counting inversions, offline orthogonal range counting, and related problems. In: Discrete Algorithms, pp. 161–173 (2010)
8. Choi, B.S., Van Meter, R.: An $\Theta\sqrt{n}$-depth quantum adder on a 2d NTC quantum computer architecture. J. Emerg. Technol. Comput. Syst. **8**(3), 1–22 (2012)
9. Devitt, S.J., Fowler, A.G., et al.: Architectural design for a topological cluster state quantum computer. New J. Phys. **11**(8), 083032 (2009)

10. Ding, J., Yamashita, S.: Exact synthesis of nearest neighbor compliant quantum circuits in 2d architecture and its application to large-scale circuits. IEEE Trans. Comput. Aided Des. Integr. Circ. Syst. **39**, 1045–1058 (2019)
11. DiVincenzo, D.P.: IBM: the physical implementation of quantum computation. Fortschr. der Phys. **48**(9–11), 771–783 (2000)
12. DiVincenzo, D.P., Solgun, F.: Multi-qubit parity measurement in circuit quantum electrodynamics. New J. Phys. **15**(7), 075001 (2013)
13. Dueck, G.W., Pathak, A., et al.: Optimization of Circuits for IBM's five-qubit Quantum Computers, pp. 680–684 (2018)
14. Fowler, A.G., et al.: Implementation of Shor's Algorithm on a Linear Nearest Neighbour Qubit Array. arXiv:quant-ph/0402196 (2004)
15. Fowler, A.G., et al.: Quantum error correction on linear nearest neighbor qubit arrays. Phys. Rev. A **69**(4), 042314 (2004)
16. Garey, M.R., Johnson, D.S.: Computers and intractability; a guide to the theory of NP-completeness. W. H. Freeman & Co., New York, NY, USA (1979)
17. Große, D., et al.: Exact multiple-control toffoli network synthesis with SAT techniques. Comput. Aided Des. Integr. Circ. Syst. **28**(5), 703–715 (2009)
18. Grover, L.K.: Quantum mechanics helps in searching for a needle in a haystack. Phys. Rev. Lett. **79**(2), 325–328 (1997)
19. Hattori, W., Yamashita, S.: Quantum circuit optimization by changing the gate order for 2d nearest neighbor architectures. In: Kari, J., Ulidowski, I. (eds.) RC 2018. LNCS, vol. 11106, pp. 228–243. Springer, Cham (2018). https://doi.org/10.1007/978-3-319-99498-7_16
20. Herrera-Martí, D.A., et al.: A photonic implementation for the topological cluster state quantum computer. Phys. Rev. A **82**(3), 032332 (2010)
21. Hirata, Y., et al.: An efficient conversion of quantum circuits to a linear nearest neighbor architecture. Quant. Inf. Comput. **11**(1&2), 25 (2011)
22. Itoko, T., et al.: Quantum circuit compilers using gate commutation rules. In: Design Automation, pp. 191–196. ACM Press, Tokyo, Japan (2019)
23. Jerrum, M.R.: The complexity of finding minimum-length generator sequences. Theor. Comput. Sci. **36**, 25 (1985)
24. Jones, N.C., et al.: Layered architecture for quantum computing. Phys. Rev. X **2**(3), 031007 (2012)
25. Kawahara, J., et al.: The time complexity of the token swapping problem and its parallel variants. In: WALCOM: Algorithms and Computation, pp. 448–459 (2017)
26. Knuth, D.E.: The Art of Computer Programming, Volume 3: Sorting and Searching, Second Edition, vol. 3, 2nd edn. (1974)
27. Kole, A., et al.: A heuristic for linear nearest neighbor realization of quantum circuits by SWAP gate insertion using$N$-gate lookahead. IEEE J. Emerg. Sel. Top. Circ. Syst. **6**(1), 62–72 (2016)
28. Kole, A., et al.: A new heuristic for N - dimensional nearest neighbor realization of a quantum circuit. IEEE Trans. Comput. Aided Des. Integr. Circ. Syst. **37**(1), 182–192 (2018)
29. Kole, A., et al.: Towards a cost metric for nearest neighbor constraints in reversible circuits. Rev. Comput. **9138**, 273–278 (2015)
30. Kumph, M., Brownnutt, M., Blatt, R.: Two-dimensional arrays of radio-frequency ion traps with addressable interactions. New J. Phys. **13**(7), 073043 (2011)
31. Lin, C., et al.: PAQCS: physical design-aware fault-tolerant quantum circuit synthesis. IEEE Trans. Very Large Scale Int. Syst. **23**(7), 1221–1234 (2015)
32. Linke, N.M., et al.: Experimental comparison of two quantum computing architectures. Proc. Natl. Acad. Sci. U.S.A. **114**(13), 3305–3310 (2017)

33. Markov, I.L., Saeedi, M.: Constant-Optimized Quantum Circuits for Modular Multiplication and Exponentiation. arXiv:1202.6614 [quant-ph] (2012)
34. Maslov, D., et al.: Quantum circuit simplification using templates. In: Design, Automation and Test in Europe, pp. 1208–1213. IEEE, Munich, Germany (2005)
35. Matsuo, A., Yamashita, S.: Changing the gate order for optimal LNN conversion. In: De Vos, A., Wille, R. (eds.) RC 2011. LNCS, vol. 7165, pp. 89–101. Springer, Heidelberg (2012). https://doi.org/10.1007/978-3-642-29517-1_8
36. Nickerson, N.H., et al.: Topological quantum computing with a very noisy network and local error rates approaching one percent. Nat. Commun. **4**(1), 1756 (2013)
37. Nielsen, M.A., et al.: Quantum computation and quantum information. Am. J. Phys. **70**(5), 558–559 (2002)
38. Ohliger, M., Eisert, J.: Efficient measurement-based quantum computing with continuous-variable systems. Phys. Rev. A **85**(6), 062318 (2012)
39. Pedram, M., Shafaei, A.: Layout optimization for quantum circuits with linear nearest neighbor architectures. IEEE Circ. Syst. Mag. **16**(2), 62–74 (2016)
40. Pham, P., Svore, K.M.: A 2d nearest-neighbor quantum architecture for factoring in polylogarithmic depth. arXiv:1207.6655 [quant-ph] (2012)
41. Saeedi, M., et al.: Synthesis of quantum circuits for linear nearest neighbor architectures. Quant. Inf. Process. **10**(3), 355–377 (2011)
42. Shafaei, A., et al.: Optimization of quantum circuits for interaction distance in linear nearest neighbor architectures. In: Design Automation Conference, pp. 1–6 (2013)
43. Shafaei, A., et al.: Qubit placement to minimize communication overhead in 2d quantum architectures. In: Design Automation Conference, pp. 495–500 (2014)
44. Shor, P.: Algorithms for quantum computation: discrete logarithms and factoring. In: Foundations of Computer Science, pp. 124–134 (1994)
45. Siraichi, M.Y., et al.: Qubit allocation. In: Code Generation and Optimization, pp. 113–125. ACM, New York, NY, USA (2018)
46. Takahashi, Y., et al.: The Quantum Fourier transform on a linear nearest neighbor architecture. Quant. Info. Comput. **7**(4), 383–391 (2007)
47. Versluis, R., et al.: Scalable quantum circuit and control for a superconducting surface code. Phys. Rev. Appl. **8**(3), 034021 (2017)
48. Wille, R., et al.: Mapping Quantum Circuits to IBM QX Architectures Using the Minimal Number of SWAP and H Operations. In: Design Automation Conference, pp. 1–6. (2019)
49. Wille, R., et al.: RevLib: an online resource for reversible functions and reversible circuits. In: Multiple Valued Logic, pp. 220–225 (2008)
50. Wille, R., et al.: Look-ahead schemes for nearest neighbor optimization of 1d and 2d quantum circuits. In: Design Automation Conference, pp. 292–297 (2016)
51. Wille, R., et al.: Considering nearest neighbor constraints of quantum circuits at the reversible circuit level. Quant. Inf. Process. **13**(2), 185–199 (2014)
52. Wille, R., et al.: Exact reordering of circuit lines for NN quantum architectures. IEEE Comput. Aided Des. Integr. Circ. Syst. **33**(12), 1818–1831 (2014)
53. Yao, N.Y., et al.: Quantum logic between remote quantum registers. Phys. Rev. A **87**(2), 022306 (2013)

54. Zulehner, A., Bauer, H., Wille, R.: Evaluating the flexibility of A* for mapping quantum circuits. In: Thomsen, M.K., Soeken, M. (eds.) RC 2019. LNCS, vol. 11497, pp. 171–190. Springer, Cham (2019). https://doi.org/10.1007/978-3-030-21500-2_11
55. Zulehner, A., et al.: An efficient methodology for mapping quantum circuits to the IBM QX architectures. IEEE Trans. Comput. Aided Des. Integr. Circ. Syst. **38**(7), 1226–1236 (2019)

Translating Constraints into QUBOs for the Quadratic Knapsack Problem

Tariq Bontekoe[1], Frank Phillipson[1,2(✉)], and Ward van der Schoot[1]

[1] TNO Department Applied Cryptography and Quantum Algorithms, The Hague, The Netherlands
{frank.phillipson,ward.vanderschoot}@tno.nl

[2] School of Business and Economics, Maastricht University, Maastricht, The Netherlands

Abstract. One of the first fields where quantum computing will likely show its use is optimisation. Many optimisation problems naturally arise in a quadratic manner, such as the quadratic knapsack problem. The current state of quantum computers requires these problems to be formulated as a quadratic unconstrained binary optimisation problem, or QUBO. Constrained quadratic binary optimisation can be translated into QUBOs by translating the constraint. However, this translation can be made in several ways, which can have a large impact on the performance when solving the QUBO. We show six different formulations for the quadratic knapsack problem and compare their performance using simulated annealing. The best performance is obtained by a formulation that uses no auxiliary variables for modelling the inequality constraint.

Keywords: quadratic knapsack problem · quadratic unconstrained binary optimisation problem · quantum computing · simulated annealing

1 Introduction

Over the past decades quantum computing research has made an impressive growth. In less than 25 years, quantum computers have evolved from laboratory experiments to public-access devices which are already being used in practice [8,19]. While there is currently no advantage in using quantum devices over their classical counterparts, their development suggests that this may soon be otherwise. Experts believe that quantum computers will first show its use in the fields of chemistry, machine learning and optimisation, as indicated by [21]. In this work, we will focus on the latter.

Within the field of optimisation, the *Knapsack Problem* is well-known. In short, the problem entails of selecting a subset of items with the highest total gain, such that the weight or costs of that subset is below a certain limit. In mathematical terms we can define it as follows. We have a list of N objects, labelled by indices i, where the weight of each object is given by w_i, and its

J. Mikyška et al. (Eds.): ICCS 2023, LNCS 14077, pp. 90–107, 2023.
https://doi.org/10.1007/978-3-031-36030-5_8

value given by c_i. We have a knapsack which can only carry weight W. If x_i is a binary variable denoting whether (denoted by 1) or not (denoted by 0) object i is contained in the knapsack, the total weight and gain of the knapsack are: $\mathcal{W} = \sum_{i=1}^{N} w_i x_i$ and $\mathcal{C} = \sum_{i=1}^{N} c_i x_i$, respectively. We wish to maximise within the limitations of our knapsack, which gives the optimisation problem $\max \mathcal{C}$ such that $\mathcal{W} \leq W$.

Over the years, various adaptations to the basic knapsack problem have been proposed. In [1] and [4], they give an overview of such adaptations, such as multiple knapsacks, special constraints, fractional items, multiple dimensions and many more. This work studies the adaptation called the Quadratic Knapsack Problem (QKP) as proposed by [9]. The knapsack problem above is extended by adding a value c_{ij} for every two nodes i and j. This value equals the extra gain we obtain if both object i and j are in the knapsack. Using the notation $c_{ii} = c_i$, the total gain in case of the QKP then equals: $\mathcal{C} = \sum_{i=1}^{N} \sum_{j=1}^{N} c_{ij} x_i x_j$, which is a problem of quadratic nature. The problem is known to be (strongly) NP-hard as shown in [12]. Applications of the QKP can be found in the field of telecommunication, logistics, production and more [20].

Numerous methods have been proposed to solve the QKP since its introduction in 1980. From early on, various methods based upon the Langrangian methods were proposed, for example by Billionnet et al. [2] and Caprara et al. [5]. These methods work by decomposing the QKP in smaller instances, after which a branch-and-bound algorithm is used to find the final solution. A similar approach is given by Pisinger et al. [20], which reduces the dimension of the problem by aggresive reduction instead of decomposition, after which a branch-and-bound algorithm is used to find the final solution as well. Other methods take a different approach in which they linearise the QKP, after which the method ends with a branch-and-bound algorithm again. For example, Billionnet and Soutif [3] proposed three ways of linearising the QKP using Mixed-Integer Programs, while Rodrigues et al. [23] obtained linearisation by replacing quadratic terms by linear constraints. More recently, Schauer [24] showed that a greedy algorithm achieves asymptotically the same results on the instances usually used in QKP works. The most recent work comes from Fomeni et al. [6], which presented a cut-and-branch algorithm which combines the concepts of cutting-planes and branch-and-bound.

With the rise of quantum computing, it is interesting to wonder how quantum devices perform at solving the QKP. Especially the quadratic nature of the QKP is of particular interest for quantum computing, as currently most quantum computing methods for solving optimisation problems require problems to be of quadratic nature. While we do not expect quantum computing to efficiently give best-case solutions for all NP-hard problems, it does seem to help in finding better approximate solutions and finding them faster [21], as shown in, for example, [13,16,19].

To apply quantum computing to quadratic optimisation problems, we usually require them to be formulated as a Quadratic Unconstrained Binary Optimisation problem (QUBO), or Ising problem. At the time of writing, there are two popular quantum techniques that can solve QUBOs, namely the so-called

Quantum Approximate Optimisation Algorithm (QAOA) on gate-based quantum devices, as proposed by [7], and Quantum Annealing, which runs on a different form of quantum device, namely quantum annealers [11]. While these methods solve such QUBOs in a generic, problem-independent way, the formulating of such QUBOs is different for each problem. It should be noted that there can be different ways of formulating a problem as a QUBO.

For many famous optimisation problems at least one QUBO formulation is known, examples of which can be found in [10,14]. The second overview also shows a QUBO formulation for the QKP, which is used in the thesis from [15]. On the contrary, the number of optimisation problems for which multiple QUBO formulations are present in literature is much lower. Even if QUBO formulations solve the same problem, they can be different both mathematically and performance-wise. That is why for practical purposes, it can be interesting to consider different QUBO formulations for the same problem.

In this work, we list various QUBO formulations for the QKP, either from literature or constructed by the authors. In addition, we compare their performance in practice by using the classical method of Simulated Annealing, which is similar to quantum annealing. While different mathematically equivalent formulations for general QUBOs have been studied before [22], we take a different approach. We specifically consider how the translation of the constraint into the QUBO influences its performance. To the our best knowledge, we are the first to make a comparison of the performance of different QUBO formulations for the QKP based on different ways of incorporating the constraint into the QUBO. While our work focuses on the QKP, it has implications for other optimisation problems as well. Particularly, we show that it can be very beneficial to consider other or multiple ways of translating constraints into a QUBO.

We have structured our work as follows. We start with background information on quantum and simulated annealing, a definition of a QUBO and different QUBO formulations for the QKP in Sect. 2. Then, we describe our approach at comparing these formulations and list the corresponding results in Sect. 3. In Sect. 4 the conclusions and directions for further research are presented.

2 Background

In this section we explain how quantum and simulated annealing works and define the general QUBO formulation. Subsequently, we explain how to formulate the Quadratic Knapsack Problem as a QUBO and present five additional QUBO formulations of the QKP. All of these formulations use a different technique to include the weight constraint into the objective function.

2.1 Quantum and Simulated Annealing

Quantum annealing is a quantum computing optimisation process specifically suitable for finding minimal solutions of objective functions with many local minima. It is designed for objective functions which have a certain number of binary decision variables and it works by mapping each of the decision variables to one or multiple qubits on the quantum annealer. Each of the basis states of the qubits then correspond to a possible assignment of the decision variables. QUBO problems are perfectly suitable for being solved in a quantum annealer.

A quantum annealer finds minima in the following way. It starts out in an equal superposition of all possible states, after which it lets the qubits evolve under a problem-specific Hamiltonian. This Hamiltonian yields a certain energy landscape, in which minima of the energy landscape corresponds to minimal solutions of the original objective function. By evolving this system for a suitable time, the quantum state ends up near a minimum of the energy landscape, after which a measurement likely results into a local minimum of the objective function with high probability. The shape of the energy landscape directly influences the ability of the system to evolve towards lower local minima. In general, a smoother energy landscape results in lower minima and hence has a better performance.

In this work we do not work with quantum annealing, but with simulated annealing instead. This which can be seen as the classical alternative of quantum annealing, but it does not simulate quantum annealing. Instead, it simulates the annealing process found in metallurgy. Still, simulated annealing is suitable for the same family of problems as quantum annealing. Just like quantum annealing, simulated annealing is a method which walks along the energy landscape and tries to find local minima. Where quantum annealing achieves this by running a quantum system under a specific Hamiltonian, simulated annealing uses a temperature parameter. At each time step, the simulated annealing solver chooses a candidate solution which directly neighbours the current solution. When the temperature parameter is higher, the system is more likely to accept worse solutions, allowing it to explore a wide range of solutions. By gradually decreasing the temperature, this becomes less likely because of which the system is likely to settle in one of the local minima at the end of the process.

2.2 QUBO Definition

A QUBO is an optimisation problem of the form $\min y = x^t Q x$, where $x \in \{0,1\}^n$ are binary decision variables and Q is an $n \times n$ coefficient matrix. The term y is called the *objective function.* Note that this expression contains linear as well as quadratic terms, as $x_i^2 = x_i$ for decision variables $x_i \in \{0,1\}$. Many NP-hard problems can be written as a QUBO [10].

While some NP-hard problems naturally arise in this form, most, including the QKP, contain constraints as well. If these constraints are linear, they can be transformed into a quadratic *penalty function* and added to the objective function. This should be done so that minimising the penalty function corresponds to satisfying the constraints. In this way we can write quadratic problems with linear constraints as a QUBO as well. Linear inequality contraints can be included as a quadratic penalty function as well by considering them as many equality constraints together.

For example, consider a general quadratic objective function $x^t Q x$ with linear constraint $Ax = b$, corresponding to the problem,

$$\min y = x^t Q x, \text{ subject to } Ax = b. \tag{1}$$

We bring the constraint into the objective function by adding a penalty term:

$$\min y = x^t Qx + \lambda(Ax - b)^t(Ax - b) = x^t Qx + x^t Rx + d = x^t Px, \qquad (2)$$

where $P = Q + R$ and the matrix R and constant d follow from the matrix multiplication. Note that the term d can be neglected, as it is constant.

The term λ is called the *weight* of the penalty function, or the *penalty value.* This parameter controls both the importance of the constraint, as well as the performance of the resulting QUBO. On the one hand, larger values of λ correspond to a higher likelihood that the constraint is met in minimal solutions. On the other hand, larger values of λ also reduce the performance of the resulting QUBO in practice. The choice of suitable weight is a study in itself and depends largely on the application. While the theoretical value to enforce the constraint can be computed for each use case, this usually results in too large weights, which influence the energy landscape too much. Each time, a deliberate consideration has to be made between performance and whether the constraints are satisfied.

2.3 Original QUBO Formulation for the QKP

To model the QKP as a QUBO, we start with the formulation from Sect. 1:

$$\max \mathcal{C} = \sum_{i=1}^{N}\sum_{j=1}^{N} c_{ij}x_i x_j \quad \text{s.t.} \quad \mathcal{W} = \sum_{i=1}^{N} w_i x_i \leq W. \qquad (3)$$

If we assume the constraint in Eq. (1) to be an equality first, we can create the following objective and penalty function respectively, with penalty weight λ:

$$\sum_{j=1}^{N}\sum_{i=1}^{N} c_{ij}x_i x_j \quad \text{and} \quad \lambda\left(W - \sum_{i=1}^{N} w_i x_i\right)^2. \qquad (4)$$

To derive the inequality constraint we have to introduce auxiliary variables to fill up the inequality to an equality:

$$H_B = \lambda\left(W - \sum_{i=1}^{N} w_i x_i - \sum_{j=1}^{M} 2^{j-1} y_j\right)^2,$$

where y_j are $M = \lceil \log_2(W + 1) \rceil$ auxiliary binary variables. This yields a binary sum which can yield all numbers up to $\mathcal{W}$. This results in the following QUBO:

$$-\sum_{i=1}^{N}\sum_{j=1}^{N} c_{ij}x_i x_j + \lambda\left(W - \sum_{i=1}^{N} w_i x_i - \sum_{k=1}^{M} 2^{k-1} y_k\right)^2. \qquad (5)$$

This QUBO formulation was originally posed by [10].

2.4 Alternative QUBO Formulations for the QKP

We now list five alternative QUBO formulations for the QKP. These QUBO formulations are devised by the authors and are, to the best of our knowledge, not listed in literature yet. We call the formulation of Eq. (5) **Type 1**. All these formulations can easily be shown to be solve the same QKP problem. For each type, we will indicate what the difference is between the different formulations and mention in which way they encode the constraint.

Type 2:

$$-\sum_{i=1}^{N}\sum_{j=1}^{N} c_{ij}x_ix_j + \lambda\left(\left(W+1-2^{M-1}\right)y_M + \sum_{k=1}^{M-1} 2^{k-1}y_k - \sum_{i=1}^{N} w_ix_i\right)^2. \quad (6)$$

In contrast to Type 1, we use the slack variables to encode the remaining capacity instead of encoding the total weight of the items. To achieve this, an offset of $2^{M-1}-1$ is introduced and again a set of binary auxiliary variables. In this case again $M = \lceil \log_2(W+1) \rceil$ auxiliary binary variables are needed.

Type 3:

$$-\sum_{i=1}^{N}\sum_{j=1}^{N} c_{ij}x_ix_j + \lambda\left(W - \sum_{i=1}^{N} w_ix_i - \sum_{k=1}^{M} (k-1)\,y_k\right)^2. \quad (7)$$

This QUBO formulation is similar to Type 1, but in this definition a one-hot encoding is used instead of a binary one. Here, $M = \max_{i=1}^{n} w_i$ auxiliary variables are needed, as any solution with a difference between total weight and capacity larger than M is clearly suboptimal, since one can add any item without violating the capacity constraint. Note that this QUBO formulation only works when all c_{ij} are non-negative, when any of the c_{ij} are non-positive, one should set $M = W + 1$. Also note that the one-hot encoding is not strictly enforced.

Type 4:

$$-\sum_{i=1}^{N}\sum_{j=1}^{N} c_{ij}x_ix_j + \lambda\left(\sum_{k=1}^{M} (W-k+1)\,y_k - \sum_{i=1}^{N} w_ix_i\right)^2. \quad (8)$$

This formulation is similar to Type 2, however we use a one-hot encoding instead of a binary encoding. Here also, $M = \max_{i=1}^{n} w_i$ auxiliary variables are needed, using the same trick. Again, this trick requires that all c_{ij} are non-negative and if not, we require $M = W + 1$ auxiliary variables. Note that again the one-hot encoding is not strictly enforced.

Type 5:

$$-\sum_{i=1}^{N}\sum_{j=1}^{N} c_{ij}x_ix_j + \lambda\left(W - W_{\text{offset}} - \sum_{i=1}^{N} w_ix_i\right)^2. \quad (9)$$

This type requires no auxiliary variables. The goal of this QUBO is to get the total weight $\sum_{i=1}^{N} w_i x_i$ as close to the capacity W as possible. To achieve this, we introduce a variable W_{offset}, which measures the offset from the capacity. The QUBO then has the goal to get the capacity as close to the fictional capacity $W_{\text{fictional}} := W - W_{\text{offset}}$ as possible. Note that while this formulation does have the advantage that it does not have any auxiliary binary variables, this comes at the cost of having an extra free parameter W_{offset}.

Type 6:

$$-\sum_{i=1}^{N}\sum_{j=1}^{N} c_{ij}x_i x_j + \lambda_1 \left(W - \sum_{i=1}^{N} w_i x_i - \sum_{k=1}^{M} (k-1)\, y_k\right)^2 + \lambda_2 \left(\sum_{k=1}^{M} y_k - 1\right)^2. \tag{10}$$

This type is an extension to type 3. By adding an extra penalty term with weight λ_2 we enforce the one-hot encoding in type 3. The original penalty term has weight λ_1. Again, we need $M = \max_{i=1}^{n} w_i$ auxiliary variables if all c_{ij} are non-negative and $M = W + 1$ if not.

3 Benchmark of QUBO Formulations

In this section, we compare the performance of the QUBO formulations defined in the previous section. First, we discuss the QKP instances used to asses the performance, and then reason which penalty values we used across the different formulations. Finally, we discuss how we implemented these instances using Simulated Annealing and present the performance of each QUBO formulation.

3.1 Problem Instances

We use the existing QKP instances from [3]. Each QKP instance is labelled as N_D_S, which corresponds to an instance with the following parameters: the QKP considers N objects, the matrix $C = (c_{ij})$ has density D, and the seed or index of the instance is denoted by S. For each of the following combinations of N and D, there are 10 different instances, namely $D = 25\%, 50\%$ for $N = 100, 200, 300$ and $D = 50\%, 100\%$ for $N = 100, 200$. For each instance, the cost values c_{ij} and weight values w_i lie in the interval $[0, 100]$.

3.2 Penalty Values

We now discuss the process of determining penalty values for our different formulations. To theoretically enforce the constraint, we require penalty values in the order of magnitude of 10 000. However, our experience suggests that these values do not work with annealing. As these values are much larger than the QUBO entries, this would influence the energy landscape too much and likely result in a bad performance. Instead, we believe that penalty values between 1 and 10 will be more suitable. In addition, in [17], they suggest

$$\lambda = \text{strength} \cdot N \cdot D,$$

with a strength of 0.1. In our case this results in values from 2.5 to 20. That is why we have decided to focus on penalty values between 1 and 20.

To determine the optimal penalty value for each formulation, we use simulated annealing to determine an approximation of where the optimal penalty value should lie for each of the given QUBO formulations. For these experiments, we try all integers from 1 to 10 as well as 15 and 20 as potential penalty values. For QUBO Type 6 we take λ_1 and λ_2 to be equal.

From these initial experiments, it follows that the seed S has no significant influence on the optimal penalty value, which is therefore discarded in the final selection for the best penalty value. By considering 3 or 4 different seeds per (size, density) instance, we determined the computed optimal values for each (size, density) instance and QUBO type combination. Optimality is determined by using the penalty value that results in the highest area-under-the-curve value (see Sect. 3.4 for an explanation). It turned out that the optimal penalty value is independent of the QUBO type. However, the different instances show quite a variety of optimal penalty values, ranging from 3 to 20. The resulting penalty values can be found in Table 6.

For Type 5, we chose the offset variable as follows. As we believe that local minima will likely lie close to the constraints, we choose offset variables which are close to 0. To fully gauge the performance of this somewhat special formulation, we consider it a total of six times, each with a different offset from 0 up to 5.

3.3 Approach

We now gauge the performance of the various QUBO formulations by finding solutions with simulated annealing (SA). Specifically, we use the D-Wave implementation[1] with its default parameters. For each QUBO formulation with corresponding penalty values, we run the following steps a total of 200 000 times:

1. Pick a random initial value for the simulated annealing algorithm.
2. Run simulated annealing for our QUBO formulation from this initial value.
3. Check whether the result satisfies the constraints. If so, we add it to our results. If not, we discard it.

Note that we repeat the above steps 200 000 times to limit the considerable amount of randomness involved in applying simulated annealing. We call each of the 200 000 different runs an *SA sample*. Note that due to discarding of the results not meeting the constraints, it is likely that for each QUBO formulation there are less than 200 000 SA samples considered in the results below.

---

[1] https://docs.ocean.dwavesys.com/projects/neal/en/latest/reference/sampler.html.

3.4 Results

In this section we discuss the results for each QUBO instance. We consider three different metrics to measure the performance of the formulations, namely, the best solution value that is found; the number of times the optimal[2] value is found; and the area-under-the-curve (AUC) value of the curve that plots the highest solution value (y-axis) found within the number of SA samples (x-axis) so far. The y-axis is normalised against the optimal solution values of the respective instances as defined in [18]. More details can be found in below.

The first two metrics give an intuition on how likely it is that a given formulation returns an optimal solution. However, it does not say anything about the quality of other solutions. The AUC value tells us how many simulating annealing steps one needs to get a good solution. In particular, the AUC value will be high if a value close to the optimum is found within a reasonable number of steps. If either of those are not the case it will be notably lower.

Best solution. The first performance metric is the best solution value found after 200 000 SA samples. We gauge this best solution value by comparing it to the optimum as a percentage of this optimum. This allows us to compare the formulations over all problem instances. We also compare our best solutions with those found by the QUBO formulation solved with SA in [15] to see how our QUBO formulations perform against the solutions there. These results are summarised in Tables 1 and 2.

Surprisingly, some of the solutions found in [15] return higher objective values than the optimal value. It is suspected that these values belong to infeasible solutions, since the optimal values found in [18] are widely assumed to be correct. Despite this, it is interesting to see that, for most instances, the solutions of all our QUBO formulations seem to be better than the SA solutions of [15].

Additionally, we find that formulations of Type 1, 2, and 4 seem to perform worse. For Type 3, 5, and 6, we see similar performances, where one does not clearly outperform the other. Although, we do observe that Type 5 performs slightly better with higher offsets, especially for the larger instances.

Optimal Solution Fraction. This metric measures the number of times an optimal solution is found. The results are shown in Table 3. We see that for most instances - especially those with a higher number of objects - no optimal value is found for all of the QUBO formulations. However, when an optimal value is found by any QUBO type, then usually the Type 5 formulation also finds the optimal value. In addition, the Type 5 formulations usually find the optimal value most often, specifically the formulations with higher offset value.

Area Under the Curve. The area-under-the-curve value represents the area under the curve of which the graph which depicts the best solution found over a certain number of SA samples. In this diagram, the x-axis denotes the number of SA samples that have been taken from 500 up to and including 20 000, and

[2] Optimal meaning true optimal solution for each instance (from [18]), and thus not necessarily the best value we found.

Table 1. Optimal objective value per instance, accompanied by the relative best found objective value as a percentage of the optimal value for the QUBO formulations over 200,000 SA samples. The highest and lowest percentage per problem instance are displayed in bold and underline respectively.

Problem instance	Optimal value	[15]	Type 1	Type 2	Type 3	Type 4	Type 5					Type 6
							Offset 0	*Offset 1*	*Offset 2*	*Offset 3*	*Offset 4*	
100_25_1	18558	93.6%	97.8%	97.4%	97.7%	84.3%	97.8%	97.7%	97.6%	97.5%	98.3%	**98.8%**
100_25_2	56525	75.5%	85.1%	88.4%	97.3%	98.1%	98.7%	98.8%	98.8%	**99.1%**	98.7%	97.1%
100_25_3	3752	92.1%	98.6%	98.6%	98.6%	0.0%	98.6%	98.6%	**99.1%**	**99.1%**	**99.1%**	98.6%
100_25_4	50382	72.6%	93.7%	91.6%	94.9%	97.4%	97.0%	97.4%	**98.1%**	97.7%	97.4%	95.6%
100_25_5	61494	66.7%	81.5%	79.9%	97.4%	**100.0%**	**100.0%**	**100.0%**	**100.0%**	**100.0%**	**100.0%**	97.4%
100_25_6	36360	62.1%	97.2%	98.4%	97.4%	97.2%	97.3%	97.6%	98.0%	**99.3%**	98.5%	97.5%
100_25_7	14657	86.9%	99.1%	99.3%	99.5%	0.0%	99.5%	98.6%	99.5%	99.3%	**99.7%**	98.9%
100_25_8	20452	92.1%	99.3%	97.9%	99.3%	85.5%	98.0%	98.7%	97.3%	97.8%	97.4%	**100.0%**
100_25_9	35438	81.1%	94.9%	96.4%	95.2%	94.6%	**97.5%**	94.8%	95.7%	95.8%	95.7%	94.3%
100_25_10	24930	89.1%	96.6%	96.1%	94.8%	95.5%	**97.1%**	96.2%	95.7%	96.0%	96.6%	96.3%
100_50_1	83742	87.4%	97.9%	97.2%	**99.0%**	97.7%	97.8%	97.8%	98.3%	98.1%	98.5%	98.4%
100_50_2	104856	64.2%	96.2%	96.8%	**99.6%**	99.0%	99.1%	99.5%	99.1%	99.0%	99.4%	99.4%
100_50_3	34006	96.9%	99.2%	**99.9%**	**99.9%**	0.0%	99.2%	99.5%	99.8%	99.8%	99.6%	99.5%
100_50_4	105996	60.2%	91.8%	90.1%	98.6%	98.8%	98.9%	99.2%	99.1%	**99.3%**	99.2%	99.0%
100_50_5	56464	83.3%	98.4%	98.2%	**99.7%**	98.6%	98.2%	99.2%	99.0%	98.8%	99.3%	98.7%
100_50_6	16083	97.0%	**100.0%**	**100.0%**	**100.0%**	0.0%	**100.0%**	**100.0%**	**100.0%**	**100.0%**	**100.0%**	**100.0%**
100_50_7	52819	87.4%	96.6%	96.5%	97.3%	96.2%	97.1%	97.3%	97.1%	97.1%	97.9%	**98.3%**
100_50_8	54246	93.4%	97.9%	98.0%	98.1%	98.0%	97.0%	97.5%	97.5%	97.6%	98.4%	**98.5%**
100_50_9	68974	87.5%	96.9%	97.0%	97.2%	96.9%	96.3%	96.9%	97.3%	96.9%	97.1%	**97.3%**
100_50_10	88634	61.4%	96.8%	97.1%	98.2%	98.4%	98.7%	98.7%	98.9%	98.7%	98.9%	**99.2%**
100_75_1	189137	62.1%	83.0%	82.1%	**100.0%**	**100.0%**	**100.0%**	**100.0%**	**100.0%**	**100.0%**	**100.0%**	**100.0%**
100_75_2	95074	95.5%	**98.9%**	97.5%	98.4%	97.5%	98.5%	97.9%	98.6%	97.9%	97.9%	98.4%
100_75_3	62098	95.0%	98.4%	97.5%	97.5%	95.3%	97.5%	**99.9%**	97.7%	98.4%	97.7%	99.7%
100_75_4	72245	95.7%	**99.7%**	98.7%	99.2%	98.3%	98.5%	98.6%	98.7%	98.8%	99.3%	99.2%
100_75_5	27616	99.0%	99.7%	99.7%	99.7%	0.0%	99.7%	99.9%	**100.0%**	**100.0%**	**100.0%**	99.7%
100_75_6	145273	65.9%	97.5%	97.5%	99.2%	99.2%	98.9%	99.4%	99.4%	**99.7%**	98.9%	99.3%
100_75_7	110979	96.1%	98.3%	98.1%	**98.9%**	98.1%	98.7%	97.9%	98.1%	98.0%	97.9%	98.4%
100_75_8	19570	97.7%	**100.0%**	99.8%	**100.0%**	0.0%	**100.0%**	99.7%	99.2%	99.2%	98.6%	99.4%
100_75_9	104341	87.1%	98.5%	97.7%	98.6%	97.5%	97.7%	98.0%	97.7%	97.6%	**99.0%**	98.8%
100_75_10	143740	98.9%	97.7%	96.7%	99.5%	99.2%	99.4%	99.5%	99.3%	99.4%	99.1%	**99.6%**
100_100_1	81978	**100.1%**	99.1%	100.0%	100.0%	91.8%	99.1%	97.5%	100.0%	99.8%	97.6%	99.9%
100_100_2	190424	86.5%	97.3%	97.2%	97.6%	**98.1%**	97.7%	98.1%	97.9%	97.9%	97.9%	97.6%
100_100_3	225434	76.7%	87.0%	91.9%	99.5%	99.8%	99.8%	99.9%	99.8%	99.5%	**100.0%**	99.7%
100_100_4	63028	**193.6%**	97.4%	97.7%	97.6%	0.0%	97.3%	97.6%	97.2%	99.7%	97.7%	98.0%
100_100_5	230076	72.9%	85.7%	87.6%	97.9%	99.6%	**99.9%**	99.8%	99.8%	99.7%	99.9%	98.8%
100_100_6	74358	99.0%	96.7%	95.9%	**99.7%**	91.8%	96.2%	96.4%	96.9%	97.0%	96.7%	97.4%
100_100_7	10330	**103.7%**	75.6%	73.6%	88.5%	0.0%	0.0%	97.1%	97.1%	98.6%	100.0%	88.5%
100_100_8	62582	**99.7%**	96.3%	96.4%	96.6%	74.9%	96.6%	96.2%	96.1%	97.0%	96.1%	96.6%
100_100_9	232754	76.5%	87.1%	84.0%	99.7%	100.0%	**100.0%**	100.0%	**100.0%**	**100.0%**	100.0%	99.6%
100_100_10	193262	77.6%	97.6%	96.8%	98.9%	99.0%	98.9%	99.1%	99.1%	99.3%	**99.3%**	99.0%
200_25_1	204441	74.8%	89.2%	91.3%	98.5%	98.9%	**98.9%**	98.8%	98.4%	98.7%	98.6%	98.6%
200_25_2	239573	53.8%	79.5%	81.8%	99.7%	99.8%	99.8%	99.8%	**99.9%**	99.9%	99.8%	99.8%
200_25_3	245463	54.4%	74.1%	77.2%	99.7%	99.8%	**100.0%**	100.0%	99.9%	100.0%	100.0%	99.7%
200_25_4	222361	60.8%	83.5%	80.1%	**99.2%**	98.8%	98.9%	98.9%	98.9%	99.1%	98.9%	99.1%
200_25_5	187324	65.9%	93.7%	93.2%	98.0%	98.3%	98.3%	98.8%	98.5%	98.4%	98.3%	**99.1%**
200_25_6	80351	79.6%	94.8%	94.3%	95.5%	95.1%	95.4%	95.7%	96.4%	96.6%	**96.6%**	96.1%
200_25_7	59036	73.3%	**98.6%**	98.1%	97.4%	0.0%	98.2%	98.6%	98.4%	98.4%	98.2%	98.0%
200_25_8	149433	65.5%	97.0%	97.4%	**98.3%**	97.9%	97.7%	98.1%	97.8%	98.1%	97.7%	98.3%
200_25_9	49366	67.9%	97.9%	97.4%	**99.2%**	0.0%	97.6%	98.4%	98.8%	98.4%	99.0%	97.8%
200_25_10	48459	76.6%	98.1%	99.0%	98.4%	0.0%	98.4%	98.5%	98.9%	**99.3%**	98.7%	98.4%

(*continued*)

Table 1. (*continued*)

Problem instance	Optimal value	[15]	Type 1	Type 2	Type 3	Type 4	Type 5					Type 6
							Offset 0	*Offset 1*	*Offset 2*	*Offset 3*	*Offset 4*	
200_50_1	372097	84.9%	90.3%	91.7%	96.9%	96.7%	**97.1%**	96.8%	97.0%	96.8%	96.8%	96.5%
200_50_2	211130	75.2%	94.3%	94.5%	94.6%	94.1%	94.7%	94.0%	93.6%	**94.9%**	94.1%	94.7%
200_50_3	227185	73.4%	95.0%	94.3%	95.0%	95.5%	94.8%	**95.7%**	95.2%	95.6%	95.0%	95.5%
200_50_4	228572	70.2%	93.4%	94.1%	94.5%	94.1%	94.0%	94.8%	94.1%	**95.1%**	94.2%	94.5%
200_50_5	479651	77.3%	71.6%	73.3%	98.6%	99.8%	99.7%	99.8%	99.7%	99.8%	**99.9%**	99.2%
200_50_6	426777	76.5%	83.3%	78.0%	98.4%	98.5%	98.6%	98.6%	98.6%	98.6%	**98.7%**	98.7%
200_50_7	220890	86.6%	93.3%	93.6%	**94.5%**	93.1%	93.5%	93.5%	94.4%	93.6%	94.1%	94.0%
200_50_8	317952	49.7%	94.7%	94.2%	96.2%	96.2%	95.9%	96.1%	95.9%	96.0%	95.8%	**96.3%**
200_50_9	104936	91.2%	96.4%	95.1%	95.5%	0.0%	94.1%	95.2%	96.3%	**97.5%**	96.2%	96.1%
200_50_10	284751	81.5%	95.0%	95.1%	95.8%	95.5%	95.0%	95.7%	95.4%	95.5%	**95.9%**	95.7%
200_75_1	442894	59.5%	93.6%	93.9%	94.9%	94.2%	95.4%	94.7%	94.7%	**95.5%**	94.3%	95.0%
200_75_2	286643	77.6%	89.5%	89.2%	91.2%	90.0%	89.8%	91.3%	89.8%	89.8%	**91.3%**	91.2%
200_75_3	61924	**101.7%**	98.1%	97.7%	96.7%	0.0%	92.9%	99.8%	99.8%	100.0%	99.6%	96.2%
200_75_4	128351	**93.2%**	82.3%	86.4%	85.9%	0.0%	85.7%	85.0%	85.7%	83.8%	86.3%	86.6%
200_75_5	137885	93.5%	92.6%	93.8%	95.3%	0.0%	94.3%	95.6%	95.3%	**96.4%**	94.6%	95.4%
200_75_6	229631	73.1%	93.6%	94.5%	94.3%	85.8%	94.6%	93.4%	93.1%	94.8%	94.6%	**95.5%**
200_75_7	269887	80.5%	90.4%	92.2%	**93.1%**	91.3%	90.2%	92.4%	92.7%	91.6%	92.3%	91.9%
200_75_8	600858	68.7%	87.8%	86.3%	97.0%	97.3%	97.2%	97.3%	97.4%	98.0%	97.2%	**98.1%**
200_75_9	516771	84.0%	91.4%	92.4%	94.8%	95.0%	95.6%	95.1%	95.0%	95.1%	**95.7%**	95.5%
200_75_10	142694	94.7%	91.5%	92.7%	91.8%	0.0%	91.3%	92.5%	**95.2%**	92.3%	91.8%	93.7%

the y-axis denotes the best solution over all valid solutions given by this number of SA samples. All y-values are normalised against the optimal solution value for the respective problem instance. In other words, if the value on the y-axis equals 1.0000 for some number n of SA samples, we know that at least 1 of those n SA samples was an optimal solution. Therefore, the y-values lie in the range $[0, 1]$.

Since SA is a random algorithm, its behaviour can differ each run, influencing the AUC value significantly. To mitigate this, the curve used for computation of the AUC values is obtained by averaging this curve over 10 different runs of 200 000 SA samples. An AUC value of 1.0000 thus implies that for each of the 10 runs a valid solution with optimal value was already found in the first 500 SA samples.

The AUC values for each combination of instance and QUBO Type is given in Tables 4 and 5. We see that the Type 1, 2, and 4 (almost) never have the highest AUC value. Moreover, we observe that quite often they even have an AUC value that is significantly worse than that of the other formulations. Especially the Type 4 formulation fails to find any valid solutions at all for quite some instances, corresponding to an AUC value of 0.0000.

Table 2. Optimal objective value per instance, accompanied by the relative best found objective value as a percentage of the optimal value for the QUBO formulations over 200,000 SA samples. The highest and lowest percentage per problem instance are displayed in bold and underline respectively.

Problem instance	Optimal value	[15]	Type 1	Type 2	Type 3	Type 4	Type 5					Type 6
							Offset 0	*Offset 1*	*Offset 2*	*Offset 3*	*Offset 4*	
200_100_1	937149	79.1%	69.5%	68.4%	98.5%	99.7%	99.7%	99.7%	99.7%	99.8%	**99.8%**	98.7%
200_100_2	303058	86.8%	92.8%	93.6%	94.3%	0.0%	92.9%	**94.4%**	93.6%	92.6%	92.4%	92.7%
200_100_3	29367	**103.4%**	83.5%	83.4%	93.1%	0.0%	0.0%	100.0%	100.0%	100.0%	100.0%	92.7%
200_100_4	100838	**101.2%**	96.9%	96.4%	97.4%	0.0%	96.8%	96.5%	97.2%	97.5%	97.2%	97.8%
200_100_5	786635	94.4%	81.5%	85.2%	96.1%	**96.6%**	96.5%	96.6%	96.5%	96.5%	96.6%	96.3%
200_100_6	41171	**102.9%**	86.8%	86.8%	91.7%	0.0%	0.0%	100.0%	100.0%	100.0%	100.0%	91.7%
200_100_7	701094	87.2%	91.1%	90.3%	95.6%	95.1%	95.1%	95.0%	95.2%	95.6%	95.3%	**95.8%**
200_100_8	782443	78.2%	90.4%	86.2%	97.5%	97.5%	97.1%	97.5%	**97.5%**	97.4%	97.4%	97.2%
200_100_9	628992	87.1%	93.3%	94.4%	**95.3%**	95.1%	94.6%	94.8%	94.9%	95.1%	95.1%	95.1%
200_100_10	378442	86.5%	91.6%	92.0%	93.0%	91.5%	91.3%	91.4%	91.6%	91.9%	**93.4%**	93.2%
300_25_1	29140	89.7%	98.6%	98.8%	98.3%	0.0%	91.6%	99.1%	99.6%	99.2%	**99.7%**	98.1%
300_25_2	281990	68.2%	90.0%	88.4%	89.0%	89.0%	89.3%	89.5%	89.2%	89.0%	**90.3%**	89.3%
300_25_3	231075	86.7%	88.4%	87.6%	89.1%	89.4%	89.4%	**89.8%**	87.7%	88.5%	89.4%	88.7%
300_25_4	444759	77.8%	78.1%	78.3%	93.1%	94.7%	95.2%	94.6%	**95.2%**	94.8%	94.8%	93.9%
300_25_5	14988	**101.0%**	92.0%	92.8%	93.4%	0.0%	0.0%	0.0%	100.0%	100.0%	100.0%	93.6%
300_25_6	269782	76.0%	88.5%	87.7%	**89.3%**	87.7%	86.9%	88.3%	87.6%	88.0%	88.2%	87.6%
300_25_7	485263	85.6%	73.9%	74.2%	93.8%	96.3%	96.5%	96.5%	**96.7%**	96.4%	96.4%	94.6%
300_25_8	9343	**102.4%**	88.6%	88.7%	90.6%	0.0%	0.0%	0.0%	94.4%	100.0%	100.0%	90.3%
300_25_9	250761	58.5%	88.5%	87.3%	89.1%	88.5%	88.8%	90.1%	88.9%	88.9%	89.0%	**90.4%**
300_25_10	383377	71.2%	87.2%	85.9%	90.4%	91.5%	91.2%	90.8%	91.1%	**91.7%**	91.4%	90.5%
300_50_1	513379	80.3%	89.2%	89.2%	91.1%	89.5%	89.8%	90.2%	89.7%	90.3%	89.8%	**91.2%**
300_50_2	105543	75.2%	92.3%	90.5%	92.5%	0.0%	89.9%	91.4%	92.4%	**94.4%**	92.1%	91.8%
300_50_3	875788	76.8%	79.9%	78.8%	94.7%	95.7%	95.7%	95.0%	**95.7%**	95.3%	95.6%	95.0%
300_50_4	307124	74.4%	89.7%	89.9%	91.3%	0.0%	90.9%	**92.9%**	90.2%	90.8%	91.0%	91.4%
300_50_5	727820	84.5%	90.5%	89.7%	**94.3%**	93.3%	93.4%	94.0%	93.2%	94.0%	94.1%	93.8%
300_50_6	734053	75.4%	90.5%	89.9%	94.2%	93.7%	94.3%	93.6%	**94.5%**	94.2%	94.1%	93.9%
300_50_7	43595	**101.0%**	97.8%	99.6%	99.8%	0.0%	97.9%	99.8%	99.4%	99.7%	99.2%	98.9%
300_50_8	767977	80.0%	88.8%	89.6%	93.7%	94.3%	**94.5%**	94.5%	94.4%	94.3%	94.3%	93.8%
300_50_9	761351	58.8%	90.0%	90.3%	93.9%	94.1%	94.3%	94.3%	**94.9%**	94.5%	94.6%	94.0%
300_50_10	996070	81.5%	69.5%	71.6%	97.2%	**98.0%**	97.8%	97.7%	97.7%	97.8%	97.8%	96.4%

Furthermore, we see that QUBO Type 3 and 6 almost never have the worst AUC value and often have a relatively high AUC value. For most instances the AUC values of these two formulations are close together. Therefore, this metric does not give a clear indication which formulation performs best.

When looking at the Type 5 formulation we see that it quite often has the best AUC value, especially for offsets with value 3 and 4. Specifically, for the instances with 300 items we see that this formulation is performing relatively well when compared to the other QUBO formulations.

Table 3. Number of times the optimal value was obtained out of 200 000 SA samples per problem instance and QUBO formulation. The 85 instances for which no formulation found the optimal value are left out.

Problem instance	Type 1	Type 2	Type 3	Type 4	Type 5					Type 6
					Offset 0	*Offset 1*	*Offset 2*	*Offset 3*	*Offset 4*	
100_25_5	0	0	0	2	4	4	4	6	8	0
100_25_8	0	0	0	0	0	0	0	0	0	1
100_50_6	163	172	238	0	97	98	67	38	11	235
100_75_1	0	0	133	8756	12946	12716	12938	13575	14012	244
100_75_5	0	0	0	0	0	0	74	18	16	0
100_75_8	2	0	1	0	1	0	0	0	0	0
100_100_7	0	0	0	0	0	0	0	0	29609	0
100_100_9	0	0	0	0	2	0	1	1	0	0
200_25_3	0	0	0	0	1	0	0	0	0	0
200_75_3	0	0	0	0	0	0	0	1	0	0
200_100_3	0	0	0	0	0	313	154	35	28	0
200_100_6	0	0	0	0	0	153	139	121	204	0
200_100_10	0	0	0	0	0	0	0	0	0	0
300_25_5	0	0	0	0	0	0	58	57	73	0
300_25_8	0	0	0	0	0	0	0	4739	3384	0

Table 4. Area-under-the-curve values per problem instance and QUBO type, averaged over 10 runs of 200 000 samples each. The highest and lowest AUC values are depicted in bold and underline respectively.

Problem instance	**Type 1**	**Type 2**	**Type 3**	**Type 4**	**Type 5**					**Type 6**
					Offset 0	*Offset 1*	*Offset 2*	*Offset 3*	*Offset 4*	
100_25_1	0.9590	0.9600	**0.9676**	0.2935	0.9566	0.9575	0.9572	0.9596	0.9640	0.9649
100_25_2	0.8208	0.8089	0.9487	0.9756	0.9761	0.9784	0.9795	0.9781	**0.9796**	0.9516
100_25_3	0.9706	0.9759	0.9788	0.0000	0.8446	0.9853	**0.9907**	0.9907	0.9907	0.9816
100_25_4	0.8850	0.8764	0.9349	0.9627	0.9648	0.9645	0.9666	**0.9672**	0.9663	0.9391
100_25_5	0.7707	0.7643	0.9532	0.9961	0.9976	0.9970	0.9974	**0.9979**	0.9975	0.9584
100_25_6	0.9585	0.9627	0.9582	0.9605	0.9619	0.9621	0.9651	**0.9695**	0.9686	0.9637
100_25_7	0.9738	0.9747	0.9825	0.0000	0.9696	0.9711	0.9753	0.9755	**0.9826**	0.9773
100_25_8	0.9611	0.9609	**0.9730**	0.3519	0.9569	0.9611	0.9615	0.9589	0.9616	0.9719
100_25_9	0.9302	0.9307	0.9193	0.9280	0.9351	0.9343	0.9393	0.9379	**0.9405**	0.9251
100_25_10	0.9412	0.9409	0.9355	0.9365	0.9413	0.9406	0.9419	0.9417	**0.9435**	0.9394
100_50_1	0.9643	0.9631	**0.9780**	0.9711	0.9706	0.9706	0.9718	0.9711	0.9709	0.9752
100_50_2	0.9174	0.9267	0.9879	0.9844	0.9838	0.9873	0.9846	0.9839	0.9847	**0.9880**
100_50_3	0.9780	0.9801	0.9853	0.0000	0.9514	0.9828	0.9869	**0.9891**	0.9890	0.9862
100_50_4	0.8702	0.8736	0.9780	0.9810	0.9810	0.9814	0.9814	**0.9819**	0.9816	0.9796
100_50_5	0.9720	0.9699	**0.9804**	0.9707	0.9387	0.9728	0.9761	0.9766	0.9794	0.9740
100_50_6	0.9998	0.9998	0.9999	0.0000	0.9943	0.9997	0.9996	0.9989	0.9968	**0.9999**

(*continued*)

Table 4. (*continued*)

Problem instance	Type 1	Type 2	Type 3	Type 4	Type 5					Type 6
					Offset 0	*Offset 1*	*Offset 2*	*Offset 3*	*Offset 4*	
100_50_7	0.9552	0.9550	0.9610	0.9478	0.9548	0.9568	0.9571	0.9554	0.9567	**0.9635**
100_50_8	0.9602	0.9642	0.9695	0.9601	0.9576	0.9582	0.9610	0.9615	0.9630	**0.9705**
100_50_9	0.9517	0.9511	**0.9611**	0.9526	0.9546	0.9574	0.9591	0.9553	0.9581	0.9601
100_50_10	0.9484	0.9486	0.9759	0.9753	0.9756	0.9759	0.9782	0.9771	**0.9783**	0.9751
100_75_1	0.7781	0.7785	0.9996	**1.0000**	**1.0000**	**1.0000**	**1.0000**	**1.0000**	**1.0000**	0.9998
100_75_2	0.9666	0.9654	**0.9742**	0.9665	0.9684	0.9680	0.9688	0.9688	0.9686	0.9720
100_75_3	0.9626	0.9621	0.9661	0.9244	0.9503	**0.9673**	0.9667	0.9667	0.9671	0.9665
100_75_4	0.9719	0.9711	**0.9831**	0.9689	0.9705	0.9738	0.9710	0.9722	0.9734	0.9820
100_75_5	0.9913	0.9890	0.9953	0.0000	0.9801	0.9892	**0.9995**	0.9992	0.9989	0.9954
100_75_6	0.9576	0.9581	0.9820	0.9787	0.9796	0.9806	0.9806	0.9804	0.9791	**0.9849**
100_75_7	0.9710	0.9688	**0.9769**	0.9709	0.9704	0.9704	0.9718	0.9708	0.9704	0.9753
100_75_8	0.9790	0.9800	0.9816	0.0000	0.9800	0.9744	0.9757	0.9722	0.9742	**0.9823**
100_75_9	0.9659	0.9632	0.9712	0.9669	0.9668	0.9687	0.9677	0.9680	0.9691	**0.9727**
100_75_10	0.9530	0.9418	0.9869	0.9840	0.9838	0.9829	0.9838	0.9842	0.9842	**0.9874**
100_100_1	0.9680	0.9711	**0.9818**	0.7725	0.9409	0.9636	0.9724	0.9694	0.9691	0.9794
100_100_2	0.9244	0.9250	0.9704	0.9741	0.9732	0.9753	0.9742	0.9747	**0.9754**	0.9707
100_100_3	0.8171	0.8217	0.9806	0.9861	0.9886	0.9872	0.9874	0.9861	**0.9899**	0.9815
100_100_4	0.9646	0.9618	**0.9696**	0.0000	0.9594	0.9585	0.9626	0.9663	0.9660	0.9695
100_100_5	0.8220	0.8206	0.9749	0.9902	0.9928	0.9916	0.9928	**0.9931**	0.9921	0.9761
100_100_6	0.9424	0.9341	**0.9595**	0.7809	0.9336	0.9431	0.9472	0.9531	0.9543	0.9579
100_100_7	0.7007	0.7198	0.8540	0.0000	0.0000	0.9459	0.9679	0.9859	**1.0000**	0.8550
100_100_8	0.9366	0.9422	0.9527	0.0509	0.9325	0.9352	0.9410	0.9394	0.9432	**0.9539**
100_100_9	0.7982	0.7882	0.9872	0.9977	0.9984	0.9981	**0.9988**	0.9984	0.9984	0.9880
100_100_10	0.9332	0.9362	0.9752	0.9749	0.9774	0.9780	0.9769	0.9791	**0.9799**	0.9742
200_25_1	0.8532	0.8488	0.9797	0.9804	0.9805	0.9820	0.9808	0.9812	**0.9822**	0.9790
200_25_2	0.7707	0.7745	0.9946	0.9958	**0.9964**	0.9963	0.9960	0.9962	0.9959	0.9948
200_25_3	0.7158	0.7158	0.9939	0.9962	0.9971	**0.9973**	0.9965	0.9971	0.9972	0.9942
200_25_4	0.7709	0.7643	0.9870	0.9840	0.9853	0.9849	0.9861	0.9857	0.9854	**0.9876**
200_25_5	0.9060	0.8995	0.9769	0.9767	0.9778	0.9771	0.9778	**0.9786**	0.9781	0.9781
200_25_6	0.9274	0.9254	0.9394	0.9218	0.6330	0.9088	0.9389	0.9508	**0.9561**	0.9405
200_25_7	0.9543	0.9583	0.9627	0.0000	0.9209	0.9625	0.9650	**0.9693**	0.9688	0.9638
200_25_8	0.9613	0.9643	0.9726	0.9681	0.9689	0.9722	0.9708	0.9729	0.9724	**0.9737**
200_25_9	0.9572	0.9566	0.9625	0.0000	0.9518	0.9657	0.9727	0.9732	**0.9766**	0.9662
200_25_10	0.9570	0.9564	0.9672	0.0000	0.9054	0.9636	0.9680	0.9730	**0.9748**	0.9649
200_50_1	0.8774	0.8848	0.9588	0.9599	0.9603	0.9612	**0.9629**	0.9628	0.9613	0.9579
200_50_2	0.9210	0.9215	0.9355	0.9277	0.9221	0.9251	0.9262	0.9292	0.9295	**0.9357**
200_50_3	0.9274	0.9270	0.9376	0.9374	0.9192	0.9285	0.9373	0.9374	0.9389	**0.9389**
200_50_4	0.9198	0.9212	**0.9356**	0.9301	0.9229	0.9279	0.9306	0.9328	0.9323	0.9346
200_50_5	0.6817	0.6877	0.9825	0.9957	0.9949	0.9956	0.9956	0.9957	**0.9961**	0.9845
200_50_6	0.7690	0.7576	0.9792	0.9819	0.9825	0.9828	0.9826	0.9828	**0.9832**	0.9795
200_50_7	0.9169	0.9161	**0.9317**	0.9216	0.9166	0.9248	0.9264	0.9263	0.9277	0.9308
200_50_8	0.9295	0.9286	0.9519	0.9496	0.9500	0.9495	0.9489	0.9513	0.9510	**0.9527**
200_50_9	0.9350	0.9295	0.9406	0.0000	0.8862	0.9334	0.9380	**0.9502**	0.9475	0.9437
200_50_10	0.9369	0.9365	**0.9499**	0.9458	0.9409	0.9435	0.9453	0.9447	0.9467	0.9488

Table 5. Area-under-the-curve values per problem instance and QUBO type, averaged over 10 runs of 200 000 samples each. The highest and lowest AUC values are depicted in bold and underline respectively.

Problem instance	Type 1	Type 2	Type 3	Type 4	Type 5					Type 6
					Offset 0	*Offset 1*	*Offset 2*	*Offset 3*	*Offset 4*	
200_75_1	0.9202	0.9207	**0.9366**	0.9343	0.9304	0.9335	0.9359	0.9364	0.9360	0.9364
200_75_2	0.8790	0.8806	**0.8955**	0.8858	0.8772	0.8855	0.8854	0.8887	0.8922	0.8926
200_75_3	0.9206	0.9128	0.9470	0.0000	0.2275	0.9754	0.9898	**0.9925**	0.9898	0.9412
200_75_4	0.8020	0.8188	**0.8345**	0.0000	0.8098	0.8134	0.8223	0.8219	0.8284	0.8342
200_75_5	0.8961	0.8951	0.9268	0.0000	0.6068	0.9182	0.9234	**0.9305**	0.9301	0.9232
200_75_6	0.9134	0.9197	0.9230	0.2653	0.9000	0.9137	0.9219	0.9250	0.9238	**0.9277**
200_75_7	0.8911	0.8938	**0.9097**	0.8806	0.8765	0.9009	0.9048	0.9009	0.9053	0.9080
200_75_8	0.8310	0.8200	0.9622	0.9688	0.9674	0.9675	0.9690	**0.9693**	0.9684	0.9664
200_75_9	0.8944	0.8989	0.9417	0.9431	0.9415	0.9426	0.9428	0.9450	**0.9461**	0.9432
200_75_10	0.8788	0.8876	0.9042	0.0000	0.8682	0.8824	0.8980	0.8987	0.9014	**0.9074**
200_100_1	0.6648	0.6667	0.9687	0.9939	0.9951	0.9950	0.9950	**0.9958**	0.9953	0.9741
200_100_2	0.8924	0.8952	**0.9260**	0.0000	0.8112	0.8997	0.9121	0.9085	0.9100	0.9215
200_100_3	0.8065	0.8125	0.9010	0.0000	0.0000	**1.0000**	0.9999	0.9989	0.9990	0.8961
200_100_4	0.9355	0.9345	**0.9572**	0.0000	0.8877	0.9299	0.9466	0.9524	0.9532	0.9560
200_100_5	0.7928	0.7943	0.9492	0.9599	**0.9602**	0.9599	0.9593	0.9596	0.9596	0.9534
200_100_6	0.7891	0.8300	0.9119	0.0000	0.0000	0.9179	0.9974	0.9978	**0.9995**	0.9116
200_100_7	0.8763	0.8778	0.9442	0.9443	0.9398	0.9426	**0.9458**	0.9447	0.9456	0.9441
200_100_8	0.8491	0.8335	0.9608	0.9665	0.9623	0.9669	0.9674	0.9668	**0.9682**	0.9629
200_100_9	0.9186	0.9168	0.9420	**0.9443**	0.9388	0.9417	0.9418	0.9423	0.9436	0.9420
200_100_10	0.8979	0.8997	0.9169	0.9001	0.8896	0.8967	0.9027	0.9051	0.9105	**0.9173**
300_25_1	0.9236	0.9168	0.9428	0.0000	0.5000	0.9690	0.9829	0.9823	**0.9835**	0.9329
300_25_2	0.8722	0.8741	0.8736	0.8778	0.8766	0.8802	0.8797	0.8820	**0.8831**	0.8768
300_25_3	0.8647	0.8635	0.8716	0.8708	0.8657	0.8660	0.8680	0.8742	**0.8754**	0.8719
300_25_4	0.7521	0.7482	0.9180	0.9385	0.9393	0.9401	0.9406	0.9405	**0.9412**	0.9200
300_25_5	0.8345	0.8228	0.9212	0.0000	0.0000	0.0000	0.9989	0.9996	**0.9997**	0.9197
300_25_6	0.8610	0.8575	0.8656	0.8614	0.8570	0.8616	0.8656	**0.8675**	0.8664	0.8649
300_25_7	0.6921	0.6928	0.9283	0.9570	0.9576	0.9585	0.9588	0.9589	**0.9591**	0.9350
300_25_8	0.6386	0.5824	0.8923	0.0000	0.0000	0.0000	0.9232	**1.0000**	**1.0000**	0.8843
300_25_9	0.8649	0.8614	0.8741	0.8732	0.8630	0.8702	0.8726	0.8754	**0.8794**	0.8775
300_25_10	0.8376	0.8254	0.8914	0.9036	0.8984	0.9000	0.9034	0.9065	**0.9067**	0.8934
300_50_1	0.8784	0.8812	0.8936	0.8858	0.8790	0.8845	0.8850	0.8851	0.8868	**0.8937**
300_50_2	0.8682	0.8687	0.9036	0.0000	0.8336	0.8882	0.9000	**0.9072**	0.9050	0.8933
300_50_3	0.7516	0.7485	0.9368	0.9467	0.9464	0.9473	0.9472	0.9471	**0.9482**	0.9394
300_50_4	0.8768	0.8708	**0.8985**	0.0000	0.7979	0.8632	0.8811	0.8883	0.8914	0.8951
300_50_5	0.8849	0.8779	0.9268	0.9259	0.9232	0.9264	0.9265	**0.9286**	0.9283	0.9263
300_50_6	0.8706	0.8669	0.9279	0.9315	0.9306	0.9305	**0.9337**	0.9329	0.9330	0.9295
300_50_7	0.9329	0.9325	0.9478	0.0000	0.6653	0.9840	0.9856	0.9863	**0.9865**	0.9479
300_50_8	0.8501	0.8513	0.9297	0.9328	0.9332	**0.9351**	0.9341	0.9340	0.9346	0.9320
300_50_9	0.8711	0.8649	0.9306	0.9353	0.9329	0.9348	0.9364	0.9365	**0.9377**	0.9327
300_50_10	0.6713	0.6690	0.9558	0.9736	0.9735	0.9734	0.9738	**0.9744**	0.9735	0.9589

Table 6. Penalty values for each problem instance.

100_25	**100_50**	**100_75**	**100_100**	**200_25**	**200_50**	**200_75**	**200_100**	**300_25**	**300_50**
3	4	5	10	3	8	15	20	10	15

4 Conclusion

In this paper, we defined six different QUBO formulations for the Quadratic Knapsack Problem and compared their performance when solved by simulated annealing. Specifically, they were compared on how often an optimal value was found, what the best value found was and how many SA steps are required until the formulation delivers an optimal or good solution.

First of all, we saw that SA is in most cases able to find solutions that are close to optimal. For some QUBO formulations and instances optimal solutions were found. We also found that our QUBO formulations mostly performed better than those found in [15]. This shows that all QUBO formulations are capable of solving different instances of the QKP at a high level.

However, the performance of the 6 different formulations is still very different. First of all, we see that Type 1, 2 and 4 are outperformed by Type 3, 5 and 6. Of these last three, Type 3 and 6 perform similarly, while in turn Type 5 outperforms both of them. Generally, the Type 5 formulation with higher offsets (such as 3 and 4) shows the best performance. Note that all three metrics suggest this and that we can hence quite confidently draw these conclusions.

The better performance of the Type 5 formulation could be explained by the fact that it is quite different in nature than the other formulations. The other 5 formulations require auxiliary variables to incorporate the capacity constraint, while Type 5 accomplishes this by introducing an offset variable. In general, more variables means a larger energy landscape, which makes finding a solution harder. In the case of the QKP, this indeed seems to be the case.

However, it should be mentioned that Type 5 also brings a disadvantage. While Type 5 manages to remove the need for auxiliary variables, it does introduce a new variable called the offset variable. For each use case, research needs to be performed to gauge what a suitable value for this offset variable is. This makes it less directly implementable than the other QUBO formulations.

Even though our conclusions are based on many different results, there are some comments which can be made about these results. Firstly, we should note that our results mainly consider one family of QKP problems. It will be interesting to see whether our results could be reproduced for different families of QKP problems. Secondly, we should mention that we only considered a small set of potential penalty values and offset variables. It could very well be that these values benefit one QUBO formulation more than the other. That is why we think that repeating our assessment for a wider variety of penalty values would allow for a fairer comparison between the different formulations.

While this work focuses on the QKP, it also has implications for other optimisation and QUBO problems. First of all, we show that the QUBO formulation significantly influences the performance of the QUBO. It is hence advisory for other QUBO problems to consider alternative formulations as well. In addition, we show that the introduction of an offset variable might be more beneficial than introducing auxiliary variables, a method which is currently not so widely used. This also opens the door to research new, different techniques for generating QUBOs to see whether they perform even better in practice.

Another avenue for future research would be to test our QUBO formulations on quantum annealing devices. As quantum annealing devices currently have limited resources and hence limited applicability, it is likely that performance varies significantly over the different QUBO formulations. It might even turn out that different QUBO formulations are preferred by quantum annealing devices than the ones preferred by the simulated annealing solver. Since quantum annealing devices are promising great applicability in the (near) future, choosing suitable QUBO formulations could become an important are of research to enhance to applicability of quantum annealers.

References

1. Assi, M., Haraty, R.A.: A survey of the knapsack problem. In: 2018 International Arab Conference on Information Technology (ACIT), pp. 1–6. IEEE (2018)
2. Billionnet, A.: Éric Soutif: An exact method based on lagrangian decomposition for the 0–1 quadratic knapsack problem. EJOR **157**(3), 565–575 (2004)
3. Billionnet, A., Soutif, É.: Using a mixed integer programming tool for solving the 0–1 quadratic knapsack problem. INFORMS J. Comput. **16**(2), 188–197 (2004)
4. Cacchiani, V., Iori, M., Locatelli, A., Martello, S.: Knapsack problems-an overview of recent advances. part ii: Multiple, multidimensional, and quadratic knapsack problems. Comput. Operations Res. 105693 (2022)
5. Caprara, A., Pisinger, D., Toth, P.: Exact solution of the quadratic knapsack problem. INFORMS J. Comput. **11**(2), 125–137 (1999)
6. Djeumou Fomeni, F., Kaparis, K., Letchford, A.N.: A cut-and-branch algorithm for the quadratic knapsack problem. Discret. Optim. **44**, 100579 (2022)
7. Farhi, E., Goldstone, J., Gutmann, S.: A quantum approximate optimization algorithm. arXiv preprint arXiv:1411.4028 (2014)
8. Feld, S., et al.: A hybrid solution method for the capacitated vehicle routing problem using a quantum annealer. Frontiers in ICT **6**, 13 (2019)
9. Gallo, G., Hammer, P.L., Simeone, B.: Quadratic knapsack problems. In: Combinatorial Optimization, pp. 132–149. Springer (1980). https://doi.org/10.1007/BFb0120892
10. Glover, F., et al.: Quantum bridge analytics I: a tutorial on formulating and using QUBO models. In: Annals of Operations Research, pp. 1–43 (2022)
11. Hauke, P., Katzgraber, H.G., Lechner, W., Nishimori, H., Oliver, W.D.: Perspectives of quantum annealing: Methods and implementations. Rep. Prog. Phys. **83**(5), 054401 (2020)
12. Kellerer, H., Pferschy, U., Pisinger, D.: Multidimensional knapsack problems. In: Knapsack Problems, pp. 235–283. Springer (2004). https://doi.org/10.1007/978-3-540-24777-7_9
13. Van der Linde, S., et al.: Hybrid classical-quantum computing in geophysical inverse problems: The case of quantum annealing for residual statics estimation. In: Sixth EAGE High Performance Computing Workshop, vol. 2022, pp. 1–5. EAGE Publications BV (2022)
14. Lucas, A.: Ising formulations of many NP problems. Front. Phys. 5 (2014)
15. Mohamed, M.: Quantum Annealing: Research and Applications. Master's thesis, University of Waterloo (2021)
16. Neukart, F., Compostella, G., Seidel, C., Von Dollen, D., Yarkoni, S., Parney, B.: Traffic flow optimization using a quantum annealer. Frontiers in ICT **4**, 29 (2017)

17. Parizy, M., Togawa, N.: Analysis and acceleration of the quadratic knapsack problem on an ising machine. IEICE Trans. Fundam. Electron. Commun. Comput. Sci. **104**(11), 1526–1535 (2021)
18. Patvardhan, C., Bansal, S., Srivastav, A.: Solving the 0–1 quadratic knapsack problem with a competitive quantum inspired evolutionary algorithm. J. Comput. Appl. Math. **285**, 86–99 (2015)
19. Phillipson, F., Bhatia, H.S.: Portfolio optimisation using the d-wave quantum annealer. In: Paszynski, M., Kranzlmüller, D., Krzhizhanovskaya, V.V., Dongarra, J.J., Sloot, P.M.A. (eds.) ICCS 2021. LNCS, vol. 12747, pp. 45–59. Springer, Cham (2021). https://doi.org/10.1007/978-3-030-77980-1_4
20. Pisinger, D.: The quadratic knapsack problem-a survey. Discret. Appl. Math. **155**(5), 623–648 (2007)
21. Preskill, J.: Quantum computing in the NISQ era and beyond. Quantum **2**, 79 (2018)
22. Punnen, A.P., Pandey, P., Friesen, M.: Representations of quadratic combinatorial optimization problems: A case study using quadratic set covering and quadratic knapsack problems. Comput. Operations Res. **112**, 104769 (2019)
23. Rodrigues, C.D., Quadri, D., Michelon, P., Gueye, S.: 0–1 quadratic knapsack problems: An exact approach based on a t-linearization. SIAM J. Optim. **22**(4), 1449–1468 (2012)
24. Schauer, J.: Asymptotic behavior of the quadratic knapsack problem. Eur. J. Oper. Res. **255**(2), 357–363 (2016)

Qubit: The Game. Teaching Quantum Computing through a Game-Based Approach

Daniel Escanez-Exposito(✉), Javier Correa-Marichal, and Pino Caballero-Gil

Department of Computer Engineering and Systems, University of La Laguna, 38200 Tenerife, Spain
{jescanez,pcaballe}@ull.edu.es, jcorreamarichal@gmail.com

Abstract. Quantum computing is a promising and rapidly growing interdisciplinary field that attracts researchers from science and engineering. Based on the hypothesis that traditional teaching is insufficient to prepare people for their introduction to this field, this paper presents *Qubit: The Game*, an innovative board game to promote both the motivation to learn quantum computing and the understanding of several essential concepts of that field. The reasons for the choice of game type, design and mechanics, and the followed methodology are described in detail here. This paper also includes a preliminary study to determine the effect of the proposed game on the perception, interest and basic knowledge of quantum computing in a group of high school students. The study findings reveal that the designed game is a powerful tool to foster interest and teach essential concepts of a subject as difficult as quantum computing, which can be of great help in introducing more complex concepts.

Keywords: Quantum Computing · Education · Game-Based Learning.

1 Introduction

Quantum computing is an exciting discipline that will become increasingly relevant due to its potential applicability in many areas. One of those areas is cybersecurity, which must be reconstructed in the face of the threat posed by the possible availability of quantum computers that make it possible to break the cryptographic algorithms on which the main current secure communication technologies are based. However, quantum computing is also a conceptually challenging discipline, as for most people, everything related to quantum concepts is perceived as an abstract and challenging field that involves difficult mathematical formalisms and is incompatible with a conception based on the classical model. This work aims to help people adapt to a world where quantum computing will be essential.

J. Mikyška et al. (Eds.): ICCS 2023, LNCS 14077, pp. 108–123, 2023.
https://doi.org/10.1007/978-3-031-36030-5_9

Research in didactics of mathematics and physics shows that the majority of students present difficulties in the acquisition of complex concepts such as those related to quantum mechanics. In fact, numerous studies conclude that students need to feel engaged and interested in the subject for learning to take place [1]. In addition, it is also known that for learning to be meaningful and authentic, it must be built piece by piece, since learning occurs as the student processes and interprets the meaning of each new information they receive. For this reason, games to teach science have been used for more than a century in order to take advantage of the opportunity they offer to integrate cognitive, affective and social aspects with learning [2]. In fact, the main conclusions of some studies carried out on learning based on games, in comparison with the one based on traditional methods, can be summarized as follows: it promotes greater student interest, it can consume more time, and it can result in a greater understanding of a concept and greater retention.

This work proposes *Qubit: The Game*, an innovative game as a way to introduce, in an entertaining way, essential concepts related to quantum states and transitions between states. It is a game that can be played by two or more players, and that is based on three types of cards: qubit, quantum gate and projection, with completely new mechanics. The object of the game is to obtain a binary sequence that is chosen at the beginning as the target key.

A preliminary study on the effect of the proposed game was carried out with secondary school students, and the conclusions were promising. That study is part of a more comprehensive analysis that is being done with larger groups of players and broader surveys to look at different results that could perhaps lead to improvements to the game.

This paper is organized as follows. In Section 2 some related works are mentioned. Section 3 introduces the quantum computing concepts covered in the proposed game. Section 4 is devoted entirely to a detailed description of the proposed game, including design, mechanics and quick start. Sect. 5 describes the pilot experience on the game that was carried out with a group of students, highlighting some findings of the study and a brief discussion. Finally, Sect. 6 of Conclusions closes the paper.

2 Related Works

In the literature we can find different proposals to improve the teaching of quantum concepts, including tools based on interactive simulations and visualization and some quantum-based games, although in most cases, these game proposals are digital. In this work, however, the decision was made to design a board game that must be played in person, in groups, in order to favor the integration of cognitive, affective and social aspects with learning. For the design of the cards, the authors drew on extensive experience with the IBM's Qiskit [3] tool, one of whose greatest strengths in visualization.

Among the efforts to improve the teaching of quantum computing, the publications [4–6] and [7] stand out. On the one hand, the work [4] gives an overview of online learning and teaching materials for a first course in university quantum mechanics, including research-based interactive simulations. On the other hand, the paper [5] is a recent paper on the teaching of a graduate-level quantum computing course and quantum computing modules in high schools, which develops problems to be solved on IBM's quantum computing simulator. Besides, the report [6] describes an experience teaching an undergraduate course on quantum computing using a practical, software-driven programming-oriented approach. Finally, the work [7] proposes an approach for secondary school students, including instructional materials to enhance the educational and cultural potential of quantum computing. Unlike those four works, the research presented in this paper is not focused on the design of modules for curricular or extracurricular courses, but on an educational proposal based on a game that can be used in any of these courses, or as a simple leisure activity.

Various proposals for tools supported by interactive simulations and visualization to teach quantum mechanics can be found in the bibliography [8–11]. First, the work [8] explores how interactive animations and simulations can enhance student understanding of quantum mechanics and quantum information theory. In the remaining papers, several specific tools are proposed for quantum visualization, including PhET [9], QuILT [10] and QuVIS. Note that the objective of this work is different from the one of the last five works mentioned because the focus is not on teaching quantum mechanics through interactive simulations or visualization tools, but rather on introducing essential concepts of quantum computing through a board game.

Among the closest bibliographical references to this work, two games proposed to teach quantum computing stand out: Quantum Odyssey [12] and Entanglion [13]. On the one hand, Odyssey is a software-assisted visual game for learning how to create new quantum algorithms and optimize them for quantum computers. On the other hand, Entanglion is a board game whose main objective is to introduce the fundamental concepts of quantum computing. *Qubit: The Game* is a card game, with an introductory focus for all types of audiences. In fact, the objective is not only educational but also to promote attraction towards the subject, and both objectives can be considered fulfilled according to the feedback received from the study carried out with young people.

3 Quantum Concepts

3.1 Qubit

Traditionally in computing, the smallest unit of information is called a bit. At any time, a bit can be in only one of two possible states, normally represented as 0 and 1. On the contrary, in quantum computing the basic unit of information is the quantum bit or qubit, which can be found in one of the infinite possible combinations of 0 and 1, in what is known as a superposition state [14]. This concept is normally represented with respect to the computational basis ($\{|0\rangle, |1\rangle\}$) by Eq. 1, with complex coefficients ($\alpha, \beta \in \mathbb{C}$), whose sum of the squares of their norms is equal to one ($||\alpha||^2 + ||\beta||^2 = 1$).

$$|\psi\rangle = \alpha|0\rangle + \beta|1\rangle \tag{1}$$

Thus, the state $|0\rangle$ is the one whose $\alpha = 1$ and whose $\beta = 0$. The opposite case occurs for the $|1\rangle$ state.

3.2 Superposition

A state in superposition is one that cannot be represented by a single component with respect to a basis. Two superposition states with respect to the computational basis are illustrated in Eq. 2, 3.

$$|+\rangle = \frac{1}{\sqrt{2}}(|0\rangle + |1\rangle) \tag{2}$$

$$|-\rangle = \frac{1}{\sqrt{2}}(|0\rangle - |1\rangle) \tag{3}$$

In the two cases shown in Eq. 2, 3, the states lie in an equiprobable superposition of the states $|0\rangle$ and $|1\rangle$, given that $||\alpha||^2 = ||\beta||^2 = \frac{1}{2}$.

However, the superposition with respect to one basis can be seen as a basic state on another basis. For example, an alternative way of defining qubits is with respect to the Hadamard basis, formed by the states $|+\rangle$ and $|-\rangle$. In this case, these vectors that make up the basis are in basic states, and the rest are in a superposition state. For example, with respect to the Hadamard basis, the states $|0\rangle$ and $|1\rangle$ are in an equiprobable superposition as illustrated in Eq. 4, 5.

$$|0\rangle = \frac{1}{\sqrt{2}}(|+\rangle + |-\rangle) \tag{4}$$

$$|1\rangle = \frac{1}{\sqrt{2}}(|+\rangle - |-\rangle) \tag{5}$$

3.3 Bloch Sphere

To better visualize the concept of superposition, the intuitive visual representation of the qubit, called the Bloch sphere, can be used. It is important to note that with this object, the goal is to represent two complex numbers (each represented by two real numbers) in three-dimensional space. This leaves one dimension missing, which is achieved by considering two opposite states on the sphere as orthogonal states (see Fig. 1).

Thus, to represent the state of a qubit, a sphere and an arrow inside it pointing to a point on its surface are needed. The higher the sphere points, the more likely it is to measure the $|0\rangle$ state; and the further down, the more likely to get the $|1\rangle$ state. The same is true for left and right, and the states $|+\rangle$ and $|-\rangle$, respectively.

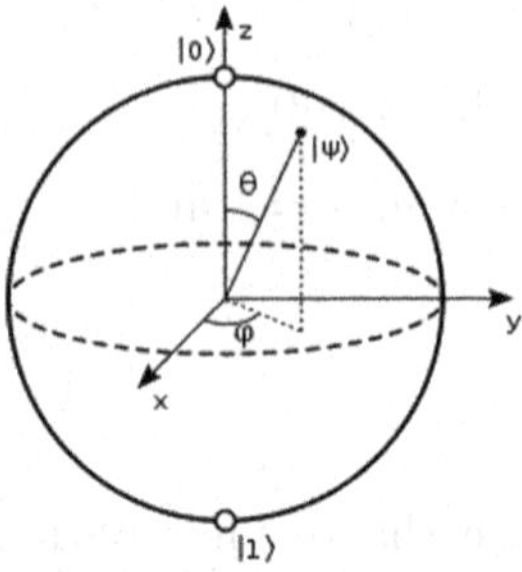

Fig. 1. Bloch Sphere

3.4 Quantum Gates

A quantum logic gate is a transformation that can be applied on the states of a small number of qubits. In the case of gates applied to a single qubit, they can be visualized as rotations in the state arrow of the qubit on the Bloch sphere. The gates that are dealt with in the proposed game are the Pauli gates X and Z, the Hadamard gate H, and the $SWAP$ gate.

Pauli-X This gate applies to a single qubit and can be represented as a rotation of π radians about the X axis on the Bloch sphere. This is due to the fact that the amplitudes of the qubit state are exchanged with respect to the computational base (see Eq. 6).

$$\alpha|0\rangle + \beta|1\rangle \xrightarrow{G_X} \beta|0\rangle + \alpha|1\rangle \tag{6}$$

In this way, for each one of the two states of the computational basis, the Pauli-X gate acts as a classical NOT gate (see Eq. 7).

$$|0\rangle \xleftrightarrow{G_X} |1\rangle \tag{7}$$

Pauli-Z. Analogous to the Pauli-X gate, the Pauli-Z gate represents a rotation of π radians to the state of the input qubit about the Z axis on the Bloch sphere. Thus, this gate applied to the state of a qubit performs the exchange of their amplitudes with respect to the Hadamard basis (see Eq. 8).

$$\alpha|+\rangle + \beta|-\rangle \xrightarrow{G_Z} \beta|+\rangle + \alpha|-\rangle \tag{8}$$

Therefore, the Pauli-Z gate also acts as a classical NOT gate for each of the two states of the Hadamard basis (see Eq. 9).

$$|+\rangle \xleftrightarrow{G_Z} |-\rangle \tag{9}$$

Hadamard (H). The Hadamard gate is one of the most used quantum gates, since it allows to go from a basic state to a superposition state. Thus, it can be used to transform a qubit in the computational basis and get its analog in the Hadamard basis (see Eq. 10). This gate represents a rotation of π radians around the diagonal X+Z axis of the Bloch sphere.

$$\begin{aligned} |0\rangle &\xleftrightarrow{G_H} |+\rangle \\ |1\rangle &\xleftrightarrow{G_H} |-\rangle \end{aligned} \tag{10}$$

SWAP. This gate is applied to the states of two input qubits, swapping their values (see Eq. 11).

$$|\psi_a\rangle|\psi_b\rangle \xleftrightarrow{G_{SWAP}} |\psi_b\rangle|\psi_a\rangle \tag{11}$$

3.5 Projection

In most quantum algorithms, qubits go through numerous quantum gates until a point is reached where it is interesting to observe the result. However, it is not possible to determine the quantum state directly. Instead, when measuring a qubit, a projection of its state vector along a given axis is obtained. Specifically, forcing the state of a qubit to collapse is equivalent to randomly projecting it onto one of the axes of the Bloch sphere. The states of qubits that are already on an axis, when projected on the corresponding basis, are measured with probability 1 in the real value of the state. However, when measured on an axis with respect to which the state is in a state of equiprobable superposition, it is expected that half of the time it is projected in one direction and half in the opposite direction. In this way, for example, when projecting onto the Z axis any of the states $|+\rangle$ or $|-\rangle$ becomes one of the two states $|0\rangle$ and $|1\rangle$ with probability 0.5. The same would happen to an arbitrary state $|?\rangle$ that was in a superposition state with respect to that base (see Eq. 12).

$$\left.\begin{matrix} |+\rangle \\ |-\rangle \\ |?\rangle \end{matrix}\right\} \xrightarrow{P_Z} \left\{\begin{matrix} |0\rangle \\ |1\rangle \end{matrix}\right. \tag{12}$$

A situation analogous to the one explained above occurs with the projection on the X axis (see Eq. 13).

$$\left.\begin{matrix} |0\rangle \\ |1\rangle \\ |?\rangle \end{matrix}\right\} \xrightarrow{P_X} \left\{\begin{matrix} |+\rangle \\ |-\rangle \end{matrix}\right. \tag{13}$$

3.6 Entanglement

Entanglement in quantum computing represents a correlation between the indeterminate states of two or more qubits. This effect has no classical analogue, so it can be hard to imagine. To try to visualize it, the existence of two entangled coins can be imagine so that tossing both coins at the same time will always yield exactly the same result. This effect does not depend on the distance between the coins. It is important to emphasize that there is no communication of any kind between the entangled objects. On the contrary, the result of the identical measurement of two entangled qubits a and b is a direct consequence of the fact that both share the same global state defined in Eq. 14.

$$|\psi\rangle = \frac{1}{\sqrt{2}}(|0_a 0_b\rangle + |1_a 1_b\rangle) \tag{14}$$

Therefore, two entangled qubits that are measured with respect to a basis in which their states are in superposition remain jointly indeterminate, since it is possible to obtain any of the two possible values with a 50% probability, although when performing the measurement (if the entanglement follows Eq. 14) the same value is obtained with a 100% probability.

3.7 Decoherence

Quantum decoherence is the process by which quantum systems lose the properties that characterize them, collapsing by themselves to a classical system. In quantum experiments, this happens quite often, so normally quantum systems are made as independent of the environment as possible. However, on the other hand, absolute control of the system is desirable to fulfill the computational objective, which implies the need to initialize the system to a known state, manipulate the system through well-defined transformations, etc. That is why decoherence reduction is a broad field of study in hardware implementations of quantum processors.

4 *Qubit: The Game*

4.1 General Description

Objective of the Game. *Qubit: The Game* is a card game whose objective is to be the first player to form a binary key. This chain is randomly selected at the beginning of the game. The objective of the game includes the case in which said chain is a subsequence of the chain formed with all the cards on the player's board. Eq. 15 shows the language containing all possible sequences of cards played on the board that win the game, with $key \in (0|1)^*$ being the target binary chain.

$$L = \{w : (0|1| + |-)^* \text{ key } (0|1| + |-)^*\} \tag{15}$$

The game is designed to be played between two and six players, with four being the ideal number for a standard game. The duration of the games will depend on the experience of the players with the game, since the first games can be a little slower, so that they will lighten up as more knowledge is gained about the rules and mechanics of the game. Up to four players, it is recommended to set the target chain size to 3 or 4, depending on whether you want a short game (10-20 min) or a medium game (15-30 min). For games with five players or more, an objective key of length 3 is recommended, to achieve intermediate or even long games (+ 30 min).

Actions. There are 2 types of possible actions:

- Draw a card from the draw piles.
- Play a card that is in your hand.

It is important to perform each action individually, fully completing one action before beginning the next. This is important as the order in which actions are performed greatly affects how the game unfolds. More details about the actions can be found in Sect. 4.3.

Turns. After choosing the target key, the starting player is randomly chosen. In the first round, each player performs three actions on their turn. Each time a player ends their turn, the turn passes to the next player in clockwise order. This dynamic is repeated until a round is completed, where all the players have already played their first turn of three actions. From this moment, the players must roll a dice to know how many actions they must perform per turn. The result of the dice defines how many actions must be performed as follows. Values one, two, and three grant 1 action; four and five provide 2 actions; and six sets 3 actions. In this way the probabilities are balanced, reducing the duration of the turns and favoring more dynamic games.

4.2 Design

The layout of the playing area is made up of each player's hand cards, a tabletop section for each player to place their played cards, three draw piles, and four resource piles.

Each player's hand is made up of the cards that are currently playable. Some of the played cards remain on the player's side of the table where the goal of obtaining the key is attempted before the opponents.

The three draw piles are described below:

- Qubit: Drawing from this pile is expected to obtain a qubit in an undefined state ($|?\rangle$), which should be interpreted as a qubit in a state on the Y axis ($|?\rangle \in \{|i\rangle, |-i\rangle\}$). Therefore, when projecting one of these on either axis (X or Z), the probability is equal for the two possible values. For the simplicity of the game, the application of gates on these qubits is not allowed.
- Gate: In this pile are the letters referring to the operations to be applied on the qubits. The following types can be expected: X, Z, H and $SWAP$ (see Eq. 7, 9, 10 and 11).

- Projection: In this pile are the cards that allow a qubit to be projected against a fixed axis. There are two kinds of projections:
  1. With respect to the Z axis: To apply this card to a qubit that is not currently on this axis, a coin is tossed and, depending on the result, the qubit becomes $|0\rangle$ or $|1\rangle$. In particular, heads indicates the conversion to the value $|0\rangle$ and tails to the value $|1\rangle$ (see Eq. 12).
  2. Regarding the X axis: The same applies for this other axis, where heads indicates the value $|+\rangle$ and tails $|-\rangle$ (see Eq. 13).

In any of these piles, a special type of cards, called event cards, surprise the player. To play an event card, simply follow the instructions given in the text. In addition, some of these cards have a lightning bolt mark indicating that they are "instant" cards that must be played immediately, without counting actions consumed for that player's turn.

In each resource pile there are 16 cards of each value ($|0\rangle$, $|1\rangle$, $|+\rangle$, or $|-\rangle$). They cannot be stolen arbitrarily. To take them, another card must indicate it. For example, an event card may indicate that a $|0\rangle$ should be put into the player's hand; or by applying an X gate to a $|0\rangle$ qubit, discard both cards and put a $|1\rangle$ card in their place; etc.

4.3 Mechanics

It might seem that the most effective way to achieve the objective of the game is for the player to use their cards in their favor to achieve victory. However, this is usually not the only or the best path. Gates, projections, and events can be applied to the played qubits of the player playing them, or to the played qubits of their opponents. A card that can be useless applied to a player's cards can significantly slow down their opponent's progress in the game. That is why fun competition and coalition mechanics can be generated.

The order in which the cards are played is also extremely important. When playing a qubit, it is placed on the player's side of the table; placing, as desired, as far to the right or left as possible with respect to the qubits already played. Once a qubit card is placed on the table, it cannot change its position (unless an event card indicates so). In this way, if when projecting a qubit the desired value is not obtained, it must be corrected using gates and/or projections. Special emphasis is placed on the fact that the application of gates and projections are not commutative operations. Projecting first and then applying a gate may not be the same as applying the same gate and then the same projection, since projecting involves changing what is being observed.

There is an event card with which it is possible to entangle two indefinitely played qubits. Once interleaved, when either of the two is projected about any axis, both are defined with the same result value. Furthermore, it is possible to entangle more than two qubits among themselves, entangling a new one with others that were already entangled.

It should be noted that events are definitively discarded once they are played, unlike the rest of the cards that can appear more than once. In this way, an analogy with decoherence is made, since there are fewer and fewer events, it is as if the state of the game tends to be more and more classic. As time passes, fewer events occur, fewer surprises, less randomness, and the game becomes less quantum. This also balances the game if the game lasts too long, since without events it becomes a much more strategic game.

4.4 Quick Start

The following is a detailed description of an example of game development. To keep the example game short, the key length is set to 2. Subsequently, an arbitrary random string of this length is chosen to be considered as the target key, presumed to be "10". Assume two players: Alice and Bob. Both, on the first turn, have 3 actions.

1. Alice starts by performing her first action: draw a card from the qubit pile, and get the expected qubit card in $|?\rangle$ state. In her second action, she decides to draw a gate card. In this case, she is surprised by an event card, which tells her that she has two additional actions. She draws a card from this pile again and gets a Hadamard gate. Next, she decides to draw a card from the projection pile, getting one on the Z-axis. She has one action left, in which she decides to play her qubit card by placing it on her side of the table. Now Alice's turn is up, so it is Bob's turn.
2. Bob decides to draw a qubit card, obtaining an event that forces him to pick a qubit in $|0\rangle$ state. Note that this card stays in Bob's hand, if he wished to play it he would have to consume another action. However, he prefers to draw another card from the same pile, and unluckily another event occurs that consumes the rest of the remaining actions in that turn.
3. Now that both players have played a full turn, at the beginning of each turn the dice must be rolled to decide the number of actions per turn to be performed. Alice rolls the dice and gets a 3, so she has an action that she spends taking a qubit card. She gets another qubit in $|?\rangle$ state.
4. Bob rolls the dice and gets a 6, so he has 3 actions. He uses the first one to draw a gate card and gets a quantum entanglement card. Now he draws another card from the qubit pile, getting a qubit in $|?\rangle$ state. Finally he places his qubit in state $|0\rangle$ on his side of the table.
5. Alice rolls the dice again and gets a 4, so she gets 2 actions. She uses both to place her two qubits in state $|?\rangle$ on her side of the table.
6. Bob rolls and gets a 5, so he gets 2 actions. The first of them is to place his qubit in state $|?\rangle$ to the left of the qubit already played in state $|0\rangle$. The last action is used in a move that can be considered a really good one: applying the entanglement card to the two undefined qubits of Alice. This way, when she measures one of them, the other one will collapse to exactly the same state and this can be a problem since the key is "10".
7. However, what could be a very good move will give the victory this turn to Alice. She rolls the dice and gets a 6, getting 3 actions. The first of these she consumes by drawing a gate card, with such good luck that she gets the

X gate. At this point, she already knows she will win the game. Alice then applies projection card Z to any of her qubits. To do so, she flips a coin: it comes up tails, so both qubits are defined in the $|1\rangle$ state. With her last action, she flips her second qubit, getting the winning key "10".

In this small example, some of the basic mechanics of the game have been put to the test, showing how the randomness present in the game can quickly generate victory, making the game very dynamic. All this without losing the high strategic component, since knowing the possibilities of the results of randomness, it is up to the player whether to choose safer or riskier situations.

5 Pilot Study

5.1 Survey Design

A first study on the effect of the proposed game was carried out with a sample of 19 students from a group of 1st year of high school of a secondary school in Fuerteventura (Canary Islands, Spain), who came to the University of La Laguna to participate in this study. They were administered a survey in different phases, producing data that was collected for use in statistical tests to draw conclusions about the potential of the game. In particular, the methodology followed was as described below. In order to determine each student's level of achievement, the student was given a simple pre-test on attitude, interest, and knowledge about the topic in question. Then they were exposed to a half-hour talk on the basics of quantum computing and the game, given by one of the authors of this paper. New questions specifically about the game were administered right after the talk. Then, after playing the game for an hour or so, the entire test was administered to the same students so that both test and retest reliability can be used, since scores from both tests can be correlated.

The ten questions included in the survey were the following:

Q1 Do you know anything about quantum computing?
Q2 Are you interested in quantum computing?
Q3 Does quantum computing seem easy to you?
Q4 In the future, would you like to work in an area close to quantum computing?
Q5 Does the game seem easy to you?
Q6 Do you find the game useful to understand quantum computing concepts?
Q7 Would you like to have the game?
Q8 Mark the 3 concepts that you find most interesting:

- ☐ Bloch sphere
- ☐ qubit
- ☐ quantum gate
- ☐ projection
- ☐ entanglement

Q9 When applying an H gate card to a played card of qubit $|0\rangle$, what is left?

- ☐ $|0\rangle$
- ☐ $|1\rangle$
- ☐ $|+\rangle$
- ☐ $|-\rangle$

Q10 What do you get when applying a projection card on the Z axis on a played qubit card $|+\rangle$?

- ☐ The qubit card $|1\rangle$
- ☐ One of the qubit cards $|0\rangle$ or $|+\rangle$, randomly on coin toss
- ☐ One of the qubit cards $|0\rangle$ or $|1\rangle$, randomly on coin toss
- ☐ The qubit chart $|+\rangle$, staying as it was

The first seven questions had to be answered with a 5-point Likert scale from 1 (not at all) to 5 (extremely), while the last ones were multiple choice (question Q8) and single (questions Q9 and Q10).

All questions were answered twice: the first four were answered in the pre-test before the talk and at the end after playing, while questions Q5 to Q10 were answered right after the talk and at the end after playing.

5.2 Findings

To verify the achievement of the objective of the game on the promotion of motivation to learn quantum computing, the first hypothesis raised in the study of the data obtained with the first four questions of the survey is:

1. There is no significant difference in students' attitudes toward quantum computing before and after exposure to the game.

With questions Q5 to Q7 of the survey, another specific hypothesis about the game is analyzed through this null hypothesis:

2. There is no significant difference in students' interest in playing the game before and after exposure to it.

Finally, the understanding of some essential concepts of quantum computing is checked with the last two questions, which allows testing the hypothesis:

3. There is no significant difference in students' basic knowledge of quantum computing before and after exposure to the game.

In this study, paired two-sample Student's t-tests have been used because a single sample of individuals answered the survey twice. In particular, null hypothesis tests that both means before and after the game are equal for each question are carried out with a significance level of 0.05. Therefore, whenever a p-value less than 0.05 is obtained, the result is considered statistically significant and the corresponding null hypothesis is rejected.

The first analysis of students' responses to question Q1 yields a first clear result as the median response for this question before the game is 1 while after the game is 3. Indeed, Table 1 shows the results of the t-test on the responses to this question Q1, and the conclusion is that there is a significant difference in the self-perception of individuals about their own knowledge of quantum computing before and after the game.

Table 1. Student's t-test on question 1.

	Q1	Q1 retest
Mean	1,263157895	3,526315789
Variance	0,204678363	0,929824561
Pearson correlation coefficient	0,046917565	
df	18	
t stat	-9,433414271	
$P(T \leq t)$ one-tail	1,08727E-08	
t critical one-tail	1,734063607	
$P(T \leq t)$ two-tail	2,17455E-08	
t critical two-tail	2,10092204	

A similar analysis with question Q2 yields a p-value of 0,034935624, which also confirms that significant difference in responses related to interest in quantum computing between pre and post game, showing that after the game the individuals are more interested in the subject than before.

The analysis of variance of the perception of the difficulty of the topic before and after playing, shown with question Q3, produces a p-value of 0,001386775. Thus, the result shows that there is a significant difference between that students' perception before and after playing.

Regarding question Q4, the obtained p-value is 0,124235321, so in the analysis of variance of interest in working in quantum computing, it is not possible to conclude a significant difference in the attitude of the students. This is the only question in which the conclusion of the t-test is not conclusive with respect to the rejection of the null hypothesis.

As mentioned above, the results of the t-tests applied to these questions Q1 to Q4 can be used to support the rejection of hypothesis 1, so it is concluded that there is a significant difference in students' attitudes toward quantum computing before and after exposure to the game.

With question Q5, the results of the two-sample Student's t-test include a p-value of 2,83965E-05, so the conclusion is that the game is perceived as easy.

The t-test analysis of question Q6 yields a p-value of 0,015972972, so students perceive the game as useful for understanding quantum computing concepts.

Finally, from question Q7 of the survey a p-value of 0,003947853 is obtained, so the students are interested in having the game.

Thus, the specific hypothesis 2 about the game is also rejected, concluding that the students are clearly interested in playing the game. In this sense, it is worth including here some verbatim comments included by the students in their responses, such as: "I really like the game because it is fun and interesting" and "The game is cool :-)".

In question Q8, it was observed that after the game the greatest interest shown before playing in the concepts of qubit, quantum gate and entanglement was maintained, compared to less interest in the concepts of Bloch sphere and projection.

Finally, questions Q9 and Q10, answered twice, after the talk and after playing, allow us to confirm the usefulness of the game for knowledge acquisition regarding essential concepts of quantum computing. On the one hand, in question Q9 about the Hadamard gate, 26.3% of answers were incorrect before playing, while there was unanimity in the correct answer after playing. On the other hand, the question Q10 on projection received 31.6% incorrect answers before playing the game, and only 10% incorrect answers after the game. Therefore, from both questions it can be concluded that the hypothesis 3 is rejected, since there is a significant difference in the basic knowledge of quantum computing of the students before and after the exposure to the game.

5.3 Discussion

In general, this research has had as a general objective to verify the hypothesis that the use by teachers of methods based on gamification contributes to a great extent to maintain and motivate students' interest in learning complex subjects in general. In fact, this preliminary study clearly confirms that the use of a card game environment with interaction between students leads to an improvement in performance and a positive attitude towards learning complex a priori subjects such as quantum computing, so the game can be considered a useful tool to teach quantum computing to non-specialists.

Although most of the findings obtained from this preliminary analysis correspond to what is indicated by intuition, it is reassuring to see statistical evidence that the game produces the expected effect. However, it is clear that a larger study with a greater variety of individuals, questions and aspects is needed. In particular, as this was the first public release of the game, the main interest was to get a quick, high-level first impression of the game. However, in subsequent experiments, the Game Experience Questionnaire [15] will be used. Besides, another pending task is the study of the effects of the different event cards on the probability of winning the game.

An additional aspect that is worth looking at in future studies and versions of the game is the minimum level or age at which the game could be introduced. In fact, it would be interesting to analyze whether introducing these concepts at an early age would foster an interest in science and computing. Another interesting fact is that it seems that the use of games could be effective in improving performance and attracting female students to science and engineering [16], therefore, as an extension of this work, a specific study will be carried out in this regard.

6 Conclusions

This work has presented *Qubit: The Game*, which is an innovative card game whose main objective is to promote interest in quantum computing as well as the understanding of several essential concepts of the field. The preliminary study carried out with a group of high school students has made it possible to determine statistically the achievement of the objectives of the game. However, there are still several open lines in this work. It is planned to release *Qubit: The Game* as an open source project so that it can be freely used by educators, students, and card game enthusiasts. Another potential future work is the implementation of a simulator with AI players to study possible improvements of the game.

Acknowledgements. This research has been supported by the Cybersecurity Chair of the University of La Laguna and the Eureka CELTIC-NEXT project C2020/2-2 IMMINENCE funded by the Centro para el Desarrollo Tecnológico Industrial (CDTI).

References

1. Newman, F.M., Wehlage, G.G., Secada, W.G., Marks, H.M., Gamorman, A.: Authentic pedagogy standards that boost student performance. Issues in restructuring schools. Issues report No. 8. Madison, W.I. Center one organize and restructuring of schools, ED 39091 (1995)
2. Randel, J.M., Morris, B.A., Wetzel, C.D., Whitehill, B.V.: The effectiveness of games for educational purposes: A review of recent research. Simulation & gaming **23**(3), 261–276 (1992)
3. Qiskit. https://qiskit.org (2023)
4. Kohnle, A., Bozhinova, I., Browne, D., Everitt, M., Fomins, A., Kok, P., Kulaitis, G., Prokopas, M., Raine, D., Swinbank, E.: A new introductory quantum mechanics curriculum. European Journal of physics **35**(1), 015001 (2013)
5. Tappert, C.C., Frank, R.I., Barabasi, I., Leider, A.M., Evans, D., Westfall, L.: Experience Teaching Quantum Computing. In: Annual Meeting Annual Meeting of the Association Supporting Computer Users in Education (2019)
6. Mykhailova, M., Svore, K.M.: Teaching quantum computing through a practical software-driven approach: Experience report. In: Proceedings of the 51st ACM technical symposium on computer science education, pp. 1019–1025 (2020)
7. Satanassi, S., Ercolessi, E., Levrini, O.: Designing and implementing materials on quantum computing for secondary school students: The case of teleportation. Physical Review Physics Education Research **18**(1), 010122 (2022)
8. Singh, C.: Helping students learn quantum mechanics for quantum computing. In: AIP Conference Proceedings, vol. 883(1), pp. 42–45. American Institute of Physics (2007)
9. McKagan, S.B., Perkins, K.K., Dubson, M., Malley, C., Reid, S., LeMaster, R., Wieman, C.E.: Developing and researching PhET simulations for teaching quantum mechanics. American Journal of Physics **76**(4), 406–417 (2008)
10. Singh, C.: Interactive learning tutorials on quantum mechanics. American Journal of Physics **76**(4), 400–405 (2008)
11. Kohnle, A., Baily, C., Campbell, A., Korolkova, N., Paetkau, M.J.: Enhancing student learning of two-level quantum systems with interactive simulations. American Journal of Physics **83**(6), 560–566 (2015)
12. Nita, L., Chancellor, N., Smith, L. M., Cramman, H., Dost, G.: Inclusive learning for quantum computing: supporting the aims of quantum literacy using the puzzle game Quantum Odyssey. arXiv preprint arXiv:2106.07077 (2021)

13. Weisz, J. D., Ashoori, M., Ashktorab, Z.: Entanglion: A board game for teaching the principles of quantum computing. In: Proceedings of the 2018 Annual Symposium on Computer-Human Interaction in Play, pp. 523–534 (2018)
14. McMahon, D.: Quantum computing explained. John Wiley & Sons (2007)
15. IJsselsteijn, W.A., de Kort, Y.A.W., Poels, K.: The Game Experience Questionnaire. Technische Universiteit Eindhoven. (2013)
16. Aremu, A.: Strategies for Improving the Performance of Female in Mathematics. African Journal of Educational Research. **5**(1), 77–85 (1999)

Software Aided Approach for Constrained Optimization Based on QAOA Modifications

Tomasz Lamża[1,2(✉)], Justyna Zawalska[1,2], Mariusz Sterzel[2], and Katarzyna Rycerz[1,2]

[1] Institute of Computer Science, AGH, al. Mickiewicza 30, 30-059 Krakow, Poland
{tomasz.lamza,justyna.zawalska}@cyfronet.pl, kzajac@agh.edu.pl

[2] Academic Computer Center Cyfronet AGH, ul. Nawojki 11, 30-950 Krakow, Poland

Abstract. We present two variants of the QAOA modification for solving constrained combinatorial problems. The results presented in this paper were obtained using the QHyper framework, which we developed specifically for this purpose. More specifically, we use the created framework to compare the QAOA results with its two modifications, namely: Weight-Free QAOA (WF-QAOA) and Hyper QAOA (H-QAOA). Additionally, we compare the Basin-hopping global optimization method for subsequent sampling of the initial points for the proposed QAOA modifications with a simple Random Search. The results obtained for the Knapsack Problem indicate that the proposed solution outperforms the original QAOA algorithm and can be promising for QUBO, where adjusting the relative importance of the cost function and the constraints is a significant challenge.

Keywords: QAOA · Constrained Optimization · Hyperparameters · QUBO · Penalty

1 Introduction

Recently, using and developing quantum algorithms for solving combinatorial problems is becoming a popular subject in the research field of quantum computation. Examples of the main findings in this area include variational algorithms for gate-based quantum devices, such as the Quantum Approximate Optimization Algorithm (QAOA) [4], or quantum annealing solvers realized by D-Wave devices[1]. However, successful usage of such algorithms usually requires careful setting of their initial parameters, which is quite crucial yet nontrivial task. Furthermore, the required formulation of the objective function as Quantum Unconstrained Binary Optimization (QUBO) in most cases requires cautious setting of weights between the cost function and the constraints. This is also not obvious; therefore, the weights often become additional hyperparameters of the problem. There are also attempts to efficiently search for such parameters and hyperparameters [1,13,15] but research in this direction is in its early development stage. In

[1] https://www.dwavesys.com/.

J. Mikyška et al. (Eds.): ICCS 2023, LNCS 14077, pp. 124–137, 2023.
https://doi.org/10.1007/978-3-031-36030-5_10

this paper, we propose two variants of the QAOA modification that improve optimization for constrained problems. The first modification (Weight-Free QAOA or WF-QAOA) is based on the definition of the new weight-free observable for the QAOA-based variational algorithm. Furthermore, this allows extending the variational parameter set with the QUBO weights in the second modification (Hyper QAOA or H-QAOA). When comparing the performance of the proposed variants, we examined two ways of sampling the initial points for the variational algorithms: simple Random Search and global optimizer Basin-hopping [18]. Additionally, to facilitate conducting a variety of planned experiments, we introduce the architecture of the QHyper software framework as a modular tool for researchers. The results show that the proposed variants perform better than the regular QAOA algorithm. Random Search with WF-QAOA produced the best results when the number of variational algorithm launches was small. However, Basin Hopping variants performed better with a larger number of the algorithm launches, particularly when using multiple initial points. Although we present the results using the Knapsack Problem, the proposed solution can be used for any combinatorial problem transformed to QUBO.

This paper is organized as follows: in Sect. 2 we introduce important preliminaries and in Sect. 3 we describe related work. The proposed variants of the QAOA modification are shown in Sect. 4. The QHyper experiment framework is presented in Sect. 5. Section 6 discusses different sampling methods based on global optimizers used in our experiments. The results are presented in Sect. 7 and conclusions can be found in Sect. 8.

2 Preliminaries

In this section, we provide a concise overview of the key aspects related to quantum solutions for combinatorial problems necessary to enhance the the readability of the paper.

2.1 Combinatorial Problem Formulation Models

In the Constrained Quadratic Model[2] (CQM), the cost function $C_f(\boldsymbol{x})$ can be represented using an $n \times n$ cost matrix $\boldsymbol{C}$ where

$$C_f(\boldsymbol{x}) = \boldsymbol{x}^T \boldsymbol{C} \boldsymbol{x} = \sum_i c_{ii} x_i + \sum_{i<j} c_{ij} x_i x_j + const. \tag{1}$$

Similarly, the constraints are given by functions $G_f^{(k)}(\boldsymbol{x})$ represented by an $n \times n$ cost matrix $\boldsymbol{G}^{(k)}$ where

$$G_f^{(k)}(\boldsymbol{x}) = \boldsymbol{x}^T \boldsymbol{G}^{(k)} \boldsymbol{x} = \sum_i g_{ii}^{(k)} x_i + \sum_{i<j} g_{ij}^{(k)} x_i x_j + const \circ 0 \tag{2}$$

[2] https://docs.ocean.dwavesys.com/en/stable/concepts/cqm.html.

where $k = 1, \ldots, M$ and M is the number of constraints. The variables $\{x_i\}_{i=1,\ldots,N}$ can be binary or integer and the matrices $\boldsymbol{C}$, $\boldsymbol{G}^{(k)}$ are real-valued. The symbol $\circ$ denotes a comparison operator $\{\geq, \leq, =\}$.

To use quantum solvers, the CQM problem must be transformed into the forms of (1) and (2), where all $\{x_i\}_{i=1,\ldots,N}$ are binary. Additionally, all functions have to be combined into Quantum Unconstrained Binary Optimization. In the first step, this is done by transforming the constraints $G_f^{(k)}(\boldsymbol{x})\circ 0$ into the equality constraints $K_f^{(k)}(\boldsymbol{x}) = 0$ where $\forall \boldsymbol{x}, K_f^{(k)}(\boldsymbol{x}) \geq 0$. Next, by adding weighted constraints and the cost function together, the objective function is obtained in the form

$$f_{QUBO}(\boldsymbol{x}) = \alpha_0 C_f(\boldsymbol{x}) + \sum_{k=1}^{M} \alpha_k K_f^{(k)}(\boldsymbol{x}), \tag{3}$$

where α_0 is the weight of the cost function and α_k is the weight of the k-th constraint, $\alpha_i > 0$.

2.2 Quantum Approximate Optimization Algorithm

The QAOA is a variational combinatorial optimization algorithm for gate-based quantum devices. To apply the algorithm, the objective function $f_{QUBO}(\boldsymbol{x})$ is translated into the cost Hamiltonian H_C

$$H_C = \alpha_0 H_{C_f} + \sum_{k=1}^{M} \alpha_k H_{K_f^{(k)}}, \tag{4}$$

where H_{C_f} and $H_{K_f^{(k)}}$ are the Hamiltonians that encode the cost function and the constraints, respectively. In addition, a mixing Hamiltonian H_M that is not commuting with H_C is required. A common choice is $H_M = \sum_{i=1}^{N} X_i$, where X_i is the Pauli-X gate. The QAOA ansatz consists of p alternating layers U_C and U_M

$$U_C(\gamma) = e^{-i\gamma H_C}, \tag{5}$$

$$U_M(\beta) = e^{-i\beta H_M}, \tag{6}$$

where γ, β are adjustable parameters. The algorithm uses a chosen classical optimizer to minimize the expectation value

$$F_p(\boldsymbol{\gamma}, \boldsymbol{\beta}) = \langle \boldsymbol{\gamma}, \boldsymbol{\beta} | \, H_C \, | \boldsymbol{\gamma}, \boldsymbol{\beta} \rangle, \tag{7}$$

of the quantum state prepared with a quantum device (or its simulator):

$$|\boldsymbol{\gamma}, \boldsymbol{\beta}\rangle = U_M(\beta_p) U_C(\gamma_p) \cdots U_M(\beta_1) U_C(\gamma_1) \, |+\rangle^{\otimes n}, \tag{8}$$

where $|+\rangle^{\otimes n}$ is a uniform superposition of n qubits. The optimization is performed with respect to $2p$ parameters $(\boldsymbol{\gamma}, \boldsymbol{\beta}) \in [0, 2\pi]^p \times [0, \pi]^p$.

2.3 Knapsack Problem

The Knapsack Problem is a combinatorial optimization problem where given a set of items, each of a certain weight and value, the aim is to determine which items should be packed to the knapsack to maximize the total value of selected items while not exceeding the knapsack's weight limit. The problem can be represented as a QUBO [9]

$$f_{QUBO\_KP}(\boldsymbol{x}, \boldsymbol{y}) = -\alpha_0 \sum_{i=1}^{N} c_i x_i + \alpha_1 (1 - \sum_{i=1}^{W} y_i)^2 + \alpha_1 (\sum_{i=1}^{W} i y_i - \sum_{i=1}^{N} w_i x_i)^2, \quad (9)$$

where N is the number of items available, W is the maximum weight of the knapsack, c_i and w_i are the value and weight of the item i, respectively. $\boldsymbol{x} = [x_i]_N$ is a Boolean vector, where $x_i = 1$ if and only if the item i was selected to be inserted into the knapsack. $\boldsymbol{y} = [y_i]_W$ is a one-hot vector where $y_i = 1$ if and only if the weight of the knapsack is equal to i. In this paper, we focus on the version from [9], where the weight for both of the constraints is the same. That is, α_0 is the weight of the cost function and α_1 is the weight of the constraints.

3 Related Work

The original version of the QAOA [4] focuses on solving unconstrained binary optimization problems. However, constrained problems can also be solved by this algorithm. There are two main methods to incorporate constraints into the QAOA [8]. The first of them, called the *hard* constraints, is based on designing a Quantum Approximate Optimization Ansatz with a special mixing Hamiltonian that prevents finding unfeasible solutions [7]. Examples of such mixing Hamiltonians include XY mixers (and their variations) [5,8,19], Grover mixers [2], and problem-specific mixers based on the use of quantum machine learning [14].

Another option — called *soft* constraints — is incorporating the constraints in the objective function (in the form of QUBO) and using the standard X mixing Hamiltonian. In this case, setting optimal weights between the cost function and the constraints that will allow obtaining feasible results is a major challenge. There are several methods available to select appropriate weights. These include ways of static settings of weights using the information taken from the objective function, such as the maximum QUBO coefficient or the maximum absolute value of the difference achieved by flipping one QUBO variable [1]. There are also efforts of using Monte Carlo-based iterative improvement of distribution from which hyperparameters are sampled, such as the Cross Entropy Method [15] or Tree-structured Parzen Estimator [13]. Monte Carlo methods are also an inspiration for global optimizers such as SHGO [3] or Basin-hopping [18]. As these methods originated from effort towards finding the minimum energy structure for molecules [12], their usefulness in the context of Quantum Variational Algorithms is a promising approach [6,10].

To the best of our knowledge, the QAOA modification proposed in this paper (see Sect. 4) was never considered in the literature. Additionally, although most of the libraries for gate-based quantum computation provide an implementation of variational algorithms such as QAOA and VQE [16], performing experiments that join global optimizers with QAOA variants requires additional tedious effort from the researchers. The presented QHyper system (see Sect. 5) is intended to fill this gap.

4 Proposed QAOA Modifications for Constrained Problems

In this section, we propose two QAOA modifications for QUBO-based constrained problems. The obtained experiment results for the Knapsack Problem described in Sect. 7 indicated that presented variants seem to be a promising alternative for the QAOA.

4.1 Weight-Free Hamiltonian

A constrained problem presented in the Hamiltonian form (see Eq. 4) as a diagonal matrix requires appropriate weights ($\boldsymbol{\alpha}$) setting so that the lower energy states are better solutions than higher energy ones. In this paper, we propose the formulation of an alternative weight-free diagonal Hamiltonian that satisfies the condition of appropriate order of energies. The new Hamiltonian is used for the expectation value estimation (see Eq. 7). The formulation is valid for problems with cost function satisfying $C_f(\boldsymbol{x}) \geq 0$ for all $\boldsymbol{x}$, where $\boldsymbol{x} = [x_i]_N$ is a binary vector that encodes the solution fulfilling constraints (see Eq. 3). This is true for numerous combinatorial problems including the Knapsack Problem. Under this assumption, we define the Hamiltonian H_{wf} as a diagonal matrix of the order 2^N, namely

$$H_{wf}[\text{decimal}(\boldsymbol{x}), \text{decimal}(\boldsymbol{x})] = \begin{cases} -C_f(\boldsymbol{x}) & \text{if } \boldsymbol{x} \text{ satisfies constraints} \\ 0 & \text{otherwise} \end{cases} \tag{10}$$

Algorithm 1. Estimating Expectation Value with Weight-Free Hamiltonian

Require: results - vector of binary results from ansatz measurements in shots, shots_num - number of shots

for $i = 1 \rightarrow$ shots_num **do**
    sol $\leftarrow$ results[i] ▷ taking binary vector of solution i
    if sol satisfies constraints **then**
        score $\leftarrow -C_f(sol)$ ▷ equivalent to $\langle sol|\, H_{wf}\, |sol\rangle$
    else
        score $\leftarrow 0$ ▷ equivalent to $\langle sol|\, H_{wf}\, |sol\rangle$
    end if
    sum$+=$ score
end for
exp_value $= \frac{\text{sum}}{\text{shots\_num}}$ ▷ estimating expectation value of the measured state
return exp_value

The main idea of the QAOA modification is to use an ansatz based on its original Hamiltonian (see Eq. 4) while changing the operator for estimating the expectation value. This is illustrated in Fig. 1b in comparison to the regular QAOA in Fig. 1a. The proposition can also be seen as a variant of the VQE algorithm valid due to the variational principle [16].

Since the weight-free Hamiltonian is not used for the ansatz building, there is also no need to present the full formula of the new Hamiltonian as the Ising model. We require only a simple algorithmic way of calculating eigenstates for a fixed number of eigenvectors that were measured to estimate the expectation value. This is presented as Algorithm 1.

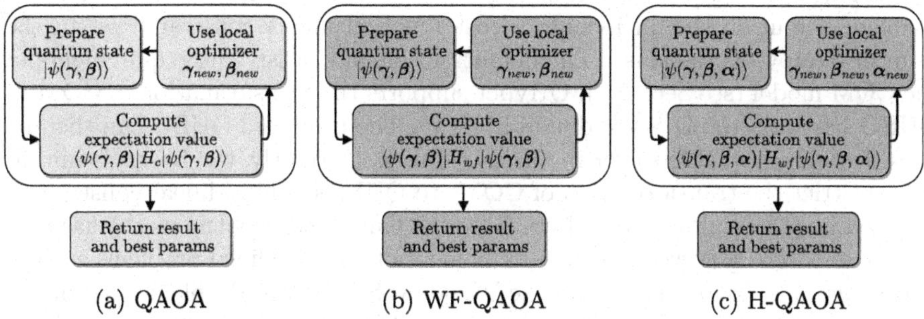

Fig. 1. Comparison of execution schemes for regular QAOA(a), Weight-free QAOA(b) and Hyper QAOA(c)

4.2 Alternate Variational Parameter Set

For ansätze built with weighted Hamiltonians (see Eq. 4), the measurement result depends not only on the choice of the $\boldsymbol{\gamma}, \boldsymbol{\beta}$ angles but also on the Hamiltonian weights – hyperparameters $\boldsymbol{\alpha}$. Based on Eq. 4 and Eq. 5 it can be observed that the U_C layer that encodes the objective function of the optimization problem is parameterized by both the angle γ_l and the weights $\boldsymbol{\alpha}$

$$\begin{aligned} \forall l \in 1, \ldots, p \quad \widehat{U_C}(\boldsymbol{\alpha}, \gamma_l) &= exp(-i \cdot \gamma_l (\alpha_0 H_{C_f} + \sum_{k=1}^{M} \alpha_k H_{K_f^{(k)}})) \\ &= exp(-i \cdot 1(\gamma_l \alpha_0 H_{C_f} + \sum_{k=1}^{M} \gamma_l \alpha_k H_{K_f^{(k)}})) \\ &= \widehat{U_C}(\gamma_l \boldsymbol{\alpha}, 1). \end{aligned} \tag{11}$$

The vector of hyperparameters $\boldsymbol{\alpha}$ must be a proper multiplier for all of the p angles $\boldsymbol{\gamma}$. As a result, $\boldsymbol{\alpha}$ can be treated as an additional variational parameter together with $\boldsymbol{\gamma}$ and $\boldsymbol{\beta}$. Therefore, we propose the Hyper QAOA (H-QAOA) extension using such parameters set with a weight-free Hamiltonian as an observable (see Fig. 1c).

5 QHyper Experiment Framework

The experiments results presented in this paper were performed using the QHyper framework designed as a software support tool for researchers working with the problem of setting hyperparameters. The architecture of the system is shown in Fig. 2. Below, we present the main components of the system.

Supported Combinatorial Optimization Problems. QHyper allows defining various optimization problems that can be solved by integrated solvers. The problem definition should include a cost function and a constraint list in the form of Sympy expressions[3]. The definitions should satisfy the requirements of the CQM model (see Sect. 2.1). QHyper supports the transformation of CQM to QUBO based on the D-Wave dimod library[4]. The produced QUBO can then be passed to any available solver in our framework. Currently, the library's limitations restrict the transformation of CQMs to QUBOs only to linear constraints. Inspired by the results from [15] we chose the Knapsack Problem as the use case for our research. However, the system includes other predefined problems such as multistage calculations planning (the Workflow Scheduling Problem [17]) or the Traveling Salesman Problem [9]. Custom problem definitions can also be added.

Supported Solvers. Currently, as shown in Fig. 2, the system is integrated with different types of solvers. Apart from the regular QAOA solver (see Sect. 2.2), modifications proposed in Sect. 4 are provided, namely: WF-QAOA described in Sect. 4.1 and H-QAOA with both modifications from Sect. 4. The solvers are implemented using the PennyLane[5] library. Additionally, the system also provides an overlay to the D-Wave Leap CQM Solver[6]. Another integrated solver is the purely classical solution Gurobi[7] that is used as a convenient reference method. The modular structure of the system enables effortless integration of additional solvers.

Global Optimizers. We added the possibility of different sampling of the initial parameters set by using a predefined set of global optimizers. Currently, QHyper supports the following: Basin-hopping, the Cross Entropy Method (CEM), and the simple parallel optimizer based on a Random Search. The details of the possible variants used in this paper are presented in Sect. 6

[3] https://docs.sympy.org/latest/tutorials/intro-tutorial/intro.html.

[4] https://docs.ocean.dwavesys.com/en/stable/docs_dimod/reference/generated/dimod.cqm_to_bqm.html.

[5] https://pennylane.ai/.

[6] https://docs.dwavesys.com/docs/latest/doc_leap_hybrid.html).

[7] https://www.gurobi.com/.

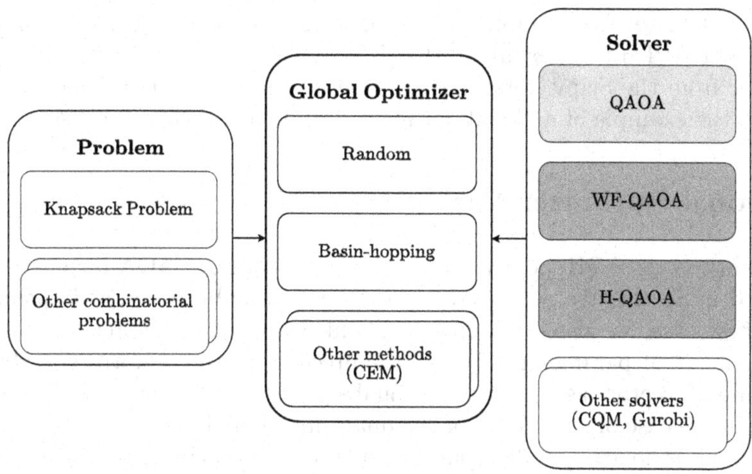

Fig. 2. QHyper architecture.

```
# 1. Creating the Knapsack Problem instance
knapsack = KnapsackProblem(max_weight=1, items=[(1, 2), (1, 1)])
# 2. Specyfing the solver configuration
solver_config = {
    # Choosing the type of the local optimizer
    'optimizer': {
        'type': 'scipy',
        'maxfun': 200
    },
    # Choosing the type of the parameterized quantum circuit
    'pqc': {
        'type': 'hqaoa',
        'layers': 5,
    },
}
# 3. Specyfing the initial parameters configuration
params_config = {
    'angles': [[0.5]*5, [1]*5],
    'hyper_args': [1, 2.5, 2.5],
}
# 4. Creating the variational algorithm (VQA)
vqa = VQA(knapsack, config=solver_config)
# 5. Creating the Random global optimizer
random = Random(processes=5, number_of_samples=100,
        bounds=[[1, 10], [1, 10], [1, 10]])
# 6. Running the solver with the Random hyperoptimizer
best_params = vqa.solve(params_cofing, random)}
```

Listing 1.1. Sample usage of QHyper for solving a Knapsack Problem

Example Usage. Code snippet in the Listing. 1.1 shows the sample usage of the QHyper API. In the example, the H-QAOA solver is used with the default optimizer from the Scipy library (Broyden-Fletcher-Goldfarb-Shanno — BFGS [11]). As the example of a global optimizer Random Search was chosen.

6 Global Optimizer Variants

In this work we used QHyper to compare three variants of QAOA-inspired solvers described in Sect. 4. We employed two different probabilistic global optimizers. In the first case, we used a simple Random Search (RS) approach, where an N-element set of parameter vectors is sampled from the uniform distribution. Next, for each vector, a separate variational algorithm is launched. In the QHyper system, in the case of simulation of the quantum algorithm on the classical HPC machine, this is done in parallel (parameter study approach). After the execution, the best result is chosen. The detailed pseudocode is shown as Algorithm 2.

Algorithm 2. Random Search

Require: n - samples number, **a** - vector of lower bounds, **b** - vector of upper bounds
$x_1, ..., x_n \sim \mathcal{U}_{[a,b]}$
$v_1, ..., v_n \leftarrow$ variational_algorithm(x_1), ..., variational_algorithm(x_n) ▷ This step can be done in parallel
return (x_i, v_i), where $v_i = min(v_1, ..., v_n)$

Algorithm 3. Basin-hopping

Require: x_0 - initial guess, niter - number of Basin-hopping iterations, stepsize - maximum step size for use in the random displacement, **a** - vector of lower bounds, **b** - vector of upper bound
$v\_x_0 \leftarrow$ variational_algorithm(x_0)
for $i = 1 \rightarrow$ niter **do**
    $x \leftarrow$ random_step(x_0) ▷ The step is chosen uniformly in the region from x_0-stepsize to x_0+stepsize, in each dimension
    while $x \notin (\boldsymbol{a}, \boldsymbol{b})$ **or** x doesn't meet all criteria **do**
        $x \leftarrow$ random_step(x_0)
    end while
    $v\_x \leftarrow$ variational_algorithm(x)
    if $v\_x < v\_x_0$ **then**
        $x_0 \leftarrow x$
        $v\_x_0 \leftarrow v\_x$
    end if
end for
return (x_0, v_x_0)

In the second case, single sampling with improvement was used on the example of the Basin-hopping algorithm [18]. A single parameter vector is sampled with the uniform distribution. Next, the variational algorithm is run and its output is accepted if the result is better. Finally, the next parameter vector is chosen uniformly in the region around the last accepted result. The process is repeated N times and, after that, the best result is chosen. The detailed pseudocode of the Basin-hopping algorithm is shown as Algorithm 3.

7 Experiment Results

All experiments were performed on a Knapsack Problem instance with three items, each with a weight of 1 and values of 2, 2, and 1. The maximum knapsack weight capacity was 2, so the maximum possible value was 4. Hence the number of binary variables is equal to 5. To compare the results of all the presented approaches, the same local minimizer was used (the Scipy implementation of L-BFGS-B). The sampling range for angles $\gamma, \beta \in [0, 2\pi]$, and for weights $\alpha_1, \alpha_2 \in [1, 10]$, where $\alpha_1, \alpha_2 \in \mathbb{R}$. We set the number of layers in each variational algorithm to $p = 5$.

Random Search tests were performed by sampling 10.000 initial points (i.e., variational parameter initial sets) and then calculating the results of Algorithm 2 for each point. In particular, as we wanted to check the performance against the number of variational algorithm (QAOA, WS-WAOA or H-QAOA) launches, the results were obtained as follows (n – number of launches of a particular QAOA-based variant):

- shuffle set of 10.000 points,
- split initial set into $10.000/n$ groups, each containing n points,
- for each group, find the minimum value in this group,
- calculate mean and standard deviation for the $10.000/n$ minimum values.

Tests for Basin-hopping (see detailed parameters in Table 1) were performed with different numbers of starting points, keeping the number of variational algorithm launches the same as for Random Search. For Basin-hopping the choice of the initial point is important as it determines the next steps of the algorithm (see *random_step()* in Algorithm 3). Therefore, in this paper, we also compare the results of splitting the global optimizer run into several independent trials with different initial points. These configurations were: (10, 1), (10, 5), (10, 10), (10, 20), (10, 25), (10, 50), (10, 100), (50, 1), (50, 2), (50, 4), (50, 5), (50, 10), (50, 20), (100, 1), (100, 2), (100, 5), (100, 10), (200, 1), (200, 5), (250, 1), (250, 2), (250, 4), (500, 1), (500, 2), (1000, 1), where the first number indicates the number of iterations per initial point, and the second one indicates the number of initial points. Each configuration was tested 10 times and then the mean and

Table 1. Basin-hopping parameters. N is the number of iterations per init point. The description of each parameter can be found in the SciPy documentation.

Basin-hopping parameters				
x_0	niter	T	stepsize	interval
init point	N	1	0.5	50

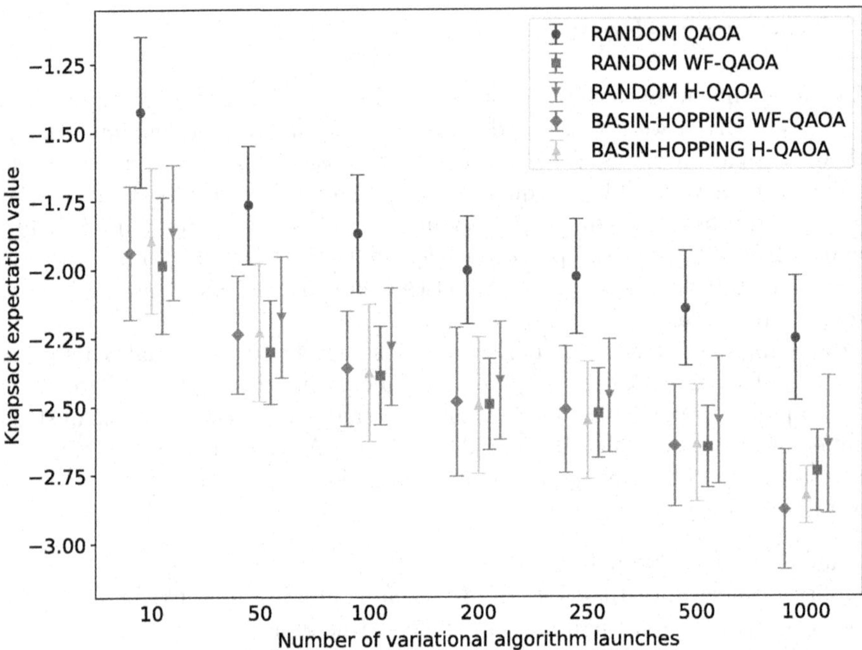

Fig. 3. Negated knapsack expectation value calculated by Algorithm 1 with standard deviation (standard error, normalized by $N-1$). For Basin-hopping only the best configuration for each number of launches was chosen (the exact ratio of initial points and number of iterations can be found in Tab. 2). The lower the expectation value, the better the quality of the results.

standard deviation were calculated. All calculations were performed on Ares Supercomputer (ACC Cyfronet AGH) with an Intel(R) Xeon(R) Platinum 8268 CPU @ 2.90GHz.

Figure 3 and the corresponding Tab. 2 indicate that with increasing number of launches of the QAOA-based variants, the results improve, which is expected. We can observe that all the proposed modifications performed better than the

Table 2. The detailed results from Fig. 3 with Basin-hopping configuration of number of iterations (iters) and number of initial points (init).

Number of launches	Basin-hopping						Random Search		
	H-QAOA			WF-QAOA					
	iters	inits	result	iters	inits	result	H-QAOA	WF-QAOA	QAOA
10	10	1	-1.89±0.26	10	1	-1.93±0.24	-1.87±0.24	-1.99±0.24	-1.42±0.27
50	10	5	-2.21±0.25	10	5	-2.23±0.21	-2.17±0.21	-2.29±0.20	-1.76±0.22
100	10	10	-2.37±0.25	10	10	-2.36±0.21	-2.29±0.22	-2.39±0.17	-1.89±0.20
200	10	20	-2.49±0.25	200	1	-2.48±0.27	-2.39±0.22	-2.51±0.17	-1.99±0.21
250	10	25	-2.56v0.22	50	5	-2.50±0.19	-2.43±0.22	-2.52±0.17	-2.02±0.20
500	250	2	-2.69±0.16	10	50	-2.63±0.23	-2.55±0.22	-2.62±0.16	-2.13±0.22
1000	10	100	-2.80±0.17	200	5	-2.85±0.24	-2.65±0.22	-2.71±0.17	-2.24±0.22

regular QAOA. For lower number of variational algorithm launches, the best results were achieved by Random Search with WF-QAOA. On the contrary, Basin Hopping variants took advantage of higher number of launches, especially with more initial points (see Tab. 2 for details) and performed slightly better. In general, the difference between results of Basin-hopping H-QAOA and WF-QAOA extensions is very small and requires further investigation.

8 Summary and Future Work

In this paper, we proposed two variants of QAOA modifications dedicated to constrained combinatorial problems. The first one is based on using weight-free Hamiltonian as the observable. The second method further extends this idea by adding QUBO weights to the set of variational parameters. We used the presented QHyper framework to compare proposed modifications with the original algorithm on the example of the Knapsack Problem. Additionally, we compare the sampling of the initial variational parameters using two global optimizers: a simple Random Search and a more sophisticated algorithm — Basin-hopping. All of the proposed modifications outperform the original QAOA algorithm. Due to the flexibility of the QHyper API, we were able to quickly evaluate various approaches with different configurations. Additionally, the modular architecture of QHyper should allow for extension to repeat experiments for a new problem.

Our experiments demonstrate that using the proposed weight-free Hamiltonian (see Sect. 4) to calculate the expectation value of the problem cost function can have a significant impact on the outcomes. This is feasible in gate-based variational algorithms, where the Hamiltonian used for the ansatz problem encoding does not have to be identical to the actual observable. What is more, estimating the expectation value can be performed using a classical algorithm based on a fixed number of measurement results (see Algorithm 1). On the contrary, the solution cannot be easily transformed to quantum annealers, where the Ising formulation of the problem Hamiltonian is required.

The next step will be to further investigate the impact of the Basin-hopping method on the results as well as the usefulness of other global optimizers such as the CEM. We also plan to try out the presented approach on larger Knapsack Problem instances and for different problems like the Traveling Salesman Problem or the Workflow Scheduling Problem.

Acknowledgements. This work was supported (in part) by the EuroHPC PL infrastructure funded at the Smart Growth Operational Programme (2014-2020), Measure 4.2 under the grant agreement no. POIR.04.02.00-00-D014/20-00. We gratefully acknowledge Poland's high-performance computing infrastructure PLGrid (HPC Centers: ACK Cyfronet AGH) for providing computer facilities and support within computational grant no. PLG/2023/016301. The authors are grateful for support from the subvention of Polish Ministry of Education and Science assigned to AGH University of Science and Technology (Faculty of Computer Science, Electronics and Telecommunications).

Code available at: github.com/qc-lab/QHyper.

References

1. Ayodele, M.: Penalty weights in QUBO formulations: permutation problems. In: Pérez Cáceres, L., Verel, S. (eds.) EvoCOP 2022. LNCS, vol. 13222, pp. 159–174. Springer, Cham (2022). https://doi.org/10.1007/978-3-031-04148-8_11
2. Bärtschi, A., Eidenbenz, S.: Grover Mixers for QAOA: shifting complexity from mixer design to state preparation. In: 2020 IEEE International Conference on Quantum Computing and Engineering (QCE), pp. 72–82 (Oct 2020). https://doi.org/10.1109/QCE49297.2020.00020, arXiv:2006.00354 [quant-ph]
3. Endres, S.C., Sandrock, C., Focke, W.W.: A simplicial homology algorithm for Lipschitz optimisation. J. Global Optim. **72**(2), 181–217 (2018). https://doi.org/10.1007/s10898-018-0645-y, http://link.springer.com/10.1007/s10898-018-0645-y
4. Farhi, E., Goldstone, J., Gutmann, S.: A Quantum Approximate Optimization Algorithm, arXiv:1411.4028 (Nov 2014), [quant-ph]
5. Fuchs, F.G., Lye, K.O., Nilsen, H.M., Stasik, A.J., Sartor, G.: Constrained mixers for the quantum approximate optimization algorithm. Algorithms **15**(6), 202 (2022). https://doi.org/10.3390/a15060202, arXiv:2203.06095 [quant-ph]
6. Golden, J., Bärtschi, A., Eidenbenz, S., O'Malley, D.: Evidence for Super-Polynomial Advantage of QAOA over Unstructured Search (Feb 2022), arXiv: 2202.00648 [quant-ph]
7. Hadfield, S., Wang, Z., O'Gorman, B., Rieffel, E.G., Venturelli, D., Biswas, R.: From the quantum approximate optimization algorithm to a quantum alternating operator ansatz. Algorithms **12**(2), 34 (2019). https://doi.org/10.3390/a12020034, arXiv:1709.03489 [quant-ph]
8. Hadfield, S., Wang, Z., Rieffel, E.G., O'Gorman, B., Venturelli, D., Biswas, R.: Quantum approximate optimization with hard and soft constraints. In: Proceedings of the Second International Workshop on Post Moores Era Supercomputing,

Denver CO USA, pp. 15–21. ACM (Nov 2017). https://doi.org/10.1145/3149526.3149530, https://dl.acm.org/doi/10.1145/3149526.3149530

9. Lucas, A.: Ising formulations of many NP problems. Front. Phys. **2** (2014). https://doi.org/10.3389/fphy.2014.00005, arXiv:1302.5843 [cond-mat, physics:quant-ph]
10. Mesman, K., Al-Ars, Z., Möller, M.: QPack: Quantum Approximate Optimization Algorithms as universal benchmark for quantum computers (Apr 2022), arXiv:2103.17193 [quant-ph]
11. Nocedal, J., Wright, S.J.: Numerical Optimization. Springer Series in Operations Research and Financial Engineering. Springer, New York (2006). https://doi.org/10.1007/978-0-387-40065-5, http://link.springer.com/10.1007/978-0-387-40065-5
12. Olson, B., Hashmi, I., Molloy, K., Shehu, A.: Basin hopping as a general and versatile optimization framework for the characterization of biological macromolecules. Adv. Artif. Intell. **2012**, 1–19 (2012). https://doi.org/10.1155/2012/674832, https://www.hindawi.com/journals/aai/2012/674832/
13. Parizy, M., Kakuko, N., Togawa, N.: Fast Hyperparameter Tuning for Ising Machines (Nov 2022), arXiv:2211.15869 [cs]
14. Radha, S.K.: Quantum constraint learning for quantum approximate optimization algorithm (Dec 2021). arXiv:2105.06770 [physics, physics:quant-ph]
15. Roch, C., Impertro, A., Phan, T., Gabor, T., Feld, S., Linnhoff-Popien, C.: Cross Entropy Hyperparameter Optimization for Constrained Problem Hamiltonians Applied to QAOA (Aug 2020). arXiv:2003.05292 [quant-ph]
16. Tilly, J., et al.: The Variational Quantum Eigensolver: A review of methods and best practices. Phys. Reports **986**, 1–128 (2022). https://doi.org/10.1016/j.physrep.2022.08.003, https://linkinghub.elsevier.com/retrieve/pii/S0370157322003118
17. Tomasiewicz, D., Pawlik, M., Malawski, M., Rycerz, K.: Foundations for workflow application scheduling on D-wave system. In: Krzhizhanovskaya, V.V., et al. (eds.) ICCS 2020. LNCS, vol. 12142, pp. 516–530. Springer, Cham (2020). https://doi.org/10.1007/978-3-030-50433-5_40
18. Wales, D.J., Doye, J.P.K.: Global optimization by basin-hopping and the lowest energy structures of lennard-jones clusters containing up to 110 atoms. J. Phys. Chem. A **101**(28), 5111–5116 (1997). https://doi.org/10.1021/jp970984n, https://pubs.acs.org/doi/10.1021/jp970984n
19. Wang, Z., Rubin, N.C., Dominy, J.M., Rieffel, E.G.: XY-mixers: analytical and numerical results for QAOA. Phys. Rev. A **101**(1), 012320 (2020). https://doi.org/10.1103/PhysRevA.101.012320, arXiv:1904.09314 [quant-ph]

GCS-Q: Quantum Graph Coalition Structure Generation

Supreeth Mysore Venkatesh[2], Antonio Macaluso[1(✉)], and Matthias Klusch[1]

[1] German Research Center for Artificial Intelligence (DFKI), Saarbruecken, Germany
{antonio.macaluso,matthias.klusch}@dfki.de
[2] University of Saarland, Saarbruecken, Germany
s8sumyso@stud.uni-saarland.de

Abstract. The problem of generating an optimal coalition structure for a given coalition game of rational agents is to find a partition that maximizes their social welfare and known to be NP-hard. Though there are algorithmic solutions with high computational complexity available for this combinatorial optimization problem, it is unknown whether quantum-supported solutions may outperform classical algorithms.

In this paper, we propose a novel quantum-supported solution for coalition structure generation in Induced Subgraph Games (ISGs). Our hybrid classical-quantum algorithm, called GCS-Q, iteratively splits a given n-agent graph game into two nonempty subsets in order to obtain a coalition structure with a higher coalition value. The GCS-Q solves the optimal split problem $\mathcal{O}(n)$ times, exploring $\mathcal{O}(2^n)$ partitions at each step. In particular, the optimal split problem is reformulated as a QUBO and executed on a quantum annealer, which is capable of providing the solution in linear time with respect to n. We show that GCS-Q outperforms the currently best classical and quantum solvers for coalition structure generation in ISGs with its runtime in the order of n^2 and an expected approximation ratio of 93% on standard benchmark datasets.

Keywords: Coalition Formation · Quantum Computing · Quantum Artificial Intelligence · Multi-Agent Systems

1 Introduction

One major challenge of rational cooperation in multi-agent systems is to solve the coalition structure generation (CSG) problem. Given a coalition game (A, v) with a set A of n agents and a characteristic function $v : \mathcal{P}(A) \to \mathbb{R}$ for coalition values $v(C)$ for all non-empty coalitions C in A, the problem is to find a coalition structure CS^* of A that maximizes the social welfare $\sum_{C \in CS^*} v(C)$. This combinatorial optimization (partitioning) problem is known to be NP-complete [1] with an exponential number of possible coalition structures. Several solution methods for the CSG problem exist, such as the currently best solver BOSS [5] with run-time complexity of $O(3^n)$, that avoid exploring the complete search space to find an optimal solution. In the following, we focus on the CSG problem

J. Mikyška et al. (Eds.): ICCS 2023, LNCS 14077, pp. 138–152, 2023.
https://doi.org/10.1007/978-3-031-36030-5_11

for Induced Subgraph Games. In this case, the coalition game is induced by an undirected weighted graph where the agents are denoted as nodes and the coalition values are the sum of the weights of edges between coalition members in the graph. However, the problem remains NP-complete [2] and therefore intractable for large values of n in practice.

One open question is whether the usage of quantum computational means may contribute to solve this problem faster than it is possible with the currently best state-of-the-art solvers. To this end, we developed a novel quantum-supported solution, called GCS-Q, for solving the coalition structure generation problem for Induced Subgraph Games. In particular, the GCS-Q leverages classical and quantum computation for an approximate, anytime solution of the problem, and is inspired by divisive hierarchical clustering. The GCS-Q starts with the grand coalition and iteratively splits it up until the coalitions are singleton sets such that it builds the hierarchy in $n-1$ steps for a coalition game with n agents and, at each step, explores $\mathcal{O}(2^n)$ partitions. However, the GCS-Q identifies the optimal split for a given coalition using a quadratic unconstrained binary optimization (QUBO) problem formulation, which can be solved experimentally on a real quantum annealing device in linear time. As a result, the overall time complexity of the GCS-Q is in the order of n^2, significantly outperforming the currently approximate classical state-of-the-art solvers. We conduct our comparative performance evaluation of the GCS-Q with selected classical baselines on standard benchmark data using a D-Wave 2000Q quantum annealer and sufficient worst-case approximation ratio.

The remainder of the paper is structured as follows: The problem of coalition structure generation for graph games is formalized in Sect. 2, followed by a discussion of related work in Sect. 3. Our quantum-supported solution GCS-Q is described in Sect. 4 and its performance comparatively evaluated in Sect. 5. Section 6 concludes with a summary of achievements and future work.

2 Problem Formulation

As mentioned above, the coalition structure generation problem is to find a partition or coalition structure of a given set of rational agents whose value or social welfare is maximal for a given coalition game. We focus on the *Induced Subgraph Games (ISGs)* that are based on graph-restricted games, i.e., coalition games induced by connected, undirected weighted graphs (cf. Definition 1) [17].

Definition 1. *An Induced Subgraph Game (A, v) is induced by a connected, undirected, weighted graph $G(V, w)$: The set $V = \{i\}_{i \in \{1..n\}}$ of nodes in G represents the set $A = \{a_1, a_2, ...a_n\}$ of n agents, and the real valued weights $w_{i,j}$ of edges (i, j) in w denote the synergies of cooperation or joint utilities of agents $a_i, a_j \in A$ in feasible coalitions $C \subseteq A$. A coalition C is feasible if and only if it induces a connected subgraph of G. The coalition value $v(C)$ of a feasible coalition C is*

$$v(C) = \sum_{(i,j)\in w,\ i,j\in C} w_{i,j}. \tag{1}$$

For given ISG (A, v)*, coalition structures* CS *are partitions of* A *into mutually disjoint, feasible coalitions* C*. The coalition structure generation problem for a given ISG is to find the optimal coalition structure* CS^* *with maximal coalition value (or social welfare) for* (A, v)*:*

$$CS^* = \arg\max_{CS} \sum_{C\in CS} v(C) \tag{2}$$

In ISGs, the coalition values depend only on the pairwise interactions between agents represented in the graph, but the set of possible solutions is not restricted and the problem remains NP-complete [2]. However, if $G(V, w)$ is a connected graph and the edge weights are all positives, i.e., $(w_{i,j}) \in \mathbb{R}^+$, the grand coalition $g_c = A$ is always the optimal coalition structure [1].

In this paper, we refer to the CSG problem in ISG as the *ISG problem* and we assume a fully connected graph with positive and negative edge weights.

3 Related Works

There are two broad classes of solutions for solving a CSG problem. *Exact* methods operate to find the global optimum by exploring a large number of possible coalition structures. Contrarily, *approximate* methods reformulate the original problem to find near-optimal solutions. Any algorithmic solution (exact or approximate) can be characterized based on the minimum time required to provide the solution. Anytime optimal algorithms (e.g., IP [13]) generate an initial set of possible solutions within a bound from the optimal and then improve their quality iteratively. The downside is that these algorithms might end up searching the entire space of all possible coalition structures, which translates into a worst-case time complexity of $\mathcal{O}(n^n)$ for n agents. Nevertheless, such methods allow intermediate solutions during execution, which can be critical for many real-world scenarios. An alternative approach consists of using Dynamic Programming [12], which avoids exploring the entire solution space without losing the guarantees of finding the optimal coalition structure. However, these approaches must be executed entirely to obtain the final solution. Nonetheless, algorithms within this category represent state-of-the-art in terms of worst-case time complexity ($\mathcal{O}(3^n)$) when considering any generic coalition game with no prior knowledge of the nature of the characteristic function [5,11]. The method DyCE (Dynamic programming for optimal connected Coalition structure Evaluation) [18] is based on IDP [12] and exact but not anytime. Its memory requirements are of the order of $\mathcal{O}(2^n)$ with a reported limitation of application to up to 30 agents in practice.

Approximate methods such as C-link (Coalition-link) [6] solve the CSG problem based on agglomerative clustering. C-link considers only the values of coalitions up to size two as interaction scores between the two agents by discarding other coalition values in the input. C-Link scales as $\mathcal{O}(n^3)$ with an estimated worst-case approximation ratio of 80% on custom datasets.

In the context of ISGs, different approaches assume specific graph structures to improve the runtime for the final solution. For example, CFSS [4] is an anytime solver for graph games that uses a branch-and-bound search technique in the solution space. CFSS has been used for comparative evaluation with the algorithms mentioned above and has shown excellent results for sparse graphs. However, the worst-case computational complexity of CFSS is still of the order $\mathcal{O}(n^n)$. Lately, the idea of applying *k-constrained Graph Clustering (KGC)* formulated as *Integer Linear Programming (ILP)* has been investigated [3]. This approach allows efficient implementation when dealing with sparse graphs.

Recently, the first general hybrid quantum-classical algorithm for solving any generic coalition game has been formulated. The algorithm, named BILP-Q [16], is suitable for gate-based quantum computing and quantum annealing and shows a complexity of the order of $\mathcal{O}(2^n)$ in the case of constrained CSG. However, BILP-Q requires the number of logical qubits to be exponential in n, which is a significant limitation considering near-term quantum technology.

4 Methods

In this section, we present GCS-Q (*Quantum-supported solution for Graph-restricted Coalition Structure generation*), a novel quantum-supported anytime approximate algorithm for CSG in the context of ISGs. GCS-Q starts from the grand coalition as the initial coalition structure and recursively performs a split to find the best bipartition of the agents based on the graph induced by the coalition game. This approach's primary source of complexity is given by finding the optimal split at each step, which is NP-hard. However, properly formulating this problem and delegating it to a quantum annealer allows for obtaining a quantum-supported algorithm that runs quadratically in the number of agents and can outperform state-of-the-art classical and quantum solutions.

4.1 Optimal Split

Given a coalition game (A, v), where $A = \{a_1, a_2, ...a_n\}$ is a set of agents of size $|A| = n$ and $v : \mathcal{P}(A) \rightarrow \mathbb{R}$ is a characteristic function, a split $\{C, \overline{C}\}$ is defined as the bipartition of A into two disjoint subsets $C, \overline{C}$.

Definition 2. *Finding the optimal split of a coalition game (A, v) is an optimization problem of the form:*

$$\arg\max_{C,\overline{C}} \; v(\{C, \overline{C}\}) = v(C) + v(\overline{C}) \tag{3}$$

$$s.\ t.\ C \cup \overline{C} = A \;\; and \;\; C \cap \overline{C} = \emptyset. \tag{4}$$

The exhaustive enumeration of all possible splits for a coalition game of n agents is of the order $\mathcal{O}(2^n)$.

For a given ISG, the optimal split into two mutually disjoint sets translates into the minimum cut problem of the graph underlying the coalition game.

Definition 3. *Let be $G(A, w)$ a weighted undirected graph. A cut is a partition of the vertices into two sets C and $\overline{C}$ such that $\overline{C} = A - C$. The value of a cut $\delta(S)$, where $S = \{C, \overline{C}\}$, is defined as the sum of the edge weights connecting the nodes of the two sets C and $\overline{C}$:*

$$\delta(C, \overline{C}) = \sum_{i \in C, j \in \overline{C}} w_{i,j}. \tag{5}$$

The weighted minimum cut (min-cut) is an optimization problem that aims to find a cut with minimum value:

$$\textit{min-cut}\ (G) = \arg\min_{S} \delta(S). \tag{6}$$

Thus, we establish the equivalence between the optimal split (cf. Definition 2) and *min-cut* (cf. Definition 3) in the context of ISGs.

Lemma 1. *Given an ISG (A, v) with an underlying graph $G(A, w)$, finding the optimal split for (A, v) is equivalent to solving the min-cut for $G(A, w)$.*

Proof. The value of a coalition A for a given coalition game can be considered constant and calculated as the sum of all the edge weights between the agents into the coalition:

$$v(A) = \sum_{i,j \in A} w_{i,j}. \tag{7}$$

A cut $\delta(S)$ on $G(A, w)$, where $S = \{C, \overline{C}\}$, produces two independent sets of nodes representing two separate coalitions (cf. Definition 2). The sum of the cut edge weights gives the value of $\delta(S)$. Furthermore, the value of the split generated by $\delta(S)$ is given by the sum of the remaining interactions between agents within the two coalitions C and $\overline{C}$. Thus, the following equivalence applies:

$$v(S) = v(A) - \delta(S), \tag{8}$$

where $S = \{C, \overline{C}\}$, $\forall C, \overline{C} \subseteq A$, s.t. $\overline{C} = A - C$. As a consequence, we can write the value of a cut as the difference between the value of A and S:

$$\delta(S) = v(A) - v(S). \tag{9}$$

Let S be the partition that minimizes $\delta(S)$, i.e., $\delta(S) \leq \delta(S') \forall S'$, then the following inequality always holds:

$$v(A) - \delta(S) \geq v(A) - \delta(S') \implies v(S) \geq v(S'). \tag{10}$$

Therefore, finding the partition S which provides minimum value for $\delta(S)$ is equivalent to finding the optimal split $S = \{C, \overline{C}\}$ which maximizes $v(S)$.

This work considers the generic case for ISGs, assuming a complete graph (i.e., fully connected) with positive and negative weights. In this case, the *min-cut* problem is proven to be NP-hard [7] and requires an exponential number of steps with respect to the number of input nodes/agents [8,9].

4.2 QUBO Formulation for Optimal Split

In this section, the *min-cut* problem (or, equivalently, the optimal split problem) is reformulated as QUBO suitable to be executed on a real quantum annealer.

Let $G(A, w)$ be a weighted undirected graph. The *min-cut* (cf. Definition 3) can be formulated as a quadratic objective function of binary variables $\{x_i\}_{i=1,\dots,n}$:

$$\arg\min_x \sum_{i=1}^{n} w_i x_i + \sum_{1 \leq i < j \leq n} w_{i,j} x_i (1 - x_j) = x^t W x, \tag{11}$$

where $w_{i,j}$ is the edge weight connecting the nodes i and j, w_i is a self-loop on node i, and the value x_i indicates the membership of one of the two disjointed sets generated by the cut. Therefore, the value of the binary string solution allows differentiating the vertices belonging to the subsets C or $\overline{C}$, as follows:

$$\forall x_i \in A, x_i = \begin{cases} 1 & \text{if } x_i \in C \\ 0 & \text{if } x_i \in \overline{C} \end{cases} \tag{12}$$

where $C \cap \overline{C} = \emptyset$ and $C \cup \overline{C} = A$. Given the QUBO formulation in Eq. (11), we can easily define the corresponding optimization problem of an Ising Hamiltonian with the assignment $x_i \rightarrow (1 - Z_i)/2$, where Z_i is the Pauli-Z operator.

4.3 GCS-Q Algorithm

The algorithm follows the strategy of divisive hierarchical clustering (DHC) and applies it in the context of ISGs. Traditional DHC algorithms create a sequence of partitions of a set of n elements, such that an optimality criterion that considers the separation between groups is maximized. In particular, DHCs start with all the elements in a unique group and generate a bipartition to minimize the distance between intra-cluster elements at each step. The number of possible bipartitions of a set of n elements is $\mathcal{O}(2^n)$, and the complete exploration is highly time-consuming. For this reason, standard DHC methods split up the groups according to a distance matrix that explores n^2 different possibilities at each step. This approach allows obtaining a method that scales polynomially ($\mathcal{O}(n^3)$) in the number of elements by drastically reducing the number of partitions examined.

The idea behind GCS-Q is to adopt the top-down strategy of DHC for solving any ISG problem. The algorithm starts initializing the current optimal coalition structure CS^* with the grand coalition g_c where all the agents belong to a single coalition, i.e., $CS^* = \{g_c\}$. Thus, the optimal split problem for g_c is formulated in terms of *min-cut* (Sect. 4.2) and solved using a quantum annealer. In particular, the annealer evaluates all possible bipartitions of g_c and provides the binary encoding of the bipartition $\{C, \overline{C}\}$ that maximizes the characteristic function (Eq. (3)). Then, the coalition value of g_c is compared with the value of the coalition structure comprising C and $\overline{C}$, i.e.,

$$v(CS^*) = v(\{g_c\}) > v(\{C, \overline{C}\}) = v(C) + v(\overline{C}). \tag{13}$$

If the splitting produces a lower coalition value, g_c is returned as the optimal coalition structure and the algorithm stops. Otherwise, if the inequality (13) does not hold, the coalition structure $\{C, \overline{C}\}$ is assigned to CS^*. The second step consists of the optimal bipartition for each coalition in the current coalition structure, and the splitting is decided based on the criterion of Eq. (13). This approach allows to generate, at each step and for each coalition, a partition of the agents that has a higher coalition value for the characteristic function. This process continues until none of the coalitions in CS^* can be split to produce a bipartition with a higher coalition value.

Notice that the condition in Eq. (13) provides an automatic stopping criterion for GCS-Q: the algorithm stops if there is no advantage in splitting the coalitions in CS^*. This is a key difference between standard DHC algorithms that always start from a single group and end up with singletons. Furthermore, the additive nature of the coalition value function in Eq. (1) for ISGs allows acting independently on the coalitions to maximize the value of the coalition structure. The pseudocode of GCS-Q is shown in Algorithm 1.

Algorithm 1. GCS-Q Algorithm

Input: Coalition game (A, v), with underlying graph $G(A, w)$ where $|A| = n$ and $w : A \times A \to \mathbb{R}$
Output: Optimal coalition structure CS^*

$CS^* \leftarrow \{\}$ ▷ initialize CS^* with empty list
$queue \leftarrow g_c$ ▷ initialize *queue* with grand coalition g_c
while $queue \neq \emptyset$ **do**
  $S \leftarrow queue.pop$ ▷ Fetch a coalition from *queue*
  Begin Optimal Split problem
    Create a weight matrix W from edges in S ▷ Eq. (11)
    Define the Ising Hamiltonian $\mathcal{H}$ for W
    Solve Ising Hamiltonian $\mathcal{H}$ on a quantum annealer
    Decode binary string to get C and $\overline{C}$
  End Optimal Split problem
  if $\overline{C} = \emptyset$ **then** ▷ no further splitting of S is inefficient
    add C to CS^*
  else ▷ try further splitting C and $\overline{C}$
    add C to *queue*
    add $\overline{C}$ to *queue*
  end if
end while
Return CS^*

4.4 Discussion

In terms of runtime, the execution of GCS-Q using the classical computation would be $\mathcal{O}(n2^n)$. In particular, given an n-agent ISG, in the case of a superadditive game, the algorithm needs to solve n times the *min-cut* problem for a fully connected graph with positive and negative edge weights, which is NP-hard [9].

Nevertheless, we experimentally show that delegating the problem of finding the best bipartition on quantum annealing allows achieving a runtime that scales in the order of n^2. This represents a significant improvement with respect to state-of-the-art quantum and classical algorithms.

The top-down strategy adopted by GCS-Q allows for obtaining several convenient properties: *i)* for an n-agent ISG, GCS-Q always converges in at most n steps. In fact, the algorithm proceeds top-down, starting from a coalition structure containing all the agents into a single coalition (the grand coalition), and it terminates (in the worst case) with a coalition structure containing the singletons. Therefore, the hierarchy is built in $n-1$ steps; *ii)* The proposed algorithm is *anytime*: the splitting in terms of *min-cut* ensures that each coalition is split into disjoint sets, which correspond to two separate coalitions. This approach automatically considers the underlying graph of the coalition game while guaranteeing a valid solution at each time step. *iii)* for superadditive games, GCS-Q always returns the best coalition structure (the grand coalition). Thanks to the top-down approach, the algorithm is initialized with the grand coalition, which has no subpartitions with higher coalition value; *iv)* GCS-Q is an approximate solver which explores a larger portion of the solution space compared to existing clustering-based approaches, such as C-Link [6]. Standard hierarchical clustering algorithms (agglomerative or divisive) examine n^2 bipartitions at each step, leading to a time complexity cubic in the number of elements at the cost of drastically reducing the number of partitions explored. In contrast, GCS-Q evaluates all possible 2^n bipartitions and selects the optimal one. This translates into exploring $\mathcal{O}(2^n)$ possible configurations for the coalition structure at each step. This approach is usually avoided because extremely time-consuming. However, the reformulation as a QUBO allows leveraging a quantum annealer that experimentally shows to solve the problem with an average runtime in the order of n (Sect. 5.2). Figure 1 provides a small example of how the GCS-Q works.

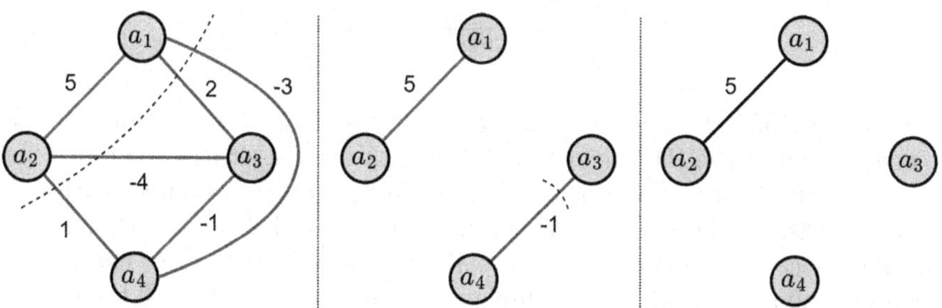

Fig. 1. Graphical representation of how the GCS-Q algorithm works. The figure represents the set of agents alongside the two steps of GCS-Q. The blue nodes indicate the vertices affected by the current min-cut. The green nodes indicate the nodes identified as belonging to the best possible coalition, and further splitting is inefficient. The red edges are the ones that are cut. The green edge denotes the ones which remain after the current optimal cut. Notice that, at each step, the algorithm returns a valid sub-optimal coalition structure. (Color figure online)

5 Evaluation

In this section, we comparatively evaluate the GCS-Q in terms of approximation ratio and runtime on synthetic benchmark datasets.

5.1 Experimental Settings

Dataset Generation. The standard approach to evaluate algorithms for cooperative games is to generate the values of the coalition function from several probability distributions. In case of ISGs, the datasets refer to the interaction score associated with the pairs of agents. Given n agents, a real value is assigned for each pair of agents, i.e., a total $\binom{n}{2}$ values (unlike standard CSGs where 2^n values are independently generated). The coalitions of size greater than two are assigned according to Eq. (1)[1]. Specifically, the edge weights of the graph game $G(A, w)$ are drawn from the following two distributions:

$$w_{i,j} \sim \mathcal{L}(\mu_1, b) \qquad\qquad w_{i,j} \sim \mathcal{N}(\mu_2, \sigma)$$

where $\mathcal{L}$, $\mathcal{N}$ are the *Laplace* and *Normal* distribution respectively, with $\mu_1 = \mu_2 = 0$ and $\sigma = b = 5$. The parameters are chosen to guarantee positive and negative values for $w_{i,j}$.

Metrics. One of the requirements to estimate apriori the approximation error of a CSG algorithm is to assume the value of any coalition to be positive [14]. Since we consider more generic coalition games where the value of a coalition can also be negative, to assess the approximation ratio of GCS-Q we evaluate its performance on a standard benchmark dataset. Given a n-agent coalition game (A, v), the approximation error is defined as the relative error between the value of the *optimal* coalition structure $CS^{(\text{opt})}$ and the value of a *near-optimal* solution $CS^{(\text{n-opt})}$ returned by a selected approximate solver:

$$e_r = \frac{|v(CS^{(\text{opt})}) - v(CS^{(\text{n-opt})})|}{v(CS^{(\text{opt})})} \in [0, 1]. \tag{14}$$

Implementation. To find the optimal coalition structure, we run a python implementation of the IDP algorithm [12], which always returns the best possible coalition structure. However, this algorithm runs out of resources with more than 20 agents. Furthermore, rather than running the GCS-Q directly on a real quantum device prone to errors, we evaluate its approximation error using classical computation. In particular, we implement the Algorithm 1, with the optimal split problem solved using a brute force strategy that explores all possible splits. We refer to this approach as GCS-Q$^{(c)}$. In addition, we implement GCS-Q by leveraging the D-Wave 2000Q to solve the sub-task of the optimal split problem. We refer to this approach as GCS-Q$^{(q)}$.

[1] Notice that to better fit standard graph games studied in the literature, we omit self-loop, which would require sampling other n values. However, the GCS-Q formulation is suitable for dealing with graphs containing self-loop.

5.2 Results

GCS-Q Runtime. In order to estimate the runtime of the quantum annealing in solving the optimal split problem, we generate 64 ISGs, each with a different number of agents for both distributions (from 2 to 65). Thus, the time-to-solution is recorded for five different runs to obtain an estimation (alongside a variability measure) of the runtime. Results are reported in Fig. 2.

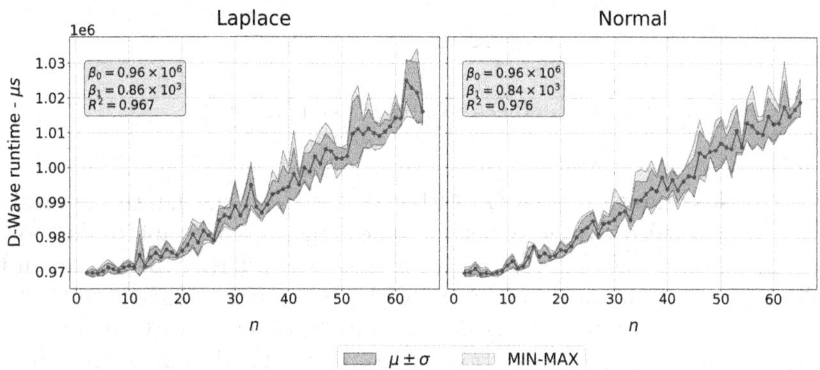

Fig. 2. Runtime of the D-Wave 2000Q when solving the optimal split problem. The same QUBO problem is solved 5 times for each number of agents. The blue line is the average runtime of the 5 experiments. The yellow shaded area represents the maximum and minimum runtimes. The green shaded area is calculated considering the mean and the standard deviation of the runtimes for each problem instance. The runtimes are reported in microseconds (μs). (Color figure online)

The order of runtime growth when increasing the number of agents is linear. To confirm this hypothesis, we estimate a linear regression model of the form $T = \beta_0 + \beta_1 n$ where T is the runtime of the quantum annealing, n is the number of agents, and β_0, β_1 are the parameters to be estimated from data. To assess the quality of the linear fitting, we calculate the coefficient of determination R^2, which is equal to 1 in the case of a perfect deterministic linear function between n and T. A value of 97% for both distributions indicates that the relationship between T and n increases linearly. Thus, we can conclude that the average-case complexity of the quantum annealing solution for solving the optimal split problem is linear in the number of agents, i.e., $\Theta(n)$.

Furthermore, to test the efficiency of GCS-Q$^{(q)}$ against GCS-Q$^{(c)}$, we run them on the two benchmark datasets described in Sect. 5.1 considering games up to 27 agents. Results in terms of runtime are depicted in Fig. 3. As expected, GCS-Q$^{(q)}$ runs polynomially in the number of agents, providing a practical quantum advantage over its classical counterpart.

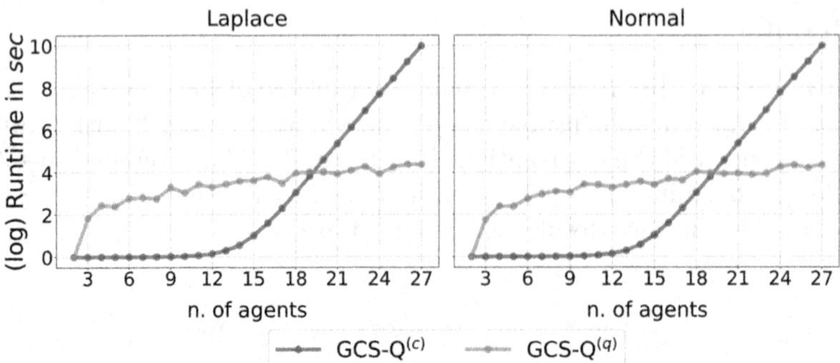

Fig. 3. Runtime of GCS-Q$^{(c)}$ and GCS-Q$^{(q)}$.

Quality Assessment of GCS-Q. Although the advantage in terms of runtime, the solution provided by the quantum annealing degrades rapidly due to the limited precision of the quantum device in use. We perform experiments on the two distributions mentioned in Sect. 5.1 with the number of agents up to 20. In particular, we compare the quality of the solutions obtained with GCS-Q$^{(c)}$ and GCS-Q$^{(q)}$ with best possible coalition structure calculated running IDP. Results are shown in Fig. 4.

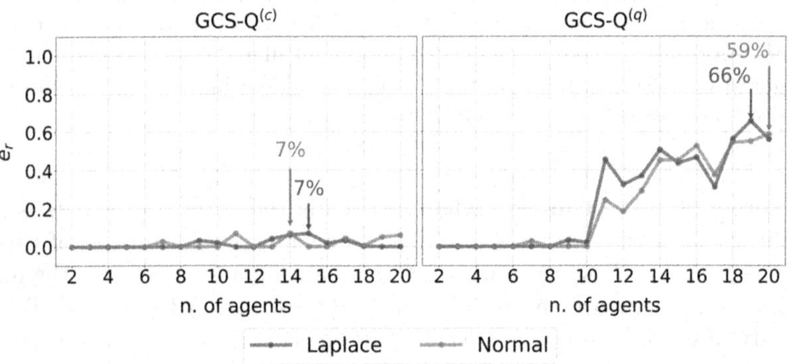

Fig. 4. Assessment of GSC-Q quality in terms of approximation error calculated on the two benchmark datasets described in Sect. 5.1.

For both distributions, the GCS-Q$^{(c)}$ has a worst-case approximation ratio of 7%. However, the GCS-Q$^{(q)}$ produces the expected solution of the Algorithm 1 only for coalition games of size up to 10. In particular, the approximation error suddenly increases for problems with more than 10 agents, which results in a worst-case approximation error of 66% and 59%). The deterioration is also observed for experiments up to 27 agents. In this case, we consider as baseline $CS^{(\text{opt})}$ the coalition structure returned by the GCS-Q$^{(c)}$ and as approximate solution $CS^{(\text{n-opt})}$ the one returned by GCS-Q$^{(q)}$. The results of the quality assessments are reported in Fig. 5. The tendency for a decrease in performance is due to the quality of the quantum annealer when solving the optimal split problem. In fact, with $n \geq 11$ the error is cascaded through further executions of the algorithm and the final solution is far from optimality.

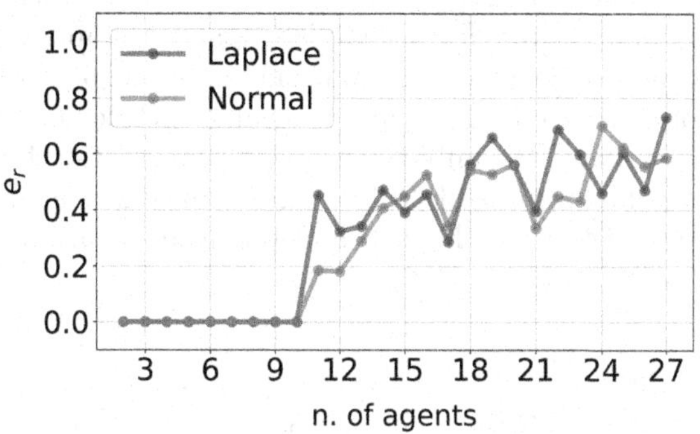

Fig. 5. Relative approximation error GCS-Q$^{(q)}$ algorithm using as baseline the GCS-Q$^{(c)}$.

The limitations of the D-Wave 2000Q have already been emphasized for specific optimization tasks [19]. Nonetheless, the latest generation of D-Wave QPUs, named *Advantage*, outperforms D-Wave 2000Q for any problem size. Furthermore, *Advantage* systems can solve larger problems with up to 120 logical qubits. In some cases, not only the *Advantage* system can find better-quality solutions but it also can find same quality solutions faster [15][2]. However, each problem Hamiltonian must be explicitly studied on the quantum device in use. In this regard, several technical optimization strategies are still possible on the D-Wave 2000Q. For instance, adopting different embedding strategies or leveraging hybrid quantum annealing computation that can deliver better quality solutions for large-size problems at the cost of worsening the runtime (for more details, see D-Wave documentation[3]).

5.3 Performance Analysis

In the case of subadditive games, the optimal coalition structure is given by the singletons. This is the worst-case scenario for GCS-Q, which needs to solve the optimal split problem n times. Therefore, considering a linear runtime of the quantum annealing (Sect. 5.2), we obtain a quantum-supported solver which scales quadratically with respect to the number of agents n.

[2] A comparison between Advantage with the D-Wave 2000Q is reported [10].
[3] https://docs.dwavesys.com/docs/latest/index.html.

Thus, we compare this runtime with state-of-the-art classical and quantum solutions when considering ISGs with an underlying fully connected graph. Since ISGs are a special case of general coalition games, we also consider the classic solvers for CSG. The best classical exact solvers for any generic CSG problem are represented by methods based on IDP [12], such as BOSS [5] and DyCE [18], having worst-case complexity of $\mathcal{O}(3^n)$. In the context of ISGs, the CFSS [4] and KGC algorithms [3] have shown excellent results with sparse graphs, but the worst-case complexity for a complete graph remains $\mathcal{O}(n^n)$. The best approximate solution in terms of the runtime is C-link [6], which is based on hierarchical agglomerative clustering and has cubic complexity in the number of agents, i.e., $\mathcal{O}(n^3)$. Finally, the only general quantum solution for CSG is BILP-Q [16], which showed a runtime of $\mathcal{O}(2^n)$ using quantum annealing A graphical comparison of state-of-the-art classical and quantum solutions is provided in Fig. 6.

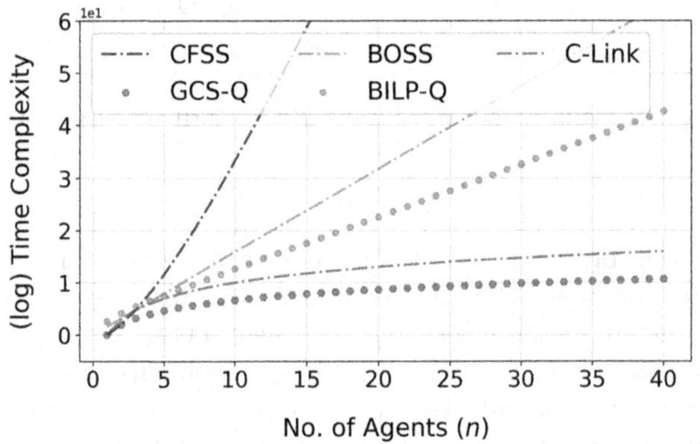

Fig. 6. Cost complexity as a function of the number of agents n. Classical solutions are indicated with blue lines, while quantum solutions are in green. (Color figure online)

The ability of GCS-Q to be executed with a runtime quadratic in the number of agents makes it the best solver for ISGs.

6 Conclusion and Future Work

In this work, we proposed GCS-Q, the first quantum-supported solution for coalition structure generation in Induced Subgraph Games. The key idea is to partition the graph underlying the coalition game into two subsets iteratively in order to obtain a coalition structure with a better coalition value. By delegating the task of finding the optimal split to a quantum annealer, we obtain a solver capable of running faster than the state-of-the-art solutions (quantum and classical). Given a n-agent coalition game, the ability of the D-Wave 2000Q quantum device to solve the optimal split problem in linear time with respect to n allows for conveying an overall runtime that scales in the order of n^2. Furthermore, by exploring all possible partitions of a given coalition, the GCS-Q examines a larger portion of the solution space compared with other approximate solvers, such as C-Link [6]. In fact, this latter adopts a bottom-up strategy and never considers the global distribution of the agents. Another important feature of GCS-Q is the ability to provide sub-optimal solutions during its execution, which makes it an anytime solver.

In addition, we provided a practical implementation of GCS-Q and evaluated its performance on standard benchmark datasets. Specifically, we generated coalition games with fully connected graphs, sampling the edge weights of the graph underlying the coalition games from two distributions (*Laplace* and *Normal*). We implemented two variants of the GCS-Q (Algorithm 1): GCS-$Q^{(c)}$ is executed entirely on a classical computer and served to estimate the expected approximation ratio (93%) for the benchmark datasets. The second approach, named GCS-$Q^{(q)}$, leverages the D-Wave 2000Q for solving the sub-task of the optimal split. As expected, the GCS-$Q^{(q)}$ scales polynomially in the number of agents and exponentially faster than GCS-$Q^{(c)}$. However, when calculating the quality of the solutions, the performance deteriorates due to the limitations of the quantum hardware in use. For this reason, the main challenge to tackle in the near future is the investigation of alternative embedding strategies or the adoption of hybrid quantum-classical solvers proposed by D-Wave.

Another natural follow-up is the execution of GCS-Q on better quantum hardware, such as the D-Wave *Advantage*. This latest generation of quantum devices has outperformed the D-Wave 2000Q in terms of both quality of solutions obtained and runtime in several combinatorial optimization problems.

In conclusion, we showed the feasibility and the benefit of adopting quantum computation in multi-agent systems with a novel quantum-supported solution suitable for solving practical real-world AI problems.

Acknowledgments. This work has been funded by the German Ministry for Education and Research (BMB+F) in the project QAI2-QAICO under grant 13N15586.

Code Availability. All code to generate the data, figures, analyses and additional technical details on the experiments are available at https://github.com/supreethmv/GCS-Q.

References

1. Aziz, H., De Keijzer, B.: Complexity of coalition structure generation. arXiv preprint arXiv:1101.1007 (2011)
2. Bachrach, Y., Kohli, P., Kolmogorov, V., Zadimoghaddam, M.: Optimal coalition structure generation in cooperative graph games. In: Twenty-Seventh AAAI Conference on Artificial Intelligence (2013)
3. Bistaffa, F., Chalkiadakis, G., Farinelli, A.: Efficient coalition structure generation via approximately equivalent induced subgraph games. IEEE Trans. Cybern. (2021)

4. Bistaffa, F., Farinelli, A., Cerquides, J., Rodriguez-Aguilar, J.A., Ramchurn, S.D.: Anytime coalition structure generation on synergy graphs. In: 13th International Conference on Autonomous Agents and Multi-agent Systems, 5–9 May 2014 (2014)
5. Changder, N., Aknine, S., Ramchurn, S.D., Dutta, A.: BOSS: a bi-directional search technique for optimal coalition structure generation with minimal overlapping. In: Proceedings of the AAAI Conference on Artificial Intelligence, vol. 35, pp. 15765–15766 (2021)
6. Farinelli, A., Bicego, M., Ramchurn, S., Zuchelli, M.: C-link: a hierarchical clustering approach to large-scale near-optimal coalition formation. In: 23rd International Joint Conference on Artificial Intelligence (IJCAI 2013), 3–9 August 2013 (2013)
7. Garey, M.R., Johnson, D.S.: Computers and Intractability, vol. 174. Freeman, San Francisco (1979)
8. González-Arangüena, E., Manuel, C.M., del Pozo, M.: Values of games with weighted graphs. Eur. J. Oper. Res. **243**(1), 248–257 (2015)
9. Mccormick, S.T., Rao, M., Rinaldi, G.: When is min cut with negative edges easy to solve? Easy and difficult objective functions for max cut. Consiglio Nazionale Delle Ricerche (CNR) (2000)
10. McGeoch, C., Farré, P.: The D-wave advantage system: an overview. D-Wave Systems Inc., Burnaby, BC, Canada, Technical report (2020)
11. Rahwan, T., Jennings, N.R.: Coalition structure generation: dynamic programming meets anytime optimization. In: AAAI Conference on Artificial Intelligence (2008)
12. Rahwan, T., Jennings, N.R.: An improved dynamic programming algorithm for coalition structure generation. In: Proceedings of the 7th International Joint Conference on Autonomous Agents and Multiagent Systems, vol. 3, pp. 1417–1420 (2008)
13. Rahwan, T., Ramchurn, S.D., Dang, V.D., Giovannucci, A., Jennings, N.R.: Anytime optimal coalition structure generation. In: AAAI, vol. 7, pp. 1184–1190 (2007)
14. Sandholm, T., Larson, K., Andersson, M., Shehory, O., Tohmé, F.: Coalition structure generation with worst case guarantees. Artif. Intell. **111**(1–2), 209–238 (1999)
15. Tasseff, B., et al.: On the emerging potential of quantum annealing hardware for combinatorial optimization. arXiv preprint arXiv:2210.04291 (2022)
16. Venkatesh, S.M., Macaluso, A., Klusch, M.: BILP-Q: quantum coalition structure generation. In: Proceedings of the 19th ACM International Conference on Computing Frontiers, pp. 189–192 (2022)
17. Voice, T., Polukarov, M., Jennings, N.R.: Coalition structure generation over graphs. J. Artif. Intell. Res. **45**, 165–196 (2012)
18. Voice, T., Ramchurn, S.D., Jennings, N.R.: On coalition formation with sparse synergies. In: Proceedings of the 11th International Conference on Autonomous Agents and Multiagent Systems, vol. 1, pp. 223–230. International Foundation for Autonomous Agents and Multiagent Systems (2012)
19. Willsch, D., et al.: Benchmarking advantage and D-wave 2000Q quantum annealers with exact cover problems. Quantum Inf. Process. **21**(4), 1–22 (2022)

Learning QUBO Models for Quantum Annealing: A Constraint-Based Approach

Florian Richoux[1,3(✉)], Jean-François Baffier[2,3], and Philippe Codognet[3]

[1] AIST, Tokyo, Japan
florian@richoux.fr
[2] IIJ Research Lab, Tokyo, Japan
jf@baffier.fr
[3] JFLI, CNRS/Sorbonne University/University of Tokyo, Tokyo, Japan
codognet@is.s.u-tokyo.ac.jp

Abstract. Quantum Annealing is an optimization process taking advantage of quantum tunneling to search for the global optimum of an optimization problem, although, being a heuristic method, there is no guarantee to find the global optimum. Optimization problems solved by a Quantum Annealer machine are modeled as Quadratic Unconstrained Binary Optimization (QUBO) problems. Combinatorial optimization problems, where variables take discrete values and the optimization is under constraints, can also be modeled as QUBO problems to benefit from Quantum Annealing power. However, defining quadratic penalty functions representing constraints within the QUBO framework can be a complex task. In this paper, we propose a method to learn from data constraint representations as a combination of patterns we isolated in Q matrices modeling optimization problems and their constraint penalty functions. We actually model this learning problem as a combinatorial optimization problem itself. We propose two experimental protocols to illustrate the strengths of our method: its scalability, where correct pattern combinations learned over data from a small constraint instance scale to large instances of the same constraint, and its robustness, where correct pattern combinations can be learned over very scarce data, composed of about 10 training elements only.

Keywords: Quantum Annealing · QUBO · Machine Learning · Constrained Optimization Problems · Constraint Satisfaction Problems

1 Introduction

As Quantum Computing is getting more real with the effective development of quantum processors, one can distinguish two approaches: the *gate-based paradigm*, in which the main industrial players such as IBM, Google, Intel and many start-up companies (IonQ, Rigetti, IQM, Pasqal, ...) have developed systems with up to a few hundreds of qubits, and the *adiabatic computation paradigm*, in which companies like D-Wave Systems have developed systems with thousands of qubits.

J. Mikyška et al. (Eds.): ICCS 2023, LNCS 14077, pp. 153–167, 2023.
https://doi.org/10.1007/978-3-031-36030-5_12

Quantum Annealing (QA) is an instance of adiabatic computation that is interesting in the current Noisy Intermediate-Scale Quantum (NISQ) era, and it has been applied in the field of combinatorial optimization. Indeed, combinatorial problems can be modeled as Quadratic Unconstrained Binary Optimization (QUBO) as input language and solved by QA systems.

More complex Constrained Optimization and Constraint Satisfaction Problems, coming from the Constraint Programming or Operations Research domains, are now being tackled with QUBO modeling and QA solving [6], although the size of the current QA machines still prevent experiments on large instances. Nevertheless, an interesting issue which is appearing is the modeling of complex problems in the constraint-based approach, and, although it is easy to add penalties into the objective function to represent simple constraints appearing in the problem, it could be difficult to express in a QUBO formulation the penalties corresponding to complex constraints as found in the Constraint Programming paradigm. Indeed, if some constraints are such as the *one-hot* or the *permutation (two-way one-hot)* constraints are easy to represent in QUBO, as we illustrate in Sect. 3, this is not the case for more general constraints. Many classical combinatorial problems are usually modeled with integer variables and constraints over those integer variables. Thus, to transform those models into QUBO, one has first to encode integer variables by binary variables (this is not difficult) and then to transform the constraints over integer values as penalties over binary variables, which may not be obvious at all. Therefore, an approach that would automatically create the QUBO penalties corresponding to integer constraints would be valuable. We can do that by learning the QUBO matrix representation corresponding to a constraint from the solution and non-solution candidates.

Learning constraints from data has been explored in different directions. Paulus et al. [18] proposes to integrate a combinatorial optimization module directly into a neural network as a layer, learning both the constraints and their costs from data. Another approach is proposed by Kumar et al. [14], where the constraints and the objective function of Mixed-Integer Linear Programs are learned from data. However, unlike our work, these two papers deal with linear constraints only, and they both learn constraints on a fixed number of variables. In comparison, our method can handle linear and non-linear constraints, and it learns a constraint representation that is independent of the number of variables. This work is inspired from [20], where a method to learn error functions representing constraints is proposed. This paper describes a model, named Interpretable Compositional Network, to learn error functions as an interpretable composition of elementary operations, in such a way that the learned compositions are independent of the number of the target constraint. The main difference with this current paper is that error functions are not necessarily quadratic but must verify a structured property upon error values implying a hierarchy among non-solution candidates, which is not necessary in the present QUBO setting.

2 Quantum Annealing and QUBO

Quantum Annealing (QA) has been proposed as a concrete form of adiabatic computation more than two decades ago by Kadowaki et al. [12] and Farhi et al. [8], and takes advantage of the physical phenomenon of *quantum tunneling*, allowing to traverse energy barriers in the energy landscape as long as they are not too large [19,23]. QA has gained momentum in the last decade with the development of special hardware based on QA, such as the quantum computers of D-Wave Systems [4,16] and, more recently, the so-called "quantum-inspired" systems which are realized with classical (non-quantum) electronics by Fujitsu [1], Hitachi [24], Toshiba [11] or Fixstars Amplify [15]. These systems are sometimes referred to as *Ising Machines*, as they can solve problems stated as Hamiltonians in the Ising model and are aimed to solve a large class of combinatorial problems [17,22], including industrial applications [25].

Interestingly, such a formulation is equivalent to the modeling in Quadratic Unconstrained Binary Optimization, a formalism whose roots go back to pseudo-binary optimization in the late 60's and which has been proposed as a simple but powerful modeling language for combinatorial problems about 15 years ago [2]. QUBO is now seen as a general modeling language for a variety of combinatorial problems [10,13]. For these reasons, QUBO has become the standard input language for Ising machines.

Simply put, a QUBO problem is given by a vector of n binary variables $x_1, \cdots, x_n$ and a quadratic expression over $x_1, \cdots, x_n$ that has to be minimized, which, without loss of generality, is of the form $\sum_{i \leq j} q_{ij} x_i x_j$. Therefore, a QUBO problem is determined by a vector x of n binary decision variables and an upper triangular $n \times n$ square matrix Q with coefficients q_{ij}. The QUBO problem can thus be written: minimize $y = x^T Q x$, where x^T is the transpose of x.

Moreover, in order to use QUBO to model *Constrained Optimization Problems* (COP) from the field of Operations Research, e.g., the well-known Traveling Salesman problem (TSP) or Quadratic Assignment Problem (QAP), one has to find a way to represent constraint expressions in QUBO models. This can be done by using *penalties* and adding them in the objective function to minimize, that is, as quadratic expressions whose value is minimal when the constraint is satisfied. An easy way to formulate such a penalty is to create a quadratic expression which has value 0 if the constraint is satisfied and a positive value otherwise, representing somehow the degree of violation of the constraint.

Although this works fine for simple constraints, defining the penalties corresponding to complex constraints can, however, become complicated. This is the starting point of our work investigating an automatic manner to generate QUBO penalties corresponding to complex constraints.

3 Motivating Example

3.1 Basic Example

Consider the problem of coloring the n nodes of a graph with k colors such that no adjacent nodes have the same color. The classical way to model such a problem in QUBO is to consider, for each node i of the graph, k binary variables x_{ij}, $j \in \{1, \ldots, k\}$ such that $x_{ij} = 1$ if the node i has the color j, and $x_{ij} = 0$ otherwise. Then, one has to devise an objective function that will be minimal when adjacent

$$\underset{Q_c}{\begin{pmatrix} -1 & 2 & 2 & 0 & 0 & 0 & 0 & 0 & 0 \\ 0 & -1 & 2 & 0 & 0 & 0 & 0 & 0 & 0 \\ 0 & 0 & -1 & 0 & 0 & 0 & 0 & 0 & 0 \\ 0 & 0 & 0 & -1 & 2 & 2 & 0 & 0 & 0 \\ 0 & 0 & 0 & 0 & -1 & 2 & 0 & 0 & 0 \\ 0 & 0 & 0 & 0 & 0 & -1 & 0 & 0 & 0 \\ 0 & 0 & 0 & 0 & 0 & 0 & -1 & 2 & 2 \\ 0 & 0 & 0 & 0 & 0 & 0 & 0 & -1 & 2 \\ 0 & 0 & 0 & 0 & 0 & 0 & 0 & 0 & -1 \end{pmatrix}} \qquad \underset{Q'_c}{\begin{pmatrix} -1 & 1 & 1 & 1 & 0 & 0 & 1 & 0 & 0 \\ 0 & -1 & 1 & 0 & 1 & 0 & 0 & 1 & 0 \\ 0 & 0 & -1 & 0 & 0 & 1 & 0 & 0 & 1 \\ 0 & 0 & 0 & -1 & 1 & 1 & 1 & 0 & 0 \\ 0 & 0 & 0 & 0 & -1 & 1 & 0 & 1 & 0 \\ 0 & 0 & 0 & 0 & 0 & -1 & 0 & 0 & 1 \\ 0 & 0 & 0 & 0 & 0 & 0 & -1 & 1 & 1 \\ 0 & 0 & 0 & 0 & 0 & 0 & 0 & -1 & 1 \\ 0 & 0 & 0 & 0 & 0 & 0 & 0 & 0 & -1 \end{pmatrix}}$$

Fig. 1. QUBO matrices corresponding to the penalties in Examples 1 and 2

nodes have different colors. But the problem variables are also subject to the constraints that each node has only a single color, i.e., that $\sum_1^k x_{ij} = 1$.

This is the well-known *one-hot* constraint: among k Boolean variables, exactly one has to be equal to 1 and others have to be equal to 0. It is classically used for the representation of integer variables with a domain of size n by n Boolean variables, although other encoding such as domain-wall or unary are possible [5,7].

Let us remark that $\sum_{j=1}^k x_{ij} = 1 \iff (\sum_{j=1}^k x_{ij} - 1)^2 = 0$, and by developing this expression, a quadratic penalty expression is obtained:

$$2\sum_{i=1}^{n}\sum_{j<j'} x_{ij}x_{ij'} - \sum_{i=1}^{n}\sum_{j=1}^{k} x_{ij}$$

This penalty has to be added to the QUBO objective function of the original problem in order to enforce the original *one-hot* constraint of the initial problem.

Example 1. Consider a QUBO model with 9 binary variables x_{ij}, $i \in \{1,2,3\}, j \in \{1,2,3\}$, and an objective function f to minimize over x_{ij}, subject to the three one-hot constraints: $\sum_{j=1}^3 x_{1j} = 1,\ \sum_{j=1}^3 x_{2j} = 1,\ \sum_{j=1}^3 x_{3j} = 1$.

The 9×9 QUBO matrix Q can be decomposed as the sum $Q = Q_o + Q_c$, with Q_o being the 9×9 matrix corresponding to the objective function f, and Q_c the 9×9 matrix corresponding to the three one-hot constraints. The Q_c matrix representing the penalty is depicted in Fig. 1.

The above transformation is straightforward, and the corresponding QUBO penalty is easy to derive mathematically from the initial constraint (as is the QUBO matrix), but this might not always the case.

More interestingly, we can see that the 9×9 matrix corresponding to the penalty part of the QUBO model (i.e., the constraint part of the initial problem) shows a particular structure with 3×3 submatrix around the diagonal. Could the penalty part of the QUBO matrix representing constraints coming from integer problems be seen as a combination of basic patterns such as submatrices? If so, could we *learn* automatically such matrix representation? We will see that the answer to both questions is "yes".

3.2 A More Complex Example

Let us consider now a slightly more complex example: *permutation* constraints. Many classical combinatorial optimization problems, such as the well-known Traveling Salesman Problem (TSP) and the Quadratic Assignment Problem (QAP), or Constraint Satisfaction problems such as the N-queens and Magic Square puzzles are usually modeled with a vector of integer decision variables which are subject to the constraint that each feasible solution forms a permutation. In the Constraint Programming community, such a *permutation constraint* is a special case of the AllDifferent constraint, which has been the subject of large literature and various solving techniques [9].

To enforce that n integer variables x_i with values in $\{1,\ldots,n\}$ represent a permutation, we need to enforce that each value $j \in \{1,\ldots,n\}$ is assigned once and only once. When translated to QUBO, with n binary variables x_{ij}, $j \in \{1,\ldots,n\}$, encoding an integer variable x_i of the original problem formulation, this amounts to the so-called *two-way one-hot* constraints: $2 \times n$ one-hot constraints with one set of n constraints corresponding to each of the n variables x_i stating that it can have only one value k and one set of n constraints for each of the n values k stating that it can be assigned to only one variable x_i.

These $2 \times n$ one-hot constraints are as follows:

$$\forall i \in \{1,\ldots,n\},\ \sum_{j=1}^{n} x_{ij} = 1 \quad \forall j \in \{1,\ldots,n\},\ \sum_{i=1}^{n} x_{ij} = 1$$

Adding all corresponding penalty expressions together and simplifying the quadratic expression gives the following penalty for the permutation constraint:

$$\sum_{i=1}^{n}\sum_{j<j'} x_{ij}x_{ij'} + \sum_{j=1}^{n}\sum_{i<i'} x_{ij}x_{i'j} - \sum_{i=1}^{n}\sum_{j=1}^{n} x_{ij}$$

Example 2. Consider a combinatorial problem on 3 integer variables x_1, x_2, x_3, with values in $\{1, 2, 3\}$ subject to a permutation constraint and an objective function f to minimize over x_i.

If each original integer variable is encoded by 3 binary variables with one-hot encoding, the corresponding QUBO model will have 9 binary variables x_{ij}, $i \in \{1, 2, 3\}, j \in \{1, 2, 3\}$, and the 9×9 QUBO matrix can be decomposed as the sum $Q' = Q'_o + Q'_c$, with Q'_o corresponding to the translation of the objective function f to minimize over x_{ij}, and Q'_c corresponding to the permutation constraint. The Q'_c matrix representing the penalty is depicted in Fig. 1.

This matrix is somewhat different from the one in Example 1, but a similar pattern of 3×3 submatrices around the diagonal can be observed, together with patterns on the lines.

4 Method Design

The main contribution of this paper is to propose a method to automatically learn from data a pattern composition representing a Q matrix, such that Q corresponds to a target constraint c. This method is directly inspired from the method proposed in [20], to learn error functions from data in an interpretable and scalable fashion. We call such a pattern composition a **Q matrix representation**. The data is the training set obtained from an instance of c, *i.e.*, the constraint c over a fixed number of variables taking their values over domains of a fixed size.

Let's consider a constraint c and an instance ι of c. A **candidate** of ι is an assignment of all variables composing ι. A candidate is said to be **positive** if it satisfies c, and **negative** otherwise. We denote by S the set of all possible tuples (x, y), where x is a candidate of the constraint instance ι, and $y \in \{0, 1\}$ describes if x is a positive or a negative candidate, such that

$$y = \begin{cases} 1 & \text{if x is a positive candidate} \\ 0 & \text{if x is a negative candidate.} \end{cases}$$

Giving some positive and negative candidates of instance ι of a constraint c, our goal is to learn a pattern composition representing a Q matrix corresponding to c, *i.e.*, Q must verify the following property:

$$\forall (x, y) \in S, x^T Qx \text{ is minimal iff } y = 1 \tag{1}$$

Variables of discrete constraints take their value from a domain composed of integers. However, QUBO problems are considering binary variables only. As written in the introduction, there are several ways to convert constraint variables into QUBO variables: unary expansion, binary expansion, etc. In this work, we will only consider one-hot encoding: a constraint variable x_i over a k-ary domain will be encoded by a k-dimensional binary vector, such that the j-th element x_{ij} of this vector is set to true if and only if $x_i = j$ holds.

To represent a discrete constraint, Q can be composed of integers only. If we aim to represent a constraint over n variables x_i taking their value in a k-ary domain, then, due to the one-hot encoding of the variables, Q is an upper triangular matrix of size $nk \times nk$. Our method is based on learning a correct combination of submatrix patterns. The one-hot constraint is systematically added into this pattern combination.

Q being an upper triangular matrix, we consider two kinds of submatrices: $k \times k$ triangle submatrices containing the diagonal of Q and representing the properties of a variable x_i, and $k \times k$ square submatrices representing properties between two variables x_i and x_j, with $i \neq j$.

Our method considers 14 square and 3 triangle submatrix patterns, depicted in Fig. 2. Square submatrix patterns can be composed by summing their elements, however, we consider some square submatrix patterns to be mutually exclusive. Indeed, it would make not sense for instance to enforce both the properties $x_i = x_j$ and $x_i \neq x_j$ at the same time. Square patterns 12, 13, 14, and triangle pattern 3 take some parameters. It is important to keep in mind that submatrices and their patterns are on binary variables.

We can divide square submatrix patterns into 3 categories:

- Comparison patterns from Square 1 to 6, representing some comparison properties between x_i and x_j. For instance, $x_i \neq x_j$ (Fig. 2a).
- Position patterns from Square 7 to 11, encoding properties such that the values of the i-th and j-th variables x_i and x_j depend on their respective position i and j. For instance, favoring $x_i = i$ and $x_j = j$ (Fig. 2g).
- Complex patterns for Square 12, 13 and 14. For instance, the repel property described in the next paragraph (Fig. 2l).

Due to the page limitation, we chose not to explain all patterns. Instead, we focus here on the less trivial ones. For Square pattern 14 (Fig. 2n) and triangle pattern 3 (Fig. 2q), b_{x_i} represents the coefficient of the variable x_i in a linear combination $\sum b_{x_i} x_i = a$. Square patterns 12 (Fig. 2l) and 13 (Fig. 2m), respectively called *repel* and *attract*, take a parameter p. These patterns look like a diagonal magnetic field with an intense value p in its center (repel) or on its borders (attract), which decays towards the borders (repel) or the center (attract). This is illustrated in Fig. 2 with $p = 3$. The intuitive idea behind these patterns is that variables x_i and x_j are repelling or attracting each others, until their values are separated by at least a distance p, such that $|x_i - x_j| > p$ holds (repel), or until they are closed enough, below a $k - p$ threshold, such that $|x_i - x_j| < k - p$ holds (attract).

Consider the function m determining if a candidate x is such that $x^T Q x$ is minimal or not:

$$m(x, Q) = \begin{cases} 1 & \text{if } x^T Q x \text{ is minimal,} \\ 0 & \text{otherwise.} \end{cases}$$

Learning a correct combination of submatrix patterns from a training set $X \subseteq S$ is a machine learning problem, but it can also be tackled as a combinatorial problem. We modeled this as a Constrained Optimization Problem (COP) described in Table 1.

We have 14 binary variables to describe which square patterns are combined to give a global pattern for square submatrices, and one variable over a ternary domain indicating which pattern is selected for triangle submatrices. Thus, all square submatrices together will share the same pattern, as well as all triangle submatrices. Although our method still works if we decide to combine different patterns for each submatrix individually, this way of doing has the advantage of learning the Q matrix representation quicker. Moreover, it is motivated by the fact that natural constraints in Constraint Programming usually apply the same property over all variables in their scope (for instance, all variables must be assigned to a different value). If it is not the case, then the target constraint can certainly be decomposed into a series of smaller constraints. For instance, let's assume we work with the constraint $c(x_1, x_2, x_3) :=$ "$x_1 + x_2 + x_3 = 5$ such that $x_1 < x_2$ holds". Here, x_1 and x_2 have a property $x_1 < x_2$ that is not shared by x_3. But the constraint c can be decomposed into two constraints

$$\begin{pmatrix}1&0&0&0&0\\0&1&0&0&0\\0&0&1&0&0\\0&0&0&1&0\\0&0&0&0&1\end{pmatrix}$$

(a) Square 1 $x_i \neq x_j$

$$\begin{pmatrix}0&1&1&1&1\\1&0&1&1&1\\1&1&0&1&1\\1&1&1&0&1\\1&1&1&1&0\end{pmatrix}$$

(b) Square 2 $x_i = x_j$

$$\begin{pmatrix}0&0&0&0&0\\1&0&0&0&0\\1&1&0&0&0\\1&1&1&0&0\\1&1&1&1&0\end{pmatrix}$$

(c) Square 3 $x_i \leq x_j$

$$\begin{pmatrix}1&0&0&0&0\\1&1&0&0&0\\1&1&1&0&0\\1&1&1&1&0\\1&1&1&1&1\end{pmatrix}$$

(d) Square 4 $x_i < x_j$

$$\begin{pmatrix}0&1&1&1&1\\0&0&1&1&1\\0&0&0&1&1\\0&0&0&0&1\\0&0&0&0&0\end{pmatrix}$$

(e) Square 5 $x_i \geq x_j$

$$\begin{pmatrix}1&1&1&1&1\\0&1&1&1&1\\0&0&1&1&1\\0&0&0&1&1\\0&0&0&0&1\end{pmatrix}$$

(f) Square 6 $x_i > x_j$

$$\begin{pmatrix}0&0&0&-1&0\\-1&-1&-1&-1&-1\\0&0&0&-1&0\\0&0&0&-1&0\\0&0&0&-1&0\end{pmatrix}$$

(g) Square 7 Favor $x_i = i$

$$\begin{pmatrix}0&0&0&1&0\\1&1&1&1&1\\0&0&0&1&0\\0&0&0&1&0\\0&0&0&1&0\end{pmatrix}$$

(h) Square 8 Avoid $x_i = i$

$$\begin{pmatrix}-1&0&-1&0&-1\\0&0&0&0&0\\-1&0&-1&0&-1\\0&0&0&0&0\\-1&0&-1&0&-1\end{pmatrix}$$

(i) Square 9 Favor $x_i, x_j \neq \{i, j\}$

$$\begin{pmatrix}1&0&1&0&1\\0&0&0&0&0\\1&0&1&0&1\\0&0&0&0&0\\1&0&1&0&1\end{pmatrix}$$

(j) Square 10 Avoid $x_i, x_j \neq \{i, j\}$

$$\begin{pmatrix}0&0&0&1&0\\1&1&1&-1&1\\0&0&0&1&0\\0&0&0&1&0\\0&0&0&1&0\end{pmatrix}$$

(k) Square 11 Swap x_i, x_j

$$\begin{pmatrix}3&2&1&0&0\\2&3&2&1&0\\1&2&3&2&1\\0&1&2&3&2\\0&0&1&2&3\end{pmatrix}$$

(l) Square 12 Repel

$$\begin{pmatrix}0&0&1&2&3\\0&0&0&1&2\\1&0&0&0&1\\2&1&0&0&0\\3&2&1&0&0\end{pmatrix}$$

(m) Square 13 Attract

$$\begin{pmatrix}2b_{x_i}b_{x_j}&2b_{x_i}b_{x_{j+1}}&2b_{x_i}b_{x_{j+2}}\\2b_{x_{i+1}}b_{x_j}&2b_{x_{i+1}}b_{x_{j+1}}&2b_{x_{i+1}}b_{x_{j+2}}\\2b_{x_{i+2}}b_{x_j}&2b_{x_{i+2}}b_{x_{j+1}}&2b_{x_{i+2}}b_{x_{j+2}}\end{pmatrix}$$

(n) Square 14 Linear combination

$$\begin{pmatrix}0&0&0&0&0\\&0&0&0&0\\&&0&0&0\\&&&0&0\\&&&&0\end{pmatrix}$$

(o) Triangle 1 Neutral

$$\begin{pmatrix}-1&0&0&0&0\\&-1&0&0&0\\&&-1&0&0\\&&&-1&0\\&&&&-1\end{pmatrix}$$

(p) Triangle 2 Favor x_i

$$\begin{pmatrix}-(2a-b_{x_i})b_{x_i}&2b_{x_i}b_{x_{i+1}}&2b_{x_i}b_{x_{i+2}}\\&-(2a-b_{x_{i+1}})b_{x_{i+1}}&2b_{x_{i+1}}b_{x_{i+2}}\\&&-(2a-b_{x_{i+2}})b_{x_{i+2}}\end{pmatrix}$$

(q) Triangle 3 Linear combination

Fig. 2. Submatrix patterns used in our method, with their property. Simple examples are displayed with $k = 5$; complex examples (n) and (q) with $k = 3$.

$c_1(x_1, x_2, x_3) := x_1 + x_2 + x_3 = 5$ and $c_2(x_1, x_2) := x_1 < x_2$, where all variables in their scope share the same properties.

Like expressed above, we forbid some square pattern combinations: We cannot have more than one square pattern from Square 1 to Square 6 in the combination, and couples of patterns Square 7 – Square 8, Square 9 – Square 10, and Square 12 – Square 13 are mutually exclusive. The 4 first constraints in Table 1 are here to forbid such combinations.

Table 1. COP model to learn Q from data $X \subseteq S$

Variables	$v_{f_1}, \ldots, v_{f_{14}}, v_h$ One variable v_{f_i} for each possible square pattern i, a unique variable v_h for the triangle pattern
Domains	$D_{f_i} = \{0, 1\}$, with $1 \leq i \leq 14$ $D_h = \{1, 2, 3\}$
Constraints	$\sum_{i=1}^{6} v_{f_i} \leq 1$ $v_{f_7} + v_{f_8} \leq 1$ $v_{f_9} + v_{f_{10}} \leq 1$ $v_{f_{12}} + v_{f_{13}} \leq 1$ $\sum_{(x,y) \in X} \lvert m(x, Q) - y \rvert = 0$
Objective function	$\min \sum_{i=1}^{14} v_{f_i}$, minimizing the number of (square) submatrix patterns in the composition

The fifth constraint in our COP model makes sure the learned Q matrix representation can correctly handle all positive and negative candidates from the training set X.

One strength of this model is its independence regarding the size of the target constraint instance: whatever the size of the data, *i.e.*, the number of variables in candidates or the size of the domains, our model will still be composed of 15 variables to express any pattern combinations describing any constraints. This makes our model scalable, allowing the learning of a Q matrix representation over a small instance of a constraint c that is valid for all instance sizes of the same constraint, as shown in Experiment 1.

5 Experiments

To show the versatility of our method, we tested it on five different constraints: AllDifferent, Ordered, LinearSum, NoOverlap1D, and Channel. Following the XCSP^3-core specifications[1] [3], those global constraints belong to four major constraint families: Comparison (AllDifferent and Ordered), Counting/Summing (LinearSum), Packing/Scheduling (NoOverlap1D) and Connection (Channel). These constraints are also among the 25 most popular and common constraints [3]. We give a brief description of those five constraints below:

- **AllDifferent** ensures that variables must all be assigned to different values.
- **Ordered** ensures that an assignment of n variables $(x_1, \ldots, x_n)$ must be ordered, given a total order. In this paper, we choose the total order $\leq$. Thus, for all indices $i, j \in \{1, n\}$, $i < j$ implies $x_i \leq x_j$.
- **LinearSum** ensures that the equation $x_1 + x_2 + \ldots + x_n = p$ holds, with the parameter p a given integer.
- **NoOverlap1D** considers variables as tasks, starting from a certain time (their value) and each with a given length p (their parameter). The constraint

[1] see also http://xcsp.org/specifications.

ensures that no tasks are overlapping, *i.e.*, for all indices $i, j \in \{1, n\}$ with n the number of variables, we have $x_i + p_i \leq x_j$ or $x_j + p_j \leq x_i$. To have a simpler code, we have considered in our system that all tasks have the same length p.
- **Channel** ensures that the i-th variable x_i assigned to j with $j \neq i$ implies that x_j is assigned to i. In other words, Channel accepts all permutations of the vector $(1, \ldots, n)$ such that each variable has been swapped with another variable at most once.

5.1 Experimental Protocols

We set up two different experimental protocols to show the scalability, the robustness, and more globally, the efficiency of our method.

Like presented in Sect. 4, learning pattern compositions of Q matrices from data is handled as a combinatorial optimization problem. To model and solve this problem, we use the framework GHOST [21] which runs a stochastic local search solver to solve problems. Due to this stochastic solving, all learning and testing have been done 100 times, but over the same pre-computed training sets, to not let the randomness of sampled sets impact the results in some way. Training and test sets that are too large to be complete have been pre-computed using Latin hypercube sampling to have a good diversity among drawn candidates. These sets have been generated such that they contain an equal number of positive and negative candidates. GHOST's solver requires a timeout, such that it tries to improve the best solution found until it reaches the given timeout. We set it to 1 s. We did not fine-tune the solver parameters, running our experiments with their default values. In addition to our two experiments, we did 100 runs for each training set of Experiment 1 and 2 disabling the objective function, to see how fast our method can find a correct Q matrix representation. Indeed, without an objective function, GHOST's solver does not take into account the timeout and halts as soon as it finds a solution satisfying all constraints in the COP model.

We have hold-out test sets of assignments from larger dimensions to evaluate the quality of our learned Q matrix representations. We did not re-run batches of experiments to keep the ones with the best results, as it should always be the case with such experimental protocols.

All experiments have been done on a computer with a Core i9 9900 CPU and 32 GB of RAM, running on Ubuntu 22.04.2. Programs have been compiled with GCC with the 03 optimization option. Our entire system, its C++ source code, and experimental setups are accessible on Zenodo[2] and GitHub.

Experiment 1: Scalability. One of the key-points of our method is that we can learn the pattern composition of a Q matrix from data about a small instance of a constraint c, *i.e.*, over few variables and small domains, which gives us the blueprint to build Q matrices handling large instances of c.

[2] https://doi.org/10.5281/zenodo.7800168.

In this experiment, we learn Q matrix representations upon complete training sets composed of all possible tuples (x, y), with x a candidate of the target constraint instance and y the binary value indicating if x is a solution of the constraint or not, as explained at the beginning of Sect. 4. These small, complete training sets are built considering constraint instances with 4 variables over domains of size 4, giving complete training sets with 256 candidates, except for the constraint NoOverlap1D for which such an instance size does not make sense. The training set for NoOverlap1D takes into account 3 variables over domains of size 7 (and with a parameter $p = 2$ for the length of each task), leading to a training set with $7^3 = 343$ candidates. The parameter p of the LinearSum constraint instance was fixed to 10, giving the constraint $\sum_{i=1}^{4} x_i = 10$, with $x_i \in \{1, \ldots, 4\}, 1 \leq i \leq 4$.

To test both the scalability of our method, we test the learned matrix representations over significantly larger constraint instances, with 30 variables over domains of size 30. Indeed, a learned composition can be represented by a vector of 15 elements, one for each variable of our COP model, independently of the size of the Q matrix and the constraint instance it represents. Those test sets containing too many candidates ($30^{30} \simeq 2 \times 10^{44}$) to be fully tested or even fully generated, we build them in order to get exactly 10,000 build positive and 10,000 drawn negative candidates, giving 20,000 unique candidates. Once again, we need to make an exception for NoOverlap1D due to the nature of this constraint. Its test set is then composed of 10,000 built positive and 10,000 drawn negative candidates of a constraint instance with 20 variables and domains of 160 elements (with $p = 6$), giving a space of $160^{20} \simeq 1.2 \times 10^{44}$ candidates. Q matrices built for these test sets are then of size 900×900, except for NoOverlap1D which has a Q matrix of size 3200×3200, going to the limit of what current Quantum Annealing machines can handle nowadays. All test sets are pre-generated, *i.e.*, sampled once and kept for all experiments.

Experiment 2: Robustness. In Experiment 1, we learn pattern compositions of Q matrices to represent constraint instances over complete training sets, *i.e.*, composed of all possible candidates. However, providing such complete sets to learn Q matrix representations may not always be convenient. To test the robustness of our method, we learn Q matrix representations over training sets of constraint instances that are too large to be completely generated in a reasonable time: 12 variables over domains of 12 elements ($12^{12} \simeq 9 \times 10^{12}$). Instead, we sample 5 positive and 5 negative candidates only in such spaces, then we test the learned Q matrix representations on the same test sets than for Experiment 1. For NoOverlap1D, we sample 5 positive and 5 negative candidates of an instance with 8 variables over domains of 35 elements (and $p = 3$), leading to a space composed of $35^8 \simeq 2.2 \times 10^{12}$ candidates.

In addition, we also tested how restricted training sets can be until we observe significant efficiency drops. For this, we repeated this experimental protocol with training sets composed of 2, 4, 6, 8, 10 and 12 candidates, each time with half

Table 2. Success rates over 100 runs on test sets. Timeout is fixed at 1s for each run.

Experiment	AllDifferent	Ordered	LinearSum	NoOverlap1D	Channel
1	100	100	100	100	100
2	100	100	100	97	100

positive and half negative candidates. Like test sets, these training sets are pre-generated.

5.2 Experimental Results

Table 2 shows the number of times, over 100 runs and for each target constraint, a Q matrix representation has been learned and corresponds to a correct Q matrix, *i.e.*, satisfying the property of Eq. 1 over tuples (x, y) of the constraint test set. This first line corresponds to the results of Experiment 1, demonstrating that our method perfectly scales, learning Q matrix representation over small training sets of about 300 candidates and successfully tested over 20,000 randomly drawn candidates from a huge candidate space of about 2×10^{44} candidates. The second line is the results of Experiment 2. It shows that our method is able to learn correct Q matrix representations from very sparse data: here, we learn Q matrix representation from 5 positive and 5 negative candidates, randomly drawn from candidate spaces of about 9×10^{12} candidates. All learned Q matrices shown themselves to be correct on test sets, except 3 times for NoOverlap1D. Indeed, for these 3 times, the solver has been trapped into some local optimum, finding a correct pattern composition for the training set but containing unnecessary patterns, that lead to incorrectly handled candidates in the NoOverlap1D test set. More specifically, almost all positive candidates from the NoOverlap1D test set were incorrectly handled and considered as negative candidates. Having false negatives is indeed more frequent than false positive because a positive candidate x_p with any values of $x_p^T Q x_p$ above the expected minimal value will be considered to be a negative candidate, but mistaking a negative candidate x_n as a positive one requires having $x_n^T Q x_n$ to be equals to the expected minimal value. Having a value $x_n^T Q x_n$ below the expected minimal value in the test set is unlikely because it would imply that all positive candidates would be considered as false negatives.

These three incorrectly learned Q matrix representations for NoOverlap1D is not a serious problem in practice: users can get around this problem either by considering more candidates in the training set, or setting a timeout longer than 1 s to let the solver more time to find an optimal solution, or finally in the worst case, running again the learning. Since learning a Q matrix representation only takes one second in our current setup, users can easily afford re-learning a Q matrix representation if the first learning outputs an incorrect one.

Table 3. Success rate over 100 runs on test sets, regarding the size of the incomplete training sets. Timeout is fixed at 1 s for each run.

Nb candidates	AllDifferent	Ordered	LinearSum	NoOverlap1D	Channel
12	100	100	100	92	100
10	100	100	100	96	100
8	37	100	100	97	100
6	33	100	100	92	10
4	27	100	100	39	10
2	0	0	19	8	0

$$\begin{pmatrix}
-1 & 2 & 2 & 2 & 0 & -1 & -1 & -1 & 0 & -1 & -1 & -1 & 0 & -1 & -1 & -1 \\
0 & -1 & 2 & 2 & -1 & 0 & 1 & 1 & 1 & -1 & -1 & -1 & 1 & -1 & -1 & -1 \\
0 & 0 & -1 & 2 & 1 & -1 & -1 & -1 & -1 & 1 & 0 & 1 & 1 & -1 & -1 & -1 \\
0 & 0 & 0 & -1 & 1 & -1 & -1 & -1 & 1 & -1 & -1 & -1 & -1 & 1 & 1 & 0 \\
0 & 0 & 0 & 0 & -1 & 2 & 2 & 2 & -1 & 1 & -1 & -1 & -1 & 1 & -1 & -1 \\
0 & 0 & 0 & 0 & 0 & -1 & 2 & 2 & -1 & 0 & -1 & -1 & -1 & 0 & -1 & -1 \\
0 & 0 & 0 & 0 & 0 & 0 & -1 & 2 & 1 & -1 & 0 & 1 & -1 & 1 & -1 & -1 \\
0 & 0 & 0 & 0 & 0 & 0 & 0 & -1 & -1 & 1 & -1 & -1 & 1 & -1 & 1 & 0 \\
0 & 0 & 0 & 0 & 0 & 0 & 0 & 0 & -1 & 2 & 2 & 2 & -1 & -1 & 1 & -1 \\
0 & 0 & 0 & 0 & 0 & 0 & 0 & 0 & 0 & -1 & 2 & 2 & -1 & -1 & 1 & -1 \\
0 & 0 & 0 & 0 & 0 & 0 & 0 & 0 & 0 & 0 & -1 & 2 & -1 & -1 & 0 & -1 \\
0 & 0 & 0 & 0 & 0 & 0 & 0 & 0 & 0 & 0 & 0 & -1 & 1 & 1 & -1 & 0 \\
0 & 0 & 0 & 0 & 0 & 0 & 0 & 0 & 0 & 0 & 0 & 0 & -1 & 2 & 2 & 2 \\
0 & 0 & 0 & 0 & 0 & 0 & 0 & 0 & 0 & 0 & 0 & 0 & 0 & -1 & 2 & 2 \\
0 & 0 & 0 & 0 & 0 & 0 & 0 & 0 & 0 & 0 & 0 & 0 & 0 & 0 & -1 & 2 \\
0 & 0 & 0 & 0 & 0 & 0 & 0 & 0 & 0 & 0 & 0 & 0 & 0 & 0 & 0 & -1
\end{pmatrix}$$

Learned Q matrix for the Channel constraint

$$\begin{pmatrix} -1 & 0 & 0 & 0 \\ & -1 & 0 & 0 \\ & & -1 & 0 \\ & & & -1 \end{pmatrix}$$

Triangle 2
Favor x_i

$$\begin{pmatrix} 0 & 0 & -1 & 0 \\ -1 & -1 & -1 & -1 \\ 0 & 0 & -1 & 0 \\ 0 & 0 & -1 & 0 \end{pmatrix}$$

Square 7
Favor $x_i = i$

$$\begin{pmatrix} -1 & 0 & -1 & 0 \\ 0 & 0 & 0 & 0 \\ -1 & 0 & -1 & 0 \\ 0 & 0 & 0 & 0 \end{pmatrix}$$

Square 9
Favor
$x_i, x_j \neq \{i, j\}$

$$\begin{pmatrix} 0 & 0 & 1 & 0 \\ 1 & 1 & -1 & 1 \\ 0 & 0 & 1 & 0 \\ 0 & 0 & 1 & 0 \end{pmatrix}$$

Square 11
Swap x_i, x_j

Fig. 3. Q matrix for Channel and patterns in the learned composition.

In Experiment 2, we learned Q matrix representation from training sets composed of 5 positive and 5 negative candidates. We tested how many candidates were needed to correctly learn pattern compositions for our different constraints. Table 3 sums up these trials, showing the success rate over 100 runs of representation learning with training sets of size n, with $\frac{n}{2}$ positive and $\frac{n}{2}$ negative candidates. Unsurprisingly, we can see no correct representations can be learned with one positive and one negative candidate only. However, four balanced candidates is sufficient to perfectly learn the Q matrix representation of the constraints Ordered and LinearSum.

Figure 3 illustrates a Q matrix obtained with our method, for the Channel constraint with 4 variables over domains of size 4, and the patterns that are taking part of the learned combination. The one-hot constraint is not depicted as a pattern in this figure, but it is always implicitly added in the combination.

6 Conclusion

In this paper, we presented a method to learn a pattern composition representing the penalty part of a QUBO matrix, which can be difficult to do manually for complex constraints. We showed that this can be done with a limited set of examples and, via two experimental protocols, that our method has excellent scalability and robustness properties.

The current limitations of our method is the incapacity to handle ancillary variables. Indeed, some constraints require additional binary variables that do not directly represent integer variables from the target constraint, but are necessary to model some interactions between integer variables. Adding automatically the right number of ancillary variables to handle constraints like Element or NValues would be a natural extension of our work.

A second limitation is the lack of input size-dependent patterns, such as the size of domains, for instance. Such patterns would be greatly helpful to model some constraints like Minimum, seeing their satisfaction depending on some fixed values, without having a combinatorial explosion of the number of patterns that could severely hurt the learning efficiency. However, defining such generic and input size-dependent patterns might be challenging.

References

1. Aramon, M., Rosenberg, G., Valiante, E., Miyazawa, T., Tamura, H., Katzgraber, H.G.: Physics-inspired optimization for quadratic unconstrained problems using a digital annealer. Front. Phys. **7**, 48 (2019)
2. Boros, E., Hammer, P.L., Tavares, G.: Local search heuristics for quadratic unconstrained binary optimization (QUBO). J. Heuristics **13**(2), 99–132 (2007)
3. Boussemart, F., Lecoutre, C., Audemard, G., Piette, C.: XCSP3-core: a format for representing constraint satisfaction/optimization problems. arXiv abs/2009.00514 (2020)
4. Bunyk, P.I., et al.: Architectural considerations in the design of a superconducting quantum annealing processor. IEEE Trans. Appl. Supercond. **24**(4), 1–10 (2014)
5. Chancellor, N.: Domain wall encoding of discrete variables for quantum annealing and QAOA. Quantum Sci. Technol. **4**, 045004 (2019)
6. Codognet, P.: Constraint solving by quantum annealing. In: Silla, F., Marques, O. (eds.) ICPP Workshops 2021: 50th International Conference on Parallel Processing, USA, 9–12 August 2021, pp. 25:1–25:10. ACM (2021)
7. Codognet, P.: Domain-wall/unary encoding in QUBO for permutation problems. In: 2022 IEEE International Conference on Quantum Computing and Engineering (QCE), pp. 167–173 (2022)
8. Farhi, E., Goldstone, J., Gutmann, S., Lapan, J., Lundgren, A., Preda, D.: A quantum adiabatic evolution algorithm applied to random instances of an NP-complete problem. Science **292**(5516), 472–475 (2001)
9. Gent, I.P., Miguel, I., Nightingale, P.: Generalised arc consistency for the AllDifferent constraint: an empirical survey. Artif. Intell. **172**(18), 1973–2000 (2008)
10. Glover, F.W., Kochenberger, G.A., Du, Y.: Quantum bridge analytics I: a tutorial on formulating and using QUBO models. 4OR **17**(4), 335–371 (2019)

11. Goto, H., Tatsumura, K., Dixon, A.R.: Combinatorial optimization by simulating adiabatic bifurcations in nonlinear Hamiltonian systems. Sci. Adv. **5**(4) (2019)
12. Kadowaki, T., Nishimori, H.: Quantum annealing in the transverse Ising model. Phys. Rev. E **58**, 5355–5363 (1998)
13. Kochenberger, G., et al.: The unconstrained binary quadratic programming problem: a survey. J. Comb. Optim. **28**(1), 58–81 (2014). https://doi.org/10.1007/s10878-014-9734-0
14. Kumar, M., Kolb, S., De Raedt, L., Teso, S.: Learning mixed-integer linear programs from contextual examples. arXiv e-prints abs/2107.07136, pp. 1–11 (2021)
15. Matsuda, Y.: Research and development of common software platform for Ising machines. In: 2020 IEICE General Conference (2020)
16. McGeoch, C.C., Harris, R., Reinhardt, S.P., Bunyk, P.I.: Practical annealing-based quantum computing. Computer **52**(6), 38–46 (2019)
17. Mohseni, N., McMahon, P.L., Byrnes, T.: Ising machines as hardware solvers of combinatorial optimization problems. Nat. Rev. Phys. **4**(6), 363–379 (2022)
18. Paulus, A., Rolínek, M., Musil, V., Amos, B., Martius, G.: CombOptNet: fit the right NP-hard problem by learning integer programming constraints. In: Proceedings of the 38th International Conference on Machine Learning (ICML 2021), pp. 8443–8453. PMLR, Online (2021)
19. Rajak, A., Suzuki, S., Dutta, A., Chakrabarti, B.K.: Quantum annealing: an overview. Philos. Trans. R. Soc. A: Math. Phys. Eng. Sci. **381**(2241) (2022)
20. Richoux, F., Baffier, J.F.: Automatic error function learning with interpretable compositional networks. Ann. Math. Artif. Intell. 1–35 (2023). Springer
21. Richoux, F., Uriarte, A., Baffier, J.F.: GHOST: a combinatorial optimization framework for real-time problems. IEEE Trans. Comput. Intell. AI Games **8**(4), 377–388 (2016)
22. Tanahashi, K., Takayanagi, S., Motohashi, T., Tanaka, S.: Application of Ising machines and a software development for Ising machines. J. Phys. Soc. Jpn. **88**(6), 061010 (2019)
23. Tanaka, S., Tamura, R., Chakrabarti, B.K.: Quantum Spin Glasses, Annealing and Computation, 1st edn. Cambridge University Press, USA (2017)
24. Yamaoka, M., Okuyama, T., Hayashi, M., Yoshimura, C., Takemoto, T.: CMOS annealing machine: an in-memory computing accelerator to process combinatorial optimization problems. In: IEEE Custom Integrated Circuits Conference, Austin, TX, USA, pp. 1–8. IEEE (2019)
25. Yarkoni, S., Raponi, E., Bäck, T., Schmitt, S.: Quantum annealing for industry applications: introduction and review. Rep. Prog. Phys. **85**(10), 104001 (2022)

TAQOS: A Benchmark Protocol for Quantum Optimization Systems

Valentin Gilbert(✉), Stéphane Louise, and Renaud Sirdey

Université Paris-Saclay, CEA-List, 91120 Palaiseau, France
{valentin.gilbert,stephane.louise,renaud.sirdey}@cea.fr

Abstract. The growing availability of quantum computers raises questions about their ability to solve concrete problems. Existing benchmark protocols still lack problem diversity and attempt to summarize quantum advantage in a single metric that measures the quality of found solutions. Unfortunately, the solution quality metric is insufficient for measuring quantum algorithm performance and should be presented along with time and instances coverage metrics. This paper aims to establish the TAQOS protocol to perform a Tight Analysis of Quantum Optimization Systems. The combination of metrics considered by this protocol helps to identify problems and instances liable to produce quantum advantage on Noisy-Intermediate Scale Quantum (NISQ) devices for useful applications. The methodology used for the benchmark process is detailed and an illustrative short case study on the Max-Cut problem is provided.

Keywords: Benchmark protocol · Quantum computing · QAOA · Quantum annealing

1 Introduction

Quantum manufacturers are currently building chips with several hundred qubits for circuit-based quantum computers and thousands of qubits for quantum annealers. As the NISQ era [20] begins, it remains unclear whether noisy quantum computers will have useful applications in the near term, since quantum error correction codes still require too much qubits to be efficient. Defining whether a quantum algorithm could bring a quantum advantage on a specific task is far from straightforward, as the full quantum stack usually involves complex classic and quantum processing where each subpart constitutes a full research domain. One relevant class of problems that may be subject to quantum advantage are optimization problems that naturally map on Adiabatic Quantum Optimization (AQO) systems. Hybrid quantum algorithms also provide an interesting option to solve optimization problems, especially using the Quantum Approximate Optimization Algorithm (QAOA) [8]. This algorithm exhibits a robust behavior under noisy regime [11,23] and encouraging theoretical bounds of convergence have been proven for specific problems at fixed depth [4,8]. The plethora of optimization problems being developed and benchmarked using Quantum Annealing (QA) and the QAOA requires a rigorous methodology to report the

J. Mikyška et al. (Eds.): ICCS 2023, LNCS 14077, pp. 168–176, 2023.
https://doi.org/10.1007/978-3-031-36030-5_13

performance of these heuristics. To this end, this paper introduces one methodology termed the TAQOS protocol, which performs a Tight Analysis of Quantum Optimization Systems performance. This protocol defines the guidelines and properties of an application-based competitive benchmark. The instances and the source code are available at [10].

1.1 Related Work

Protocols used to benchmark the performance of classical heuristics appeared in the 1970 s s and provide useful guidelines to produce high-quality classical computer benchmarks. Several best practices and guidelines for evaluating computer performance exist in the current literature [15]. One important approach is found in [3], which splits the performance study into two types of benchmarks. The first type is *competitive benchmark* which aims to directly and quantitatively compare the performance of different algorithms. Whereas the second type, named *descriptive benchmark*, is used to analyze and understand the factors that impact algorithm performances. While *competitive benchmark* should be composed of fast-to-compute unbiased metrics to compare algorithm performance, *descriptive benchmark* can be composed of more complex metrics serving a better understanding of the algorithmic behavior.

As quantum annealers have improved (e.g., D-Wave systems [1]), the scientific community has begun to evaluate their performance against advanced classical heuristics. T. Albash et al. [2] showed that the scaling advantage of QA could outperform well-known classical heuristics such as simulated annealing. Several studies on specific Ising models, such as Spin-Glass and Sherrington-Kirkpatrick models, have shown that quantum annealers could perform better than classical methods on specific cases [12,19].

Quantum circuit performance evaluation started with randomized benchmarking methods of single-qubit gates circuits. This protocol, presented by E. Knill in [13], was then extended to multi-qubits gates circuits in [17]. Both protocols are scalable as they are strictly based on circuits only using Clifford gates, producing an output distribution that can be known efficiently with a classical computer.

Other studies have tried to define a set of metrics to measure the potential of quantum circuits. The Quantum Volume [6] evaluates the maximum size of a square circuit that can run reliably on a given quantum chip. The Volumetric Benchmark [5] extends this method to rectangle circuits. Both metrics provide insights about the volume (width and depth) that can run reliably on a chip. The precise and costly evaluation of the output distribution (based on Heavy Output Generation) classifies both metric use into *descriptive benchmark*. These metrics do not report on the fidelity nor quality of the output of specific application circuits and are not scalable for such uses.

A scalable competitive metric, called the Q-score, has recently emerged to evaluate solutions to the Max-Cut problem [18]. This metric is the first attempt to design a hardware-independent way to measure quantum performance. Fellous et al. [9] introduced a methodology named Metric-Noise-Resource (MNR) to evaluate the ratio between energy consumption and quality of the solution provided by an error-corrected quantum computer. MNR is the first methodology to estimate energy consumption with the launching of the Quantum Energy Initiative.

Finally, several frameworks have been developed to benchmark applications, such as the QEC-D framework [16] and the QASM Bench [14]. These frameworks provide sets of metrics to test applications but are still dedicated to perform *descriptive benchmark* of small instances by studying the fidelity of the output distribution of the circuit.

2 TAQOS Benchmark Protocol

The TAQOS protocol aims to establish a fair benchmarking protocol to compare quantum algorithms, such as AQO and the QAOA, with classical algorithms. Figure 1 shows the workflow of the two quantum heuristics. Each dotted box is an abstract description of a computational task. One can perform a factorial study by testing several implementations of a single dotted box and letting the rest of the workflow unchanged. Each of the two quantum heuristics exhibits at least one critical task proven NP-Hard: the Quadratic Unconstrained Binary Optimization (QUBO) problem mapping for AQO (task #2) and the transpilation of the circuit to the hardware topology for the QAOA (task #8). The methods used to select optimized hyper-parameters (task #3, #7, #8 and #10) for the execution of quantum algorithms should be specified. For example, the

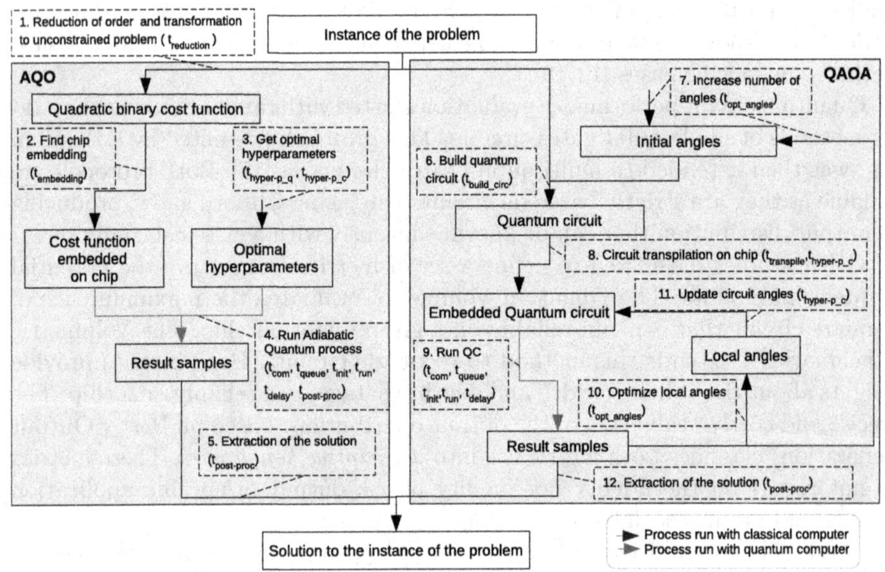

Fig. 1. Workflow of quantum optimization methods: AQO and the QAOA. Each dotted line box defines a processing action with run time variables t_x involved.

selection of the chain strength and the unembedding method should be documented for experiments on AQO using D-Wave systems. For the QAOA, the local and global optimization procedure to get an appropriate set of angles should be detailed with processing time spent and termination conditions. For the fairness of benchmark studies, each experiment should analyze the computation time corresponding to each dotted box.

2.1 Metrics

Competitive benchmark metrics must be scalable, hardware-independent, and efficiently computed. In addition, the metrics must be comparable to the results obtained by state-of-the-art optimization methods. The TAQOS protocol considers three different quantities to evaluate the performance of a quantum solver: the quality of the result, the time to get the solution, and the energy spent by the computer to get the result. A fourth metric evaluates the robustness of the quantum heuristic and computes the coverage of the set of instances.

Let $\mathcal{P} = \{P_1, P_2, ..., P_{\mathrm{N}}\}$ be a set of combinatorial problems and $\mathcal{I}$ the set of all possible instances associated to the problem P_n. Each instance $I_i \in \mathcal{I}$ has a set of solutions S. Let S_{q} be the subset of solutions found by a quantum computer. The objective function c evaluates the quality of a solution s. Considering a maximization problem, the best solution obtained by a quantum computer has the cost $c_{\mathrm{q}}^* = \max_{s_i \in S_{\mathrm{q}}} c(s_i)$ and is denoted s_{q}^*. Let $c_{\mathrm{c}}*$ be the cost associated with the best solution obtained by a classical heuristic and c_{ref} the cost of reference (for example, the best-known solution to a specific problem). The function r_{ref} evaluates the quality of a solution s_i as a ratio of the reference cost:

$$r_{\mathrm{ref}}(c(s_i), c_{\mathrm{ref}}) = \frac{c_{\mathrm{ref}} - c(s_i)}{c_{\mathrm{ref}}} \tag{1}$$

A negative ratio implies that the quality of the solution s_i is better than the solution of reference. Let the wall clock time associated with the quantum (classical) computation be t_{q} (t_{c}) and the energy consumption be e_{q} (e_{c}). The following set of inequalities defines a definitive quantum advantage over classical computation for a specific instance:

$$\begin{aligned} r_{\mathrm{ref}}(c_{\mathrm{q}}^*, c_{\mathrm{ref}}) &\leq r_{\mathrm{ref}}(c_{\mathrm{c}}^*, c_{\mathrm{ref}}) \\ t_{\mathrm{q}}^* &\leq t_{\mathrm{c}}^* \\ e_{\mathrm{q}}^* &\leq e_{\mathrm{c}}^* \end{aligned} \tag{2}$$

Quality Metric. The benchmark of NP-Hard problems requires the definition of an efficiently calculated quality metric. We opt to measure the quality of the solution following the recommendations of R. S. Barr et al. [3], taking the Best-Known Solution (BKS) as the solution of reference of cost c_{ref}. $r_\epsilon(\mathcal{I})$ computes the fraction of instances for which the ratio r_{ref} is less than ϵ. We define slices with $\epsilon \in [0, 1]$, e.g., within 1, 5 or 10% to optimality, to detect sets of instances amenable to produce close-to-optimal results:

$$r_\epsilon(\mathcal{I}) = \frac{|\{I_i \in \mathcal{I} \text{ with } r_{\text{ref}}(c_{\text{q}}*, c_{\text{ref}}) < \epsilon\}|}{|\mathcal{I}|} \tag{3}$$

$r_0(\mathcal{I})$ outputs the ratio of solutions that are better than the solution of reference.

Wall Clock Time Metric. Wall clock time metric should include the whole processing time from the problem formulation to the solution extraction (see Fig. 1). The device setting time of hyper-parameter that require quantum (classical) processing is denoted $t_{\text{hyper}-\text{p}_\text{q}}$ ($t_{\text{hyper}-\text{p}_\text{c}}$). The wall clock processing time of AQO is defined in Eq. 4.

$$\begin{aligned} t_{\text{quantum}} &= (t_{\text{init}} + t_{\text{run}} + t_{\text{delay}}) \times nb_{\text{shots}} + t_{\text{hyper}-\text{p}_\text{q}} \\ t_{\text{classical}} &= t_{\text{reduction}} + t_{\text{embedding}} + t_{\text{hyper}-\text{p}_\text{c}} + t_{\text{com}} + t_{\text{queue}} + t_{\text{post}-\text{proc}} \\ t_{\text{AQO\_wall\_clock}} &= t_{\text{quantum}} + t_{\text{classical}} \end{aligned} \tag{4}$$

The wall clock processing time of the QAOA with local and global optimization of angles can be specified as:

$$\begin{aligned} t_{\text{local\_quantum}} &= (t_{\text{init}} + t_{\text{run}} + t_{\text{delay}}) \times nb_{\text{shots}} \times nb_{\text{local\_opt}} \\ t_{\text{local\_classical}} &= nb_{\text{local\_opt}} \times (t_{\text{com}} + t_{\text{queue}} + t_{\text{hyper}-\text{p}_\text{c}} + t_{\text{opt\_angles}}) \\ t_{\text{quantum}} &= nb_{\text{global\_opt}} \times t_{\text{local\_quantum}} \\ t_{\text{classical}} &= nb_{\text{global\_opt}} \times (t_{\text{build\_circ}} + t_{\text{compile}} + t_{\text{local\_classical}}) + t_{\text{post}-\text{proc}} \\ t_{\text{QAOA\_wall\_clock}} &= t_{\text{quantum}} + t_{\text{classical}} \end{aligned} \tag{5}$$

Energy Consumption Metric. Quantum computers are deemed less energy-consuming than supercomputers. However, their power consumption is presently not disclosed with enough precision by quantum hardware manufacturers. At this stage, we therefore let energy consumption metrics as perspectives.

Coverage metric. The last metric evaluates the coverage of the set $\mathcal{I}$. Classical studies based on Algorithm Selection Problem [21] demonstrated an existing link between the instance structure and the relative performance of specific heuristics [7,22]. Combined with the quality metric, the coverage metric evaluates the robustness of the heuristic. We follow the work of I. Dunning [7] and compute a set of metrics specific to one optimization problem (e.g., the density of an instance for a problem based on graphs). The coverage c_ϵ of a metric f is an interval at fixed ϵ:

$$c_\epsilon(f, I_i) = [f(I_i) - \epsilon, f(I_i) + \epsilon] \cap [0, 1] \tag{6}$$

The whole coverage of a metric f on a set of instances $\mathcal{I}$ is:

$$C_\epsilon(f, \mathcal{I}) = \bigcup_{I_i \in \mathcal{I}} c_\epsilon(f, I_i) \tag{7}$$

These four metrics define the building blocks of the TAQOS protocol. An illustration of their use is presented in the next section.

2.2 Use Case on the Max-Cut Problem

This section presents a case study of the TAQOS protocol on the Max-Cut problem. The Max-Cut formulation is very close to the Ising model problem, easily mapped on existing qubit interconnects. Moreover, the classical community has studied this problem well, with several open-source implementations of heuristics (e.g., the MQLib [7]). Let $\mathcal{G} \stackrel{\text{def}}{=} (\mathcal{V}, \mathcal{E})$ the graph with a set of vertices $\mathcal{V}$ and a set of edges $\mathcal{E}$. The maximum cut of a graph is the partition of its vertices into two subsets S and T such that the number of edges shared by S and T is maximum. The cost function to be maximized is $C(\mathcal{G}) = -\sum_{i,j \in \mathcal{E}} \omega_{ij} s_i s_j$ with $s_i, s_j = \pm 1$. The problem is turned into a minimization problem by changing the sign of ω_{ij}.

Our instances of the Max-Cut problem are generated from the topology of four D-Wave systems. Random ω_{ij} coefficients are drawn from the set $\{+1, -1\}$ with same probability. Each instance is strongly favorable to D-Wave systems as it perfectly maps the topology of the quantum chip. However, the generated Ising Spin-glass problem is still hard to solve for classical heuristics. Results are presented in Table 1. The benchmark is done on 30 instances for each graph, considering D-Wave solutions as reference solutions. The annealing time is set to 100 µs and the sampling is done over 256 shots (nb_{shots}). We did not tune the gauge

Table 1. Performance comparison between quantum and classical algorithms used to solve the Max-Cut problem on four different graphs tailored for D-Wave's quantum chips topology. Results are averaged over 30 instances for each graph. Green cells underline best classical runs for each time frame: {1, 10, 100}.

Quantum solvers	Wall clock time (s)	Chimera graph DW_2000Q $\lvert\mathcal{V}\rvert : 2041$ $\lvert\mathcal{E}\rvert : 5974$				Pegasus graph Adv4.1 $\lvert\mathcal{V}\rvert : 5621$ $\lvert\mathcal{E}\rvert : 40279$				Pegasus graph Adv6.1 $\lvert\mathcal{V}\rvert : 5616$ $\lvert\mathcal{E}\rvert : 40135$				Zephyr graph Adv2 $\lvert\mathcal{V}\rvert : 563$ $\lvert\mathcal{E}\rvert : 4790$			
DW2000Q	1.43	c_{ref}				/				/				/			
Adv4.1	2.90	/				c_{ref}				/				/			
Adv6.1	2.88	/				/				c_{ref}				/			
Adv2	1.18	/				/				/				c_{ref}			
Classical		r_0	$r_{0.01}$	$r_{0.05}$	$r_{0.1}$	r_0	$r_{0.01}$	$r_{0.05}$	$r_{0.1}$	r_0	$r_{0.01}$	$r_{0.05}$	$r_{0.1}$	r_0	$r_{0.01}$	$r_{0.05}$	$r_{0.1}$
Random	1	0	0	0	0	0	0	0	0	0	0	0	0	0	0	0	0.03
DUARTE	1	0	0	0.97	1	0	0	0.9	1	0	0	0.7	1	0.1	0.3	1	1
2005	10	0	0	1	1	0	0	1	1	0	0	1	1	0.23	0.47	1	1
	100	0	0	1	1	0	0	1	1	0	0	1	1	0.4	0.6	1	1
FESTA	1	0	0	0	1	0	0	0	0.37	0	0	0	0.53	0	0.17	0.93	1
2002	10	0	0	0.9	1	0	0	0	1	0	0	0	0.9	0.3	0.5	1	1
GPR	100	0	0	1	1	0	0	0	1	0	0	0	1	0.3	0.53	1	1
FESTA	1	0	0	0	1	0	0	0.03	1	0	0	0	1	0	0.07	0.97	1
2002	10	0	0	0	1	0	0	0.03	1	0	0	0.07	1	0.1	0.3	1	1
GVNS	100	0	0	0	1	0	0	0.33	1	0	0	0.23	1	0.17	0.5	1	1
FESTA	1	0	0	0.03	1	0	0	0.03	1	0	0	0	1	0.07	0.27	1	1
2002	10	0	0	1	1	0	0	0.17	1	0	0	0.2	1	0.3	0.6	1	1
GVNSPR	100	0	0	1	1	0	0	0.5	1	0	0	0.37	1	0.4	0.67	1	1

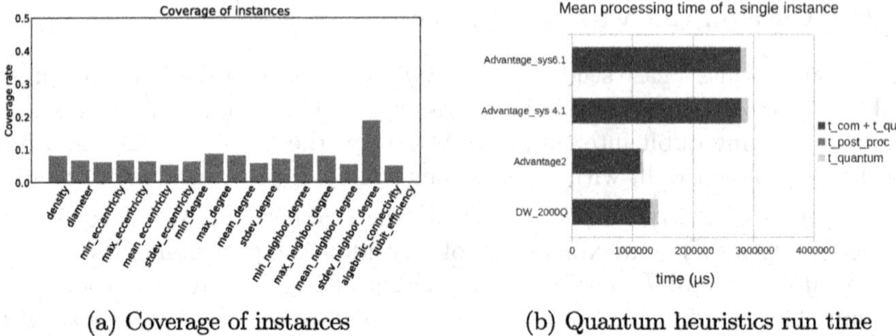

(a) Coverage of instances (b) Quantum heuristics run time

Fig. 2. (a) Shows the coverage rate of the set of evaluated instances. Coverage rates are computed from normalized graph metrics such as density, diameter, eccentricity, etc. The last metric measures the qubit mapping efficiency. The minimum, maximum, mean and standard deviation are available to study the distribution of these metrics. The coverage rate corresponds to the total length of intervals in $C_{0.05}(f, I)$. (b) Details the run time of quantum heuristics. Each time is averaged over the 30 instances. $t_{\text{reduction}}$, $t_{\text{embedding}}$ and $t_{\text{hyper-p}_\text{q}}$ are set to 0 as the study does not require any of the corresponding computational task.

inversion or pausing times. The D-Wave's performance is compared with algorithms from the MQLib [7] that constitute state-of-the-art methods used to solve the Max-Cut problem. Each classical algorithm is run over three time periods (1 s, 10 s, 100 s) on a single processor Intel® Core™ i7-6600U 2.6GHz. The metric c_ϵ is measured for $\epsilon \in \{0, 0.01, 0.05, 0.1\}$. For large graphs (i.e., Chimera and Pegasus) D-Wave annealers constantly outperform classical heuristics, even with less run time. The classical heuristics perform well on smaller graph (Zephyr) and outperform some reference solutions found by D-Wave, even with less run time. However, the competitive performance of the D-Wave systems must be interpreted considering the coverage rate of tested instances, shown in the Fig. 2a. These four graphs cover a very small range of graph-specific coverage metrics (less than 10% for almost every metric with $\epsilon = 0.05$). The run time of D-Wave systems is low because the set of instances, owing to their topology, avoids time-consuming operations such as reduction, embedding and hyper-parameter settings (see Fig. 2b). This use case shows the importance of being transparent about experiments done on quantum devices. The topology of instances strongly impacts the quality of the results returned by quantum devices. The coverage metric quantifies its robustness and can be used to identify classes of instances producing high-quality results on quantum devices.

3 Conclusion

This paper has introduced the TAQOS benchmark methodology, which fairly compares classical and quantum heuristics performance. TAQOS is a scalable framework of metrics that analyses the trade-offs between quality and robustness. It constitutes a competitive methodology to benchmark hybrid algorithms such as AQO and the QAOA. It uses field-proven metrics to compare quantum to classical results obtained with existing benchmark methodologies. This paper illustrated the application of the TAQOS protocol on the Max-Cut problem in a favorable context for D-Wave systems and showed that performance reports should consider instances set coverage to avoid misleading conclusions. The use case illustrates the ability of TAQOS to gauge the fairness of quantum optimization experiments. In particular, this allows us to separate the experiments favorable to some quantum hardware from the more generic experiments that would manifest a real and robust quantum advantage. Future studies will be done on other optimization problems (especially Higher Order Binary Optimization problems and the TSP). This future work will provide insight into problems and instance properties that might benefit from a quantum advantage.

References

1. D-wave systems, d-wave ocean sdk, release 6.0.1 (2022). https://github.com/dwavesystems/dwave-ocean-sdk
2. Albash, T., Lidar, D.A.: Demonstration of a scaling advantage for a quantum annealer over simulated annealing. Phys. Rev. X **8**(3), 031016 (2018)
3. Barr, R.S., Golden, B.L., Kelly, J.P., Resende, M.G.C., Stewart, W.R.: Designing and reporting on computational experiments with heuristic methods. J. Heuristics **1**(1), 9–32 (1995)
4. Basso, J., et al.: The quantum approximate optimization algorithm at high depth for maxcut on large-girth regular graphs and the sherrington-kirkpatrick model. arXiv preprint arXiv:2110.14206 (2021)
5. Blume-Kohout, R., Young, K.C.: A volumetric framework for quantum computer benchmarks. Quantum **4**, 362 (2020)
6. Cross, A.W., Bishop, L.S., Sheldon, S., Nation, P.D., Gambetta, J.M.: Validating quantum computers using randomized model circuits. Phys. Rev. A **100**, 032328 (2019)
7. Dunning, I., Gupta, S., Silberholz, J.: What works best when? a systematic evaluation of heuristics for max-cut and QUBO. INFORMS J. Comput. **30**(3) (2018)
8. Farhi, E., Goldstone, J., Gutmann, S.: A quantum approximate optimization algorithm. arXiv preprint arXiv:1411.4028 (2014)
9. Fellous Asiani, M., et al.: Optimizing resource efficiencies for scalable full-stack quantum computers. arXiv preprint arXiv:2209.05469 (2022)
10. Gilbert, V.: Taqos (2023). https://github.com/CEA-LIST/Quantum-Benchmark-CEA-LIST/tree/main/TAQOS
11. Harrigan, M.P., et al.: Quantum approximate optimization of non-planar graph problems on a planar superconducting processor. Nat. Phys. **17**(3), 332–336 (2021)
12. Hen, I., Job, J., Albash, T., Rønnow, T.F., et al.: Probing for quantum speedup in spin-glass problems with planted solutions. Phys. Rev. A **92**, 042325 (2015)
13. Knill, E., et al.: Randomized benchmarking of quantum gates. Phys. Rev. A **77**, 012307 (2008)
14. Li, A., Stein, S., Krishnamoorthy, S., Ang, J.: Qasmbench: A low-level qasm benchmark suite for nisq evaluation and simulation. arXiv preprint arXiv:2005.13018 (2020)

15. Lilja, D.J.: Measuring computer performance: a practitioner's guide. Cambridge University Press (2005)
16. Lubinski, T., Coffrin, C., McGeoch, C., Sathe, et al.: Optimization applications as quantum performance benchmarks. arXiv preprint arXiv:2302.02278 (2023)
17. Magesan, E., Gambetta, J.M., Emerson, J.: Scalable and robust randomized benchmarking of quantum processes. Phys. Rev. Lett. **106**, 180504 (2011)
18. Martiel, S., Ayral, T., Allouche, C.: Benchmarking quantum coprocessors in an application-centric, hardware-agnostic, and scalable way. IEEE Trans. Quantum Eng. **2**, 1–11 (2021)
19. Oshiyama, H., Ohzeki, M.: Benchmark of quantum-inspired heuristic solvers for quadratic unconstrained binary optimization. Scient. Reports **12**(1) (2022)
20. Preskill, J.: Quantum computing in the nisq era and beyond. Quantum **2**, 79 (2018)
21. Rice, J.R.: The algorithm selection problem. In: Advances in Computers, pp. 65–118. Elsevier (1976)
22. Smith-Miles, K., Baatar, D., Wreford, B., Lewis, R.: Towards objective measures of algorithm performance across instance space. Comput. Oper. Res. **45**, 12–24 (2014)
23. Xue, C., Chen, Z.Y., Wu, Y.C., Guo, G.P.: Effects of quantum noise on quantum approximate optimization algorithm. Chin. Phys. Lett. **38**(3), 030302 (2021)

Enabling Non-linear Quantum Operations Through Variational Quantum Splines

Matteo Antonio Inajetovic[1], Filippo Orazi[1(✉)], Antonio Macaluso[2], Stefano Lodi[1], and Claudio Sartori[1]

[1] University of Bologna, Bologna, Italy
matteo.inajetovic@studio.unibo.it,
{filippo.orazi2,stefano.lodi,claudio.sartori}@unibo.it
[2] German Research Center for Artificial Intelligence (DFKI), Saarbruecken, Germany
antonio.macaluso@dfki.de

Abstract. One of the major issues for building a complete quantum neural network is the implementation of non-linear activation functions in a quantum computer. In fact, the postulates of quantum mechanics impose only unitary transformations on quantum states, which is a severe limitation for quantum machine learning algorithms. Recently, the idea of QSplines has been proposed to approximate non-linear quantum activation functions by means of the HHL. However, QSplines rely on a problem formulation to be represented as a block diagonal matrix and need a fault-tolerant quantum computer to be correctly implemented.

This work proposes two novel methods for approximating non-linear quantum activation functions using variational quantum algorithms. Firstly, we develop the variational QSplines (VQSplines) that allow overcoming the highly demanding requirements of the original QSplines and approximating non-linear functions using near-term quantum computers. Secondly, we propose a novel formulation for QSplines, the Generalized QSplines (GQSplines), which provide a more flexible representation of the problem and are suitable to be embedded in existing quantum neural network architectures. As a third meaningful contribution, we implement VQSplines and GQSplines using Pennylane to show the effectiveness of the proposed approaches in approximating typical non-linear activation functions in a quantum computer.

Keywords: Quantum Machine Learning · Quantum Neural Networks · Quantum Computing

1 Introduction

Quantum computers are machines that leverage the properties of quantum mechanics to store and process information. Although a potential quantum advantage has already been shown in different domains, such as quantum chemistry [1], multi-agent systems [2,3], it is still unclear whether quantum computation can be used efficiently in machine learning (ML).

J. Mikyška et al. (Eds.): ICCS 2023, LNCS 14077, pp. 177–192, 2023.
https://doi.org/10.1007/978-3-031-36030-5_14

The majority of the approaches proposed in the field of Quantum Machine Learning (QML) relies on using hybrid quantum-classical optimization to train parameterized quantum circuits to perform typical ML tasks. Although these techniques represent the most promising attempt to leverage near-term quantum technology, it is still unclear whether they can outperform classical algorithms.

One class of QML algorithms gaining momentum in recent years is Quantum Neural Networks (QNN) which try to emulate the behavior of classical neural networks using parametrized quantum circuits. The key feature of classical neural networks is the ability to capture complex patterns by applying multiple non-linear activation functions. On the other side, QNN models utilize quantum kernels to explicitly map data into a high-dimensional space and learn the complex patterns in data. Although this approach has shown promising results, it is not a credible alternative capable of providing a robust quantum advantage over classical methods. An alternative strategy to unlock the full potential of quantum computing in ML is to implement classical neural networks using the properties of quantum computing as computational resources to obtain a robust speed-up. However, the postulates of quantum mechanics forbid non-unitary operations on quantum states, which is a strong limitation for QML algorithms. Thus, the ability to approximate non-linear activation functions in a quantum computer is essential to obtain a quantum advantage of QML algorithms over their classical counterpart. In this paper, we propose two novel approaches for quantum activation functions based on QSplines [4], which leverage hybrid quantum-classical optimization and are suitable for near-term quantum computation.

The remainder of the paper is structured as follows: a brief discussion of related work is provided in Sect. 2. Section 3 gives a highlight of the contribution of the paper while Sect. 4 contains the detailed methodological approach. In Sect. 5 the results of the experiments are presented and then discussed in Sect. 6. The paper concludes with a summary of achievements and future work.

2 Related Works

Recently, several attempts have been made to provide a routine for quantum activation functions and overcome the constraint of the unitarity of quantum operations. The quantum Splines (QSplines) [4] rely on classical B-Spline regression models that aim to find the optimal set of parameters to minimize the Residual Sum of Squares in a ridge regression problem where observed variables are augmented with polynomials. However, the QSplines suffer from several drawbacks and limitations: the use of the HHL [5] as a subroutine imposes running the algorithm on a perfectly error-corrected quantum computer, and it is not suitable to be executed on near-term quantum devices. Moreover, the output of the QSplines requires a post-processing step to obtain the value of the non-linear function which is stored in the quantum state. Furthermore, QSplines require ad-hoc formulation for the basis expansion matrix in terms of a diagonal block matrix that is hardly generalizable.

Other works on quantum activation functions rely on repeat-until-success technique [6]. In this case, the most significant limitation is that the input must be in the range $[0, \frac{\pi}{2}]$, which is a severe constraint for real-world problems. Recently, the problem of non-linear approximation has been considered by means of a Quantum Perceptron [7]. The proposed quantum algorithm produces the output of a non-linear activation function leveraging an iterative computation of all the powers of the inner product up to an order d. The main drawback of this approach is the number of qubits required, which depends linearly on the number of input features and the order of the polynomial. in particular, given an n-dimensional feature vector and a degree of the polynomial d, the idea of the Quantum Perceptron requires $n + d$ qubits. Experimentally, this approach shows good results with a degree of polynomial $3 \leq d \leq 10$ for 1 dimensional feature vector. Finally, parametrized quantum circuits have been used to achieve non-linear approximation by means of hybrid quantum-classical optimization [8]. However, this approach relies on utilizing multiple copies of variational quantum states to treat nonlinearities, and its use in the context of quantum neural networks is unclear. Furthermore, the experiments approximate Schrödinger equation only and do not address the estimation of typical activation functions.

3 Contribution

This work proposes two alternative approaches for quantum activation functions: Variational QSpline (VQsplines) and Generalized Quantum Splines (GQSplines). The VQSplines adopt the problem formulation of the QSplines [4], translating it in the context of hybrid quantum-classical computation using the VQLS as a quantum routine for matrix inversion and uses the quantum dot product to calculate the value of the non-linear activation function. The advantage of this approach is twofold: on one side, the use of the HHL is avoided allowing the VQSplines to be executed using near-term quantum technology. On the other side, the use quantum dot product provides a quantum state encoding the value of the activation function without requiring a post-processing step.

The second main methodological contribution is a novel formulation for QSplines, the GQSpline, that relies on a more flexible problem formulation in terms of basis expansion and allows obtaining an end-to-end quantum routine that can approximate any non-linear activation function. Importantly, the GQSplines can be adopted as a sub-routine in existing quantum neural networks encoding into the amplitudes of a quantum states the value of the non-linear function.

Importantly, the proposed algorithms encode data (basis expansion of the input features and the value of the non-linear activation function) into the amplitudes of quantum states. This means that the qubit complexity (i.e., the number of qubits required) scales logarithmically with respect to the input size which is a significant improvement with respect to existing approaches whose number of qubits is extremely demanding.

4 Methods

In this section, the methodological contributions are presented. We describe two different approaches for quantum splines: the *VQSplines* adopts the piece-wise formulation of QSplines but replaces the HHL with the Variational Quantum Linear Solver (VQLS) [9]. As a consequence, we obtain a quantum algorithm suitable for NISQ devices capable of estimating any non-linear function. Second, we propose the *GQSplines* model, a novel formulation of QSplines that allows us to obtain an end-to-end quantum algorithm for non-linear approximation suitable for existing quantum neural networks.

4.1 Preliminaries

Quantum Splines
Spline functions are smoothing methods for modeling the relationships between variables, typically adopted either as a visual aid in data exploration or for estimation purposes [10]. The underlying idea is to use linear models in which the input features are augmented with the *basis expansions*. Technically, splines are constructed by dividing the sample data into sub-intervals delimited by break-points, also referred to as knots. A fixed degree polynomial is then fitted in each of the segments, thus resulting in piecewise polynomial regression.

While the formulation in terms of truncated basis functions is conceptually simple, its numerical and computational properties are not very attractive. For this reason, in practice, the *B-splines* parametrization [11] is adopted. This generates a block design matrix where the *sparsity* is constant and depends on the degree of the polynomial fitted in each local interval. Given a sequence of knots $\xi_1, \xi_2, \cdots, \xi_T$, we fit a line in each interval $[\xi_k, \xi_{k+1}]_{k=1,\cdots,T-1}$ without derivability constraints.

$$\tilde{\boldsymbol{y}} = \boldsymbol{S}\boldsymbol{\beta} \rightarrow \begin{pmatrix} \tilde{y}_1 \\ \tilde{y}_2 \\ \cdots \\ \tilde{y}_K \end{pmatrix} = \begin{pmatrix} S_1 & 0 & \cdots & 0 \\ 0 & S_2 & \cdots & 0 \\ \cdots & \cdots & \cdots & \cdots \\ 0 & 0 & \cdots & S_K \end{pmatrix} \begin{pmatrix} \beta_1 \\ \beta_2 \\ \cdots \\ \beta_K \end{pmatrix}, \tag{1}$$

where $\tilde{y_k}$ contains the function evaluations in ξ_k and ξ_{k+1}, β_ks are the spline coefficients and $\boldsymbol{S}_{(2K)\times(2K)}$ is a block diagonal matrix with each block S_k that represents the basis expansions in the k-th interval. Therefore, solving the linear system in Eq. (1) allows computing the splines coefficients, which serve to approximate non-linear functions encoded in the vector $\tilde{\boldsymbol{y}}$.

The idea of the Quantum Splines (*QSplines*) [4] is to adopt the B-spline formulation in the context of quantum computation to approximate non-linear functions. In particular, the computation of the QSplines is performed in three steps. First, the HHL computes the spline coefficients for the k-th interval encoded into the quantum state $|\beta_k\rangle$. Second, $|\beta_k\rangle$ interacts with the quantum state $|x_k\rangle$ encoding the input in the k-th interval via quantum interference through the swap-test [12]. This allows generating the state $|f_k\rangle$ which encodes the estimate of the non-linear function evaluated in x_k; Third, $|f_k\rangle$ is measured and post-processed to obtain y_k.

Although QSplines allow overcoming the limitation of unitary operation on quantum states, their applicability is very limited since the use of the HHL as a subroutine requires a massive number of error-corrected qubits to be executed. Furthermore, in the current formulation, the final quantum state is obtained using the swap test, whose results are not directly encoded in the amplitude and need a post-processing step to be calculated. All these factors forbid the adoption of QSplines in current models of quantum neural networks, which run on a limited set of noisy qubits.

Variational Quantum Linear Solver
An alternative quantum algorithm to solve a linear system of equations is the Variational Quantum Linear Solver (*VQLS*) [9]. Specifically, the VQLS solves a linear system of equations using a variational hybrid quantum-classical approach which is suitable for near-term quantum devices. Given a matrix A and a state vector $|b\rangle$, the VQLS prepares the state $|x\rangle$ such that:

$$A|0\rangle = A|x\rangle \propto |b\rangle. \tag{2}$$

The matrix A is defined as a linear combination of unitary matrices A_l:

$$A = \sum_{l=0}^{L} A_l c_l \tag{3}$$

where c_l are complex numbers. With this input, the VQLS generates the state $|x\rangle$ employing an ansatz for the gate sequence $V(\theta)$ such that $|x(\theta)\rangle = V(\theta)|0\rangle$. The parameters θ are input to a quantum computer, which prepares $|\theta\rangle$ and runs an efficient quantum circuit that estimates a cost function $C(\theta)$. The value of $C(\theta)$ from the quantum computer is returned to the classical computer which then adjusts θ (via a classical optimization algorithm) in an attempt to reduce the cost. This process is iterated many times until one reaches a termination condition of form $C(\theta) < \gamma$, at which point we say that $\theta = \theta_{\text{opt}}$.

Once the cost function is minimized, the Ansatz $V(\theta_{opt})$ with the optimal parameters θ_{opt} prepares the normalized state $|x(\theta_{\text{opt}})\rangle$ which approximates $|x\rangle$:

$$V(\theta_{\text{opt}})|0\rangle = |x(\theta_{\text{opt}})\rangle. \tag{4}$$

Notice that this description of the problem is coherent with the original VQLS. However, in the case of QSplines, the linear system of equation in Eq.(1) represents the target solution (the B-spline coefficients) as $|\beta\rangle$, the A_l circuit encodes the matrix S, and $|y\rangle$ replaces $|b\rangle$.

4.2 VQSplines: Variational Algorithm for QSplines

The idea behind the *VQSplines* is to implement the QSplines in the context of hybrid quantum-classical computation and provide an efficient method to estimate non-linear functions using near-term quantum computers. In particular, the VQSplines replace the HHL with the VQLS and adopt the quantum inner product [13] to generate a quantum state encoding the value of the non-linear target function, avoiding the use of the swap-test.

In practice, the computation of the VQSplines is performed in two steps. First, the VQLS estimates the spline coefficients for the k-th interval:

$$S_k|\beta'_k\rangle = |y'_k\rangle \xRightarrow{VQLS} |\beta'_k\rangle = \beta'_{k,0}|0\rangle + \beta'_{k,1}|1\rangle \approx S_k^{-1}|y'_k\rangle \tag{5}$$

In order to obtain the quantum state $|\beta'_k\rangle$, each 2×2 linear system requires the training of a parametrized quantum circuit which needs three different quantum gates: a unitary $V(\theta_k)$ implemented with a Pauli rotation gate R_y that allows generating a quantum state whose amplitudes encode the B-spline coefficients, once the optimal set of parameters $\theta_{k,\text{opt}}$ is obtained. Notice that, as for the QSplines, the entire problem is decomposed into K 2×2 linear systems, and each set of coefficients can be encoded using a single qubit; the second set of quantum gates are the unitaries A_l, which encode the information related to the S_k matrix. In particular, the matrix S_k is decomposed by means of a linear combination of known unitary matrices as follows:

$$S_k = \begin{bmatrix} 1 & a \\ 1 & b \end{bmatrix} = \sum_{l=0}^{3} A_l c_l = Ic_0 + Xc_1 + Zc_2 + R_y(3\pi)c_3 \tag{6}$$

where the coefficients of the linear combination are initialized as follows:

$$c_0 = (b+1)/2; \quad c_1 = (a+1)/2; \quad c_2 = (1-b)/2; \quad c_3 = (a-1)/2;$$

Third, the unitary U encodes the classical vector y_k into the amplitude of $|y'_k\rangle$ [14]. The QSplines model for the k-th interval is depicted in Fig. 1.

As a result of the optimization process, we obtain a variational circuit that creates a quantum state encoding the solution of the linear system:

$$V(\theta_{k,opt})|0\rangle = |\beta'_k\rangle = \beta'_{k,0}|0\rangle + \beta'_{k,1}|1\rangle \tag{7}$$

The second step of the VQSplines computes the inner product between the basis expansion of the input $|x'_k\rangle$ and the B-splines coefficients $|\beta'_k\rangle$. To this end, the quantum inner product [13] generates a quantum state whose amplitudes encode the value of the non-linear function:

$$B^\dagger V(\theta_{k,opt})|0\rangle_n = a_o|0\rangle_n + \sum_{i=1}^{N-1} a_i|i\rangle_n \tag{8}$$

$$a_0 = \langle 0|B^\dagger V(\theta_{k,opt})|0\rangle = \langle x'_k|BB^\dagger|\beta'_k\rangle = \langle x'_k||\beta'_k\rangle = \hat{y}_k \tag{9}$$

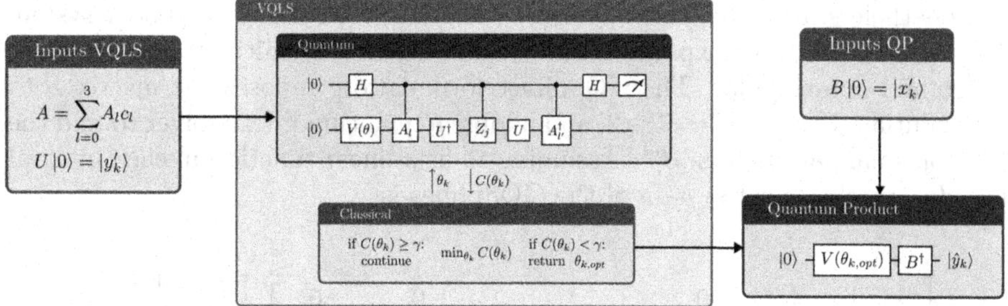

Fig. 1. The VQSplines architecture. The quantum part of the VQLS is optimized classically and then $\theta_{k,opt}$ is used in the quantum inner product to compute $|\hat{y}_k\rangle$

where B is the amplitude encoding quantum routine [14] of the basis expansion of the input x_k. The quantum circuit of the non-linear target function for the k-th interval is built as shown in Fig. 1 (*Quantum Inner product*).
These steps are repeated for each sub-interval and the solutions of the quantum system in Eq. (1) are obtained. Therefore, as for QSplines, once $|\beta'\rangle$ is known, the VQSplines require $\mathcal{O}(K)$ repetitions of the quantum inner product circuit to calculate a spline function with K knots.

Importantly, the use of the quantum inner product allows one to directly obtain the value of the target function as amplitudes of a quantum state and does not require any post-processing as for QSplines.

4.3 GQSplines: Generalized Quantum Splines

The original formulation of the QSplines and VQSplines relies on a block diagonal B-Spline matrix which allows decomposing the entire problem into K sub-problems that can be solved independently. However, this approach does not allow the estimation of a non-linear function in a full quantum manner, which is a requirement for embedding a quantum activation function into a quantum neural network. For this reason, we propose the *Generalized Quantum Splines* (GQSplines), a novel approach that relies on a more general formulation of the B-splines and allows estimating the value of a non-linear function with a single quantum circuit without problem decomposition.

Given the recursive definition of B-Spline [11] and the related basis expansion with knots list $\xi = [\xi_1, .., \xi_i, \xi_{i+1}, .., \xi_T]$, a non linear function f can be estimated using the observed values $Y = \{y_1, .., y_K\}$ given the inputs $X = \{x_1, .., x_K\}$. In this case, the linear system of equations describing the relation between the estimates of the activation function Y, the matrix S and the spline coefficients β is the following:

$$Y_{K\times 1} = S_{K\times K}\beta_{K\times 1} \implies \begin{bmatrix} y_1 \\ y_2 \\ \vdots \\ y_K \end{bmatrix} = \begin{bmatrix} B_{1,d}(x_1) & \dots & B_{l,d}(x_1) \\ B_{1,d}(x_2) & \dots & B_{l,d}(x_2) \\ \vdots & \ddots & \vdots \\ B_{1,d}(x_K) & \dots & B_{l,d}(x_K) \end{bmatrix} \begin{bmatrix} \beta_1 \\ \beta_2 \\ \vdots \\ \beta_K \end{bmatrix}, \qquad (10)$$

where d is the degree of the B-spline, T is number of knots and $l = T - d - 1$.

Nonetheless, to adopt the VQLS (or the HHL) for solving the linear system of equations, the basis expansion matrix S has to be hermitian and, therefore, square and non-singular. The GQSplines formulation imposes the matrix S to be Hermitian, i.e., $K = l = T - 2$, and adopts a quantum linear solver to find the set of optimal parameters $|\beta\rangle$[1]. Assuming to fit a linear function in each interval (i.e., $d = 1$), the linear system of the GQSplines is:

$$\begin{bmatrix} y_1 \\ y_2 \\ y_3 \\ y_4 \\ \vdots \\ y_{k-1} \\ y_K \end{bmatrix} = \begin{bmatrix} 1 & 0 & \dots & \dots & 0 & 0 \\ 0 & 1-x_2 & x_2 & \dots & \dots & 0 \\ \dots & 0 & 1-x_3 & x_3 & \dots & \dots \\ \dots & \dots & 0 & 1-x_4 & \dots & \dots \\ \dots & \dots & \dots & \dots & \dots & \dots \\ 0 & \dots & \dots & \dots & 1-x_{K-1} & x_{K-1} \\ 0 & 0 & \dots & \dots & 0 & 1 \end{bmatrix} \begin{bmatrix} \beta_1 \\ \beta_2 \\ \beta_3 \\ \beta_4 \\ \vdots \\ \beta_{K-1} \\ \beta_K \end{bmatrix}. \tag{11}$$

which leads to the following quantum linear system of equations:

$$S|\beta\rangle = |Y\rangle \tag{12}$$

where $|\beta\rangle$ and $|Y\rangle$ are two quantum states that encode in their amplitudes the vectors Y and β. Importantly, the normalization constraint of quantum states $|\beta\rangle$ and $|Y\rangle$ imposes rescaling the values of the target variable such that $\langle Y|Y\rangle = \langle\beta|\beta\rangle = 1$.

With a particular focus on the VQLS as a quantum linear solver, the size of the linear system in Eq. (12) is K which gives us the number of qubits required to find the optimal coefficients. Precisely, since the VQLS requires the vectors to be encoded as quantum state, the number of qubits scale logarithmically in the number of knots K, which is directly related to the quality of the fitting of the curve (the higher is K, the better the fitting is). This exponential scaling with respect to the number of inputs is a significant improvement when compared to other quantum approaches for quantum activation functions [7] whose number of qubits scales linearly with the size of the inputs.

Given the linear system in Eq. (10), the GQSplines proceed in two steps. Firstly, the coefficients $|\beta'\rangle$ are generated using the VQLS (or the HHL):

$$|\beta'\rangle = VQLS(S, Y) = \sum_{i=1}^{K} \beta_i |i\rangle. \tag{13}$$

Thus, quantum state $|\beta'\rangle$ interacts via interference with $|x'\rangle$ representing the basis expansion of x by means of the quantum inner product, as for VQSplines.

[1] It is also possible to define the Hermitian matrix H from S as $H = \begin{pmatrix} 0 & S \\ S^\dagger & 0 \end{pmatrix}$.

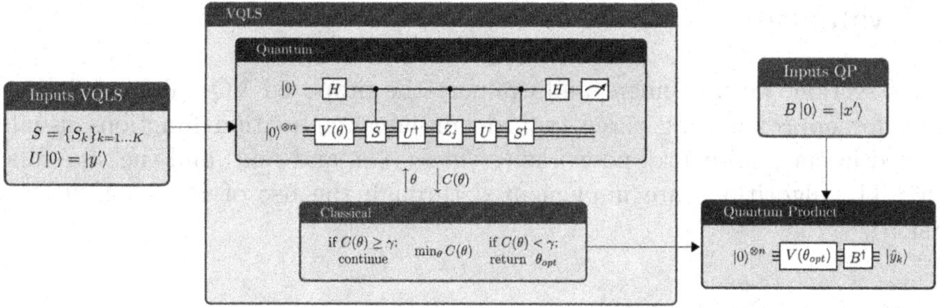

Fig. 2. GQSplines architecture. The quantum part of the VQLS is optimized classically to obtain θ_{opt} that is then used in the quantum inner product to compute $|\hat{y}_k\rangle$. All $y'_k \in y$ can be approximated with the results of a single optimization of the VQLS.

In the case of GQSplines, the VQLS circuit requires $n+1$ qubits, where $K = 2^n$ is the number of knots. Therefore, the Hadamard test [15] employs n qubits to implement the operators ($V(\theta)$,A and U) and one for the ancilla qubit. Furthermore, if $d = 1$, we end up with a diagonal block matrix which allows decomposing the matrix S (Eq. (3)) in terms of quantum gates in such a way to define the quantum circuit of the VQLS efficiently. Specifically, we can act independently on each interval to define the k-th matrix decomposition as follows:

$$S_k = \begin{bmatrix} 1-a & a \\ 0 & 1-b \end{bmatrix} = \sum_{l=0}^{3} A_l c_{k,l} = I c_{k,0} + X c_{k,1} + Z c_{k,2} + R_y(3\pi) c_{k,3} \quad (14)$$

where the coefficients of the linear combination are computed as:

$$c_0 = 1 - a/2 - b/2; \qquad c_1 = a/2; \qquad c_2 = (b-a)/2; \qquad c_3 = a/2$$

This method allows the S matrix to be efficiently decomposed and obtain the linear combination of quantum gates required for the VQLS ([9]). The operator U is realized by amplitude encoding state preparation [14]. Still, the normalization of Y is required. With this new formulation, we are able to solve a singular linear system and encode all the spline coefficients β in a unique quantum state by implementing the Ansatz only once. Subsequently, this state is used to compute the inner product with the k-th row of the matrix S (encoded through the routines B_i) and return the $\hat{y}_k$ estimates describing $\hat{Y}$. The workflow of GQSplines is depicted in Figure 2.

5 Evaluation

In this section, we implement and evaluate the proposed VQSplines and GQSplines for approximating three typical non-linear activation functions usually adopted in classical neural networks (*sigmoid*, *elu*, and *relu*) and the *sine* function[2]. The algorithms are implemented through the use of Pennylane[3](0.27.0 version).

5.1 Experimental Settings

Data Generation. We consider *sigmoid*, *elu*, and *relu* activation functions and the *sine* function to showcase the non-linearity of our models. Equation (15) shows the formulation of the functions tested for VQSplines, which is coherent with the experimental setting of original QSplines [4]:

$$\text{ReLu}(x) = max(0, x) \qquad \text{Elu}(x) = \begin{cases} x \text{ if } x > 0 \\ 0.3(e^x - 1) \text{ otherwise} \end{cases}$$
$$\text{Sigmoid}(x) = \frac{1}{1 + e^{-4x}} \qquad f_{\text{sin}}(x) = sin(\pi x) \tag{15}$$

In the case of GQSplines, the input features and the value of the non-linear activation function are encoded into the amplitude of a quantum state. In this respect, the same functions are normalized in the interval [0, 1]. Thus, for the VQSplines the approximation of *Elu*, *ReLu*, and *Sigmoid* is carried out according to the experiments of the original QSplines paper, while *sin* allows for comparison of other approaches [7].

Metrics. As discussed in the previous sections, different models are tested using different renormalizations. In order to perform a fair comparison between all methods, we consider the Normalized Root Mean Squared Errors (NRMSE) are defined as:

$$\mathbf{NRMSE} = \frac{\sqrt{N^{-1} \sum_{i=0}^{N} (\hat{y}_i - y_i)^2}}{y_{max} - y_{min}}. \tag{16}$$

The NMRSE allows for a robust comparison that investigates the quality of the fitting independently of the scale of the target variable and the number of points used to approximate it.

[2] Though the primary objective of QSplines is to embed non-linearity into quantum neural networks, they can serve to approximate other types of non-linearity.
[3] https://pennylane.ai/.

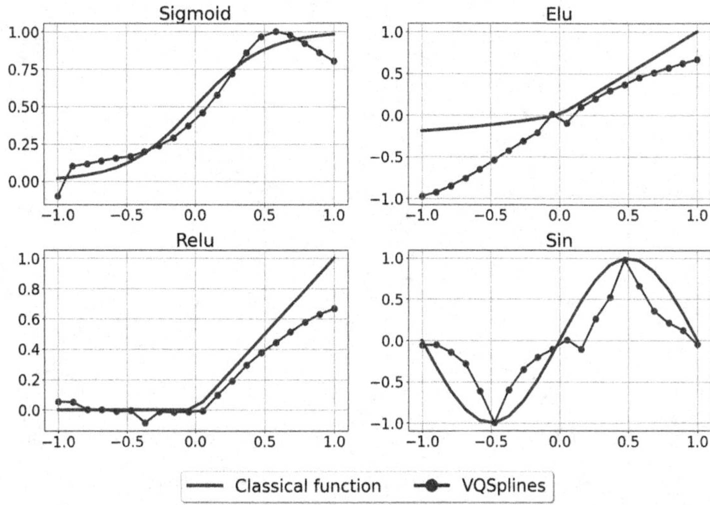

Fig. 3. VQSpline experimental results. The model is able to approximate the function and obtain non-linearity. Each point j is computed as a product between $|x_j\rangle$ and the spline coefficient $|\beta'_j\rangle$

5.2 Results

We test the VQSplines by training the VQLS and computing the quantum inner product to produce the final estimates $\hat{Y}$ for curve described in Sect. 5.1. The results are shown in Fig. 3.

We can see that in all the cases, the VQSplines can capture the non-linearity of the curves as expected. For *relu* and *sigmoid*, we have a good approximation, while the same cannot be observed for the *elu* and *sine*. In particular, the approximation of the curves deteriorates at the boundaries of the activation functions with input close to -1 or 1.

Considering the results of the GQSplines, the experiments are performed on the four functions normalized into the interval $[0, 1]$. For each curve, we show the fitting results in Fig. 4, using 16 knots and 4 qubits. In this case, the approximation quality seems to be sensitively better with respect to the VQSplines. However, while increasing the number of knots allows a better fitting, it implies using a larger quantum state which flatters the curve since the vectors $|Y\rangle$ and $|\beta\rangle$ are normalized to 1.

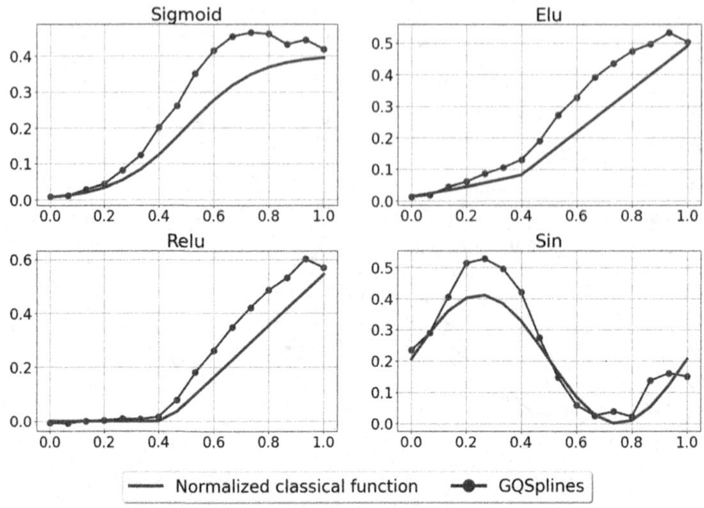

Fig. 4. GQSplines experimental results. The model is able to approximate and emulate the trend of all 4 normalized functions. Each point is computed as the product of the j-th row of S and the B-spline coefficients $|\beta'\rangle$.

We can observe that there is a slight tendency of the GQSplines to overestimate the target function. Nevertheless, the method can capture the non-linearity of the curves while approximating their, especially in the case of the activation function. These results are achieved using only 4 qubits which is a significant improvement with respect to the experiments of other proposed approaches in the literature.

In order to make a fair comparison between QSplines, VQSplines and GQSplines, we calculate the NRMSE for the four curves and report the results in Table 1. We can observe that both proposed methods outperform the original QSplines. Although VQSplines perform better in terms of NRMSE than GQSplines for the *relu* and *sigmoid*, the differences between the two methods are minimal. The same happens when looking at the approximation of *elu* and *sigmoid*, where the GQSplines perform better.

Table 1. NRMSE on each function for the proposed models and the baseline. The best approximation for each function is highlighted, and we can see that our model provides a considerable increase in performance with respect to QSplines.

Model	Knots	Elu	Relu	Sigmoid	Sin
Qsplines	20	0.4874	0.5240	0.1589	—
GQSpline	16	**0.0126**	0.0111	0.0156	**0.0099**
VQSpline	20	0.1278	**0.0069**	**0.0067**	0.0677

6 Discussion

GQSplines and VQSplines allow estimating the value of non-linear functions by means of parameterized quantum circuits. Both methods outperform the original proposal of QSplines in terms of fitting and require a significantly less number of qubits with respect to other existing approaches for quantum activation functions [6,7].

In the case of VQSplines, the quality of the estimates in each interval depends to the specific condition number of linear systems, that for activation functions increases as one moves towards the extremes of the function. In fact, if the matrix S in Eq. (12) is ill-conditioned (i.e., it has a high condition number) the quality of the solution provided by a quantum solver is negatively affected. In particular, the larger the condition number is, the higher the probability of obtaining numerical errors in solving the linear system [9] since ill-conditioned systems converge slowly [16]. This implies that subsystems where the condition number is high lead to worse estimated coefficients.

The GQSplines algorithm uses a problem formulation that allows obtaining an end-to-end quantum algorithm representing the spline coefficients with a single linear system of equations. In this case, results are more stable, the shape of the curve is well-approximated, and the non-linearity is well-captured. However, the limitation is that the curves must be normalized, and the normalization depends on the number of Knots which defines the size of the linear system. To tackle this problem, one can consider adopting amplitude amplification [17] to stretch the shape of the curve and increase the amplitudes to correct the normalization effects. Nonetheless, increasing the number of Knots (which scales logarithmically with the number of qubits of the VQLS) allows to build smaller intervals and potentially improves the fitting quality. This is a significant improvement with respect to existing quantum approaches [7] where the input scale linearly with the number of qubits.

Furthermore, to use of GQSplines as a method for quantum activation functions requires the ability to calculate the target variable $\hat{y_j}$ from a generic input x_j. In this regard, we have to devise a mapping circuit M that creates a vector representing the basis expansion of x_j. Then, in order to obtain $\hat{y}_j$ we have to apply the dot-product between the basis expansion of x_j and the β coefficients, both encoded as a quantum state. As before, the output $\hat{y}_j$ will be encoded by the quantum state calculated by the following circuit:

$$|x_j'\rangle \otimes |0\rangle^{\otimes(n-1)} \text{ — } \boxed{M(T,[\xi_i,\xi_{i+1}])} \text{ — } \boxed{V^\dagger(\theta_{opt})} \text{ — } \quad \hat{y}_j|0\rangle + \Sigma_{i=1}^{k}\alpha_i|i\rangle \quad (17)$$

Note that the only difference between this approach and the one described for the GQSplines is that here the Ansatz ($V^\dagger$) is transposed and applied in reverse order with respect to the encoding of the basis expansion of x_j. This procedure allows the proposed GQSplines method to produce a non-linear function f for a given input x_j which approximates the target variable y. This approach can be adopted in existing quantum neural networks such as the quantum Single Layer Perceptron [18,19] or the more general MAQA Framework [20], which require an end-to-end quantum routine storing the input-output of the non-linear activation function into the amplitudes of a quantum state.

Table 2 summarizes the properties of QSplines [4] and the two proposed methods. The VQSplines overcome the issue of performing the post-processing step and replace the HHL with the VQLS. However, they still require an ad-hoc problem decomposition with a diagonal block matrix. The GQSplines overcome this issue by defining a single linear system of equations representing the splines function.

Table 2. Comparison between QSplines [4], VQSplines, and GQSplines. The table describes whether the approach allows obtaining an end-to-end quantum routine for non-linear approximation (*end-to-end*), the quantum linear solver in use (VQLS [9] or HHL [5]), and whether the results are directly encoded into a quantum state or not depending on the use either the swap-test [21] (not accessible) or the quantum inner product [13] (accessible). This dichotomy is described by the column *Post.*

Method	End-to-end	Linear Problem Solver	Quantum Inner Product	Post
QSplines	×	HHL	Swap Test	×
VQSplines	×	VQLS	Inner Product	✓
GQSplines	✓	VQLS	Inner Product	✓

7 Conclusion

Quantum Machine Learning (QML) has recently attracted ever-increasing attention and promises to impact various applications by leveraging quantum computational power and novel algorithmic models, such as Variational Algorithms. However, the field is still in its infancy, and its practical benefits need further investigation. One of the major issues in building a complete quantum neural network is the limitation of unitary, and therefore linear, operations on quantum states. In this work, we move toward a non-linear approximation of quantum activation functions using parametrized quantum circuits. In particular, we showed that it is possible to circumvent the constraint of unitarity in quantum computation by presenting an efficient version of the QSplines, whose implementation falls within the context of fault-tolerant quantum computation. The proposed methods do not require a fault-tolerant quantum subroutine (such as HHL) and allow the approximation of non-linear functions using near-term quantum technology.

The contribution of this paper is twofold. The first part proposes the Variation Quantum Splines (VQSplines), an implementation of the fault-tolerant

QSplines [4] in the context of hybrid quantum-classical computation. Additionally, Generalized QSplines (GQSplines) are formulated and discussed. The benefit of this new formulation lies in the ability to be generalizable with respect to the structure of the spline matrix. The GQSplines adopt a new basis expansion matrix formulation, avoid the problem decomposition, and allow for tackling the problem of the matrix inversion in an end-to-end manner, with one single linear system and the number of qubits that scales logarithmically with the number of knots. Furthermore, the GQSplines are more efficient with respect to the number of qubits required with respect to existing quantum approaches for quantum activation functions. Experiments showed that both methods outperform the QSplines and can efficiently capture the non-linearity of typical activation functions adopted in classical neural networks.

Future work will be dedicated to embedding the GQSplines as a subroutine in existing quantum neural networks to leverage the properties of quantum computing in typical machine learning tasks where the non-linearity is crucial.

Acknowledgments. This work has been partially funded by the German Ministry for Education and Research (BMB+F) in the project QAI2-QAICO under grant 13N15586.

Code Availability. All code to generate the data, figures, analyses, and additional details on the experiments are available at https://github.com/inajetovic/Variational-Quantum-Splines.

References

1. Yuan, X.: A quantum-computing advantage for chemistry. Science **369**(6507), 1054–1055 (2020)
2. Venkatesh, S.M., Macaluso, A., Klusch, M.: BILP-Q: quantum coalition structure generation. In: Proceedings of the 19th ACM International Conference on Computing Frontiers, pp. 189–192 (2022)
3. Venkatesh, S.M., Macaluso, A., Klusch, M.: GCS-Q: quantum graph coalition structure generation. arXiv preprint arXiv:2212.11372 (2022)
4. Macaluso, A., Clissa, L., Lodi, S., Sartori, C.: Quantum splines for non-linear approximations. In: Proceedings of the 17th ACM International Conference on Computing Frontiers, CF 2020, pp. 249–252, New York, USA, Association for Computing Machinery (2020)
5. Harrow, A.W., Hassidim, A., Lloyd, S.: Quantum algorithm for linear systems of equations. Phys. Rev. Lett. **103**(15), 10 (2009)
6. Cao, Y., Guerreschi, G.G., Aspuru-Guzik, A.: Quantum neuron: an elementary building block for machine learning on quantum computers
7. Maronese, M., Destri, C., Prati, E.: Quantum activation functions for quantum neural networks (2022)
8. Lubasch, M., Joo, J., Moinier, P., Kiffner, M., Jaksch, D.: Variational quantum algorithms for nonlinear problems. Phys. Rev. A **101**, 010301 (2020)
9. Bravo-Prieto, C., LaRose, R., Cincio, M.L., Coles, P.J.: Variational quantum linear solver, Cerezo, Yigit Subasi (2019)
10. Hastie, T., Tibshirani, R., Friedman, J.: The Elements of Statistical Learning. SSS, Springer, New York (2009). https://doi.org/10.1007/978-0-387-84858-7

11. de Boor, C.: A Practical Guide to Splines. Springer Verlag, New York (1978)
12. Buhrman, H., Cleve, R., Watrous, J., de Wolf, R.: Quantum fingerprinting. Phys. Rev. Lett. **87**, 167902 (2001)
13. Markov, V., Stefanski, C., Rao, A., Gonciulea, C.: A generalized quantum inner product and applications to financial engineering. arXiv preprint arXiv:2201.09845 (2022)
14. Mottonen, M., Vartiainen, J.J.: Decompositions of general quantum gates. Ch. 7 in Trends in Quantum Computing Research, NOVA Publishers, New York, 2006 (2005)
15. Cleve, R., Ekert, A., Macchiavello, C., Mosca, M.: Quantum algorithms revisited. Proc. R. Soc. Lond. Ser. A: Math. Phys. Eng. Sci. **454**(1969), 339–354 (1998)
16. Rice, J.R.: A theory of condition. SIAM J. Numer. Anal. **3**(2), 287–310 (1966)
17. Brassard, G., Hoyer, P., Mosca, M., Tapp, A.: Quantum amplitude amplification and estimation. Quantum Comput. Quantum Inf. **305**, 53–74 (2000)
18. Macaluso, A., Clissa, L., Lodi, S., Sartori, C.: A variational algorithm for quantum neural networks. In: Krzhizhanovskaya, V.V. (ed.) ICCS 2020. LNCS, vol. 12142, pp. 591–604. Springer, Cham (2020). https://doi.org/10.1007/978-3-030-50433-5_45
19. Macaluso, A., Orazi, F., Klusch, M., Lodi, S., Sartori, C.: A variational algorithm for quantum single layer perceptron. In: , et al. Machine Learning, Optimization, and Data Science. LOD 2022. Lecture Notes in Computer Science. vol. 13811. Springer, Cham (2023). https://doi.org/10.1007/978-3-031-25891-6_26
20. Macaluso, A., Klusch, M., Lodi, S., et al.: MAQA: a quantum framework for supervised learning. Quantum Inf. Process. **22**, 159 (2023). https://doi.org/10.1007/s11128-023-03901-w
21. Barenco, A., Berthiaume, A., Deutsch, D., Ekert, A., Jozsa, R., Macchiavello, C.: Stabilization of quantum computations by symmetrization. SIAM J. Comput. **26**(5), 1541–1557 (1997)

Exploring the Capabilities of Quantum Support Vector Machines for Image Classification on the MNIST Benchmark

Mateusz Slysz[1,2(✉)], Krzysztof Kurowski[1], Grzegorz Waligóra[2], and Jan Węglarz[2]

[1] Poznań Supercomputing and Networking Center, IBCH PAS, Poznań, Poland
{mslysz,krzysztof.kurowski}@man.poznan.pl

[2] Institute of Computing Science Poznań, Poznań University of Technology, Poznań, Poland
{grzegorz.waligora,jan.weglarz}@cs.put.poznan.pl

Abstract. Quantum computing is a rapidly growing field of science with many potential applications. One such field is machine learning applied in many areas of science and industry. Machine learning approaches can be enhanced using quantum algorithms and work effectively, as demonstrated in this paper. We present our experimental attempts to explore Quantum Support Vector Machine (QSVM) capabilities and test their performance on the collected well-known images of handwritten digits for image classification called the MNIST benchmark. A variational quantum circuit was adopted to build the quantum kernel matrix and successfully applied to the classical SVM algorithm. The proposed model obtained relatively high accuracy, up to 99%, tested on noiseless quantum simulators. Finally, we performed computational experiments on real and recently setup IBM Quantum systems and achieved promising results of around 80% accuracy, demonstrating and discussing the QSVM applicability and possible future improvements.

Keywords: Quantum Support Vector Machine · Quantum Kernel Alignment · Image Classification · MNIST Benchmark

1 Introduction

Quantum computing is a growing field of science that uses quantum phenomena and brings much hope for more efficiency than classical computing. It has been proven theoretically that some classically intractable problems can be solved more effectively using quantum approaches. Examples of well-known quantum algorithms include Shor's algorithm for factorizing large numbers [19], or Grover's algorithm for quickly searching through unsorted data sets [8]. However, we are currently in the Noisy Intermediate Scale Quantum (NISQ) era [16], which, due to hardware implementation difficulties, is characterized by quantum devices with limited capacity and accuracy. While attempts to create fault-tolerant quantum computers are underway, researchers are looking for

J. Mikyška et al. (Eds.): ICCS 2023, LNCS 14077, pp. 193–200, 2023.
https://doi.org/10.1007/978-3-031-36030-5_15

different areas in which they could exploit the capabilities of existing quantum devices for computation. Examples of such fields include discrete optimization [6], simulation of quantum systems [11], and machine learning [5,13,18], also specifically for image classification purposes [1,15,20].

The aim of this paper is to demonstrate the potential advantages of existing NISQ devices used in a machine learning image classification problem. Using a gate-based quantum device, we run experiments with the Support Vector Machine (SVM) algorithm [3] with quantum kernels to classify a benchmark dataset.

2 Theoretical Background

2.1 Introduction to Quantum Computing

In a nutshell, quantum computers use qubits instead of bits to encode information. Qubits are two-level quantum variables and are subject to follow principles of quantum mechanics, such as superposition and entanglement. Superposition is a feature of quantum systems that allows particles to be in many states simultaneously (in this case pseudo binary states $|0\rangle$ and $|1\rangle$ are used), and entanglement is a property that binds quantum variables together in a way that allows to transfer more information than would be classically possible. These quantum effects can be exploited by cleverly designed algorithms to achieve faster and better results than it would be possible with any classical machines [14].

2.2 Support Vector Machine

A classical Support Vector Machine is a machine learning model used for data classification and regression. SVM parameters w and b are trained to find an optimal separating hyperplane f to classify data points x into classes (nominally $y = \pm 1$):

$$f(x) = \text{sign}\,(w \cdot x + b) \tag{1}$$

This problem of maximizing the margin between decision classes resolves to an optimization problem:

$$\begin{aligned} &\min_{w} \frac{1}{2}\|w\|^2 \\ \text{subject to} \quad &y_i\,(w \cdot x_i + b) \geq 1 \end{aligned} \tag{2}$$

However, classes in most datasets are not linearly separable, so SVMs must be able to tackle this problem. It is done by using kernel functions to map data to some high-dimensional feature space, in which it is possible to separate them with a hyperplane. A kernel function K is, in fact, a dot product between feature maps ϕ. For a pair of points, x_i, x_j it is given by

$$K(x_i, x_j) = \langle \phi(x_i)| \cdot |\phi(x_j)\rangle \tag{3}$$

and, for the sake of our optimization problem, should keep the similarity measure between data points as close as possible to the original feature space.

After finding the optimal parameter values, a new example $\tilde{x}$ is classified by the model following

$$f(\tilde{x}) = \text{sign}\left(\sum_i y_i \alpha_i K(x_i, \tilde{x}) + b\right) \tag{4}$$

This so-called kernel trick is a very useful tool and allows achieving good results on complicated datasets, without losing the ability of generalization. However, selecting a kernel is a difficult task and different kernel functions (e.g. linear, polynomial, Radial Based Functions, etc.) and have different strengths and weaknesses, making it worthwhile to look for more and more mapping functions that are better suited to specific applications.

2.3 Quantum Kernel

For this reason, we experimented with a new type of kernel which is computed using a quantum computer. Based on [7,9] we encode the kernel function as a readout of a parametrised quantum circuit. The data x is mapped to an n-qubit quantum feature state through a unitary operator $U(x)$:

$$\phi(x) = U(x)|0^n\rangle\langle 0^n|U^\dagger(x) \tag{5}$$

From Eqs. (3) and (5) it follows that the kernel function of two variables x_i and x_j is of the form of

$$K(x_i, x_j) = \|\langle 0^n|U^\dagger(x_i)U(x_j)|0^n\rangle\|^2 \tag{6}$$

which can be estimated as the readout from a quantum circuit implementing $U^\dagger(x_i)U(x_j)$. Since the characteristics of the quantum circuit representing the kernel function must depend not only on the input data x, but also on the parameters describing the feature map, we must take into consideration the more general definition of the feature map, which can be described as:

$$\phi(x) = D_x|\psi\rangle\langle\psi|D_x^\dagger \tag{7}$$

where $|\psi\rangle$ is some fiducial quantum state, corresponding to the feature map, and D_x is some data-dependent unitary operator. We can impose a condition in which the $|\psi\rangle$ state value is the effect of applying an operator V_λ on the initial state $|0^n\rangle$:

$$|\psi\rangle = V_\lambda|0^n\rangle \tag{8}$$

The unitary operator $U(x)$ is now a combination of V_λ and D_x, so the kernel function can be described as

$$K(x_i, x_j) = \|\langle 0^n|V_\lambda^\dagger D_{x_i}^\dagger D_{x_j} V_\lambda|0^n\rangle\|^2 \tag{9}$$

This can be represented in the form of a quantum variational circuit, which was designed based on [7]. For the data encoding D block, $R_X(\theta)$ and $R_Z(\theta)$ gates were chosen to represent data points, with angles θ corresponding to the numerical values of variables. Due to the fact that R_X and R_Z are orthogonal state rotations, it is possible to encode two data features on a single qubit - one for each axis. In the variational V block $R_Y(\lambda)$ gates were used and λ parameters are subject to training, alongside with CZ gates to create entanglement between qubits in the quantum circuit.

3 Computational Experiments

To test and compare the behaviour of the QSVM algorithm with different parameters, we decided to apply it for image classification. We tested the algorithm on images from the well-known MNIST dataset, containing images of handwritten digits at 28 × 28 resolution [12]. The data was filtered only to have two classes - '0's and '1's, as the basic version of the SVM algorithm only supports binary classification.

Since the limitations of currently available quantum computers and simulators did not allow the algorithm to fit a full-sized data set with a feature space of 784, reducing the number of features was necessary. Since the data is image-based, it was possible to use image scaling algorithms such as *resize* from Python *scikit-image* library [22]. We obtained images with smaller resolutions of 4 × 4, 6 × 6 and 8 × 8. After scaling, we got data points of dimensionality 16, 36 and 64, respectively, which correspond to the size of a quantum circuit equal to 8, 18 and 32 qubits (Fig. 1).

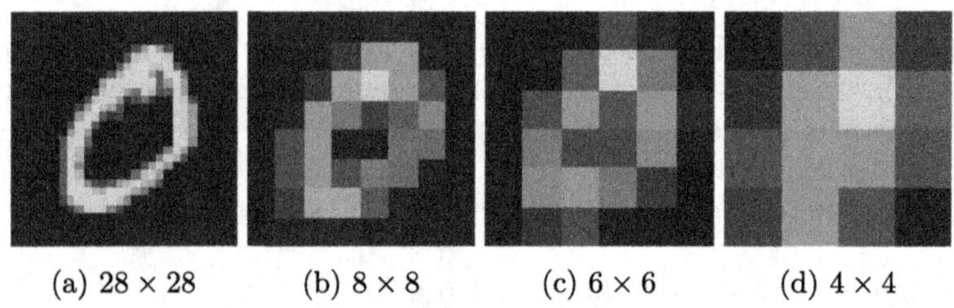

(a) 28 × 28 (b) 8 × 8 (c) 6 × 6 (d) 4 × 4

Fig. 1. A grid of original size and resized images of a digit '0' in different resolutions.

The quantum circuit was designed so that CZ gates that create an entangled state connect each qubit with its two neighbours. This method allowed to create a maximum number of connections between neighbouring variables, as the original data represents pixels, thus having a spatial interpretation. An example of the circuit with randomly chosen starting parameters λ and two data points from the training set x_1 and x_2 encoded is shown in Fig. 2.

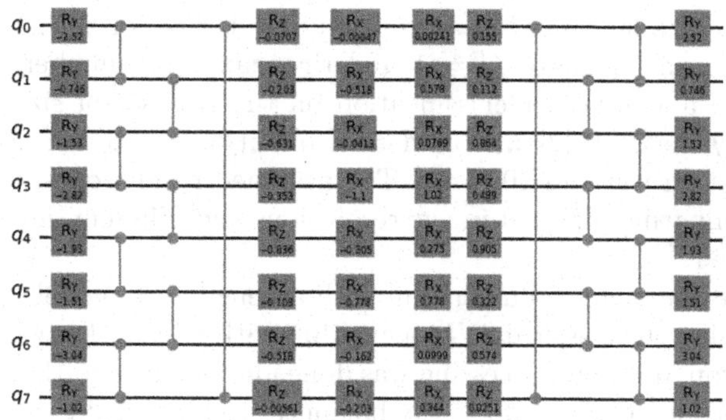

Fig. 2. An example quantum circuit for 2 digits from the training set.

Using this representation, we could run our experiments using Qiskit [17] and IBM Quantum services for all the resolutions on IBM *qasm_simulator*, which supports the simulation of quantum circuits up to 32 qubits. However, due to the connectivity limitations of actual quantum devices, it has been much more difficult to fit such circuits on quantum processors. We were limited to running experiments using 4×4 resolution images on available 27-qubit quantum computers (e.g. *ibmq_montreal* with 128 Quantum Volume) and on 127-qubit *ibm_washington* [10].

The training process was performed using the Simultaneous Perturbation Stochastic Approximation (SPSA) [21] algorithm to find the optimal set of λ parameters and solve the kernel alignment problem [4]. Based on averaged readouts from 1024 samples of the circuit executions, it was possible to construct an approximated kernel matrix that best described the similarity between data points in the dataset. The matrix dimensions were the same as the training set size and was equal to 20. Figure 3 shows an example kernel matrix.

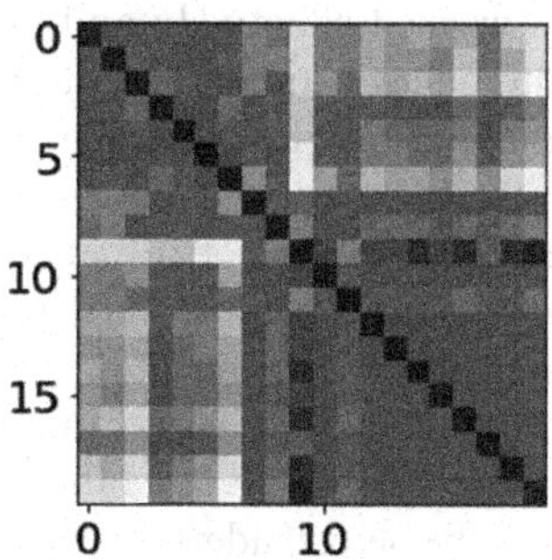

Fig. 3. Example kernel matrix of size 20×20 computed on a NISQ device.

This matrix was later used as a pre-computed kernel for the SVM algorithm with a soft margin [3] factor $C = 1$ and the maximum number of the SPSA iterations $= 10$ and used for classification on the test set of size 10. The *balanced accuracy* measure [2] was used as a quality indicator of the algorithm. Each configuration was run 10 times. The averaged results obtained from a set of readouts comparing different image resolutions and different devices were presented in Table 1.

To compare the quantum algorithm with a benchmark model, we conducted experiments using the classical SVM algorithm with a linear kernel. To maintain a fair comparison, data preprocessing was done identically, and the instance sizes were also the same. The results of the balanced accuracy for the classical SVM for 3 resolutions - 4×4, 6×6 and 8×8 are also shown in the Table 1.

Table 1. Accuracy of QSVM algorithm on the MNIST dataset, comparing different instance sizes and computers.

Computer	Number of qubits	Instance size	Balanced Accuracy [%]
classical	-	4×4	89.36
		6×6	97.71
		8×8	99.38
qasm_simulator	32	4×4	79.83
		6×6	95.79
		8×8	99.29
ibmq_montreal	27	4×4	77.68
ibm_washington	127	4×4	81.03

The results of the quantum algorithm are slightly worse than those of the classical SVM classifier. However, the differences, especially between experiments on larger instances - 6×6 and 8×8 - are relatively small, which puts the conclusion that the quantum algorithm worked effectively and performed the classification task well. Slightly larger differences can be seen on an 4×4 instance size, but in this case still the vast majority of instances in the test set were classified accurately.

Another interesting comparison is the performance of the quantum computer simulator with the actual quantum machines. For the same configuration, the results are very close to each other, which shows that the simulation process is consistent with the performance of the actual quantum computers. The fact that it was possible to successfully run the machine learning algorithm fully on a quantum processor can already be considered a promising result of the study.

In addition, a trend analogous to classical machine learning algorithms can be seen on quantum machines, where increasing the size of the instance significantly affects the quality of the results. Thus, it can be concluded that in the near future, with the growth of quantum processors, the performance of this and other quantum machine learning algorithms will increase.

4 Conclusions

Quantum Support Vector Machine is a quantum algorithm incorporating quantum computing into the machine learning landscape. So far, the proposed model has been discussed theoretically and tested on relatively simple tasks, which only contain a few variables and is relatively simple to solve by state-of-the-art machine learning techniques. We have demonstrated experimentally in this paper that the algorithm works well on the more complex image classification problem on the MNIST dataset.

Admittedly, it was necessary to perform appropriate preprocessing to reduce the resolution of the original images to fit them on still limited capabilities offered by a quantum simulator and IBM quantum devices. Nevertheless, the experimental results obtained show that relatively high classification accuracy is possible. It was also possible to run instances on real quantum NISQ devices from IBM, utilising IBM flagship quantum computers - the 127-qubit *ibm_washington* and *ibmq_montreal*, receiving consistent results.

Additionally, we noticed an impact on the overall QSVM performance, as the selected quantum devices provide a specific topology and connectivity among superconducting qubits. Although recent research suggests adapting a model to the interconnection network topology of a given quantum processor [9], we decided to stay with a more dense structure of entanglement connections due to the spatial nature of the data set to be processed. Still, it naturally requires further inventions, especially in the context of new NISQ devices with denser connectivity among qubits.

Further experiments with QSVM and other quantum machine learning algorithms can be conducted to study the impact that different entanglement strategies may have on the qualitative results of the considered algorithm. Also, other ideas can be used to further improve the quality of the results, for example using different classical preprocessing algorithms to shrink dataset size, such as PCA and its variations, as well as the usage of error correction and mitigation techniques. As quantum technologies quickly evolve, quantum computers will have more and better quality qubits, which will be more densely connected soon. Consequently, we can run experiments for much larger input datasets by applying the QSVM-based approach discussed in this paper.

References

1. Al-Ogbi, S., Ashour, A., Felemban, M.: Quantum image classification on NISQ devices. In: 2022 14th International Conference on Computational Intelligence and Communication Networks (CICN), pp. 1–7 (2022). https://doi.org/10.1109/CICN56167.2022.10008259
2. Brodersen, K.H., Ong, C.S., Stephan, K.E., Buhmann, J.M.: The balanced accuracy and its posterior distribution. In: 2010 20th International Conference on Pattern Recognition, pp. 3121–3124. IEEE (2010)
3. Cortes, C., Vapnik, V.: Support-vector networks. Mach. Learn. **20**, 273–297 (1995)

4. Cristianini, N., Shawe-Taylor, J., Elisseeff, A., Kandola, J.: On kernel-target alignment. In: Advances in Neural Information Processing Systems, vol. 14 (2001)
5. Dawid, A., et al.: Modern applications of machine learning in quantum sciences. arXiv preprint arXiv:2204.04198 (2022)
6. Farhi, E., Goldstone, J., Gutmann, S.: A quantum approximate optimization algorithm. arXiv preprint arXiv:1411.4028 (2014)
7. Glick, J.R., et al.: Covariant quantum kernels for data with group structure. arXiv preprint arXiv:2105.03406 (2021)
8. Grover, L.K.: A fast quantum mechanical algorithm for database search. In: Proceedings of the Twenty-Eighth Annual ACM Symposium on Theory of Computing, pp. 212–219 (1996)
9. Havlíček, V., et al.: Supervised learning with quantum-enhanced feature spaces. Nature **567**(7747), 209–212 (2019)
10. IBM: IBM Quantum – quantum-computing.ibm.com. https://quantum-computing.ibm.com/. Accessed 03 Feb 2023
11. Kandala, A., et al.: Hardware-efficient variational quantum eigensolver for small molecules and quantum magnets. Nature **549**(7671), 242–246 (2017)
12. LeCun, Y., Bottou, L., Bengio, Y., Haffner, P.: Gradient-based learning applied to document recognition. Proc. IEEE **86**(11), 2278–2324 (1998)
13. Lloyd, S., Schuld, M., Ijaz, A., Izaac, J., Killoran, N.: Quantum embeddings for machine learning. arXiv preprint arXiv:2001.03622 (2020)
14. Nielsen, M.A., Chuang, I.: Quantum computation and quantum information (2002)
15. Park, S., Park, D.K., Rhee, J.K.K.: Variational quantum approximate support vector machine with inference transfer. Sci. Rep. **13**(1), 3288 (2023)
16. Preskill, J.: Quantum computing in the NISQ era and beyond. Quantum **2**, 79 (2018)
17. Qiskit contributors: Qiskit: an open-source framework for quantum computing (2023). https://doi.org/10.5281/zenodo.2573505
18. Schuld, M., Killoran, N.: Quantum machine learning in feature Hilbert spaces. Phys. Rev. Lett. **122**(4), 040504 (2019)
19. Shor, P.W.: Polynomial-time algorithms for prime factorization and discrete logarithms on a quantum computer. SIAM Rev. **41**(2), 303–332 (1999)
20. Singh, G., Kaur, M., Singh, M., Kumar, Y., et al.: Implementation of quantum support vector machine algorithm using a benchmarking dataset. Indian J. Pure Appl. Phys. (IJPAP) **60**(5), 407–414 (2022)
21. Spall, J.C.: An overview of the simultaneous perturbation method for efficient optimization. J. Hopkins APL Tech. Dig. **19**(4), 482–492 (1998)
22. Van der Walt, S., et al.: Rescale, resize, and downscale—skimage v0.19.2 docs – scikit-image.org. https://scikit-image.org/docs/stable/auto_examples/transform/plot_rescale.html (2014). Accessed 03 Feb 2023

Determination of the Lower Bounds of the Goal Function for a Single-Machine Scheduling Problem on D-Wave Quantum Annealer

Wojciech Bożejko[1(✉)], Jarosław Pempera[1], Mariusz Uchroński[1,2], and Mieczysław Wodecki[3]

[1] Department of Control Systems and Mechatronics, Wrocław University of Science and Technology, Janiszewskiego 11-17, 50-372 Wrocław, Poland
{wojciech.bozejko,jaroslaw.pempera,mariusz.uchronski}@pwr.edu.pl

[2] Wroclaw Centre for Networking and Supercomputing, Wybrzeże Wyspiańskiego 27, 50-370 Wrocław, Poland

[3] Department of Telecommunications and Teleinformatics, Wrocław University of Science and Technology, Wybrzeże Wyspiańskiego 27, 50-370 Wrocław, Poland
mieczyslaw.wodecki@pwr.edu.pl

Abstract. The fundamental problem of using metaheuristics and almost all other approximation methods for difficult discrete optimization problems is the lack of knowledge regarding the quality of the obtained solution. In this paper, we propose a methodology for efficiently estimating the quality of such approaches by rapidly – and practically in constant time – generating good lower bounds on the optimal value of the objective function using a quantum machine, which can be an excellent benchmark for comparing approximate algorithms. Another natural application is to use the proposed approach in the construction of exact algorithms based on the Branch and Bound method to obtain real optimal solutions.

Keywords: Quantum Annealing · Lower Bound · Scheduling

1 Introduction

The concept of quantum computing and computers was independently introduced in the early 1980s. Since then, it has soaked up very significant developments in theory and, most importantly, in the last 20 years, in machines implementing quantum computing paradigms. Currently, the two leading types of quantum machines are quantum gate-based computers, developed mainly by IBM and Google, and adiabatic quantum computing (AQC), developed by D-Wave and NEC. In the gate-based model, calculations are performed by applying unitarity gates to quantum bits (i.e. qubits), whose states can be read out at the end of the calculation. In contrast, in AQC, in particular quantum annealing, a

J. Mikyška et al. (Eds.): ICCS 2023, LNCS 14077, pp. 201–208, 2023.
https://doi.org/10.1007/978-3-031-36030-5_16

starting state of the system modeled in hardware on multiple qubits is prepared as the ground state of the Hamiltonian encoding the solution to the desired optimization problem, to which adiabatic evolution is then applied, aiming at the minimal-energy state of the whole system. Most importantly, it is shown that the AQC is polynomially equivalent to a universal gate-based quantum computer, since any quantum circuit can be represented as a time-dependent Hamiltonian with at most polynomial charge [1].

There are quite a few descriptions in the literature of transforming classical NP-hard combinatorial optimization problems into forms suitable for quantum annealers [3]. These can be represented in Ising form using a $-1, 1$ basis (representing spins), or as a quadratic unconstrained binary optimization (QUBO) problem using a binary basis. These two forms are equivalent. This makes it easy to solve difficult discrete optimization problems – with some (unknown) – approximation. However, there is so far no description in the literature of methods that can quickly indicate the error of such an approximation. In this paper, we try to fill this research gap by proposing the idea of determining a lower bound on the value of the objective function of an optimization problem by solving with quantum annealing a dual problem resulting from Lagrange relaxation.

2 Formulation of the Problem

We will present the method of constructing a lower bound on the D-Wave quantum machine using the example of the NP-hard single-machine Total Weighted Tardiness Problem (TWTP), denoted in the literature by $1||\sum w_iT_i$. There is given a *set of tasks* $\mathcal{J} = \{1, 2, ..., n\}$, which must without interruption be executed on a single-machine. The start of the tasks begins at time 0. At any time, a machine can execute at most one task. The following are associated with each task $i \in \mathcal{J}$: *execution time* p_i, *critical line* d_i, and *weight of penalty function* w_i. For a fixed order of execution of tasks on the machine, let S_i be the starting moment and $C_i = S_i + p_i$ the ending moment of the execution of task $i \in \mathcal{J}$. Then, *delay* $T_i = \max\{0, C_i - d_i\}$, and *cost of tardiness (penalty)* $f_i(C_i) = w_i \cdot T_i$. The TWTP problem considered in this paper consists in determining the execution schedule of the machine described by S_i, C_i, $i \in \mathcal{J}$ with a minimal *total cost* $\sum_{i=1}^{n} f_i(C_i) = \sum_{i=1}^{n} w_iT_i$.

The task execution schedule described by the sequences S_i, C_i, $i \in \mathcal{J}$ is feasible if the following constraints are met:

$$S_i + p_i \le S_j \vee S_j + p_j \le S_i,\ i \neq j,\ i, j = 1, 2, \ldots, n, \tag{1}$$

$$S_i \ge 0,\ \ C_i = S_i + p_i,\ \ i = 1, 2, \ldots, n. \tag{2}$$

The single-machine problem of minimizing the sum of delay costs formulated above is NP-hard. Optimal algorithms for solving the problem based on the methods of dynamic programming, i.e. on Lagrange relaxation and branch and bound, are described in the works by (Potts [6], and Wodecki [12]). These algorithms are time consuming, thus in practice, small-scale examples can be solved on classical computers with their help. These are mainly metaheuristics that have been widely used since the 1990s: tabu search (Bożejko et al. [4], Uchroński [10]), dynamic programming (Rostami et al. [9]), simulated annealing (Potts and Van Wassenhove [7]). Extensive reviews of the literature on scheduling problems with due dates was also presented by Adamu and Adewumi [2]. The literature also deals with single-machine scheduling problems with uncertain execution times or desired completion dates: Rajba and Wodecki [8], Bożejko et al. [5].

3 Determining the Lower Bound on the D-Wave Quantum Machine

The calculation of the lower bound of the objective function will be performed in two steps. In step one, for a quantum computer, using Lagrange relaxation we will define a dual optimization problem that will be maximized on a QPU. In step two, using a classical CPU, the exact value of the lower bound will be determined based on the results obtained in step one.

Let us consider a certain optimization problem having the following property: its solution (in the sense of value) is always less than or equal to the optimal one. A relaxed version of the problem considered in this paper using the Lagrange function has this property. The relaxation will be governed by the non-overlapping constraint (i.e. their decouplability), the inequality of the (1).

For simplicity of notation, let us assume that the tasks are executed in the natural order of π, $\pi(i) = i$. The TWTP problem under consideration can be written in the form of an optimization task:

$$\min_{S} \sum_{i=1}^{n} w_i T_i \tag{3}$$

subject to

$$S_i + p_i - S_j \leq K(1 - y_{ij}),\ j = i+1, \ldots, n,\ i = 1, \ldots, n, \tag{4}$$

$$S_j + p_j - S_i \leq K y_{ij},\ j = i+1, \ldots, n,\ i = 1, \ldots, n, \tag{5}$$

$$y_{ij} \in \{0, 1\},\ j = i+1, \ldots, n,\ i = 1, \ldots, n, \tag{6}$$

$$S_i \geq 0,\ i = 1, \ldots, n, \tag{7}$$

where K is some sufficiently large number. In turn, y_{ij} is a binary variable equal to 1 if the task i precedes j and 0 otherwise. The Lagrange function with multipliers u_{ij} and v_{ij}, $i, j = 1, 2, \ldots, n$ takes for the vector $S = (S_1, S_2, \ldots, S_n)$ and the matrix $y = [y_{ij}]_{n \times n}$ the form:

$$L(S, y, u, v) = \sum_{i=1}^{n} w_i T_i + \sum_{i=1}^{n} \sum_{j=i+1}^{n} u_{ij}(S_i + p_i - S_j - K(1 - y_{ij}))$$

$$+ \sum_{i=1}^{n} \sum_{j=i+1}^{n} v_{ij}(S_j + p_j - S_i - K y_{ij})$$

Transforming this expression we obtain

$$L(S,y,u,v) = \sum_{i=1}^{n} L_i(S_i,u,v) + K\sum_{i=1}^{n}\sum_{j=i+1}^{n} Q_{ij}(y_{ij},u,v) + V(u,v). \quad (8)$$

where

$$L_i(S_i,u,v) = w_i T_i + \alpha_i S_i, \;\; \alpha_i = \sum_{j=i+1}^{n}(u_{ij}-v_{ij}) + \sum_{j=1}^{i-1}(v_{ji}-u_{ji}),$$

$$Q_{ij}(y_{ij},u,v) = (u_{ij}-v_{ij})y_{ij}, \;\; V(u,v) = \sum_{i=1}^{n} p_i\left(\sum_{j=1}^{i-1} v_{ji} + \sum_{j=i+1}^{n} u_{ij}\right).$$

Let us note that if S^* is an optimal solution to the TWTP problem, then for *any non-negative* $u, v \geq 0$ there is a

$$\sum_{j=1}^{n} w_j T_j \geq \sum_{j=1}^{n} w_j T_j + \sum_{i=1}^{n}\sum_{j=i+1}^{n} u_{ij}(S_i^* + p_i - S_j^* - K(1-y_{ij}))$$

$$+\sum_{i=1}^{n}\sum_{j=i+1}^{n} v_{ij}(S_i^* + p_i - S_j^* - Ky_{ij}) \geq \min_{S}\min_{y} L(S,y,u,v).$$

Therefore, when looking for a good lower bound, one should compute

$$LB = \max_{u,v}\min_{S,y} L(S,y,u,v) = \max_{u,v}\left(\sum_{i=1}^{n}\min_{0\leq S_i\leq T-p_i} L_i(S_i,u,v)\right.$$

$$\left. +K\sum_{i=1}^{n}\sum_{j=i+1}^{n}\min_{y} Q_{ij}(y_{ij},u,v) + V(u,v)\right) \quad (9)$$

whereby the maximization with respect to u and v can be approximate, while that with respect to S and y is exact.

Determination of Lower Bound (Step 1) on a D-Wave Quantum Annealer. Let us note that the lower bound (9) can be written as a minimization of the opposite (minus) value, with constraints:

$$LB = -\min_{u,v,S,y}\left[-\left(\sum_{i=1}^{n} L_i(S_i,u,v) + K\sum_{i=1}^{n}\sum_{j=i+1}^{n} Q_{ij}(y_{ij},u,v) + V(u,v)\right)\right] \quad (10)$$

s.t.

$$L_i(S_i,u,v) \leq L_i(0,u,v), \; i = 1,2,\ldots,n, \quad (11)$$

$$L_i(S_i, u, v) \leq L_i(1, u, v),\ i = 1, 2, \ldots, n, \tag{12}$$

$$\vdots$$

$$L_i(S_i, u, v) \leq L_i(T - p_i, u, v),\ i = 1, 2, \ldots, n, \tag{13}$$

and

$$Q_{ij}(y_{ij}, u, v) \leq Q_{ij}(0, u, v),\ i, j = 1, 2, \ldots, n, \tag{14}$$

$$Q_{ij}(y_{ij}, u, v) \leq Q_{ij}(1, u, v),\ i, j = 1, 2, \ldots, n, \tag{15}$$

where each of the constraints (11)–(13) of the form $L_i(S_i, u, v) \leq L_i(t, u, v)$, $i = 1, 2, \ldots, n$, $n = 0, 1, \ldots, T - p_i$, where $L_i(S_i, u, v) = w_i T_i + \alpha_i S_i$ is technically written in the D-Wave machine program as one of two constraints – each of (11)–(13) is encoded as expressed in the Algorithm 1, since in the constraints of the QUBO model, there cannot be a function *maximum* resulting from the formula to calculate the delay for a task i starting at time t equalling $T_i(t) = \max\{0, t + p_i - d_i\}$.

Algorithm 1: Adding S minimalization constraints to the QUBO model

for $i = 1, 2, \ldots, n$ **do**
  for $t = 0, 1, 2, \ldots, T - p_i$ **do**
    if $(t + p_i - d_i > 0)$ **then**
      Add constraint $L_i(S_i, u, v) \leq w_i \cdot (t + p_i - d_i) + \alpha_j \cdot t$
    else
      Add constraint $L_i(S_i, u, v) \leq w_i \cdot 0 + \alpha_i \cdot t$

The task formulated in this way can already be directly implemented on a D-Wave machine since all constraints, as well as the objective function, are linear. The difficulty is the possible suboptimality of the resulting quantum annealing vector S and binary matrix y with respect to the formulation (9).

4 Experimental Research

To verify the effectiveness of the proposed method of determining the lower bound, computational experiments were carried out on the quantum algorithm implemented on the D-WAVE quantum annealer and the algorithm determining the lower bound on a classical silicon computer with an i7-12700H 2.30 GHz processor. The research was carried out on 30 instances divided into three groups of 10 instances each. Instance groups differ in the number of tasks. A full set of test instances can be found in [11].

Table 1 presents the results of experimental research, in particular, in column 1 there is LB^Q determined by the quantum algorithm, while in column 3 there is LB^{CPU} determined by the classical algorithm. Columns 2 and 4 show the time of quantum computations and computations on a classical computer, respectively. In addition, column 5 includes the acceleration of calculations and the relative difference of the LB value as a measure of the quality assessment of the generated solutions (column 6), determined as $Quality = \frac{LB^Q - LB^{CPU}}{LB^{CPU}}$. Analyzing the results presented in the Table, we can conclude that in a significant number of instances, the LB determined by the quantum annealer is significantly greater than the LB determined on a classical computer. The LB value determined by the annealer is not lower for all instances, with the LB value determined on the CPU, and in 26 out of 30 instances it is better. For the instance $wt7\_70$ LB^Q is nearly 200 times better than LB^{CPU}. The Quantity value occurs on average 17 times for the $n = 5$ group, 8 times for the group $n = 6$ and 92 times for the group $n = 7$. Comparing the calculation time of a quantum exponent and a classical computer, we can conclude that the time of quantum calculations is from 6 to nearly 140 times shorter than the time of calculations on a classical computer. The advantage of quantum computing increases as the number of tasks increases. For the $n = 5$ group, it is on average 9 times lower, while for the $n = 8$ group, it is 97 times lower on average (Fig. 1).

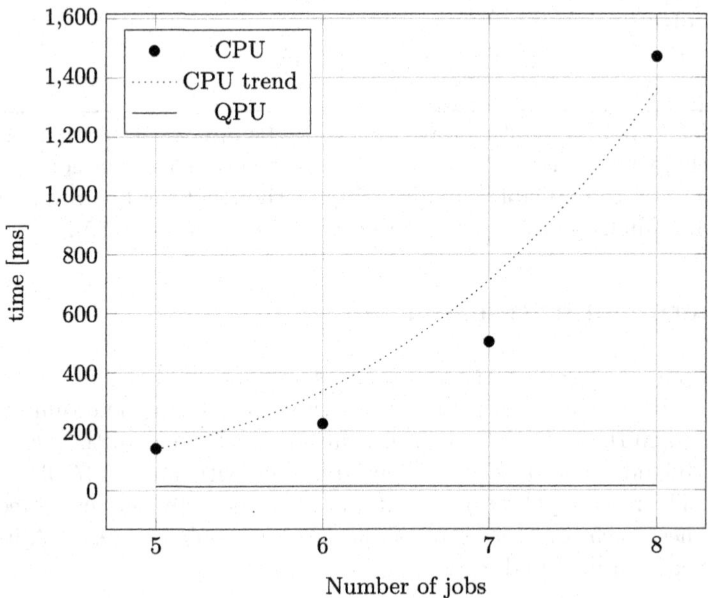

Fig. 1. Computation time of LB calculations on quantum processor QPU and silicon processor CPU

Table 1. The results of experiments.

example	LB^Q	$Time^Q$	LB^{CPU}	$TIME^{CPU}$	SPEED-UP	Quality
wt5_40	423	15	0	183	12,20	
wt5_41	2153	15	456	140	9,33	4,72
wt5_42	1657	15	300	103	6,87	5,52
wt5_43	1001	15	10	148	9,87	100,1
wt5_44	1588	15	116	115	7,67	13,69
wt5_45	2099	15	0	187	12,47	
wt5_46	1791	15	604	116	7,73	2,97
wt5_47	2443	15	783	147	9,80	3,12
wt5_48	3353	15	1138	123	8,20	2,95
wt5_49	1578	15	358	100	6,67	4,41
wt6_70	469	15	0	202	13,47	
wt6_71	3328	15	385	241	16,07	8,64
wt6_72	3563	15	290	359	23,93	12,29
wt6_73	2630	15	421	178	11,87	6,25
wt6_74	3216	15	612	312	20,80	5,25
wt6_75	1280	15	0	324	21,60	—
wt6_76	0	15	0	261	17,40	—
wt6_77	8	15	0	242	16,13	—
wt6_78	0	15	0	299	19,93	—
wt6_79	16	15	0	186	12,40	—
wt7_70	3049	15	15	450	30,00	203,27
wt7_71	3635	15	317	582	38,80	11,47
wt7_72	1395	15	0	282	18,80	—
wt7_73	3806	15	62	451	30,07	61,39
wt7_74	3117	15	0	420	28,00	—
wt7_75	2840	15	0	238	15,87	—
wt7_76	0	15	0	605	40,33	—
wt7_77	64	15	0	436	29,07	—
wt7_78	0	15	0	302	20,13	—
wt7_79	12	15	0	381	25,40	—
wt8_80	100	15	0	1407	93,80	—
wt8_81	1271	15	0	1278	85,20	—
wt8_82	992	15	0	1249	83,27	—
wt8_83	662	15	0	1576	105,07	—
wt8_84	292	15	0	945	63,00	—
wt8_85	481	15	0	1682	112,13	—
wt8_86	3522	15	0	2053	136,87	—
wt8_87	1961	15	0	1127	75,13	—
wt8_88	5529	15	0	1774	118,27	—
wt8_89	2333	15	0	1512	100,80	—

5 Summary

This paper presents an algorithm for determining the lower bound on the value of the objective function for the TWTP problem implemented on a D-Wave quantum computer. The presented approach can be adapted to estimate the value of the optimal solution of other NP-hard discrete optimization problems, such as the commutator problem or multi-machine problems (e.g. job shop). A natural direction for further research will be to apply the proposed method for determining lower bounds on a quantum machine, together with the (natural) determination of upper bounds by simply solving the problem formulated as QUBO, also on a QPU, to the construction of an exact algorithm based on the Branch and Bound method. This will allow – against the intuition associated with the probabilistic nature of computation on QPUs – to the generation of truly optimal solutions.

References

1. Aharonov, D., Dam, W., Kempe, J., Landau, Z., Lloyd, S., Regev, O.: Adiabatic quantum computation is equivalent to standard quantum computation. SIAM Rev. **50**(4), 755–787 (2008)
2. Adamu, M.O., Adewumi, A.O.: A survey of single-machine scheduling to minimize weighted number of tardy jobs. J. Ind. Manage. Optim. **10**, 219–241 (2013)
3. Bożejko, W., Pempera, J., Uchroński, M., Wodecki, M.: Distributed quantum annealing on D-wave for the single-machine total weighted tardiness scheduling problem. In: Groen, D., de Mulatier, C., Paszynski, M., Krzhizhanovskaya, V.V., Dongarra, J.J., Sloot, P.M.A. (eds.) Computational Science - ICCS 2022. LNCS, vol. 13353, pp. 171–178. Springer, Cham (2022). https://doi.org/10.1007/978-3-031-08760-8_15
4. Bożejki, W., Grabowski, J., Wodecki, M.: Block approach-tabu search algorithm for single-machine total weighted tardiness problem. Comput. Ind. Eng. **50**, 1–14 (2006)
5. Bożejko, W., Rajba, P., Wodecki, M.: Stable scheduling of single-machine with probabilistic parameters. Bull. Pol. Acad. Sci. Tech. Sci. **65**, 219–231 (2017)
6. Potts, C.N., Van Wassenhove, L.N.: A branch and bound algorithm for the total weighted tardiness problem. Oper. Res. **33**, 177–181 (1985)
7. Potts, C.N., Van Wassenhove, L.N.: Single-machine tardiness sequencing heuristics. IIE Trans. **23**, 346–354 (1991)
8. Rajba, P., Wodecki, M.: Stability of scheduling with random processing times on one machine. Applicationes Mathematicae **39**, 169–183 (2012)
9. Rostami, S., Creemers, S., Leus, R.: Precedence theorems and dynamic programming for the single-machine weighted tardiness problem. Eur. J. Oper. Res. **272**, 43–49 (2019)
10. Uchroński, M.: Parallel algorithm with blocks for a single-machine total weighted tardiness scheduling problem. Appl. Sci. **11**(5), 2069 (2021)
11. Uchroński, M.: Test instances for a single-machine total weighted tardiness scheduling problem. https://zasobynauki.pl/zasoby/74584
12. Wodecki, W.: A branch-and-bound parallel algorithm for single-machine total weighted tardiness problem. Int. J. Adv. Manuf. Technol. **37**, 996–1004 (2008)

Simulating Sparse and Shallow Gaussian Boson Sampling

Zoltán Kolarovszki[1,2(✉)], Ágoston Kaposi[1,2], Tamás Kozsik[2], and Zoltán Zimborás[1,2,3]

[1] Quantum Computing and Quantum Information Research Group, Department of Computer Sciences, Wigner Research Centre for Physics, Budapest, Hungary
{kolarovszki.zoltan,kaposi.agoston,zimboras.zoltan}@wigner.hu
[2] Faculty of Informatics, Eötvös Loránd University, Budapest, Hungary
kto@inf.elte.hu
[3] Algorithmiq Ltd., Helsinki, Finland

Abstract. Gaussian Boson Sampling (GBS) is one of the most popular quantum supremacy protocols as it does not require universal control over the quantum system, which favors current photonic experimental platforms and there is strong theoretical evidence for its computational hardness. However, over the years, several algorithms have been proposed trying to increase the performance of classically simulating GBS assuming certain constraints, e.g., a low number of photons or shallow interferometers. Most existing improvements of the classical simulation of GBS provide a performance increase regarding the probability calculation, leaving the sampling algorithm itself untouched. This paper provides an asymptotically better sampling algorithm in the case of low squeezing and shallow circuits.

Keywords: Quantum Computing · Quantum Computer Simulation · Quantum Advantage

1 Introduction

In the last decade, photonic quantum computing has gained more relevance in the quantum computing world due to its apparent scalability and the recent demonstration of photonic quantum advantage [17]. In particular, in recent experiments, the so-called Gaussian Boson Sampling (GBS) scheme was used in recent experiments to demonstrate that photonic quantum devices are capable of solving a classically hard task [16]. This sampling schemes has also been demonstrated to have several special applications, e.g., in graph problems [2] and quantum chemistry [9]. Therefore, trying to improve algorithms for simulating GBS has gained attraction.

Simulation of GBS is also a necessary step in creating a photonic quantum computer, since quantum computer manufacturers need to compare experimental data with a simulation for assessing inaccuracies. Moreover, an efficient simulation algorithm enables researchers to be able to perform various numerical experiments on a classical computer testing new quantum protocols.

J. Mikyška et al. (Eds.): ICCS 2023, LNCS 14077, pp. 209–223, 2023.
https://doi.org/10.1007/978-3-031-36030-5_17

In its essence, GBS means the mode-wise photon detection of a multimode Gaussian state [8,10]. The scheme can be simulated by calculating the probabilities of the photon detection events using the displacement vector and covariance matrix of the Gaussian state. However, the probabilities turn out to be classically hard (#P-hard) to compute in general and there is also theoretical evidence that sampling from this distribution is hard [7], hence an existing photonic quantum computer has an advantage over classical computers.

One feature that makes the direct simulation of GBS hard is the size of the event space. The number of photon detection events exponentially increases in the number of modes, repeating the calculation of the probabilities many times. To counter this problem, one can introduce a mode-by-mode sampling algorithm which does not calculate the probabilities for all the possible events, but only for a certain subset of the sample space, hence reducing the complexity of the calculation [12].

The complexity of the (loop) hafnian is at the heart of the classical simulation of GBS. The state-of-the-art algorithm for calculating this quantity is the power trace method introduced by Björklund et al. [4]. Additionally, one could achieve a considerable speedup by modifying the original mode-by-mode sampling algorithm, described by Quesada et al. [12]. In their paper, they achieved a quadratic speedup over the original particle-resolved GBS algorithm, using the fact that the formula of (loop) hafnian for Gaussian pure states simplifies [14].

Since the inception of GBS, several algorithms have been proposed to simulate certain edge cases more efficiently [3,11]. Important edge cases are, e.g., when the Gaussian states have relatively low average particle numbers (sparse circuit) or when the interferometer used in the production of the Gaussian state is shallow. In the sparse and shallow case, a faster algorithm for calculating the (loop) hafnian has been already proposed [11], but a modification of the sampling algorithm in this scenario has not been considered before. Therefore, we propose an algorithm for simulating GBS for sparse and shallow photonic Gaussian circuits.

The structure of the paper is as follows: in Sect. 2 the basics of the GBS scheme are introduced. In Sect. 3 the proposed sampling algorithm is derived and introduced. Lastly, in Sect. 4, the complexity of the algorithm is calculated in the case of non-displaced threshold GBS.

2 Setup

In this section, basic familiarity with quantum optics and photonic Gaussian states is assumed. For reference, the reader may visit Refs. [1,15].

A d-mode Gaussian state ρ can be completely characterized by its displacement vector $\bar{r} = \operatorname{tr}[\hat{r}\rho]$ and covariance matrix $\sigma = \frac{1}{2}\operatorname{tr}[\{\hat{r}-\bar{r}, \hat{r}-\bar{r}\}\rho]$, where $\hat{r} = (x_1, \dots, x_d, p_1, \dots, p_d)$ is the vector containing the x_i, p_i quadrature operators, and $\{A, B\}$ denotes the anticommutator of the operators A and B .

Gaussian states can be similarly characterized by their complex displacement vector and their Q-function covariance matrix

$$\gamma = \mathrm{tr}\left[\xi\rho\right], \tag{1}$$

$$\Sigma = \frac{1}{2}\left(\mathrm{tr}\left[\{(\xi-\gamma),(\xi-\gamma)^\dagger\}\rho\right] + \mathbb{1}\right), \tag{2}$$

where $\xi = (a_1, \ldots, a_d, a_1^\dagger, \ldots, a_d^\dagger)$ is the vector containing the creation and annihilation operators.

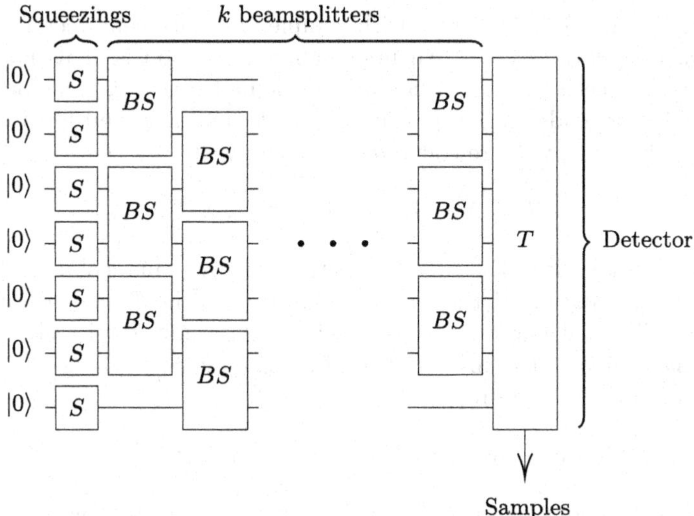

Fig. 1. The basic setup of GBS without displacement. On the vacuum state, a column of squeezings is applied with different squeezing parameters, and then k column of beamsplitters with alternating starting positions depending on the parity of the column, as demonstrated by the diagram.

In the most usual variant of the GBS scheme one performs a mode-wise photon number detection measurement, which is a projective measurement described by the projections

$$P_n = |n\rangle\langle n| \qquad (n \text{ photon}, n \in \mathbb{N}_0). \tag{3}$$

One can easily specialize towards a coarser detection scheme, the so called *threshold detection*. The threshold detection measurement for one mode is defined by projections

$$\begin{aligned} Q_0 &= P_0 = |0\rangle\langle 0| && (0 \text{ photon}), \\ Q_1 &= \sum_{n=1}^{\infty} P_n = \mathbb{1} - Q_0 && (\geq 1 \text{ photons, click}). \end{aligned} \tag{4}$$

On multiple modes, the tensor product of the one-mode projection elements define measurement projections

$$\{R_{s_1} \otimes \cdots \otimes R_{s_d}\}_{(s_1,\ldots,s_d)\in I^d} =: \{R_S\}_{S\in I^d}\,, \tag{5}$$

where I could either denote $\{0,1\}$ or $\mathbb{N}_0 = \{0,1,2,\ldots\}$ and R_S could denote either P_S or Q_S, and d is the number of modes. Given a quantum state ρ, the probability of detecting the sample $S \in I^d$ is

$$p(S) = \operatorname{tr}\left[\rho R_S\right]. \tag{6}$$

Consider a d-mode Gaussian state with complex displacement vector γ and Q-function covariance matrix Σ. An important observation is that in GBS the probability of detecting a sample $S \in I^d$ can be multiplicatively factorized to a function which depends on $S, \operatorname{br}_S \gamma, \operatorname{br}_S \Sigma^{-1}$ only, and a second function which depends purely on Σ, γ. More concretely, one can write

$$p(S) = N(\gamma,\Sigma) f\left(S, \operatorname{br}_S \gamma, \operatorname{br}_S \Sigma^{-1}\right), \tag{7}$$

where br_S denotes the block-wise repetition of rows and columns by S (0 entry means elimination of the row or column); for the formal definition see Appendix A. The calculation of f is usually a classically hard task, and N is just a normalization factor, independent of the sample S, fulfilling the following block matrix factorization property:

$$N(\gamma_1 \oplus_b \gamma_2, \Sigma_1 \oplus_b \Sigma_2) = N(\gamma_1,\Sigma_1)N(\gamma_2,\Sigma_2). \tag{8}$$

Probability expressions of the form of Eq. (7) are motivated by several GBS-related sampling schemes, where the function f contains the (loop) torontonian [5] or (loop) hafnian [4]. As a simple example, for a Gaussian state 0 complex displacement vector and Q-function covariance matrix Σ, the threshold detection probability can be calculated by the formula

$$p(S) = \frac{\operatorname{tor}\left(\operatorname{br}_S O\right)}{\sqrt{\det\left(\Sigma\right)}}, \tag{9}$$

where $O = \mathbb{1} - \Sigma^{-1}$ and tor denotes the torontonian [13]. Naively, to sample from the probability distribution one would need to calculate the probability for all elements in the sample space $\Omega = \{S \in I^d\}$. Since $|\Omega| = |I|^d$ scales exponentially in the number of modes d, the sampling algorithm constructed in this fashion may be computationally demanding in general. Instead, one can opt for a sampling method where knowledge of all the probabilities is not necessary. For the simulation of photonic quantum computing, the mode-by-mode sampling method is used extensively, which is introduced by Quesada et al. [13] for the case of non-displaced threshold GBS.

3 Classical Simulation of Sparse and Shallow GBS

In this section, we would like to introduce a classical algorithm which may have better performance than previous algorithms in the case of low particle numbers (sparse) and shallow interferometers. A key observation is that the (loop) torontonian and (loop) hafnian functions factorize over block direct sum due to Theorem 3 in Appendix B. The following question arises: can the mode-by-mode algorithm be improved using Theorem 3? The answer turns out to be positive, as demonstrated in this section.

Consider a d-mode photonic quantum system according to Fig. 1 with k beamsplitter columns. Let $r = (r_1, \dots, r_d) \in \mathbb{R}_+^d$ squeezing parameters corresponding to the squeezing gates and $U \in U(d)$ unitary matrix corresponding to the interferometer consisting of the beamsplitters. Then the Q-function covariance matrix of the state before threshold detection is

$$\Sigma = \begin{bmatrix} U & \\ & U^* \end{bmatrix} \begin{bmatrix} \cosh^2 r & \cosh r \sinh r \\ \cosh r \sinh r & \cosh^2 r \end{bmatrix} \begin{bmatrix} U^\dagger & \\ & U^T \end{bmatrix}. \tag{10}$$

The unitary $U \in U(d)$ corresponding to the columns of beamsplitters can get "banded" if the number of beamsplitter columns k is smaller than d. To properly formulate this, we need the definition of bandwidth:

Definition 1 (Set of k-bandwidth matrices). *The set of k-bandwidth matrices is*

$$\mathrm{Band}_k(\mathbb{C}^{n\times n}) :=$$
$$\left\{ A \in \mathbb{C}^{n\times n} \mid \forall i \in [n] : \begin{cases} \text{if } i < n-k : \forall a \in [n-i-k] : A_{i,i+k+a} = 0 \\ \text{if } i > k : \forall a \in [i-k] : A_{i,i-k-a} = 0 \end{cases} \right\}. \tag{11}$$

For convenience, let us define a function which assigns the minimal number of matching row and column eliminations to decompose the "overestimated" matrix to direct sums of smaller matrices.

Definition 2. *Let $n \in \mathbb{N}_+$ and let $A \in \mathbb{C}^{n\times n}$. The function* band $: \mathbb{C}^{n\times n} \to \mathbb{N}$ *is defined as*

$$A \mapsto \min_{k\in[n]} \left\{ k : A \in \mathrm{Band}_k(\mathbb{C}^{n\times n}) \right\} \tag{12}$$

Consider a d-mode circuit of $m \leq d/2$ columns of beamsplitters, and denote the interferometer of each column by $U_i \in U(d), 1 \leq i \leq m$. One can easily show that the bandwidth of $U := U_1 \dots U_m$ is just m, i.e. $\mathrm{band}(U_1 \dots U_m) = m$, and hence $\mathrm{band}(\Sigma) = 2m =: k$. Therefore, for circuits where $\mathrm{band}(\Sigma) < d$, one can have occurrences during the sampling procedure when the reduction bitstring S has 0-substrings of length k and using Theorem 3, the probabilities would factorize. More concretely, define a bitstring S by concatenating two arbitrary bitstring S_1, S_2 by k 0s, i.e. let $S = S_1 0^k S_2$. Moreover, let $n = |S|$ and consider a matrix $M \in \mathbb{C}^{2n \times 2n}$ which can be decomposed into $A, B, C, D \in \mathbb{C}^{n \times n}$ blocks, i.e.

$$M = \begin{bmatrix} A & B \\ C & D \end{bmatrix} \tag{13}$$

such that $\mathrm{band}(A) = \mathrm{band}(B) = \mathrm{band}(C) = \mathrm{band}(D) = k$. Then we can write

$$\mathrm{br}_S\, M = \begin{bmatrix} A_1 \oplus A_2 & B_1 \oplus B_2 \\ C_1 \oplus C_2 & D_1 \oplus D_2 \end{bmatrix} =: M_1 \oplus_b M_2, \tag{14}$$

where - again - the bandwidth of every submatrix is k. In Theorem 3 it has already been shown, that the part of the probability which depends on the sample S factorizes over block direct sums. A natural question would be whether one could use this formula in the mode-by-mode sampling algorithm. One could immediately note that factorizing is not a trivial matter, since in the algorithm, the Q-function covariance matrix is block reduced for each iteration by the first n modes. In each iteration indexed by n, the $(\mathrm{r}_{1^n 0^{d-n}}\, \Sigma)^{-1}$ matrix is calculated, which is different for each iteration. This matrix then needs to be reduced again by the previous samples to calculate probability, according to Eq. (7). If one could interchange the inversion on Σ and the first reduction by $1^n 0^{d-n}$, Theorem 3 could be trivially used to factorize in the algorithm whenever a 0^k substring appears in the sample, but it is not entirely obvious if such interchange of operations is permitted. Luckily, the following theorem asserts that this is indeed the case:

Theorem 1. *Let $\Sigma \in \mathbb{C}^{2d \times 2d}$ be the Q-function covariance matrix for a pure Gaussian state in which the interferometer matrix $U \in U(d)$ is m-bandwidth and $r = (r_1, \dots, r_d) \in \mathbb{R}^d_{\geq 0}$. Let $k = 2m$, $k \leq n \leq d$, $S = 1^{(n-k)} 0^k$ and $F = 1^n 0^{(d-n)}$. Then*

$$(\mathrm{br}_S \circ \mathrm{inv} \circ \mathrm{br}_F)\, \Sigma = (\mathrm{br}_S \circ \mathrm{br}_F \circ \mathrm{inv})\, \Sigma = \mathrm{br}_S\, \mathrm{br}_F \left(\Sigma^{-1}\right). \tag{15}$$

Proof. By direct computation one gets

$$\mathrm{br}_S\, \mathrm{br}_F \left(\Sigma^{-1}\right) = \begin{bmatrix} \mathbb{1} & -\mathrm{r}_S\, \mathrm{r}_F\, B \\ -\mathrm{r}_S\, \mathrm{r}_F\, B^* & \mathbb{1} \end{bmatrix}, \qquad \Sigma^{-1} = \begin{bmatrix} \mathbb{1} & -B \\ -B^* & \mathbb{1} \end{bmatrix}, \tag{16}$$

where $B := U \tanh(r) U^T \in \mathbb{C}^{d \times d}$, which is equivalent to the result of Proposition 4 from Appendix C. □

Let $\Sigma \in \mathbb{C}^{2d \times 2d}$ have k-bandwidth submatrices as before. Let S_1 and S_2 be arbitrary bitstrings, and let $n := |S_1| + |S_2| + k \leq d$. Then

$$\mathrm{br}_{S_1 0^k S_2}\, \Sigma^{-1} = \mathrm{br}_{S_1 0^k S_2}\, \Sigma^{-1} = \mathrm{br}_{S_1 0^k 0^{|S_2|}}\, \Sigma^{-1} \oplus_b \mathrm{br}_{0^{|S_1|} 0^k S_2}\, \Sigma^{-1}, \tag{17}$$

where using Theorem 1 we can write

$$\mathrm{br}_{S_1 0^k 0^{|S_2|}}\, \Sigma^{-1} = (\mathrm{br}_{S_1 0^k} \circ \mathrm{inv} \circ \mathrm{br}_{1^{|S_1|} 1^k 0^{|S_2|}}) \Sigma, \tag{18}$$

$$\mathrm{br}_{0^{|S_1|} 0^k S_2}\, \Sigma^{-1} = (\mathrm{br}_{0^k S_2} \circ \mathrm{inv} \circ \mathrm{br}_{0^{|S_1|} 1^k 1^{|S_2|}})\, \Sigma, \tag{19}$$

$$\begin{aligned} \mathrm{br}_{S_1 0^k S_2}\, \Sigma^{-1} = {} & (\mathrm{br}_{S_1 0^k} \circ \mathrm{inv} \circ \mathrm{br}_{1^{|S_1|} 1^k 0^{|S_2|}})\, \Sigma \\ & \oplus_b (\mathrm{br}_{0^k S_2} \circ \mathrm{inv} \circ \mathrm{br}_{0^{|S_1|} 1^k 1^{|S_2|}})\, \Sigma. \end{aligned} \tag{20}$$

Putting it all together, it is apparent that we can cut the mode-by-mode sampling at a 0^k pattern. Finally, using Theorem 3 we conclude that

$$\begin{aligned} &f(S_1 0^k S_2, \mathrm{br}_{S_1 0^k S_2}\,\gamma, \mathrm{br}_{S_1 0^k S_2}\,\Sigma^{-1}) \\ &\quad = f(S_1 0^k, \mathrm{br}_{S_1 0^k}\,\mathrm{br}_{1^{|S_1|}1^k 0^{|S_2|}}\,\gamma, \mathrm{br}_{S_1 0^k}(\mathrm{br}_{1^{|S_1|}1^k 0^{|S_2|}}\,\Sigma^{-1})) \\ &\quad \times f(0^k S_2, \mathrm{br}_{0^k S_2}\,\mathrm{br}_{0^{|S_1|}1^k 1^{|S_2|}}\,\gamma, \mathrm{br}_{0^k S_2}(\mathrm{br}_{0^{|S_1|}1^k 1^{|S_2|}}\,\Sigma^{-1})), \end{aligned} \tag{21}$$

which means that when at least k zeroes are encountered, the Q-function covariance matrix Σ will be factorized during the sampling algorithm. More concretely, the terms in the product are independent of S_1 and S_2 respectively, so during the sampling algorithm where the S_2 sample is varied for calculating the probability distribution, the first term dependent on S_1 will not change. Furthermore, one can also omit to calculate the normalization factor N and keep track of the probability from the previous sampling for calculating conditional probability.

In particular, for the case of non-displaced Gaussian Threshold Boson Sampling, one can write

$$\begin{aligned} (\mathrm{tor} \circ \mathrm{br}_{S_1 0^k S_2})(\mathbb{1} - \Sigma^{-1}) &= \mathrm{tor}\,\mathrm{br}_{S_1 0^k}\left(\mathbb{1} - (\mathrm{br}_{1^{|S_1|}1^k 0^{|S_2|}}\,\Sigma)^{-1}\right) \\ &\quad \times \mathrm{tor}\,\mathrm{br}_{0^k S_2}\left(\mathbb{1} - (\mathrm{br}_{0^{|S_1|}1^k 1^{|S_2|}}\,\Sigma)^{-1}\right), \end{aligned} \tag{22}$$

and using all these insights, one can build an algorithm which could be seen in Algorithm 1. In this algorithm, when a 0^k substring appears in the samples, the input of the torontonian function can be factorized, and it is sufficient to consider only a subsystem which starts from the beginning of the 0^k substring in the sample. The non-displaced threshold GBS is emphasized because one could give an upper bound for its average complexity, discussed in the following section.

Algorithm 1. Proposed non-displaced threshold GBS algorithm for small-depth interferometer

Require: Q-function covariance matrix Σ, bandwidth k
- $d \leftarrow \dim \Sigma/2$
- $S, A \leftarrow []$ ▷ Empty list for samples and accumulator
- **while** $|S| + |A| < d$ **do**
  - $n \leftarrow |A| + 1$
  - **if** $|S| = 0$ **then**
    - $F \leftarrow 1^n 0^{d-n}$
    - $S_0 \leftarrow S + [0]$
    - $S_1 \leftarrow S + [1]$
  - **else**
    - $F \leftarrow 0^{|S|-k} 1^{k+n} 0^{d-n-|S|}$
    - $S_0 \leftarrow 0^k + S + [0]$
    - $S_1 \leftarrow 0^k + S + [1]$
  - **end if**
  - $w_0 \leftarrow N\ \mathrm{tor}(\mathrm{br}_{S_0}(\mathbb{1} - \mathrm{br}_F(\Sigma)^{-1}))$ ▷ 0-detection weight
  - $w_1 \leftarrow N\ \mathrm{tor}(\mathrm{br}_{S_1}(\mathbb{1} - \mathrm{br}_F(\Sigma)^{-1}))$ ▷ 1-detection weight
  - $c \leftarrow$ choose from $(0, 1)$ with weights (w_0, w_1)
  - $S \leftarrow S + [c]$
  - **if** last k elements in S are all 0s **then**
    - $A \leftarrow A + S$
    - $S \leftarrow []$ ▷ Sample is emptied
  - **end if**
- **end while**

4 Complexity of the Classical Algorithm for Non-displaced Gaussian Threshold Boson Sampling

According to the proposed Algorithm 1, the size of the matrices serving as inputs of the torontonian can be reduced, since the sample S is emptied when a 0^k bitstring is encountered in the sample. To give an upper bound for the complexity of the algorithm, one is required to find a lower bound for the vacuum detection probability, given a set of squeezing parameters.

Theorem 2 (Lower bound for vacuum probability). *Let G_r^d be the set of d-mode pure non-displaced Gaussian states with squeezing parameters $r = (r_1, \dots, r_d)$, and $S \in P([d])$ a set containing mode numbers. Then*

$$\inf_{\rho \in G_r^d} p_\rho^S(0) \geq \cosh^{-2|S|} \max_{i \in [d]} r_i, \tag{23}$$

where $p_\rho(S = 0)$ is the probability of detecting vacuum on modes defined by S.

Proof. For a non-displaced Gaussian state, the vacuum probability can be calculated by $p_\rho^S(0) = \det \left[\mathrm{br}_S\, \Sigma_\rho\right]^{-\frac{1}{2}}$, where Σ_ρ is the Q-function covariance matrix corresponding to ρ. Hence, one needs to calculate

$$\inf_{\rho \in G_r^d} p_\rho^S(0) = \sup_{U \in U(d)} \det \left[\mathrm{br}_S\, \Sigma\right]^{-\frac{1}{2}}.$$

According to Eq. (10), for pure Gaussian states the Q-function covariance matrices only depend on the squeezing parameters r and the interferometer U, i.e., $\Sigma = \Sigma(r, U)$. Since the squeezing parameters r are fixed, one can restate the previous optimization over states as an optimization over unitary matrices as

$$\sup_{\rho \in G_r^d} \det \left[\mathrm{br}_S\, \Sigma_\rho\right] = \sup_{U \in U(d)} \det \left[\mathrm{br}_S\, \Sigma(r, U)\right]. \tag{24}$$

If A is a positive semi-definite matrix, then $\det A \leq \left(\frac{\operatorname{Tr} A}{\dim A}\right)^{\dim A}$, from the arithmetic mean-geometric mean inequality. Using this relation, one can provide an upper bound for $\det \left[\mathrm{br}_S\, \Sigma\right]$. Let $r_{\max} = \max_{i \in [d]} r_i$ and write

$$\begin{aligned} \sup_{U \in U(d)} \det \left[\mathrm{br}_S\, \Sigma\right] &\leq \sup_{U \in U(d)} \left[\frac{1}{|S|} \sum_{s \in S} \left(\sum_{j=1}^{d} u_{sj} u_{sj}^* \cosh^2 r_j\right)\right]^{2|S|} \\ &= \cosh^{4|S|} r_{\max}. \end{aligned} \tag{25}$$

□

Using Theorem 2, one can bound the probability distribution p_ρ^S from below by an i.i.d. probability distribution p as

$$p_\rho^S(0) \geq p(0)^{|S|} = \cosh^{-2|S|} r_{\max}, \tag{26}$$

where $p(0)$ is the lower bound for the probability of detecting vacuum on the single mode S_i, and the corresponding probability distribution is

$$p(0) = \cosh^{-2} r_{\max}, \qquad p(1) = 1 - p(0), \tag{27}$$

which yields an i.i.d. probability distribution over all modes in S.

To give an estimate of the average complexity of Algorithm 1, we need a probability distribution which gives the probability of detecting n many 1s between 2 bitstrings of the form 0^k. For example, one could consider the bitstring

$$\underbrace{0000}_{k\ 0\text{s}}\underbrace{101100010011}_{n=6\ 1\text{s}}\underbrace{0000}_{k\ 0\text{s}}, \tag{28}$$

where 6 1s are found between two 0^k substrings, and in this section, we will refer to these as *clusters*. Suppose that the probability of detecting 0 is uniform and independent for all modes, and denote this probability p. To generate bitstring beginning and ending with a 1, we can write

$$G = 1T + 1\left(\sum_{i=0}^{k-1} 0^i\right) G. \tag{29}$$

This recursion could be used to generate many other quantities, e.g., probabilities, by formally replacing 1 and 0 in the formula. For this reason, we replace $1 \mapsto z(1-p)$ and $0 \mapsto p$, where we included a parameter z in 1 since we only want to count the 1s in the bitstring (the 0s do not increase the complexity). The resulting equation can be solved for $G := G(z)$ as

$$G(z) = \frac{1-p}{z^{-1} - 1 + p^k}, \tag{30}$$

which is to be interpreted as the generator for the probability distribution for detecting 1s. However, $G(z)$ is not normalized in the sense that the coefficients of $G(z)$ would not sum up to 1, therefore one has to normalize it to yield a proper probability distribution. After normalization and expansion, we acquire

$$\hat{G}(z) := \frac{G(z)}{G(1)} = \frac{p^k}{z^{-1} - 1 + p^k} = \sum_{n=1}^{\infty} z^n p_c(n), \tag{31}$$

where $p_c(n) = p^k(1-p^k)^{n-1}$ represents the probability of encountering n 1s between two 0^k substrings. With the knowledge of this probability distribution, it is a trivial matter to calculate the average number of 1s between two 2 0^k substrings and is given by $\mathbb{E}[n] = p^{-k}$. The lower bound given by Eq. (27) is a probability distribution which is uniform and independent over all modes and using $p(0) = p$ one can write

$$\mathbb{E}[n] = (\cosh \max_i r_i)^{2k}, \tag{32}$$

which gives an upper bound for the number of 1s in terms of the maximal squeezing parameter.

To calculate the complexity of Algorithm 1, one needs to use the complexity for the torontonian itself. It should be emphasized, that the underlying method of calculating the torontonian can be chosen freely. Suppose that the calculation of the torontonian has exponential complexity and write $\mathcal{C}_{\text{tor}} = N^\alpha \beta^N$, where $\alpha, \beta > 1$. One can give an upper bound to the complexity of a cluster of n 1s as

$$\mathcal{C}_{\text{cluster}}(n) = \sum_{N=1}^{n} N^\alpha \beta^N \leq n^{\alpha+1} \beta^n, \tag{33}$$

which can be used to give an upper bound for the average complexity of the cluster

$$\mathbb{E}[\mathcal{C}_{\text{cluster}}] \leq \sum_{n=1}^{\infty} p_c(n) n^{\alpha+1} \beta^n = \frac{p^k}{1-p^k} \sum_{n=1}^{\infty} n^{\alpha+1} [\beta(1-p^k)]^n, \tag{34}$$

and this expression is convergent when $p^k > 1 - \beta^{-1}$.

In summary, the complexity of the problem reduces in the proposed algorithm when a 0-substring is encountered with a certain length, which is not considered in the original threshold GBS algorithm presented in [13]. Hence, the presented sampling algorithm is faster in general for sparse and shallow circuits. It should also be noted, that the speedup presented in this section is just an illustration of the main principle of the proposed algorithm. One could simulate threshold GBS using (loop) hafnians which have lower complexity than the original algorithm [6, 14], but the main principle of the proposed algorithm is applicable in these algorithms as well.

5 Conclusion and Outlook

A modified classical algorithm has been given for simulating Gaussian Boson Sampling, which takes into account the shallowness of the circuit and the low number of particles in the system. The algorithm can be applied for the threshold and the particle-resolved GBS as well, and the average complexity of the proposed algorithm has been calculated in the case of non-displaced threshold GBS.

The proposed sampling algorithm can also be employed in Gaussian Boson Sampling with photon number resolving measurements and even using displaced states. However, calculating the average complexity needs further investigation in these cases, since the calculation is not feasible by using similar assumptions as for the Threshold Gaussian Boson Sampling using torontonian. Moreover, the case of mixed Gaussian states has not been considered during this work, and it is still an open question whether the proposed algorithm is still valid in this case. This problem has been set aside for a future project.

Acknowledgement. This research was supported by the Ministry of Culture and Innovation and the National Research, Development and Innovation Office within the Quantum Information National Laboratory of Hungary (Grant No. 2022-2.1.1-NL-2022-00004).

A Matrix Operations

There are several notations regarding matrix operations throughout the article, and this section aims to collect all of them to avoid confusion.

Definition 3 (Row reduction). *Let $A \in \mathbb{C}^{n\times m}$ and let $S \in \mathbb{N}_0^n$ be a bitstring with $k := \sum_{i=0}^{|S|} S_i$. Then the row reduction $\mathrm{rr}_S : \mathbb{C}^{n\times m} \to \mathbb{C}^{k\times m}$ is the function mapping A to a matrix formed by repeating the i-th row of A S_i many times.*

Definition 4 (Column reduction). *Let $A \in \mathbb{C}^{n\times m}$ and let $S \in \mathbb{N}_0^n$ be a bitstring with $k := \sum_{i=0}^{|S|} S_i$. Then the column reduction $\mathrm{cr}_S : \mathbb{C}^{n\times m} \to \mathbb{C}^{n\times k}$ is defined by $\mathrm{cr}_S(A) := \mathrm{rr}_S(A^T)^T$.*

Definition 5 (Reduction). *Let $A \in \mathbb{C}^{n\times n}$ and let $S \in \mathbb{N}_0^n$ be a bitstring with k 1s. Then the reduction $\mathrm{r}_S : \mathbb{C}^{n\times n} \to \mathbb{C}^{k\times k}$ is defined by $\mathrm{r}_S := \mathrm{rr}_S \circ \mathrm{cr}_S$.*

Definition 6 (Block reduction). *Block reduction can only be defined on an even-dimensional matrix $M \in \mathbb{C}^{2n\times 2n}$. Let $A, B, C, D \in \mathbb{C}^{n\times n}$ and $S \in \mathbb{N}_0^n$. Then*

$$\mathrm{br}_S \begin{bmatrix} A & B \\ C & D \end{bmatrix} = \begin{bmatrix} \mathrm{r}_S A & \mathrm{r}_S B \\ \mathrm{r}_S C & \mathrm{r}_S D \end{bmatrix}. \tag{35}$$

Definition 7 (Block direct sum). *Let $n, m \in \mathbb{N}^+$, $A_1, B_1, C_1, D_1 \in \mathbb{C}^{n\times n}$ and $A_2, B_2, C_2, D_2 \in \mathbb{C}^{m\times m}$. Then the block direct sum $\oplus_b : \mathbb{C}^{2n\times 2n} \times \mathbb{C}^{2m\times 2m} \to \mathbb{C}^{2(n+m)\times 2(n+m)}$ is defined as*

$$\begin{bmatrix} A_1 & B_1 \\ C_1 & D_1 \end{bmatrix} \oplus_b \begin{bmatrix} A_2 & B_2 \\ C_2 & D_2 \end{bmatrix} = \begin{bmatrix} A_1 \oplus A_2 & B_1 \oplus B_2 \\ C_1 \oplus C_2 & D_1 \oplus D_2 \end{bmatrix}. \tag{36}$$

B Factorizing Probabilities over Block Direct Sums

Theorem 3 (Factorization of probabilities). *Consider a 1-mode POVM $\{R_i\}_{i\in I}$ for some index set $I = \{0, 1\}$ or $I = \mathbb{N}_0$, where R_i is either P_i or Q_i described by Eq. (3) and Eq. (4). A d-mode Gaussian state with complex displacement vector γ and Q-function covariance matrix Σ. Suppose that the probability of detecting $S \in I^d$ is*

$$\mathrm{tr}\left[\rho\left(P_{S_1} \otimes \dots P_{S_d}\right)\right] = f\left(S, \mathrm{br}_S\,\gamma, \mathrm{br}_S\,\Sigma^{-1}\right) N(\gamma, \Sigma), \tag{37}$$

where the normalization N has the block matrix factorization property, i.e.

$$N(\gamma_1 \oplus_b \gamma_2, \Sigma_1 \oplus_b \Sigma_2) = N(\gamma_1, \Sigma_1) N(\gamma_2, \Sigma_2). \tag{38}$$

Then suppose ρ_1, ρ_2 are Gaussian states over d_1 and d_2 modes with γ_1 and γ_2 complex displacement vectors and Σ_1, Σ_2 Q-function covariance matrices respectively, and consider $S \in I^{d_1}$, $T \in I^{d_2}$. Then f also has the block matrix factorization property in the following sense:

$$\begin{aligned} &f(S \times T, \mathrm{br}_{S\times T}(\gamma_1 \oplus_b \gamma_2), \mathrm{br}_{S\times T}(\Sigma_1^{-1} \oplus_b \Sigma_2^{-1})) \\ &\quad = f(S, \mathrm{br}_S\, \gamma_1, \mathrm{br}_S\, \Sigma_1^{-1}) f(T, \mathrm{br}_T\, \gamma_2, \mathrm{br}_T\, \Sigma_2^{-1}). \end{aligned} \tag{39}$$

Proof. By direct calculation,

$$\begin{aligned} &f(S \times T, \mathrm{br}_{S\times T}(\gamma_1 \oplus \gamma_2), \mathrm{br}_{S\times T}(\Sigma_1^{-1} \oplus \Sigma_2^{-1})) \\ &\quad = \frac{\mathrm{tr}\left[(\rho_1 \otimes \rho_2)\left(P^{(S)} \otimes P^{(T)}\right)\right]}{N(\gamma_1 \oplus_b \gamma_2, \Sigma_1 \oplus_b \Sigma_2)} = \frac{\mathrm{tr}\left[\rho_1 P^{(S)}\right] \mathrm{tr}\left[\rho_2 P^{(T)}\right]}{N(\gamma_1, \Sigma_1) N(\gamma_2, \Sigma_2)} \\ &\quad = f(S, \mathrm{br}_S\, \gamma_1, \mathrm{br}_S\, \Sigma_1^{-1}) f(T, \mathrm{br}_T\, \gamma_2, \mathrm{br}_T\, \Sigma_2^{-1}), \end{aligned} \tag{40}$$

where $P^{(V)} = P_{V_1} \otimes \cdots \otimes P_{V_d}$, for any $V \in I^d$. □

Corollary 1. *The torontonian, loop torontonian, hafnian and loop hafnian also factorize in the manner described by Theorem 3.*

C Supplementary Calculations

Proposition 1. *Let $U \in U(d)$ and $D \in \mathbb{C}^{d\times d}$ diagonal matrix, and $S = 1^n 0^{(d-n)}$ with $n < d$. Then*

$$\mathrm{r}_S\left(UDU^\dagger\right) = \mathrm{rr}_S\left(U\right) D\, \mathrm{cr}_S\left(U^\dagger\right) =: VDV^\dagger, \tag{41}$$

where $V = \mathrm{rr}_S\left(U\right) \in \mathbb{C}^{n\times d}$, and rr, cr are defined in Appendix A.

Proposition 2. *Let $U \in U(d)$ be an m-bandwidth unitary, and let $m < n < d$, $S = 1^n 0^{(d-n)}$. Then let V be defined by*

$$\mathrm{rr}_S\left(U\right) =: V = \left[W \middle| X\right]. \tag{42}$$

where $W \in \mathbb{C}^{n\times(n-m)}$, $X \in \mathbb{C}^{n\times(d-n+m)}$. Then $V^\dagger V = \mathbb{1}_{n-m} \oplus K$, where $X^\dagger X = K \in \mathbb{C}^{(d-n+m)\times(d-n+m)}$.

Proof. $U \in U(d)$ can be written using d unit vectors $\{\boldsymbol{u}_i\}_{i=1}^d, \boldsymbol{u}_i \in \mathbb{C}^d$ as $U = [\boldsymbol{u}_1, \ldots \boldsymbol{u}_d]$, where $\langle \boldsymbol{u}_i, \boldsymbol{u}_j \rangle = \delta_{ij}$ and $\langle \cdot, \cdot \rangle$ is the standard inner product over $\mathbb{C}^d$. When we reduce by $S = 1^n 0^{d-n}$, elements from $\boldsymbol{u}_i$ are removed consequently. Split up the reduced matrix as $V = [W|X]$. We know that $W^\dagger W = \mathbb{1}_{n-m}$ since i, j fulfill the conditions $i, j \leq n - m$, and the reduction only removes zero elements from $\boldsymbol{u}_i$ and $\boldsymbol{u}_j$. More concretely, one can write $\mathrm{rr}_S\, \boldsymbol{u}_i = \boldsymbol{w}_i \oplus 0_{n-i-m}$ which is still orthogonal to $\boldsymbol{u}_j$ vectors where only zeros have been trimmed. Moreover, $W^\dagger X = 0$, since for $i \leq n - m$ and $j > n - m$, the reduction by S only cancels zeros from $\boldsymbol{u}_i$, and the non-zero elements cancelled in $\boldsymbol{u}_j$ by the reduction are multiplied by zero when calculating the inner product with $\boldsymbol{u}_i$, hence the inner product of reduced vectors is equal to the inner product of original vectors in this case. Putting everything together one can conclude that $V^\dagger V = \mathbb{1}_{n-m} \oplus X^\dagger X$. □

Proposition 3. *Let $U \in U(d)$ be a unitary and let $n \in [d]$, $S = 1^n 0^{d-n}$. Let $V = \mathrm{rr}_S(U)$ and $D \in \mathbb{C}^{d \times d}$ a diagonal matrix. Then*

$$(VDV^\dagger)^{-1} = VD^{-1}V^\dagger. \tag{43}$$

Proposition 4. *Let $\Sigma \in \mathbb{C}^{2d \times 2d}$ be a Q-function covariance matrix of the form*

$$\Sigma = \begin{bmatrix} U & \\ & U^* \end{bmatrix} \begin{bmatrix} \cosh^2 r & \cosh r \sinh r \\ \cosh r \sinh r & \cosh^2 r \end{bmatrix} \begin{bmatrix} U^\dagger & \\ & U^T \end{bmatrix}, \tag{44}$$

where $U \in U(d)$ is m-bandwidth, and $k := 2m$. Let $F = 1^n 0^{d-n}$ and $S = 1^{n-k}0^k$ with $k \leq n \leq d$. Then

$$(\mathrm{br}_S \circ \mathrm{inv} \circ \mathrm{br}_F)\Sigma = \begin{bmatrix} \mathbb{1} & -\mathrm{r}_S \, \mathrm{r}_F(B) \\ -\mathrm{r}_S \, \mathrm{r}_F(B)^* & \mathbb{1} \end{bmatrix}. \tag{45}$$

where $B = U \tanh(r) U^T$.

Proof. Let us denote $\cosh^2(r) =: C$ and $\cosh(r)\sinh(r) =: S$. Then

$$\mathrm{br}_F \, \Sigma = \begin{bmatrix} VCV^\dagger & VSV^T \\ V^*SV^\dagger & V^*CV^T \end{bmatrix}, \tag{46}$$

using Proposition 1. The inverse can be divided into blocks as

$$(\mathrm{br}_F \, \Sigma)^{-1} = \begin{bmatrix} (\mathrm{br}_F \, \Sigma)^{-1}_{11} & (\mathrm{br}_F \, \Sigma)^{-1}_{12} \\ (\mathrm{br}_F \, \Sigma)^{-1}_{21} & (\mathrm{br}_F \, \Sigma)^{-1}_{22} \end{bmatrix}. \tag{47}$$

By explicit calculation, one can show that

$$(\mathrm{br}_F \, \Sigma)^{-1}_{11} = \left(V(C - SV^T(V^*CV^T)^{-1}V^*S)V^\dagger\right)^{-1}, \tag{48}$$

and using Proposition 3 we can write

$$C - SV^T(V^*CV^T)^{-1}V^*S = C - SV^TV^*C^{-1}V^TV^*S. \tag{49}$$

Moreover, using Proposition 2 we may write

$$\begin{aligned} C - SV^TV^*C^{-1}V^TV^*S &= C - S(\mathbb{1}_{n-m} \oplus K^*)C^{-1}(\mathbb{1}_{n-m} \oplus K^*)S \\ &=: \mathbb{1}_{n-m} \oplus M, \end{aligned} \tag{50}$$

where $K, M \in \mathbb{C}^{(d+m-n)\times(d+m-n)}$. We also know that

$$\left(V(\mathbb{1} \oplus M)V^\dagger\right)_{ij} = \sum_{a,b=1}^{d} V_{ia}(\mathbb{1}_{n-m} \oplus M)_{a,b} V^*_{jb}, \tag{51}$$

but since V is m-bandwidth, we know that there is no terms overlapping with m if $i, j \leq n-m-m = n-k$. Therefore it is guaranteed that $(\mathrm{br}_F \, \Sigma)^{-1}_{11} = \mathbb{1}_{n-k} \oplus E$, where $E \in \mathbb{C}^{(d+k-n)\times(d+k-n)}$, which means that $\mathrm{r}_S(\mathrm{br}_F\,\Sigma)^{-1}_{11} = \mathbb{1}_{n-k}$. Similarly, one can write

$$(\mathrm{br}_F\,\Sigma)^{-1}_{12} = -(\mathbb{1}_{n-k} \oplus E)\, VSV^T V^* C^{-1} V^T, \tag{52}$$

and after reduction one acquires

$$\mathrm{r}_S((\mathrm{br}_F\,\Sigma)^{-1}_{12}) = -\,\mathrm{r}_S\left(V(\mathrm{r}_G(\tanh r) \oplus H)V^T\right), \tag{53}$$

where $G = 1^{(n-m)}0^{(d-n+m)}$ and $H \in \mathbb{C}^{(d+m-n)\times(d+m-n)}$. Consider indices $i, j \leq n-k$. Then

$$\left(V(\mathrm{r}_G(\tanh(r)) \oplus H)V^T\right)_{ij} = -\sum_{a,b=1}^{d} V_{ia}\left(\mathrm{r}_G(\tanh(r)) \oplus H\right)_{ab} V_{bj}, \tag{54}$$

but we know that $V_{ia} = V_{bj} = 0$ for $a, b > n-m$, i.e. the values of H are irrelevant when we are computing matrix elements with indices $i, j \leq n-k$. Therefore we can just write

$$\left(V(\mathrm{r}_G(\tanh(r)) \oplus H)V^T\right)_{ij} = \left(V \tanh(r) V^T\right)_{ij}, \tag{55}$$

which essentially yields that

$$\begin{aligned}\mathrm{r}_S(\mathrm{br}_F(\Sigma)^{-1})_{12} = -\,\mathrm{r}_S\,\mathrm{r}_F(V\tanh(r)V^T) &= -\,\mathrm{r}_S\,\mathrm{r}_F(U\tanh(r)U^T)\\ &= -\,\mathrm{r}_S\,\mathrm{r}_F(B).\end{aligned} \tag{56}$$

□

References

1. Adesso, G., Ragy, S., Lee, A.R.: Continuous variable quantum information: Gaussian states and beyond. Open Syst. Inf. Dyn. **21**(01n02), 1440001 (2014). https://doi.org/10.1142/s1230161214400010
2. Arrazola, J.M., Bromley, T.R.: Using Gaussian Boson sampling to find dense subgraphs. Phys. Rev. Lett. **121**, 030503 (2018). https://doi.org/10.1103/PhysRevLett.121.030503
3. Barvinok, A.: Two algorithmic results for the traveling salesman problem. Math. Oper. Res. **21** (2001). https://doi.org/10.1287/moor.21.1.65
4. Björklund, A., Gupt, B., Quesada, N.: A faster Hafnian formula for complex matrices and its benchmarking on a supercomputer (2018). https://doi.org/10.48550/ARXIV.1805.12498
5. Bulmer, J.F.F., Paesani, S., Chadwick, R.S., Quesada, N.: Threshold detection statistics of bosonic states. Phys. Rev. A **106**(4) (2022). https://doi.org/10.1103/physreva.106.043712
6. Bulmer, J.F.F., et al.: The boundary for quantum advantage in Gaussian Boson sampling. Sci. Adv. **8**(4) (2022). https://doi.org/10.1126/sciadv.abl9236
7. Deshpande, A., et al.: Quantum computational advantage via high-dimensional Gaussian Boson sampling. Sci. Adv. **8**(1), eabi7894 (2022). https://doi.org/10.1126/sciadv.abi7894

8. Hamilton, C.S., Kruse, R., Sansoni, L., Barkhofen, S., Silberhorn, C., Jex, I.: Gaussian Boson sampling. Phys. Rev. Lett. **119**(17) (2017). https://doi.org/10.1103/physrevlett.119.170501
9. Huh, J., Guerreschi, G.G., Peropadre, B., McClean, J.R., Aspuru-Guzik, A.: Boson sampling for molecular vibronic spectra. Nat. Photonics **9**(9), 615–620 (2015). https://doi.org/10.1038/nphoton.2015.153. Springer
10. Kruse, R., Hamilton, C.S., Sansoni, L., Barkhofen, S., Silberhorn, C., Jex, I.: Detailed study of Gaussian boson sampling. Phys. Rev. A **100**(3) (2019). https://doi.org/10.1103/physreva.100.032326
11. Qi, H., Cifuentes, D., Brádler, K., Israel, R., Kalajdzievski, T., Quesada, N.: Efficient sampling from shallow Gaussian quantum-optical circuits with local interactions. Phys. Rev. A **105**, 052412 (2022). https://doi.org/10.1103/PhysRevA.105.052412
12. Quesada, N., Arrazola, J.M.: Exact simulation of Gaussian Boson sampling in polynomial space and exponential time. Phys. Rev. Res. **2**, 023005 (2020). https://doi.org/10.1103/PhysRevResearch.2.023005
13. Quesada, N., Arrazola, J.M., Killoran, N.: Gaussian Boson sampling using threshold detectors. Phys. Rev. A **98**(6) (2018). https://doi.org/10.1103/physreva.98.062322
14. Quesada, N., et al.: Quadratic speed-up for simulating Gaussian Boson sampling. PRX Quantum **3**, 010306 (2022). https://doi.org/10.1103/PRXQuantum.3.010306
15. Serafini, A.: Gaussian States of Continuous Variable Systems. CRC Press, Taylor & Francis Group (2017)
16. Zhong, H.S., et al.: Phase-programmable Gaussian Boson sampling using stimulated squeezed light. Phys. Rev. Lett. **127**, 180502 (2021). https://doi.org/10.1103/PhysRevLett.127.180502
17. Zhong, H.S., et al.: Quantum computational advantage using photons. Science **370**(6523), 1460–1463 (2020). https://doi.org/10.1126/science.abe8770

Solving Higher Order Binary Optimization Problems on NISQ Devices: Experiments and Limitations

Valentin Gilbert[1(✉)], Julien Rodriguez[1,2], Stéphane Louise[1], and Renaud Sirdey[1]

[1] Université Paris-Saclay, CEA-List, 91120 Palaiseau, France
{valentin.gilbert,julien.rodriguez,stephane.louise,renaud.sirdey}@cea.fr
[2] Université de Bordeaux, INRIA, Palaiseau, France

Abstract. With the recent availability of Noisy Intermediate-Scale Quantum devices, the potential of quantum computers to impact the field of combinatorial optimization lies in quantum variational and annealing-based methods. This paper further compares Quantum Annealing (QA) and the Quantum Approximate Optimization Algorithm (QAOA) in solving Higher Order Binary Optimization (HOBO) problems. This case study considers the hypergraph partitioning problem, which is used to generate custom HOBO problems. Our experiments show that D-Wave systems quickly reach limits solving dense HOBO problems. Although the QAOA demonstrates better performance on exact simulations, noisy simulations reveal that the gate error rate should remain under 10^{-5} to match D-Wave systems' performance, considering equal compilation overheads for both device.

Keywords: HOBO · Balanced hypergraph partitioning · Quantum computing · QAOA · Quantum annealing

1 Introduction

As we enter the Noisy Intermediate Scale Quantum (NISQ) era, companies are now building chips that control a few hundred qubits for quantum circuit models and several thousand for quantum annealers. The selection of interesting problems that run successfully on noisy quantum chips is now a key point of interest for researchers and industries alike. As quantum heuristics performance limits are easier to reveal in the higher instance density regime, which either requires qubit duplications or larger circuit depths, we use HOBO problems to generate k-local Hamiltonians of custom density. An experimental study of the impact of the HOBO formulation on the QAOA was done in [5], demonstrating that higher order formulations were favorable to the QAOA. E. Pelofske et al. [11] also compared the ability of QA and the QAOA to solve HOBOs perfectly adapted to

J. Mikyška et al. (Eds.): ICCS 2023, LNCS 14077, pp. 224–232, 2023.
https://doi.org/10.1007/978-3-031-36030-5_18

ibm_washington's graph connectivity containing cubic interaction terms, showing that current ideal QAOA execution on real hardware could not match QA results quality.

We propose another study case to experimentally evaluate bounds on the error rate that would permit QAOA to beat QA on results quality. As a use case, we generate HOBO formulations from Balanced Hypergraph Partitioning (BHP) problems, which is well-known in combinatorial optimization due to the difficulty of finding a good solution. It consists in dividing the vertices into different subsets, considering a balancing constraint while minimizing the number of hyperedges connecting the partitions. The balancing constraint acts as a global constraint and requires a strong coupling between the variables of the problem. This problem is interesting as its transformation into the Ising model gives a fully connected 2-local Hamiltonian with some k-local terms representing hyperedges. A general formulation for graph bi-partitioning using the Ising model was proposed in [9]. This formulation has been extended to graph k-partitioning in [16], with an experimental comparison between state-of-the-art partitioning methods and the quantum hybrid method *qbsolv* , which seems competitive. H. N. Djidjev et al. [6] are less optimistic and demonstrate that the advantage of the quantum annealer is still limited by the size of the quantum chip. They also underline the importance of accounting for compilation time, which can represent up to 99% of the computation run time for large instances. Recent theoretical results on the limitations of the QAOA on pure k-spin model are available in [2]. The authors show that the QAOA is subject to optimality limitations for any even $k \geq 4$ in the infinite size limit for fixed p. It sets a first theoretical bound, proving that the QAOA may encounter strong limitations in solving HOBO problems.

Our contributions are two-fold. The first one is a recursive formulation of the BHP problem as a HOBO problem. The second contribution is a performance comparison of two quantum heuristics: the QA and the QAOA. In particular, our experiments suggest that noisy QAOA will only compete with D-Wave systems on low density problems if the error rate remains under 10^{-5}.

2 Problem Formulation

The formulation of the BHP problem is an extension of a previous work based on hypergraph bi-partitioning [13]. A Hypergraph is a generalization of a graph where hyperedges can be connected to one or more vertices. Let $\mathcal{H} \overset{\text{def}}{=} (\mathcal{V}, \mathcal{E})$ the hypergraph defined from a set of vertices $\mathcal{V}$ and a set of hyperedges $\mathcal{E}$. A k-partition Π of $\mathcal{H}$ is a splitting of $\mathcal{V}$ into k vertex subsets π_i with $1 \leq i \leq k$, called parts, such that : (i) each part π_i respects the capacity constraint : $\forall i, |\pi_i| \leq \frac{|\mathcal{V}|}{k}$; (ii) all parts are pairwise disjoint : $\forall i, j \; i \neq j, \pi_i \cap \pi_j \;=\; \emptyset$; the union of all parts is equal to $\mathcal{V}$: $\bigcup_i \pi_i = \mathcal{V}$. A cut for a k-partition Π of $\mathcal{H}$ is the union of hyperedges that contain at least two vertices in different parts and the cut-size f_c is the number of cut edges. Our formulation minimizes the *min-cut* metric with a balanced constraint. Considering k the final number of partitions, at a given level of recursion, the capacity constraint for one recursion splitting the vertices v into 2 sub-parts $x_v \in \{0, 1\}$ is:

$$H_A = \left(\sum_{v \in \mathcal{V}} \omega_v x_v - \left\lceil \left\lfloor \frac{k}{2} \right\rfloor / k \times \Omega(\mathcal{V}) \right\rceil \right)^2 \tag{1}$$

where ω_v is the weight of each node v and $\Omega(\mathcal{V}) = \sum_{v \in \mathcal{V}} \omega_v$. $\Omega(\mathcal{V}) = |\mathcal{V}|$ for unweighted graphs. This expression weights the number of nodes that should appear in π_0 and π_1 according to their total weight. The second component of the cost function is used to minimize the *min-cut* metric f_c:

$$H_B = \sum_{e \in \mathcal{E}} \left(\omega_e \times \left(1 - \prod_{v \in e} x_v - \prod_{v \in e} (1 - x_v) \right) \right) \tag{2}$$

ω_e corresponds to the weight of the hyperedge e. The weight ω_e is added to f_c when $\forall v \in e, \exists v' \in e$ with $v' \neq v$ and $x_v \neq x_{v'}$. The objectives H_A and H_B are then gathered to create the final objective to minimize. Coefficients A and B are real numbers and are used to weight each objective:

$$C(x) = AH_A + BH_B \tag{3}$$

The reader can refer to the method described in the paper of Lucas et al. [9] to set the coefficients A and B. The upper formulation is only valid for a recursive k-partitioning algorithm. If the formulation was for a k-direct partitioning, it would be possible to encode vertex affectation to each partition using logarithmic k-partition encoding, as for coloring problems [15].

3 Experimental Setup

We solve HOBO problems using two quantum optimization methods: QA and the QAOA. The metric used for comparison is the energy gap Δ_E^*, which is the difference between the energy of the ground state (classically exhaustively computed) and the mean energy of the expectation value.

Our work is based on hypergraphs composed of 10 nodes and 15 hyperedges. A first set of instances is composed of k-uniform hypergraphs with $k \in \{2, 3, 4\}$. The parameter k is limited to 4 to avoid trivial solutions. For each value of k, 15 instances are randomly generated.

3.1 Setup of D-Wave Systems

D-Wave processors [3] are designed to minimize an Ising cost function H taking an input vector $s = (s_1, s_2, ..., s_n)$ with $s_i \in \{+1, -1\}$ where h_i and J_{ij} are real numbers.

$$H(s_1, s_2, ..., s_n) = -\sum_{i=1}^{n} h_i s_i - \sum_{i<j}^{n} J_{ij} s_i s_j \tag{4}$$

The translation between QUBO and Ising cost function is straightforward with a simple variable change $x_i = \frac{1-s_i}{2}$.

Experiments were done on the most recent chip *Advantage2_prototype*1.1, which produced the best results minimizing the Ising cost functions. D-Wave systems require a QUBO formulation of the initial problem. The transformation used to convert HOBO to QUBO is done using Rosenberg reduction [14] to quadratize the terms of the cost function.

We used the heuristic presented in [4] to map the QUBO on the D-Wave Quantum Process Unit. Our experiments consider average-quality embedding to avoid bias by selecting only the best embeddings over multiple tries. Majority voting is used during the post-processing phase to determine the final value of each variable. We do not use further specific processing such as spin reversal technique [12] or pausing time.

For each group of HOBO problems transformed to QUBO problems, we numerically study their optimal chain strength cs using a factor called Relative Chain Strength [17]:

$$cs = RCS \times max(\{h_i\} \cup \{J_{ij}\}) \tag{5}$$

The sampling of different values for the chain strength experimentally determines the optimal RCS factor. Figure 1 a. shows that a phase transition occurs when the RCS factor becomes sufficient, leading to a significant improvement. The duplication error rate of qubits measured when the majority vote occurs follows the same phase transition. This evaluation is repeated for each k and optimal average values RCS^* are presented in Table 1.

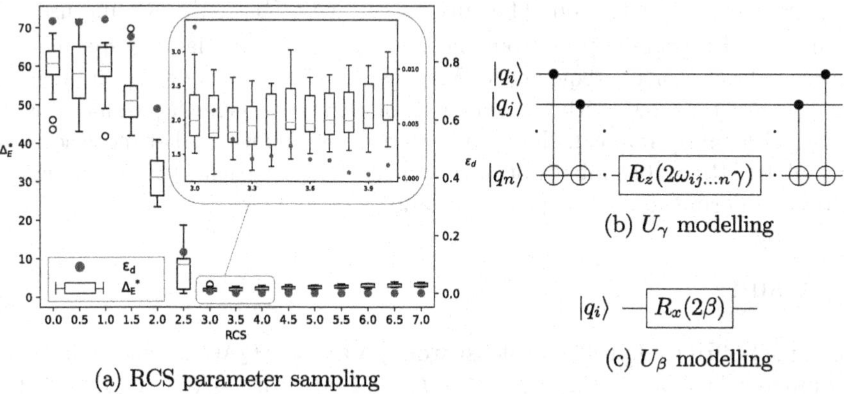

(a) RCS parameter sampling

(b) U_γ modelling

(c) U_β modelling

Fig. 1. Parameter settings of QA and the QAOA. (a) shows the RCS parameter sampling for 2-uniform instances with impact on the energy gap Δ_E^* and duplication error rate on qubits ϵ_d. (b) shows the implementation of each $\omega_{ij\ldots n}\sigma_i\sigma_j\ldots\sigma_n$ term derived from the HOBO cost function terms. (c) shows U_β implementation with β rotation around the X-axis since the domain is not restricted.

3.2 Setup of the QAOA

The QAOA [7] circuit is built from the Hamiltonian derived from the HOBO cost function. Unlike D-Wave systems, the k-local Hamiltoninan can be implemented by the QAOA without quadratization. Figure 1 b. and c. show the unitary implementation corresponding to both problem and mixing Hamiltonian. We perform perfect and noisy simulations of these quantum circuits using the IBM Qiskit library [10]. The *Aer* simulator is used to perform the simulations, which offers a nice trade-off between execution speed and quality of the results. We use Qiskit Pauli error model for noisy simulation with the same error rate ϵ for bit-flip and phase-flip. We assume that the initialization and measurement of qubits are noiseless. Noisy simulations of quantum circuits are executed on different topologies to analyze the benefits of each chip density D. We study 3 different topologies: one which is fully connected, another one based on IBM's *ibmq_guadalupe* heavy-hex topology with a cycle layout of 12 qubits ($D = 0.17$). The last topology comes from *sycamore* chip and is a grid layout of 12 qubits ($D = 0.26$). The mapping of circuits on topologies that are not fully connected requires additional SWAP gates added by the Qiskit transpiler, which generates gate depth overheads.

The QAOA experiment is done from $p = 1$ to $p = 30$. At each step $p = i$, a local optimizer is used to find the optimal set of angles $\vec{\gamma}_p^* = (\gamma_1, ..., \gamma_p)$, $\vec{\beta}_p^* = (\beta_1, ..., \beta_p)$. We use *Nelder-Mead* optimization method [8] with a maximal number of function evaluations (i.e., quantum circuit execution) set to 300. The concentration of good parameters at p-depth $(\vec{\gamma}_p^*, \vec{\beta}_p^*)$ for small values of p has been analytically proven in [1]. L. Zhou et al. [18] introduced an optimization method based on discrete sine and cosine transform that benefits from this parameter concentration. The authors call it FOURIER[q, R], and use it to initialize angles $(\gamma_{p+1}, \beta_{p+1})$ from the sets $\vec{\gamma}_p, \vec{\beta}_p$. The variable q specifies the length of the vector of frequencies. We consider the case when $q = p$, meaning that q parameter grows with p when a pair of angles is added. R parameter is the number of local optima calculated at each level p. Following their notations, we use FOURIER$[\infty, 10]$ global optimization method for each QAOA simulation. For each experiment, we set $\gamma \in [0, 2\pi]$ and $\beta \in [0, \pi]$.

4 Results

The impact of HOBO problem's density on QA and the QAOA is shown in Fig. 2. It compares D-Wave *Advantage2_prototype1.1* and the QAOA ability to find optimal solutions to HOBO problems generated from k-uniform hypergraphs. The perfect simulation of the QAOA surpasses D-Wave systems on 2-uniform (3-uniform) hypergraphs when $p = 24$ ($p = 22$). It shows that the increase in the cardinality of the hyperedges severely limits the performance of the D-Wave quantum computer, which becomes highly inefficient for cardinalities greater than 4. This performance loss is caused by the Rosenberg decomposition coupled with the qubits duplication needed for mapping the problem on D-Wave chip's topology. These two processing steps multiply the required physical qubits by 581% for 4-uniform hypergraphs for D-Wave systems (see Table 1). Comparatively, 4-uniform hypergraphs only increase QAOA depth by 155% on the cycle topology compared to a fully connected topology. The overheads difference is less important for 2-uniform hypergraphs (175% qubit overhead for D-Wave against 173% gate depth overhead on the cycle topology). One can observe that there is no significant difference in circuit depth overheads between cycle and grid layouts, meaning that the Qiskit transpiler algorithm doesn't fully take advantage of the higher connectivity of the grid layout.

Table 1. BHP problem instances description. Each set of instances is composed of 15 hypergraphs having 10 nodes and 15 edges. The (RCS^*) is calculated for each set. The first table shows the overheads of physical qubits needed by D-Wave systems. The second table shows the overheads of gate depth for each topology compared with a fully connected topology. Green cells highlight smallest overheads.

BHP instances		D-Wave system Advantage2 prototype 1.1							
	#terms	RCS^*	#var QR			#qubits			
			min	max	mean	min	max	mean	ratio
2-uniform	55	3.6	10	10	10	17	18	17.5	175%
3-uniform	55	3.0	10	10	10	17	18	17.5	175%
4-uniform	119.6	0.5	24	29	26.8	48	70	58.1	581%

BHP instances	QAOA Circuit depth $p = 1$										
	complete topology			cycle topology				grid topology			
	min	max	mean	min	max	mean	ratio	min	max	mean	ratio
2-uniform	96	126	110.6	167	220	191.3	173%	178	220	196.9	178%
3-uniform	105	150	129.2	174	234	206	159%	185	264	213.9	166%
4-uniform	281	362	323.2	438	581	499	155%	400	485	439	135%

We further study the group of 2-uniform hypergraphs and estimate a noise rate threshold of single and double qubit gates that would permit the QAOA to reach D-Wave performance. Optimal angles are considered to be known at each p-layer. The simulation is done with optimal angles $(\vec{\gamma_p^{opt}}, \vec{\beta_p^{opt}})$ found by the FOURIER$[\infty, 10]$ method on *Aer* simulator. Figure 3 shows the simulation of the QAOA considering various qubit layouts compared to D-Wave systems' best performance on the same instances. The QAOA simulation reaches the best expectation value found by D-Wave *Advantage2_prototype1.1* at $p = 30$ with $\epsilon = 10^{-5}$. Under this threshold, the QAOA becomes inefficient at $p \approx 10$ for $\epsilon = 10^{-3}$ and $p \approx 27$ for $\epsilon = 10^{-4}$. Curves on Fig. 3 a) and b) demonstrate lots of fluctuation, reminiscent of the noise impact on the optimization landscape, even when perfect angles are already known. This last experiment can be considered equally favorable to D-Wave and the QAOA, implying approximately 175% overheads for each physical implementation: 175% qubits overheads for D-Wave

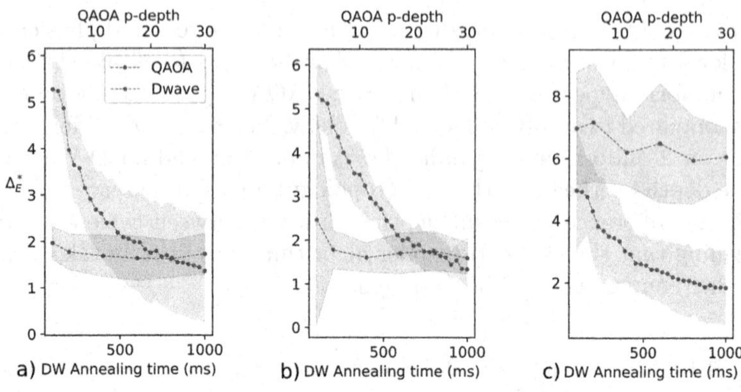

Fig. 2. Quantum heuristics performance solving the BHP problem, using the energy gap Δ_E^* as performance measure. Graphs (a), (b) and (c) respectively show the performance of D-Wave *Advantage2_prototype1.1* and QAOA on 2, 3 and 4-uniform hypergraphs. The shaded area represents the standard deviation of each curve.

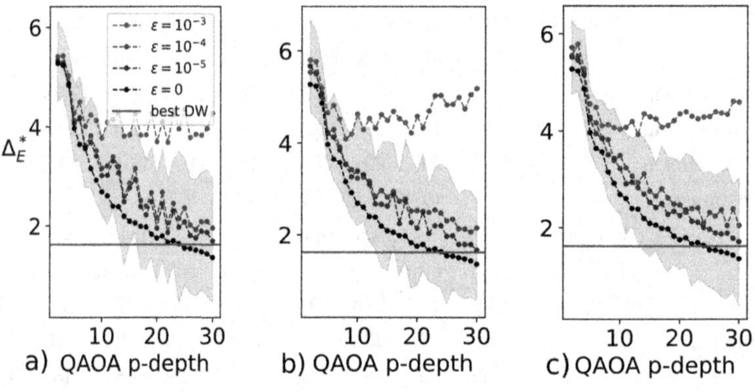

Fig. 3. Simulation of bi-partitioning 2-uniform hypergraphs using QAOA (red) and D-Wave (blue). a) , b) and c) respectively show noisy simulations on fully connected, cycle, and grid layouts. The shaded area represents the standard deviation at $\epsilon = 10^{-5}$. (Color figure online)

systems, versus 173% and 178% depth overheads for cycle and grid topologies. Considering an equal overhead produced by the compilation step, this experiment sets a first bound on noise rate to allow the QAOA to reach D-Wave best available systems, which is 10^{-5}.

5 Conclusion

This work proposes a general approach to compare the performance of D-Wave systems with the QAOA solving HOBO problems. We proposed a method to generate HOBO cost functions from BHP problems with various densities. The higher density regime, illustrated by HOBO problems with many terms, identifies the performance limitations of the QAOA and D-Wave systems. The former is limited by the noisy implementation of gates and the latter by its sparse topology. Even if the QAOA reaches D-Wave systems performance on perfect simulations for low-density problems, the variational heuristic gets rapidly stuck on noisy simulations. Our experiment suggests that a single quantum gate error rate $\epsilon < 10^{-5}$ would permit the QAOA to reach D-Wave systems performances, when the compilation overheads are the same for QA and the QAOA. This bound could be improved with more experiments and a larger set of instances. Current circuit chip designers are approaching this threshold with superconducting systems having $\epsilon \approx 10^{-3}$ and ion-based qubits systems having $\epsilon \approx 10^{-4}$ for single-qubit gates. On the other hand, dense HOBO problems represent hard instances for D-Wave systems, implying the use of quadratic reduction techniques and qubits duplications. Future work will investigate the density threshold for which the performance of the QAOA and QA crosses.

Acknowledgment. The work presented in this paper has been supported by AIDAS - AI, Data Analytics and Scalable Simulation - which is a Joint Virtual Laboratory gathering the Forschungszentrum Jülich (FZJ) and the French Alternative Energies and Atomic Energy Commission (CEA). We thank D. Vert for useful advice and fruitful discussions.

References

1. Akshay, V., Rabinovich, D., Campos, E., Biamonte, J.: Parameter concentrations in quantum approximate optimization. Phys. Rev. A **104**(1), L010401 (2021)
2. Basso, J., Gamarnik, D., et al.: Performance and limitations of the qaoa at constant levels on large sparse hypergraphs and spin glass models. arXiv preprint arXiv:2204.10306 (2022)
3. Bunyk, P., et al.: Architectural considerations in the design of a superconducting quantum annealing processor. IEEE Trans. Appl. Supercond. **24**(4), 1–10 (2014)
4. Cai, J., Macready, W.G., Roy, A.: A practical heuristic for finding graph minors. arXiv preprint arXiv:1406.2741 (2014)
5. Campbell, C., Dahl, E.: Qaoa of the highest order. In: 2022 IEEE 19th International Conference on Software Architecture Companion (ICSA-C), pp. 141–146. IEEE (2022)
6. Djidjev, H.N., Chapuis, G., Hahn, G., Rizk, G.: Efficient combinatorial optimization using quantum annealing. arXiv preprint arXiv:1801.08653 (2018)
7. Farhi, E., et al.: A quantum approximate optimization algorithm (2014)
8. Gao, F., Han, L.: Implementing the nelder-mead simplex algorithm with adaptive parameters. Comput. Optim. Appl. **51**(1), 259–277 (2012)
9. Lucas, A.: Ising formulations of many np problems. Front. Phys., 5 (2014)
10. Md Sajid, A., et al.: Qiskit: An open-source framework for quantum computing (2022). https://doi.org/10.5281/zenodo.2573505
11. Pelofske, E., Bärtschi, A., Eidenbenz, S.: Quantum annealing vs. qaoa: 127 qubit higher-order ising problems on nisq computers. arXiv preprint arXiv:2301.00520 (2023)

12. Pudenz, K.L.: Parameter setting for quantum annealers. In: 2016 IEEE High Performance Extreme Computing Conference (HPEC), pp. 1–6 (2016)
13. Rodriguez, J.: Quantum algorithms for hypergraph bi-partitioning. In: 23ème congrès annuel de la Société Française de Recherche Opérationnelle et d'Aide à la Décision. INSA Lyon, Villeurbanne - Lyon, France (Feb 2022), https://hal.archives-ouvertes.fr/hal-03595234
14. Rosenberg, I.G.: Reduction of bivalent maximization to the quadratic case. Cahiers du Centre d'Etudes de Recherche Operationnelle **17**, 71–74 (1975)
15. Tabi, Z., et al.: Quantum optimization for the graph coloring problem with space-efficient embedding. In: 2020 IEEE International Conference on Quantum Computing and Engineering (QCE), pp. 56–62. IEEE (2020)
16. Ushijima-Mwesigwa, H., Negre, C.F., Mniszewski, S.M.: Graph partitioning using quantum annealing on the d-wave system. In: Proceedings of the Second International Workshop on Post Moores Era Supercomputing, pp. 22–29 (2017)
17. Willsch, D., et al.: Benchmarking advantage and d-wave 2000q quantum annealers with exact cover problems. Quantum Inf. Process. **21**(4), 1–22 (2022)
18. Zhou, L., Wang, S.T., Choi, S., Pichler, H., Lukin, M.D.: Quantum approximate optimization algorithm: Performance, mechanism, and implementation on near-term devices. Phys. Rev. X **10**(2), 021067 (2020)

Constructing Generalized Unitary Group Designs

Ágoston Kaposi[1,2(✉)], Zoltán Kolarovszki[1,2], Adrian Solymos[2,3], Tamás Kozsik[1], and Zoltán Zimborás[1,2,4]

[1] Faculty of Informatics, Eötvös Loránd University, Budapest, Hungary
kaposiagoston@inf.elte.hu, kto@elte.hu
[2] Quantum Computing and Quantum Information Research Group, Department of Computer Sciences, Wigner Research Centre for Physics, Budapest, Hungary
{kolarovszki.zoltan,solymos.adrian,zimboras.zoltan}@wigner.hu
[3] Institute of Physics, Eötvös Loránd University, Budapest, Hungary
[4] Algorithmiq Ltd, Helsinki, Finland

Abstract. Unitary designs are essential tools in several quantum information protocols. Similarly to other design concepts, unitary designs are mainly used to facilitate averaging over a relevant space, in this case, the unitary group $U(d)$. The most appealing case is when the elements of the design form a group, which in turn is called a unitary group design. However, the application of group designs as a tool is limited by the fact that there is no trivial construction method to get even a group 2-design for arbitrary dimensions. In this paper, we present novel construction methods, based on the representation theory of the unitary group and its subgroups, that allow the building of higher-order unitary designs from group designs.

Keywords: Quantum information theory · Unitary t-designs · Representation theory

1 Introduction

Ever since their introduction, *unitary t-designs* have played a ubiquitous role in quantum information science. These finite sets of d-degree unitary operators have the property that averaging an operator over the t-fold tensor products of them equals the same type of averaging over the *entire* unitary group $U(d)$ with respect to the Haar measure. Unitary designs were proved to be useful in particular for the construction of unitary codes [13], the realization of quantum information protocols [6], the derandomization of probabilistic constructions [9], the study of SIC-POVMs [5], the detection of entanglement [2], process tomography [14], randomized benchmarking [17] and for shadow estimation [1,11].

The most well-known example of a unitary t-design is that of the multi-partite Clifford group which forms a unitary 3-design for qubit systems and a unitary 2-design for qudit systems when the local dimension d is an odd prime [12,18,19].

J. Mikyška et al. (Eds.): ICCS 2023, LNCS 14077, pp. 233–245, 2023.
https://doi.org/10.1007/978-3-031-36030-5_19

It is a well-established fact, that t-designs for $U(d)$ always exist for any t and d [15], but the actual construction of these designs is usually a mathematically challenging task. Evidently, this limits their use in concrete applications.

One of the most elegant ways of constructing them goes through representation theory. From unitary irreducible representations of finite groups, one can get a unitary 1-design, and with additional extra properties, the represented group elements can form a unitary 2-design or even a unitary 3-design. However, it has been shown that a representation of a finite group cannot be a unitary t-design for arbitrary $t \geq 4$ and $d > 2$ [3,10]. Moreover, there is no generic construction to find these so-called group 2- and 3-designs for an arbitrary dimension d.

In this paper, a generalization of the aforementioned group design construction is presented which provides methods to construct unitary 2- and possibly higher designs. Concrete examples are also provided in most cases.

The structure of the paper is as follows: Sect. 2 contains some basic definitions and statements regarding t-designs; in Sect. 3 a construction of t-designs from finite unitary subgroups is provided for $t = 2$ and 3 and some examples for the construction is presented; Sect. 4 presents a different construction with which a unitary design can be created from an orthogonal or unitary symplectic design and some examples.

2 Background and Notation

Several different definitions for unitary designs and group designs can be found in the literature [4,7,15]. The following section introduces the ones used in this paper. Most importantly, this paper only considers exact designs.

Definition 1 (t-design of a group). *Let $\mathcal{G} \subseteq U(d)$ be a compact matrix Lie group. A finite set $\mathcal{V} \subseteq \mathcal{G}$ with weight function $w : \mathcal{V} \to [0,1]$ is called a weighted t-design of the group $\mathcal{G}$ if the following equation holds for any linear transformation M on $(\mathbb{C}^d)^{\otimes t}$:*

$$\sum_{V \in \mathcal{V}} w(V)\, V^{\otimes t} M \left(V^{\otimes t}\right)^{\dagger} = \int_{g \in \mathcal{G}} g^{\otimes t} M \left(g^{\otimes t}\right)^{\dagger} \mathrm{d}g. \tag{1}$$

where the integral on the right-hand side is taken over all elements in $\mathcal{G}$ with respect to the Haar measure. The number t is called the order of the design.

Remark 1. In this definition and in the rest of the paper $\mathcal{G}$ can be naturally identified with its defining representation. Therefore, $g^{\otimes t} = \Pi(g)^{\otimes t}$, where Π is the defining representation of $\mathcal{G}$.

Definition 2 (Unitary t-design). *A t-design $\mathcal{V}$ (with weight function w) of a group $\mathcal{G}$ is called a weighted unitary t-design if $\mathcal{G} = U(d)$. If $\mathcal{V}$ forms a group, then it is called a unitary t-group or group t-design.*

Remark 2. The weight function of a t-design $\mathcal{V}$ is the constant function $w \equiv 1/|\mathcal{V}|$ unless otherwise stated.

An alternative characterization of unitary designs can be given through the representation theory of $U(d)$. Considering the t-fold tensor product of the defining representation of $U(d)$, the underlying vector space $(\mathbb{C}^d)^{\otimes t}$ splits up into the different irreducible sectors of $U(d)$ labelled by Young diagrams

$$(\mathbb{C}^d)^{\otimes t} \cong \bigoplus_{\gamma \in \Gamma} \mathcal{K}_\gamma \otimes \mathcal{H}_\gamma, \tag{2}$$

where Γ is the set of Young diagrams containing at most d rows and t number of boxes, $\mathcal{H}_\gamma$ carries the $U(d)$ irrep labelled by the Young diagram γ and $\mathcal{K}_\gamma$ is the multiplicity space (where $U(d)$ acts trivially). Let us denote by $P_\gamma = P_\gamma^{\mathcal{K}} \otimes P_\gamma^{\mathcal{H}}$ the projections corresponding to the $V_\gamma = \mathcal{K}_\gamma \otimes \mathcal{H}_\gamma$ subspaces.

Proposition 1. *A finite set $\mathcal{V} \subset U(d)$ forms a unitary t-design if and only if the following equation is true for all linear transformations M on $(\mathbb{C}^d)^{\otimes t}$:*

$$\sum_{V \in \mathcal{V}} w\left(V\right) V^{\otimes t} M \left(V^{\otimes t}\right)^\dagger = \sum_{\gamma \in \Gamma} \frac{\mathrm{Tr}_\gamma^{\mathcal{H}}(P_\gamma M P_\gamma) \otimes P_\gamma^{\mathcal{H}}}{\mathrm{Tr}(P_\gamma)}, \tag{3}$$

where we used the notation as before, and $\mathrm{Tr}_\gamma^{\mathcal{H}}$ is the partial trace over $\mathcal{H}_\gamma$ of operators supported on the subspace $\mathcal{K}_\gamma \otimes \mathcal{H}_\gamma$.

This proposition can be proven using Schur's lemma, since the left hand side of Eq. (3) commutes with all $U^{\otimes t}$ (this follows from Eq. (1)). Schur's lemma can be invoked after decomposing the tensor product of representations into irreps. If an irrep's multiplicity is one, the resulting intertwining map is simply a constant multiple of the projection to the support, the constant being given by the appropriate proportion of the M operator's trace. If the irrep has higher multiplicity the resulting multiplicity is as indicated on the right hand side of Eq. (3):

A particular set of exact t-designs (for low t) can be constructed using finite groups [8].

Proposition 2. *Let $\mathcal{G}$ be a compact matrix Lie group and $\mathcal{V} < \mathcal{G}$ a finite subgroup. $\mathcal{V}$ is a group t-design if and only if the irreducible subspaces of the defining representation of $\mathcal{G}$ at the t-fold tensor product are equivalent to the irreducible subspaces of the representations' restrictions to the subgroup $\mathcal{V}$.*

3 Constructing Higher Order Designs from Lower Ones

In this section, we want to show a construction method that creates higher-order designs from lower-order ones. The main idea is based on examining the behaviours of representations of finite groups on the relevant invariant subspaces of the t-fold tensor product of the defining representation of the unitary group $U(d)$.

The defining representation of $U(d)$ is the most straightforward one, denoted by $\Pi_{\square}$ and given by

$$\Pi_{\square}(U) = U, \qquad U \in U(d). \tag{4}$$

One can easily verify that this representation is irreducible. In contrast, the t-fold tensor product of this representation $\Pi_{\square}^{\otimes t}$ acting on $\left(\mathbb{C}^d\right)^{\otimes t}$ is reducible for $t \geq 2$. The irreducible decomposition of the t-fold tensor product can be described through Young diagrams and the Schur-Weyl duality. For this paper, one only needs to consider the 2- and 3-fold tensor products of $\Pi_{\square}$, which decompose using the Schur-Weyl duality as

$$\Pi_{\square}^{\otimes 2} \cong \Pi_{\ydiagram{2}} \oplus \Pi_{\ydiagram{1,1}}, \tag{5}$$

$$\Pi_{\square}^{\otimes 3} \cong \Pi_{\ydiagram{3}} \oplus \Pi_{\ydiagram{2,1}}^{\oplus 2} \oplus \Pi_{\ydiagram{1,1,1}}, \tag{6}$$

where the irreducible representations are labelled by their Young diagrams. Evidently Eq. (5) holds for $d > 1$ and Eq. (6) for $d > 2$. It is known from basic representation theory that the irreducible representations $\Pi_{\ydiagram{2}}$ and $\Pi_{\ydiagram{1,1}}$ are supported on the symmetric and antisymmetric subspaces, respectively, and will be referred to as such in the following.

3.1 Constructing 2-designs

A unitary 2-design can be constructed from a unitary representation of a finite group if the irreducible subspaces of the 2-fold tensor product are equivalent to the irreducible subspaces of the 2-fold tensor product of the unitary group according to Proposition 2. In general, the decomposition may not be preserved after restricting to a finite subgroup, but there may exist certain subgroups, for which the decomposition is preserved on some irreducible subspace(s). The current section investigates the possibility of creating 2-designs from such subgroups. This can be done by the construction method stated by the following theorem:

Theorem 1. *Let $d > 1$ and $H, K < U(d)$ be finite subgroups with $\Pi_{\ydiagram{2}}|_K$ and $\Pi_{\ydiagram{1,1}}|_H$ being irreducible representations, then the sets of unitaries $HK = \{hk : h \in H, k \in K\}$ and $KH = \{kh : k \in K, h \in H\}$ both form a weighted unitary 2-design with weights*

$$w_{HK}(U) = w_{KH}(U^{\dagger}) = \frac{|\{(h,k) \in H \times K : hk = U\}|}{|H|\,|K|}. \tag{7}$$

Proof. According to Proposition 3 found in the Appendix, if the statement is true for HK, the same statement is automatically true for KH as well, therefore only proof for HK is needed.

Let M be an arbitrary matrix of dimension $d^2 \times d^2$. Consider the averaging over the elements of HK, taking into account the possibly non-equal weights given by Eq. (7):

$$\sum_{V\in HK} w_{HK}(V)\, V^{\otimes 2}M(V^{\otimes 2})^{\dagger} = \frac{1}{|H|}\sum_{h\in H} h^{\otimes 2}\left(\frac{1}{|K|}\sum_{k\in K} k^{\otimes 2}M(k^{\otimes 2})^{\dagger}\right)(h^{\otimes 2})^{\dagger}. \tag{8}$$

The irrep $\Pi_{\yng(2)}|_K$ appears with multiplicity one in the irrep decomposition of the 2-fold tensor product representation of K by dimensional arguments. By Proposition 1, when performing the averaging with respect to K one acquires

$$\overline{M^K} := \frac{1}{|K|}\sum_{k\in K} k^{\otimes 2}M(k^{\otimes 2})^{\dagger} = c_{\yng(2)}P_{\yng(2)} + N, \tag{9}$$

where $c_{\yng(2)} = \mathrm{Tr}(P_{\yng(2)}M)$ and N is some operator such that $N = P_{\yng(1,1)}NP_{\yng(1,1)}$. The averaging of $\overline{M^K}$ with respect to H can be done with respect to the splitting $V_{\yng(2)} \oplus V_{\yng(1,1)}$ since the off-diagonal blocks are zero. The block corresponding to $V_{\yng(2)}$ remains the same since it commutes with any operator. On the other hand, the block corresponding to $V_{\yng(1,1)}$ after the averaging becomes $c_{\yng(1,1)}P_{\yng(1,1)}$ based on Schur's lemma since $\Pi_{\yng(1,1)}|_H$ is an irreducible representation. As a result, we get

$$\frac{1}{|H|}\sum_{h\in H} h^{\otimes 2}\overline{M^K}(h^{\otimes 2})^{\dagger} = c_{\yng(2)}P_{\yng(2)} + c_{\yng(1,1)}P_{\yng(1,1)}, \tag{10}$$

which proves the theorem by Proposition 1. □

Using the GAP system [16] we have found groups which have the property as described in Theorem 1:

Example 1. A 6-dimensional unitary 2-design can be constructed from the groups $PSU(3,3)$ and A_7. The first group, $PSU(3,3)$, has a unitary irreducible representation in 6 dimensions for which the symmetric irreducible representation on the 2-fold tensor product of the unitary group restricted to this representation of $PSU(3,3)$ remains irreducible while the antisymmetric does not. On the other hand, the alternating group on 7 elements has a 6 dimensional unitary irreducible representation that remains irreducible on the antisymmetric subspace of the 2-fold tensor product while being reducible on the symmetric subspace. As a result, we can construct a weighted unitary 2-design in 6 dimensions with weights given by Eq. (7) based on Theorem 1.

Example 2. Similarly to Example 1, a 4-dimensional unitary 2-design can be constructed from two groups found in the `SmallGroup` library of GAP. The group K which is obtained from the 6-th irreducible representation of `SmallGroup(640,21454)` and the group H which is obtained from the 2-nd irreducible representation of `SmallGroup(120,34)` behave as described in Theorem 1. This means that the representation $\Pi_{\square}^{\otimes 2}$ restricted to group K

remains irreducible on the symmetric subspace and restricted to group H remains irreducible on the antisymmetric subspace (however, they are reducible on their respective complement).

3.2 Possible Construction of Higher Designs

As in the previous construction described in Sect. 3.1 for the $t = 2$ case, a similar method is expected to work for $t = 3$ or higher-order designs. However, for $t \geq 3$, the irreducible decomposition of the t-fold tensor product of $\Pi_{\square}$ contains at least a subspace with multiplicity 2 or higher, which results in different behaviour when averaging over the unitary group, as described in Proposition 1. Luckily, the following theorem asserts that the previous construction generalizes:

Theorem 2. *Let $d > 2$ such that $d \neq 4$ and $H, K < U(d)$ finite subgroups such that for each $\gamma = \square\square\square, \begin{smallmatrix}\square\square\\\square\end{smallmatrix}, \begin{smallmatrix}\square\\\square\\\square\end{smallmatrix}$ either $\Pi_\gamma|_K$ or $\Pi_\gamma|_H$ is irreducible. Then the sets of unitaries $HK = \{hk : h \in H, k \in K\}$ and $KH = \{kh : h \in H, k \in K\}$ both form a weighted unitary 3-design with weights given by*

$$w_{HK}(U) = w_{KH}(U^\dagger) = \frac{|\{(h,k) \in H \times K : hk = U\}|}{|H|\,|K|}. \tag{11}$$

Remark 3. In case of $d = 4$ the dimensions of $\Pi_{\square\square\square}$ and $\Pi_{\begin{smallmatrix}\square\square\\\square\end{smallmatrix}}$ are equal which would result in a different condition for this Theorem which will not be discussed here.

Proof. According to Proposition 3 found in the Appendix, if the statement is true for HK, the same statement is automatically true for KH as well, therefore only proof for HK is needed.

Let M be an arbitrary matrix of dimension $d^3 \times d^3$. One can write

$$\sum_{V \in HK} w_{HK}(V) V^{\otimes 3} M \left(V^\dagger\right)^{\otimes 3} = \frac{1}{|H|} \sum_{h \in H} h^{\otimes 3} \left(\frac{1}{|K|} \sum_{k \in K} k^{\otimes 3} M \left(k^\dagger\right)^{\otimes 3} \right) \left(h^\dagger\right)^{\otimes 3}, \tag{12}$$

where the appearance of w_{HK} follows from the fact that some elements in the product of groups H and K may coincide. For brevity the following is introduced:

$$\overline{M^K} := \frac{1}{|K|} \sum_{k \in K} k^{\otimes 3} M \left(k^\dagger\right)^{\otimes 3}, \tag{13}$$

$$\overline{M^{HK}} := \frac{1}{|H|} \sum_{h \in H} h^{\otimes 3} \overline{M^K} \left(h^\dagger\right)^{\otimes 3}. \tag{14}$$

To prove the theorem, we need to show that $\overline{M^{HK}}$ takes the form of the RHS in Eq. (3) from Proposition 1. This could be demonstrated by describing all the block elements of $\overline{M^{HK}}$ given by the projections corresponding to Eq. (6). Firstly, we investigate the block in the decomposition corresponding to the subspace $V_{\ydiagram{3}}$. If the representation $\Pi_{\ydiagram{3}}|_K$ remains irreducible on the subspace $V_{\ydiagram{3}}$, one can write

$$\frac{1}{|K|}\sum_{k\in K} k^{\otimes 3} P_{\ydiagram{3}} M P_{\ydiagram{3}} (k^{\dagger})^{\otimes 3} = \operatorname{Tr}(P_{\ydiagram{3}} M) P_{\ydiagram{3}}. \tag{15}$$

Consequently, the matrix $\overline{M^K}$ on the subspace $V_{\ydiagram{3}}$ acts as an identity matrix, therefore taking the average over H has no effect here. Moreover, if the representation $\Pi_{\ydiagram{3}}|_H$ remains irreducible then the average of $P_{\ydiagram{3}}\overline{M^K}P_{\ydiagram{3}}$ over H diagonalizes the matrix. As a consequence, by taking the average over HK with weights w_{HK}, the resulting transformation on the subspace $V_{\ydiagram{3}}$ acts as a unitary 3-design. The case of block $P_{\ydiagram{1,1,1}} M P_{\ydiagram{1,1,1}}$ is analogous to the case $V_{\ydiagram{3}}$.

According to the statement, $d \neq 4$, hence $\dim(V_\gamma) \neq \dim(V_{\gamma'})$ for $\gamma \neq \gamma'$. Consider a block described by $P_\gamma M P_{\gamma'}$ where $\gamma \neq \gamma'$. Let $\dim(V_\gamma) < \dim(V_{\gamma'})$, without loss of generality. By assumption, the representation restricted to either group H or K remains irreducible on $V_{\gamma'}$. By taking the average by H or K, the considered block may become zero, since it could only be an intertwiner between two different dimensional irreducible subspaces. After taking the average by both H and K consecutively, every off-diagonal block must vanish.

If the representation $\Pi_{\ydiagram{2,1}}|_K$ remains irreducible, then after averaging over K the matrix $P_{\ydiagram{2,1}} M P_{\ydiagram{2,1}}$ is by Proposition 2 and Proposition 1:

$$P_{\ydiagram{2,1}}\overline{M^K}P_{\ydiagram{2,1}} = \frac{1}{\operatorname{Tr}(P_{\ydiagram{2,1}})}\operatorname{Tr}^{\mathcal{H}}_{\ydiagram{2,1}}(P_{\ydiagram{2,1}} M P_{\ydiagram{2,1}}) \otimes P^{\mathcal{H}}_{\ydiagram{2,1}}. \tag{16}$$

Since for all $h \in H$ the $h^{\otimes 3}$ is block diagonal on the subspace corresponding to $V_{\ydiagram{2,1}}$, the matrix in Eq. (16) commutes with it.

If the representation $\Pi_{\ydiagram{2,1}}|_H$ remains irreducible, then the same happens to the matrix $\overline{M^K}$, and due to the cyclic property of the partial trace if tracing out over $\mathcal{H}_{\ydiagram{2,1}}$ it gives the same as partial trace for the matrix M:

$$\operatorname{Tr}^{\mathcal{H}}_{\ydiagram{2,1}}\left(P_{\ydiagram{2,1}} M P_{\ydiagram{2,1}}\right) = \operatorname{Tr}^{\mathcal{H}}_{\ydiagram{2,1}}\left(P_{\ydiagram{2,1}}\overline{M^K}P_{\ydiagram{2,1}}\right). \tag{17}$$

Consequently, all blocks are the same as in Proposition 1. □

Example 3. Using the GAP system it can be shown that a 10-dimensional unitary 3-design can be constructed from two groups using Theorem 2. Let H be the 3-rd irreducible representation of the group ``(3xU5(2)).2" and $K$ the 36-th irreducible representation of the group ``2x2.M22". Then $H^{\otimes 3}|_{\ydiagram{3}}$ is irreducible and $K^{\otimes 3}|_{\ydiagram{2,1}}$ and $K^{\otimes 3}|_{\ydiagram{1,1,1}}$ are irreducible. Therefore the set of unitaries given by the product HK with weights given by Eq. (11) produces a 10-dimensional 3-design by Theorem 2.

4 Unitary 2-Designs from Orthogonal and Symplectic 2-Designs

In the previous section, we provided methods to build unitary designs from the irreducible representations of two finite unitary subgroups. Let us turn our attention to the scenario where one fixes a 2-design $\mathcal{V}$ of a subgroup $\mathcal{G}$ of the unitary group. This is then transformed resulting in a set of unitary matrices also exhibiting design properties. Using this construction, *unitary* designs can be obtained from these two sets. In particular, the orthogonal and the unitary symplectic group (also called compact symplectic group) will be considered for $\mathcal{G}$. Note that using Definition 1, orthogonal and unitary symplectic t-designs are just t-designs of the groups $\mathcal{G} = O(d)$ and $\mathcal{G} = USp(d)$, respectively.

Theorem 3. *Let $\mathcal{V} \subset O(d)$ form an orthogonal 2-design, and consider the set $\mathcal{W}_\alpha := W_\alpha \mathcal{V} W_\alpha^\dagger$, where W_α is the unitary describing the basis transformation in Eq. (18), then the set of unitaries $\mathcal{W}_\alpha \cdot \mathcal{V}$ forms a unitary 2-design.*

Proof. Let $\{|j\rangle\}_{j=0}^{d-1}$ be the basis of $\mathbb{C}^d$ with respect to which the representation of the orthogonal group is considered as real matrices. This leads to a basis on the tensor square $(\mathbb{C}^d)^{\otimes 2}$ defined by the tensor power of the elements: $\{|j\rangle \otimes |k\rangle\}_{j,k=0}^{d-1}$. Let W_α be the operation on the basis elements defined by

$$W_\alpha |j\rangle = (\tau_\alpha)^j |j\rangle \,, \tag{18}$$

where $\tau_\alpha = e^{\frac{2\pi i \alpha}{2d}}$.

Let $\Phi_\square$ denote the defining representation of the orthogonal group with respect to the basis $\{|j\rangle\}_{j=0}^{d-1}$. This can be embedded into the defining representation of the unitary group. The irreducible decomposition of $\Phi_\square^{\otimes 2}$ is the following:

$$\Phi_\square^{\otimes 2} \cong \Phi_{|\psi\rangle} \oplus \Phi_{|\psi\rangle}^c \oplus \Phi_{\boxminus}, \tag{19}$$

where $\Phi_{|\psi\rangle}$ is a 1-dimensional representation acting on the subspace spanned by $|\psi\rangle = \frac{1}{\sqrt{d}} \sum_{j=0}^{d-1} |j\rangle \otimes |j\rangle$, $\Phi_{|\psi\rangle} \oplus \Phi_{|\psi\rangle}^c$ and $\Phi_{\boxminus}$ act on the symmetric subspace $V_{\square\square}$ and on the anti-symmetric subspace $V_{\boxminus}$, respectively, where the indices are used as in Eq. (5). The projections to these subspaces are $P_{|\psi\rangle} = |\psi\rangle\langle\psi|$, $P_{\square\square} - P_{|\psi\rangle}$ and $P_{\boxminus}$, respectively.

Let us now examine the set $\mathcal{W}_\alpha = W_\alpha \mathcal{V} W_\alpha^\dagger$. This also forms an orthogonal 2-design, however in this case for the representation $\Phi'_\square$ which is unitarily equivalent to $\Phi_\square$ but the matrices are considered real with respect to the basis $\{|W_\alpha j\rangle\}_{j=0}^{d-1}$. This means that in the irrep decomposition of the two-fold tensor product, the distinguished one-dimensional subspace is spanned by $|\psi'\rangle = \frac{1}{\sqrt{d}} \sum_{j=0}^{d-1} |W_\alpha j\rangle \otimes |W_\alpha j\rangle$.

One can take the scalar product $f(\alpha) := |\langle \psi, W_\alpha \otimes W_\alpha \psi\rangle|^2$. This gives $f(0) = 1$ and $f(1) = 0$ and, by the continuity of the scalar product, for arbitrary $q \in [0,1]$ there is an $\alpha \in [0,1]$ which gives $f(\alpha) = q$. This is used later to define the value of α for a given d. The main idea of this proof is, using the fact that when $\mathcal{V}$ and $\mathcal{W}_\alpha$ form an orthogonal 2-design, that the value of α can be determined in a way that the product $\mathcal{W}_\alpha\mathcal{V}$ also forms a unitary 2-design.

Let M be an arbitrary complex matrix of dimension $d^2 \times d^2$. According to Proposition 1, averaging with $\mathcal{V}$ results in

$$\begin{aligned}\overline{M} &:= \sum_{v\in\mathcal{V}} (v\otimes v)M(v\otimes v)^\dagger \\ &= \frac{\operatorname{Tr}\left(P_{|\psi\rangle}M\right)}{\operatorname{Tr}(P_{|\psi\rangle})}P_{|\psi\rangle} + \frac{\operatorname{Tr}\left(P_{\Box\Box}M - P_{|\psi\rangle}M\right)}{\operatorname{Tr}\left(P_{\Box\Box} - P_{|\psi\rangle}\right)}\left(P_{\Box\Box} - P_{|\psi\rangle}\right) + \frac{\operatorname{Tr}\left(P_{\substack{\Box\\\Box}}M\right)}{\operatorname{Tr}(P_{\substack{\Box\\\Box}})}P_{\substack{\Box\\\Box}}.\end{aligned} \tag{20}$$

Moreover, averaging $\overline{M}$ with $\mathcal{W}_\alpha$ one acquires

$$\begin{aligned}\overline{\overline{M}} &:= \sum_{w\in\mathcal{W}_\alpha} (w\otimes w)\,\overline{M}(w\otimes w)^\dagger \\ &= \frac{\operatorname{Tr}\left(P_{|\psi'\rangle}\overline{M}\right)}{\operatorname{Tr}(P_{|\psi'\rangle})}P_{|\psi'\rangle} + \frac{\operatorname{Tr}\left(P_{\Box\Box}\overline{M} - P_{|\psi'\rangle}\overline{M}\right)}{\operatorname{Tr}(P_{\Box\Box} - P_{|\psi'\rangle})}\left(P_{\Box\Box} - P_{|\psi'\rangle}\right) + \\ &\quad + \frac{\operatorname{Tr}\left(P_{\substack{\Box\\\Box}}\overline{M}\right)}{\operatorname{Tr}(P_{\substack{\Box\\\Box}})}P_{\substack{\Box\\\Box}}.\end{aligned} \tag{21}$$

On the antisymmetric subspace, the averaging acts like a unitary design since $\operatorname{Tr}\left(P_{\substack{\Box\\\Box}}\overline{\overline{M}}\right) = \operatorname{Tr}(P_{\substack{\Box\\\Box}}\overline{M}) = \operatorname{Tr}(P_{\substack{\Box\\\Box}}M)$. However, for it to act like a unitary design on the symmetric subspace, it is required that all diagonal elements are equal to each other. This means that in the remaining part of the proof it is enough to consider only the symmetric subspace so as to get this desired property.

Let the action of $\overline{M}$ on the unitary symmetric subspace be $D := \overline{M}|_{V_{\Box\Box}} = a\cdot P_{\Box\Box} + b\cdot P_{|\psi\rangle}$ for some $a, b \in \mathbb{C}$. By averaging it with $\mathcal{W}_\alpha$ this expression gets modified to

$$\frac{\operatorname{Tr}(P_{|\psi'\rangle}D)}{\operatorname{Tr}(P_{|\psi'\rangle})}P_{|\psi'\rangle} + \frac{\operatorname{Tr}\left(\left(P_{\Box\Box} - P_{|\psi'\rangle}\right)D\right)}{\operatorname{Tr}\left(P_{\Box\Box} - P_{|\psi'\rangle}\right)}\left(P_{\Box\Box} - P_{|\psi'\rangle}\right). \tag{22}$$

One can calculate the coefficient of the first term of Eq. (22) by

$$\operatorname{Tr}(P_{|\psi'\rangle}D) = \operatorname{Tr}(P_{|\psi'\rangle}aP_{\Box\Box} + P_{|\psi'\rangle}bP_{|\psi\rangle}) = a + bq. \tag{23}$$

Analogously, the coefficient of the second term is

$$\operatorname{Tr}\left(\left(P_{\Box\Box} - P_{|\psi'\rangle}\right)D\right) = \left(\frac{d(d+1)}{2} - 1\right)a + (1-q)b. \tag{24}$$

The two coefficients in equation (22) need to be equal for $\mathcal{W}_\alpha \mathcal{V}$ to form a unitary 2-design. Therefore the following condition needs to be met:

$$\left(\frac{d(d+1)}{2} - 1\right)(a + bq) = \left(\frac{d(d+1)}{2} - 1\right)a + (1-q)b, \tag{25}$$

which simplifies to

$$q = \frac{2}{d(d+1)}. \tag{26}$$

This can be achieved independently from the a and b values. □

Using the GAP system [16] we have found groups which have the property as described in Theorem 3:

Example 4. The group `PSU(3,3)` has a 7-dimensional irreducible representation (`PSU(3,3)`$_{[3]}$ in GAP where the index denotes the third irrep) with which a 7-dimensional 2-design can be constructed using Theorem 3.

For constructing unitary 2-designs from symplectic 2-designs a similar theorem can be formulated:

Theorem 4. *Let $\mathcal{V} \subset USp(d)$ form a unitary symplectic 2-design, and consider the set $\mathcal{W}_\alpha = W_\alpha \mathcal{V} W_\alpha^\dagger$, where W_α is a unitary describing the basis transformation in Eq.* (27)*, then the set of unitaries $\mathcal{W}_\alpha \cdot \mathcal{V}$ forms a unitary 2-design.*

Proof. The proof is similar to the proof of Theorem 3 with the 1-dimensional subspace determined by $|\psi\rangle = \sum_{j=0}^{d-1} |j\rangle \otimes |j+d\rangle - |j+d\rangle \otimes |j\rangle$. The operator corresponding to the basis transformation is defined as

$$\begin{aligned} W_\alpha |j\rangle &= \tau^j |j\rangle \\ W_\alpha |j+d\rangle &= \tau^j |j+d\rangle \end{aligned} \tag{27}$$

for each $j = 0, \ldots, d-1$ and $\tau_\alpha = e^{\frac{2\pi i \alpha}{2d}}$. □

Using the GAP system [16] we have found groups which have the property as described in Theorem 4:

Example 5. The group denoted as `SmallGroup(640,21454)` in GAP has 4-dimensional irreducible representation (`SmallGroup(640,21454)`$_{[6]}$ in GAP where the index denotes the sixth irrep) with which a 4-dimensional 2-design can be constructed using Theorem 4.

Example 6. The group `PSU(3,3)` has 6-dimensional irreducible representation (`PSU(3,3)`$_{[2]}$ in GAP) with which a 6-dimensional 2-design can be constructed using Theorem 4.

5 Summary and Outlook

The current paper establishes a procedure for constructing a weighted t-design using finite groups whose representation admits an easily verifiable property and some examples are shown to use this procedure to construct 2-designs. Furthermore, a method for constructing a unitary 2-design from an orthogonal or a unitary symplectic 2-design is proposed with some examples to demonstrate the working of construction. However, there are a plethora of possible research directions regarding group representations based on these ideas. In particular, we plan to carry out a very thorough symbolic search through finite groups using GAP to identify cases, and perhaps even families of cases, when our current methods could be used successfully. We also aim to extended the basis change trick to cases beyond orthogonal and symplectic 2-designs, considering rather general examples of the splitting of the 2-fold tensor product representation of finite group families. On the more ambitious side, one of the goals could be to extend some of the results (or the ideas) in the paper to families of random circuits, which have different splitting and convergence properties in different irreducible subspaces of $U(d)$. Specifically, one may intend to study random circuits that are made of sequences of random orthogonal and random symplectic gates, and compare them with the convergence of other gate-set families.

Acknowledgement. This research was supported by the Ministry of Culture and Innovation and the National Research, Development and Innovation Office through the Quantum Information National Laboratory of Hungary (Grant No. 2022-2.1.1-NL-2022-00004) and OTKA grant No. FK 135220.

Appendix A: Symmetry of Product Designs

Proposition 3. *Let $H, K \subset U(d)$ be finite subsets invariant to the elementwise adjoint operation ($H^\dagger = H$, $K^\dagger = K$). If HK forms a t-design, then KH also forms a t-design, where $HK = \{hk : h \in H, k \in K\}$.*

Remark 4. H, K being finite subgroups of $U(d)$ is a special case.

Proof. It is easy to see from the properties of the adjoint that $(HK)^\dagger = KH$, where $(HK)^\dagger = \{(hk)^\dagger : h \in H, k \in K\}$. Starting from this observation, we now provide proof that for any t-design $\{V_i\}_{i=1}^n$ it follows that $\{V_i^\dagger\}_{i=1}^n$ is also a t-design, which then completes the proof of the theorem.

To see that the above-mentioned proposition is true, we will use the non-degeneracy of the Hilbert-Schmidt inner product $\langle A, B\rangle_{\text{HS}} = \text{Tr}(A^\dagger B)$, which implies that if $\text{Tr}(AB) = \text{Tr}(CB)$ for all $B \in \mathcal{B}(\mathbb{C}^d)$ then $A = C$. From this non-degeneracy statement it follows that given a t-design $\{V_i\}_{i=1}^n$, the set $\{V_i^\dagger\}_{i=1}^n$ is also a t-design if and only if

$$\text{Tr}\left(\frac{1}{n}\sum_i (V_i^\dagger)^{\otimes t} X V_i^{\otimes t}\, Y\right) = \text{Tr}\left(\int_{U(d)} U^{\otimes t} X (U^\dagger)^{\otimes t} \text{d}U\, Y\right) \tag{28}$$

holds for all $X, Y \in \mathcal{B}((\mathbb{C}^d)^{\otimes t})$. Using the linearity and the cyclic property of the trace one can write

$$\begin{aligned}
\operatorname{Tr}\left(\frac{1}{n}\sum_i (V_i^\dagger)^{\otimes t} X V_i^{\otimes t}\, Y\right) &= \operatorname{Tr}\left(\frac{1}{n}\sum_i V_i^{\otimes t}\, Y (V_i^\dagger)^{\otimes t} X\right) \\
&= \operatorname{Tr}\left(\int_{U(d)} U^{\otimes t} Y (U^\dagger)^{\otimes t} \mathrm{d}U\, X\right) \\
&= \operatorname{Tr}\left(\int_{U(d)} (U^\dagger)^{\otimes t} X U^{\otimes t} \mathrm{d}U\, Y\right) \\
&= \operatorname{Tr}\left(\int_{U(d)} U^{\otimes t} X (U^\dagger)^{\otimes t} \mathrm{d}U\, Y\right), \qquad (29)
\end{aligned}$$

where the last equality follows from the invariance of the Haar measure with respect to inversion, that is for any $X \in \mathcal{B}((\mathbb{C}^d)^{\otimes t})$:

$$\int_{U(d)} U^{\otimes t} X (U^\dagger)^{\otimes t} \mathrm{d}U = \int_{U(d)} (U^\dagger)^{\otimes t} X U^{\otimes t} \mathrm{d}U. \qquad (30)$$

Using this line of thought, Eq. 28 follows, which proves the theorem. □

References

1. Arienzo, M., Heinrich, M., Roth, I., Kliesch, M.: Closed-form analytic expressions for shadow estimation with brickwork circuits. arXiv preprint (2022). https://doi.org/10.48550/arXiv.2211.09835
2. Bae, J., Hiesmayr, B.C., McNulty, D.: Linking entanglement detection and state tomography via quantum 2-designs. New J. Phys. **21**(1), 013012 (2019). https://doi.org/10.1088/1367-2630/aaf8cf
3. Bannai, E., Navarro, G., Rizo, N., Tiep, P.H.: Unitary t-groups. J. Math. Soc. Japan **72**(3), 909–921 (2020). https://doi.org/10.2969/jmsj/82228222
4. Bengtsson, I., Życzkowski, K.: Geometry of quantum states: an introduction to quantum entanglement. Cambridge University Press (2017)
5. Czartowski, J., Goyeneche, D., Grassl, M., Życzkowski, K.: Isoentangled mutually unbiased bases, symmetric quantum measurements, and mixed-state designs. Phys. Rev. Lett. **124**(9), 090503 (2020). https://doi.org/10.1103/physrevlett.124.090503
6. Dankert, C., Cleve, R., Emerson, J., Livine, E.: Exact and approximate unitary 2-designs and their application to fidelity estimation. Physical Review A **80**(1) (2009). https://doi.org/10.1103/physreva.80.012304
7. Di Matteo, O.: A short introduction to unitary 2-designs. CS867/QIC890 (2014)
8. Gross, D., Audenaert, K., Eisert, J.: Evenly distributed unitaries: On the structure of unitary designs. J. Mathem. Phys. **48**(5), 052104 (2007). https://doi.org/10.1063/1.2716992
9. Gross, D., Krahmer, F., Kueng, R.: A partial derandomization of phaselift using spherical designs. J. Fourier Anal. Appl. **21**(2), 229–266 (2014). https://doi.org/10.1007/s00041-014-9361-2

10. Guralnick, R., Tiep, P.: Decompositions of small tensor powers and Larsen's conjecture. Representation Theory **9**(5), 138–208 (2005). https://doi.org/10.1090/S1088-4165-05-00192-5
11. Helsen, J., Walter, M.: Thrifty shadow estimation: re-using quantum circuits and bounding tails. arXiv preprint (2022). https://doi.org/10.48550/arXiv.2212.06240
12. Kueng, R., Gross, D.: Qubit stabilizer states are complex projective 3-designs. arXiv preprint arXiv:1510.02767 (2015). https://doi.org/10.48550/arXiv.1510.02767
13. Roy, A., Scott, A.J.: Unitary designs and codes. Designs Codes Cryptograp. **53**(1), 13–31 (2009). https://doi.org/10.1007/s10623-009-9290-2
14. Scott, A.J.: Optimizing quantum process tomography with unitary 2-designs. J. Phys. A: Mathem. Theoret. **41**(5), 055308 (2008). https://doi.org/10.1088/1751-8113/41/5/055308
15. Seymour, P., Zaslavsky, T.: Averaging sets: A generalization of mean values and spherical designs. Adv. Math. **52**(3), 213–240 (1984). https://doi.org/10.1016/0001-8708(84)90022-7
16. The GAP Group: Gap-groups, algorithms, and programming, vol. 412, p. 2 (2022), https://www.gap-system.org
17. Wallman, J.J., Flammia, S.T.: Randomized benchmarking with confidence. New J. Phys. **16**(10), 103032 (2014). https://doi.org/10.1088/1367-2630/16/10/103032
18. Webb, Z.: The Clifford group forms a unitary 3-design. Quantum Inform. Comput. **16**, 1379–1400 (2015). https://doi.org/10.26421/QIC16.15-16-8
19. Zhu, H.: Multiqubit Clifford groups are unitary 3-designs. Phys. Rev. A **96**, 062336 (2017). https://doi.org/10.1103/PhysRevA.96.062336

Multi-objective Quantum-Inspired Genetic Algorithm for Supervised Learning of Deep Classification Models

Jerzy Balicki(✉)

Faculty of Mathematics and Information Science, Warsaw University of Technology, ul. Koszykowa 75, 00-662 Warsaw, Poland
jerzy.balicki@pw.edu.pl

Abstract. Quantum decision making is an emerging field that explores how quantum computing can be used to make decisions more efficiently and effectively than classical computing. The main advantage of quantum decision making is the ability to explore multiple possible solutions to a problem simultaneously, using the principles of superposition and entanglement. A quantum-inspired genetic algorithm can improve a quality of a multi-criteria supervised learning of deep classification models. Designed classifiers can be trained by a quantum simulator with Hadamard, CNOT and rotation gates. To demonstrate advantages of the new algorithm, we analyze the Pareto-optimal classifiers for an efficient diagnosis of SARS-CoV-2 infection based on remote analysis of X-rays images with the quantum computing platform QI.

Keywords: Deep Learning Models · Genetic Algorithms · Quantum Gates

1 Introduction

The power of quantum computing comes from its ability to perform many computations in parallel. Quantum genetic algorithms leverage this capability by representing the search space in a quantum superposition of states, which allows for exploration of multiple potential solutions at once. These properties make quantum genetic algorithms a promising area of research for solving complex optimization problems, such as those encountered in machine learning. For the above reasons, we propose a quantum-inspired genetic algorithm to train deep neural networks for diagnosis of Covid-19 cases using chest radiography X-rays images. Obtained results of numerical experiments with the quantum simulator QuTech confirmed the great potential of quantum algorithms [24].

To order many important issues in this paper, related work is described in Sect. 2. Then, the issues related to using deep learning in SARS-CoV-2 are characterized in Sect. 3. Next, Sect. 4 presents some studies under quantum encoding and an evolution device for evolutionary algorithms. Finally, some experimental results are presented in Sect. 5.

J. Mikyška et al. (Eds.): ICCS 2023, LNCS 14077, pp. 246–253, 2023.
https://doi.org/10.1007/978-3-031-36030-5_20

2 Related Work

Richard Feynman introduced hypothesis that a classical computer is not able to simulate physical phenomena such as the quantum computer [11]. Moreover, Benioff confirmed that a quantum computer could meet the principles of a Turing machine [7]. It was also shown that a universal quantum computer could perform tasks impossible to solve by a universal Turing machine [10]. Quantum calculations are based on quantum gates, where qubits in the quantum register are initialized in the $|0\rangle$ state due to the z-basis. There are several quantum algorithms such as Shor's algorithm for factoring large integers [21]. Quantum Monte Carlo, quantum phase estimation, and HHL algorithm have potential applications in machine learning [1]. Grover's algorithm searches an unsorted database in time complexity $O(\sqrt{n})$, while the best-known classical algorithm needs $O(n)$ operations [12].

However, constructing a quantum processor is a challenge because of requirements that are in conflict: state preparation, long coherence times, universal gate operations and qubit readout. Processors based on a few qubits have been demonstrated using several technologies such as nuclear magnetic resonance, cold ion trap and optical systems [18]. A calculation operation is required to be completed much more quickly than the decoherence time [4].

The undoubted development of quantum computers goes hand in hand with the intensive development of artificial intelligence. One of the most interesting ideas concerns the implementation of quantum-inspired genetic algorithms [13]. In a quantum-inspired evolutionary algorithm (QEA), a quantum register is defined by a string of qubits [14]. We believe the quantum register ensures a higher population diversity than other known representations. Besides, the evolution quantum operator implemented by the quantum gate replaces the selection and mutation operators [20]. Solutions are represented probabilistically because the quantum register represents the linear superposition of all possible states. The QEA has been developed for the combinatorial optimization problems such as face verification, the knapsack problem and the Travelling Salesman Problem [6, 26]. Besides, the QMEA found solutions close to the Pareto-optimal front for multi-objective 0/1 knapsack problems [16].

Recently, a framework of genetic algorithm-based CNN has been proposed on multi-access edge computing for automated detection of COVID-19 [15]. In this context, we constructed an adaptive multi-objective quantum-based genetic algorithm (MQGA) for supervised training of deep learning models based on Convolutional Neural Networks. The MQGA works with a quantum register that provides a population of chromosomes that represents hyperparameters of CNNs. In this approach, some strategies for adaptive parameters of the algorithm can be developed, too. The MQGA improves proximity to the Pareto-optimal front and preserving diversity by employing advantages of quantum-inspired gates.

Quantum computing can be simulated in computing clouds, too. IBM offers Quantum Computing as a Service, QCaaS [7]. Alibaba and CAS provide public quantum computing services, too [2]. D-Wave launched Leap, the real-time quantum application environment. Rigetti Computing delivers public Quantum Cloud Services, QCS, where quantum processors are integrated with classical computing infrastructure and made available to user over the cloud [21]. Moreover, Amazon Quantum Solutions Lab allows developers to work with experts in quantum computing, machine learning, and high-performance computing [3]. Azure Quantum enables an access for diverse quantum software, hardware, and solutions from Microsoft and our partners [19].

3 Dataset with X-Ray Image Collection

The COVID-19 pandemic is still having a devastating impact not only on health and the economy, but also on people's sense of security, well-being and satisfaction. The dataset called COVID-19 Image Data Collection is a publicly available dataset for diagnosis using deep learning algorithms. It consists of chest X-ray and CT scan images (https://github.com/ieee8023/covid-chestxray-dataset). The dataset COVIDx is available at (https://github.com/lindawangg/COVID-Net). The set RSNA COVID-19 Detection Challenge is provided by the Radiological Society of North America (https://www.kaggle.com/c/rsna-pneumonia-detection-challenge/data). Besides, the SIRM COVID-19 CT Dataset consists of CT scan images collected by the Italian Society of Medical and Interventional Radiology available at https://www.sirm.org/en/category/articles/covid-19/. Also, the dataset COVID-CT can be used to classify COVID-19 patients and patients with other lung diseases by deep learning models. It was created by researchers from the University of San Diego and is available at https://github.com/UCSD-AI4H/COVID-CT.

The COVIDx dataset is was created by researchers from Qatar University, the University of Dhaka, and the University of Malaysia. The dataset contains over 13,000 chest X-ray images, including 5,445 COVID-19 positive images, as well as images of patients with other types of respiratory diseases and healthy individuals. It is available on various online platforms, including Kaggle, GitHub, and the COVID-19 Image Data Collection website [28].

We designed an experimental Covid-19 CXR diagnostic cloud (Covid CXR) to be made available for some clinicians to support the diagnosis of coronavirus 2. In the next stage, the Covid-19 detection application will be available for patients via the Internet. The advantage of this solution is the reduction of costs and a short waiting time for the test result (few seconds), which in turn is of key importance for improving not only the health situation, but also the economic and social situations.

When infected patients are effectively screened, they can receive immediate treatment and care, and be isolated to reduce the spread of the virus. It is worth mentioning that reverse transcriptase-polymerase chain reaction (RT-PCR) testing is currently being used to detect COVID-19 cases. However, RT-PCR tests are very time-consuming, complicated, and also require the involvement of diagnosticians, who are few in relation to the needs. The sensitivity of RT-PCR tests varies depending on the manufacturer and the batch, and the studies show relatively low precision. Some studies even suggest that radiographic data analysis could be used as a primary screening tool for COVID-19. In particular, most positive cases show bilateral abnormalities in the CXR images, including lack of transparency and interstitial abnormalities. An initial diagnosis can also be made using the mobile application [28].

4 Multi-objective Quantum Deep Learning

A multi-objective quantum-inspired genetic algorithm MQGA operates on the quantum register that represents hyperparameters of CNNs. It is made up of several layers that perform different operations on the input chest X-ray images. The basic structure consists of three types of layers. A convolutional layer applies a set of learnable filters to extract relevant features from the input image. The filters are small, square-shaped matrices that slide over the image in a specific pattern, performing a convolution operation at each position. The convolutional layer provides a set of feature maps that highlight the most important features of the input image. The pooling layer reduces the spatial dimensions of the entered feature maps. The most common type of pooling is *max pooling*, which takes the maximum value of each non-overlapping subregion of the feature map.

The fully connected layer classifies the chest X-ray image based on the pooling layer. It takes the flattened output of the previous layers and applies a set of weights to produce a final output vector. A sequence of these three layers are repeated L times. The final output layer is *softmax* layer that produces a set of probabilities of the chest X-ray image belonging to two of the possible classes. The class with the highest predicted probability is then assigned as the final output.

Training adjusts the filter weights so that the CNN can learn to recognize important features. This is done using a form of gradient descent optimization algorithm, such as stochastic gradient descent (SGD) or ADAM with the learning rate to control the updates of weights during training. We need to know the number L of weave layers with parameters related to the number of neurons in three dimensions. The size of the input image determines the size of the feature maps and there are both L filters and feature maps generated in the convolutional layers. Besides, the given size of the filters determines the size of the receptive field and the features that are extracted. We tune the stride level S (in %) that determines the amount of shift of the filter.

The padding value PV (in %) determines extra rows and columns of pixels to the input image to preserve the spatial dimensions of the output feature maps. We consider the most widely used activation function ReLU Besides, a sigmoid activation function and a tanh activation function (Hyperbolic Tangent) are used. Their limitations for gradient based algorithms, such as vanishing gradients, can be easily omitted by MQGA. Moreover, softmax is used to produce a probability distribution over two classes, with the sum of all probabilities equal to 1. Another hyperparameter of the CNN is the number of hidden layers in the network that affects the complexity of the learned features. The batch size BS determines the number of samples used in each iteration.

A qubit can exist in more than one state (a superposition) at the same moment in time and can be represented by the Bloch sphere. The qubit can be modeled as a two-layer quantum bit from the Hilbert space H_2 with the base $B = \{|0\rangle, |1\rangle\}$. The qubit may be in the "1" binary state, in the "0" state, or in any superposition of them [17]. The state x_m of the mth qubit in the Q-chromosome can be written, as follows [5]:

$$\boldsymbol{Q}_m = \boldsymbol{\alpha}_m|0\rangle \oplus \boldsymbol{\beta}_m|1\rangle, \tag{1}$$

where

α_m and β_m – the complex numbers that specify the amplitudes of the states 0 and 1, respectively;

$\oplus$– a superposition operation;
m – the index of the gene in the chromosome, $m = \overline{1, M}$.

The value $|\alpha_m|^2$ is the probability that we observe the state "0". Similarly, $|\beta_m|^2$ is the probability that state "1" is measured. The qubit is characterized by the pair (α_m, β_m) with the constraint, as below [21]:

$$|\alpha_m|^2 + |\beta_m|^2 = 1. \tag{2}$$

Dirac notation is often used to select a basis. The basis for a qubit (two dimensions) is $|0\rangle = (1, 0)$ and $|1\rangle$= (0,1). The most commonly used representation of a chromosome in a genetic algorithm is the matrix, as follows [22]:

$$Q = \begin{bmatrix} |\alpha_1| \ldots |\alpha_m| \ldots |\alpha_M| \\ |\beta_1| \ldots |\beta_m| \ldots |\beta_M| \end{bmatrix} \tag{3}$$

However, the state $Q_m = \alpha_m|0\rangle \oplus \beta_m|1\rangle$ of the mth qubit can be represented as the point on the 3D Bloch sphere (Fig. 1), as follows [23]:

$$\left|Q_m\right\rangle = \cos\frac{\theta_m}{2}\left|0\right\rangle + e^{i\phi_m}\sin\frac{\theta_m}{2}\left|1\right\rangle, m = \overline{1, M} \tag{4}$$

where $0 \le \theta_m \le \pi$ and $0 \le \phi_m \le 2\pi$.

Two angles θ_m and ϕ_m determines the localization of mth qubit on the Bloch sphere. In addition to the representation of (3), we can therefore distinguish the following other models of the chromosome [27]:

$$Q^{sign} = \begin{bmatrix} |\alpha_1| & \ldots & |\alpha_m| & \ldots & |\alpha_M| \\ sign(r_1) & \ldots & sign(r_m) & \ldots & sign(r_M) \end{bmatrix} \tag{5}$$

where r_m – the random number from the interval [−1; 1], and also it is, as below [29]:

$$Q^{vector} = [|\alpha_1|, \ldots, |\alpha_m|, \ldots, |\alpha_M|] \tag{6}$$

$$Q^{angle} = \begin{bmatrix} \theta_1 \ldots \theta_m \ldots \theta_M \\ \phi_1 \ldots \phi_m \ldots \phi_M \end{bmatrix} \tag{7}$$

There are two important criteria for deep learning. The first one is an accuracy and the second criterion is F_1-score. We determine non-dominated solutions from the current population and copy them to the archive after verification. Figure 2 shows the initialization *Q(t)* by Hadamard gates and rotation gates. Then, digital population *P(t)* is created by observation the states of *Q(t)*.

Figure 3 shows a histogram after updating the register $Q(t)$ using the rotation gates *Rx*, *Ry*, *Rz* refer to best group. Digital algorithms for calculation this probability distribution is exponentially more difficult as the number of qubits (width) and number of gate cycles (depth) raise [24].

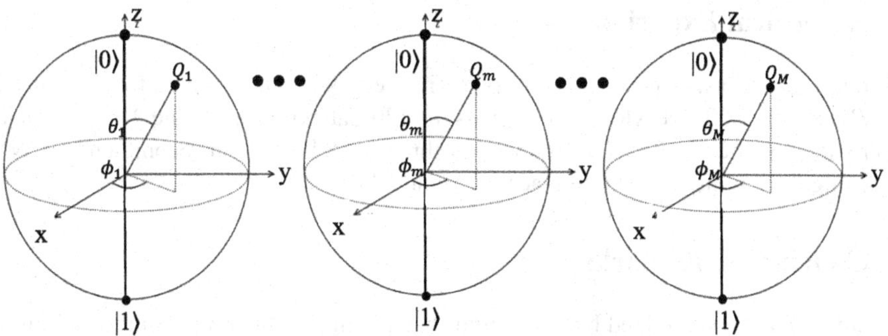

Fig. 1. The set of Bloch spheres for the quantum register Q.

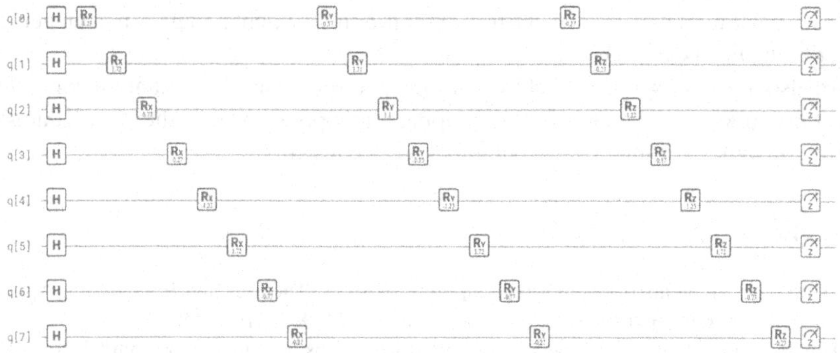

Fig. 2. A diagram of the quantum evolution circuit at Quantum Inspire platform.

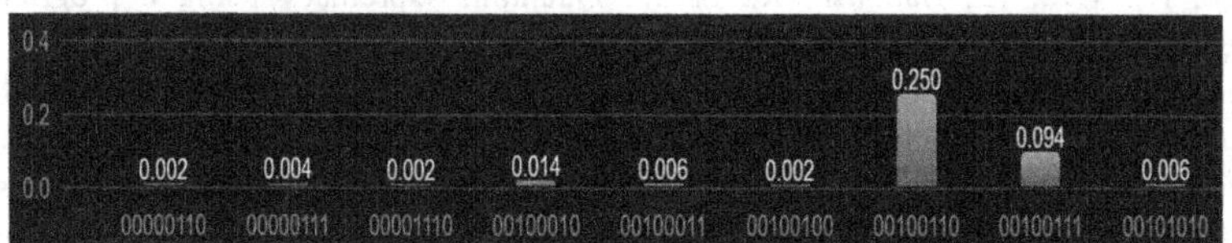

Fig. 3. Histogram after updating the population $Q(t)$.

Table 1. Comparison of Convolutional Neural Networks.

	QCNN		RESNET		SENET	
	Value	#	Value	#	Value	#
Accuracy	0.982	1	0.952	3	0.954	2
F1-score	0.984	1	0.949	2	0.948	3

5 Numerical Experiments

We consider an instance of the multi-objective seep learning instance for the dataset COVIDx with 13,975 CXR images across 13,870 patient cases. Table 1 shows characteristics for three CNNs. By comparison three models, we can recommend QCNN regarding its dominance regarding accuracy and F1-score.

6 Concluding Remarks

We presented QCNN trained by a quantum-inspired multi-objective evolutionary algorithm in an attempt to gain deeper insights into critical factors associated with COVID-19 cases, which can aid clinicians in improved screening. Development of Pareto-optimal deep learning solutions for detecting COVID-19 cases from CXR images can predict hospitalization duration which would be useful for triaging, patient population management, and individualized care planning.

We also introduces a new machine learning paradigm based on quantum computers. Our future works will be focused on developing this approach to the other Covid datasets. Besides, the other artificial neural networks will be tested.

References

1. Akama, S.: Elements of Quantum Computing. History, Theories and Engineering Applications, Springer, Cham (2015). https://doi.org/10.1007/978-3-319-08284-4
2. Alibaba Cloud Quantum Development Platform. https://damo.alibaba.com/labs/quantum. Accessed 10 Jan 2023
3. Amazon Braket. https://aws.amazon.com/braket/. Accessed 15 Jan 2023
4. Arute, F., Arya, K., Babbush, R., et al.: Quantum supremacy using a programmable superconducting processor. Nature **574**, 505–510 (2019)
5. Balicki, J.: Many-objective quantum-inspired particle swarm optimization algorithm for placement of virtual machines in smart computing cloud. Entropy **24**, 58 (2022)
6. Balicki, J., Korłub, W., Krawczyk, H., Paluszak, J.: Genetic programming with negative selection for volunteer computing system optimization. In: Proceedings of 6th International Conference on Human System Interactions (HSI), Sopot, Poland, pp. 271–278 (2013)
7. Benioff, P.: Quantum mechanical models of Turing machines that dissipate no energy. Phys. Rev. Lett. **48**, 1581–1585 (1982)
8. Bozzo-Rey, M., Loredo, R.: Introduction to the IBM Q experience and quantum computing. CASCON **2018**, 410–412 (2018)
9. DeBenedictis, E.P.: Beyond quantum supremacy. Computer **53**(2), 91–94 (2020)
10. Dicarlo, L., et al.: Demonstration of two-qubit algorithms with a superconducting quantum processor. Nature **460**(7252), 240–244 (2009)
11. Feynman, R.: Simulating physics with computers. Int. J. Theor. Phys. **21**, 467–488 (1982)
12. Grover, L.K.: A fast quantum mechanical algorithm for database search. In: Proceedings of 28th ACM Symposium on the Theory of Computing, pp. 212–220 (1996)
13. Gu, J., Gu, X., Gu, M.: A novel parallel quantum genetic algorithm for stochastic job shop scheduling. J. Math. Anal. Appl. **355**, 63–81 (2009)
14. Han, K.-H., Han, J.-H.: Genetic quantum algorithm and its application to combinatorial optimization problem. In: Proceedings of the Congress on Evolutionary Computation, vol. 2, pp. 1354–1360 (2000)

15. Hassan, M.R., Ismail, W.N., Chowdhury, A., et al.: A framework of genetic algorithm-based CNN on multi-access edge computing for automated detection of COVID-19. J. Supercomput. **78**, 10250–10274 (2022)
16. Kim, J., Kim, J.-H., Han, K.-H.: Quantum-inspired multiobjective evolutionary algorithm for multiobjective 0/1 knapsack problems. In: IEEE Congress on Evolutionary Computation Vancouver, Canada, July 16–21, pp. 9151–9156 (2006)
17. Li, R., Wu, B., Ying, M., Sun, X., Yang, G.: Quantum supremacy circuit simulation on sunway taihulight. IEEE Trans. Parallel Distributed Syst. **31**(4), 805–816 (2020)
18. Michael, I.L.C., Nielsen, A.: Quantum Computation and Quantum Information. Cambridge University Press (2011)
19. Microsoft Azure Quantum. https://azure.microsoft.com/. Accessed 15 Jan 2023
20. Narayanan, A., Moore, M.: Quantum-inspired genetic algorithms. In: Proceedings of IEEE International Conference on Evolutionary Computation, Nagoya, Japan, pp. 61–66 (1996)
21. Olivares-Sánchez, J., Casanova, J., Solano, E., Lamata L.: Measurement-based adaptation protocol with quantum reinforcement learning in a Rigetti quantum computer (2018). https://arxiv.org/abs/1811.07594
22. Pednault, E., Gunnels, J., Maslov, D., Gambetta, J.: On quantum supremacy. https://www.ibm.com/blogs/research/2019/10/on-quantum-supremacy/. Accessed 25 Nov 2022
23. Preskill, J.: Quantum computing and the entanglement frontier. In: Proceedings of the 25th Conference Physics (2011)
24. Quantum Inspire (QI) is a quantum computing platform designed and built by QuTech. https://www.quantum-inspire.com/. Accessed 25 Feb 2023
25. Shor, P.: Algorithms for quantum computation: discrete logarithms and factoring. In: Proceedings of the 35th Symposium on Foundations of Computer Science, November (1994), pp. 124–134. IEEE Press (1994)
26. Talbi, H., Batouche, M., Draao, A.: A quantum-inspired evolutionary algorithm for multiobjective image segmentation. Int. J. Math. Phys. Eng. Sci. **1**(2), 109–114 (2007)
27. Vandersypen, L.M.K., Steffen, M., Breyta, G., Yannoni, C.S., Sherwood, M.H., Chuang, I.L.: Experimental realization of Shor's quantum factoring algorithm using nuclear magnetic resonance. Nature **414**, 883–887 (2001)
28. Wang, W., et al.: Detection of SARS-CoV-2 in different types of clinical specimens. JAMA (2020)
29. Xing, H., Ji, Y., Bai, L., Liu, X., Qu, Z., Wang, X.: An adaptive-evolution-based quantum-inspired evolutionary algorithm for QoS multicasting in IP/DWDM networks. Comput. Commun. **32**, 1086–1094 (2009)

Simulations of Flow and Transport: Modeling, Algorithms and Computation

Numerical Simulation of Virus-Laden Droplets Transport from Lung to Lung by Using Eighth-Generation Airway Model

Shohei Kishi[1(✉)], Masashi Yamakawa[1], Ayato Takii[2], Tomoaki Watamura[1], Shinichi Asao[3], Seiichi Takeuchi[3], and Minsuok Kim[4]

[1] Kyoto Institute of Technology, Matsugasaki, Sakyo-Ku, Kyoto 606-8585, Japan
m2623010@edu.kit.ac.jp

[2] RIKEN Center for Computational Science, 7-1-26 Minatojima-Minami-Machi, Chuo-Ku, Kobe 650-0047, Hyogo, Japan

[3] College of Industrial Technology, 1-27-1 Amagasaki, Kobe 661-0047, Hyogo, Japan

[4] School of Mechanical, Electrical and Manufacturing Engineering, Loughborough University, Loughborough, UK

Abstract. In this study, we simulated the trajectory of virus-laden droplets from the lung of an infected person to that of the exposed person using computational fluid dynamics. As numerical models, the model of the infected person who had a bifurcated airway and that of the exposed person who had an eighth-generation airway were prepared. The volume and number of virus-laden droplets adhered to the inlet patches of the exposed person's lung were calculated to evaluate the risk of infection when the infected person was talking for 40 s. To identify the lung to which droplets adhered, we labeled the inlet patches of the exposed person's lung with 53 numbers, and then measured the volume and number of droplets on the inlet patches of lung. We also categorized the lung's 53 intake patches into five groups and calculated the overall volume and number of attached droplets for each. In addition, we parameterized the angle of the exposed person's neck to evaluate the effect of the tilting neck on the volume and number of droplets reaching the lungs. We found that the volume and number of droplets adhered to the right middle group of bronchi were remarkably smaller than the other four groups, and weakly depended on the neck angle. The volume and number of droplets adhered to the inlet patches of the lung reached the maximum values when the neck angle was 20° upward.

Keywords: Computational fluid dynamics · Respiratory organ · SARS-CoV-2

1 Introduction

Quantitative and effective methods of preventing infection are still required owing to the ongoing spread of SARS-CoV-2. Since respiratory viruses such as SARS-CoV-2 and influenza are mainly transported through the air [1], it is necessary to determine how the exposed person takes in virus-laden droplets, i.e., the route of airborne transportation. There are two major methods for identifying the routes of airborne infection: the

J. Mikyška et al. (Eds.): ICCS 2023, LNCS 14077, pp. 257–270, 2023.
https://doi.org/10.1007/978-3-031-36030-5_21

first is an experimental method using actual rooms and subjects, and the second is a simulation method using numerical calculations. As a typical example of experimental methods, Killingley et al. [2] performed SARS-CoV-2 human investigation in 2022. They intranasally inoculated human with SARS-CoV-2, and identified the dose of virus causing infection and evaluated the symptom kinetics during infection. These experimental approaches offer the advantage of collecting precise and reliable data, whereas they have the drawback of requiring a great deal of time and manpower.

On the other hand, computer simulations can efficiently reproduce this phenomenon. Actually, computational fluid dynamics (CFD) has been actively used in recent years [3–5]. Ramajo et al. [3] simulated how an urban bus's HVAC (Heat, Ventilation, Air Conditioning) system would affect the movement of virus-laden droplets. It was found that the large droplets (>200μm) were not captured by HVAC outlet; they moved more than 3m and deposited within 2 s. Small droplets (<5μm) were more sensitive to the airflow and were easily trapped by HVAC. Mariam et al. [4] also simulated the dispersion of SARS-CoV-2 in an indoor environment. They parameterized the location of the air vent and the velocity of expelled particles. They found that the airflow, during normal talking, transports virus-laden droplets from the particle generator to receptor, when there is 2m distance between them. As an example of studies focusing on the airway, Wedel et al. [5] adopted three different airway models to simulate variation in the aerosol deposition during breathing. They reported that the deposition variation in the laryngeal region depended on the shape of the airway.

Although there are numerous studies examining the interior and exterior of the human body separately, no study has examined both regions together. In addition, while many studies have parameterized the velocity of particle generation and the position of air vent as variables, there are few cases where the angle of neck is parameterized [6]. It can be thought that tilting neck alters the shape of flow pathway in the airway. Doing so, it is expected the transportation route of virus-laden droplets will be changed, which makes the differences in the volume and number of droplets adhered to the lung. Investigating this difference, we hypothesized the change of neck angle may lead to some degree of reducing the risk of infection. In this study, we parameterized the angle of the exposed person's neck and comprehensively analyzed the human body's internal and external flow fields. The movement of droplets was also calculated using the fluid analysis results to simulate how many virus-laden droplets expelled from the infected person's lung would adhere the exposed person's lung.

2 Numerical Approach

2.1 Flow Field and Heat Analysis

The flow and temperature fields were calculated employing *SCRYU/Tetra* [7]. The governing equations are the continuity equations, the incompressible Navier-Stokes equations, and the energy conservation equation.

The continuity equation is given by

$$\frac{\partial u_{a_i}}{\partial x_i} = 0. \tag{1}$$

the three-dimensional incompressible Navier-Stokes equations are defined by

$$\frac{\partial(\rho_a u_{a_i})}{\partial t} + \frac{\partial(u_{a_j}\rho_a u_{a_i})}{\partial x_j} = -\frac{\partial p}{\partial x_i} + \frac{\partial}{\partial x_j}\left\{\mu\left(\frac{\partial u_{a_i}}{\partial x_j} + \frac{\partial u_{a_j}}{\partial x_i}\right)\right\} - \rho_a g_i \beta(T_a - T_0), \quad (2)$$

where $u_{a_i}(i, j = 1, 2, 3)$ is the air velocity components in the x, y and z direction, and ρ_a is gas density, which is constant. t, p and μ denote time, pressure, viscosity, respectively. $g_i(i, j = 1, 2, 3)$ is the gravity acceleration in the x, y and z direction, and β indicates the thermal expansion coefficient. T_a and T_0 are the air temperature and reference temperature, respectively. The energy conservation equation is given by

$$\frac{\partial(\rho_a C_p T_a)}{\partial t} + \frac{\partial(u_{a_j}\rho_a C_p T_a)}{\partial x_j} = \frac{\partial}{\partial x_j}\left(K\frac{\partial T_a}{\partial x_j}\right) + \dot{q}, \quad (3)$$

where C_p is specific heat capacity. K and $\dot{q}$ denote thermal conductivity and heat flux. The SIMPLEC algorithm [8] was used to solve Eqs. (1) and (2), which couples of velocity and pressure. The second order MUSCL method [9] was applied to the convective terms of Eqs. (2) and (3). The Smagorinsky sub-grid scale (SGS) model of large eddy simulation (LES) [10] is used to accuracy capture eddies generated by turbulence.

2.2 Movement of Virus-Laden Droplets

The equation of droplet radius changing due to evaporation is defined as follows [11]:

$$\frac{dr}{dt} = -\left(1 - \frac{RH}{100}\right) \cdot \frac{De_s(T_a)}{\rho_d R_v T_a} \cdot \frac{1}{r}, \quad (4)$$

where r is droplet radius, and RH represents humidity. D, e_s and ρ_d denote the water vapor coefficient, saturated vapor pressure, and droplet density, respectively. R_v indicates the gas constant of water vapor. The equation of droplet motion is defined as [11]

$$\frac{4}{3}\pi\rho_d r^3\frac{dv_i}{dt} = \frac{4}{3}\pi\rho_d r^3 g_i + C_D(v_i) \cdot \frac{1}{2}\rho_a S\left|u_{a_i} - v_i\right|\left(u_{a_i} - v_i\right), \quad (5)$$

where $v_i(i = 1, 2, 3)$ is the droplet velocity components in the x, y and z direction, and C_D denotes the drag coefficient. S is the projected area of droplets.

3 Simulation of Droplets

3.1 Numerical Model

Figure 1(a) shows the overview of two people facing each other in the room. The room size was 3.0 m × 3.0 m × 3.0 m, and 0.15 m × 0.15 m air vent is placed at the top of the room. The distance between the mouths of two people was set at 1.0 m. Figure 1(b) indicates human models with airways for infected and exposed individuals, respectively. In this study, we made a simulation model: the talking infected person and the breathing exposed person.

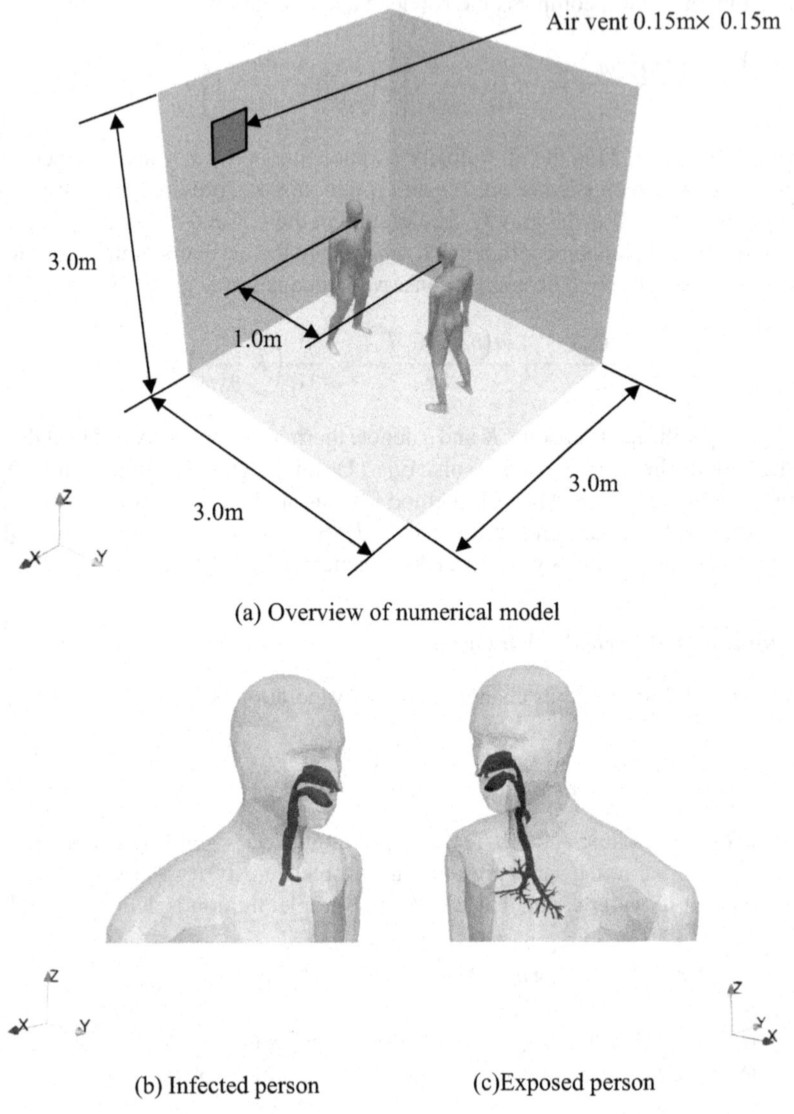

Fig. 1. Numerical model of the room and human model with the airway.

The airway of the infected person is bifurcated and its shape is created based on *Anatomy & Physiology* [12]. The airway of the exposed person is eighth-generation and is modeled from computed tomography (CT) data of adult male [13, 14]. Figure 2 shows the tilting motion of exposed person's neck. The rotation axis is placed at the neck center, and the rotation is made upward ranging $\theta = 0$ to $20°$ with $5°$ increments. The shape of the airway below the throat is independent of rotation. Figure 3 shows the computational grids of eighth-generation airway model. Computational grids were created by using

SCRYU/Tetra. The wall of the airway has three prismatic layers. The total number of computational grids inside and outside the human body was approximately 4,000,000. As a typical example, the number of tetrahedron elements is 2,423,944, pyramid elements is 2,208 and prism elements is 1,496,250 in $\theta = 0°$.

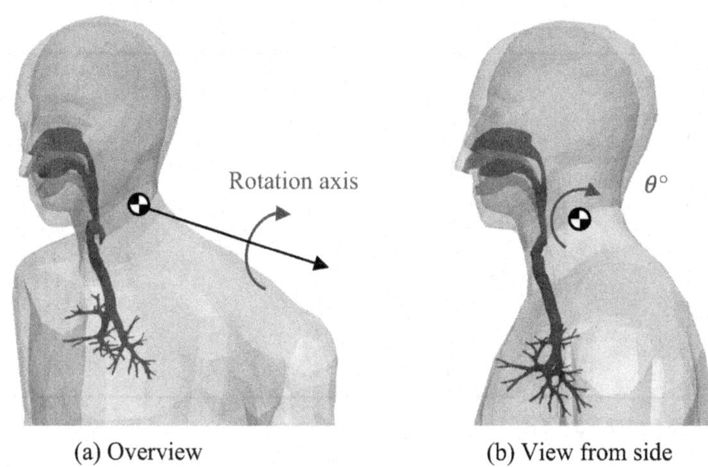

(a) Overview (b) View from side

Fig. 2. The angle of neck variation (yellow represents $\theta = 20°$).

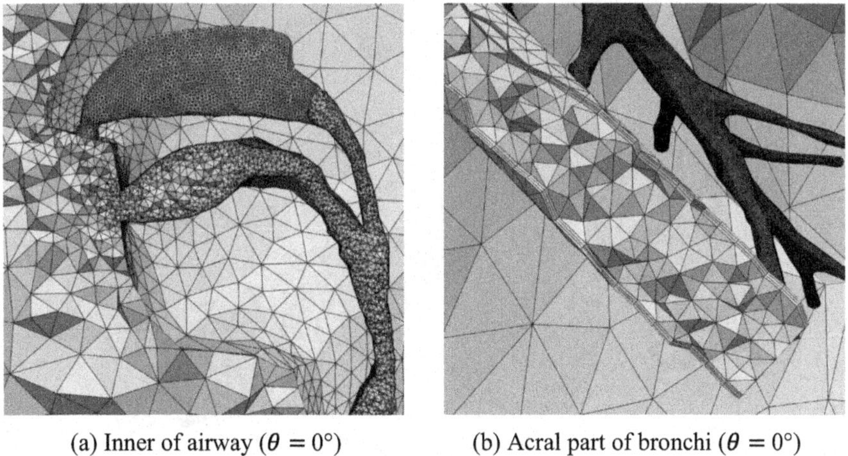

(a) Inner of airway ($\theta = 0°$) (b) Acral part of bronchi ($\theta = 0°$)

Fig. 3. Grids of eighth-generation airway.

3.2 Computational Condition for Droplets Analysis

Figure 4 indicates the frequency distribution of virus-laden droplet diameters expelled from an infected person during talking [15, 16]. Although droplets were expelled at

160 pieces/s in the literature, we released droplets at 1,600 pieces/s to achieve better accuracy. The computational time was set to 40 s by considering both the time and efficiency required to capture the trend of the phenomenon. The total number of droplets was 64,000. Figure 5 shows the initial position of generating droplets in the infected person's bronchi, which are installed equally on both sides of the airway.

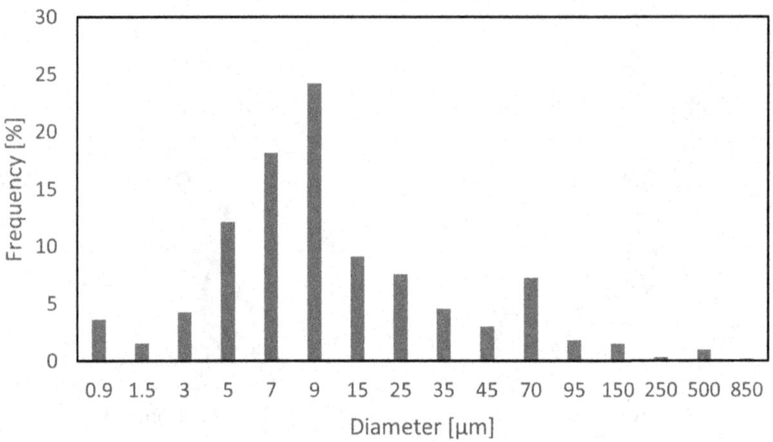

Fig. 4. Frequency distribution of virus-laden droplet diameters expelled from infected person during talking.

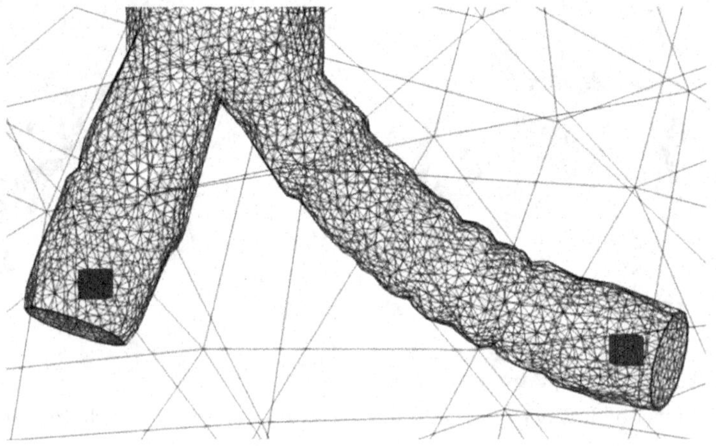

Fig. 5. Initial position of generating droplets (the acral part of infected person's bronchi).

3.3 Assessing the Risk of Infection

Figure 6(a) shows the overview of the infected person's airway and the inlet patches of the lung, and Fig. 6(b) shows the airway of the exposed person. Note that we defined the lower left bronchi in the human body as Type A and the lower right bronchi as Type B. Figures 6(c) and 6(d) display the numbers assigned to inlet patches of lung in Type A and Type B, respectively. We followed the method of grouping [17, 18]; we grouped the inlet patches of lung No. 1 to 11 as upper left of bronchi (LeftUp), No. 12 to 23 as lower left of bronchi (LeftLow), No. 24 to 34 as upper right of bronchi (RightUp), No. 35 to 38 as middle right of bronchi (RightMid) and No. 39 to 53 as lower right of bronchi (RightLow). In this study, 53 inlet patches of lungs were divided into five groups and the total volume and number of droplets adhered to each group were calculated to facilitate comparison with the literature. The greater the amount of droplets adhered to inlet patches of lungs, the higher the risk of infection is assessed in this study.

3.4 Computational Conditions for Flow and Heat Analysis

Figure 7 shows the volume flow rate during talking which was determined from Bale et al. [16] and Gupta et al. [19], and was given at the inlet surfaces of the infection person's lung in Fig. 6(a). Figure 8 shows the volume flow rate of breathing as defined by the Handbook of Physiology [20] and Ogura et al. [21], and was applied at the inlet patches of the exposed person's lung in Figs. 6(c) and 6(d). As pressure condition, the homogeneous Neumann condition was also adopted to the inlet surfaces of both lungs. The air vent in Fig. 1(a) is subjected to the boundary conditions: static pressure is 0 and velocity is the homogeneous Neumann condition. The initial room temperature is $T_r = 20\,°\mathrm{C}$, and the body temperature of infected and exposed person is $T_h = 30\,°\mathrm{C}$, considering the effect of clothes. The temperature of air expelled from the infected and the exposed person is assumed to be $T_f = 36\,°\mathrm{C}$ (Table 1).

Table 1. Computational conditions for calculation flow field and movement of droplets.

Total number of computational grids	About 400,0000
The air vent of room	Velocity: Neumann Pressure: 0
Infected person's lung inlet surface	Velocity: Talking Pressure: Neumann
Exposed person's lung inlet surface	Velocity: Breathing Pressure: Neumann
Initial room temperature	20 °C
Human body temperature	30 °C
Intake and exhaled air temperature	36 °C
Time step of flow calculation	0.0001
Time step of droplets calculation	0.00001

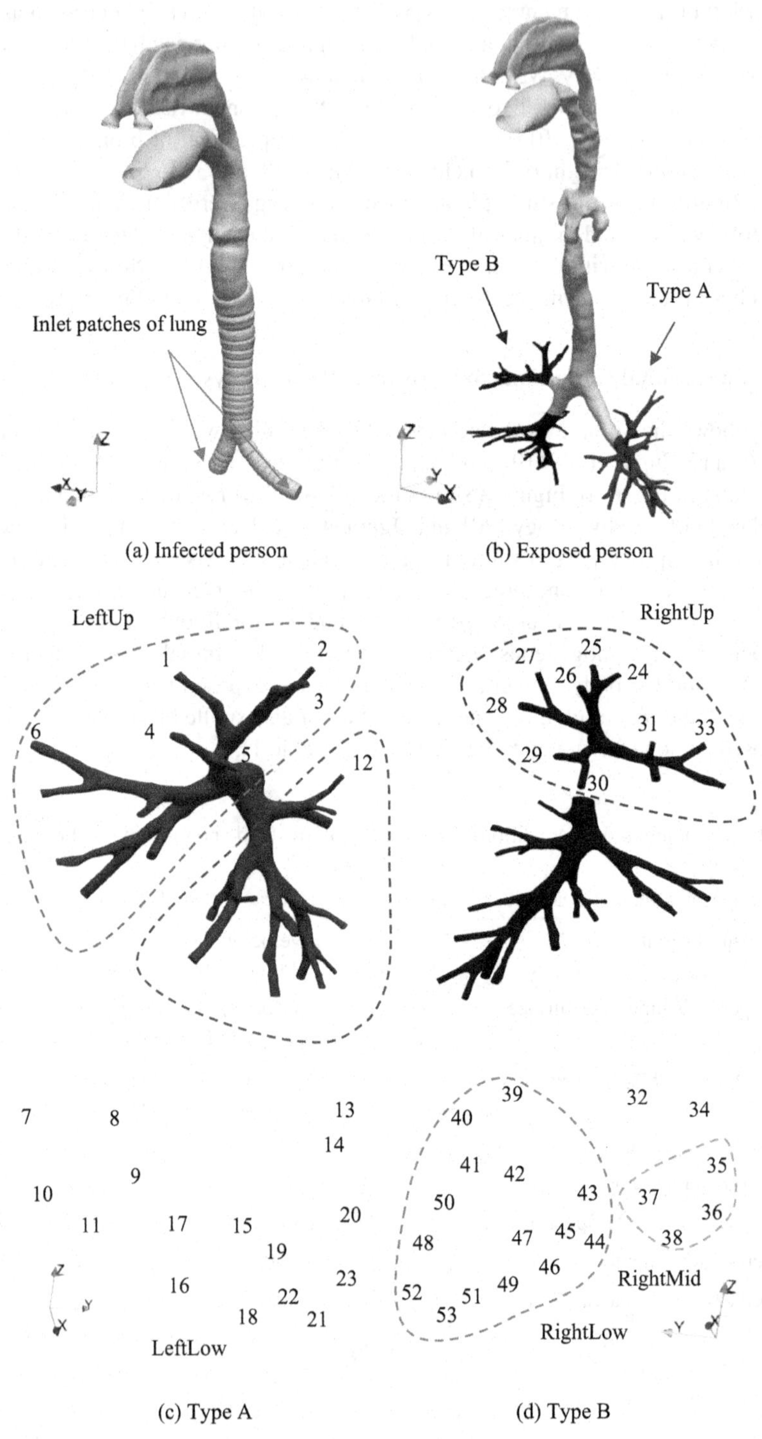

Fig. 6. Enraged view of the airway.

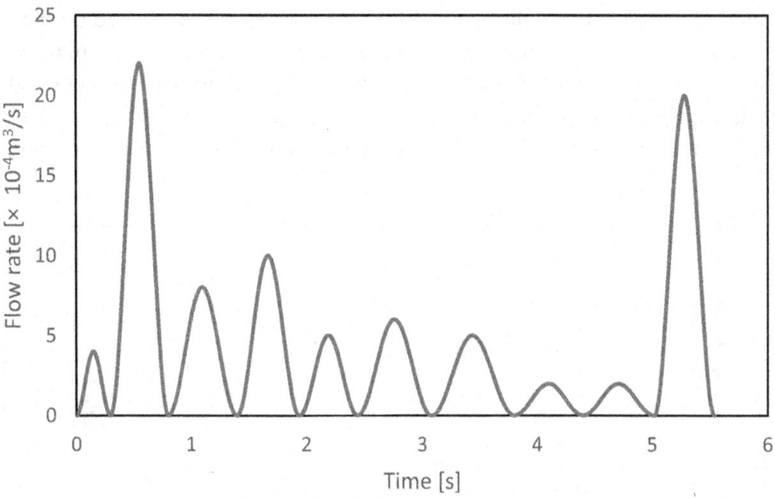

Fig. 7. The volume flow rate during talking.

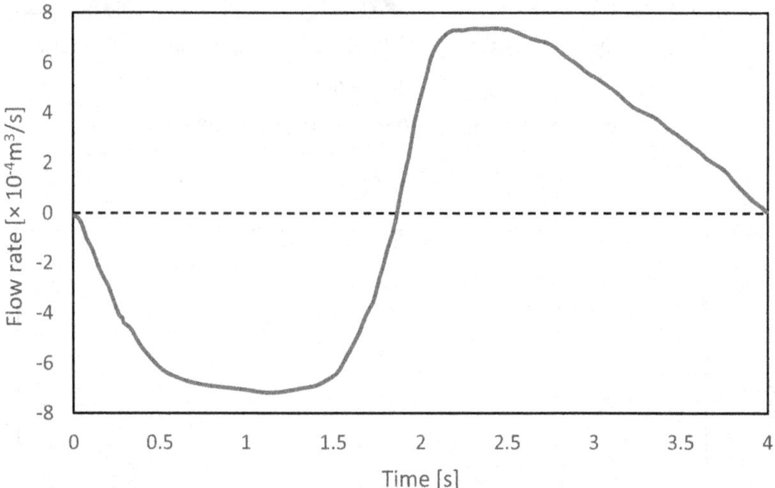

Fig. 8. The volume flow rate of breathing.

4 Results and Discussion of Droplets Analysis

Figures 9(a) and (b) represent the total volume and number of virus-laden droplets adhered to each of the five groups. The maximum volume is found at RightLow for θ of 5°, 10°, and 20°, while it takes the maximum value at LeftLow in case of $\theta = 0°$ and 15°. The number of adhered droplets is high at RightLow ($\theta = 0°$ and 20°), while it takes the maximum value at LeftLow ($\theta = 5°$, 10°, and 15°). The number of droplets adhered to RightLow ($\theta = 15°$) in Fig. 9(b) is only nine, whereas Fig. 9(a) shows the maximum volume of adhered droplets, and this suggests that the adhered droplets have large volume. Thus, there is a difference between the locations of maximum volume and those of maximum number. As an overall trend, the volume and number of adhered droplets in the RightMid group are smaller than those in the other four groups in Fig. 9. A similar trend was also found in the literature [17], and this is because the number of

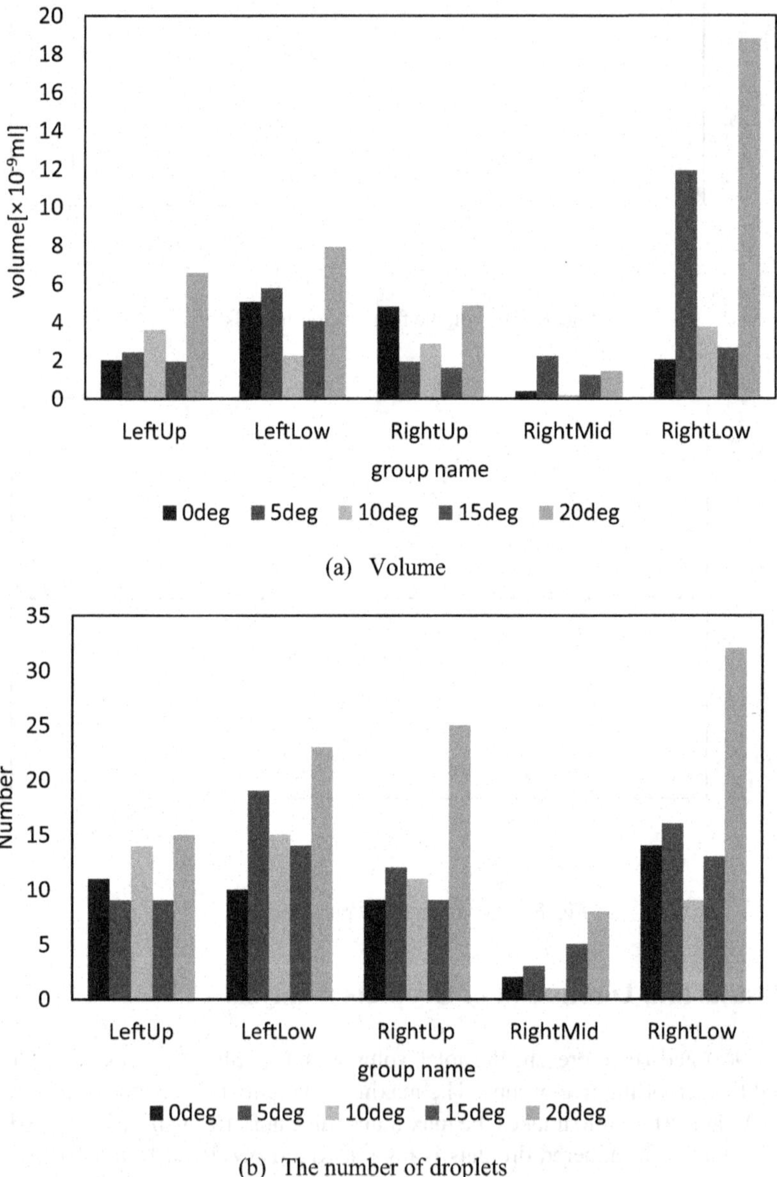

(a) Volume

(b) The number of droplets

Fig. 9. The volume and the number distribution of virus-laden droplets per bronchial group.

inlet patches originally grouped together is only four. It was also found that the volume and number of adhered droplets also took large values for each group at $\theta = 20°$. We thought that this finding was due to the difference in airflow pathways generated by tilting neck.

Figure 10 shows the dispersion of virus-laden droplets when θ is 20°. Red and green particles depict floating droplets and adhered droplets, respectively. The human model on the left side is the infected person releasing virus-laden droplets while the model on right is the exposed person who inhales the droplets by breathing. The droplets are released slightly downward because the oral cavity of infected person is curved, as shown in Fig. 6(a). Therefore, it can be seen that many droplets adhere to the torso of the exposed person. It is also confirmed that many droplets are floating over the head of the exposed person. We thought that the updrafts generated by the heat of the exposed person caused droplets to float.

Figure 11 indicates the streamline of the exposed person when the volume flow rate of inhalation took the maximum value. It can be seen that the airflow from mouth to lungs is fast at the throat with a speed of approximately 4 m/s. The exposed person in Fig. 11(b) is more likely to inhale the air located at a higher place because the mouth of exposed person is facing upward. This finding suggests that tilting the neck upward facilitates the aspiration of droplets raised by updrafts. It was also confirmed that the shape of the flow pathway in the airway of Fig. 11(a) is curved entirely, while that of Fig. 11(b) is straight from the throat to the lungs. This could be explained by tilting the neck upward, which allowed droplets inhaled into the oral cavity to fall into the lungs.

From the above description, it was concluded that tilting the neck upward led to change of airflow pathways in respiratory and room regions, which increased the number

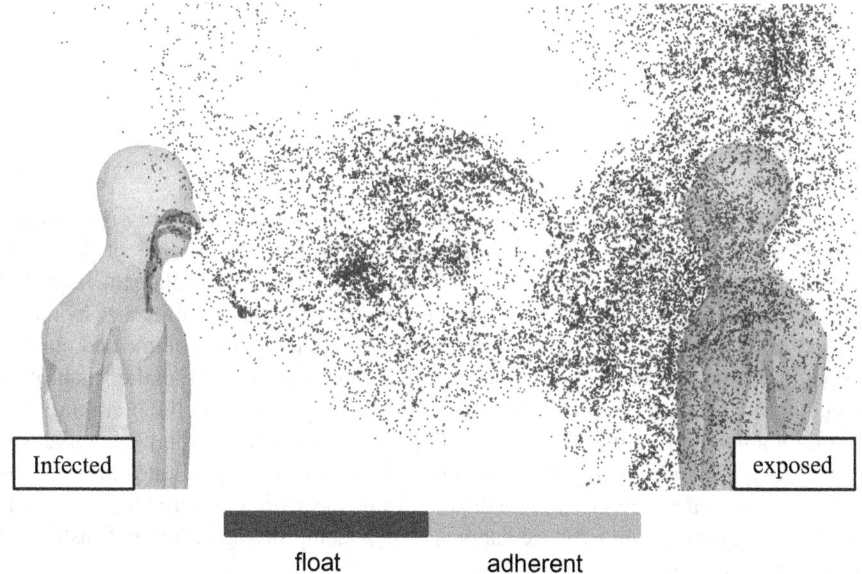

Fig. 10. Dispersion of virus-laden droplets ($\theta = 20°$, $t = 40$ s).

of droplets inhaled by the exposed person, and the volume and number of droplets adhered to the inlet patches of lungs took considerable value. Therefore, keeping the head forward could suppress the inhalation of droplets raised by the updrafts, and reduce the risk of infection.

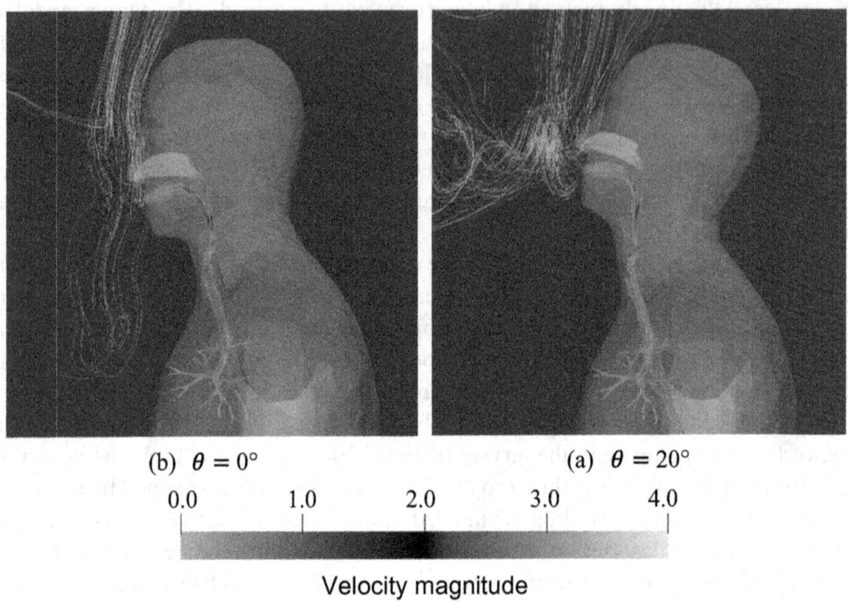

Fig. 11. Streamline of exposed person when the volume flow rate of inhalation took the maximum value ($t = 9.9$ s).

5 Conclusions

We comprehensively analyzed the flow field of the room and respiratory regions and evaluated the amount of virus-laden droplet adhered to the inlet patches of the lungs. The angle of an exposed person's neck was also parameterized to simulate the impact of the tilting neck on the risk of infection. In addition, the eighth-generation airway model was used as the airway of an exposed person. It was found that the volume and number of adhered droplets in the right middle region of the bronchi were smaller than other bronchial parts and we confirmed that our finding is consistent with the literature [17]. The total volume and number of adhered droplets increased when the neck angle was 20° upward. This finding suggests that tilting the neck upward facilitates the droplets inhaled into the oral cavity falling into the lungs, i.e., the settling of virus-laden droplets from the oral cavity to lung can be suppressed by facing the head forward instead of tilting the neck upward.

Acknowledgments. This work was supported by JST CREST Grant Number JPMJCR20H7, by JKA through its promotion funds from KEIRIN RACE, and by JSPS KAKENHI Grant Number 21K03856.

References

1. Wang, C.C. et al.: Airborne transmission of respiratory viruses. Science, Volume 373, Issue 6558 (2021)
2. Killingley, B., et al.: Safety, tolerability and viral kinetics during SARS-CoV-2 human challenge in young adults. Nat. Med. **28**, 1031–1041 (2022)
3. Ramajo, D.E., Corzo, S.: Airborne transmission risk in urban buses: A computational fluid dynamics study. Aerosol Air Qual. Res. **22**, 210334 (2022)
4. Mariam, et al.: CFD Simulation of the Airborne Transmission of COVID-19 Vectors Emitted during Respiratory Mechanisms: Revisiting the Concept of Safe Distance. ACS Omega, pp. 16876–16889 (2021)
5. Wedel, J., et al.: Anatomy matters: The role of the subject-specific respiratory tract on aerosol deposition - A CFD study. Comput. Methods Appl. Mech. Eng. 401(Part A), 115372 (2022)
6. Sugiura, N.: Supercomputer finds safest way to sit and chat while eating out. Asahi Shinbun (2020), https://www.asahi.com/ajw/articles/13817552. Accessed 25 Jan 2023
7. Watanabe, N., et al.: An 1D-3D Integrating Numerical Simulation for Engine Cooling Problem. SAE Technical Paper 2006-01-1603 (2006)
8. Doormal, V.J.P., Raithby, G.D.: Enhancements of the simple method for predicting incompressible fluid flows. Numer. Heat Transfer **7**(2), 147–163 (1984)
9. van Leer, B.: Towards the ultimate conservation difference scheme 4: A new approach to numerical convection. J. Comp. Phys. **23**, 276–299 (1977)
10. Yoshizawa, A., et al.: Computational Fluid Dynamics Series 3: Analysis of Turbulent Flows, pp. 67–84. University of Tokyo Press (1995)
11. Yamakawa, M., et al.: Computational investigation of prolonged airborne dispersion of novel coronavirus-laden droplets. J. Aerosol Sci. **155**, 105769 (2021)
12. Yamakawa, M., et al.: Influenza viral infection simulation in human respiratory tract. In: The Proceedings of 29th International Symposium on Transport Phenomena, Paper ID:13. Honolulu Hawaii (2018)
13. Kim, M., et al.: Effect of upper airway on tracheobronchial fluid dynamics. Int. J. Numer. Methods Biomed. Eng. **34**(9), e3112 (2018)
14. Collier, G.J., et al.: 3D phase contrast MRI in models of human airways: Validation of computational fluid dynamics simulations of steady inspiratory flow. J. Magn. Reson. Imaging **48**(5), 1400–1409 (2018)
15. Duguid, J.P., et al.: The size and the duration of air-carriage of respiratory droplets and droplet-nuclei. Epidemiol. Infect. **44**(6), 471–479 (1946)
16. Bale, R., et al.: Simulation of droplet dispersion in COVID-19 type pandemics on Fugaku. In: PASC'21, vol. 4, pp. 1–11 (2021)
17. Choi, J., et al.: Differences in particle deposition between members of imaging-based asthma clusters. J. Aerosol. Med. Pulm. Drug. Deliv. **32**(4), 213–223 (2019)
18. Atzeni, C., et al.: Computational fluid dynamic models as tools to predict aerosol distribution in tracheobronchial airways. Sci. Rep. **11**, 1109 (2021)
19. Gupta, J.K., et al.: Characterizing exhaled airflow from breathing and talking. Indoor Air **20**(1), 31–39 (2010)

20. Fenn, W.O., Rahn, H.: Handbook of Physiology, Section3: Respiration. American Physiological Society, Washington, DC (1965)
21. Ogura, K., et al.: Coupled simulation of Influenza virus between inside and outside the body. In: The Proceedings of the International Conference on Computational Methods, vol. 7, pp. 71–82 (2020)

Constraint Energy Minimizing Generalized Multiscale Finite Element Method for Highly Heterogeneous Compressible Flow

Leonardo A. Poveda[1], Shubin Fu[2], and Eric T. Chung[1](✉)

[1] The Chinese University of Hong Kong, Shatin, New Territories Hong Kong SAR, China
lpoveda@math.cuhk.edu.hk, tschung@math.cuhk.edu.hk
[2] Eastern Institute for Advanced Study, Ningbo, China
shubinfu@eias.ac.cn

Abstract. This work presents a Constraint Energy Minimizing Generalized Multiscale Finite Element Method (CEM-GMsFEM) for solving a single-phase compressible flow in highly heterogeneous media. To discretize this problem, we first construct a fine-grid approximation using the Finite Element Method with a backward Euler time approximation. After time discretization, we use Newton's method to handle the nonlinearity in the resulting equations. To solve the linear system efficiently, we shall use the framework of CEM-GMsFEM by constructing multiscale basis functions on a suitable coarse-grid approximation. These basis functions are given by solving a class of local energy minimization problems over the eigenspaces that contain local information on heterogeneity. In addition, oversampling techniques provide exponential decay outside the corresponding local oversampling regions. Finally, we will provide two numerical experiments on a 3D case to show the performance of the proposed approach.

Keywords: Constraint energy minimization · multiscale finite element methods · compressible flow · highly heterogeneous

1 Introduction

The phenomenon of fluid flow through heterogeneous porous materials has been studied in fields as diverse as reservoir simulation, water storage, and groundwater contamination. These problems can be prohibitively expensive to solve when applying traditional fine-scale direct techniques, primarily due to the strong heterogeneity of the geological data. Historically, the scientific community has

Eric Chung's research is partially supported by the Hong Kong RGC General Research Fund (Project number: 14304021).

J. Mikyška et al. (Eds.): ICCS 2023, LNCS 14077, pp. 271–278, 2023.
https://doi.org/10.1007/978-3-031-36030-5_22

been motivated to develop model reduction techniques. The first is the upscaling method [1], where the upscaled geological properties, such as permeability fields, are obtained by applying specific rules and then solving the problem with a mostly reduced model. The second is the multiscale method [5,8,10]; in this case, the solution of the problem is approximated by local basis functions, which are solutions of a class of local problems on the coarse mesh.

Among these multiscale methods, MsFEM and, in particular, its extension GMsFEM have achieved enormous success and are used in a wide range of practical applications [7,9]. In GMsFEM, the main idea is to use appropriately designed local spectral problems to construct the multiscale basis in GMsFEM, where multiple basis functions are allowed. Therefore, the accuracy of the multiscale solution can be tuned and controlled. In a previous work [6], significant results were obtained in the context of GMsFEM. The use of other types of multiscale methods for compressible flow, for instance, [9].

In this paper, we adopt the basic idea given in [3], which is a variation of GMsFEM based on a constraint energy minimization (CEM), for single-phase nonlinear compressible flow. This method provides a better convergence rate proportional to the size of the coarse grid. Furthermore, the CEM-GMsFEM uses the concepts of oversampling and localization [10] to compute multiscale basis functions in oversampled subregions. These basis functions are given by solving a class of local energy minimization problems over the eigenspaces that contain local information on heterogeneity.

The outline of the article is organized as follows: in Sect. 2, we briefly introduce the formulation of the model used in this work. Section 3 is devoted to constructing the offline multiscale space and framework of CEM-GMsFEM. Numerical experiments are presented in Sect. 4. Conclusions and final comments are given in Sect. 5.

2 Formulation of the Problem

We consider the following single-phase nonlinear compressible flow through a porous medium:

$$\begin{aligned} \partial_t(\phi\rho) - \nabla\cdot\left(\tfrac{\kappa}{\mu}\rho\nabla p\right) &= q, \quad \text{in } \Omega_T := \Omega\times(0,T],\\ \tfrac{\kappa}{\mu}\rho\nabla p\cdot n &= 0, \quad \text{on } \Gamma_N\times(0,T],\\ p &= p^D, \quad \text{on } \Gamma_D\times(0,T],\\ p &= p_0, \quad \text{on } \Omega\times\{t=0\}. \end{aligned} \tag{1}$$

Here, ϕ is the porosity of the medium, which is assumed to be a constant in our presentation, p is the fluid pressure we aim to seek, and μ is the constant fluid viscosity. κ denotes the permeability field that may be highly heterogeneous. Ω is the computational domain with boundary defined by $\partial\Omega_T = \Gamma_D\cup\Gamma_N$, and n is the outward unit-normal vector on $\partial\Omega_T$. The fluid density ρ is a function of the fluid pressure p as

$$\rho(p) = \rho_{\text{ref}}e^{c(p-p_{\text{ref}})}, \tag{2}$$

where ρ_{ref} is the given reference density and p_{ref} is the reference pressure.

Throughout this paper, we adopt the notation $L^2(D)$ and $H^1(D)$ to indicate the usual Sobolev spaces on subdomain $D \subset \Omega$ equipped with the norm $\|\cdot\|_{0,D}$ and $\|\cdot\|_{1,D}$ respectively. If $D = \Omega$, we omit the subscript D. In addition, we denote $\mathrm{V} := \mathrm{H}^1(D)$, $\mathrm{V}_0 := \mathrm{H}_0^1(D)$.

In the CEM-GMsFEM considered in this work, multiscale basis functions will be constructed for the pressure p. First, we introduce the notion of the two-scale grid. Then, we divide the computational domain Ω into some regular coarse blocks and denote the resulting triangulation as $\mathcal{T}^H$. We use H to represent the diameter of the coarse block $K \in \mathcal{T}^H$. Each coarse block will be further divided into a connected union of conforming fine-grid blocks across coarse-grid edges. We denote this fine-grid partition as $\mathcal{T}^h$, a refinement of $\mathcal{T}^H$ by definition. Let N^c be the number of coarse nodes, $\{x_i\}_{i=1}^{N^c}$ the set of nodes in $\mathcal{T}^H$ and $\omega_i = \bigcup\{K_j \in \mathcal{T}^H : x_i \in \overline{K}_j\}$ the neighborhood of the node x_i. In addition, given a coarse block K_i, we represent the oversampling region $K_{i,m} \subset \Omega$ obtained by enlarging K_i with m coarse grid layers, see Fig. 1. Let V_h be the space of the first-order Lagrange basis function concerning the fine-grid $\mathcal{T}^h$. Then, the finite element approximation to (1) on the fine grid is to seek

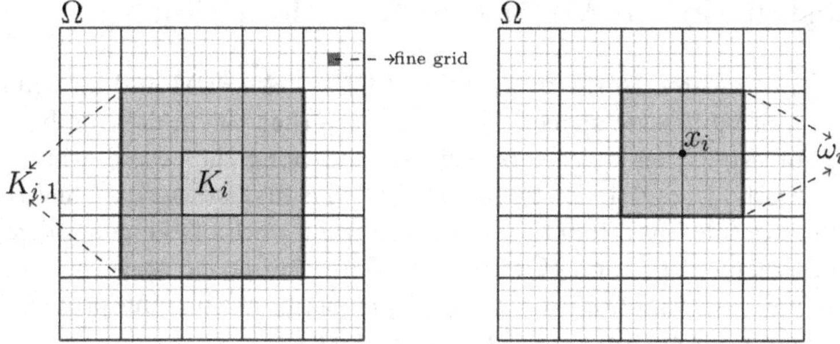

Fig. 1. Illustration of the coarse element K_i and oversampling domain $K_{i,1}$, the fine grid element and neighborhood ω_i of the node x_i.

$$(\phi \partial_t \rho(p_h), v) + \left(\tfrac{\kappa}{\mu}\rho(p_h)\nabla p_h, \nabla v\right) = (q, v), \quad \text{for each } v \in \mathrm{V}_h. \tag{3}$$

To derive the fully discrete scheme for (3), we introduce a partition of the time interval $[0, T]$ into subintervals $[t^{n-1}, t^n]$, $1 \le n \le N_t$ (N_t is an integer) and we denote the time-step size by $\Delta_t^n = t^n - t^{n-1}$. Then, using the backward Euler scheme in time, we can obtain the fully discrete scheme as follows: find p_h^n such that

$$(\phi\rho(p_h^n), v) - (\phi\rho(p_h^{n-1}), v) + \Delta_t^n \left(\tfrac{\kappa}{\mu}\rho(p_h^n)\nabla p_h^n, \nabla v\right) = \Delta_t^n (q, v), \tag{4}$$

for each $v \in \mathrm{V}_h$. Newton's method can solve the nonlinear equation (4). Specifically, let $\{\eta_i\}_{i=1}^{N^f}$ be the finite element basis functions for V_h, where

N^f is the number of fine grid nodes. We now can write $p_h^{n,k} = \sum_i p_i^{n,k}\eta_i$ and $p_h^{n-1} = \sum_i p_i^{n-1}\eta_i$, k denotes the k-th Newton iteration. Then, we can recast the nonlinear equation (4) as a residual equation system:

$$F_j^{n,k} = \left(\phi\rho\left(\sum_{i=1}^{N^f} p_i^{n,k}\eta_i\right),\eta_j\right) - \left(\phi\rho\left(\sum_{i=1}^{N^f} p_i^{n-1}\eta_i\right),\eta_j\right) \\ + \Delta_t^n\left(\frac{\kappa}{\mu}\rho\left(\sum_{i=1}^{N^f} p_i^{n,k}\eta_i\right)\sum_{i=1}^{N^f} p_i^{n,k}\nabla\eta_i,\nabla\eta_j\right) - \Delta_t^n\left(q,\eta_j\right) = 0, \tag{5}$$

for $j = 1, 2, \cdots, N^f$. To linearize the global problem, we should compute the partial derivatives of the residual equation concerning the unknown $p_i^{n,k}$, thus

$$J_{ji}^{n,k} := \frac{\phi\partial F_j^{n,k}}{\partial p_i^{n,k}} = (\phi\rho(p_h^{n,k})\eta_i,\eta_j) + \Delta_t^n\left(\frac{\kappa}{\mu}\rho(p_h^{n,k})\nabla\eta_i,\nabla\eta_j\right) \\ + \Delta_t^n\left(c\frac{\kappa}{\mu}\eta_i\rho(p_h^{n,k})\sum_i p_i^{n,k}\nabla\eta_i,\nabla\eta_j\right), \tag{6}$$

which results in a linear system that needs to solve $\mathbf{J}^{n,k}\boldsymbol{\delta}_{p^{n,k}} = -\mathbf{F}^{n,k}$, where $\mathbf{J}^{n,k} := [J_{ji}^{n,k}]_{i,j=1}^{N^f}$ represents the Jacobi matrix, $\mathbf{F}^{n,k} := [F_j^{n,k}]_{j=1}^{N^f}$ is the residual and $p^{n,k+1} = p^{n,k} + \boldsymbol{\delta}_{p^{n,k}}$.

3 Construction of Multiscale Basis Function

This section is devoted to the framework of CEM-GMsFEM and introduces the construction of multiscale spaces. We emphasize that the multiscale basis functions and corresponding spaces are defined concerning the coarse grid $\mathcal{T}^H$. The multiscale method consists of two stages. In the first stage, we construct the auxiliary multiscale basis function using the framework of the generalized multiscale finite element method (GMsFEM) [5]. In the second stage, we construct the multiscale basis function by solving some energy-minimizing problem in the coarse oversampling region $K_{i,m}$ with $m \geq 1$, see [3].

We construct auxiliary multiscale basis functions by solving the spectral problem for each coarse element K_i. We make use of the space V restricted to the coarse element K_i, *i.e.*, $\mathrm{V}(K_i) := \mathrm{V}\big|_{K_i}$. Then, we solve the following local eigenvalue problems: find $(\lambda_j^{(i)}, \varphi_j^{(i)}) \in \mathbb{R} \times \mathrm{V}(K_i)$ such that

$$a_i(\varphi_j^{(i)}, w) = \lambda_j^{(i)} s_i(\varphi_j^{(i)}, w), \quad \text{for each } w \in \mathrm{V}(K_i), \tag{7}$$

where $a_i(v,w) := \int_{K_i} \kappa\rho(p_0)\nabla v \cdot \nabla w dx$, and $s_i(v,w) := \int_{K_i} \widetilde{\kappa} v w dx$, in which $\widetilde{\kappa} = \rho(p_0)\kappa\sum_{i=1}^{N^c} |\nabla\chi_i|^2$, N^c is the total number of neighborhoods, p_0 is the initial pressure p and $\{\chi_i\}$ are the partitions of unity function of ω_i [2].

We assume that the eigenfunctions satisfy the normalized condition $s_i(\varphi_j^{(i)}, \varphi_j^{(i)}) = 1$. The eigenvalues are ordered ascendingly, *i.e.*, $\lambda_1^{(i)} \leq \lambda_2^{(i)} \leq \cdots$. Then, we can use the first L_i eigenfunctions to construct the local auxiliary multiscale $\mathrm{V}_{\text{aux}}(K_i) := \text{span}\{\varphi_j^{(i)} : 1 \leq j \leq L_i\}$. Then, the global auxiliary space V_{aux} is defined by using these local auxiliary spaces $\mathrm{V}_{\text{aux}} = \bigoplus_{i=1}^{N^c} \mathrm{V}_{\text{aux}}(K_i)$.

The inner product and s-norm of the global auxiliary multiscale spaces are defined respectively by $s(v,w)=\sum_{i=1}^{N^c} s_i(v,w)$, $\|v\|_s := \sqrt{s(v,v)}$.

To construct the CEM-GMsFEM basis functions, we use the following definition from [3].

Definition 1 **($\varphi_j^{(i)}$-orthogonality).** *Given a function $\varphi_j^{(i)} \in \mathrm{V}_{\text{aux}}$, if a function $\psi \in \mathrm{V}$ satisfies*

$$s(\psi,\varphi_j^{(i)})=1, \quad s(\psi,\varphi_{j'}^{(i')})=0, \quad \text{if } j' \neq j \text{ or } i' \neq i,$$

then, we say that is $\varphi_j^{(i)}$-orthogonal where $s(v,w)=\sum_{i=1}^{N} s_i(v,w)$.

We define the operator $\pi : \mathrm{V} \to \mathrm{V}_{\text{aux}}$ by $\pi(v)=\sum_{i=1}^{N}\sum_{j=1}^{L_i} s_i(v,\varphi_j^{(i)})\varphi_j^{(i)}$, for each $v \in \mathrm{V}$, and the null space of the operator π is defined by $\widetilde{\mathrm{V}}=\{v \in \mathrm{V} : \pi(v)=0\}$.

We will now construct the multiscale basis functions. For each coarse block K_i, we define the oversampled subdomain $K_{i,m} \subset \Omega$ by enlarging K_i with an arbitrary number of coarse grid layers $m \geq 1$. Let $\mathrm{V}_0(K_{i,m}) := \mathrm{H}_0^1(K_{i,m})$; we solve the following minimization problems: find multiscale basis function $\psi_{j,\text{ms}}^{(i)} \in \mathrm{V}_0(K_{i,m})$

$$\psi_{j,\text{ms}}^{(i)} = \operatorname{argmin}\{a(\psi,\psi) : \psi \in \mathrm{V}_0(K_{i,m}),\ \psi \text{ is } \varphi_j^{(i)}\text{-orthogonal}\}. \tag{8}$$

Then, the CEM-GMsFEM space is defined by

$$\mathrm{V}_{\text{ms}} = \operatorname{span}\{\psi_{j,\text{ms}}^{(i)} : 1 \leq j \leq L_i, 1 \leq i \leq N\}.$$

The minimization problem (8) is implicit; we can redefine it into an explicit form by using a Lagrange multiplier. Then, the problem (8) can be written as the following problem: find $\psi_{j,\text{ms}}^{(i)} \in \mathrm{V}_0(K_{i,m})$, $\lambda \in \mathrm{V}_{\text{aux}}^{(i)}(K_i)$ such that

$$\begin{cases} a(\psi_{j,\text{ms}}^{(i)},\eta)+s(\eta,\lambda) & =0, \quad \text{for all } \eta \in \mathrm{V}(K_{i,m}), \\ s(\psi_{j,\text{ms}}^{(i)}-\varphi_j^{(i)},\nu) & =0, \quad \text{for all } \nu \in \mathrm{V}_{\text{aux}}^{(i)}(K_{i,m}), \end{cases}$$

where $\mathrm{V}_{\text{aux}}^{(i)}(K_{i,m})$ is the union of all local auxiliary spaces for $K_i \subset K_{i,m}$. Note that one can solve the above continuous problem numerically on a fine-scale grid. Thus, given the above space and by using the backward Euler scheme, the full-discrete formulation reads as follows: find $p_{\text{ms}} \in \mathrm{V}_{\text{ms}}$ such that

$$(\phi\rho(p_{\text{ms}}^n),v)-(\phi\rho(p_{\text{ms}}^{n-1}),v)+\Delta_t^n\left(\tfrac{\kappa}{\mu}\rho(p_{\text{ms}}^n)\nabla p_{\text{ms}}^n,\nabla v\right)=\Delta_t^n(q,v), \tag{9}$$

for all $v \in \mathrm{V}_{\text{ms}}$. The local multiscale basis construction is motivated by the global basis construction defined below.

4 Numerical Results

In this section, we present two representative examples to confirm the performance of the CEM-GMsFEM. We use the backward Euler scheme for the time discretization and Newton's method for the nonlinear equation. All computations are performed by using the software `MatLab`. Firstly, we consider two high-contrast permeability fields for each experiment. Long channels and inclusions form these fields, the blank regions have values of 10^5 millidarcys, while other regions have values of 10^9 millidarcys (see, for instance, Fig. 2). In all numerical tests, we let viscosity $\mu = 5\,\mathrm{cP}$, porosity $\phi = 500$, fluid compressibility $c = 1.0 \times 10^{-8} 1/\,\mathrm{Pa}$, the reference pressure $p_{\text{ref}} = 2.00 \times 10^7\,\mathrm{Pa}$, and the reference density $\rho_{\text{ref}} = 850\,\mathrm{kg/m^3}$.

We consider the first experiment a full zero Neumann boundary condition, with an initial pressure field p_0 with value $2.16 \times 10^7\,\mathrm{Pa}$. Four vertical injectors are placed in the corners, and one sink is in the middle of the domain to drive the flow. We set the fine grid resolution of 64^3, with fine grid size of $h = 20\,\mathrm{m}$, the coarse grid resolution of 8^3, with $H = 8h$. The parameter Δ_t^n is 7 days, and $T = 20\Delta_t^n (= 140$ days) is the total simulation time. For the CEM-GMsFEM, we use 4 basis function and 4 oversampling layers. It is clear that the number of bases efficiently improves the accuracy of the CEM-GMsFEM [3]; in this case, the relative L^2 error is 1.7396E-03 and H^1 error is 3.8180E-01. The dimension of the coarse system is 4916 ($= 729 \times$ number of basis functions); note that the dimension of the fine-scale system is 274625. We compare the pressure profiles with singular source and zero Neumann boundary conditions in Fig. 2.

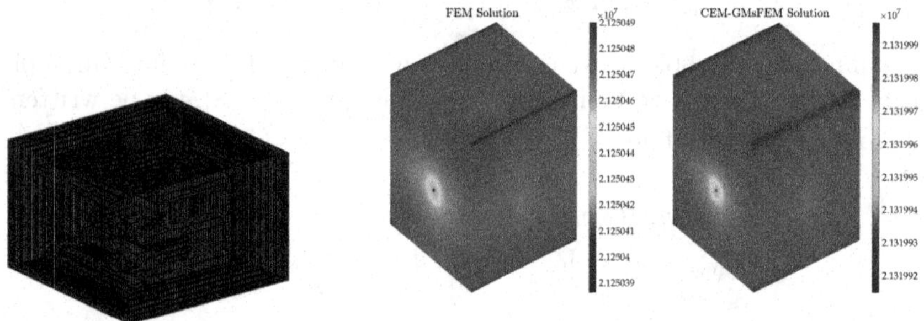

Fig. 2. Experiment with full-zero Neumann boundary condition. High-contrast permeability field (left), fine-scale reference solution (middle), and CEM-GMsFEM solution (right) with 4 basis function and 4 oversampling layers at $T = 20\Delta_t^n$.

In the second experiment, we consider a zero Neumann boundary condition and nonzero Dirichlet boundary condition [11]. We impose zero Neumann condition on boundaries of planes xy and xz and let $p = 2.16 \times 10^7\,\mathrm{Pa}$ in the first yz plane and $p = 2.00 \times 10^7\,\mathrm{Pa}$ in the last yz plane for all time instants, no additional source is imposed. The pressure difference will drive the flow, and the initial field p_0 linearly decreases along the x axis and is fixed in the yz plane. Table 1 shows that numerical results use 4 basis functions on each coarse block with different coarse grid sizes ($H = 4h, 8h$ and $16h$). In Table 1, ε_0 and ε_1 denote the relative L^2 and energy error estimate between the reference solution and CEM-GMsFEM solution.

In Fig. 3, we depict the numerical solution profiles with a fine grid resolution of 32^3 and coarse grid resolution of 8^3 at $T = 140$ day, which have a good agreement. For this case, by using 4 basis functions and the coarse grid size $H = 8h$, the CEM-GMsFEM uses 4 oversampling layers, and the relative error $\varepsilon_0 = 2.8514\text{E-}04$, while $\varepsilon_1 = 3.8213\text{E-}01$.

Table 1. Numerical result with different numbers of oversampling layers (m) for the second experiment with full zero Neumann and nonzero Dirichlet boundary condition.

Number basis	H	Number oversampling layers m	ε_0	ε_1
4	$4h$	3	2.3493E-03	5.6700E-01
4	$8h$	4	2.8514E-04	3.8213E-01
4	$16h$	5	1.3302E-04	1.6102E-01

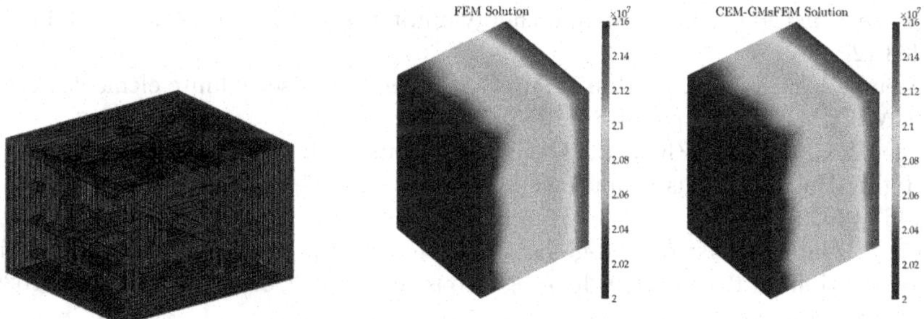

Fig. 3. Experiment with combined boundary condition. High-contrast permeability field (left), fine-scale reference solution (middle), and CEM-GMsFEM solution (right) with 4 basis function and 4 number of oversampling layers at $T = 20\Delta_t^n$.

5 Conclusions and Future Directions

We have presented CEM-GMsFEM for solving the highly heterogeneous nonlinear single-phase compressible flow in this work. For CEM-GMsFEM, the first step is constructing the additional space by solving spectral problems. The second step is based on constraint energy minimization and oversampling. So, we construct multiscale basis functions for pressure. Two representative 3D examples have been presented to verify the efficiency and accuracy of the proposed method. The convergence depends on the coarse mesh size and the decay of eigenvalues of local spectral problems. In addition, the CEM-GMsFEM is shown to have a second-order convergence rate in the L^2-norm and a first-order convergence rate in the energy norm concerning the coarse grid size.

In some applications, a future challenge remains to boost the performance of the coarse-grid simulation, especially when the source term is singular; one may need to further improve the accuracy of the approximation without additional mesh refinement. In these cases, one needs to enrich the multiscale space by adding more basis functions in the online stage [4]. These new basis functions are based on the oversampling technique and the information on local residuals. Moreover, an adaptive enrichment algorithm will be presented to reduce error in some regions with large residuals.

References

1. Arbogast, T., Xiao, H.: A multiscale mortar mixed space based on homogenization for heterogeneous elliptic problems. SIAM J. Numer. Anal. **51**(1), 377–399 (2013)
2. Babuška, I., Melenk, J.M.: The partition of unity method. Internat. J. Numer. Methods Engrg. **40**(4), 727–758 (1997)
3. Chung, E.T., Efendiev, Y., Leung, W.T.: Constraint energy minimizing generalized multiscale finite element method. Comput. Methods Appl. Mech. Engrg. **339**, 298–319 (2018)
4. Chung, E.T., Efendiev, Y., Leung, W.T.: Fast online generalized multiscale finite element method using constraint energy minimization. J. Comput. Phys. **355**, 450–463 (2018)
5. Efendiev, Y., Galvis, J., Hou, T.Y.: Generalized multiscale finite element methods (GMsFEM). J. Comput. Phys. **251**, 116–135 (2013)
6. Fu, S., Chung, E., Zhao, L.: Generalized Multiscale Finite Element Method for Highly Heterogeneous Compressible Flow. Multiscale Model. Simul. **20**(4), 1437–1467 (2022)
7. Fu, S., Chung, E.T.: A local-global multiscale mortar mixed finite element method for multiphase transport in heterogeneous media. J. Comput. Phys. **399**, 108906 (2019)
8. Hou, T.Y., Wu, X.H.: A multiscale finite element method for elliptic problems in composite materials and porous media. J. Comput. Phys. **134**(1), 169–189 (1997)
9. Kim, M.Y., Park, E.J., Thomas, S.G., Wheeler, M.F.: A multiscale mortar mixed finite element method for slightly compressible flows in porous media. J. Korean Math. Soc. **44**(5), 1103–1119 (2007)
10. Må lqvist, A., Peterseim, D.: Localization of elliptic multiscale problems. Math. Comp. **83**(290), 2583–2603 (2014)
11. Wang, Y., Hajibeygi, H., Tchelepi, H.A.: Algebraic multiscale solver for flow in heterogeneous porous media. J. Comput. Phys. **259**, 284–303 (2014)

Numerical Simulation of Propeller Hydrodynamics Using the Open Source Software

Andrey Britov[1(✉)], Sofya Yarikova[1], Andrey Epikhin[1], Stepan Elistratov[1,2], and Qin Zhang[3]

[1] Ivannikov Institute for System Programming of the RAS, Moscow, Russia
291299ab@gmail.com
[2] Shirshov Institute of Oceanology of the RAS, Moscow, Russia
[3] Ocean University of China, Qingdao, China

Abstract. The paper presents the results of numerical simulation of the propeller Ka4-70 using the actuator line model in the OpenFOAM, AMReX and Nek5000 open-source software. The modifications of the tools for wind farm simulation for these packages are carried out. Features of these implementation are described. For numerical calculations the LES and IDDES turbulence models are used. A comparison of the computational costs and accuracy of flow structures are made for the actuator line model using different methods and the arbitrary mesh interface approach. The actuator line model provides force characteristics and flow structures with good enough accuracy.

Keywords: propeller · thrust forces · wake dynamics · OpenFOAM · AMReX · Nek5000

1 Introduction

Enterprises engaged in design and development of various propellers are constantly facing the problems of their numerical modeling and characterization. These problems can be solved using different numerical approaches [1,2]. In mid-90's sliding mesh interfaces method was developed [3]. This approach turned out to be applicable in many of scientific studies related to rotational motion of solid bodies. Especially wide sliding mesh techniques are being applied in propellers development. Steijl et al. [4] proposed a method for the study of helicopter rotor-fuselage interaction by third-order sliding mesh on block-structured mesh. This technique was successfully applied in LES simulations of tidal-stream turbines, showing great agreement with experimental power and thrust predictions [5]. Ramirez et al. [6] presented a new technique to maintain the high-order stencil across the sliding interfaces. Those solutions are presented in the vast majority of CFD-toolboxes, in particular in OpenFOAM [7] via AMI and GGI techniques [8–10]. However, further investigations showed that using such techniques demand a lot of computational resources [11]. In case of propeller work simulation it turns out that it's enough

J. Mikyška et al. (Eds.): ICCS 2023, LNCS 14077, pp. 279–291, 2023.
https://doi.org/10.1007/978-3-031-36030-5_23

to replace the propeller's geometry with a simplified model saving computational resources but providing fine accuracy of calculations. In order to reduce computational time, several researchers used panel method [12] or the so-called hybrid models which combine CFD solver and blade element model (BEM) [13]. In this type of modelling the aerodynamic forces applied to a blade do not result directly from CFD, but are calculated separately using inflow data and blade geometry. These calculations are carried out in parallel with the numerical simulation. The blade forces are calculated at each iteration and are implemented as source terms in the flow. There are three hybrid models depending on the distribution of the source terms: actuator disk, actuator line and actuator surface [14].

The simplest hybrid model is the actuator disk that replaces the wind turbine rotor with a thin disk volume. Typically the disk has a diameter equal to that of the rotor and its blade forces affected on the fluid are replaced by equivalent source terms. This model related to wind turbines was introduced in by Sørensen & Myken in [15] using axisymmetric Euler solver. In this study the rotor is replaced with equivalent constant intensity sources. This intensity is calculated from the wind turbine thrust. Actuator disk model neglects separate blade geometry details. However, geometry and viscous flow around blades are not defined. The occupied swept area of the rotor is replaced with distributed source terms instead [16]. Three-dimensional calculations using actuator disk were presented by Amara et al. [17] in cases of isolated and clustered wind turbines. Although, for detailed representation of near wake or blade tip vortices, a three-dimensional model for each blade must be used.

Sørensen in [18] proposed so-called actuator line model (ALM). In accordance with it a blade is discretized by finite set of points (elements). Drag and lift forces are calculated for each element taking into account local attack angle and relative velocity. Blade elements themselves are defined by aerodynamic and geometric characteristics. Aerodynamic performances corresponds the foil, made by intersection of blade in the point where the element is situated. Geometrical parameters are derived from the geometry of analyzed foil. For each blade element, relative flow velocity and angle of attack are computed using fluid velocity and element radial one. The comparison of the actuator Line model with the experimental data reveals the effectiveness of this model for wind turbines' power characteristic calculations. The reliability of the model for representing near and far wakes was proved in [19,20].

More complete and complicated hybrid model is the actuator surface model one applied by Dobrev & Massouh [21] and Shen [22] et al. The main advantage of this model is more physically realistic force distribution along the blade. In the actuator surface model the blade geometry is represented by a surface formed by chord lengths distribution at different radial locations of a blade [23]. In fact the actuator surface model considered as an extend actuator line model where blade are no more represented as thin lines but expand along the chords to the surfaces immersed in the flow. According to [13] the distribution of the calculated forces on blade chords improved structure of induced velocities near the wake. Rotating blade effect is represented by pressure and velocity discontinuity, which related to a circulation around the aerodynamic foil. Thus, velocity and pressure gradients created by the actuator surface becomes very close to a real turbine rotation case improving the initial conditions for the wake.

Based on computational resources and calculation accuracy, it's expedient to carry out further investigations using actuator line model.

2 Numerical Method

2.1 Governing Equations

The complex flow fields around the propeller are obtained by solving the 3D incompressible Navier-Stokes equations. For the incompressible single-phase Newtonian turbulent flows, the filtered mass and momentum equations are the following:

$$\begin{cases} \nabla \cdot \boldsymbol{u} = 0, & \text{(1a)} \\ \frac{\partial \boldsymbol{u}}{\partial t} + (\boldsymbol{u} \cdot \nabla)\boldsymbol{u} = -\frac{1}{p} + \nu \nabla^2 \boldsymbol{u} + \boldsymbol{F}_{turb}. & \text{(1b)} \end{cases}$$

Where $\boldsymbol{u}$ is a three-dimensional velocity vector, t - time, p - pressure, ∇ - del operator, ν - coefficient of kinematic viscosity, $\boldsymbol{F}_{turb}$ - source term.

Simple explanation of actuator line model is presented on Fig. 1. Propellers blades are replacing by lines consisting from finite number of elements.

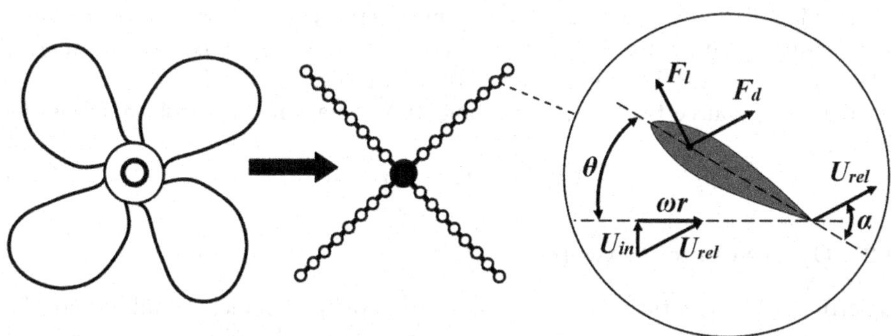

Fig. 1: Actuator line model scheme

Force and moment characteristics for each actuator line element are calculated via the following formulas [24]:

$$F_l = \frac{1}{2} A_{elem} C_l(\alpha) |\boldsymbol{u}_{rel}|^2, \tag{2}$$

$$F_d = \frac{1}{2} A_{elem} C_d(\alpha) |\boldsymbol{u}_{rel}|^2, \tag{3}$$

$$M = \frac{1}{2} A_{elem} r C_m(\alpha) |\boldsymbol{u}_{rel}|^2, \tag{4}$$

$$\boldsymbol{u}_{rel} = \boldsymbol{u}_{in} + \boldsymbol{w}r. \tag{5}$$

Where A_{elem} - element square, $C_l(\alpha)$ - lift coefficient, $C_d(\alpha)$ - drag coefficient, $C_m(\alpha)$ - pitching moment coefficient, $\boldsymbol{u}_{in}$ - inflow velocity vector, ω - angular velocity of the turbine, r - radius of element.

For each element drag force, lift force and momentum are calculating respectively to local velocity, angle of attack (α) and force coefficient. Every element has aerodynamic and geometric characteristics (chord, twist (θ)). Aerodynamic performances are corresponding to foil in section, where element is placed, assuming that lift and drag coefficients are known. Geometrical characteristics is defined by geometry of corresponding foil.

After the force on the actuator line element from the flow being calculated, it is then projected back onto the flow field as a source term in the momentum equation. To avoid instability due to steep gradients, the source term is tapered from its maximum value away from the element location by means of a spherical Gaussian function [25]:

$$\eta = \frac{1}{\epsilon^3 \pi^{\frac{3}{2}}} \exp\left[-\left(\frac{|\boldsymbol{r}|}{\epsilon} \right)^2 \right], \tag{6}$$

$$\epsilon_{mesh} = 2 C_{mesh} \Delta x, \tag{7}$$

$$\Delta x = \sqrt[3]{V_{cell}}. \tag{8}$$

where r is a distance from the actuator line element quarter-chord location, ϵ - regularization parameter, V_{cell} - cell volume, Δx - cell length, C_{mesh} - coefficient taking into account the unevenness of the cell faces.

Thus, the source term F_{turb} in momentum equation can be obtained as the following:

$$\boldsymbol{F}_{turb} = (\boldsymbol{F}_l + \boldsymbol{F}_d) \otimes \eta. \tag{9}$$

2.2 Open-Source Software

OpenFOAM. The turbinesFoam library is used for numerical simulation of the propeller hydrodynamics [24,26]. This library was developed to model wind and marine hydro-kinetic turbines in OpenFOAM using the actuator line method, which was written as an extension library, using the fvOptions functionality for adding source terms to equations at run-time. This allows the ALM to be added to many of the standard solvers included in OpenFOAM without modification. In the original turbinesFoam library the propeller rotation speed is calculated using parameterized coefficient tip speed ratio and inflow velocity. The library is modified in order to being able to run the propeller hydrodynamics simulation in a generator mode (zero flow velocity). If the inflow velocity is different from zero, the original version of the turbinesFoam library is used. Otherwise, rotational speed is set by user. The ability to continue the calculation after its suspension is also added. For the numerical simulation of propeller hydrodynamics PIMPLE algorithm is used. The algorithm is implemented in pimpleFoam solver in OpenFOAM. The IDDES turbulence model is used [27]. The LES approach is used in the free stream area and the Spalart - Allmaras turbulence model [28] is used near the hub wall. The filteredLinearM difference scheme proposed in [29] is applied for convective flux discretisation.

AMReX. The AMR-Wind [30] library is used to investigate propeller performances based on AMReX framework [31,32]. AMR-Wind is a parallel, block-structured, finite volume method, incompressible flow solver for wind turbine and wind farm simulations specialized for efficiency and scalability. The solver is built on top of the AMReX library which provides the mesh data structures, performance portable parallel algorithms compatible with different GPU architectures, linear solvers which are a combination of geometric and algebraic multigrid solvers. AMReX supports the development of adaptive mesh refinement (AMR) algorithms for solving systems of partial differential equations, in simple or complex geometries. AMR reduces computational costs and the amount of memory compared to a uniform mesh, while maintaining accurate descriptions of different physical processes in complex multi-physics algorithms. AMReX provides support for both explicit and implicit mesh discretization algorithms. Summing up, AMR-wind achiving the following advantages: an open, well-documented implementation of the state-of-the-art computational models for modeling propellers flow physics at various fidelity's. The numerical simulation is based on the LES approach and actuator line method, which used forces obtained from the OpenFOAM solution.

Nek5000. Nek5000 is an open-source spectrum-element based method [33]. The method unites advantages of finite-element methods and those of spectral element ones. The joining of the solutions in each element through the element edges is made via 'overlap' of the finite element solutions in adjacent element (Lagrangian joining) which allows to calculate solution with 8 order accuracy. The method finds the solution as a Legendre polynomial series on a grid consisting of Gauss-Lobbato-Legendre points. The velocity field obtained can be legally interpolated to a user grid due to high order of the solution. The feature is that the elements used might not be orthogonal, that allows it to be applied to a wider problem class. The main disadvantage is the inability to set pointwise force because of absence of element in sense of which finite element use it (Nek's element are really huge in actuator-line step scales), which make smoothing an obligation. Another one follows the high-order and precision of the method and is a gross time of the calculation (about a week for a case).

3 Numerical Setup

The rotation of propeller Ka4-70 [34] from Wageningen series with propeller diameter $D = 0.1$ m is considered. Rotational speed of propeller is 500 RPM. Inflow velocity is equal to 0.05 m/s.

The simulations are performed inside a domain (height is 10D, wide is 6D and full length equal to 2D + L, where L is local refinement length equal to 3D or 10D) presented in the Fig. 2. It is similar to size that used in Ocean University of China (OUC) numerical simulation [35]. For 10D distance simulation an open tank is proposed.

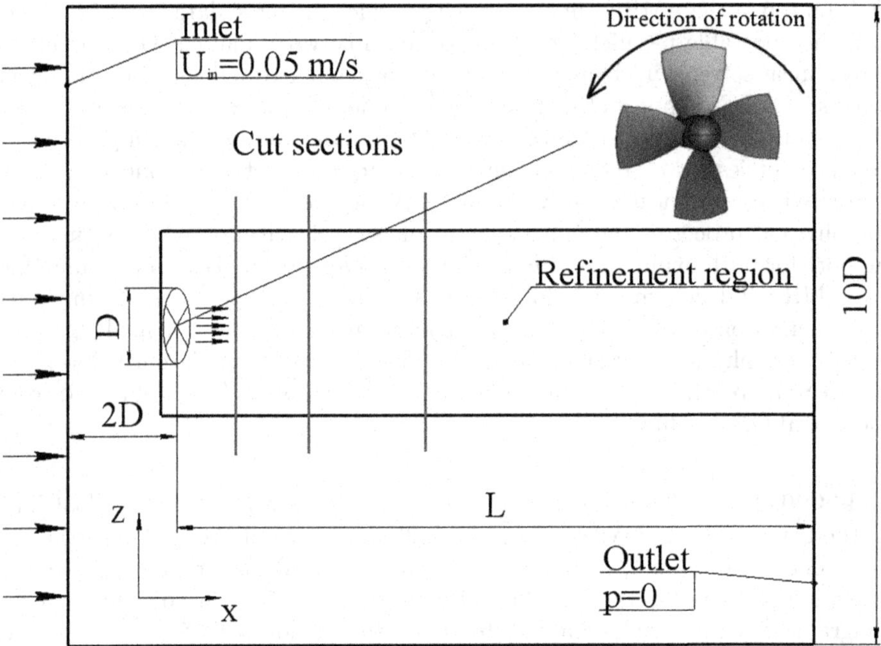

Fig. 2: Computational domain scheme with initial and boundary conditions

Number of actuator line elements is 16. One of the main complexities of ALM applying is the determination of the aerodynamic coefficients of an element. The most accurate method to solve the problem is to capture the foils by cylindrical intersections with subsequent numerical simulation, but this approach is too time consuming. An alternative to this method is using a foil database, such as airfoiltools.com [36], where are drag and lift coefficients for airfoils that are geometrically similar to those obtained by blade model dissecting. Disadvantage of the foil database resources is the lack of data for high angles of attack. For eliminate this problem it's proposed to use the Viterna method of extrapolation [37]. The main time costs are associated with the preparation of data for each airfoil. Further, they can be reduced by using simpler methods for aerodynamic characteristics determination [38].

An example of the lift coefficient dependence from angle of attack obtained using Viterna extrapolation method and a comparison of the real foil and a similar one from airfoiltools.com are presented on Fig. 3.

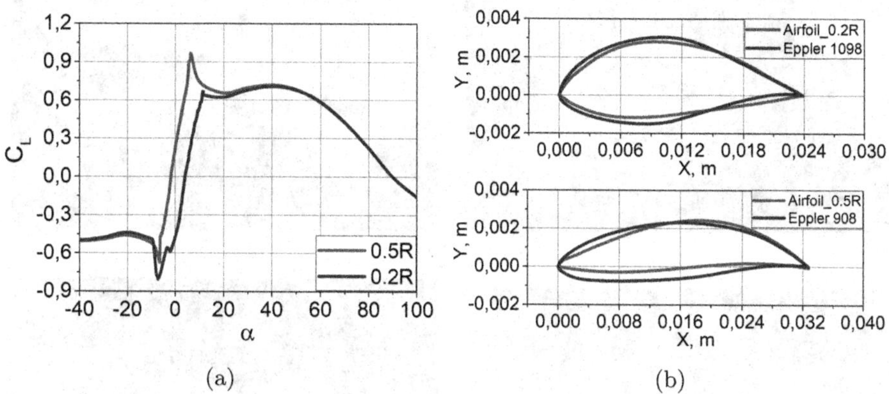

Fig. 3: Blade intersection: a) lift coefficient from angle of attack b) comparison between real airfoil and similar from airfoiltools.com

The main goal of the actuator line is to replace real screw with the mass force in the right part of Navier-Stocks equation. For the AMReX and Nek5000 simulation the set of 16 pointwise forces per blade obtained from the OpenFOAM solution are used. To convert them into the mass forces they are smoothed by gaussian bell by formula:

$$f(x,y,z) = \sum_i f_i N(\epsilon) exp\left(-\frac{(x-x_i)^2 + (y-y_i)^2 + (z-z_i)^2}{2\epsilon^2}\right), \qquad (10)$$

where $N(\epsilon)$ is a norm of this bell. Here (x_i, y_i, z_i) is a point where the current poinwise force is applied. Hence the screw rotates, these points are also rotate around it's center. Additionally, we must sum the forces by both points and blades. Finally, this force is added into Navier-Stocks equation.

4 Results and Discussions

Figures 4 and 5 show mean velocity flow structures behind the propeller for different distances. Comparison of the results obtain from different CFD-toolboxes with the implementation of hybrid models and the arbitrary mesh interface method [35] is presented on Figs. 6 and 7.

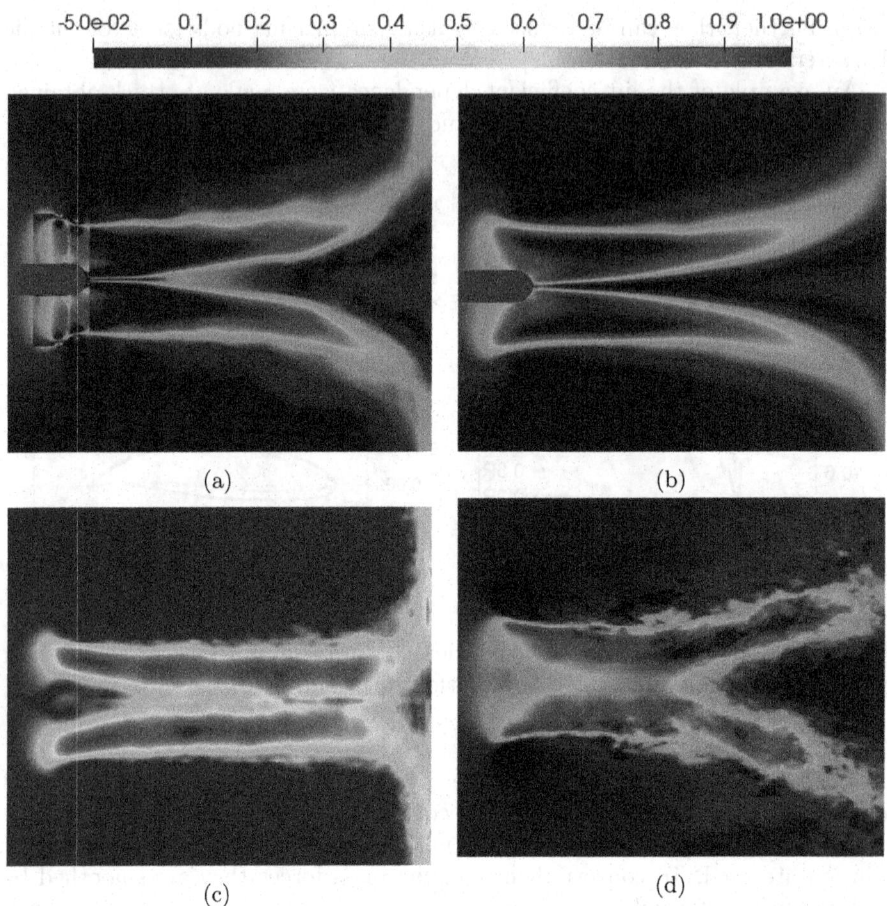

Fig. 4: Mean velocity field on 3D: a) OUC numerical simulation; OpenFOAM; c) Nek5000; d) AMReX

Flow structure generated by propeller has four areas: near-wake, transition, far-wake regions and near-wall depending from degree of destabilization. The near-wake region is close to the propeller disk (0.5D), there is an area with maximum values of velocity. Beside, this is the region, where hub's vortexes starting to affect on the flow. It should be noted that the actuator model cannot reproduce the effect of a rigid body on the flow, so propeller hub must be included in the grid. The area where the vortexes lose their original morphology and interact with outer flow is the transition area (1D). Downstream, velocity profile lose its charge, vortexes are broken down into small-scale, disordered turbulence in the far-wake region (2D and more). For cases with 3D distance between propeller and wall, far wake region transforms into the near-wall region, where propellers wake vortex structure interacting the wall.

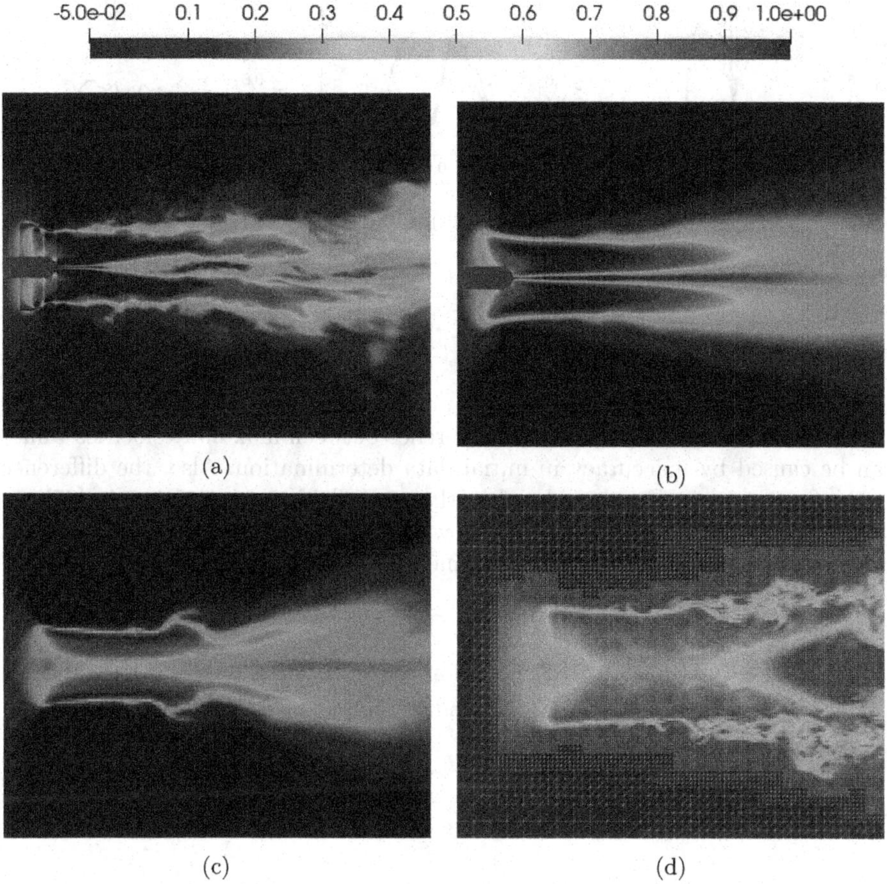

-5.0e-02 0.1 0.2 0.3 0.4 0.5 0.6 0.7 0.8 0.9 1.0e+00

(a) (b)

(c) (d)

Fig. 5: Mean velocity field on 10D: a) OUC numerical simulation; b) OpenFOAM; c) AMReX; d) Adaptive mesh refinement

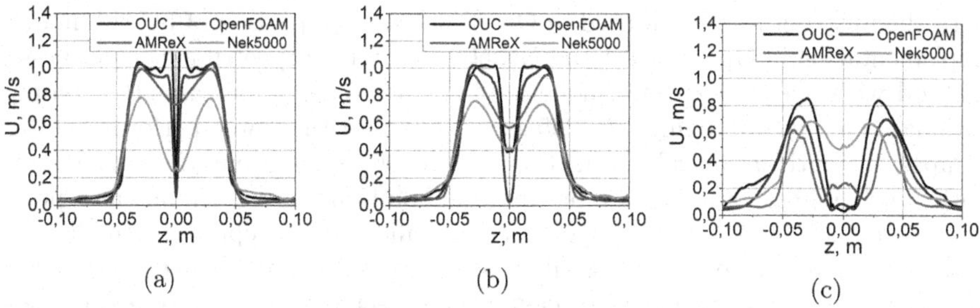

Fig. 6: Comparison of mean velocity at distance: a) 0.5D; b) 1D; c) 2D from propeller

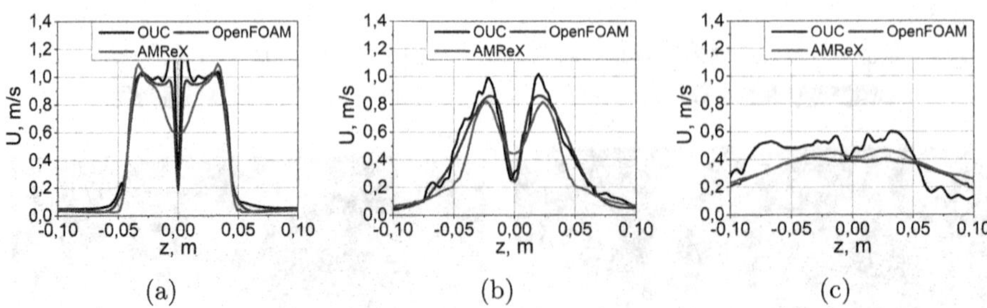

Fig. 7: Comparison of mean velocity at distance: a) 0.5D; b) 1D; c) 2D from propeller

The obtained results show that the actuator line model represents propeller hydrodynamics correctly. The main difference between maximal velocities values can be caused by inaccuracy in initial data determination. Also, the difference in the flow structure can be related to the small field averaging time in [35].

Different simulation approaches show the similarity flow structures. For all calculations 36 cores are used. Table 1 includes main calculation characteristics.

Table 1: Calculation characteristics.

	AMR-Wind	turbinesFOAM	Nek5000
t_{iter}, sec	6	15	3.5
number of cores	36	36	36
number of cells	15 000 000	10 000 000	10 000

To compare the efficiency of CFD software it's necessary to convert obtained calculation characteristics into the calculation time per core per computational cell. Thus, applying of actuator line model on block-structured mesh in AMReX software led to a reduction of computational time by about 4 times compared with the unstructured mesh used in OpenFOAM. Comparison of hybrid models implementation in OpenFOAM and Nek5000 show that spectrum-element based method with actuator surface model and 8-order scheme increases the computational time by approximately 235 times. Despite the lower performance (in comparison with finite volume methods) Nek5000 carries out the calculations with much higher order and represent the solution as the set of basis polynomials rather than the set of point values, allowing to reproduce small-scale turbulence that can be investigated in post-processing after being interpolated on a fine grid. Such small vortices cannot be detected by finite volume methods because they are limited by their mesh step. Another advantage of Nek5000 is the ability of accurate calculation of spatial derivative, which can be necessary for further problem development (e.g., sedimentation problem). The main cause of the divergence between wake structure obtained via OpenFOAM and Nek5000 must be insufficient grid resolution in vertical direction. That's why only in a horizontal one wakes are similar. Mesh refining in this direction must yield a more proper results, but requires additional computational resources.

5 Conclusion

Investigation of applying actuator line model for numerical simulation of propeller Ka4-70 are carried out. The finite volume method (OpenFOAM), finite volume method on block-adapted mesh (AMReX) and spectrum-element based method (Nek5000) are compared. The results show that the application of actuator line model saves computing resources and reproduces the characteristics and hydrodynamics of the propeller with sufficient accuracy. For modelling of instantaneous characteristics in the far field it's advisable to use adaptive mesh approach. However, for detailed representation of near wake Nek5000 using is more efficiency, while for determining mean fields and force and moment characteristics it's preferably to use OpenFOAM.

Acknowledgements. The reported study was funded by Russian Foundation for Basic Research (RFBR, Proj. No. 21-57-53019) and National Natural Science Foundation of China (NSFC, Proj. No. 52111530047).

References

1. Bachler, G., Schiffermüller, H., Bregant, A.: A parallel fully implicit sliding mesh method for industrial CFD applications. In: Jenssen, C.B., et al. (eds.) Parallel Computational Fluid Dynamics 2000, pp. 501–508. North-Holland, Amsterdam (2001). https://doi.org/10.1016/B978-044450673-3/50129-9
2. Paulo, A.S.F.S., Tsoutsanis, P., Antoniadis, A.F.: Simple multiple reference frame for high-order solution of hovering rotors with and without ground effect. Aerosp. Sci. Technol. **111**, 106518 (2021). https://doi.org/10.1016/j.ast.2021.106518
3. Rai, M.M.: A conservative treatment of zonal boundaries for Euler equation calculations. J. Comput. Phys. **62**(2), 472–503 (1986). https://doi.org/10.1016/0021-9991(86)90141-5
4. Steijl, R., Barakos, G.: Sliding mesh algorithm for CFD analysis of helicopter rotor- fuselage aerodynamics. Int. J. Num. Methods Fluids **58**(5), 527–549 (2008). https://doi.org/10.1002/fld.1757
5. McNaughton, J., Afgan, I., Apsley, D.D., Rolfo, S., Stallard, T., Stansby, P.K.: A simple sliding-mesh interface procedure and its application to the CFD simulation of a tidal-stream turbine. Int. J. Num. Methods Fluids **74**(4), 250–269 (2014). https://doi.org/10.1002/fld.3849
6. Ramírez, L., Foulquié, C., Nogueira, X., Khelladi, S., Chassaing, J.-C., Colominas, I.: New high-resolution-preserving sliding mesh techniques for higherorder finite volume schemes. Comput. Fluids **118**, 114–130 (2015). https://doi.org/10.1016/j.compfluid.2015.06.008
7. Jasak, H., Jemcov, A., Tukovic, Z.: OpenFOAM: A c++ library for complex physics simulations (2013)

8. Chandar, D., Gopalan, H.: Comparative analysis of the arbitrary mesh interface(AMI) and overset methods for dynamic body motions in OpenFOAM. (2016). https://doi.org/10.2514/6.2016-3324
9. Vilfayeau, S., Pesci, C., Ferraris, S., Heather, A., Roesler, F.: Improvement of arbitrary mesh interface (AMI) algorithm for external aerodynamic simulation with rotating wheels. Fourth international conference in numerical and experimental aerodynamics of, road vehicles and trains (Aerovehicles 4), Berlin, Germany, August 23–25 (2021)
10. Nuernberg, M., Tao, L.: Three dimensional tidal turbine array simulations using OpenFOAM with dynamic mesh. Ocean Eng. **147**, 629–646 (2018). https://doi.org/10.1016/j.oceaneng.2017.10.053
11. Daaou Nedjari, H., Guerri, O., Saighi, M.: Full rotor modelling and generalized actuator disc for wind turbine wake investigation. Energy Rep. **6**, 232–255 (2020). https://doi.org/10.1016/j.egyr.2019.10.041. Technologies and Materials for Renewable Energy, Environment and Sustainability
12. Baltazar, J.M., Rijpkema, D., Falcão de Campos, J., Bosschers, J.: Prediction of the open-water performance of ducted propellers with a panel method. J. Marine Sci. Eng. **6**(1) (2018). https://doi.org/10.3390/jmse6010027
13. Vermeer, L.J., Sørensen, J.N., Crespo, A.: Wind turbine wake aerodynamics. Prog. Aerosp. Sci. **39**(6), 467–510 (2003). https://doi.org/10.1016/S0376-0421(03)00078-2
14. Amer, E., Dobrev, I., Massouh, F.: Determination of wind turbine far wake using actuator disk (2014)
15. Sørensen, J.N., Myken, A.: Unsteady actuator disc model for horizontal axis wind turbines. J. Wind Eng. Indust. Aerodyn. **39**(1), 139–149 (1992). https://doi.org/10.1016/0167-6105(92)90540-Q
16. Martínez Tossas, L., Leonardi, S., Churchfield, M., Moriarty, P.: A comparison of actuator disk and actuator line wind turbine models and best practices for their use (2012). https://doi.org/10.2514/6.2012-900
17. Ammara, I., Leclerc, C., Masson, C.: A viscous three-dimensional differential/actuator-disk method for the aerodynamic analysis of wind farms. J. Solar Energy Eng. Trans. ASME - J. Sol. Energy Eng. **124** (2002). https://doi.org/10.1115/1.1510870
18. Sorensen, J., Shen, W.Z.: Numerical modeling of wind turbine wakes. J. Fluids Eng. **124**, 393 (2002). https://doi.org/10.1115/1.1471361
19. Troldborg, N., Sørensen, J., Mikkelsen, R.: Numerical simulations of wake characteristics of a wind turbine in uniform flow. Wind Energy **13**, 86–99 (2010). https://doi.org/10.1002/we.345
20. Lynch, C.E., Prosser, D.T., Smith, M.J.: An efficient actuating blade model for unsteady rotating system wake simulations. Comput. Fluids **92**, 138–150 (2014). https://doi.org/10.1016/j.compfluid.2013.12.014
21. Dobrev, I., Massouh, F., Rapin, M.: Actuator surface hybrid model. J. Phys. Conf. Ser. **75**(1), 012019 (2007). https://doi.org/10.1088/1742-6596/75/1/012019
22. Shen, W.Z., Zhang, J.: The actuator surface model: A new navier-stokes based model for rotor computations. J. Solar Energy Eng. Trans. ASME - J. Sol. Energy Eng. **131** (2009). https://doi.org/10.1115/1.3027502
23. Yang, X., Sotiropoulos, F.: A new class of actuator surface models for wind turbines (2018)
24. Bachant, P., Goude, A., Wosnik, M.: Actuator line modeling of vertical-axis turbines. arXiv preprint arXiv:1605.01449 (2016)

25. Troldborg, N.: Actuator line modeling of wind turbine wakes (2009)
26. turbinesFoam library. https://github.com/turbinesFoam/turbinesFoam. Accessed 17 Apr 2023
27. Gritskevich, M.S., Garbaruk, A., Schütze, J., Menter, F.R.: Development of DDES and IDDES formulations for the k-! shear stress transport model. Flow Turbul. Combust. **88**, 431–449 (2012)
28. Spalart, P., Allmaras, S.: A one-equation turbulence model for aerodynamic flows. AIAA **439** (1992). https://doi.org/10.2514/6.1992-439
29. Epikhin, A.S.: Numerical schemes and hybrid approach for the simulation of unsteady turbulent flows. Mathemat. Model. Comput. Simulat. **11**(6), 1019–1031 (2019). https://doi.org/10.1134/S2070048219060024
30. AMR-Wind Solver. https://github.com/Exawind/amr-wind. Accessed 14 Apr 2023
31. Zhang, W., et al.: Amrex: A framework for blockstructured adaptive mesh refinement. J. Open Source Softw. **4**, 1370 (2019). https://doi.org/10.21105/joss.01370
32. AMReX Software. https://github.com/AMReX-Codes/amrex. Accessed 01 Apr 2023
33. Nek5000 Software. https://github.com/Nek5000. Accessed 31 Jan 2023
34. Kuiper, G.: The Wageningen Propeller Series. MARIN Publication. Maritime Research Institute, Netherlands (1992)
35. Wang, M., QingXu, Zhang, Q., Epikhin, A., Liang, B.: Comparative analysis of non/ductedpropeller under the influence of vertical wall. In: 2022 Ivannikov Ispras Open Conference (ISPRAS), pp. 124–129 (2022). https://doi.org/10.1109/ISPRAS57371.2022.10076863
36. Airfoiltools. http://airfoiltools.com/. Accessed 04 Mar 2023
37. Viterna, L., Janetzke, D.: Theoretical and experimental power from large horizontal-axis wind turbines. NASA Technical Memorandum (1982)
38. Petrov, A.G., Sukhov, A.D., Sibgatullin, I.N., Britov, A.D.: Analytical and numerical methods for Zhukovsky airfoils aerodynamics coefficients. In: 2022 Ivannikov Ispras Open Conference (ISPRAS), pp. 62–64 (2022). https://doi.org/10.1109/ISPRAS57371.2022.10076854

Numerical Simulation of Supersonic Jet Noise Using Open Source Software

Andrey Epikhin[1,2,3] and Ivan But[1,2(✉)]

[1] Ivannikov Institute for System Programming of the RAS, 109004 Moscow, Russia
ivan.but@ispras.ru
[2] Keldysh Institute of Applied Mathematics of the RAS, 125047 Moscow, Russia
[3] Bauman Moscow State Technical University, 105005 Moscow, Russia

Abstract. The paper is devoted to the study of various numerical algorithms for calculating the flow and acoustics characteristics of supersonic jets implemented in open source software. The ideally expanded supersonic jet with parameters $M = 2.1$, $Re = 70000$ is considered. A comparison of various approaches implemented in the OpenFOAM and block-structured adaptive mesh refinement framework of AMReX is conducted. Numerical algorithms for compressible gas flow implemented in pimpleCentralFoam, QGDFoam and CNS solvers are considered. Acoustic noise are calculated using the Ffowcs Williams and Hawkings analogy implemented in the libAcoustics library. Cross-validation comparison of the flow fields and acoustic characteristics is carried out.

Keywords: Aeroacoustics · Noise · Jet · Compressible flow · Quasi-gas dynamic equations · OpenFOAM · AMReX

1 Introduction

The relevance of the research topic is determined by the prevalence of jet streams in nature and technology. Laminar jets are quite rare in nature, therefore, in the future, more attention was paid to both theoretical and experimental works on turbulent jets [1].

There are two main approaches for studying jet flows: full-scale experiments [2,3] and numerical simulations [4,5]. Numerical simulations are more cost-effective, making them more popular. There are various methods used in numerical simulations, such as high-order accuracy methods and the Galerkin method [6], Godunov-type approximation methods (Kurganov-Tadmor, Rusanov, HLLC, etc.), hybrid approach [8], and algorithms based on regularized (quasi-gasdynamic) equations of gas dynamics [9]. Each method has its own advantages and limitations, and choosing the right method depends on the specific goals of the study. For instance, Godunov-type approximation methods are

Supported by Moscow Center of Fundamental and Applied Mathematics, Agreement with the Ministry of Science and Higher Education of the Russian Federation, No. 075 – 15 – 2022 – 283.

J. Mikyška et al. (Eds.): ICCS 2023, LNCS 14077, pp. 292–302, 2023.
https://doi.org/10.1007/978-3-031-36030-5_24

limited in their applicability to Mach numbers larger 1, making it difficult to use them for subsonic flows. These methods are implemented in various computing packages. It should be noted that despite the convergence and stability of the solution when using the Kurganov-Tadmor scheme, the approach is dissipative, as a result of which it is necessary to use a finer grid, which leads to high costs for RAM and causes difficulties with a large amount of stored data. Therefore, there is a need to optimize the grid and adapt it during the calculation; within the framework of this approach, it is possible to single out the open library AMReX [10,11]. However, it is important to consider the accuracy and computational costs of each method before deciding which one to use.

Currently, reducing the noise levels from supersonic jets is a major concern in industries such as combustion chamber design, jet engines, and pollution control. There are several sources of noise in supersonic jets [12–16], including large-scale turbulence, small-scale turbulence, broadband noise from the interaction of shockwaves and hydrodynamic instabilities (Mach waves), and narrowband noise from resonant flow regimes between shockwaves and hydrodynamic instabilities (Screech tone). Understanding the interaction between the high-speed flow, instabilities, and the environment is crucial in studying the acoustic noise of trans- and supersonic jets. To accurately predict noise in the far field, integral analogies for solving the Ffowcs-Williams and Hawkings equations can be used [19,20]. In this work, this method was first implemented in the AMReX package for predicting the noise level, the results of this calculation were compared with the results for the libAcoustics library implemented in OpenFOAM.

In conclusion, the study of turbulent free jets has been an important area of research for many years due to its prevalence in nature and technology. The goal of reducing the noise level from supersonic jets remains an important area of study and research, with various approaches being taken to address this issue, including full-scale experiments, numerical simulations, and integral analogies. The choice of approach depends on a number of factors, including the accuracy of the method, computational costs, and the ability to study the acoustic noise level from the jet. Despite the progress made in this field, there is still much to be learned about turbulent free jets and their associated noise, and ongoing research continues to be conducted in this area.

2 Mathematical Model and Numerical Method

For the calculation, a mathematical model is used, including an assembly for a compressible flow:

$$\frac{\partial \rho}{\partial t} + \nabla \cdot \mathbf{j_m} = 0, \tag{1}$$

$$\frac{\partial \rho \mathbf{U}}{\partial t} + \nabla \cdot (\mathbf{j_m} \otimes \mathbf{U}) + \nabla p = \nabla \cdot \hat{\sigma}, \tag{2}$$

$$\frac{\partial \rho E}{\partial t} + \nabla \cdot (\mathbf{j_m} E) + \nabla \cdot \mathbf{q} = \nabla \cdot (\hat{\sigma} \mathbf{U}), \tag{3}$$

$$p = \rho RT, \tag{4}$$

where ρ - density; $\mathbf{j_m}$ - mass flux density; $\mathbf{U}$ - velocity vector; $E = e + |\mathbf{U}|/2$ - total energy, e - specific internal gas energy; $\hat{\sigma}$ - viscous stress tensor, $\mathbf{q}$ - heat flux. For all solvers, the mathematical model described above is used, but the numerical algorithm for solving them is different and has some peculiarities:

for pimpleCentralFoam [8,21] and AMReX CNS solver [10,11]:

$$\mathbf{j_m} = \rho\mathbf{U},\ \hat{\sigma} = \hat{\sigma}_{NS} = \mu[(\nabla \otimes \mathbf{U}) + (\nabla \otimes \mathbf{U})^T]; \tag{5}$$

for QGDSolver [9,22,23]:

$$\mathbf{j_m} = \rho(\mathbf{U} - \mathbf{w}),\ \hat{\sigma} = \hat{\sigma}_{NS} + \hat{\sigma}_{QGD},\ \mathbf{q} = \mathbf{q}_{NS} + \mathbf{q}_{QGD}. \tag{6}$$

2.1 OpenFOAM Software, HybridCentralSolvers

The pimpleCentralFoam solver [21] is used for numerical simulation. This solver uses the operator splitting technique for the system of partial differential equations describing the low-speed motion of the fluid. For high-speed flows, the explicit Godunov-type methods are used. Two approaches were merged in the single hybrid method, proposed and developed by Kraposhin, for the simulation of flows in a wide range of Mach numbers [8]. Within this approach, the standard techniques for temporal derivatives, diffusion, and source terms are mixed with the KT/KNP fluxes for the convective terms. The KT/KNP convective fluxes are formulated for the unknown fields from the new time layer, yielding to the implicit approximation of a convection-diffusion equation. The modified PIMPLE algorithm is employed to couple pressure, velocity, and density. More details about the code, including the governing equations, can be found in the paper [8].

2.2 OpenFOAM Software, QGDSolver

The QGDFoam solver [22] is used in the study, in which a numerical algorithm for solving regulated quasi-gasdynamic (QGD) equations is implemented. Being an extension of the classical system of Navier-Stokes equations, QGD systems contain additional terms that are proportional to the small coefficient τ, which has the dimension of time [9]. When the parameter τ tends to zero, the QGD system of equations transitions to the system of Navier-Stokes equations. In dimensionless form, the value of τ is proportional to the Knudsen number. For density gases, the value of τ is too small to use its direct value, since it does not provide the required stability of the numerical algorithm. In this case, the role of the free path in the numerical algorithm can be played by the computational grid step in space:

$$\tau = \alpha_{QGD}\frac{\Delta_h}{a},$$

where $\alpha_{QGD} \in [\,0, 1]$ is a constant, which is the tuning parameter of the numerical QGD algorithm, Δ_h is the size of the calculation cell, a is the speed of sound of a mixture of gases. When solving problems with high numbers Ma and Re, the introduced dissipation with the help of τ-terms is not enough, and therefore an additional viscosity is introduced into the system in the form of a coefficient in the viscous stress tensor σ: $\mu \rightarrow \mu + p\tau Sc_{QGD}$, where Sc_{QGD} is a scheme parameter that ensures its stability at high values of the local Ma number. As mentioned earlier, the variables w, σ_{QGD}, q_{QGD} - quasi-gasdynamic parameters that depend on τ. More details about the code, including the governing equations, can be found in paper [9,23].

2.3 AMReX Software, CNS Solver

AMReX [10] is a C++ framework that supports the development of block-structured adaptive mesh refinement algorithms for solving partial differential equations (PDE) systems with complex boundary conditions for current and new numerical method architectures. The flow solver is implemented within the block-structured adaptive mesh refinement (AMR) framework of AMReX [18]. To solve the Navier-Stokes equations, finite-volume schemes of the second order of accuracy are used. The second-order Runge-Kutta method is used for the temporal discretization. This software package allows use block-orthogonal grids and automatic grid adaptation according to the selected parameter, while the equations on the new adaptation layer follow internal time. To solve the system of conservation equations, the third-order least squares method is used to calculate the velocity gradients and the CNS solver. The libAcoustics library is used together with AMReX CNS solver for sound pressure prediction of the jet, the calculation results are validated on experimental data and are presented in Sect. 4.

2.4 LibAcoustics Library

The Farassat 1A [19] formulation implemented in the libAcoustics library developed by the authors [24,25] is used. This analogy is used to define the far field noise generated by an acoustic source moving through a gas. This library was verified on the problems of calculating noise from a monopoly and dipole source. And validation was carried out on a number of inkjet tasks, the results are published in papers [26,27]. In this study, this library has been adapted for use in conjunction with the AMReX package. The following formula is used to calculate the sound pressure level (SPL):

$$SPL\,(dB) = 20log_{10}\left(\frac{p_{rms}}{p_{ref}}\right),$$

where p_{rms} is the RMS sound pressure, p_{ref} is the reference sound pressure. The developed library based on the OpenFOAM package is in the public domain [24] and can be compiled independently of any modules of the main package and the type of solvers.

3 Computational Setup

The problem of the high-speed free air jets flow from a round tube with a nozzle exit diameter $D = 0.01m$ into a space flooded with air is considered. The initial data correspond to the values of the parameters $M = U_j/a = U_j/\sqrt{kRT_j} = 2.1$, $Re = \rho_j U_j D/\mu_j = 70000$, where $U_j = 526m/s$ - jet exit velocity, $T_j = 156K$ - jet exit temperature, ρ_j - jet exit density, $k = 1.4$ - isentropic expansion factor, $p_j = 5066Pa$ - jet exit pressure, $p_c = p_j$ - chamber pressure, $p_a = 101325Pa$ - ambient pressure, $T_0 = 294K$ - constant stagnation temperature. The flow parameters and geometry correspond to the experimental study carried out by Troutt [3]. The initial condition for the velocity at the inlet is set given by the equation:

$$U(r) = 0.5 \cdot U_j \cdot \left[1 + tanh\left(10 \cdot \left(1 - \frac{2r}{D}\right)\right)\right].$$

The computational domain is a rectangular parallelepiped, in which the outlet boundary is removed by $100D$, and the side boundaries are removed by $20D$. The inlet boundary corresponded to the round nozzle exit and coincides with the origin of coordinates (see Fig. 1).

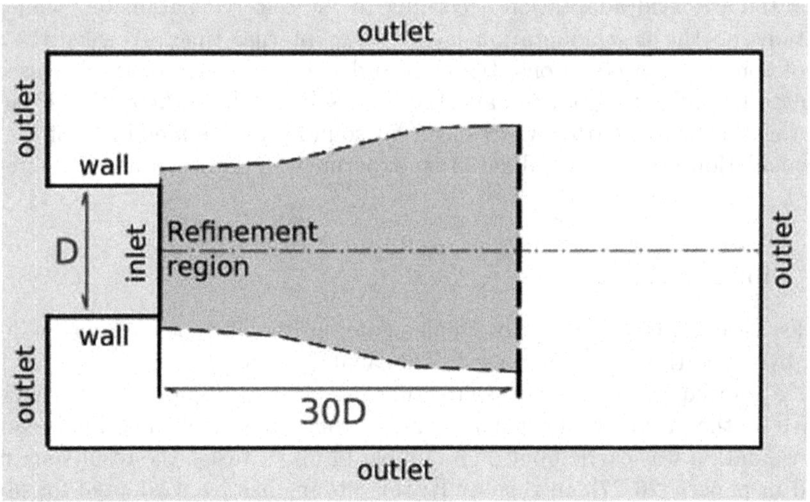

Fig. 1. Computational domain geometry.

To use solvers based on OpenFOAM, the computational grid is additionally refined by $30D$ downstream. Based on the recommendations presented in [7,26], in this area of refinement, a computational grid with a resolution of 32 cells per diameter (CPD) is used. However, during the evaluation calculations, it is found that at a moderate Reynolds number and such a grid resolution, the hybrid solver poorly reproduces hydrodynamic instabilities. As a result, for hybridCentralSolvers, a computational grid is made with additional local refinement along the jet axis with a resolution of $60CPD$ and a total number of cells of the order of 38 million (see Fig. 2). In the AMReX package, adaptive mesh refinement is performed according to the local Reynolds number. The mesh resolution in the region of the jet core is $32CPD$. The maximum number of cells in the calculation process is about 45 million.

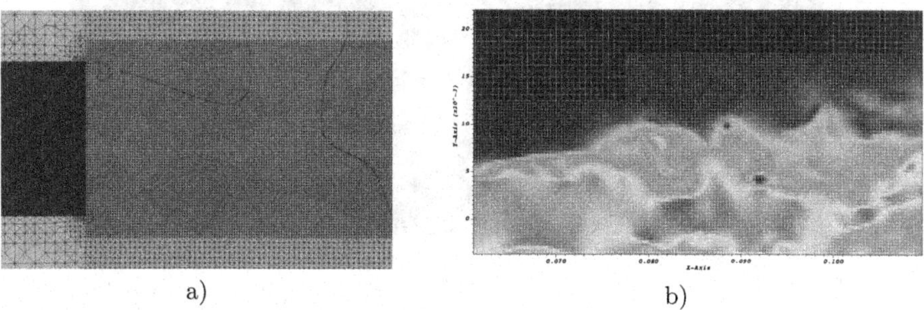

a) b)

Fig. 2. A fragment of the computational grid in: (a) OpenFOAM; (b) AMReX

The virtual microphones are located at a distance $R = 40D$, the microphone position angle is set from 15 to 90 degrees. The pimpleCentarlFoam solver used the vanLeer scheme. In the QGDFoam solver, the following tuning parameters are used, which are defined in [26]: $\alpha_{QGD} = 0.15$, $Sc_{QGD} = 0$. For CNS, the criterion for mesh refinement by the local Reynolds number.

4 Results and Discussion

Figure 3 shows that the hybrid pimpleCentralFoam solver is more dissipative, the QGD algorithm and the CNS solver more correctly reproduce the process of formation and propagation of hydrodynamic instabilities. So it makes sense to compare the ability of these two solvers to describe both pressure waves and temperature distribution in the jet (Fig. 4).

According to the Fig. 4a for QGDFoam solver and Fig. 4b for AMReX CNS solver for dimensionless pressure and temperature the CNS solver better describes the propagation of pressure waves in space due to adaptive mesh refinement. While in the QGDFoam solver, pressure waves are immediately attenuated in the coarse mesh region when propagating beyond the open control surfaces surrounding the jet flow. Based on the recommendations [28,29] on choosing the shape of the control surface for calculating the acoustic pressure from jet flows, an open surface is constructed, which is schematically shown in Fig. 4. Figure 5 shows the axial distribution of the time-averaged Mach number and comparison of noise with experimental data.

The Figs. 3 and 5a show that due to the use of a structured grid and a higher order solver based on the AMReX package, it is possible to obtain a good match with the QGD algorithm which used mesh with more refinement. In this case, all three solvers give good agreement with the acoustic characteristics of the jet. The highest levels of generated noise occur at angles around 30°.

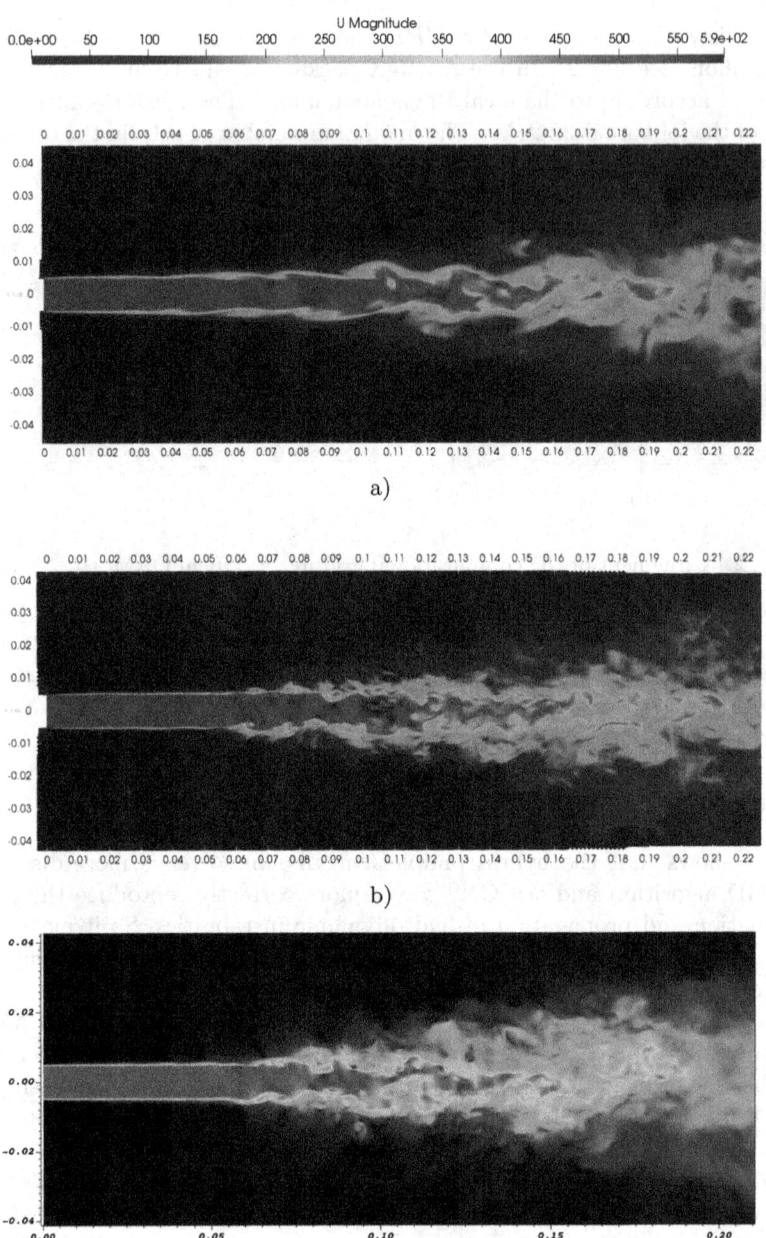

Fig. 3. Instantaneous jet velocity distribution at M=2.1, Re=70000: (a) pimpleCentralFoamSolver; (b) QGDFoam solver (c) CNS solver

Fig. 4. Dimensionless fields of pressure and temperature at $M = 2.1$, $Re = 70000$, $T_0 = 156K$, $P_0 = 5066Pa$; green line - schematic showing the open control surfaces surrounding the jet flow: (a) QGDFoam solver (b) AMReX CNS solver (Color figure online)

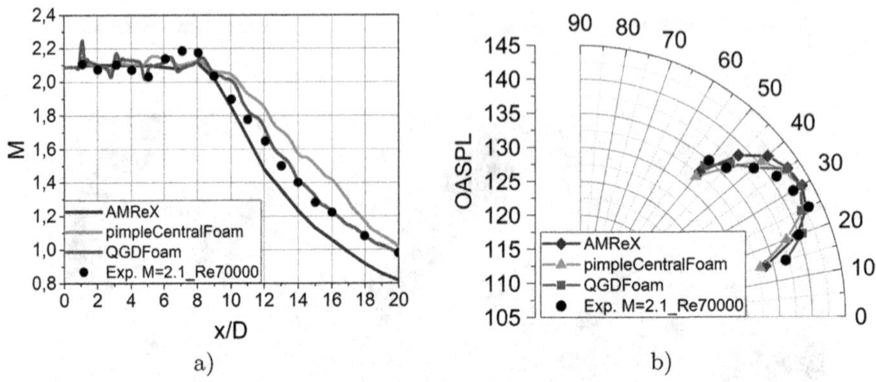

Fig. 5. Flow characteristics: (a) Axial distribution of the mean Mach number; (b) Sound pressure level directivity distributions

5 Conclusion

In modern research, numerical simulation of physical problems plays an important role, as a result of which the choice of an appropriate algorithm for numerical simulation is of paramount importance. Numerical studies of the applicability of various numerical algorithms for calculating supersonic flows implemented in various open-source solvers such as pimpleCentralFoam, QGDFoam, AMReX CNS solver are carried out. Cross-validation is performed on the example of calculating the ideally-expanded viscous gas jet and the acoustic noise generated by it at a $M = 2.1$ and a $Re = 70000$. For the first time, the libAcoustics library is used together with AMReX software for sound pressure propagation and further comparison with the results obtained in other solvers are conducted.

The space-time fields of gas-dynamic parameters of the near field and the acoustic pressure in the far field are defined. The analysis of the obtained data made it possible to determine the settings of numerical algorithms for solving such a class of the problems. All calculation results are obtained on the same grid resolution greater than $32CPD$. The results for the hybrid approach match the experimental data worse than the QGD algorithm and the solver based on block adaptive technology. The hybrid solver is more dissipative than the other considered algorithms and requires a more detailed grid in the region of the jet core in order to correctly reproduce hydrodynamic instabilities. The QGD algorithm, with regularization parameters $Sc = 0$, $\alpha_{QGD} = 0.15$ makes it possible to accurately describe the flow structure and acoustic characteristics, but due to the implementation features, it requires more computational time. An approach based on a block-structured adaptive grid makes it possible to obtain a more accurate result with the same grid resolution in the region of the jet core. This is due to the higher order of approximation scheme that can be used on the structured grids. Therefore, the CNS solver would be reasonable to use in such cases where it is necessary to calculate the propagation of turbulent jets over long distances.

Acknowledgements. This work was supported by Moscow Center of Fundamental and Applied Mathematics, Agreement with the Ministry of Science and Higher Education of the Russian Federation, grant number $075-15-2022-283$.

References

1. Ginevsky, A.S.: Theory of turbulent jets and traces: Integral methods of calculation. Engineering (1969)
2. Stromberg, J.L., McLaughlin, D.K., Troutt, T.R.: Flow field and acoustic properties of a Mach number 0.9 jet at a low Reynolds number. J. Sound Vibr. **72**(2), 159–176 (1980)
3. Troutt, T.R., McLaughlin, D.K.: Experiments on the flow and acoustic properties of a moderate-Reynolds-number supersonic jet. J. Fluid Mech. **116**, 123–156 (1982)
4. Biswas, S., Qiao, L.: A numerical investigation of ignition of ultra-lean premixed H/air mixtures by pre-chamber supersonic hot jet. SAE Int. J. Eng. **10**(5), 2231–2247 (2017)
5. Li, X.R., et al.: Acoustic feedback loops for screech tones of underexpanded free round jets at different modes. J. Fluid Mechan. **902**, A17 (2020)
6. Galerkin, B.G.: On electrical circuits for the approximate solution of the Laplace equation. Vestnik Inzh. **19**, 897–908 (1915). (In Russian)
7. Epikhin, A., Kraposhin, M., Vatutin, K.: The numerical simulation of compressible jet at low Reynolds number using OpenFOAM. E3S Web Conf. **128** (2019)
8. Kraposhin, M.V., Banholzer, M., Pfitzner, M., Marchevsky, I.K.: A hybrid pressure-based solver for non ideal single-phase fluid flows at all speeds. Int. J Numer. Methods Fluids **88**(2), 79–99 (2018)
9. Elizarova, T.G.: Quasi-gas Dynamic Equations, Springer, Berlin (2009). https://doi.org/10.1007/978-3-642-00292-2
10. AMReX Guided Tutorials. https://amrex-codes.github.io/amrex/tutorials. Accessed 3 Feb 2023
11. AMReX CNS Flow Solver. https://github.com/AMReX-Codes/amrex/tree/development/Tests/EB/CNS. Accessed 3 Feb 2023
12. Baars, W.J., Tinney, C.E., Murray, N.E., Jansen, B.J., Panickar, P.: The effect of heat on turbulent mixing noise in supersonic jets. In: AIAA Paper, pp. 2011–1029 (2011)
13. Tam, C.K.W., Viswanathan, K., Ahuja, K.K., Panda, J.: The sources of jet noise: Experimental evidence. J. Fluid Mech. **615**, 253–292 (2008)
14. Tam, C.K.W., Shen, H., Raman, G.: Screech tones of supersonic jets from bevelled rectangular nozzles. AIAA J. **35**(7), 1119–1125 (1997)
15. Tam, C.K.W., Burton, D.E.: Sound generated by instability waves of supersonic flows. Part 2. Axisymmetric jets. J. Fluid Mech. **138**, 273–295 (1984)
16. Tam, C.K.W.: Mach wave radiation from high-speed jets. AIAA J. **47**(10), 2440–2448 (1984)
17. Arnold, D.N., Brezzi, F., Cockburn, B., Marini, L.: Unified analysis of discontinuous Galerkin methods for elliptic problems. SIAM J. Numer. Anal. **39**(5), 1749–1779 (2002)

18. Natarajan, M., et al.: A moving embedded boundary approach for the compressible Navier-Stokes equations in a block-structured adaptive refinement framework. J. Comput. Phys. (2022)
19. Brentner, K.S., Farassat, F.: An analytical comparison of the acoustic analogy and Kirchhoff formulations formoving surfaces. AIAA J. **36**, 1379–1386 (1998)
20. Brès G., A., Pérot, F., Freed, D.: A Ffowcs Williams-Hawkings solver for lattice Boltzmann based computational aeroacoustics. In: AIAA Paper, pp. 2010–3711 (2010)
21. hybridCentralSolvers. https://github.com/unicfdlab/hybridCentralSolvers. Accessed 3 Feb 2023
22. QGDSolvers. https://github.com/unicfdlab/QGDsolver. Accessed 3 Feb 2023
23. Kraposhin, M.V., Smirnova, E.V., Elizarova, T.G., Istomina, M.A.: Development of a new OpenFOAM solver using regularized gas dynamic equations. Comput. Fluids **166**, 163–175 (2018)
24. LibAcoustics Library. https://github.com/unicfdlab/libAcoustics. Accessed 3 Feb 2023
25. Epikhin, A., Evdokimov, I., Kraposhin, M., Kalugin, M., Strijhak, S.: Development of a dynamic library for computational aeroacoustics applications using the OpenFOAM open source package. Procedia Comput. Sci. **66**, 150–157 (2015)
26. Epikhin, A., Kraposhin, M.: Prediction of the free jet noise using quasi-gas dynamic equations and acoustic analogy. In: Krzhizhanovskaya, V.V., et al. (eds.) ICCS 2020. LNCS, vol. 12143, pp. 217–227. Springer, Cham (2020). https://doi.org/10.1007/978-3-030-50436-6_16
27. Melnikova, V.G., Epikhin, A.S.. Kraposhin, M.V.: The Eulerian-Lagrangian approach for the numerical investigation of an acoustic field generated by a high-speed gas-droplet flow. Fluids (2021)
28. Uzun, A., Lyrintzis, A.S., Blaisdell, G.A.: Coupling of integral acoustics methods with LES for jet noise prediction. In: AIAA Paper, pp. 4982–5001 (2004)
29. Shur, M., Spalart, P., Strelets, M.: Noise prediction for increasingly complex jets. Part I: Methods and tests. Int. J. Aeroacoust. **4**, 213–246 (2005)

DNS of Thermocapillary Migration of a Bi-dispersed Suspension of Droplets

Néstor Balcázar-Arciniega(✉), Joaquim Rigola, and Assensi Oliva

Heat and Mass Transfer Technological Center (CTTC), Universitat Politècnica de Catalunya-BarcelonaTech (UPC), Colom 11, 08222 Terrassa, Barcelona, Spain
nestor.balcazar@upc.edu, nestorbalcazar@yahoo.es

Abstract. The multiple markers unstructured conservative level-set method for two-phase flow with variable surface tension is applied in the Direct Numerical Simulation of thermocapillary-migration of a bi-dispersed suspension of droplets. Surface tension is a function of temperature on the interface. Consequently, the called Marangoni stresses induced by temperature gradients on the interface lead to a coupling of the momentum transport equation with the thermal energy transport equation. The finite-volume method on three-dimensional collocated unstructured meshes discretizes the transport equations. Interface capturing is carried out by the unstructured conservative level-set method. The multiple marker approach avoids fluid particles' numerical and potentially unphysical coalescence. The classical fractional-step projection method solves the pressure-velocity coupling. Unstructured flux limiters schemes solve the convective term of transport equations. Adaptive mesh refinement is incorporated to optimize computational resources. Verifications, validations and numerical findings are reported.

Keywords: Unstructured Conservative Level-Set Method · Unstructured Flux-Limiters · Finite-Volume Method · Unstructured Meshes · Adaptive Mesh Refinement · Variable Surface Tension · Thermocapillarity

1 Introduction

Interfacial phenomena induced by variable surface tension, e.g., thermocapillary or surfactants, are frequent in nature and industry. Diverse engineering systems, from nuclear reactors to unit operations and chemical reactors, from combustion engines to wastewater treatment plants, entail bubbles or droplets inside another fluid phase with complex interfacial physics. This work focuses on the Marangoni migration (Thermocapillarity) of droplets, an interfacial phenomenon induced by a nonuniform temperature distribution on the fluid interface. Because the surface tension is a function of the temperature,

The principal author, N. Balcázar-Arciniega, as a Serra-Húnter Lecturer (UPC-LE8027), acknowledges the Catalan Government for the financial support through this program. The computing time granted by the RES (IM-2023-1-0003, IM-2022-2-0009, IM-2021-3-0013, IM-2020-2-0002) and PRACE 14th Call (2016153612) on the supercomputer MareNostrum IV based in Barcelona - Spain, is thankfully acknowledged. The financial support of the MINECO - Spain (PID2020-115837RB-100) is acknowledged.

J. Mikyška et al. (Eds.): ICCS 2023, LNCS 14077, pp. 303–317, 2023.
https://doi.org/10.1007/978-3-031-36030-5_25

surface tension gradients arise. Consequently, shear stresses on the interface induce the migration of droplets in the direction of the temperature gradient. Beyond its scientific importance, thermocapillary migration is essential in microgravity environments [39] and micro-devices [21].

Experimental research on bubble swarms or suspension of droplets with complex interfacial physics, e.g., thermocapillary, is constrained by optical access. In contrast, analytical methods can be applied only for particular cases with a substantial simplification of physics. Consequently, the development of computational methods [34,45] for complex multiphase flows is well justified. In this framework, many methods have been designed for Direct Numerical Simulation (DNS) of gas-liquid multiphase flow [34,45]. For instance, level-set (LS) [24,33,42], Volume of Fluid (VoF) [25,34,36,46], coupled VoF-LS [10,40,41], conservative level-set (CLS) [9,15,32], and front-tracking (FT) [44,47]. Although a similar idea is shared in designing these methods, their numerical implementations on structured or unstructured meshes present significative differences [6,9,10,15].

Further efforts to extend the aforementioned interface capturing/tracking methods for two-phase flows with variable surface tension have been reported, e.g., thermocapillary effects. For instance, Balcazar et al. [13] reported a level-set model for thermocapillary migration of individual and multiple droplets. [30,31] researched the thermocapillary migration of multiple deformable droplets by using front-tracking simulations. [28,37] reported DNS of thermal Marangoni effects at deformable interfaces based on the volume-of-fluid method. [49,50] reported front-tracking simulations of an isolated spherical drop in thermocapillary migration for low and high Marangoni numbers. Two- and three-dimensional level-set simulations of thermocapillary migration of droplets were reported by [19,51]. A front-tracking method for insoluble and soluble surfactants was introduced by [29]. The previous works have reported remarkable numerical and physical findings. Nevertheless, many other configurations and flow conditions have to be explored yet. Consequently, this research is a systematic effort toward designing computational methods for two-phase flow with complex interface physics, i.e., thermocapillary-driven two-phase flow, using the unstructured conservative level-set (UCLS) method proposed by Balcazar et al. [4,6–9,13,15,17]. Contributions of this work include the incorporation of novel adaptive mesh refinement and unstructured conservative level-set method for thermocapillary migration of droplets. In addition, the DNS of the thermocapillary-driven motion of a bi-dispersed suspension of droplets is presented on three-dimensional fixed meshes.

The organization of this paper is described as follows: The mathematical formulation and numerical methods are introduced in Sect. 2. Section 3 reports numerical experiments. Conclusions are outlined in Sect. 4.

2 Mathematical Formulation and Numerical Methods

2.1 One Fluid Formulation for Incompressible Two-Phase Flow

The one-fluid formulation [34,46] solves the Navier-Stokes equations for the dispersed phase (Ω_d) and continuous phase (Ω_c), using the multi-marker UCLS approach [15,17]:

$$\frac{\partial}{\partial t}(\rho \mathbf{v}) + \nabla \cdot (\rho \mathbf{v}\mathbf{v}) = -\nabla p + \nabla \cdot \mu (\nabla \mathbf{v}) + \nabla \cdot \mu (\nabla \mathbf{v})^T + (\rho - \rho_0)\mathbf{g} + \mathbf{f}_\sigma, \quad (1)$$

$$\nabla \cdot \mathbf{v} = 0, \quad (2)$$

where p is the pressure, $\mathbf{v}$ is the fluid velocity, $\mathbf{g}$ denotes the gravitational acceleration, ρ is the fluid density, μ is the dynamic viscosity, $\mathbf{f}_\sigma$ refers to the surface tension force per unit volume concentrated on the interface (Γ), subscripts d and c refer to the dispersed and continuous phases. Density and viscosity are constant. Nevertheless, a jump discontinuity arises on the interface: $\mu = \mu_c H_c + \mu_d H_d,\ \ \rho = \rho_c H_c + \rho_d H_d$. The Heaviside step function (H_c) is one in Ω_c and zero in Ω_d. On the other hand, $H_d = 1 - H_c$. In case periodic boundary conditions are set in the y-axis (parallel to $\mathbf{g}$), a force $-\rho_0 \mathbf{g}$ should be included in Eq. (1) [5,6,15]. In that case, $\rho_0 = V_\Omega^{-1} \int_\Omega (\rho_c H_c + \rho_d H_d)\, dV$.

2.2 The Multi-marker UCLS Method

The Unstructured Conservative Level-Set (UCLS) approach proposed by Balcazar et al. [9,15] is used for interface capturing in the framework of the finite volume method. Furthermore, the multiple markers UCLS method [4–6,13,15,17] is adopted to circumvent the numerical coalescence of fluid particles. In this context, a modified level-set function [9,13,15] represents each marker, $\phi_i = \frac{1}{2}\left(\tanh\left(\frac{d_i}{2\varepsilon}\right) + 1\right)$, where ε sets the thickness of the interface profile, and d_i is a signed distance function [33,42]. The ith UCLS advection equation is computed in the conservative form:

$$\frac{\partial \phi_i}{\partial t} + \nabla \cdot \phi_i \mathbf{v} = 0, \ \ i = \{1, 2, ..., N_m - 1, N_m\}, \quad (3)$$

N_m is the number of UCLS markers, which equals the number of bubbles or droplets. To maintain a constant and sharp UCLS profile, a re-initialization equation [9] is solved:

$$\frac{\partial \phi_i}{\partial \tau} + \nabla \cdot \phi_i (1 - \phi_i) \mathbf{n}_i^0 = \nabla \cdot \varepsilon \nabla \phi_i, \ \ i = \{1, 2, ..., N_m - 1, N_m\}. \quad (4)$$

Here, $\mathbf{n}_i^0$ denotes the interface normal unit vector evaluated at $\tau = 0$. At the control volume Ω_P, $\varepsilon_P = 0.5(h_P)^\alpha$ with $\alpha = 0.9$, h_P refers to the local grid size [9,13, 15]. Equation (4) is computed for the pseudo-time τ up to the steady state. Interface curvatures κ_i and normal vectors $\mathbf{n}_i$ are computed as follows [6,9,15]: $\kappa_i = -\nabla \cdot \mathbf{n}_i$, $\mathbf{n}_i = \frac{\nabla \phi_i}{||\nabla \phi_i||}$.

2.3 Marangoni Force

The Continuous Surface Force (CSF) model [18] is adopted for computing the surface tension force ($\mathbf{f}_\sigma$, Eq. (1)). This model has been extended to the multiple marker UCLS approach by Balcazar et al. [5,6,13,15,17]:

$$\mathbf{f}_\sigma = \sum_{i=1}^{N_m} (\mathbf{f}_{\sigma,i}^{(n)} + \mathbf{f}_{\sigma,i}^{(t)}). \quad (5)$$

Here, the interface tangential component $\mathbf{f}_{\sigma,i}^{(t)}$, is the so-called Marangoni force [22], defined as follows:

$$\begin{aligned}\mathbf{f}_{\sigma,i}^{(t)} &= \nabla_{\Gamma_i}\sigma(T)\delta_{\Gamma,i}^{s} = (\nabla\sigma(T) - \mathbf{n}_i(\mathbf{n}_i \cdot \nabla\sigma(T)))\delta_{\Gamma,i}^{s} \\ &= (\nabla\sigma(T) - \mathbf{n}_i(\mathbf{n}_i \cdot \nabla\sigma(T)))||\nabla\phi_i||. \end{aligned} \tag{6}$$

The regularized Dirac delta function $\delta_{\Gamma,i}^{s} =||\nabla\phi_i||$ [5,9,15,17] is concentrated on the interface. Note that the tangential component of the gradient operator is $\nabla_{\Gamma_i} = \nabla - \mathbf{n}_i(\mathbf{n}_i \cdot \nabla)$. Furthermore, $\sigma = \sigma(T)$ denotes the equation of state for the surface tension coefficient. On the other hand, the normal component of the surface tension force, $\mathbf{f}_{\sigma,i}^{(n)}$, is calculated as follows:

$$\mathbf{f}_{\sigma,i}^{(n)} = \sigma\kappa_i\mathbf{n}_i\delta_{\Gamma,i}^{s}, = \sigma\kappa_i\mathbf{n}_i||\nabla\phi_i||, = \sigma\kappa_i\nabla\phi_i. \tag{7}$$

This force is perpendicular to the interface (Γ_i), whereas k_i is the curvature.

2.4 Equation of State $\sigma = \sigma(T)$ and Energy Equation

The properties of the vapour and liquid phases for a specific fluid become similar as its critical temperature is reached. Indeed, an increment of temperature reduces the surface tension, i.e., $\partial\sigma(T)/\partial T = \sigma_T$ with $\sigma_T < 0$. For most fluids, a linear equation of state for surface tension is expected: $\sigma = \sigma(T) = \sigma_0 + \sigma_T(T - T_0)$, with $\sigma_0 = \sigma(T_0)$. As a consequence, the Marangoni force (Eq. (6)) for linear $\sigma(T)$ is written as:

$$\mathbf{f}_{\sigma,i}^{(t)} = \mathbf{f}_{\sigma,i}^{(t)}(T) = (\sigma_T\nabla T - \sigma_T\mathbf{n}_i(\mathbf{n}_i \cdot \nabla T))||\nabla\phi_i||. \tag{8}$$

The following energy transport equation [13] address the evolution of the temperature:

$$\rho c_p\left(\frac{\partial T}{\partial t} + \nabla \cdot (\mathbf{v}T)\right) = \nabla \cdot (\lambda\nabla T). \tag{9}$$

Here, $\lambda = \lambda_d H_d + \lambda_c H_c$ is the thermal conductivity, and $c_p = c_{p,d}H_d + c_{p,c}H_c$ is the specific heat capacity. Subindex d and c denotes the dispersed and continuous phases.

2.5 Regularization of Physical Properties

This research employs a regularization of physical properties $\{\mu, \lambda, \rho c_p\}$ proposed by [13] for thermocapillary migration of droplets: $\rho = \rho_d H_d^s + \rho_c H_c^s$, $\mu^{-1} = \mu_d^{-1}H_d^s + \mu_c^{-1}H_c^s$, $(\rho c_p) = (\rho c_p)_d H_d^s + (\rho c_p)_c H_c^s$, $\lambda^{-1} = \lambda_d^{-1}H_d^s + \lambda_c^{-1}H_c^s$. Further details can be found in our previous research [6, 13, 15].

2.6 Numerical Methods

Transport equations are discretized by the finite-volume method on 3D collocated unstructured meshes [15]. The convective term of level-set advection equation (Eq. (3)), momentum transport equation (Eq. (1)) and energy transport equation (Eq. (9)) is explicitly calculated at the cell-faces by the unstructured flux-limiter schemes proposed by Balcazar et al. [9,15]. Consequently, the convective term in the cell Ω_P is written as follows $(\nabla \cdot \beta\psi\mathbf{v})_P = V_P^{-1}\sum_f \beta_f\psi_f(\mathbf{v}_f \cdot \mathbf{A}_f)$. Here, $\mathbf{A}_f = ||\mathbf{A}_f||\mathbf{e}_f$ is the area vector, subindex f refers to the cell-faces, $\mathbf{e}_f$ is a unit-vector pointing outside the local cell Ω_P, and V_P denotes the volume of Ω_P. Furthermore,

$$\psi_f = \psi_{C_p} + \frac{1}{2}\mathrm{L}(\theta_f)(\psi_{D_p} - \psi_{C_p}), \tag{10}$$

where $\theta_f = (\psi_{C_p} - \psi_{U_p})/(\psi_{D_p} - \psi_{C_p})$ is the monitor variable and $L(\theta_f)$ is the flux limiter function. Furthermore, subindex C_p denotes the upwind point, subindex U_p refers to the far-upwind point, and subindex D_p denotes the downwind point, according to the stencil proposed for the UCLS method [15]. Multiple flux limiters have been implemented on the UCLS solver [15], some of them are remarked [23,27,43]:

$$L(\theta_f) \equiv \begin{cases} \max\{0, \min\{2\theta_f, 1\}, \min\{2, \theta_f\}\} & \text{SUPERBEE,} \\ \max\{0, \min\{4\theta_f, 0.75 + 0.25\theta_f, 2\}\} & \text{SMART,} \\ (\theta_f + |\theta_f|)/(1 + |\theta_f|) & \text{VANLEER,} \\ 0 & \text{UPWIND,} \\ 1 & \text{CD,} \end{cases} \tag{11}$$

SUPERBEE limiter is used unless otherwise stated. The finite-volume discretization of the compressive term in the re-initialization equation (Eq. (4)) is performed at the cell Ω_P as proposed by [15]: $(\nabla \cdot \phi_i(1-\phi_i)\mathbf{n}_i^0)_P = \frac{1}{V_P}\sum_f (\phi_i(1-\phi_i))_f \mathbf{n}_{i,f}^0 \cdot \mathbf{A}_f$. In addition, $(\phi_i(1-\phi_i))_f$ and $\mathbf{n}_{i,f}^0$ are approximated by linear interpolation.

Discretization of the diffusive term in transport equations is performed by the central difference scheme [15]. On the other hand, linear interpolation (with a weighting factor of 0.5) [15] is used to approximate the cell-face values unless otherwise stated. The weighted least-squares method [7,9,15,17] computes the gradients at the cell centroids. The pressure-velocity coupling is solved with the fractional-step projection method [20, 34,46]. Indeed, a predictor velocity ($\mathbf{v}_P^*$) is calculated in the first step:

$$\frac{\rho_P \mathbf{v}_P^* - \rho_P^0 \mathbf{v}_P^0}{\Delta t} = \mathbf{C}_{\mathbf{v},P}^0 + \mathbf{D}_{\mathbf{v},P}^0 + (\rho_P - \rho_0)\mathbf{g} + \mathbf{f}_{\sigma,P}, \tag{12}$$

Here, subindex P denotes the local control volume (Ω_P), superindex 0 refers to the previous time-step, $\mathbf{C}_\mathbf{v} = -\nabla \cdot (\rho\mathbf{v}\mathbf{v})$, and $\mathbf{D}_\mathbf{v} = \nabla \cdot \mu\nabla\mathbf{v} + \nabla \cdot \mu(\nabla\mathbf{v})^T$. After imposing the incompressibility constraint ($(\nabla \cdot \mathbf{v})_P = 0$) in the corrector step (Eq. (14)), a Poisson equation for the pressure field is obtained:

$$\left(\nabla \cdot \left(\frac{\Delta t}{\rho}\nabla p\right)\right)_P = (\nabla \cdot \mathbf{v}^*)_P, \quad \mathbf{e}_{\partial\Omega} \cdot \nabla p|_{\partial\Omega} = 0. \tag{13}$$

A linear system results from the finite volume approximation of Eq. (13), which is computed by a preconditioned (Jacobi pre-conditioner) conjugate gradient method [26,48]. Here $\partial\Omega$ refers to the boundary of Ω, excluding regions with periodic conditions where information of the corresponding periodic nodes is used [6,15]. In the next step, an updated velocity ($\mathbf{v}_P$) is calculated as follows:

$$\frac{\rho_P \mathbf{v}_P - \rho_P \mathbf{v}_P^*}{\Delta t} = -(\nabla p)_P. \tag{14}$$

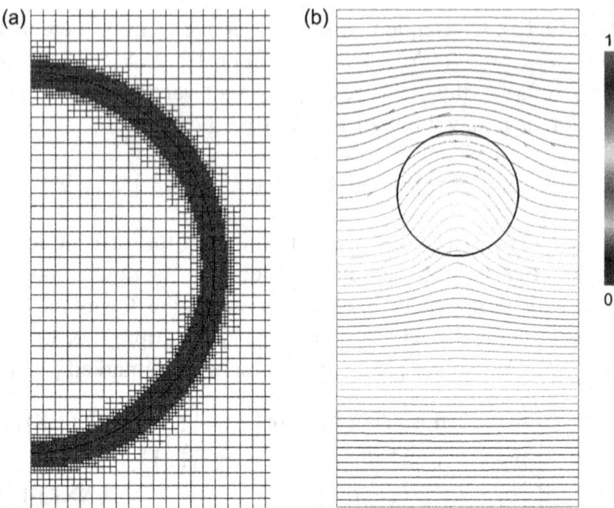

Fig. 1. Thermocapillary migration of a single droplet, $\mathbf{g} = \mathbf{0}$. (a) Adaptive mesh refinement (AMR) around the interface. The maximum grid size is $h_{max} = L_x/60$, and the minimum grid size is $h_{min} = h_{max}/2^4$. (b) Temperature isocontours

Furthermore, to avoid pressure-velocity decoupling on collocated meshes [35] and to fulfill the incompressibility constraint, a cell-face velocity $\mathbf{v}_f$ is interpolated [13, 15]. Consistently with $\mathbf{v}_f$, the volume flux $(\mathbf{v}_f \cdot \mathbf{A}_f)$, normal velocity $(\mathbf{v}_f \cdot \mathbf{e}_f)$ or a compatible variable is used to calculate the convective term of transport equations (see appendix A of [13] for example). The reader is referred to [7,15,17] for an example of a global algorithm for complex interfacial physics and further details on the finite-volume discretization.

3 Numerical Experiments

Validations, verifications and extensions of the UCLS method [5,7,9,10,15] have been systematically reported, including: the gravity-driven motion of single bubbles [2,3, 5,6,9], Thermocapillary migration of single and multiple droplets on fixed unstructured meshes [13,14], falling droplets [8], gravity-driven bubbly flows [5,11,15–17], the bouncing collision of a droplet against a fluid-fluid interface [11], binary droplet collision [11], deformation of droplets under shear stresses [10], primary atomization of a liquid [38], mass transfer in bubble swarms [4,15–17], and saturated liquid-vapour phase change [7]. A comparison of the UCLS method [6,9,13] and unstructured coupled VoF-LS method [10] is performed in [8]. Consequently, this research is a further systematic step in thermocapillary-driven two-phase flows.

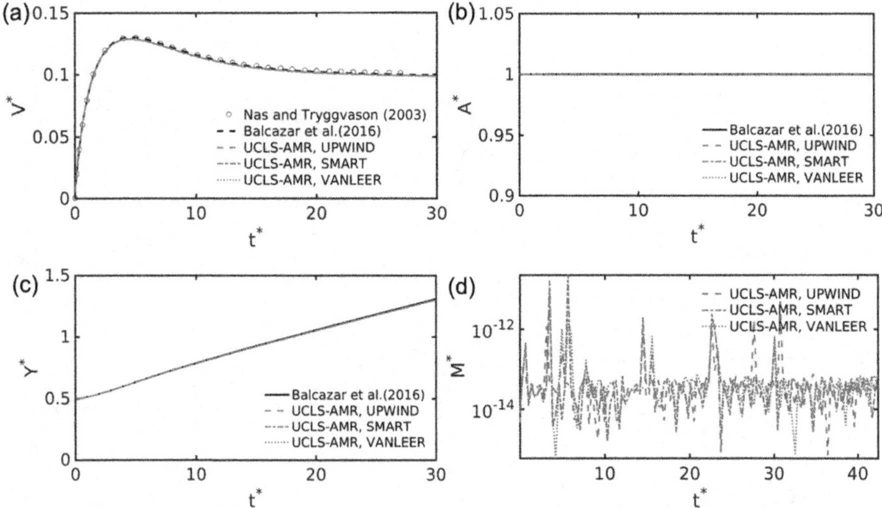

Fig. 2. Effect of flux-limiter schemes (UPWIND, SMART, VANLEER) on thermocapillary migration of a single droplet, $\mathbf{g} = \mathbf{0}$. Comparison of present simulations (AMR-UCLS) against front tracking simulations of Nas and Tryggvason (2003) [31], and conservative level-set simulations of Balcazar et al. (2016) [12] on fixed meshes. Here $t^* = t/t_r$. (a) Migration velocity $V^* = (\mathbf{e}_y \cdot \mathbf{v})U_r^{-1}$. (b) Dimensionless droplet surface $A^* = A(t)/A(0)$, $A(t) = \int_\Omega ||\nabla\phi||dV$. (c) Position of the droplet centre on the y-axis $Y^* = y/L_x$. (d) Mass conservation $M^* = (M(t) - M(0))/M(0)$, $M(t) = \int_\Omega H_d^s dV$

The following dimensionless numbers characterize the thermocapillary migration of droplets:

$$\text{Ma} = \frac{|\sigma_T|||\nabla T_\infty||d^2\rho_c c_{p,c}}{4\mu_c\lambda_c}, \text{Re} = \frac{|\sigma_T|||\nabla T_\infty||d^2\rho_c}{4\mu_c^2},$$
$$\text{Ca} = \frac{|\sigma_T|||\nabla T_\infty||d}{2\sigma_0}, \eta_\beta = \frac{\beta_c}{\beta_d}. \tag{15}$$

where Ca is the capillary number, Ma is the Marangoni number, Re is the Reynolds number, η_β denotes the physical property ratio, $\beta = \{\rho, \mu, \lambda, c_p\}$, $\nabla T_\infty = ((T_h - T_c)/L_y)\mathbf{e}_y$, T_h denotes the temperature at the top boundary (hot), and T_c denotes the temperature at the bottom boundary (cold), as depicted in Fig. 1b. On the other hand, $T_r = ||\nabla T_\infty||(0.5d)$ defines the reference temperature, $U_r = |\sigma_T|||\nabla T_\infty||(0.5d)/\mu_c$ is the reference velocity, and $t_r = 0.5d/U_r$ is the reference time.

3.1 Thermocapillary Migration of a Single Droplet

This test case was reported by [31] in the framework of the front-tracking method. Furthermore, it has been successfully computed through the UCLS method on fixed unstructured meshes by Balcazar et al. [13]. In what follows the UCLS method

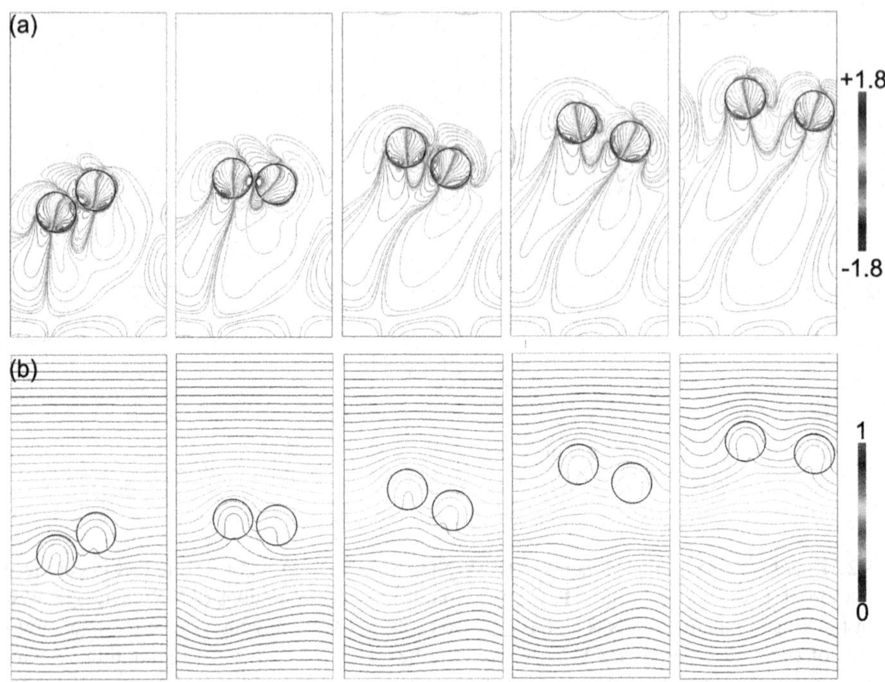

Fig. 3. Thermocapillary interaction of two droplets $\mathbf{g} = \mathbf{0}$. $Re = 60$, $Ma = 60$, $Ca = 0.0416\bar{6}$, $\eta_\rho = \eta_\mu = \eta_{c_p} = \eta_\lambda = 2$. Uniform hexahedral mesh with grid size $h = d/48$. (a) Vorticity contours $((\nabla \times \mathbf{v}) \cdot \mathbf{e}_z)$. (b) Temperature contours

[9,13,15,17] for two-phase flow with variable surface tension is coupled to a hexahedral Adaptive Mesh Refinement strategy (AMR) [3]. The hexahedral AMR technique was introduced by [1] for single-phase turbulent flows. In a further step, [3] extended and optimized this technique for rising bubbles at high Reynolds numbers (wobbling-regime) in the framework of the UCLS two-phase flow solver proposed by Balcazar et al. [8,9,15].

The computational setup consists of a rectangle domain (Ω) extending $L_x = 4d$ in the x direction and $L_y = 8d$ in the y direction, whereas d is the droplet diameter. As an initial condition, the droplet centroid is located above the bottom wall at a distance d. The top and bottom walls are no-slip boundaries with temperature T_h and $T_c < T_h$, respectively. On the other hand, the periodic boundary condition is applied to lateral boundaries (x-axis). The material property ratios η_ρ, η_μ, η_{c_p} and η_λ are set to 0.5, whereas the dimensionless parameters are chosen as $\mathrm{Re} = 5$, $\mathrm{Ma} = 20$, and $\mathrm{Ca} = 0.016\bar{6}$. Figure 1 shows an instantaneous result for the temperature field induced by thermocapillary migration and details on the AMR applied to the droplet interface. Figure 2 illustrates that numerical results computed by the AMR-UCLS method are in close agreement with those reported by [31] and [13] on fixed meshes. Moreover, Fig. 2d shows excellent mass conservation of fluid phases.

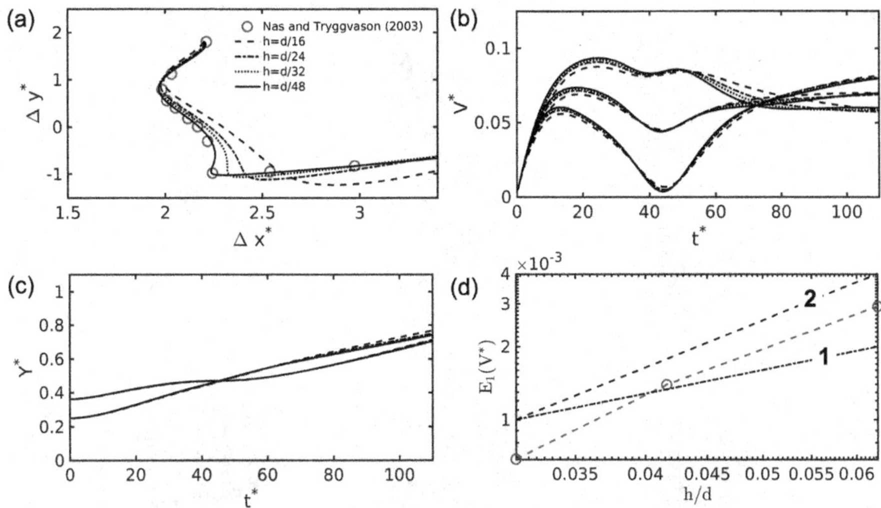

Fig. 4. Thermocapillary interaction of two droplets, with $\mathbf{g} = \mathbf{0}$, $Ma = 60$, $Re = 60$, $Ca = 0.0416\bar{6}$, $\eta_\rho = 2$, $\eta_\mu = 2$, $\eta_{c_p} = 2$, $\eta_\lambda = 2$. Lines denote the present numerical results with grid size $h = \{d/16, d/24, d/32, d/48\}$. Red symbols depict the front-tracking simulation reported by [31]. (a) Vertical versus horizontal separation distance. (b) Migration velocity. (c) Position of the droplet centre on the y-axis $Y^* = y/L_y$. (d) Order of convergence: The red line denotes present simulations. Black lines for first-order and second-order convergence. $E_1(V^*) = N^{-1}\sum_{i=1}^{N} ||V_i^* - V_{i,ref}^*||$, $V_{i,ref}^*$ refers to numerical results for the finest mesh $h = d/48$

3.2 Thermocapillary Interaction of Two Droplets

This test case was reported by [31] in the framework of the front-tracking method. Here, this case is computed by the UCLS method on fixed meshes. The computational setup consists of a rectangle domain (Ω), which extends $L_x = 4\,d$ on the x-axis, $L_y = 8\,d$ on the y-axis, and $L_z = h$, where h is the grid size. Ω is discretized by $\{8192, 18432, 32768, 73728\}$ uniform hexahedral cells. Accordingly, the grid sizes are $h = \{d/16, d/24, d/32, d/48\}$. The material properties ratios are $\eta_\rho = 2$, $\eta_\mu = 2$, $\eta_\lambda = 2$, $\eta_{c_p} = 2$. Further dimensionless parameters are set to $\mathrm{Ma} = 60$, $\mathrm{Ca} = 0.041\bar{6}$, and $\mathrm{Re} = 60$. The initial droplet centroids are $(x/d, y/d) = (0.95, 2.0)$, and $(x/d, y/d) = (2.05, 2.9)$. The droplets are circular cylinders of diameter d at $t = 0$. Concerning the boundary conditions, periodic conditions are set in the x direction. On the other hand, the top and bottom walls are no-slip boundaries. Temperatures at the top and bottom boundaries (y-axis) are T_t and T_b, respectively, with $T_t > T_b$.

Figure 3 depicts the isotherms and vorticity contours ($\mathbf{e}_z \cdot \nabla \times \mathbf{v}$) for thermocapillary interaction of the two droplets. Additionally, Fig. 4a depicts the vertical separation distance of droplets against its horizontal separation distance. As the grid is refined, numerical results using the UCLS method converge toward the front-tracking simulation reported by [31]. Figure 4b illustrates the migration velocity V^*, whereas grid convergence (second order) is demonstrated in Fig. 4d. An acceleration and deceleration period is experienced by both droplets, which is evidenced by the overshoot in

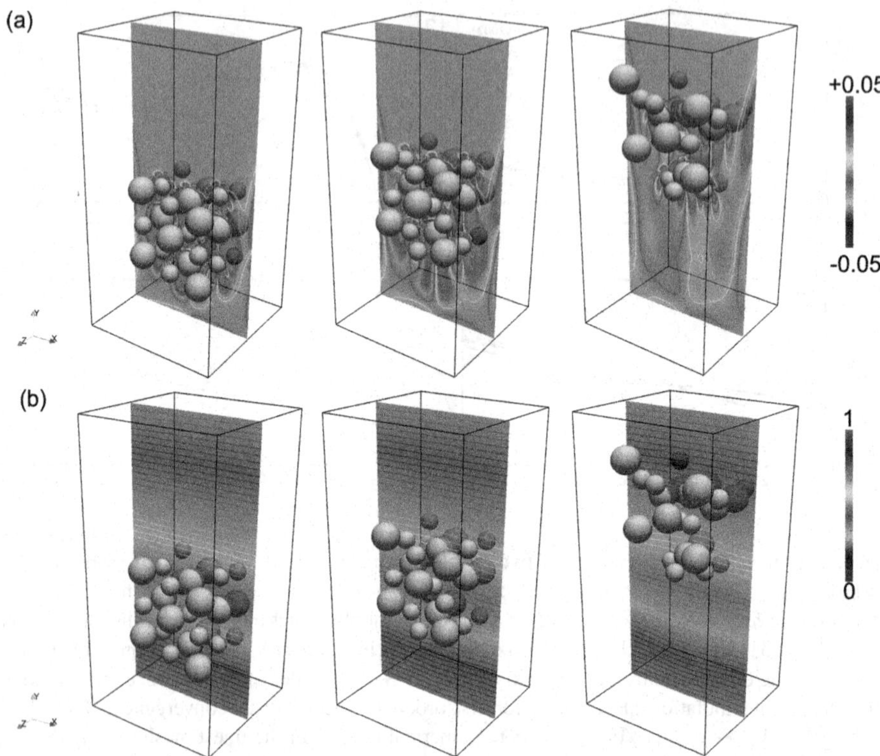

Fig. 5. Thermocapillary interaction of a bi-dispersed suspension of 27 droplets ($\mathbf{g} = \mathbf{0}$), distributed as 13 droplets with diameter d, and 14 droplets with diameter d^*. Re $= 80$, Ma $= 10$, Ca $= 0.0416\bar{6}$, $\eta_\rho = \eta_\mu = \eta_{c_p} = \eta_\lambda = 2$, $d/d^* = 1.5$. Uniform hexahedral mesh with grid size $h = d/48$, equivalent to 27648000 control volumes. Simulation performed on 1536 CPU cores. (a) Vorticity contours ($(\nabla \times \mathbf{v}) \cdot \mathbf{e}_z$) at $t^* = t/t_r = \{16.4, 32.8, 65.7\}$. (b) Temperature contours.

their migration velocities. After that, one of the droplets reaches a quasi-steady state ($t^* \geq 60$), whereas the other presents a new acceleration stage ($t^* \geq 40$).

3.3 Thermocapillary Migration of a Bi-dispersed Suspension of Droplets

The multi-marker UCLS method performs the DNS of thermocapillary migration of a bi-dispersed suspension of droplets. Ω is a rectangular channel of size $L_y = 10.66\,d$ in the y axis, and $(L_x, L_z) = (5.33\,d, 5.33\,d)$ on the plane $x - z$. Ω is discretized by 27648000 uniform hexahedral control volumes ($240 \times 240 \times 480$ grid points), with grid size $h = d/45$. At $t = 0$, 27 droplets are set near the bottom boundary, following a random pattern (Fig. 5a). The suspension of droplets consists of 13 droplets of diameter d and 14 droplets of diameter d^*. The diameter ratio is $d/d^* = 1.5$. Dimensionless numbers are defined concerning the droplet diameter d, as outlined in Fig. 5 and Fig. 6. The fluids are initially quiescent, and the temperature increases linearly from the

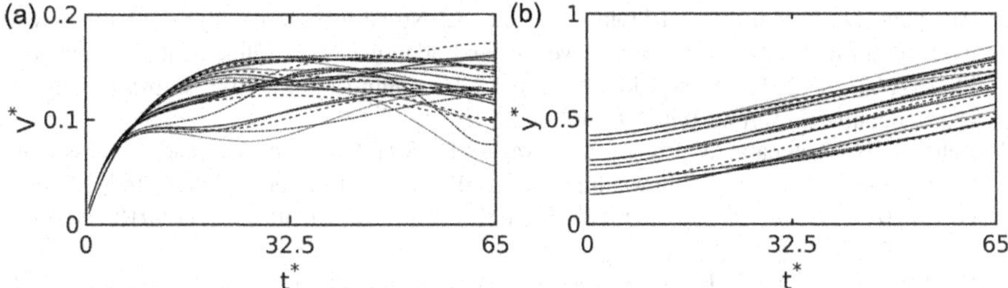

Fig. 6. Thermocapillary interaction of a bi-dispersed suspension of 27 droplets $\mathbf{g} = \mathbf{0}$, distributed as 13 droplets with diameter d, and 14 droplets with diameter d^*. Re $= 80$, Ma $= 10$, Ca $= 0.0416\bar{6}$, $\eta_\rho = \eta_\mu = \eta_{c_p} = \eta_\lambda = 2$, $d/d^* = 1.5$. Uniform hexahedral mesh with grid size $h = d/48$, equivalent to 27648000 control volumes. Simulation performed on 1536 CPU cores. (a) Dimensionless migration velocity $V^* = \mathbf{e}_y \cdot \mathbf{v}_{c,i}/U_r$, $\mathbf{v}_{c,i}$ is the droplet velocity. (b) Dimensionless vertical position, $y^* = \mathbf{e}_y \cdot \mathbf{x}_{c,i}/L_y$, $\mathbf{x}_{c,i}$ is the droplet centroid.

bottom wall to the top wall. The adiabatic boundary condition is applied to lateral walls. On the other hand, the temperature at the top and bottom boundaries are T_h and T_c, respectively, with $T_h > T_c$. The no-slip boundary condition applies to all the boundaries. Figure 5 show instantaneous vorticity and temperature contours as the droplets migrate to the hot wall. Figure 6 depicts the migration velocity of each droplet and the vertical position of droplet centroids.

4 Conclusions

The multi-marker UCLS method for two-phase flow with variable surface tension has been applied to the thermocapillary migration of droplets. Validations and verifications include the Marangoni migration of a single droplet using AMR and the interaction of two droplets on fixed meshes. The unstructured flux-limiters schemes proposed by Balcazar et al. [9,15], to discretize the convective term of transport equations in the framework of the UCLS method, minimize the so-called numerical diffusion and avoid numerical oscillations at the interface. Altogether, numerical schemes lead to a robust and accurate numerical method for complex thermocapillary-driven two-phase flow on 3D collocated unstructured meshes.

References

1. Antepara, O., Lehmkuhl, O., Borrell, R., Chiva, J., Oliva, A.: Parallel adaptive mesh refinement for large-eddy simulations of turbulent flows. Comput. Fluids **110**, 48–61 (2015). https://doi.org/10.1016/j.compfluid.2014.09.050. https://linkinghub.elsevier.com/retrieve/pii/S0045793014003958
2. Antepara, O., Balcázar, N., Oliva, A.: Tetrahedral adaptive mesh refinement for two-phase flows using conservative level-set method. Int. J. Numer. Meth. Fluids **93**(2), 481–503 (2021). https://doi.org/10.1002/fld.4893. https://onlinelibrary.wiley.com/doi/10.1002/fld.4893

3. Antepara, O., Balcázar, N., Rigola, J., Oliva, A.: Numerical study of rising bubbles with path instability using conservative level-set and adaptive mesh refinement. Comput. Fluids **187**, 83–97 (2019). https://doi.org/10.1016/j.compfluid.2019.04.013. https://linkinghub.elsevier.com/retrieve/pii/S0045793018306297
4. Balcázar, N., Antepara, O., Rigola, J., Oliva, A.: DNS of drag-force and reactive mass transfer in gravity-driven bubbly flows. In: García-Villalba, M., Kuerten, H., Salvetti, M.V. (eds.) DLES 2019. ES, vol. 27, pp. 119–125. Springer, Cham (2020). https://doi.org/10.1007/978-3-030-42822-8_16
5. Balcazar, N., Castro, J., Rigola, J., Oliva, A.: DNS of the wall effect on the motion of bubble swarms. Procedia Comput. Sci. **108**, 2008–2017 (2017). https://doi.org/10.1016/j.procs.2017.05.076. https://linkinghub.elsevier.com/retrieve/pii/S1877050917306142
6. Balcazar, N., Lehmkuhl, O., Jofre, L., Oliva, A.: Level-set simulations of buoyancy-driven motion of single and multiple bubbles. Int. J. Heat Fluid Flow **56** (2015). https://doi.org/10.1016/j.ijheatfluidflow.2015.07.004
7. Balcazar, N., Rigola, J., Oliva, A.: Unstructured level-set method for saturated liquid-vapor phase change. In: Fluid Dynamics and Transport Phenomena, WCCM-ECCOMAS 2020, vol. 600, pp. 1–12 (2021). https://doi.org/10.23967/wccm-eccomas.2020.352. https://www.scipedia.com/public/Balcazar_et_al_2021a
8. Balcazar, N., Castro, J., Chiva, J., Oliva, A.: DNS of falling droplets in a vertical channel. Int. J. Comput. Meth. Exp. Meas. **6**(2), 398–410 (2017). https://doi.org/10.2495/CMEM-V6-N2-398-410. http://www.witpress.com/doi/journals/CMEM-V6-N2-398-410
9. Balcazar, N., Jofre, L., Lehmkuhl, O., Castro, J., Rigola, J.: A finite-volume/level-set method for simulating two-phase flows on unstructured grids. Int. J. Multiph. Flow **64**, 55–72 (2014). https://doi.org/10.1016/j.ijmultiphaseflow.2014.04.008. https://linkinghub.elsevier.com/retrieve/pii/S030193221400072X
10. Balcazar, N., Lehmkuhl, O., Jofre, L., Rigola, J., Oliva, A.: A coupled volume-of-fluid/level-set method for simulation of two-phase flows on unstructured meshes. Comput. Fluids **124**, 12–29 (2016). https://doi.org/10.1016/j.compfluid.2015.10.005. https://linkinghub.elsevier.com/retrieve/pii/S0045793015003394
11. Balcazar, N., Lehmkuhl, O., Rigola, J., Oliva, A.: A multiple marker level-set method for simulation of deformable fluid particles. Int. J. Multiph. Flow **74**, 125–142 (2015). https://doi.org/10.1016/j.ijmultiphaseflow.2015.04.009. https://linkinghub.elsevier.com/retrieve/pii/S0301932215001019
12. Balcazar, N., Oliva, A., Rigola, J.: A level-set method for thermal motion of bubbles and droplets. J. Phys. Conf. Ser. **745**, 032113 (2016). https://doi.org/10.1088/1742-6596/745/3/032113
13. Balcazar, N., Rigola, J., Castro, J., Oliva, A.: A level-set model for thermocapillary motion of deformable fluid particles. Int. J. Heat Fluid Flow **62**, 324–343 (2016). https://doi.org/10.1016/j.ijheatfluidflow.2016.09.015. https://linkinghub.elsevier.com/retrieve/pii/S0142727X16301266
14. Balcazar Arciniega, N., Rigola, J., Oliva, A.: A level-set model for two-phase flow with variable surface tension: thermocapillary and surfactants. In: 8th European Congress on Computational Methods in Applied Sciences and Engineering (CIMNE) (2022). https://doi.org/10.23967/eccomas.2022.011. https://www.scipedia.com/public/Balcazar_Arciniega_et_al_2022a
15. Balcazar-Arciniega, N., Antepara, O., Rigola, J., Oliva, A.: A level-set model for mass transfer in bubbly flows. Int. J. Heat Mass Transf. **138**, 335–356 (2019). https://doi.org/10.1016/j.ijheatmasstransfer.2019.04.008
16. Balcázar-Arciniega, N., Rigola, J., Oliva, A.: DNS of mass transfer from bubbles rising in a vertical channel. In: Rodrigues, J.M.F., et al. (eds.) ICCS 2019. LNCS, vol. 11539, pp. 596–610. Springer, Cham (2019). https://doi.org/10.1007/978-3-030-22747-0_45

17. Balcazár-Arciniega, N., Rigola, J., Oliva, A.: DNS of mass transfer in bi-dispersed bubble swarms. In: Groen, D., de Mulatier, C., Paszynski, M., Krzhizhanovskaya, V.V., Dongarra, J.J., Sloot, P.M.A. (eds.) Computational Science – ICCS 2022. LNCS, vol. 13353, pp. pp 284–296. Springer, Cham (2022). https://doi.org/10.1007/978-3-031-08760-8_24
18. Brackbill, J.U., Kothe, D.B., Zemach, C.: A continuum method for modeling surface tension. J. Comput. Phys. **100**(2), 335–354 (1992). https://doi.org/10.1016/0021-9991(92)90240-Y. https://linkinghub.elsevier.com/retrieve/pii/002199919290240Y
19. Brady, P.T., Herrmann, M., Lopez, J.M.: Confined thermocapillary motion of a three-dimensional deformable drop. Phys. Fluids **23**(2), 022101 (2011). https://doi.org/10.1063/1.3529442. http://aip.scitation.org/doi/10.1063/1.3529442
20. Chorin, A.J.: Numerical solution of the Navier-Stokes equations. Math. Comput. **22**(104), 745 (1968). https://doi.org/10.2307/2004575. https://www.jstor.org/stable/2004575?origin=crossref
21. Darhuber, A.A., Troian, S.M.: Principles of microfluidic actuation by modulation of surface stresses. Ann. Rev. Fluid Mech. **37**(1), 425–455 (2005). https://doi.org/10.1146/annurev.fluid.36.050802.122052. https://www.annualreviews.org/doi/10.1146/annurev.fluid.36.050802.122052
22. Deen, W.: Analysis of Transport Phenomena. Oxford University Press (2011)
23. Gaskell, P.H., Lau, A.K.C.: Curvature-compensated convective transport: SMART, a new boundedness- preserving transport algorithm. Int. J. Numer. Meth. Fluids **8**(6), 617–641 (1988). https://doi.org/10.1002/fld.1650080602. http://doi.wiley.com/10.1002/fld.1650080602
24. Gibou, F., Fedkiw, R., Osher, S.: A review of level-set methods and some recent applications. J. Comput. Phys. **353**, 82–109 (2018). https://doi.org/10.1016/j.jcp.2017.10.006. https://linkinghub.elsevier.com/retrieve/pii/S0021999117307441
25. Hirt, C., Nichols, B.: Volume of fluid (VOF) method for the dynamics of free boundaries. J. Computat. Phys. **39**(1), 201–225 (1981). https://doi.org/10.1016/0021-9991(81)90145-5. https://linkinghub.elsevier.com/retrieve/pii/0021999181901455
26. Karniadakis, G.E., Kirby II, R.M.: Parallel Scientific Computing in C++ and MPI. Cambridge University Press (2003). https://doi.org/10.1017/CBO9780511812583. https://www.cambridge.org/core/product/identifier/9780511812583/type/book
27. LeVeque, R.J.: Finite Volume Methods for Hyperbolic Problems (2002). https://doi.org/10.1017/cbo9780511791253
28. Ma, C., Bothe, D.: Direct numerical simulation of thermocapillary flow based on the Volume of Fluid method. Int. J. Multiph. Flow **37**(9), 1045–1058 (2011). https://doi.org/10.1016/j.ijmultiphaseflow.2011.06.005. https://linkinghub.elsevier.com/retrieve/pii/S0301932211001273
29. Muradoglu, M., Tryggvason, G.: A front-tracking method for computation of interfacial flows with soluble surfactants. J. Comput. Phys. **227**(4), 2238–2262 (2008). https://doi.org/10.1016/j.jcp.2007.10.003. https://linkinghub.elsevier.com/retrieve/pii/S002199910700438X
30. Nas, S., Muradoglu, M., Tryggvason, G.: Pattern formation of drops in thermocapillary migration. Int. J. Heat Mass Transfer **49**(13–14), 2265–2276 (2006). https://doi.org/10.1016/j.ijheatmasstransfer.2005.12.009. https://linkinghub.elsevier.com/retrieve/pii/S0017931006000093
31. Nas, S., Tryggvason, G.: Thermocapillary interaction of two bubbles or drops. Int. J. Multiph. Flow **29**(7), 1117–1135 (2003). https://doi.org/10.1016/S0301-9322(03)00084-3. https://linkinghub.elsevier.com/retrieve/pii/S0301932203000843
32. Olsson, E., Kreiss, G.: A conservative level set method for two phase flow. J. Comput. Phys. **210**(1), 225–246 (2005). https://doi.org/10.1016/j.jcp.2005.04.007. https://linkinghub.elsevier.com/retrieve/pii/S0021999105002184

33. Osher, S., Sethian, J.A.: Fronts propagating with curvature-dependent speed: algorithms based on Hamilton-Jacobi formulations. J. Comput. Phys. **79**(1), 12–49 (1988). https://doi.org/10.1016/0021-9991(88)90002-2. https://linkinghub.elsevier.com/retrieve/pii/0021999188900022
34. Prosperetti, A., Tryggvason, G.: Computational Methods for Multiphase Flow. Cambridge University Press, Cambridge (2007). https://doi.org/10.1017/CBO9780511607486
35. Rhie, C.M., Chow, W.L.: Numerical study of the turbulent flow past an airfoil with trailing edge separation. AIAA J. **21**(11), 1525–1532 (1983). https://doi.org/10.2514/3.8284. https://arc.aiaa.org/doi/10.2514/3.8284
36. Rider, W.J., Kothe, D.B.: Reconstructing volume tracking. J. Comput. Phys. **141**(2), 112–152 (1998). https://doi.org/10.1006/jcph.1998.5906. https://linkinghub.elsevier.com/retrieve/pii/S002199919895906X
37. Samareh, B., Mostaghimi, J., Moreau, C.: Thermocapillary migration of a deformable droplet. Int. J. Heat Mass Transf. **73**, 616–626 (2014). https://doi.org/10.1016/j.ijheatmasstransfer.2014.02.022
38. Schillaci, E., Antepara, O., Balcázar, N., Serrano, J.R., Oliva, A.: A numerical study of liquid atomization regimes by means of conservative level-set simulations. Comput. Fluids **179**, 137–149 (2019). https://doi.org/10.1016/j.compfluid.2018.10.017. https://linkinghub.elsevier.com/retrieve/pii/S0045793018307801
39. Subramanian, R., Balasubramaniam, R., Clark, N.: Motion of bubbles and drops in reduced gravity. Appl. Mech. Rev. **55**(3), B56–B57 (2002). https://doi.org/10.1115/1.1470685. https://asmedigitalcollection.asme.org/appliedmechanicsreviews/article/55/3/B56/456871/Motion-of-Bubbles-and-Drops-in-Reduced-Gravity
40. Sun, D., Tao, W.: A coupled volume-of-fluid and level set (VOSET) method for computing incompressible two-phase flows. Int. J. Heat Mass Transf. **53**(4), 645–655 (2010). https://doi.org/10.1016/j.ijheatmasstransfer.2009.10.030. https://linkinghub.elsevier.com/retrieve/pii/S0017931009005717
41. Sussman, M., Puckett, E.G.: A coupled level set and volume-of-fluid method for computing 3D and axisymmetric incompressible two-phase flows. J. Comput. Phys. **162**(2), 301–337 (aug 2000). https://doi.org/10.1006/jcph.2000.6537. https://linkinghub.elsevier.com/retrieve/pii/S0021999100965379
42. Sussman, M., Smereka, P., Osher, S.: A level set approach for computing solutions to incompressible two-phase flow. J. Comput. Phys. **114**(1), 146–159 (1994). https://doi.org/10.1006/jcph.1994.1155. https://linkinghub.elsevier.com/retrieve/pii/S0021999184711557
43. Sweby, P.K.: High resolution schemes using flux limiters for hyperbolic conservation laws. SIAM J. Numer. Anal. **21**(5), 995–1011 (1984). https://doi.org/10.1137/0721062. http://epubs.siam.org/doi/10.1137/0721062
44. Tryggvason, G., et al.: A front-tracking method for the computations of multiphase flow. J. Comput. Phys. **169**(2), 708–759 (2001). https://doi.org/10.1006/jcph.2001.6726. https://linkinghub.elsevier.com/retrieve/pii/S0021999101967269
45. Tryggvason, G., Scardovelli, R., Zaleski, S.: The volume-of-fluid method. In: Direct Numerical Simulations of Gas-Liquid Multiphase Flows, pp. 95–132. Cambridge University Press, January 2001. https://doi.org/10.1017/CBO9780511975264.006. https://www.cambridge.org/core/product/identifier/CBO9780511975264A041/type/book_part
46. Tryggvason, G., Scardovelli, R., Zaleski, S.: Direct Numerical Simulations of Gas-Liquid Multiphase Flows (2011). https://doi.org/10.1017/CBO9780511975264
47. Unverdi, S.O., Tryggvason, G.: A front-tracking method for viscous, incompressible, multi-fluid flows. J. Comput. Phys. **100**(1), 25–37 (1992). https://doi.org/10.1016/0021-9991(92)90307-K. https://linkinghub.elsevier.com/retrieve/pii/002199919290307K

48. Van der Vorst, H.A., Dekker, K.: Conjugate gradient type methods and preconditioning. J. Comput. Appl. Math. **24**(1–2), 73–87 (1988). https://doi.org/10.1016/0377-0427(88)90344-5. https://linkinghub.elsevier.com/retrieve/pii/0377042788903445
49. Yin, Z., Chang, L., Hu, W., Li, Q., Wang, H.: Numerical simulations on thermocapillary migrations of nondeformable droplets with large Marangoni numbers. Phys. Fluids **24**(9), 092101 (2012). https://doi.org/10.1063/1.4752028. http://aip.scitation.org/doi/10.1063/1.4752028
50. Yin, Z., Li, Q.: Thermocapillary migration and interaction of drops: two non-merging drops in an aligned arrangement. J. Fluid Mech. **766**, 436–467 (2015). https://doi.org/10.1017/jfm.2015.10. https://www.cambridge.org/core/product/identifier/S0022112015000105/type/journal_article
51. Zhao, J.F., Li, Z.D., Li, H.X., Li, J.: Thermocapillary migration of deformable bubbles at moderate to large Marangoni number in microgravity. Microgravity Sci. Technol. **22**(3), 295–303 (2010). https://doi.org/10.1007/s12217-010-9193-x

Unstructured Conservative Level-Set (UCLS) Simulations of Film Boiling Heat Transfer

Néstor Balcázar-Arciniega(✉), Joaquim Rigola, and Assensi Oliva

Heat and Mass Transfer Technological Centre, Universitat Politècnica de Catalunya-BarcelonaTech, Colom 11, 08222 Terrassa, Barcelona, Spain
nestor.balcazar@upc.edu, nestorbalcazar@yahoo.es

Abstract. A novel unstructured conservative level-set method for film boiling is introduced. The finite-volume discretization of transport equations is performed on collocated unstructured grids. Mass transfer is driven by thermal phase change and computed with the temperature gradient in the liquid and vapour phases at the interface. The fractional-step projection method is used for solving the pressure-velocity coupling. The convective term of transport equations is discretized by unstructured flux-limiter schemes to avoid numerical oscillations around the interface and minimize numerical diffusion. The central difference scheme discretizes diffusive terms. Verification and validation for film boiling on a flat surface are performed.

Keywords: Unstructured Conservative Level-Set Method · Unstructured Flux-Limiters · Finite-Volume Method · Unstructured Meshes · Vapor-Liquid Phase Change · Film Boiling

1 Introduction

Liquid-vapour phase change, e.g., boiling and condensation, is frequent in nature and industrial applications. Multiple engineering devices, from steam generators to cooling towers in nuclear and conventional thermal power plants, from unit operations to chemical reactors in the chemical processing industry, entail bubbles or droplets generated by liquid-vapour phase change, i.e., boiling, condensation and evaporation. Although empirical correlations have been proposed to perform predictions in boiling heat transfer, the interaction between fluid mechanics and transport phenomena in liquid-vapour phase change still needs to be better understood.

Research in Liquid-vapour phase change follows three paths: i) Experimental measurements with the appropriate visualization and instrumentation systems, ii) theoretical methods based on the analytical solutions of mathematical models with substantial

The main author, N. Balcázar-Arciniega, as a Serra-Húnter Lecturer (UPC-LE8027), acknowledges the Catalan Government for the financial support through this program. Computing time granted by the RES (IM-2023-1-0003, IM-2022-2-0009) is acknowledged. The authors acknowledge the financial support of the MINECO, Spain (PID2020-115837RB-100).

J. Mikyška et al. (Eds.): ICCS 2023, LNCS 14077, pp. 318–331, 2023.
https://doi.org/10.1007/978-3-031-36030-5_26

simplifications of physics, and iii) numerical methods and their computational implementation. Complexities of liquid-vapour phase change constrain the analytical methods [1]. On the other hand, optical access can limit experimental measurements. Indeed, computational methods can often be the only mechanism to explore boiling heat transfer. Three computational methods are remarked: Euler-Euler method (E-E) or two-fluid model [47], Euler-Lagrange method (E-L) [47] and Direct Numerical Simulation (DNS) [47,67]. The E-E method formulates the continuous and dispersed phases as a fully interpenetrating continuum. The E-L method uses the Eulerian framework to solve the continuous phase, whereas the Lagrangian approach tracks the position and velocity of the fluid particles (bubbles or droplets). Finally, DNS solves all scales of fluid flow and interfaces without physics simplifications. Supercomputer advances have empowered the DNS as a practical approach for designing numerical experiments of liquid-vapour phase change.

Multiple methods have been reported for DNS of two-phase flows: front-tracking (FT) [65,69], level-set (LS) [26,45,61], Volume of Fluid (VoF) [32,50,67], coupled VoF-LS [11,59,60], and conservative level-set (CLS) [10,16,44], as examples. Further extensions of these methods include liquid-vapour phase change. For instance, [31,40,46,68,70,71] report VoF implementations of phase change. [51,55–58] propose extensions of LS methods to boiling heat transfer. Hybrid LS method and ghost-fluid approach [27] were reported by [28,42,63]. Coupled VoF-LS methods for phase change were reported by [43,53,64]. [23,24,33,34,48,66] inform a FT method for boiling. Although significant numerical and physical findings have been reported in previous efforts, most proposed methods employ structured and Cartesian meshes. Therefore, multiple flow conditions and engineering interest configurations must be explored. On the other hand, to the authors' knowledge, the robustness and accuracy of the unstructured conservative level-set (UCLS) method [8,10,14,16,18] to tackle film boiling heat transfer are still to be proven. Indeed, this work is a systematic step to develop numerical methods for complex interface physics in the framework of the UCLS method proposed by Balcazar et al. [8,10,14,16,18].

This research is organized as follows: Sect. 2 reviews the mathematical formulation. Section 2.4 presents the numerical methodology of the UCLS method on collocated unstructured meshes. Section 3 gives validations and numerical experiments of film boiling heat transfer. Finally, Sect. 4 report the conclusions.

2 Mathematical Formulation and Numerical Methods

2.1 Incompressible Two-Phase Flow with Phase Change

The Navier-Stokes equations for the vapour phase (Ω_v) and liquid phase (Ω_l) are presented in the framework of the so-called one fluid formulation [47,65,67]:

$$\frac{\partial}{\partial t}(\rho \mathbf{v}) + \nabla \cdot (\rho \mathbf{v}\mathbf{v}) = -\nabla p + \nabla \cdot \mu (\nabla \mathbf{v}) + \nabla \cdot \mu (\nabla \mathbf{v})^T + (\rho - \rho_0)\mathbf{g} + \mathbf{f}_\sigma, \quad (1)$$

Here p is the pressure, $\mathbf{v}$ denotes the fluid velocity, μ refers to the dynamic viscosity, ρ refers to the fluid density, $\mathbf{g}$ is the gravity, $\mathbf{f}_\sigma$ denotes the surface tension force per unit volume, acting on the interface (Γ). If periodic boundary conditions are applied along the vertical axis (parallel to $\mathbf{g}$), the acceleration of the flow field in the direction of $\mathbf{g}$ should be avoided. Accordingly Eq. (1) incorporates the force $-\rho_0 \mathbf{g}$ [6,7,16], where $\rho_0 = V_\Omega^{-1} \int_\Omega (\rho_d H_d + \rho_c H_c)\, dV$. Otherwise, $\rho_0 = 0$.

Physical properties are constant at each fluid phase. Nevertheless, a jump discontinuity arises at the interface. Consequently, $\rho = \rho_l H_l + \rho_v H_v$, and $\mu = \mu_l H_l + \mu_v H_v$. Subscripts l and v refer to the liquid and vapour phases. H_v denotes the Heaviside step function, one in Ω_v and zero elsewhere. Moreover, $H_l = 1 - H_v$.

The mass conservation equation for the vapour phase and liquid phase are written as follows [8]:

$$\frac{\partial}{\partial t} H_v + \nabla \cdot (H_v \mathbf{v}) = \frac{\dot{m}_{lv}}{\rho_v} \delta_\Gamma, \ \frac{\partial}{\partial t} H_l + \nabla \cdot (H_l \mathbf{v}) = -\frac{\dot{m}_{lv}}{\rho_l} \delta_\Gamma, \tag{2}$$

where δ_Γ is the Dirac delta function concentrated at Γ, $\dot{m}_{lv}$ is the mass transfer rate promoted by the liquid-vapour phase change. As a consequence:

$$\nabla \cdot \mathbf{v} = \left(\frac{1}{\rho_v} - \frac{1}{\rho_l} \right) \dot{m}_{lv} \delta_\Gamma. \tag{3}$$

Equation (3) poses an incompressible constraint in Ω, excluding the interface region (Γ).

2.2 Unstructured Conservative Level-Set (UCLS) Method

Interface capturing on collocated unstructured meshes is performed by the Unstructured Conservative Level-Set (UCLS) method proposed by Balcázar et al. [10,16]. An implicit function, $\phi = \frac{1}{2}\left(\tanh(\frac{d}{2\varepsilon}) + 1\right)$, represents the interface. Here, the parameter ε sets the interface thickness, and d is a signed distance function [45]. Furthermore, $\varepsilon_P = 0.5(h_P)^\alpha$ at the local cell Ω_P, h_P is the local grid size [16], $\alpha = 0.9$ unless otherwise stated. The Heaviside step function (H_l) introduced in Eq. (2) can be regularized by the UCLS function (ϕ). Therefore, $H_l^s = 1 - H_v^s = \phi$. And, consequently, Eq. (2) leads to an interface advection equation with phase change,

$$\frac{\partial \phi}{\partial t} + \nabla \cdot \phi \mathbf{v} = -\frac{1}{\rho_l} \dot{m}_{lv} \delta_\Gamma^s. \tag{4}$$

Here, $\delta_\Gamma^s = ||\nabla \phi||$ is the regularized Dirac delta function [8,10,15,16]. The UCLS profile is kept constant and sharp by solving the re-initialisation equation [10],

$$\frac{\partial \phi}{\partial \tau} + \nabla \cdot \phi(1-\phi)\mathbf{n}|_{\tau=0} = \nabla \cdot \varepsilon \nabla \phi, \tag{5}$$

where $\mathbf{n}|_{\tau=0}$ refers to the normal unit vector evaluated at $\tau = 0$. Equation (5) is advanced in the pseudo-time (τ) upon arriving at the steady state. Interface curvature (κ) and normal unit vector ($\mathbf{n}$) are computed as follows: $\mathbf{n}(\phi) = \nabla \phi_i ||\nabla \phi_i||^{-1}$ and $\kappa(\phi) = -\nabla \cdot \mathbf{n}$. Furthermore, the Continuous Surface Force (CSF) model [20] is employed to compute the surface tension force in the framework of the UCLS method [9,10,14,16, 18]. Consequently, $\mathbf{f}_\sigma = \sigma \kappa \mathbf{n} \delta_\Gamma^s = \sigma \kappa(\phi) \mathbf{n}(\phi) ||\nabla \phi|| = \sigma \kappa(\phi) \nabla \phi$, where σ is the surface tension coefficient. Finally, the smoothed Heaviside step function (H_l^s and H_v^s) regularize physical properties. Indeed, $\mu = \mu_l H_l^s + \mu_v H_v^s$ and $\rho = \rho_l H_l^s + \rho_v H_v^s$. Consistently with Eq. (4), $H_l^s = \phi$ and $H_v^s = 1 - H_l^s$.

2.3 Liquid-Vapour Phase Change and Energy Equation

The following thermal energy equation computes the temperature on Ω_v:

$$\frac{\partial T}{\partial t} + \nabla \cdot (\mathbf{v}T) = \frac{1}{\rho c_p} \nabla \cdot (\lambda \nabla T), \tag{6}$$

where $c_p = c_{p,v}$ is the specific heat capacity at constant pressure, $\lambda = \lambda_v$ is the thermal conductivity, and $\rho = \rho_v$ is the density of the vapour phase. The vapour-liquid interface is at the saturation temperature (T_{sat}) [54], $T(\mathbf{x}, t) = T_{sat}$ on Ω_l. The mass transfer ($\dot{m}_{lv}$) driven by the liquid-vapour phase change is computed on the interface [8]: $\dot{m}_{lv} = h_{lv}^{-1} (\lambda_v (\nabla T \cdot \mathbf{n})_v - \lambda_l (\nabla T \cdot \mathbf{n})_l)$. Here, h_{lv} refers to the heat of vaporization. On the other hand, $T(\mathbf{x}, t) = T_{sat}$ in Ω_l implies that $(\nabla T \cdot \mathbf{n})_l = 0$. Finally, $(\nabla T \cdot \mathbf{n})_v = \Theta_{n,v}$ is extended on $\phi_{cut} < \phi < 1$, by solving a transport equation proposed in the framework of the UCLS method by [8]: $\frac{\partial}{\partial \tau'} \Theta_{n,v} + \mathbf{n} \cdot \nabla \Theta_{n,v} = 0$. In this research $\phi_{cut} = 0.3$.

2.4 Finite-Volume Method: Unstructured Flux-Limiters

The finite-volume method discretizes transport equations for 3D collocated unstructured meshes [16]. The convective term of energy equation (Eq. (6)), level-set advection equation (Eq. (4)), and momentum transport equation (Eq. (1)), is explicitly computed through unstructured flux-limiter schemes proposed by Balcazar et al. [10, 16]. Accordingly, the convective term in the local cell Ω_P is discretized as follows: $(\nabla \cdot \beta \psi \mathbf{v})_P = V_P^{-1} \sum_f \beta_f \psi_f (\mathbf{v}_f \cdot \mathbf{A}_f)$. Here, $\mathbf{A}_f = ||\mathbf{A}_f|| \mathbf{e}_f$ is the area vector. The unit vector $\mathbf{e}_f$, perpendicular to the face f, points outside the local cell Ω_P, V_P is the volume of Ω_P, and sub-index f refers to the cell faces. On the other hand,

$$\psi_f = \psi_{C_p} + \frac{1}{2} \mathrm{L}(\theta_f)(\psi_{D_p} - \psi_{C_p}), \tag{7}$$

where $\theta_f = (\psi_{C_p} - \psi_{U_p})/(\psi_{D_p} - \psi_{C_p})$ is a monitor variable, and $L(\theta_f)$ is the flux limiter function. Consistently with [10, 16], subindex C_p refers to the upwind point, subindex U_p denotes the far-upwind point, and subindex D_p denotes the downwind point. Multiple flux limiters functions are implemented [25, 38, 39, 41, 62]:

$$L(\theta_f) \equiv \begin{cases} \max\{0, \min\{2\theta_f, 1\}, \min\{2, \theta_f\}\} & \text{SUPERBEE,} \\ (\theta_f + |\theta_f|)/(1 + |\theta_f|) & \text{VANLEER,} \\ \max\{0, \min\{4\theta_f, 0.75 + 0.25\theta_f, 2\}\} & \text{SMART,} \\ 1 & \text{CD,} \\ 0 & \text{UPWIND.} \end{cases} \tag{8}$$

A SUPERBEE flux limiter unless otherwise stated. The compressive term (Eq. (5)) is discretized as proposed by [16]: $(\nabla \cdot \phi_i (1 - \phi_i) \mathbf{n}_i^0)_P = \frac{1}{V_P} \sum_f (\phi_i (1 - \phi_i))_f \mathbf{n}_{i,f}^0 \cdot \mathbf{A}_f$. Here, $(\phi_i (1 - \phi_i))_f$ and $\mathbf{n}_{i,f}^0$ are linearly interpolated. The central difference scheme discretizes the diffusive term of transport equations [16]. Linear interpolation (arithmetic or distance weighted) [16] approximates the cell-face values. The weighted least-squares method [8, 10, 16, 18] evaluates the gradients in Ω_P. The pressure-velocity coupling is solved by the fractional-step method [21, 39, 47, 67]. First, a predictor velocity ($\mathbf{v}_P^*$) is calculated:

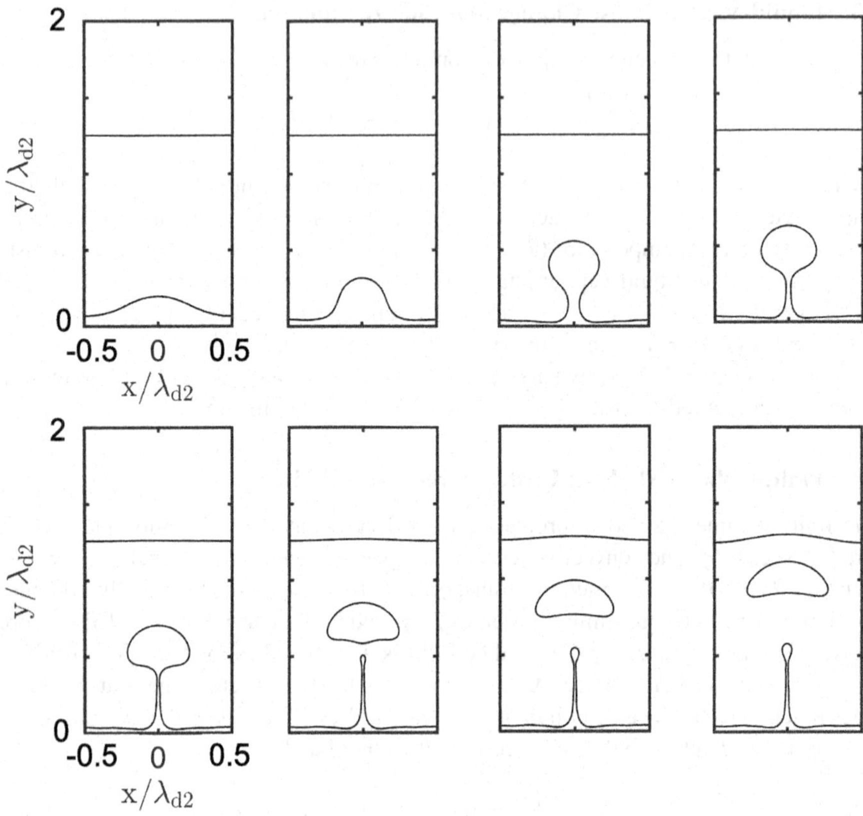

Fig. 1. Evolution of the liquid-vapour interface during film boiling. The dimensionless numbers are Gr $=$ 17.85, Pr $=$ 4.2 and Ja $=$ 0.064. Physical properties ratios are $\mu_v/\mu_l = 0.386$, $\rho_v/\rho_l = 0.209$, $\lambda_v/\lambda_l = 0.281$, and $c_{p,v}/c_{p,l} = 1.830$. The grid size is $h = \lambda_{d2}300^{-1}$. $t^* = t/t_s = \{1.02, 3.06, 5.09, 6.11, 7.13, 8.15, 9.17, 10.19\}$.

$$\frac{\rho_P \mathbf{v}_P^* - \rho_P^0 \mathbf{v}_P^0}{\Delta t} = \mathbf{C}_{\mathbf{v},P}^0 + \mathbf{D}_{\mathbf{v},P}^0 + (\rho_P - \rho_0)\mathbf{g} + \sigma\kappa_P(\nabla\phi)_P, \tag{9}$$

where $\mathbf{C}_\mathbf{v} = -\nabla \cdot (\rho\mathbf{v}\mathbf{v})$ and $\mathbf{D}_\mathbf{v} = \nabla \cdot \mu\nabla\mathbf{v} + \nabla \cdot \mu(\nabla\mathbf{v})^T$. The superindex 0 refers to the previous time step. The incompressibility constraint with phase change (Eq. (3)) is applied to the corrector-step (Eq. (11)). Consequently, the following Poisson equation for the pressure arises:

$$\left(\nabla \cdot \frac{\Delta t}{\rho}\nabla p\right)_P = (\nabla \cdot \mathbf{v}^*)_P - \left(\frac{1}{\rho_v} - \frac{1}{\rho_l}\right)\dot{m}_{lv,P}\delta_{\Gamma,P}^s, \; \mathbf{e}_{\partial\Omega} \cdot \nabla p|_{\partial\Omega} = 0. \tag{10}$$

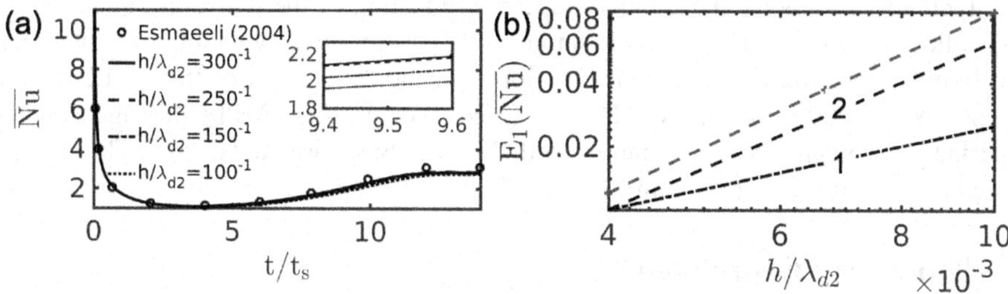

Fig. 2. Grid refinement test: (a) Time evolution of Nusselt number ($\overline{\text{Nu}}$). The dimensionless numbers are Gr $= 17.85$, Pr $= 4.2$ and Ja $= 0.064$. Physical properties ratios are $\mu_v/\mu_l = 0.386$, $\rho_v/\rho_l = 0.209$, $\lambda_v/\lambda_l = 0.281$, and $c_{p,v}/c_{p,l} = 1.830$. Benchmark results extracted from Esmaeeli and Tryggvason (2004) [23] (symbols). (b) Order of convergence: The red line denotes present simulations. Black lines for first-order (1) and second-order (2) convergence. $E_1(\overline{\text{Nu}}) = N^{-1}\sum_{i=1}^{N} ||\overline{\text{Nu}}_i - \overline{\text{Nu}}_{i,ref}||$, $\overline{\text{Nu}}_{i,ref}$ refers to numerical results for the finest mesh, N is the number of sample points $(t_i t_s^{-1}, \overline{\text{Nu}}_i)$.

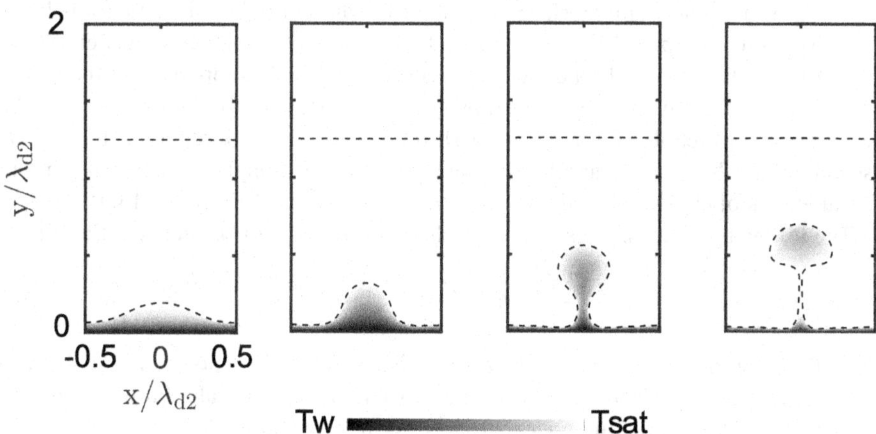

Fig. 3. Evolution of the liquid-vapour interface and temperature field during film boiling. Here, Pr $= 4.2$, Gr $= 17.85$, and Ja $= 0.064$. The ratio of thermophysical properties are $\rho_v/\rho_l = 0.209$, $\mu_v/\mu_l = 0.386$, $\lambda_v/\lambda_l = 0.281$, and $c_{p,v}/c_{p,l} = 1.830$. The grid size is $h = \lambda_{d2}300^{-1}$.

Equation (10) lead to a linear system, which is solved by the preconditioned conjugate gradient method [35]. In a further step, a corrected velocity ($\mathbf{v}_P$) is computed:

$$\frac{\rho_P \mathbf{v}_P - \rho_P \mathbf{v}_P^*}{\Delta t} = -(\nabla p)_P, \tag{11}$$

Furthermore, to avoid pressure-velocity decoupling on collocated meshes [49] and to fulfil the incompressibility constraint, a cell-face velocity $\mathbf{v}_f$ is interpolated (see [14]). The volume flux ($\mathbf{v}_f \cdot \mathbf{A}_f$), normal velocity ($\mathbf{v}_f \cdot \mathbf{e}_f$) or some equivalent variable are employed to solve the convective term of transport equations [14].

Technical details for the finite-volume discretization of the transport equations on 3D collocated unstructured meshes can be found in [16]. The new numerical methods for liquid-vapour phase change are developed in the framework of the UCLS solver proposed by Balcazar et al. [16]. The numerical code employs MPI (Message Passing Interface) for parallel communications and C++ for object-oriented design. The strong-speedup scalability is reported in [6, 16].

3 Numerical Experiments

3.1 Validations and Verifications of the UCLS Method

The UCLS method was first introduced in [10]. A main contribution was the introduction of an accurate and original formulation of unstructured flux limiters for the finite-volume discretization of the level-set advection equation on collocated unstructured meshes. Systematic validations, verifications, and extensions of the UCLS method [10, 16] include: thermocapillarity [4, 14, 18], gravity-driven rising bubbles [2,3,6,7,10], bubbly flows [6,12,16,17], binary droplet collision [12], collision of a droplet against a fluid-fluid interface [12], deformation of droplets [11], Taylor bubbles [2,29,30], falling droplets [9], atomization of gas-liquid jets [52], mass transfer in bubble swarms [5,16,17], and liquid-vapour phase change [8]. A comparison of the UCLS method [10] and coupled VoF-LS method [11] is reported in [9]. An extension of the UCLS method to tetrahedral adaptive mesh refinement has been reported in [3]. This research is a further step toward developing algorithms for complex interface physics in the framework of the UCLS method proposed by Balcazar et al. [7, 10, 12, 13, 15, 16, 18].

The following dimensionless numbers characterize the film boiling heat transfer:

$$\text{Gr} = \frac{\rho_v(\rho_l - \rho_v)||\mathbf{g}||X^3}{\mu_v^2}, \text{Pr} = \frac{\mu_v c_{p,v}}{\lambda_v}, \text{Ja} = \frac{c_{p,v}(T_w - T_{sat})}{h_{lv}}, \frac{\beta_v}{\beta_l}, \tag{12}$$

where Pr is the Prandtl number, Gr is the Grashof number, Ja denotes the Jakob number, $\beta = \{\rho, \mu, c_p, \lambda\}$. Here $X = l_s$ is a characteristic length scale. Furthermore, the computation of the Nusselt number (Nu) is performed as follows:

$$\begin{aligned} \text{Nu}(\mathbf{x}_w, t) &= \frac{X}{(T_w - T_{sat})}(\nabla T \cdot \mathbf{e}_w)(\mathbf{x}_w, t), \\ \overline{\text{Nu}}(t) &= \frac{1}{A_w}\int_{A_w} \text{Nu}(\mathbf{x}_w, t)dA, \\ \overline{\overline{\text{Nu}}} &= \frac{1}{T}\int_0^T \overline{\text{Nu}}(t)dt, \end{aligned} \tag{13}$$

where $\text{Nu}(\mathbf{x}_w, t)$ is the local Nusselt number. The unit vector $\mathbf{e}_w$, perpendicular to the bottom wall, points toward the fluids. $\overline{\text{Nu}}(t)$ refers to the space-averaged Nusselt number at the time t, A_w is the wall surface. $\overline{\overline{\text{Nu}}}$ denotes the time-averaged Nusselt number in the period T. On the other hand, $l_s = \left(\sigma||\mathbf{g}||^{-1}|\rho_l - \rho_v|^{-1}\right)^{1/2}$ is the capillary length scale, $v_s = (||\mathbf{g}||l_s)^{1/2}$ is the characteristic velocity, and $t_s = l_s/v_s$ is the characteristic time scale. The most dangerous wavelength [19] of Rayleigh-Taylor instability is defined as $\lambda_{d2} = 2\pi\sqrt{3}l_s$.

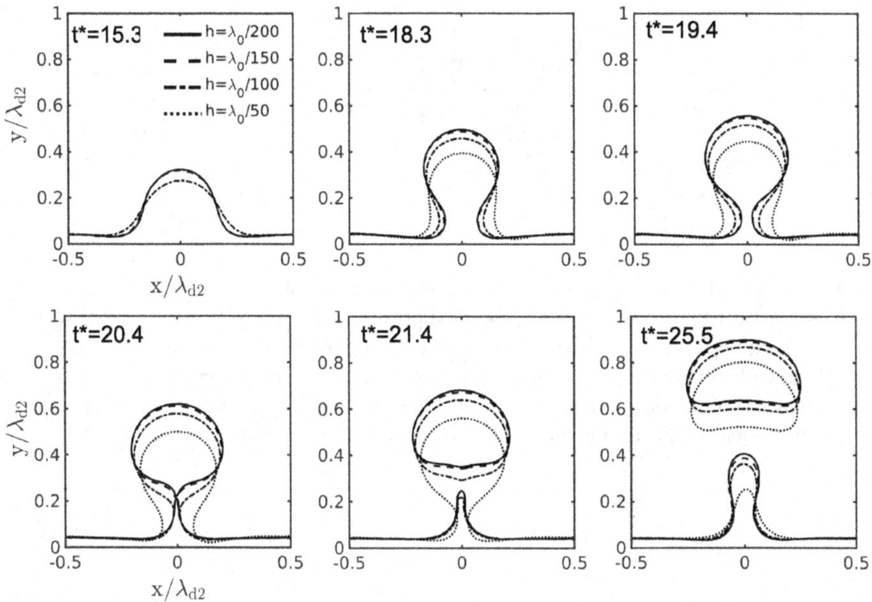

Fig. 4. Evolution of the liquid-vapour interface during film boiling. Here, Pr $= 1$, Gr $= 144.6$, and Ja $= 0.1$. The ratio of thermophysical properties are $\rho_v/\rho_l = 20^{-1}$, $\mu_v/\mu_l = 40^{-1}$, $\lambda_v/\lambda_l = 1$, and $c_{p,v}/c_{p,l} = 1$. The grid size is $h = \lambda_{d2}200^{-1}$.

3.2 Film Boiling on a Horizontal Plane

Ω is a rectangular domain $(L_x, L_y, L_z) = (\lambda_{d2}, 2\lambda_{d2}, h)$, discretized by triangular prisms with $h \approx \{\lambda_{d2}/150, \lambda_{d2}/200, \lambda_{d2}/250, \lambda_{d2}/300\}$. The initial film thickness is perturbed according to the interface shape: $y_\Gamma = y_0 + A\cos(2\pi x/\lambda_{d2})$ where $y_0 = 0.125\lambda_{d2}$, $A = 0.05\lambda_{d2}$ [23]. A vapour layer on top of the liquid layer is posed at $1.25\lambda_{d2}$. Neumann boundary conditions are set to all the boundaries except at the bottom wall. The no-slip boundary condition is applied for the bottom wall's velocity ($\mathbf{v}_w$). The temperature at the bottom wall is fixed to T_w. As initial conditions, the velocity field is zero, and the temperature field in Ω equals the saturation temperature T_{sat}. The computational setup is illustrated in Fig. 1.

A grid resolution test is depicted in Fig. 2. The time evolution of the Nusselt number ($\overline{\text{Nu}}$) is shown in Fig. 2a. The dimensionless numbers are Gr $= 17.85$, Pr $= 4.2$ and Ja $= 0.064$. Physical properties ratios are $\mu_v/\mu_l = 0.386$, $\rho_v/\rho_l = 0.209$, $\lambda_v/\lambda_l = 0.281$, and $c_{p,v}/c_{p,l} = 1.830$. Present results computed with the novel UCLS method for film boiling are in close agreement with front-tracking simulations reported by Esmaeeli and Tryggvason (2004) [23]. Figure 2b shows that the UCLS method presents second-order convergence. The red line denotes present simulations. Black lines for first-order (1) and second-order (2) convergence. The error is computed as follows: $E_1(\overline{\text{Nu}}) = N^{-1}\sum_{i=1}^{N} ||\overline{\text{Nu}}_i - \overline{\text{Nu}}_{i,ref}||$, $\overline{\text{Nu}}_{i,ref}$ refers to numerical results for the finest mesh, N is the number of sample points $(t_i t_s^{-1}, \overline{\text{Nu}}_i)$. The evolution of

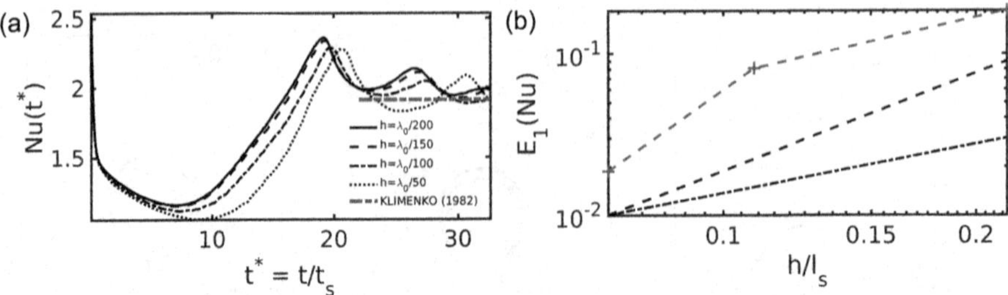

Fig. 5. Grid resolution test: (a) Time evolution of Nusselt number ($\overline{\mathrm{Nu}}$). Here, Pr $=$ 1, Gr $=$ 144.6, and Ja $= 0.1$. The ratio of thermophysical properties are $\rho_v/\rho_l = 20^{-1}$, $\mu_v/\mu_l = 40^{-1}$, $\lambda_v/\lambda_l = 1$, and $c_{p,v}/c_{p,l} = 1$. Numerical results are compared against Klimenko's correlation [36,37]. (b) Order of convergence: The red line denotes present simulations. Black lines for first-order (1) and second-order (2) convergence. $E_1(\overline{\mathrm{Nu}}) = N^{-1}\sum_{i=1}^{N}||\overline{\mathrm{Nu}}_i - \overline{\mathrm{Nu}}_{i,ref}||$, $\overline{\mathrm{Nu}}_{i,ref}$ refers to numerical results for the finest mesh, N is the number of sample points $(t_i t_s^{-1}, \overline{\mathrm{Nu}}_i)$.

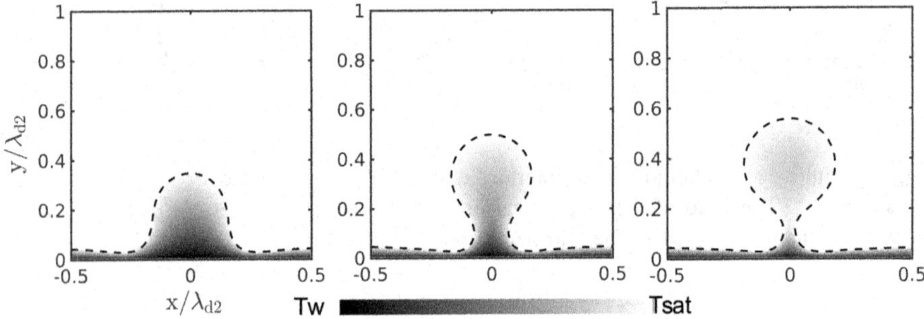

Fig. 6. Evolution of the liquid-vapour interface and temperature field during film boiling. Here, Pr $= 1$, Gr $= 144.6$, and Ja $= 0.1$. The ratio of thermophysical properties are $\rho_v/\rho_l = 20^{-1}$, $\mu_v/\mu_l = 40^{-1}$, $\lambda_v/\lambda_l = 1$, and $c_{p,v}/c_{p,l} = 1$. The grid size is $h = \lambda_{d2}200^{-1}$.

the liquid-vapour interface during film boiling is depicted in Fig. 1, and the temperature field is shown in Fig. 3.

A second case is performed on a rectangular domain $(L_x, L_y, L_z) = (\lambda_{d2}, \lambda_{d2}, h)$. Here, Pr $= 1$, Gr $= 144.6$, and Ja $= 0.1$. The ratio of thermophysical properties are $\rho_v/\rho_l = 20^{-1}$, $\mu_v/\mu_l = 40^{-1}$, $\lambda_v/\lambda_l = 1$, and $c_{p,v}/c_{p,l} = 1$. A test of grid convergence is illustrated in Fig. 5, with $h \approx \{\lambda_{d2}/50, \lambda_{d2}/100, \lambda_{d2}/150, \lambda_{d2}/200\}$. The evolution of the liquid-vapour interface during film boiling is depicted in Fig. 4, and the temperature field is shown in Fig. 6.

4 Conclusions

– The UCLS method introduced by Balcazar et al. [7,10,12,16–18] has been extended for film boiling heat transfer. The numerical model has been verified and validated against numerical results of [22] and Klimenko's correlation [36,37].

– Unstructured flux-limiters schemes proposed by Balcazar et al. [10, 11, 16] for the discretization of the convective term, in the framework of the UCLS method, improves the numerical stability and minimize the numerical diffusion in simulations of film boiling heat transfer.
– Further verifications and validations, including three-dimensional cases, will be reported in a future work.

References

1. Alexiades, V., Solomon, A.D.: Mathematical Modeling of Melting and Freezing Processes, 1st edn. CRC Press (1993). https://doi.org/10.1201/9780203749449. https://www.taylorfrancis.com/books/9781351433280
2. Antepara, O., Balcázar, N., Oliva, A.: Tetrahedral adaptive mesh refinement for two-phase flows using conservative level-set method. Int. J. Numeri. Meth. Fluids **93**, 481–503 (2021). https://doi.org/10.1002/fld.4893. https://onlinelibrary.wiley.com/doi/10.1002/fld.4893
3. Antepara, O., Balcázar, N., Rigola, J., Oliva, A.: Numerical study of rising bubbles with path instability using conservative level-set and adaptive mesh refinement. Comput. Fluids **187**, 83–97 (2019). https://doi.org/10.1016/j.compfluid.2019.04.013. https://linkinghub.elsevier.com/retrieve/pii/S0045793018306297
4. Balcazar, N., Antepara, O., Rigola, J., Oliva, A.: DNS of thermocapillary migration of deformable droplets. In: Salvetti M., Armenio V., Fröhlich J., Geurts B., Kuerten H. (eds.) Direct and Large-Eddy Simulation XI. ERCOFTAC Series, vol. 25 (2019). https://doi.org/10.1007/978-3-030-04915-7_28
5. Balcázar, N., Antepara, O., Rigola, J., Oliva, A.: DNS of drag-force and reactive mass transfer in gravity-driven bubbly flows. In: García-Villalba, M., Kuerten, H., Salvetti, M.V. (eds.) DLES 2019. ES, vol. 27, pp. 119–125. Springer, Cham (2020). https://doi.org/10.1007/978-3-030-42822-8_16
6. Balcazar, N., Castro, J., Rigola, J., Oliva, A.: DNS of the wall effect on the motion of bubble swarms. Procedia Comput. Sci. **108**, 2008–2017 (2017). https://doi.org/10.1016/j.procs.2017.05.076. https://linkinghub.elsevier.com/retrieve/pii/S1877050917306142
7. Balcazar, N., Lehmkuhl, O., Jofre, L., Oliva, A.: Level-set simulations of buoyancy-driven motion of single and multiple bubbles. Int. J. Heat Fluid Flow **56** (2015). https://doi.org/10.1016/j.ijheatfluidflow.2015.07.004
8. Balcazar, N., Rigola, J., Oliva, A.: Unstructured level-set method for saturated liquid-vapor phase change. In: Fluid Dynamics and Transport Phenomena, WCCM-ECCOMAS 2020, vol. 600, pp. 1–12 (2021). https://doi.org/10.23967/wccm-eccomas.2020.352. https://www.scipedia.com/public/Balcazar_et_al_2021a
9. Balcazar, N., Castro, J., Chiva, J., Oliva, A.: DNS of falling droplets in a vertical channel. Int. J. Comput. Meth. Exp. Meas. **6**, 398–410 (2017). https://doi.org/10.2495/CMEM-V6-N2-398-410. https://www.witpress.com/doi/journals/CMEM-V6-N2-398-410
10. Balcazar, N., Jofre, L., Lehmkuhl, O., Castro, J., Rigola, J.: A finite-volume/level-set method for simulating two-phase flows on unstructured grids. Int. J. Multiph. Flow **64**, 55–72 (2014). https://doi.org/10.1016/j.procs.2017.05.076. https://linkinghub.elsevier.com/retrieve/pii/S1877050917306142
11. Balcazar, N., Lehmkuhl, O., Jofre, L., Rigola, J., Oliva, A.: A coupled volume-of-fluid/level-set method for simulation of two-phase flows on unstructured meshes. Comput. Fluids **124**, 12–29 (2016). https://doi.org/10.1016/j.compfluid.2015.10.005. https://linkinghub.elsevier.com/retrieve/pii/S0045793015003394

12. Balcazar, N., Lehmkuhl, O., Rigola, J., Oliva, A.: A multiple marker level-set method for simulation of deformable fluid particles. Int. J. Multiph. Flow **74**, 125–142 (2015). https://doi.org/10.1016/j.ijmultiphaseflow.2015.04.009
13. Balcazar, N., Oliva, A., Rigola, J.: A level-set method for thermal motion of bubbles and droplets. J. Phys. Conf. Ser. **745**, 032113 (2016). https://doi.org/10.1088/1742-6596/745/3/032113
14. Balcazar, N., Rigola, J., Castro, J., Oliva, A.: A level-set model for thermocapillary motion of deformable fluid particles. Int. J. Heat Fluid Flow **62**, 324–343 (2016). https://doi.org/10.1016/j.ijheatfluidflow.2016.09.015. https://linkinghub.elsevier.com/retrieve/pii/S0142727X16301266
15. Balcazar-Arciniega, N., Rigola, J., Oliva, A.: A level-set model for two-phase flow with variable surface tension: thermocapillary and surfactants. In: 8th European Congress on Computational Methods in Applied Sciences and Engineering (2022). https://doi.org/10.23967/eccomas.2022.011. https://www.scipedia.com/public/Balcazar_Arciniega_et_al_2022a
16. Balcazar-Arciniega, N., Antepara, O., Rigola, J., Oliva, A.: A level-set model for mass transfer in bubbly flows. Int. J. Heat Mass Transf. **138**, 335–356 (2019). https://doi.org/10.1016/j.ijheatmasstransfer.2019.04.008
17. Balcázar-Arciniega, N., Rigola, J., Oliva, A.: DNS of mass transfer from bubbles rising in a vertical channel. In: Rodrigues, J.M.F., et al. (eds.) ICCS 2019. LNCS, vol. 11539, pp. 596–610. Springer, Cham (2019). https://doi.org/10.1007/978-3-030-22747-0_45
18. Balcázar-Arciniega, N., Rigola, J., Oliva, A.: DNS of mass transfer in bi-dispersed bubble swarms. In: Groen, D., de Mulatier, C., Paszynski, M., Krzhizhanovskaya, V.V., Dongarra, J.J., Sloot, P.M.A. (eds.) Computational Science – ICCS 2022. LNCS, vol. 13353, pp 284–296. Springer, Cham. https://doi.org/10.1007/978-3-031-08760-8_24
19. Berenson, P.J.: Film-boiling heat transfer from a horizontal surface. J. Heat Transf. **83**, 351–356 (1961). https://doi.org/10.1115/1.3682280. https://asmedigitalcollection.asme.org/heattransfer/article/83/3/351/430783/FilmBoiling-Heat-Transfer-From-a-Horizontal
20. Brackbill, J.U., Kothe, D.B., Zemach, C.: A continuum method for modeling surface tension. J. Comput. Phys. **100**, 335–354 (1992). https://doi.org/10.1016/0021-9991(92)90240-Y. https://linkinghub.elsevier.com/retrieve/pii/002199919290240Y
21. Chorin, A.J.: Numerical solution of the Navier-Stokes equations. Math. Comput. **22**, 745 (1968). https://doi.org/10.2307/2004575. https://www.jstor.org/stable/2004575?origin=crossref
22. Esmaeeli, A., Tryggvason, G.: Computations of explosive boiling in microgravity. J. Sci. Comput. (2003). https://doi.org/10.1023/A:1025347823928
23. Esmaeeli, A., Tryggvason, G.: Computations of film boiling. Part I: numerical method. Int. J. Heat Mass Transf. **47**, 5451–5461 (2004). https://doi.org/10.1016/j.ijheatmasstransfer.2004.07.027. https://linkinghub.elsevier.com/retrieve/pii/S0017931004002947
24. Esmaeeli, A., Tryggvason, G.: A front tracking method for computations of boiling in complex geometries. Int. J. Multiph. Flow **30**, 1037–1050 (2004). https://doi.org/10.1016/j.ijmultiphaseflow.2004.04.008. https://linkinghub.elsevier.com/retrieve/pii/S0301932204000576
25. Gaskell, P.H., Lau, A.K.C.: Curvature-compensated convective transport: smart, a new boundedness- preserving transport algorithm. Int. J. Numer. Meth. Fluids **8**, 617–641 (1988). https://doi.org/10.1002/fld.1650080602. https://doi.wiley.com/10.1002/fld.1650080602
26. Gibou, F., Fedkiw, R., Osher, S.: A review of level-set methods and some recent applications. J. Comput. Phys. **353**, 82–109 (2018). https://doi.org/10.1016/j.jcp.2017.10.006. https://linkinghub.elsevier.com/retrieve/pii/S0021999117307441

27. Gibou, F., Fedkiw, R.P., Cheng, L.T., Kang, M.: A second-order-accurate symmetric discretization of the Poisson equation on irregular domains. J. Comput. Phys. **176**, 205–227 (2002). https://doi.org/10.1006/jcph.2001.6977. https://linkinghub.elsevier.com/retrieve/pii/S0021999101969773
28. Gibou, F., Chen, L., Nguyen, D., Banerjee, S.: A level set based sharp interface method for the multiphase incompressible Navier-Stokes equations with phase change. J. Comput. Phys. **222**, 536–555 (2007). https://doi.org/10.1016/j.jcp.2006.07.035. https://linkinghub.elsevier.com/retrieve/pii/S0021999106003652
29. Gutiérrez, E., Balcázar, N., Bartrons, E., Rigola, J.: Numerical study of Taylor bubbles rising in a stagnant liquid using a level-set/moving-mesh method. Chem. Eng. Sci. **164**, 158–177 (2017). https://doi.org/10.1016/j.ces.2017.02.018
30. Gutiérrez, E., Favre, F., Balcázar, N., Amani, A., Rigola, J.: Numerical approach to study bubbles and drops evolving through complex geometries by using a level set - moving mesh - immersed boundary method. Chem. Eng. J. **349**, 662–682 (2018). https://doi.org/10.1016/j.cej.2018.05.110
31. Hardt, S., Wondra, F.: Evaporation model for interfacial flows based on a continuum-field representation of the source terms. J. Comput. Phys. **227**, 5871–5895 (2008). https://doi.org/10.1016/j.jcp.2008.02.020. https://linkinghub.elsevier.com/retrieve/pii/S0021999108001228
32. Hirt, C., Nichols, B.: Volume of fluid (VOF) method for the dynamics of free boundaries. J. Comput. Phys. **39**, 201–225 (1981). https://doi.org/10.1016/0021-9991(81)90145-5. https://linkinghub.elsevier.com/retrieve/pii/0021999181901455
33. Irfan, M., Muradoglu, M.: A front tracking method for direct numerical simulation of evaporation process in a multiphase system. J. Comput. Phys. **337**, 132–153 (2017). https://doi.org/10.1016/j.jcp.2017.02.036. https://linkinghub.elsevier.com/retrieve/pii/S0021999117301304
34. Juric, D., Tryggvason, G.: Computations of boiling flows. Int. J. Multiph. Flow **24**, 387–410 (1998). https://doi.org/10.1016/S0301-9322(97)00050-5. https://linkinghub.elsevier.com/retrieve/pii/S0301932297000505
35. Karniadakis, G.E., Kirby II, R.M.: Parallel Scientific Computing in C++ and MPI. Cambridge University Press, June 2003. https://doi.org/10.1017/CBO9780511812583. https://www.cambridge.org/core/product/identifier/9780511812583/type/book
36. Klimenko, V.: Film boiling on a horizontal plate - new correlation. Int. J. Heat Mass Transf. **24**, 69–79 (1 1981). https://doi.org/10.1016/0017-9310(81)90094-6
37. Klimenko, V., Shelepen, A.: Film boiling on a horizontal plate-a supplementary communication. Int. J. Heat Mass Transf. **25**, 1611–1613 (1982). https://doi.org/10.1016/0017-9310(82)90042-4
38. LeVeque, R.J.: High-resolution conservative algorithms for advection in incompressible flow. SIAM J. Numer. Anal. **33**, 627–665 (1996). https://doi.org/10.1137/0733033
39. LeVeque, R.J.: Finite Volume Methods for Hyperbolic Problems. Cambridge University Press, August 2002. https://doi.org/10.1017/CBO9780511791253. https://www.cambridge.org/core/product/identifier/9780511791253/type/book
40. Ma, C., Bothe, D.: Numerical modeling of thermocapillary two-phase flows with evaporation using a two-scalar approach for heat transfer. J. Comput. Phys. **233**, 552–573 (2013). https://doi.org/10.1016/j.jcp.2012.09.011. https://linkinghub.elsevier.com/retrieve/pii/S0021999112005426
41. Moukalled, F., Mangani, L., Darwish, M.: The Finite Volume Method in Computational Fluid Dynamics. Springer, Cham (2016). https://doi.org/10.1007/978-3-319-16874-6
42. Ningegowda, B., Ge, Z., Lupo, G., Brandt, L., Duwig, C.: A mass-preserving interface-correction level set/ghost fluid method for modeling of three-dimensional boiling flows. Int. J. Heat Mass Transf. **162**, 120382 (2020). https://doi.org/10.1016/j.ijheatmasstransfer.2020.120382. https://linkinghub.elsevier.com/retrieve/pii/S0017931020333184

43. Ningegowda, B., Premachandran, B.: A coupled level set and volume of fluid method with multi-directional advection algorithms for two-phase flows with and without phase change. Int. J. Heat Mass Transf. **79**, 532–550 (2014). https://doi.org/10.1016/j.ijheatmasstransfer.2014.08.039. https://linkinghub.elsevier.com/retrieve/pii/S0017931014007261
44. Olsson, E., Kreiss, G.: A conservative level set method for two phase flow. J. Comput. Phys. **210**, 225–246 (2005). https://doi.org/10.1016/j.jcp.2005.04.007. https://linkinghub.elsevier.com/retrieve/pii/S0021999105002184
45. Osher, S., Sethian, J.A.: Fronts propagating with curvature-dependent speed: algorithms based on Hamilton-Jacobi formulations. J. Comput. Phys. **79**, 12–49 (1988). https://doi.org/10.1016/0021-9991(88)90002-2. https://linkinghub.elsevier.com/retrieve/pii/0021999188900022
46. Perez-Raya, I., Kandlikar, S.G.: Discretization and implementation of a sharp interface model for interfacial heat and mass transfer during bubble growth. Int. J. Heat Mass Transf. **116**, 30–49 (2018). https://doi.org/10.1016/j.ijheatmasstransfer.2017.08.106. https://linkinghub.elsevier.com/retrieve/pii/S0017931017322408
47. Prosperetti, A., Tryggvason, G.: Computational Methods for Multiphase Flow. Cambridge University Press (2007). https://doi.org/10.1017/CBO9780511607486
48. Rajkotwala, A., et al.: A critical comparison of smooth and sharp interface methods for phase transition. Int. J. Multiph. Flow **120**, 103093 (2019). https://doi.org/10.1016/j.ijmultiphaseflow.2019.103093. https://linkinghub.elsevier.com/retrieve/pii/S0301932219301016
49. Rhie, C.M., Chow, W.L.: Numerical study of the turbulent flow past an airfoil with trailing edge separation. AIAA J. **21**, 1525–1532 (1983). https://doi.org/10.2514/3.8284. https://arc.aiaa.org/doi/10.2514/3.8284
50. Rider, W.J., Kothe, D.B.: Reconstructing volume tracking. J. Comput. Phys. **141**, 112–152 (1998). https://doi.org/10.1006/jcph.1998.5906. https://linkinghub.elsevier.com/retrieve/pii/S002199919895906X
51. Sato, Y., Ničeno, B.: A sharp-interface phase change model for a mass-conservative interface tracking method. J. Comput. Phys. **249**, 127–161 (2013). https://doi.org/10.1016/j.jcp.2013.04.035. https://linkinghub.elsevier.com/retrieve/pii/S0021999113003197
52. Schillaci, E., Antepara, O., Balcázar, N., Serrano, J.R., Oliva, A.: A numerical study of liquid atomization regimes by means of conservative level-set simulations. Comput. Fluids **179**, 137–149 (2019). https://doi.org/10.1016/j.compfluid.2018.10.017. https://linkinghub.elsevier.com/retrieve/pii/S0045793018307801
53. Singh, N.K., Premachandran, B.: A coupled level set and volume of fluid method on unstructured grids for the direct numerical simulations of two-phase flows including phase change. Int. J. Heat Mass Transf. **122**, 182–203 (2018). https://doi.org/10.1016/j.ijheatmasstransfer.2018.01.091. https://linkinghub.elsevier.com/retrieve/pii/S0017931017341194
54. Son, G., Dhir, V.K.: Numerical simulation of saturated film boiling on a horizontal surface. J. Heat Transf. **119**, 525–533 (1997). https://doi.org/10.1115/1.2824132. https://asmedigitalcollection.asme.org/heattransfer/article/119/3/525/415631/Numerical-Simulation-of-Saturated-Film-Boiling-on
55. Son, G., Dhir, V.K.: Numerical simulation of film boiling near critical pressures with a level set method. J. Heat Transf. **120**, 183–192 (1998). https://doi.org/10.1115/1.2830042. https://asmedigitalcollection.asme.org/heattransfer/article/120/1/183/383014/Numerical-Simulation-of-Film-Boiling-Near-Critical
56. Son, G., Dhir, V.K., Ramanujapu, N.: Dynamics and heat transfer associated with a single bubble during nucleate boiling on a horizontal surface. J. Heat Transf. **121**, 623–631 (1999). https://doi.org/10.1115/1.2826025. https://asmedigitalcollection.asme.org/heattransfer/article/121/3/623/429987/Dynamics-and-Heat-Transfer-Associated-With-a

57. Son, G., Dhir, V.K.: Numerical simulation of nucleate boiling on a horizontal surface at high heat fluxes. Int. J. Heat Mass Transf. **51**, 2566–2582 (2008). https://doi.org/10.1016/j.ijheatmasstransfer.2007.07.046. https://linkinghub.elsevier.com/retrieve/pii/S0017931007005212
58. Son, G., Dhir, V.K.: Three-dimensional simulation of saturated film boiling on a horizontal cylinder. Int. J. Heat Mass Transf. **51**, 1156–1167 (2008). https://doi.org/10.1016/j.ijheatmasstransfer.2007.04.026. https://linkinghub.elsevier.com/retrieve/pii/S0017931007003304
59. Sun, D., Tao, W.: A coupled volume-of-fluid and level set (VOSET) method for computing incompressible two-phase flows. Int. J. Heat Mass Transf. **53**, 645–655 (2010). https://doi.org/10.1016/j.ijheatmasstransfer.2009.10.030. https://linkinghub.elsevier.com/retrieve/pii/S0017931009005717
60. Sussman, M., Puckett, E.G.: A coupled level set and volume-of-fluid method for computing 3d and axisymmetric incompressible two-phase flows. J. Comput. Phys. **162**, 301–337 (2000). https://doi.org/10.1006/jcph.2000.6537. https://linkinghub.elsevier.com/retrieve/pii/S0021999100965379
61. Sussman, M., Smereka, P., Osher, S.: A level set approach for computing solutions to incompressible two-phase flow. J. Comput. Phys. **114**, 146–159 (1994). https://doi.org/10.1006/jcph.1994.1155. https://linkinghub.elsevier.com/retrieve/pii/S0021999184711557
62. Sweby, P.K.: High resolution schemes using flux limiters for hyperbolic conservation laws. SIAM J. Numer. Anal. **21**, 995–1011 (1984). https://doi.org/10.1137/0721062. https://epubs.siam.org/doi/10.1137/0721062
63. Tanguy, S., Ménard, T., Berlemont, A.: A level set method for vaporizing two-phase flows. J. Comput. Phys. **221**, 837–853 (2007). https://doi.org/10.1016/j.jcp.2006.07.003. https://linkinghub.elsevier.com/retrieve/pii/S0021999106003214
64. Tomar, G., Biswas, G., Sharma, A., Welch, S.W.J.: Multimode analysis of bubble growth in saturated film boiling. Phys. Fluids **20**, 092101 (2008). https://doi.org/10.1063/1.2976764. https://aip.scitation.org/doi/10.1063/1.2976764
65. Tryggvason, G., et al.: A front-tracking method for the computations of multiphase flow. J. Comput. Phys. **169**, 708–759 (2001). https://doi.org/10.1006/jcph.2001.6726. https://linkinghub.elsevier.com/retrieve/pii/S0021999101967269
66. Tryggvason, G., Esmaeeli, A., Al-Rawahi, N.: Direct numerical simulations of flows with phase change. Comput. Struct. **83**, 445–453 (2005). https://doi.org/10.1016/j.compstruc.2004.05.021. https://linkinghub.elsevier.com/retrieve/pii/S0045794904004158
67. Tryggvason, G., Scardovelli, R., Zaleski, S.: The volume-of-fluid method (2001). https://doi.org/10.1017/CBO9780511975264.006. https://www.cambridge.org/core/product/identifier/CBO9780511975264A041/type/book_part
68. Tsui, Y.Y., Lin, S.W.: Three-dimensional modeling of fluid dynamics and heat transfer for two-fluid or phase change flows. Int. J. Heat Mass Transf. **93**, 337–348 (2016). https://doi.org/10.1016/j.ijheatmasstransfer.2015.09.021. https://linkinghub.elsevier.com/retrieve/pii/S0017931015009643
69. Unverdi, S.O., Tryggvason, G.: A front-tracking method for viscous, incompressible, multi-fluid flows. J. Comput. Phys. **100**, 25–37 (1992). https://doi.org/10.1016/0021-9991(92)90307-K. https://linkinghub.elsevier.com/retrieve/pii/002199919290307K
70. Welch, S.W., Wilson, J.: A volume of fluid based method for fluid flows with phase change. J. Comput. Phys. **160**, 662–682 (2000). https://doi.org/10.1006/jcph.2000.6481. https://linkinghub.elsevier.com/retrieve/pii/S0021999100964817
71. Yuan, M., Yang, Y., Li, T., Hu, Z.: Numerical simulation of film boiling on a sphere with a volume of fluid interface tracking method. Int. J. Heat Mass Transf. **51**, 1646–1657 (2008). https://doi.org/10.1016/j.ijheatmasstransfer.2007.07.037. https://linkinghub.elsevier.com/retrieve/pii/S0017931007005108

Deswelling Dynamics of Chemically-Active Polyelectrolyte Gels

Bindi M. Nagda[1], Jian Du[1(✉)], Owen L. Lewis[2], and Aaron L. Fogelson[3]

[1] Florida Institute of Technology, Melbourne, FL 32901, USA
jdu@fit.edu
[2] University of New Mexico, Albuquerque, NM 87106, USA
[3] University of Utah, Salt Lake City, UT 84112, USA

Abstract. Ion-induced volume phase transitions in polyelectrolyte gels play an important role in physiological processes such as mucus storage and secretion in the gut, nerve excitation, and DNA packaging. Experiments have shown that changes in ionic composition can trigger rapid swelling and deswelling of these gels. Based on a previously developed computational model, we carry out 2D simulations of gel deswelling within an ionic bath. The dynamics of the volume phase transition are governed by the balance of chemical and mechanical forces on components of the gel. Our simulation results highlight the close connections between the patterns of deswelling, the ionic composition, and the relative magnitude of particle-particle interaction energies.

Keywords: Polyelectrolyte Gel · Deswelling · Multiphase Model · Simulation

1 Introduction

Polyelectrolyte gels are mixtures of a polymer network, a solvent and ions that are either bound to the network or dissolved in the solvent. They are prevalent in living systems, from polynucleotides like DNA molecules [1] to the glycoproteins that form a protective gastric mucus layer [2]. Experiments have shown that monovalent/divalent ion exchange between the polymer and the solvent plays a major role in inducing a volume transition in polyelectrolyte gels [3]. These volume transitions are important role in physiology, a prime example is their role in maintenance of the gastric mucus layer. Gastric mucus is a highly hydrated network consisting mainly of water mixed with mucin polymers and other electrolytes [2]. Goblet cells in the gut produce mucin and densely pack it in lipid vesicles. Inside these vesicles, high concentrations of divalent calcium ions keep the mucin network in a dehydrated state by crosslinking negatively charged groups between two polymer strands. A stimuli such as change in ionic composition can triggerff the release of these mucins into the extracellular environment, whereby they swell explosively on a timescale of milliseconds [2].

Supported by NIH research grant 1R01GM131408.

J. Mikyška et al. (Eds.): ICCS 2023, LNCS 14077, pp. 332–346, 2023.
https://doi.org/10.1007/978-3-031-36030-5_27

Pioneering work on volume transitions investigated the thermodynamic equilibrium of the gel and its dependence on parameters such as pH and temperature [4]. Later investigations of non-equilibrium transient states of swelling were based on simplifications such as a reduced model for the fluid flow, or low polymer volume fractions [5]. Recent models have addressed gel dynamics and included effects of electrochemical potentials [6]. These studies are based on the classical Flory-Huggins mixture theory [7], and do not generalize the mixing energy to account for ion/polymer chemical reactions or the network's affinity for binding with divalent cations. To address this gap, a multiphase model for polyelectrolyte gel swelling was developed with the formulation based on chemical potentials [8]. More recently, the model was rederived and extended to use force densities (i.e. potential gradients) rather than chemical potentials to reveal the underlying causes of the swelling behavior [9]. A computational method for simulating gel swelling dynamics in 2D was proposed in [10], based on a novel algorithm for evolving the dissolved ion concentrations and the electric potential originally presented in [11].

In this paper, we extend the computational investigation towards analyzing deswelling dynamics in 2D by characterizing the contributions of the forces acting on the network and solvent due to short-range interactions. In Sect. 2, we present the model equations that govern the dynamics of a two-phase gel. In Sect. 3, describe simulation results from two sets of computational experiments. We see that in one parameter regime, exposure of a low density gel to a bath of divalent calcium leads to a deswelling or condensation of the gel, with the speed and degree of deswelling controlled by the bath concentration. However, in another regime (defined by the relative size of particle-particle interaction energies), exposure to a divalent bath can lead to a qualitatively distinct form of collapse where-in a high density "ring" of network forms at the periphery of the gel sample. Finally in Sect. 4, we present a conclusion of our findings and highlight the key points from our study.

2 Model

Our model incorporates a two-fluid description of the solvent phase (s) and the network phase (n) of the gel, as well as the evolution of the dissolved and bound ion concentrations, included through binding and unbinding reactions between cations and the negatively charged network. It accounts for a number of mechanical forces (viscous, drag, pressure) and "chemical forces" (entropic, electric, short-range interactions), which depend on the spatial distribution of the network and solvent phases, the ionic species, and their respective velocities. Full details about the derivation can be found in [9]. The model equations are a coupled system of nonlinear PDEs along with two constraints - incompressibility and electroneutrality of the mixture at all points in space and time.

We denote the volume fraction of the network and solvent as θ_{n} and θ_{s}, respectively. The dynamics of the two volume fractions are governed by the advection equations

$$\frac{\partial}{\partial t}\theta_{\mathrm{n}} + \nabla \cdot (\theta_{\mathrm{n}}\mathbf{u}_{\mathrm{n}}) = 0, \qquad \frac{\partial}{\partial t}\theta_{\mathrm{s}} + \nabla \cdot (\theta_{\mathrm{s}}\mathbf{u}_{\mathrm{s}}) = 0. \tag{1}$$

Here $\mathbf{u}_{\mathrm{n}}$ and $\mathbf{u}_{\mathrm{s}}$ are the velocity of the network and solvent, respectively. These equations along with the identity $\theta_{\mathrm{n}}(\boldsymbol{x},t) + \theta_{\mathrm{s}}(\boldsymbol{x},t) = 1$ yield the volume-averaged incompressibility condition

$$\nabla \cdot (\theta_{\mathrm{n}}\mathbf{u}_{\mathrm{n}} + \theta_{\mathrm{s}}\mathbf{u}_{\mathrm{s}}) = 0. \tag{2}$$

Dissolved sodium and calcium ions can bind to and unbind from negatively charged sites on the network according to the following reactions. Here M refers to a monomer that carries one negative charge. $k_{\mathrm{j}}^{\mathrm{on}}$ and $k_{\mathrm{j}}^{\mathrm{off}}$ are the binding and unbinding rates for ion j, respectively.

$$\mathrm{M}^- + \mathrm{Na}^+ \underset{k_{\mathrm{Na}}^{\mathrm{off}}}{\overset{k_{\mathrm{Na}}^{\mathrm{on}}}{\rightleftharpoons}} \mathrm{MNa}, \quad \mathrm{M}^- + \mathrm{Ca}^{2+} \underset{k_{\mathrm{Ca}}^{\mathrm{off}}}{\overset{k_{\mathrm{Ca}}^{\mathrm{on}}}{\rightleftharpoons}} \mathrm{MCa}^+, \quad \mathrm{M}^- + \mathrm{MCa}^+ \underset{2k_{\mathrm{Ca}}^{\mathrm{off}}}{\overset{\frac{1}{2}k_{\mathrm{Ca}}^{\mathrm{on}}}{\rightleftharpoons}} \mathrm{M_2Ca}. \tag{3}$$

The concentrations (per unit solvent volume) of dissolved sodium, calcium, and chloride ions are denoted C_{Na}, C_{Ca}, and C_{Cl}, respectively. They are governed by the Nernst-Planck equations (4)–(6). The ions move by advection at the solvent velocity, diffusion, and electromigration, and participate in binding/unbinding reactions with the network. Ψ is the electrical potential scaled by $\frac{q}{k_B T}$, where q is the fundamental charge, k_B is the Boltzmann constant, and T is the temperature. The value of Ψ is determined by the electroneutrality constraint, Eq. (10). D_{Na}, D_{Ca}, and D_{Cl} are the ion diffusion coefficients.

$$\frac{\partial C_{\mathrm{Na}}}{\partial t} + \mathbf{u}_{\mathrm{s}} \cdot \nabla C_{\mathrm{Na}} = \frac{1}{\theta_{\mathrm{s}}}\nabla \cdot \Big(\theta_{\mathrm{s}} D_{\mathrm{Na}} \left(\nabla C_{\mathrm{Na}} + C_{\mathrm{Na}}\nabla\Psi\right)\Big) - k_{\mathrm{Na}}^{\mathrm{on}} m C_{\mathrm{Na}} + k_{\mathrm{Na}}^{\mathrm{off}} B_{\mathrm{MNa}}\theta_{\mathrm{s}}, \tag{4}$$

$$\frac{\partial C_{\mathrm{Ca}}}{\partial t} + \mathbf{u}_{\mathrm{s}} \cdot \nabla C_{\mathrm{Ca}} = \frac{1}{\theta_{\mathrm{s}}}\nabla \cdot \Big(\theta_{\mathrm{s}} D_{\mathrm{Ca}} \left(\nabla C_{\mathrm{Na}} + 2C_{\mathrm{Ca}}\nabla\Psi\right)\Big) - k_{\mathrm{Ca}}^{\mathrm{on}} m C_{\mathrm{Ca}} + k_{\mathrm{Ca}}^{\mathrm{off}} B_{\mathrm{MCa}}\theta_{\mathrm{s}}, \tag{5}$$

$$\frac{\partial C_{\mathrm{Cl}}}{\partial t} + \mathbf{u}_{\mathrm{s}} \cdot \nabla C_{\mathrm{Cl}} = \frac{1}{\theta_{\mathrm{s}}}\nabla \cdot \Big(\theta_{\mathrm{s}} D_{\mathrm{Cl}} \left(\nabla C_{\mathrm{Cl}} - C_{\mathrm{Cl}}\nabla\Psi\right)\Big), \tag{6}$$

The concentrations (per total volume) of ions bound to the network are denoted B_{MNa}, B_{MCa} and $B_{\mathrm{M_2Ca}}$ for sodium bound to monomer, calcium bound to a single monomer, and calcium bound to two monomers, respectively. They are affected by the ions' advection at the network velocity and their participation in binding/unbinding reactions.

$$\frac{\partial B_{\mathrm{MNa}}}{\partial t} + \nabla \cdot (\mathbf{u}_{\mathrm{n}} B_{\mathrm{MNa}}) = k_{\mathrm{Na}}^{\mathrm{on}} m C_{\mathrm{Na}}\theta_{\mathrm{s}} - k_{\mathrm{Na}}^{\mathrm{off}} B_{\mathrm{MNa}}{\theta_{\mathrm{s}}}^2, \tag{7}$$

$$\frac{\partial B_{\mathrm{MCa}}}{\partial t} + \nabla \cdot (\mathbf{u}_{\mathrm{n}} B_{\mathrm{MCa}}) = k_{\mathrm{Ca}}^{\mathrm{on}} m C_{\mathrm{Ca}}\theta_{\mathrm{s}} - k_{\mathrm{Ca}}^{\mathrm{off}} B_{\mathrm{MCa}}{\theta_{\mathrm{s}}}^2 + 2k_{\mathrm{Ca}}^{\mathrm{off}} B_{\mathrm{M_2Ca}} - \frac{1}{2}k_{\mathrm{Ca}}^{\mathrm{on}} m B_{\mathrm{MCa}}, \tag{8}$$

$$\frac{\partial B_{\mathrm{M_2Ca}}}{\partial t} + \nabla \cdot (\mathbf{u}_{\mathrm{n}} B_{\mathrm{M_2Ca}}) = -2k_{\mathrm{Ca}}^{\mathrm{off}} B_{\mathrm{M_2Ca}} + \frac{1}{2}k_{\mathrm{Ca}}^{\mathrm{on}} m B_{\mathrm{MCa}}, \tag{9}$$

In Eqs. (4)–(9), m is the concentration of unoccupied binding sites on the network, which is calculated as $m = n^{\mathrm{tot}}\theta_{\mathrm{n}} - B_{\mathrm{MNa}} - B_{\mathrm{MCa}} - 2B_{\mathrm{M_2Ca}}$ (n^{tot} is defined below.). The electroneutrality condition is that the concentration of charge is zero:

$$B_{\mathrm{MNa}} + 2B_{\mathrm{MCa}} + 2B_{\mathrm{M_2Ca}} - n^{\mathrm{tot}}\theta_{\mathrm{n}} + \theta_{\mathrm{s}}\left(C_{\mathrm{Na}} + 2C_{\mathrm{Ca}} - C_{\mathrm{Cl}}\right) = 0. \tag{10}$$

Force balance Eqs. (11)–(12) for the two-phase gel, along with the incompressibility condition (2), are solved to determine the velocities of the two phases and the pressure, p. The network and solvent viscosities are η_{n} and η_{s}, respectively, while ξ/ν_{n} is the drag coefficient for relative motion between the solvent and network. (ν_{n} is defined below.)

$$\nabla \cdot \left(\theta_{\mathrm{n}}\underline{\underline{\sigma}}^{\mathrm{n}}(\mathbf{u}_{\mathrm{n}})\right) - \frac{\xi}{\nu_{\mathrm{n}}}\theta_{\mathrm{n}}\theta_{\mathrm{s}}\left(\mathbf{u}_{\mathrm{n}} - \mathbf{u}_{\mathrm{s}}\right) - \theta_{\mathrm{n}}\nabla p + \mathbf{f}_{\mathrm{n}}^{\mathrm{E}} + \mathbf{f}_{\mathrm{n}}^{\mathrm{I}} + \mathbf{f}_{\mathrm{n}}^{\mathrm{S}} = 0, \tag{11}$$

$$\nabla \cdot \left(\theta_{\mathrm{s}}\underline{\underline{\sigma}}^{\mathrm{s}}(\mathbf{u}_{\mathrm{s}})\right) - \frac{\xi}{\nu_{\mathrm{n}}}\theta_{\mathrm{n}}\theta_{\mathrm{s}}\left(\mathbf{u}_{\mathrm{s}} - \mathbf{u}_{\mathrm{n}}\right) - \theta_{\mathrm{s}}\nabla p + \mathbf{f}_{\mathrm{s}}^{\mathrm{E}} + \mathbf{f}_{\mathrm{s}}^{\mathrm{I}} + \mathbf{f}_{\mathrm{s}}^{\mathrm{S}} = 0. \tag{12}$$

The viscous stress tensors in (11)–(12) are given by

$$\underline{\underline{\sigma}}^{\mathrm{i}} = \eta_{\mathrm{i}}(\nabla\mathbf{u}_{\mathrm{i}} + (\nabla\mathbf{u}_{\mathrm{i}})^{\mathbf{T}}) - (\eta_{\mathrm{i}}\nabla\cdot\mathbf{u}_{\mathrm{i}})\underline{\underline{I}} \qquad \mathrm{i} = \mathrm{n}, \mathrm{s}. \tag{13}$$

We solve the model Eqs. (1)–(12) numerically using the schemes described in detail in [10].

The charged nature of the ionic species induces an electric potential that exerts forces on the charged entities present in the gel. The long-range electric force densities on the network and solvent, respectively, are

$$\mathbf{f}_{\mathrm{n}}^{\mathrm{E}} = -\frac{\theta_{\mathrm{n}}}{\nu_{\mathrm{n}}}\frac{B_{\mathrm{MNa}} + 2B_{\mathrm{MCa}} + 2B_{\mathrm{M_2Ca}} - n^{\mathrm{tot}}\theta_{\mathrm{n}}}{n^{\mathrm{tot}}\theta_{\mathrm{n}}}k_B T\nabla\Psi, \tag{14}$$

$$\mathbf{f}_{\mathrm{s}}^{\mathrm{E}} = -\frac{\theta_{\mathrm{s}}}{\nu_{\mathrm{s}}}\frac{C_{\mathrm{Na}} + 2C_{\mathrm{Ca}} - C_{\mathrm{Cl}}}{s^{\mathrm{tot}}}k_B T\nabla\Psi. \tag{15}$$

Here, ν_{n} and ν_{s} are the volume of individual network and solvent molecules, respectively, while the molar concentrations of these particles in pure network and pure solvent are n^{tot} and s^{tot} (related by $\nu_{\mathrm{n}}n^{\mathrm{tot}} = \nu_{\mathrm{s}}s^{\mathrm{tot}}$). The entropic force densities on the network and solvent are

$$\mathbf{f}_{\mathrm{n}}^{\mathrm{S}} = -\frac{k_B T}{\nu_n}\left(\frac{1}{N_{\mathrm{chain}}} + \left(1 - \frac{1}{N_{\mathrm{chain}}}\right)\theta_{\mathrm{n}}\right)\nabla\theta_{\mathrm{n}}, \tag{16}$$

$$\mathbf{f}_{\mathrm{s}}^{\mathrm{S}} = -\frac{k_B T}{\nu_{\mathrm{n}}}\left(\left(\frac{1}{N_{\mathrm{chain}}} - 1\right)\theta_{\mathrm{s}}\nabla\theta_{\mathrm{s}} + \nabla\theta_{\mathrm{s}}\right). \tag{17}$$

The short-range interaction force densities on the network and solvent are

$$\mathbf{f}_{\mathrm{n}}^{\mathrm{I}} = -\frac{k_B T\theta_{\mathrm{n}}}{\nu_{\mathrm{n}}}\nabla\left(\frac{i(\alpha)}{2}{\theta_{\mathrm{s}}}^2 + \mu_{\mathrm{n}}^0(\alpha)\right), \tag{18}$$

$$\mathbf{f}_{\mathrm{s}}^{\mathrm{I}} = -\frac{k_B T \theta_{\mathrm{s}}}{\nu_{\mathrm{n}}} \nabla \left(\frac{i(\alpha)}{2} {\theta_{\mathrm{n}}}^2 + \mu_{\mathrm{s}}^0 \right), \tag{19}$$

where

$$\alpha = \frac{2B_{\mathrm{M_2Ca}}}{n^{\mathrm{tot}} \theta_{\mathrm{n}}}, \tag{20}$$

$$i(\alpha) = z(2\epsilon_{\mathrm{us}} - \epsilon_{\mathrm{uu}} - \epsilon_{\mathrm{ss}}) - 2\left(1 - \frac{1}{N_{\mathrm{chain}}}\right)(\epsilon_{\mathrm{us}} - \epsilon_{\mathrm{uu}}) - (\epsilon_{\mathrm{us}} - \epsilon_{\mathrm{uu}})\alpha, \tag{21}$$

$$\mu_{\mathrm{n}}^0(\alpha) = \epsilon_{\mathrm{uu}} \frac{z}{2} + (\epsilon_{\mathrm{pp}} - \epsilon_{\mathrm{uu}})\left(1 - \frac{1}{N_{\mathrm{chain}}}\right) + \frac{\alpha}{2}(\epsilon_{\mathrm{xx}} - \epsilon_{\mathrm{uu}}). \tag{22}$$

and

$$\mu_{\mathrm{s}}^0 = \epsilon_{\mathrm{ss}} \frac{z}{2}. \tag{23}$$

Here, α is the fraction of monomers crosslinked by doubly bound calcium ions, μ_{n}^0 and μ_{s}^0 are the standard free energies of pure network and pure solvent, and $i(\alpha)$ corresponds to the interaction parameter from Flory-Huggins mixture theory [7]. This quantity may be thought of as the change in chemical energy due to mixing. Through its dependence on α, $i(\alpha)$ changes as crosslinks are formed and broken, which represents a major novelty of this model framework.

The expressions for $i(\alpha)$, $\mu_{\mathrm{n}}^0(\alpha)$, and μ_{s}^0 are derived using lattice-based mean-field arguments [8]. In these expressions, "interaction energies" represent the energy associated with various particles occupying adjacent locations on the lattice. The energies are ϵ_{xx} and ϵ_{uu} for adjacent crosslinked and uncrosslinked network particles from different polymer chains, ϵ_{pp} for adjacent particles in the same chain, ϵ_{ss} for adjacent solvent particles, and ϵ_{us} for adjacent network and solvent particles.

Of the three chemical forces acting on the gel, the entropic force always promotes swelling/mixing and the electric force tends to resist the relative motion between the network and the solvent [10]. Therefore, in the experiments below, the deswelling/collapsing of a gel is driven by the short range interaction force. Based on Eqs. (18)–(19) and ignoring the terms that are constant in space (μ_{s}^0 and the first two terms in $\mu_{\mathrm{n}}^0(\alpha)$), we may rewrite the short range forces as:

$$\mathbf{f}_{\mathrm{n}}^{\mathrm{I}} = \frac{k_B T \theta_{\mathrm{n}}}{\nu_{\mathrm{n}}} \left(\frac{\theta_{\mathrm{s}}^2 \epsilon_1 - \epsilon_3}{2} \nabla\alpha - i(\alpha)\theta_{\mathrm{s}} \nabla\theta_{\mathrm{s}} \right) = \mathbf{f}_{\mathrm{n}}^{\mathrm{I},1} + \mathbf{f}_{\mathrm{n}}^{\mathrm{I},2} + \mathbf{f}_{\mathrm{n}}^{\mathrm{I},3}, \tag{24}$$

$$\mathbf{f}_{\mathrm{s}}^{\mathrm{I}} = \frac{k_B T \theta_{\mathrm{s}}}{\nu_{\mathrm{n}}} \left(\frac{\theta_{\mathrm{n}}^2}{2} \epsilon_1 \nabla\alpha - i(\alpha)\theta_{\mathrm{n}} \nabla\theta_{\mathrm{n}} \right) = \mathbf{f}_{\mathrm{s}}^{\mathrm{I},1} + \mathbf{f}_{\mathrm{s}}^{\mathrm{I},2}. \tag{25}$$

Here $\epsilon_1 = \epsilon_{\mathrm{us}} - \epsilon_{\mathrm{uu}}$. In this work, we assume $\epsilon_1 < 0$, which implies an uncrosslinked, network monomer "prefers" to be adjacent to a solvent particle rather than to another negatively charged monomer. Similarly, we assume $\epsilon_3 = \epsilon_{\mathrm{xx}} - \epsilon_{\mathrm{uu}} < 0$, and thus the interaction energy between network monomers is reduced by calcium crosslinking. To facilitate later discussion, we have partitioned $\mathbf{f}_{\mathrm{n}}^{\mathrm{I}}$ and $\mathbf{f}_{\mathrm{s}}^{\mathrm{I}}$ into components:

$$\mathbf{f}_{\mathrm{n}}^{\mathrm{I},1} = \frac{k_B T \theta_{\mathrm{n}} \theta_{\mathrm{s}}^2}{2\nu_{\mathrm{n}}} \epsilon_1 \nabla\alpha, \quad \mathbf{f}_{\mathrm{n}}^{\mathrm{I},2} = -\frac{k_B T \theta_{\mathrm{n}} \theta_{\mathrm{s}}}{\nu_{\mathrm{n}}} i(\alpha) \nabla\theta_{\mathrm{s}}, \quad \text{and} \quad \mathbf{f}_{\mathrm{n}}^{\mathrm{I},3} = -\frac{k_B T \theta_{\mathrm{n}}}{2\nu_{\mathrm{n}}} \epsilon_3 \nabla\alpha.$$

$$\mathbf{f}_\mathrm{s}^{\mathrm{I},1} = \frac{k_B T \theta_\mathrm{s} \theta_\mathrm{n}^2}{2\nu_\mathrm{n}} \epsilon_1 \nabla \alpha \quad \text{and} \quad \mathbf{f}_\mathrm{s}^{\mathrm{I},2} = -\frac{k_B T \theta_\mathrm{s} \theta_\mathrm{n}}{\nu_\mathrm{n}} i(\alpha) \nabla \theta_\mathrm{n}.$$

Note that $\mathbf{f}_\mathrm{n}^{\mathrm{I},2} = -\mathbf{f}_\mathrm{s}^{\mathrm{I},2}$. These forces are completely analogous to classical Flory-Huggins polymer theory. They tend to drive swelling/mixing when $i(\alpha) < 0$ and to drive deswelling when $i(\alpha) > 0$. Both $\mathbf{f}_\mathrm{n}^{\mathrm{I},1}$ and $\mathbf{f}_\mathrm{s}^{\mathrm{I},1}$ are novel to this model (they do not have analogs in Flory-Huggins theory). Both are oriented in the direction of decreasing α, driving material towards a spatial location with a lower crosslink fraction α. For a low-density network ($\theta_\mathrm{n} \ll 1$), $\mathbf{f}_\mathrm{n}^{\mathrm{I},1}$ is much larger in magnitude than $\mathbf{f}_\mathrm{s}^{\mathrm{I},1}$. Finally, $\mathbf{f}_\mathrm{n}^{\mathrm{I},3}$ tends to drive the network towards a region with higher α. Therefore, the forces on the network $\mathbf{f}_\mathrm{n}^{\mathrm{I},1}$ and $\mathbf{f}_\mathrm{n}^{\mathrm{I},3}$ play opposite roles for the same spatial distribution of α. For $\theta_\mathrm{s} \approx 1$, the direction of $\mathbf{f}_\mathrm{n}^{\mathrm{I},1} + \mathbf{f}_\mathrm{n}^{\mathrm{I},3}$ is determined by the relative magnitudes of ϵ_1 and ϵ_3 (i.e. is the energy of the lattice reduced more by replacing monomer/monomer interactions with monomer/solvent interactions or by crosslinks) . As shown in our results section, depending on the relative magnitude of the interaction energies, the ion-induced volume transition in gels may exhibit drastically different patterns.

3 Simulation Results

3.1 Model Parameters and Problem Setup

The values of model parameters, similar to those used in [9,10], are listed in Table 1. We choose the values of the interaction energies so that the interaction parameter $i(\alpha)$ in Eq. (21) is $i(\alpha) = 18\alpha - 2.76$, an increasing linear function of α. As the value of α increases from 0 to 1, $i(\alpha)$ changes from negative to positive so that the roles of the classical Flory-Huggins terms $\mathbf{f}_\mathrm{n}^{\mathrm{I},2}$ and $\mathbf{f}_\mathrm{s}^{\mathrm{I},2}$ switch from promoting swelling (mixing of the phases) to promoting deswelling (phase separation). The ion binding and unbinding rate coefficients in the table imply that calcium has a smaller dissociation constant, so calcium binding to the network is chemically preferred. The computational domain is a 2D square of size $20\ \mu\mathrm{m} \times 20\ \mu\mathrm{m}$. Both velocities are assumed to be zero along its boundaries. No flux boundary conditions are applied to each dissolved ion concentration on all boundaries. A computational grid of size 256×256 is used for all simulations. The time step is fixed at $\Delta t = 2.0 \times 10^{-7}$ s. The typical distribution of the model variables at the initial time is shown in Fig. 1. The initial profiles of the network volume fraction and bound ion distributions are set to represent a blob of gel with loosely packed network ($\theta_\mathrm{n} \approx 5\%$), immersed in a fluid solvent with a high concentration of divalent calcium ions. The region with an appreciable amount of network is referred to as the "gel"; the rest of the domain is referred to as the "bath". This setup mimics the experiments in [12]. Spatially constant concentrations of dissolved ions with zero net charge are set in the bath, with the concentration of calcium much higher than that of sodium. We denote initial bath concentrations of ions with the asterisk superscript: C_i^*. The initial conditions for the ion concentrations in the gel region are chosen so that the binding and unbinding reactions are in local chemical equilibrium [10].

Table 1. Simulation parameters.

parameter	value
network viscosity η_n	100 Poise
solvent viscosity η_s	0.01 Poise
diffusion coefficient D_j	2.5×10^{-5} cm^2/s
drag coefficient ξ/ν	2.5×10^9 g/(cm^{3}s)
$\epsilon_{us} - \epsilon_{uu}$ (ϵ_1)	−18.0
$\epsilon_{us} - \epsilon_{ss}$	11.6
$\epsilon_{xx} - \epsilon_{uu}$ (ϵ_3)	variable
$\epsilon_{pp} - \epsilon_{uu}$	0
number of monomers in a chain N	100
network charge density n^{tot}	0.1 Molar
size of monomer ν_n	1.661×10^{-8} µm^3
coordination number z	6
ion valences z_{Na}, z_{Ca}, z_{Cl}	1, 2, −1
k^{on}_{Ca}	5×10^6 M^{-1}s^{-1}
k^{off}_{Ca}	5×10^2 M^{-1}s^{-1}
k^{on}_{Na}	10^6 M^{-1}s^{-1}
k^{off}_{Na}	10^3 M^{-1}s^{-1}

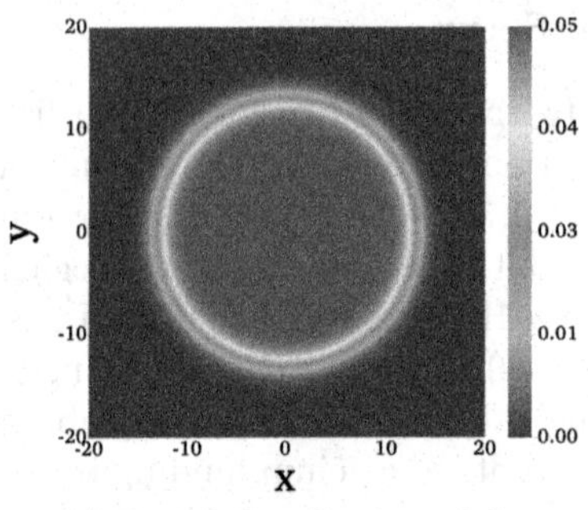

(a) Initial distribution of θ_n.

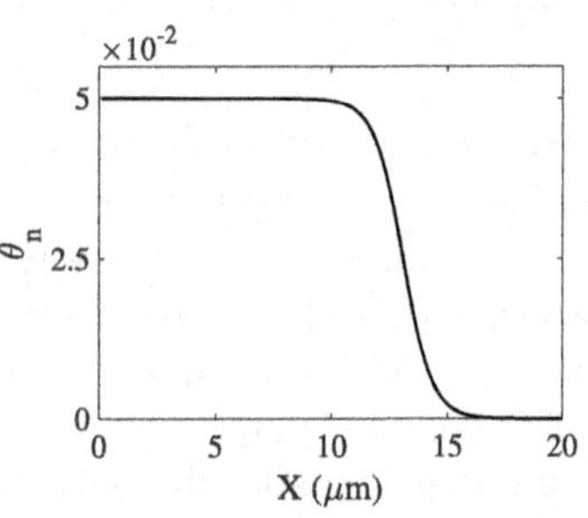

(b) Profile of θ_n at $t = 0$ along the positive x-axis.

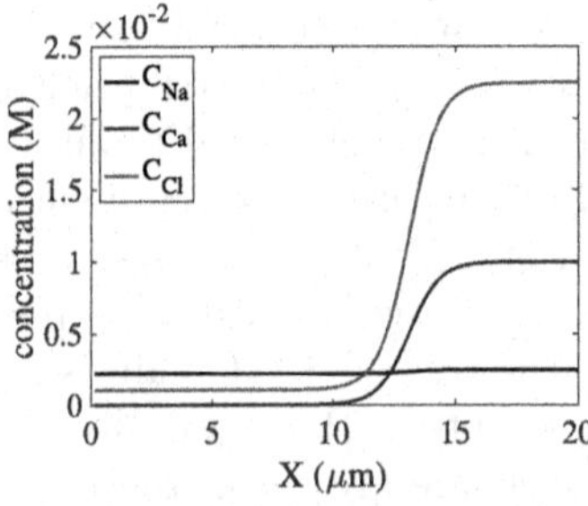

(c) Profiles of the concentrations for dissolved ions at $t = 0$ along the positive x-axis.

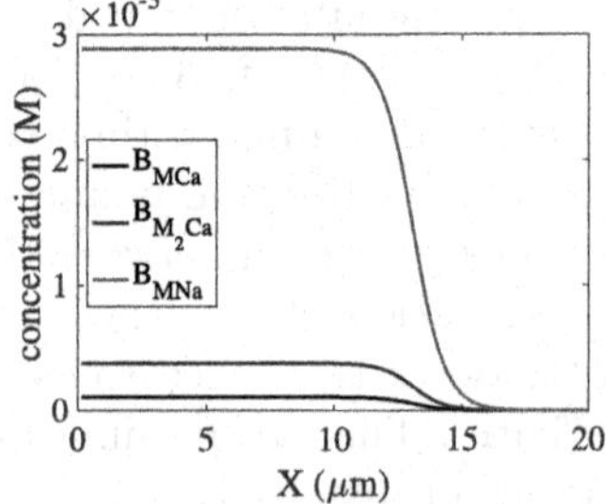

(d) Profiles of the concentrations of bound ions at $t = 0$ along the positive x-axis.

Fig. 1. The 2D computational domain and initial distributions of model variables.

For the initial profiles shown in Fig. 1, we have $\alpha \approx 0.153$ and the interaction parameter i is approximately zero. The spatially uniform profiles in the gel and in the bath are connected smoothly p a hyperbolic tangent function.

3.2 Effect of Calcium Bath Concentration on Gel Deswelling

To investigate the effect of calcium ion concentration in the bath, we carried out two simulations with $C^*_{\mathrm{Ca}} = 0.001$ M and $C^*_{\mathrm{Ca}} = 0.01$ M, respectively. The bath sodium concentration is fixed at a relatively low value of $C^*_{\mathrm{Na}} = 0.0025$ M.

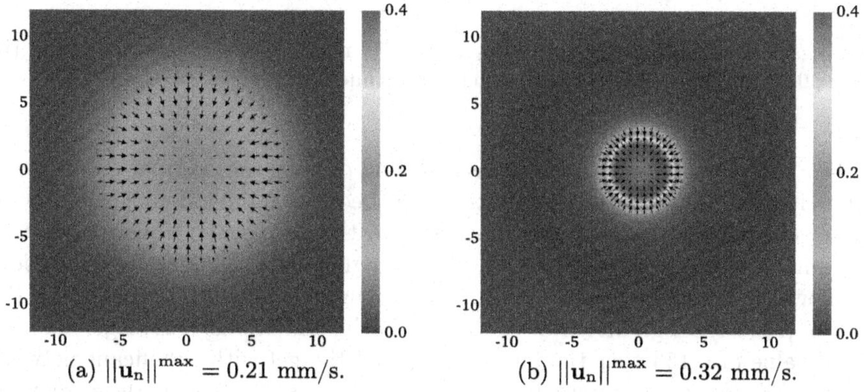

(a) $||\mathbf{u}_\mathrm{n}||^{\max} = 0.21$ mm/s. (b) $||\mathbf{u}_\mathrm{n}||^{\max} = 0.32$ mm/s.

Fig. 2. The distribution of θ_n and $\mathbf{u}_\mathrm{n}$ at t = 5.6 ms for (a) $C^*_{\mathrm{Ca}} = 0.001$ M and (b) $C^*_{\mathrm{Ca}} = 0.01$ M. $\epsilon_3 = -0.1$. The vectors have different scales.

In this subsection, $\epsilon_3 = -0.1$ so that crosslinks only slightly reduce the interaction energy between monomers. At the start of these simulations, the crosslink fraction α is higher near the edge of the gel abutting the bath where the calcium concentration is higher than in the gel's interior. As a consequence, $i(\alpha)$ is positive there, favoring deswelling. The distributions of θ_n and $\mathbf{u}_\mathrm{n}$ at t = 5.6 ms are plotted in Fig. 2. We see that as the network moves inward towards the center of the domain, the gel becomes smaller and more dense relative to its initial profile. We refer to this behavior as "collapse". A greater collapse is observed for higher C^*_{Ca}. In both simulations, the maximum magnitude of $\mathbf{u}_\mathrm{n}$ decreases with time (not shown). Plots of C_{Ca} and $\mathbf{u}_\mathrm{s}$ at the same time are shown in Fig. 3. In contrast to the network velocity, the solvent velocity exhibits complicated patterns. With the inward movement of the network, the solvent moves outward from the center of the gel. The solvent velocity near the gel's center is much larger with the higher bath calcium concentration. Due to diffusion, the distribution of C_{Ca} becomes more homogeneous with time (not shown).

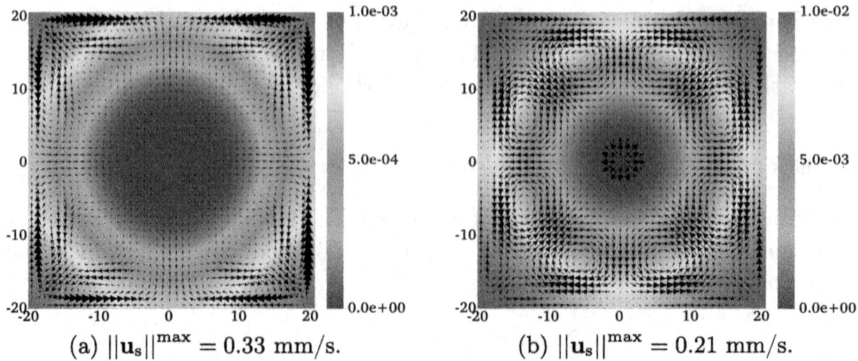

(a) $||\mathbf{u}_s||^{max} = 0.33$ mm/s. (b) $||\mathbf{u}_s||^{max} = 0.21$ mm/s.

Fig. 3. The distribution of C_{Ca} and $\mathbf{u}_{\mathrm{s}}$ at t = 5.6 ms for (a) $C^*_{\mathrm{Ca}} = 0.001$ M and (b) $C^*_{\mathrm{Ca}} = 0.01$ M. $\epsilon_3 = -0.1$. All vectors have the same scale.

In Fig. 4, we plot the distributions of θ_{n} and α along the positive x-axis. Relative to its initial profile, the distribution of θ_{n} becomes more inhomogeneous with the development of large peak values close to the gel's center. There is a more profound collapse in the simulation with a higher bath calcium concentration. As shown in Figs. 4ab, as the value of C^*_{Ca} changes from 0.001 M to 0.01 M, the peak of α increases greatly. Notice that for both simulations, α is larger than its initial value of 0.153 over the large portions of the gel with significant network volume fractions. The interaction parameter $i(\alpha)$ is positive in those regions, and so the short range force components $\mathbf{f}^{\mathrm{I},2}_{\mathrm{n}}$ and $\mathbf{f}^{\mathrm{I},2}_{\mathrm{s}}$ both promote deswelling. Because $|\epsilon_3| \ll |\epsilon_1|$, $|\mathbf{f}^{\mathrm{I},3}_{\mathrm{n}}| \ll |\mathbf{f}^{\mathrm{I},1}_{\mathrm{n}}|$, and therefore $\mathbf{f}^{\mathrm{I},3}_{\mathrm{n}}$ does not appreciably deter collapse.

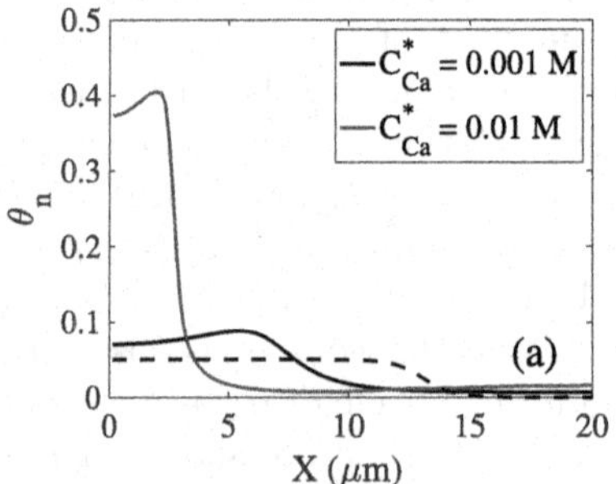

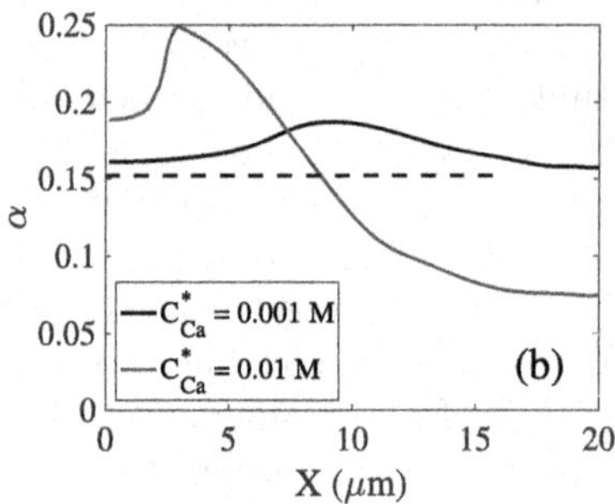

Fig. 4. 1D distributions of (a) θ_{n} and (b) α at t = 5.6 ms for $C^*_{\mathrm{Ca}} = 0.001$ M and 0.01 M. $\epsilon_3 = -0.1$. The dashed lines show the corresponding initial profiles.

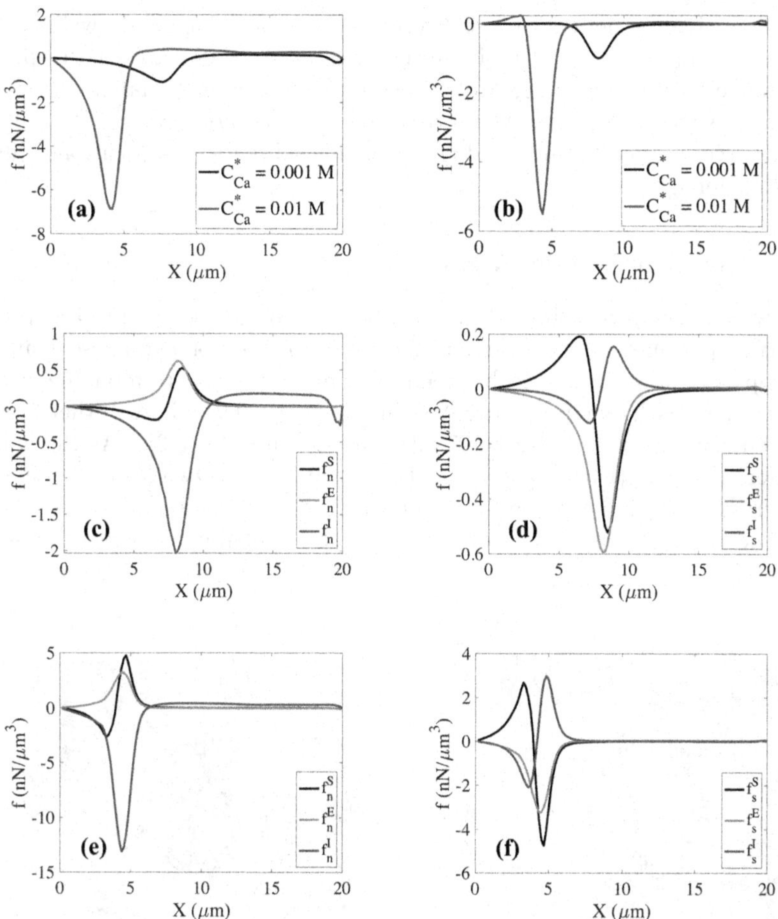

Fig. 5. Force densities along the positive **x**-axis at t = 2.8 ms. Total chemical force densities on (a) network and (b) solvent. Components of the chemical force densities for $C^*_{\mathrm{Ca}} = 0.001$ M on (c) network and (d) solvent. Components of the chemical force densities for $C^*_{\mathrm{Ca}} = 0.01$ M on (e) network and (f) solvent. (Color figure online)

To understand the roles played by different chemical forces in deswelling, we plot their 1D distributions at t = 2.8 ms in Fig. 5. Figures 5ab show that the total chemical forces on both the network and the solvent are directed towards the center of the gel and that the magnitude of the forces is much larger for the higher bath calcium concentration. A detailed look at the plots Figs. 5c–f indicates that: (1) The electric forces (green solid lines) on the network and on the solvent resist gel deswelling. The magnitude of the electric force increases with C^*_{Ca}, consistent with our previous study [9]. (2) From the center to the edge, the entropic force (black solid lines) changes direction and tends to homogenize the distribution of θ_{n} and θ_{s}. (3) The short range forces (red solid curves in c and

e) mainly drive the network towards the center, promoting gel deswelling. Since α has a sharper gradient in the simulation with $C^*_{\rm Ca} = 0.01$ M, the corresponding short range force is also much larger. The location where $\mathbf{f}^{\rm I}_{\rm n}$ changes direction is consistent with the location where α is highest (plot not shown). On the other hand, $\mathbf{f}^{\rm I}_{\rm s}$ (red solid curves in d and f) also changes direction, at locations different from that for $\mathbf{f}^{\rm I}_{\rm n}$.

3.3 Effect of Interaction Parameter

Next, we report on two simulations with $\epsilon_3 = -10$ and $\epsilon_3 = -18$. The purpose is to study parameter regimes in which crosslink formation can cause a substantial drop in interactive energy between monomers (relative to monomer/solvent mixing). The bath calcium concentration is $C^*_{Ca} = 0.01$ M. All other parameters are identical to those of the simulations shown in Sect. 3.2. As the value ϵ_3 gets more negative (with $|\epsilon_3| \leq |\epsilon_1|$), the magnitude of network force component $\mathbf{f}^{\rm I,3}_{\rm n}$ increases. With other conditions being the same, this tends to decrease the magnitude of the combined force $\mathbf{f}^{\rm I,1}_{\rm n} + \mathbf{f}^{\rm I,3}_{\rm n}$ and contributes to there being less deswelling.

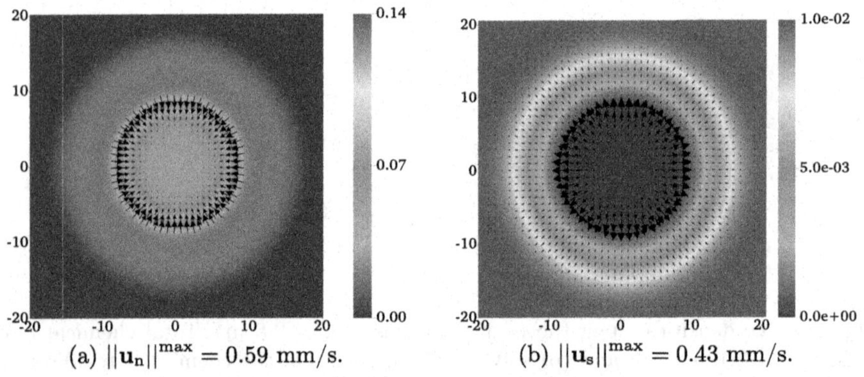

(a) $||\mathbf{u}_{\rm n}||^{\rm max} = 0.59$ mm/s. (b) $||\mathbf{u}_{\rm s}||^{\rm max} = 0.43$ mm/s.

Fig. 6. The distribution of $\theta_{\rm n}$ and $\mathbf{u}_{\rm n}$ (a) and $C_{\rm Ca}$ and $\mathbf{u}_{\rm s}$ (b) at t = 1.6 ms for $\epsilon_3 = -10$. The vectors have different scales.

For the simulation with $\epsilon_3 = -10$, the solvent tends to flow outwards, away from the gel center as the network collapses inward, similar to the previous section. The distributions of $\theta_{\rm n}$, $\mathbf{u}_{\rm n}$, $C_{\rm Ca}$ and $\mathbf{u}_{\rm s}$ at t = 1.6 ms are plotted in Fig. 6. Comparison between Fig. 6a and the results at the same time from the simulations of the previous section (not shown) indicates that the extent of deswelling is greatly reduced as the value of ϵ_3 is changed from -0.1 to -10.

As the value of ϵ_3 is further reduced to -18, network accumulation is observed along the edge of the gel and a ring pattern is formed (see Fig. 7). Concurrently, solvent is depleted from the edge of the gel, where the high density network

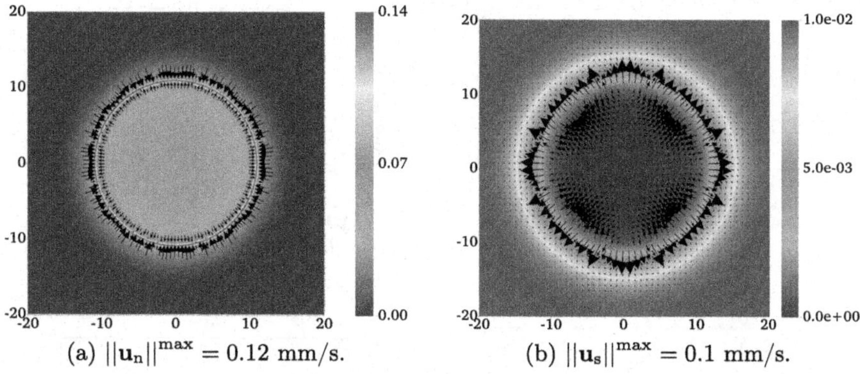

(a) $||\mathbf{u}_\mathrm{n}||^{\max} = 0.12$ mm/s. (b) $||\mathbf{u}_\mathrm{s}||^{\max} = 0.1$ mm/s.

Fig. 7. The distribution of θ_n and $\mathbf{u}_\mathrm{n}$ (a) and C_{Ca} and $\mathbf{u}_\mathrm{s}$ (b) at t = 1.6 ms for $\epsilon_3 = -18$. The vectors have different scales.

ring is formed. Solvent thus moves from this region both toward the origin and toward the domain boundary.

Figure 8 shows the 1D distributions of θ_n and α at t = 1.6 ms. When $\epsilon_3 = -18$, these quantities become strikingly inhomogeneous, with greater θ_n and α accumulation near the edge of the gel. There is very little movement of the network towards the origin when, indicated by the locations of the peak values of θ_n and α. Since α is large in a portion of the gel near its edge, $i(\alpha)$ is positive in those regions so that the short range force components $\mathbf{f}_\mathrm{n}^{\mathrm{I},2}$ and $\mathbf{f}_\mathrm{s}^{\mathrm{I},2}$ (not shown) drive more network into the ring and expel solvent from it. In the regions away from the edge of the gel where $i(\alpha)$ is negative, $\mathbf{f}_\mathrm{n}^{\mathrm{I},2}$ and $\mathbf{f}_\mathrm{s}^{\mathrm{I},2}$ promote solvent/network mixing.

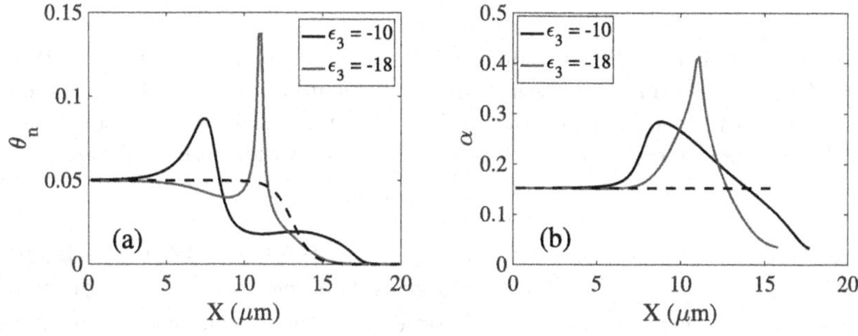

Fig. 8. 1D distributions of (a) θ_n and (b) α at t = 1.6 ms for $\epsilon_3 = -10$ and $\epsilon_3 = -18$. The black dashed lines show the initial profiles.

To further elucidate the ring formation seen for $\epsilon_3 = -18$ (and compare with the collapse seen for $\epsilon_3 = -10$), the 1D distributions of the chemical force densities at t = 0.8 ms are plotted in Fig. 9. Figure 9a shows that the total

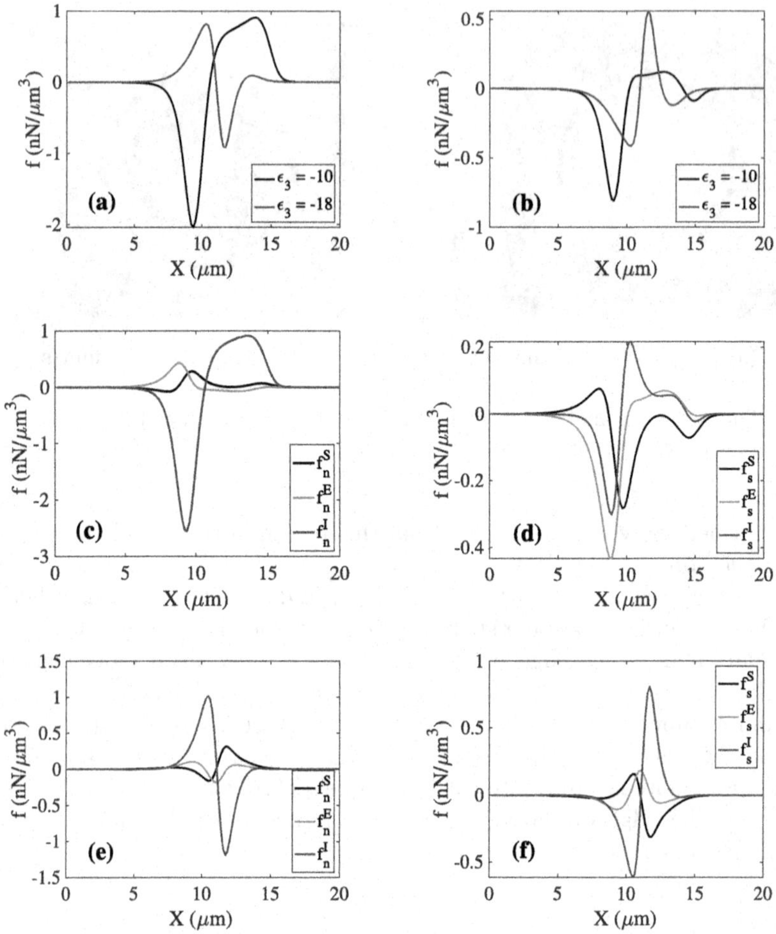

Fig. 9. 1D force density distributions at t = 0.8 ms for $\epsilon_3 = -10$ and -18. Total chemical force density in (a) network and (b) solvent. For $\epsilon_3 = -10$, components of chemical force density on (c) network and (d) solvent. For $\epsilon_3 = -18$, components of chemical force density on (e) network and (f) solvent. (Color figure online)

chemical force on the network is directed toward the edge of the gel (from both directions) when $\epsilon_3 = -18$, while the force is directed towards the center of the gel for $\epsilon_3 = -10$. For $\epsilon_3 = -18$, the force over the region where the network is being driven "outwards" is nearly equal in magnitude to the force over the region where the network is being driven "inwards". These opposing forces result in the ring formation observed in Fig. 7. At t = 0.8 ms the total chemical force on the solvent is directed away from the edge of the gel, in both directions, as shown in Fig. 9b. A detailed look at the plots c–f indicates that: (1) The electric forces (green solid lines) on the network and solvent are equal in magnitude with opposite directions. For $\epsilon_3 = -10$, the electric force resists deswelling by driving the network towards the edge of the gel and the solvent towards the center. For $\epsilon_3 = -18$, the electric force resists ring formation by driving the network away from the edge and the solvent towards the edge so the two phases can mix. (2) The entropic force tends to homogenize the distribution of θ_n (and θ_s) as indicated by the changing direction of the entropic forces (black solid lines). (3) The short range force for $\epsilon_3 = -10$ in (c) drives the network towards the center, leading to collapse. A detailed look at the individual components of the short range force for $\epsilon_3 = -18$ (plot not shown) indicates that $\mathbf{f}_n^{I,1}$ and $\mathbf{f}_n^{I,3}$ exert a nearly equal and opposite force on the network, leaving $\mathbf{f}_n^{I,2}$ to dominate. Since $i(\alpha)$ is positive over some portion of the gel and negative over another because of how α is distributed (see Fig. 8b), then $\mathbf{f}_n^{I,2}$ causes deswelling over the region where $i(\alpha) < 0$ and causes swelling over the region where $i(\alpha) > 0$. The location where $\mathbf{f}_n^{I}$ changes direction is consistent with the location where α takes its maximum in Fig. 8b. As a result, we observe accumulation of the network near the edge of the gel.

4 Conclusion

Our computational studies reveal multiple qualitatively distinct forms of deswelling which may be exhibited by a model of polyelectrolyte gel. In particular, we show that the bath ionic composition and relative magnitudes of the interaction energies play a significant role in determining the deswelling dynamics of the gel. Specifically, our results indicate that different components of the short range interaction force can drive swelling *or* deswelling of the gel. In the case where $|\epsilon_3| \ll |\epsilon_1|$, the dominant components of the short-range interaction forces drive the gel to collapse. As $|\epsilon_3|$ increases to roughly the same scale as $|\epsilon_1|$, the gel collapse is reduced significantly due to the greater competition between different component of the interaction force. It is worth noting that this form of gel collapse, and its dependence on the ionic composition of the bath has been studied in numerous theoretical and experimental systems, and is a widespread behavior of polyelectrolyte gel systems [13]. Finally, for $|\epsilon_3| = |\epsilon_1|$ we observe a more complex behavior characterized by network accumulation near the gel's edge, resulting in a ring pattern. In this case the Flory-Huggins interaction parameter, $i(\alpha)$, plays a dominant role in determining the dynamics of the gel, becoming positive as the network forms crosslinks and leading to the formation of network ring. Similar roles played by the Flory-Huggins interaction parameter were reported in experimental studies [12]. Pattern formation during gel deswelling (sometimes called spinodal decomposition) has been observed in other model systems and other geometries [14], though to our knowledge this work represents the first time it has been reported in a model which incorporates a crosslink-dependent interaction energy. Our computational results from varying the interactive energies between network and solvent particles motivate further investigation into behavior of gels with different properties.

References

1. Tongu, C., Kenmotsu, T., Yoshikawa, Y., Zinchenko, A., Chen, N., Yoshikawa, K.: Divalent cation shrinks DNA but inhibits its compaction with trivalent cation. J. Chem. Phys. **144**(20), 205101 (2016)
2. Verdugo, P.: Supramolecular dynamics of mucus. Cold Spring Harb. Perspect. Med. **2**(11), a009597 (2012)
3. Mussel, M., Basser, P.J., Horkay, F.: Ion-induced volume transition in gels and its role in biology. Gels **7**(1), 20 (2021)
4. Flory, P., J. Rehner. J.: Statistical mechanics of cross-linked polymer networks II. Swelling. J. Chem. Phys. **11**, 521–526 (1943)
5. Yamaue, T., Taniguchi, T., Doi, M.: Shrinking process of gels by stress-diffusion coupled dynamics. Theor. Phys. Supp. **138**, 4160–417 (2000)
6. Zhang, H., Dehghany, M., Hu, Y.: Kinetics of polyelectrolyte gels. J. Appl. Mech. **87**(6), 061010 (2020)
7. Doi, M.: Introduction to Polymer Physics. Oxford University Press (1996)
8. Sircar, S., Keener, J., Fogelson, A.: The effect of divalent vs. monovalent ions on the swelling of mucin-like polyelectrolyte gels: governing equations and equilibrium analysis. J. Chem. Phys. **138**, 014901 (2013)
9. Du, J., Lewis, O., Keener, J., Fogelson, A.: Modeling and simulation of the ion-binding-mediated swelling dynamics of mucin-like polyelectrolyte gels. Gels **7**(4), 244 (2021)
10. Du, J., Nagda, B., Lewis, O., Szyld, D., Fogelson, A.: A computational frame-work for the swelling dynamics of mucin-like polyelectrolyte gels. J. Non-Newton. Fluid Mech. **313**, 104989 (2023)
11. Lewis, O., Keener, J., Fogelson, A.: Electrodiffusion-mediated swelling of a two-phase gel model of gastric mucus. Gels **4**(3), 76 (2018)
12. Mussel, M., Lewis, O., Basser, P.J., Horkay, F.: Dynamic model of monovalent-divalent cation exchange in polyelectrolyte gels. Phys. Rev. Mater. **6**, 035602 (2022)
13. Horkay, F., Tasaki, I., Basser, P.: Effect of monovalent-divalent cation exchange on the swelling of polyacrylate hydrogels in physiological salt solutions. Biomacromol **2**(1), 195–199 (2001)
14. Celora, G.L., Hennessy, M.G., Münch, A., Wagner, B., Waters, S.L.: The dynamics of a collapsing polyelectrolyte gel. arXiv preprint arXiv:2105.06495 (2021)

Mathematical Modeling and Numerical Simulations for Drug Release from PLGA Particles

Yu Sun[1], Yan Li[2], and Jiangguo Liu[3(✉)]

[1] School of Advanced Materials Discovery (SAMD), Colorado State University, Fort Collins, CO 80523, USA
sundaym@rams.colostate.edu

[2] Department of Design and Merchandising and SAMD, Colorado State University, Fort Collins, CO 80523, USA
yan.li@colostate.edu

[3] Department of Mathematics and SAMD, Colorado State University, Fort Collins, CO 80523-1874, USA
liu@math.colostate.edu

Abstract. Poly lactic-co-glycolic acid (PLGA) is a copolymer that has demonstrated great potentials in development of novel drug delivery systems. This paper first discusses synthesis procedures and properties of PLGA micro/nanoparticles (MPs/NPs) and then examine mechanisms of drug release from PLGA particles. For the core-shell structure of reservoir-type PLGA MPs/NPs, diffusion through the polymeric shell is identified as the main mechanism of drug release. A time-dependent diffusion equation is used to model release from homogeneous spherical particles. Finite volume schemes are developed for the radial diffusion model. Numerical results and *in vitro* experiment data are discussed.

Keywords: Core-shell structure · Diffusion · Drug release · Finite volume schemes · Microparticles and nanoparticles · PLGA (poly lactic-co-glycolic acid)

1 Introduction

Polymers have been used in the pharmaceutical industry for several decades due to their biocompatible, biodegradable, and nontoxic properties [1,3,5,7,9,19,22]. A considerable amount of research has been devoted to the roles of poly-lactic-co-glycolic-acid (PLGA), poly-ethylene-oxide (PEO), and polyethylene glycol (PEG) in drug delivery. PLGA has shown great potentials in making drug delivery systems. It was approved by FDA as a material for therapeutic devices due to its biocompatibility and biodegradability [2,10,15,24]. After PLGA is introduced to human metabolism system, it can decompose into carbon dioxide and water, causing no harm to human bodies.

Y. Sun and J. Liu were partially supported by US National Science Foundation under grant DMS-2208590.

J. Mikyška et al. (Eds.): ICCS 2023, LNCS 14077, pp. 347–360, 2023.
https://doi.org/10.1007/978-3-031-36030-5_28

PLGA can be formed into particles at the micro- or nano-meter level and various shapes, e.g., slabs, cylinders, or spheres with therapeutic agents (drugs) encapsulated. The drugs will be released later in a controlled manner. Applications of PLGA MPs/NPs for control of bacterial infection and cures of Alzheimer's disease and breast cancer can be found in [2,10,15].

Four major mechanisms and three main release stages have been identified for drug release from PLGA micro- or nano-particles [16–18].

- Diffusion through the polymeric shell;
- Convection through the pores in the polymeric shell;
- Osmotic pumping;
- Degradation.

Diffusion is the major mechanism observed in the early stage [14].

Development of mathematical models for drug release from porous polymer systems dates back to the early 1980s, see [6] and references therein. Back then, the focus was on finding series solutions and then approximation for short-time behaviors, usually under the perfect sink condition, see [12] and references therein. Effects of variable boundary conditions and distribution of particle size on release behaviors were also investigated [12]. For diffusion-controlled release from reservoir or matrix systems, a summary of the series solutions was given in [14] for slabs, cylinders, and spheres. This approach is still very useful as demonstrated in a recent work [4] using series expansion in terms of eigenfunctions to investigate drug delivery from a multilayer spherical capsule. Finite element methods were investigated in [20,23] for diffusional drug release from complex matrix systems. All these demonstrate that mathematical models are helpful for design of new drug delivery systems [11].

The rest of this paper is organized as follows. Section 2 briefly describes a procedure for synthesis of PLGA particles and how drug, e.g., gentamicin, is encapsulated into PLGA MPs/NPs. Section 3 discusses properties of PLGA particles and commonly observed release mechanisms of drug release from PLGA MPs/NPs. Section 4 examines mathematical modeling for drug release from PLGA particles. Section 5 develops a numerical method for the radial diffusion equation that models release from spherical particles with the core-shell structure. Section 6 discusses numerical results and experimental data. The paper is concluded with some remarks in Sect. 7.

2 Synthesis of PLGA Particles and Drug Encapsulation

Various techniques have been used in preparation of PLGA MPs/NPs and encapsulation of drugs [3,7,10,22]. Gentamicin is a commonly used drug being encapsulated in PLGA particles. This is finished through a synthesis procedure based on a double emulsion evaporation method [16]. Different solutions for emulsion are prepared first. These include, for instance,

- 100 mg PLGA dissolved in 6 ml dichloromethane (DCM) as the oil phase;
- 20 mg gentamicin dissolved in water as the aqueous phase 1;
- 12% PVA solution as the aqueous phase 2.

The synthesis procedure contains 4 steps.

(i) Oil phase (PLGA/DCM) was mixed with aqueous phase 1 (gentamicin/water), yielding a primary emulsion solution;
(ii) Sonication was applied to the mixture for 3 min with a 35% amplitude;
(iii) Double-emulsion nano-droplets were formed via mixing primary emulsion solution with aqueous phase 2 (PVA/water);
(iv) The double-emulsion nano-droplet solution was allowed to evaporate for 6 h, then PLGA NPs precipitated (Figs. 1 and 2).

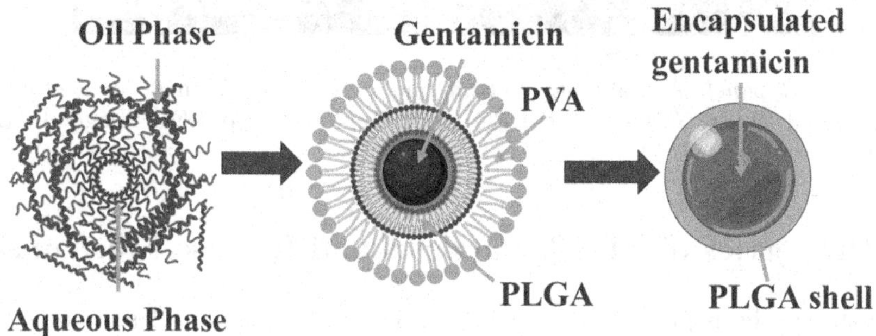

Fig. 1. An illustration of synthesis of PLGA micro-/nano-particles and encapsulation of drug gentamicin.

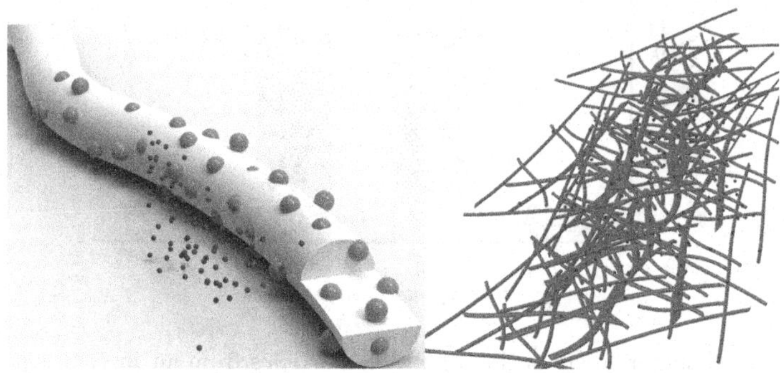

Fig. 2. An illustration of drug gentamicin being encapsulated in PLGA nanoparticles (PU-PEO nanofiber scaffolds).

Electron-spinning is an effective way to incorporate NPs into fibers. Solutions of 0.35 g PU dissolved in 10 ml DCM and 0.2 g PEO dissolved in 10 ml DCM are prepared first. Then PU and PEO solution were mixed with PLGA NP solutions for 10-min sonication to yield an emulsion solution. A scanning electron microscopy (SEM) image of gentamicin-encapsulated PLGA NPs is shown in Fig. 3. The core-shell structure and porous shells can be clearly observed.

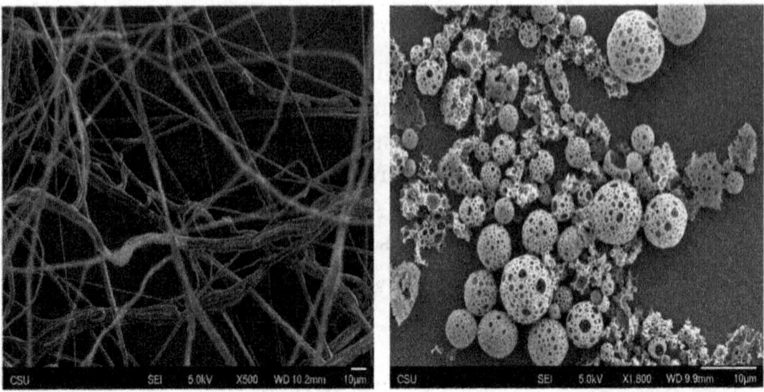

Fig. 3. *Left panel*: A scanning electron microscopy (SEM) image of gentamicin-encapsulated PLGA NPs with PU-PEO fiber; *Right panel*: An SEM image of PLGA NPs with porous shells.

3 Properties of PLGA Particles and Release Mechanisms

With the synthesis procedure discussed in Sect. 2, PLGA particles are mostly in the spherical shape, as shown in Fig. 3 right panel. Shown in Fig. 4 is a distribution of the diameters of PLGA particles from a particular experiment when the PLGA concentration was at 16.7 mg/ml.

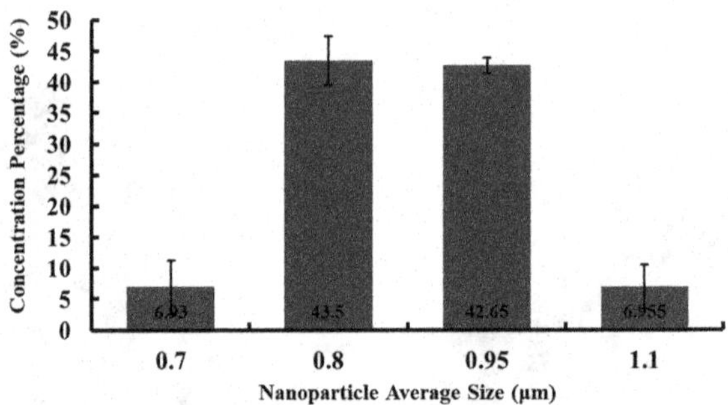

Fig. 4. Size (diameter) distribution of PLGA particles from an *in vitro* experiment.

A summary of PLGA particles average size, thickness, number of PLGA MPs/NPs and estimated loading gentamicin concentration is given in Table 1.

Major mechanisms of drug release from PLGA particles, as listed in Table 2 have been identified in the literature [16–18]. An illustration of the major mechanisms in the process of drug release from PLGA particles is also provided in Fig. 5.

Table 1. Properties of PLGA particles & drug gentamicin observed in five experiments

PLGA/DCM concentration (mg/ml)	Yield rate (%)	Average diameter (μm)	Thickness (μm)	PLGA NPs (#)	Gentamicin initial concentration (mg/ml)
13.3	68.34%	1.5	0.21	5.78×10^7	0.0768
15.0	58.89%	1.6	0.48	3.58×10^7	0.0621
16.7	96.33%	1.0	0.06	4.77×10^8	0.0177
18.3	32.12%	1.8	0.17	6.19×10^7	0.0297
20.0	38.16%	2.4	0.20	3.18×10^7	0.0224

Table 2. Summary of drug release mechanisms from PLGA particles

Mechanisms	Description
Diffusion thru polymeric shell	Early stage of release process, burst effect
Convection thru pores	Through the pores in the polymeric shell
Osmotic pumping	Due to concentration gradient
Degradation	Degradation of NPs

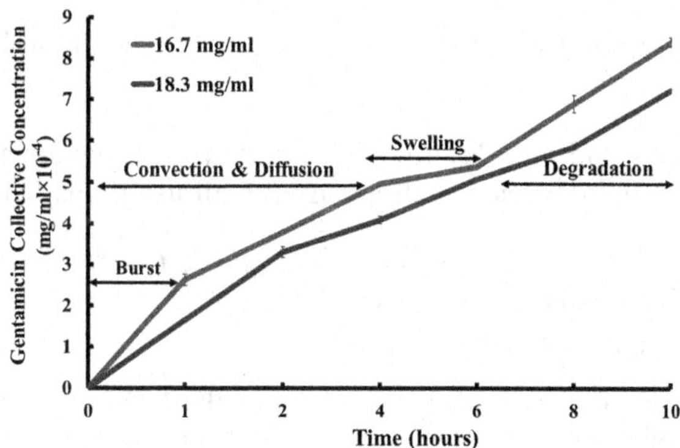

Fig. 5. An illustration of major mechanisms for drug release from PLGA particles.

4 Mathematical Modeling for Drug Release from PLGA Particles

4.1 A Model for Radial Diffusion in Polymeric Shells

We follow the approaches and major assumptions in [14]. Specifically,

(i) PLGA MPs/NPs are formed as reservoir devices for drug delivery;
(ii) The drug solution inside the core is a *constant activity resource*;
(iii) The PLGA MPs/NPs have a core-shell structure, in particular, the inner and outer radii are r_a, r_b;

(iv) The polymeric shell is porous and homogeneous, drug is diffused through this shell, the diffusivity is a positive constant D;
(v) For the early stage of the diffusion process, the drug concentration at the inner membrane (r_a) is maintained as a constant.

Shown in Fig. 6 is an illustration of the spherical core-shell structure of a typical PLGA particle.

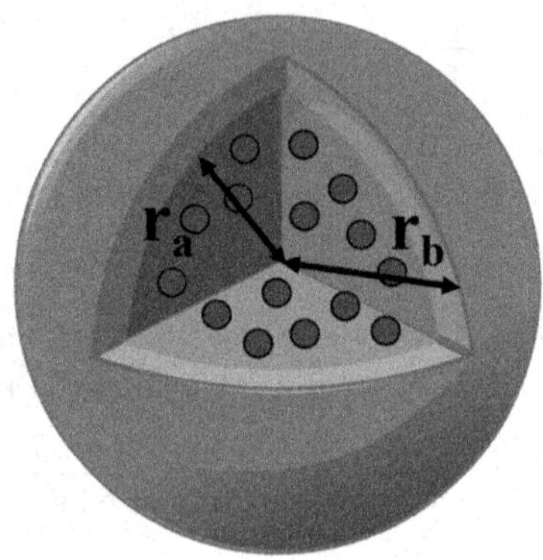

Fig. 6. An illustration of the spherical core-shell structure of typical PLGA particles.

Based on Fick's 2nd law [13,14], we write down a time-dependent diffusion equation in the radial direction with a constant diffusivity as below.

$$\begin{cases} \frac{\partial c}{\partial t} = \frac{D}{r^2}\frac{\partial}{\partial r}\left(r^2\frac{\partial c}{\partial r}\right), & r \in [r_a, r_b],\ t \in (0, T] \\ \text{Boundary conditions} \\ \text{Initial condition} \end{cases} \tag{1}$$

For the above simple model, the early work focused on finding series solutions. For a special case with a constant initial condition and a constant Dirichlet boundary condition, see [12] for a series solution. [14] summarized the classical work up to Year 2012 and provided a nice overview of various models of dosage forms and their series solutions.

As *in vitro* experiments for investigation of drug-release from PLGA particles are expensive and vary with lab conditions, the *in silico* approach with numerical simulations will offer meaningful alternative/supplementary data to experiment results. This relies on accurate and efficient numerical methods for solving the differential equation boundary initial value problems.

4.2 Other Existing Mathematical Models

There exist a good number of mathematical models for drug release from PLGA or more generally polymeric devices, depending on what mechanisms are being considered, e.g., diffusion, convection, swelling, and degradation. The popular ones are the 0th order, 1st order, Higuchi, and Korsmeyer-Peppas models, according to `ChatGPT`.

The Peppas equation shown below applies to non-swellable matrix systems in the early stage [11,12]:

$$\frac{M_t}{M_\infty} = At^\beta, \tag{2}$$

where M_t is the accumulative mass of the drug released up to time t, M_∞ is the total mass released up to time infinity, A, β are positive constants. In particular, $\beta = 0.50$ for slabs, $\beta = 0.45$ for cylindrical particles, $\beta = 0.43$ for spherical particles. It is also mentioned that, if $0.50/0.45/0.43 < \beta < 1.00$, the diffusion is anomalous, i.e., non-Fickian. This simple exponential relation was elaborated in [12] and further justified in [11].

For drug release from non-swellable polymeric devices (reservoir or matrix), a summary of mathematical models can be found in [14]. In particular, for a spherical particle of the reservoir type with a core-shell structure under the assumption of constant activity source, there holds

$$M_t = 4\pi DKc_s \frac{r_b r_a}{r_b - r_a} t, \tag{3}$$

where D is the drug diffusivity, K is a drug partition parameter, c_s is the solubility concentration, r_a, r_b are the inner and outer radii, respectively, and t is time. See [14, p. 355]. We will fit our experimental data with this model later.

5 Numerical Methods for Radial Diffusion

In this section, we develop efficient numerical schemes for solving the radial diffusion equation that models drug diffusion through the polymeric shell of sphere-shaped PLGA particles.

5.1 Node-Oriented Finite Volume Schemes for Spherical Diffusion

In this subsection, we develop numerical schemes for the spherical diffusion problem, which shares some features with the control volume method in [21]. We consider a constant radial diffusion equation with a source term

$$\frac{\partial u}{\partial t} = \frac{D}{r^2}\frac{\partial}{\partial r}\left(r^2 \frac{\partial u}{\partial r}\right) + f(r,t), \tag{4}$$

which can be rewritten as

$$r^2 \frac{\partial u}{\partial t} + \frac{\partial}{\partial r}\left(- D\frac{\partial u}{\partial r} r^2\right) = r^2 f(r,t). \tag{5}$$

Let $0 < r_a < r_b$ be the inner and outer radii of a spherical shell. Consider a uniform partition $r_a = r_1 < \cdots < r_n = r_b$ with $h = (r_b - r_a)/(n-1)$ being the mesh size. For node r_i, we consider the control volume $[r_i - \frac{h}{2}, r_i + \frac{h}{2}]$. This will be a half volume when $i = 1$ or $i = n$.

Integrating Eq. (5) LHS 1st term over the control volume, we obtain

$$\int_{r_{i-\frac{1}{2}}}^{r_{i+\frac{1}{2}}} \frac{\partial u}{\partial t} r^2 dr = \frac{\partial}{\partial t}\Big(\int_{r_{i-\frac{1}{2}}}^{r_i} u\, r^2 dr + \int_{r_i}^{r_{i+\frac{1}{2}}} u\, r^2 dr\Big). \tag{6}$$

Consider the integrand $w(r) = r^2\, u(r,t)$ in the above integrals. We approximate its derivatives in the half volumes $[r_{i-\frac{1}{2}}, r_i]$ and $[r_i, r_{i+\frac{1}{2}}]$, respectively, to obtain

$$w'(r_i) \approx \frac{r_i^2 u_i - r_{i-1}^2 u_{i-1}}{h}, \qquad w'(r_i) \approx \frac{r_{i+1}^2 u_{i+1} - r_i^2 u_i}{h}. \tag{7}$$

The Taylor expansion in these two half volumes takes the forms shown below.

$$\begin{aligned} w(r) = r^2\, u = r_i^2 u_i + (r - r_i)\tfrac{r_i^2 u_i - r_{i-1}^2 u_{i-1}}{h} + \mathcal{O}(h^2), \\ w(r) = r^2\, u = r_i^2 u_i + (r - r_i)\tfrac{r_{i+1}^2 u_{i+1} - r_i^2 u_i}{h} + \mathcal{O}(h^2). \end{aligned} \tag{8}$$

Plugging these back into Eq. (6) yields

$$\int_{r_{i-\frac{1}{2}}}^{r_{i+\frac{1}{2}}} \frac{\partial u}{\partial t} r^2 dr = h\frac{\partial}{\partial t}\Big(\frac{1}{8} r_{i-1}^2 u_{i-1} + \frac{6}{8} r_i^2 u_i + \frac{1}{8} r_{i+1}^2 u_{i+1} + \mathcal{O}(h^2)\Big). \tag{9}$$

Similar formulas can be established for the half volumes related to $i = 1, n$.

Now we check the diffusion term on the LHS of Eq. (5). The integral is treated by the Fundamental Theorem of Calculus and cancellation.

$$\begin{aligned} \textstyle\int_{r_{i-\frac{1}{2}}}^{r_{i+\frac{1}{2}}} D\frac{\partial}{\partial r}\Big(r^2\frac{\partial u}{\partial r}\Big)dr &= D\textstyle\int_{r_{i-\frac{1}{2}}}^{r_i} \frac{\partial}{\partial r}\Big(r^2\frac{\partial u}{\partial r}\Big)dr + D\textstyle\int_{r_i}^{r_{i+\frac{1}{2}}} \frac{\partial}{\partial r}\Big(r^2\frac{\partial u}{\partial r}\Big)dr \\ &= D\, r_{i+\frac{1}{2}}^2 \frac{\partial u}{\partial r}\Big|_{r_{i+\frac{1}{2}}} - D\, r_{i-\frac{1}{2}}^2 \frac{\partial u}{\partial r}\Big|_{r_{i-\frac{1}{2}}} \\ &= \tfrac{D}{h}\Big(r_{i-\frac{1}{2}}^2 u_{i-1} - \Big(r_{i-\frac{1}{2}}^2 + r_{i+\frac{1}{2}}^2\Big)u_i + r_{i+\frac{1}{2}}^2 u_{i+1}\Big) + \mathcal{O}(h^2). \end{aligned} \tag{10}$$

As for the two special cases $i = 1, n$, one has, respectively,

$$\begin{aligned} \textstyle\int_{r_1}^{r_{\frac{3}{2}}} D\frac{\partial}{\partial r}\Big(r^2\frac{\partial u}{\partial r}\Big) &\approx \tfrac{D}{h}\Big(-(r_{\frac{3}{2}}^2 - r_1^2)u_1 + (r_{\frac{3}{2}}^2 - r_1^2)u_2\Big) \\ \textstyle\int_{r_{n-\frac{1}{2}}}^{r_n} D\frac{\partial}{\partial r}\Big(r^2\frac{\partial u}{\partial r}\Big) &\approx \tfrac{D}{h}\Big(-(r_n^2 - r_{n-\frac{1}{2}}^2)u_{n-1} + (r_n^2 - r_{n-\frac{1}{2}}^2)u_n\Big). \end{aligned} \tag{11}$$

For the RHS of Eq. (5), we use the midpoint quadrature to obtain

$$\int_{r_{i-\frac{1}{2}}}^{r_{i+\frac{1}{2}}} f r^2 dr = h r_i^2 f_i + \mathcal{O}(h^3), \quad 2 \le i \le (n-1). \tag{12}$$

For the special cases $i = 1$ and $i = n$, we have

$$\int_{r_1}^{r_{\frac{3}{2}}} f r^2 dr \approx \frac{h}{2} r_1^2 f_1, \quad \int_{r_{n-\frac{1}{2}}}^{r_n} f r^2 dr \approx \frac{h}{2} r_n^2 f_n. \tag{13}$$

Let $\mathbf{v}(t) = \left[v_1(t), v_2(t), \cdots, v_n(t)\right]^T$ be the dim-n column vector consisting of the numerical solution nodal values. We obtain an ODE system

$$\mathbf{M}\,\frac{d\mathbf{v}}{dt} + \mathbf{S}\,\mathbf{v} = \mathbf{g}, \tag{14}$$

where $\mathbf{S}$ and $\mathbf{M}$ are the stiffness and mass matrices, respectively.

$$\mathbf{S} = \frac{D}{h}\begin{bmatrix} r_{\frac{3}{2}}^2 - r_1^2 & -(r_{\frac{3}{2}}^2 - r_1^2) & \cdots & \cdots & \cdots \\ \cdots & \cdots & \cdots & \cdots & \cdots \\ \cdots & -r_{i-\frac{1}{2}}^2 & r_{i-\frac{1}{2}}^2 + r_{i+\frac{1}{2}}^2 & -r_{i+\frac{1}{2}}^2 & \cdots \\ \cdots & \cdots & \cdots & \cdots & \cdots \\ \cdots & \cdots & \cdots & r_n^2 - r_{n-\frac{1}{2}}^2 & -(r_n^2 - r_{n-\frac{1}{2}}^2) \end{bmatrix}, \tag{15}$$

$$\mathbf{M} = \frac{h}{8}\begin{bmatrix} 3r_1^2 & r_2^2 & 0 & 0 & \cdots & 0 & 0 & 0 \\ r_1^2 & 6r_2^2 & r_3^2 & 0 & \cdots & 0 & 0 & 0 \\ 0 & r_2^2 & 6r_3^2 & r_4^2 & \cdots & 0 & 0 & 0 \\ \cdots & \cdots & \cdots & \cdots & \cdots & \cdots & \cdots & \cdots \\ 0 & 0 & 0 & 0 & \cdots & r_{n-2}^2 & 6r_{n-1}^2 & r_n^2 \\ 0 & 0 & 0 & 0 & \cdots & 0 & r_{n-1}^2 & 3r_n^2 \end{bmatrix}. \tag{16}$$

Note that vector $\mathbf{g}$ corresponds to the source term.

The linear ODE system in (14) can be solved using the implicit Euler or Crank-Nicolson methods. Let $\Delta t = T/m$ and $0 = t_0 < \cdots < t_{j-1} < t_j < \cdots < t_m = T$ be a uniform temporal partition. Let $\mathbf{v}^{(j)}$ be the fully discretized numerical solution at time t_j. Then the implicit Euler discretization yields

$$\mathbf{M}\frac{\mathbf{v}^{(j+1)} - \mathbf{v}^{(j)}}{\Delta t} + \mathbf{S}\,\mathbf{v}^{(j+1)} = \mathbf{g}^{(j+1)}, \tag{17}$$

and hence

$$(\mathbf{M} + \Delta t\,\mathbf{S})\mathbf{v}^{(j+1)} = \mathbf{M}\,\mathbf{v}^{(j)} + \Delta t\,\mathbf{g}, \quad j = 0, 1, \cdots, (m-1). \tag{18}$$

After modification by boundary conditions, the above linear system can be solved by a standard linear solver. This offers a well-defined time-marching scheme.

For better approximation accuracy of the numerical solution, we could also use the Crank-Nicolson discretization

$$\mathbf{M}\frac{\mathbf{v}^{(j+1)} - \mathbf{v}^{(j)}}{\Delta t} + \mathbf{S}\frac{1}{2}\left(\mathbf{v}^{(j+1)} + \mathbf{v}^{(j)}\right) = \frac{1}{2}\left(\mathbf{g}^{(j+1)} + \mathbf{g}^{(j)}\right), \tag{19}$$

which leads to another time-marching scheme,

$$\left(\mathbf{M} + \frac{\Delta t}{2}\mathbf{S}\right)\mathbf{v}^{(j+1)} = \left(\mathbf{M} - \frac{\Delta t}{2}\mathbf{S}\right)\mathbf{v}^{(j)} + \frac{\Delta t}{2}\left(\mathbf{g}^{(j+1)} + \mathbf{g}^{(j)}\right), \tag{20}$$

for $j = 0, 1, \cdots, (m-1)$.

5.2 Computation of Mass and Fluxes

For the above numerical schemes, we can compute (mass) fluxes at the two radial ends r_a and r_b, which correspond to the interior and exterior shells for the core-shell structure of spherical PLGA MPs/NPs.

At a typical time moment $t_j (j = 0, 1, \cdots, m)$, the numerical solution $v^{(j)}(r)$ is a continuous piecewise linear polynomial determined by its nodal values. The numerical total mass in the radial interval $[r_a, r_b]$ is, by direct calculations,

$$\begin{aligned} M_j &= \sum_{i=1}^{n-1} \int_{r_i}^{r_{i+1}} v^{(j)}(r) r^2 dr = \int_{r_i}^{r_{i+1}} \left(v_i^{(j)} \frac{r_{i+1}-r}{h} + v_{i+1}^{(j)} \frac{r-r_i}{h} \right) r^2 dr \\ &= \sum_{i=1}^{n-1} \left(v_i^{(j)} r_{i+1} - v_{i+1}^{(j)} r_i \right) \frac{1}{3h} (r_{i+1}^3 - r_i^3) \\ &+ \sum_{i=1}^{n-1} \left(v_{i+1}^{(j)} - v_i^{(j)} \right) \frac{1}{4h} (r_{i+1}^4 - r_i^4), \quad j = 0, 1, \cdots, m. \end{aligned} \tag{21}$$

The outflow fluxes at r_b and r_a, (for $j = 1, 2, \cdots, m$) are respectively,

$$\begin{aligned} F_j^b &= -\frac{D}{2} \left(\frac{v_{N_r}^{(j)} - v_{N_r-1}^{(j)}}{h} + \frac{v_{N_r}^{(j-1)} - v_{N_r-1}^{(j-1)}}{h} \right) r_b^2 \, \Delta t, \\ F_j^a &= \frac{D}{2} \left(\frac{v_2^{(j)} - v_1^{(j)}}{h} + \frac{v_2^{(j-1)} - v_1^{(j-1)}}{h} \right) r_a^2 \, \Delta t. \end{aligned} \tag{22}$$

Since there is no source in this case, mass conservation for the numerical solution takes the following form

$$M_j - M_{j-1} = F_j^a + F_j^b, \quad j = 1, \cdots, m. \tag{23}$$

5.3 Nondimensionalization

The numerical solvers discussed in the previous subsection could produce simulation results for release from single PLGA particle once the model parameters, e.g., diffusivity D, the inner and outer radii r_a, r_b, are provided. For units of parameters or variables, we follow [4]. See Table 3.

Table 3. Units of parameters/variables for spherical diffusion

Variables	Units	Meaning
D	$\mathrm{cm}^2\,\mathrm{s}^{-1}$	Diffusion coefficient
c	mg/ml	Concentration
r_a, r_b, r	cm	Radial positions
t	s	Time

For better performance of numerical simulations, we introduce new dimensionless parameters and variables $\hat{D}, \hat{r}, \hat{t}, \hat{c}$ so that

$$\widehat{D} = \frac{D}{D_0}, \qquad \widehat{t} = \frac{t}{T_0}, \qquad \widehat{r} = \frac{r}{r_b}, \qquad \widehat{c} = \frac{c}{C_0}, \tag{24}$$

where D_0, T_0, C_0 are chosen according to experiment data. Let $H_0 = (r_b - r_a)/r_b$ be the thickness of the polymeric shell. Then the nondimensionalized diffusion problem takes the form

$$\begin{cases} \frac{\partial \widehat{c}}{\partial \widehat{t}} = \frac{T_0 D_0}{r_b^2} \frac{\widehat{D}}{\widehat{r}^2} \frac{\partial}{\partial \widehat{r}} \left(\widehat{r}^2 \frac{\partial \widehat{c}}{\partial \widehat{r}} \right), \quad \widehat{r} \in [1 - H_0, 1], \ \widehat{t} \in (0, 1], \\ \text{Boundary conditions,} \\ \text{Initial condition.} \end{cases} \tag{25}$$

The source term does not exist for drug release from PLGA particles.

6 Experimental and Numerical Results

Note that *in vitro* experiment data vary. For our experiments gentamicin encapsulated in PLGA particles, three sets of radii are recorded as follows.

- Case 1: $r_a = 0.7\,\mu\text{m} = 7.0 * 10^{-5}\,\text{cm}$, $r_b = 0.8\,\mu\text{m} = 8.0 * 10^{-5}\,\text{cm}$;
- Case 2: $r_a = 0.8\,\mu\text{m} = 8.0 * 10^{-5}\,\text{cm}$, $r_b = 1.0\,\mu\text{m} = 1.0 * 10^{-5}\,\text{cm}$;
- Case 3: $r_a = 1.8\,\mu\text{m} = 1.8 * 10^{-4}\,\text{cm}$, $r_b = 2.0\,\mu\text{m} = 2.0 * 10^{-4}\,\text{cm}$.

This indicates the relative thickness $H_0 = (r_b - r_a)/r_b$ is in the range 0.10–0.20. The observation time T_0 could be chosen as 1 hr–2 hrs, namely, 3600–7200 s, to focus on the early (burst) stage.

The diffusivity parameter D is difficult to obtain. An empirical formula was proposed in [13] based on data fitting and knowledge from series solutions of similar problems. For the *in silico* approach, one may consider a baseline value for D_0 and vary it up or down for several magnitudes in numerical simulations.

For our five *in vitro* experiments with gentamicin encapsulated in PLGA particles, we record the accumulative percentage of gentamicin released versus time. As shown in Table 4, about 20% of total gentamicin was released after about 2 h; about 60% of total gentamicin was released after about 10 h.

Table 4. Accumulative percentage of gentamicin release in five experiments

PLGA/DCM (mg/ml) concentration	13.30	15.0	16.7	18.3	20.0
Time (h)	Accumulative % of gentamicin released				
1	11.54%	12.86%	15.77%	9.92%	9.41%
2	21.71%	22.29%	22.73%	19.82%	17.92%
4	26.03%	28.15%	29.81%	24.56%	23.51%
6	31.04%	31.68%	32.25%	30.47%	30.88%
8	36.11%	37.99%	41.43%	35.09%	33.58%
10	44.34%	46.11%	50.46%	43.37%	42.97%

The raw data suggests that we should focus on the 1st hour for a potential stage of burst. Following the discussion in [14, Fig. 4], we fit the model "reservoir devices with constant activity source" (Eq. (7) therein) using our experimental data for the 1st 60 min with 2-min sampling. Linear regression yields (Fig. 7)

$$M_t = 4.2 * 10^{-3}\, t. \tag{26}$$

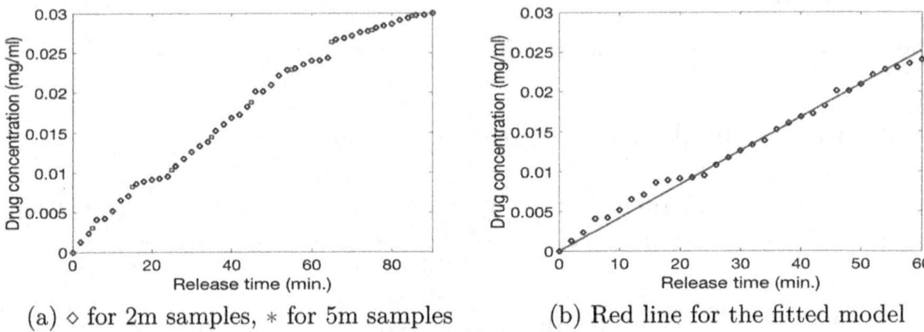

(a) $\diamond$ for 2m samples, $*$ for 5m samples (b) Red line for the fitted model

Fig. 7. Experimental data fitting for gentamicin release from PLGA particles. (a) Raw data showing the accumulative concentration of released drug during the 1st 90 min; (b) The red line represents the fitting model $M_t = 4.2{*}10^{-3}\, t$ for $t \in [0, 60]$ min. (Color figure online)

7 Concluding Remarks

This paper investigates mathematical modeling and numerical simulations for drug release from PLGA particles as our exploratory efforts in this interdisciplinary approach. Based on examination of data obtained from *in vitro* experiments, our focus is placed on modeling the early stage of release with assumptions of *no swelling of the porous shell*, *homogeneity of the shell*, and *the perfect sink condition*, a radial direction diffusion equation is adopted. Finite volume schemes have been developed for this model. `Matlab` code modules along with *in silico* results will be incorporated in our code package `DarcyLite` [8].

Due to the pores on the polymeric shells of PLGA MPs/NPs, a time-dependent convection-diffusion equation would be more suitable for modeling the drug release process. This is currently under our investigation and will be reported in our future work.

As drug release proceeds, the porous shell is subject to swelling and eventually degrades. Mathematical modeling and numerical simulations for degradation are more challenging. These are under our investigation also.

Simulations of drug release from PLGA MPs/NPs using Molecular Dynamics (MD) seem too expensive and hence will not be our pursuit. Our goal is to develop efficient simulation code modules that can run on laptop or low-configuration desktop computers. These code modules will be based on the mathematical equations that characterize transport through porous media.

References

1. Busatto, C., Pesoa, J., Helbling, I., Luna, J., Estenoz, D.: Effect of particle size, polydispersity and polymer degradation on progesterone release from PLGA microparticles: Experimental and mathematical modeling. Int. J. Pharm. **536**, 360–369 (2018)
2. Delbreil, P., Rabanel, J.M., Banquy, X., Brambilla, D.: Therapeutic nanotechnologies for Alzheimer's disease: a critical analysis of recent trends and findings. Adv. Drug Deliv. Rev. **187**, 114397 (2022)
3. Ghitman, J., Biru, E.I., Stan, R., Iovu, H.: Review of hybrid PLGA nanoparticles: future of smart drug delivery and theranostics medicine. Mater. Dsgn. **193**, 108805 (2020)
4. Jain, A., McGinty, S., Pontrelli, G., Zhou, L.: Theoretical model for diffusion-reaction based drug delivery from a multilayer spherical capsule. Int. J. Heat Mass Trans. **183**, 122072 (2022)
5. Jusu, S.M., et al.: Drug-encapsulated blend of PLGA-PEG microspheres: in vitro and in vivo study of the effects of localized/targeted drug delivery on the treatment of triple-negative breast cancer. Sci. Rep. **10**, 14188 (2020)
6. Korsmqer, R.W., Gumy, R., Doelker, E., Buri, P., Peppas, N.A.: Mechanisms of solute release from porous hydrophilic polymers. Int. J. Pharm. **15**, 25–35 (1983)
7. Lagreca, E., Onesto, V., Di Natale, C., La Manna, S., Netti, P.A., Vecchione, R.: Recent advances in the formulation of PLGA microparticles for controlled drug delivery. Prog. Biomater. **9**(4), 153–174 (2020). https://doi.org/10.1007/s40204-020-00139-y
8. Liu, J., Sadre-Marandi, F., Wang, Z.: DarcyLite: a Matlab toolbox for Darcy flow computation. Proc. Comput. Sci. **80**, 1301–1312 (2016)
9. Naghipoor, J., Rabczuk, T.: A mechanistic model for drug release from PLGA-based drug eluting stent: a computational study. Comput. Bio. Med. **90**, 15–22 (2017)
10. Operti, M.C., Bernhardt, A., Grimm, S., Engel, A., Figdor, C.G., Tagit, O.: PLGA-based nanomedicines manufacturing: technologies overview and challenges in industrial scale-up. Int. J. Pharm. **605**, 120807 (2021)
11. Peppas, N.A., Narasimhan, B.: Mathematical models in drug delivery: how modeling has shaped the way we design new drug delivery systems. J. Con. Rel. **190**, 75–81 (2014)
12. Ritger, P.L., Peppas, N.A.: A simple equation for description of solute release I. Fickian and non-Fickian release from non-swellable devices in the form of slabs, spheres, cylinders or discs. J. Con. Rel. **5**, 23–36 (1987)
13. Siepmann, J., Elkharraz, K., Siepmann, F., Klose, D.: How autocatalysis accelerates drug release from PLGA-based microparticles: a quantitative treatment. Biomacromol **6**, 2312–2319 (2005)
14. Siepmann, J., Siepmann, F.: Modeling of diffusion controlled drug delivery. J. Con. Rel. **161**, 351–362 (2012)
15. Su, Y., et al.: PLGA-based biodegradable microspheres in drug delivery: recent advances in research and application. Drug Deliv. **28**, 1397–1418 (2021)
16. Sun, Yu., Bhattacharjee, A., Reynolds, M., Li, Y.V.: Synthesis and characterizations of gentamicin-loaded poly-lactic-co-glycolic (PLGA) nanoparticles. J. Nanopart. Res. **23**(8), 1–15 (2021). https://doi.org/10.1007/s11051-021-05293-3
17. Tamani, F., Bassand, C., Hamoudi, M.C., Siepmann, F., Siepmann, J.: Mechanistic explanation of the (up to) 3 release phases of PLGA microparticles: Monolithic dispersions studied at lower temperatures. Int. J. Pharm. **596**, 120220 (2021)

18. Tamani, F., et al.: Mechanistic explanation of the (up to) 3 release phases of PLGA microparticles: Diprophylline dispersions. Int. J. Pharm. **572**, 118819 (2019)
19. Tamani, F.: Towards a better understanding of the drug release mechanisms in PLGA microparticles. Ph.D. thesis, Université de Lille, France (2019)
20. Wu, X., Zhou, Y.: Finite element analysis of diffusional drug release from complex matrix systems. II. Factors influencing release kinetics. J. Con. Rel. **51**, 57–71 (1998)
21. Zeng, Y., et al.: Efficient conservative numerical schemes for 1d nonlinear spherical diffusion equations with applications in battery modling. J. Electrochem. Soc. **160**(9), A1565–A1571 (2013)
22. Zhou, J., Zhai, Y., Xu, J., Zhou, T., Cen, L.: Microfluidic preparation of PLGA composite microspheres with mesoporous silica nanoparticles for finely manipulated drug release. Int. J. Pharm. **593**, 120173 (2021)
23. Zhou, Y., Wu, X.: Finite element analysis of diffusional drug release from complex matrix systems. I. Complex geometries and composite structures. J. Con. Rel. **49**, 277–288 (1997)
24. Zhu, X., Braatz, R.D.: Modeling and analysis of drug-eluting stents with biodegradable PLGA coating: Consequences on intravascular drug delivery. J. Biomech. Eng. **136**, 111004 (2014)

Molecular Dynamics Study of Hydrogen Dissolution and Diffusion in Different Nonmetallic Pipe Materials

Dukui Zheng[1], Jingfa Li[2(✉)], Bo Yu[2], Zhiqiang Huang[1], Yindi Zhang[1], Yafan Yang[3], Dongxu Han[2], and Jianli Li[2]

[1] College of Petroleum Engineering, Yangtze University, Wuhan 430100, China
[2] School of Mechanical Engineering and Hydrogen Energy Research Center, Beijing Institute of Petrochemical Technology, Beijing 102617, China
lijingfa@bipt.edu.cn
[3] State Key Laboratory for Geomechanics and Deep Underground Engineering, China University of Mining and Technology, Xuzhou 221116, China

Abstract. The nonmetallic pipes can effectively avoid the hydrogen embrittlement of metal pipes when transporting hydrogen. However, due to the characters of the nonmetal materials, there will be a large degree of gas permeation when conveying hydrogen by nonmetallic pipes. To select suitable nonmetal pipe materials, the solubility, diffusivity, and permeability of hydrogen in PE, PVC and PVDF amorphous polymers are investigated and compared by molecular dynamics simulations at 270–310 K and 0.1–0.7 MPa, providing guidance for the construction of nonmetallic hydrogen transportation pipes. Simulation results indicate that the solubility coefficients of hydrogen in PE and PVDF rise with the increasing temperature, but show an opposite trend in PVC. Both the diffusion and permeability coefficients increase with the rise of temperature. In a small range of pressure variation, the influence of pressure on diffusion and permeation characteristics is ignorable. Among the three studied amorphous polymers, the permeability coefficient of hydrogen in PE is the largest and that in PVDF is the smallest. In addition, the diffusion of hydrogen molecules in the polymer conforms to the hopping mechanism.

Keywords: Hydrogen · Nonmetallic materials · Dissolution · Diffusion · Permeation · Molecular dynamics

1 Introduction

It is an important measure to reduce carbon emissions to substitute hydrogen for partial natural gas for combustion and utilization. However, the safety problems such as hydrogen embrittlement of metal pipes may occur when transporting hydrogen [1]. To ensure the safety of hydrogen pipeline transportation, nonmetallic pipes such as PE are usually used in engineering practice. However, due to the difference of material properties, nonmetallic pipes demonstrate a greater degree of gas permeation than metal

J. Mikyška et al. (Eds.): ICCS 2023, LNCS 14077, pp. 361–368, 2023.
https://doi.org/10.1007/978-3-031-36030-5_29

pipes, resulting in energy waste and safety issues. Therefore, it is of great significance to select suitable nonmetallic pipe materials to reduce the permeation of hydrogen. In this study, the solubility, diffusivity, and permeability of hydrogen in nonmetallic pipe materials (PE, PVC and PVDF) are analyzed and compared under the working temperature (270–310 K) and pressure (0.1–0.7 MPa) of urban nonmetallic pipes by Giant Canonical Monte Carlo (GCMC) and Molecular dynamics (MD) methods, to provide reference for the construction of hydrogen transportation pipelines and material selection of urban nonmetallic pipes.

2 Molecular Dynamics Model

First, the PE, PVC and PVDF single chains with a polymerization degree of 100 as well as the H_2 molecule are constructed by the all-atomic model. Then, 1000 steps of geometric optimization are performed. Finally, 10 three-dimensional periodic amorphous cells of PE, PVC and PVDF consisting of five polymer chains are constructed, respectively. 30 amorphous cells are optimized with 10000 steps, and the three amorphous PE, PVC and PVDF cells with the lowest energy are selected for further study. After the geometric optimization, further geometric optimization and energy reduction of these nine selected amorphous cells need to be carried out through the following dynamic operations. (1) The NVT ensemble is used for 500 ps relaxation at 300 K. (2) Annealing: the amorphous cells are heated from 300 K to 600 K with a step of 60 K, and then cooled from 600 K to 300 K with the same temperature step. (3) The NPT ensemble is used to relax the amorphous cells for 500 ps at 300 K and 1 GPa. (4) The NPT ensemble is used to balance for 1000 ps at 298 K and 0.1 MPa. The Polymer Consistent Force Field (PCFF) is adopted in MD simulations and the cutoff distance is set as 12.5 Å.

3 Molecular Dynamics Simulation

3.1 Dissolution Simulation

Based on the GCMC principle, the Metropolis method is used to insert, delete, and translate H_2 molecules in amorphous cells, and the probabilities of occurrence are assumed to be 40%, 40% and 20%, respectively. The adsorption isotherms of H_2 in PE, PVC and PVDF at 270–310 K can be obtained and fitted by the Henry model, as shown in Eq. (1). The solubility coefficient can be calculated by taking the limit slope when the fugacity f tends to 0 for the adsorption isotherms, as shown in Eq. (2). At least five independent dissolution simulations are conducted for three amorphous cells of each material to reduce the errors. Finally, the solubility coefficients of H_2 in amorphous PE, PVC and PVDF are taken as the average values of 15 dissolution simulations results.

$$C = K_H \cdot f \tag{1}$$

$$S = \lim_{f \to 0} \frac{C}{f} = K_H \tag{2}$$

Here, C is the dissolved concentration of H_2 molecules in polymer, cm^3(STP)/cm^3; K_H is the Henry constant; f is the fugacity, MPa; S is the solubility coefficient, which represents thermodynamic characteristics between H_2 molecules and polymers, cm^3(STP)/(cm^3·MPa).

3.2 Diffusion Simulation

Firstly, the H_2 molecules are adsorbed by amorphous cells at different temperatures and fugacity based on GCMC, in which the fugacity is calculated by the Peng-Robinson equation of state (EoS). Then, the amorphous cells containing H_2 molecules are balanced by 2000 ps NPT. Finally, the 10 ns NVT dynamic operation is carried out on the amorphous PE, PVC and PVDF cells that adsorbed H_2 molecules to calculate the diffusion coefficients of H_2. Through above operations, the Mean Square Displacement (MSD)-time (t) figures of H_2 molecules in amorphous PE, PVC and PVDF cells at different temperatures (270–310 K) and pressures (0.1–0.7 MPa) can be obtained. The diffusion coefficients then can be calculated by analyzing the MSD-t figures. At first, the part with the slope of 1 in the log(MSD)-log(t) figure is found, which represents the normal diffusion of H_2 molecules. Then, in the MSD-t figure, the slope a is calculated by linear fitting of the first function for this part, and finally the diffusion coefficient can be calculated by Eq. (3) [2]. In each working condition, at least 10 independent NVT simulations are conducted for three amorphous cells of each material to eliminate the influence of error. Finally, the diffusion coefficients of H_2 in amorphous PE, PVC and PVDF cells are taken as the average values of 30 NVT simulation results.

$$D = \frac{a}{6} \tag{3}$$

Here, D is the diffusion coefficient, cm^2/s; a is the slope of normal diffusion part in the MSD-t figure.

3.3 Permeation Simulation

The "dissolution-diffusion" theory can be used to describe the permeation process of H_2 molecules in polymers, as illustrated in Fig. 1. Firstly, the H_2 molecules are dissolved in the polymer. Subsequently, H_2 molecules diffuse through the polymer, and finally H_2 molecules are desorbed to escape from the polymer. Therefore, the permeation of H_2 in PE, PVC and PVDF can be divided into the stage of dissolution and diffusion, and their relationship can be described by Eq. (4) [3].

$$P = SD \tag{4}$$

Here, P is the permeability coefficient, cm^3(STP)·cm/(cm^2·s·MPa); S is the solubility coefficient, cm^3(STP)/(cm^3·MPa); D is the diffusion coefficient, cm^2/s.

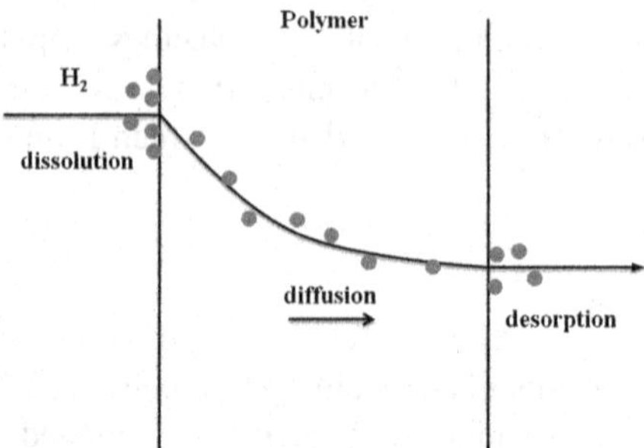

Fig. 1. Permeation process of H_2 in polymers.

4 Results Analysis and Discussion

4.1 Influences of Temperature on the Solubility Coefficient

The adsorption isotherms of H_2 in amorphous PE, PVC and PVDF cells at 270–310 K are displayed in Fig. 2. The solubility coefficients of H_2 can be obtained by fitting and calculating the data in Fig. 2 using Eqs. (1) and (2), and the results are shown in Fig. 3. It indicates that the solubility coefficients of H_2 in PE and PVDF increase with the rise of temperature, while the solubility coefficient in PVC shows an opposite trend. The reason can be attributed to that the material properties are different. At 270–310 K, the PE and PVDF are in the rubbery state, while PVC is at the glassy state. For PE and PVDF, the pores in the cell become larger with the increasing temperature, and the H_2 molecules can be adsorbed easier. For PVC, the free volume for H_2 dissolution is in the "frozen" state, so the size of pores basically does not change with the increase of temperature. Meanwhile, the density of H_2 decreases with the rise of temperature, which weakens the adsorption and results in the phenomenon that the solubility coefficient decreases with the rising temperature. Figure 3 also shows that the solubility coefficient of H_2 in PE is the largest. Additionally, the solubility coefficient of H_2 in PVC is larger than that in PVDF when the temperature is lower than 300 K.

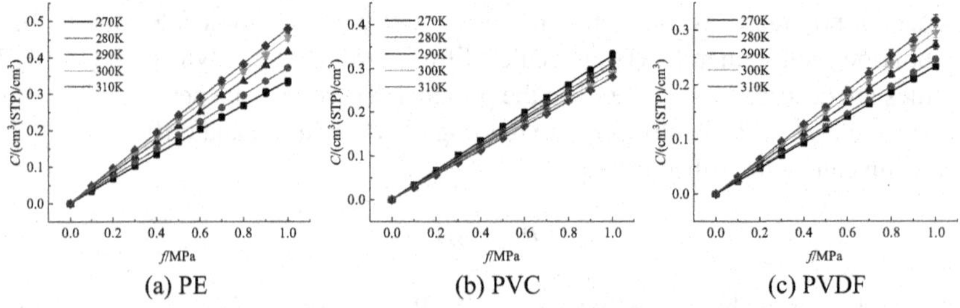

Fig. 2. Adsorption isotherms of H_2 in amorphous PE, PVC and PVDF cells at 270–310 K.

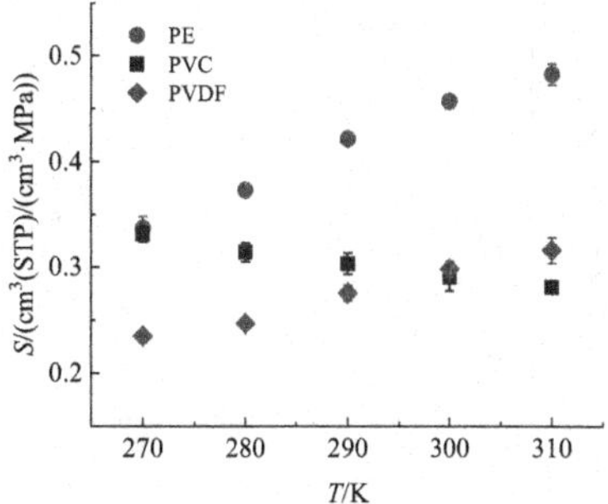

Fig. 3. Solubility coefficients of H_2 in amorphous PE, PVC and PVDF cells at 270–310 K.

4.2 Influences of Temperature on the Diffusion Coefficient

Through the 10 ns NVT simulation and the processing of MSD-*t* figures, the diffusion coefficients of H_2 in amorphous PE, PVC and PVDF cells at 270–310 K can be obtained, as shown in Fig. 4. It exhibits that the diffusion coefficients of H_2 increase with the rise of temperature, this is because the kinetic energy of H_2 molecules and the kinetic ability of polymer chains increase with the rising temperature, which makes the H_2 molecules diffuse easier. Furthermore, the diffusion coefficient of H_2 in amorphous PE cell is the largest and that in amorphous PVDF cell is the smallest at 270–310 K.

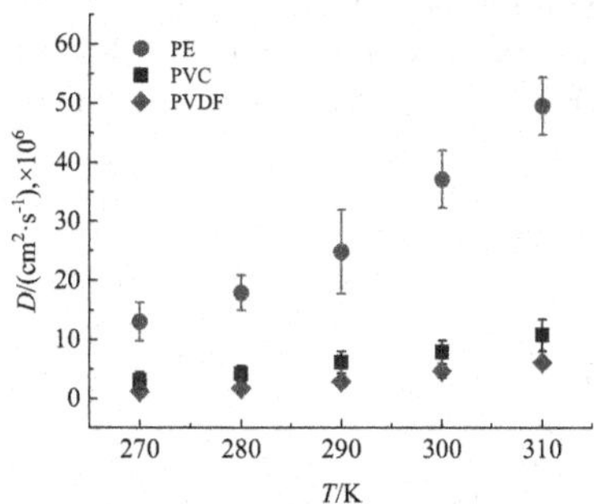

Fig. 4. Diffusion coefficients of H_2 in amorphous PE, PVC and PVDF cells at 270–310 K and 0.1 MPa.

4.3 Influences of Pressure on the Diffusion Coefficient

In addition to studying the effect of temperature on the diffusion coefficient, the effect of pressure (0.1–0.7 MPa) on the diffusion coefficient is also analyzed, as shown in Fig. 5. It demonstrates that the pressure change has a tiny effect on the diffusion coefficient, because the change of studied pressure is small in this work, resulting in that the change of diffusion coefficient is smaller than the error fluctuation. Therefore, it is difficult to observe obvious changes of diffusion coefficient when the range of pressure change is tiny.

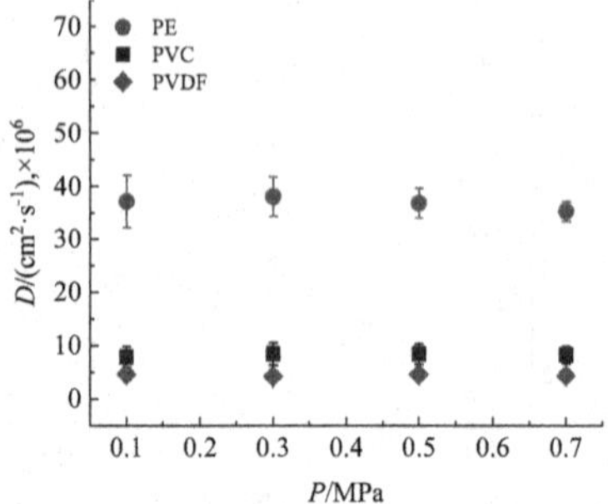

Fig. 5. Diffusion coefficients of H_2 in amorphous PE, PVC and PVDF cells at 0.1–0.7 MPa and 300 K.

4.4 Influences of Temperature and Pressure on the Permeability Coefficient

The data in Fig. 3 and 4 are calculated by Eq. (4) to investigate the influence of temperature (270–310 K) on permeability coefficients of H_2 in amorphous PE, PVC and PVDF cells, as presented in Fig. 6. It displays that the permeability coefficients increase with the rise of temperature. For PE and PVDF, this is because their solubility and diffusion coefficients of H_2 increase with the rising temperature. For PVC, the solubility coefficients decrease while the diffusion coefficients increase with the rise of temperature, indicating that the increase extent of the diffusion coefficient is larger than the decrease extent of the solubility coefficient with the increasing temperature, finally resulting in that the permeability coefficient increases with the rise of temperature. Moreover, the order of permeability coefficient of H_2 from large to small is PE, PVC and PVDF, respectively.

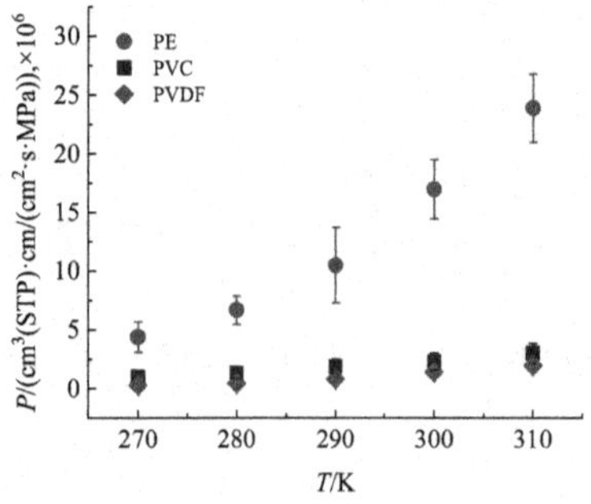

Fig. 6. Permeability coefficients of H_2 in amorphous PE, PVC and PVDF cells at 270–310 K and 0.1 MPa.

4.5 Diffusion Mechanism

Figure 7 exhibits the diffusion trajectories of H_2 molecules in amorphous PE, PVC and PVDF cells at 300 K and 0.1 MPa. It can be clearly seen that the diffusion range of H_2 in PE is the largest, and that in PVDF is the smallest, which is consistent with the results shown in Fig. 4. That is, the diffusion coefficient of H_2 in amorphous PE is the largest, and that in amorphous PVDF is the smallest.

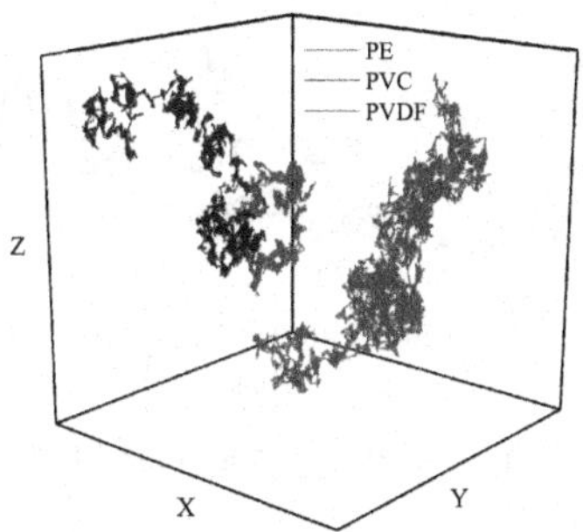

Fig. 7. Diffusion trajectories of H_2 molecules in amorphous PE, PVC and PVDF cells at 300 K and 0.1 MPa.

The diffusion distances of H_2 in amorphous PE, PVC and PVDF cells are obtained by calculating the data in Fig. 7 by using Eq. (5), as shown in Fig. 8. It indicates that the diffusion distance of H_2 molecules is the largest in PE, the second in PVC, and the smallest in PVDF, which is consistent with the diffusion coefficients in Fig. 4 and the diffusion ranges in Fig. 7. The diffusion distances of H_2 in these three materials increase with the rise of temperature because the temperature increases the kinetic energy and motion ability of H_2 molecules, and then H_2 molecules diffuse easier. In addition, it is also found in Fig. 8 that H_2 molecules vibrate in the polymer pores for a long time during the diffusion process, as shown in the black circle, and then quickly hop to other polymer pores in a short time, as marked in the purple circle, indicating the law of "long time vibration + short time hopping" as a whole.

$$D_d = \sqrt{(x_t - x_0)^2 + (y_t - y_0)^2 + (z_t - z_0)^2} \quad (5)$$

Here, D_d is the diffusion distance of H_2 molecule, Å; x_t, y_t, z_t are the x, y, z coordinates of H_2 molecule at any time t; x_0, y_0, z_0 are the x, y, z coordinates of H_2 molecule at the initial time.

5 Conclusions

The dissolution, diffusion and permeation characteristics of H_2 in amorphous PE, PVC and PVDF materials at 270–310 K and 0.1–0.7 MPa are analyzed and compared by Grand Canonical Monte Carlo and molecular dynamics simulations in this study. The following main conclusions are summarized.

(1) The solubility coefficient of H_2 decreases with the increasing temperature in PVC, but shows an opposite trend in PE and PVDF. The diffusivity and permeation coefficients of H_2 increase with the rise of temperature. The pressure has slight influence on the permeability coefficients of H_2 in PE, PVC and PVDF when the range of pressure variation is small.

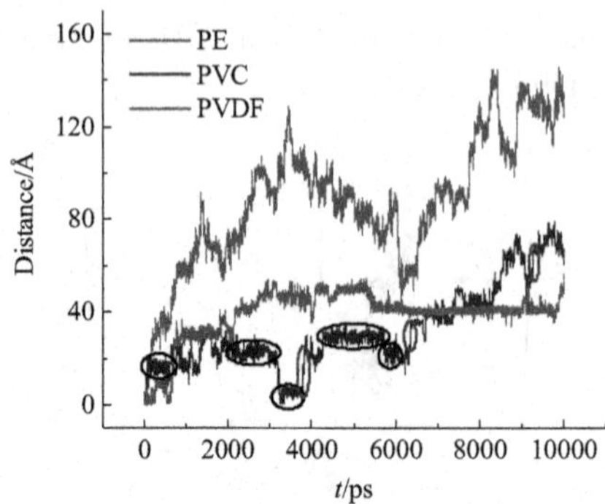

Fig. 8. Diffusion distances of H_2 molecules in amorphous PE, PVC and PVDF cells at 300 K and 0.1 MPa.

(2) Among the amorphous PE, PVC and PVDF, the permeability coefficients of H_2 in PE are the largest and that in PVDF are the smallest. The diffusion of H_2 molecules in amorphous PE, PVC and PVDF accords with the "hopping" mechanism. The H_2 molecules vibrate in the pores for a long time and then hop to adjacent pores to complete the diffusion for a short time.

Acknowledgements. This study is supported by the National Key R&D Program of China (No. 2021YFB4001601), the State Key Laboratory of Engines, Tianjin University (No. K2022-02), and the Hubei Provincial Department of Science and Technology Project: Natural Gas Blending Hydrogen Transmission Technology (No. 2022EJD031).

References

1. Li, J.F., Su, Y., Zhang, H.: Research progresses on pipeline transportation of hydrogen-blended natural gas. Nat. Gas. Ind. **41**(04), 137–152 (2021)
2. Yang, Y.F., Nair, A.K.N., Sun, S.Y.: Adsorption and diffusion of methane and carbon dioxide in amorphous regions of cross-linked polyethylene: a molecular simulation study. Ind. Eng. Chem. Res. **58**, 8426–8436 (2019)
3. Zheng, D.K., Li, J.F., Liu, B.: Molecular dynamics investigations into the hydrogen permeation mechanism of polyethylene pipeline material. J. Mol. Liq. **368**, 120773 (2022)

Component-wise and Unconditionally Energy-Stable VT Flash Calculation

Xiaoyu Feng[1], Meng-Huo Chen[2], Yuanqing Wu[3], Shuyu Sun[1(✉)], and Tao Zhang[1]

[1] Computational Transport Phenomena Laboratory (CTPL), Division of Physical Sciences and Engineering (PSE), King Abdullah University of Science and Technology (KAUST), Thuwal 23955 -6900, Saudi Arabia
shuyu.sun@kaust.edu.com

[2] Department of Mathematics, National Chung Cheng University, Chiayi 62102, Taiwan

[3] College of Mathematics and Statistics, Shenzhen University, Shenzhen 518060, Guangdong, China

Abstract. Flash calculations of the hydrocarbon mixture are essential for determining how the mixture phase behaves, which will ultimately affect subsurface flow and transport. In this paper, a novel numerical scheme is proposed for calculating the two-phase equilibrium of Peng-Robinson (PR) fluid at constant volume, temperature, and moles, namely the volume-temperature (VT) flash framework based on the dynamic model. Since the dynamic model is based on the energy dissipation law and the Onsager's reciprocal principle, we proposed a linear energy-stable scheme with the help of the convex-concave splitting technique, the energy factorization approach, and the component-wise iteration framework. The scheme eventually results in a fully explicit algorithm, and it avoids the challenges of solving non-linear systems and other difficulties in the traditional flash calculation methods. This scheme inherits the original energy stability and significantly reduces the implementation burden. It also achieves convergence unconditionally, even with a huge time step. Numerical experiments are carried out to illustrate its accuracy.

Keywords: Energy-stable scheme · Component-wise framework · Peng-Robinson fluid · Dynamic modeling · Volume-temperature (VT) flash

1 Introduction

A number of interrelated physical processes, including the multi-phases, the multi-components, and the phase equilibrium, widely exist in the subsurface reservoir, making it a complicated system [1,2,10]. Under different natural conditions and working circumstances, the characteristics of the fluid mixtures have a significant effect on the flow and compositions transfer. The flash calculation based on the equation of state (EoS) is the foundation for measuring the properties of the mixture. The Peng-Robinson (PR) EoS is widely acknowledged as the most trustworthy and precise one to describe the characteristics of hydrocarbons.

J. Mikyška et al. (Eds.): ICCS 2023, LNCS 14077, pp. 369–383, 2023.
https://doi.org/10.1007/978-3-031-36030-5_30

More people and organizations have recently turned their attention to unconventional energy sources like shale gas or methane hydrate due to the exhaustion of traditional energy sources and the advancement of petroleum industry techniques [8–10]. In the tight formations, the capillary effect is more important for flash calculations and modeling than it is in a traditional reservoir. Also, modeling unconventional reservoirs does not want the main variables, such as pressure, to be as sensitive as they are in the PT flash framework. When it comes to the phase equilibrium problems for associating fluids, like H_2O, the VT flash process, which is based on Helmholtz free energy, is more preferred because the association contribution resulting from hydrogen bonding interactions was initially defined in terms of this kind of energy.

Even though the VT flash calculation and the PT flash calculation have different structures, they both follow the principle of energy minimization. The Helmholtz free energy is minimized in the VT flash calculation, while the Gibbs free energy is minimized in the PT flash calculation [6]. Both types of flash calculations generate a set of nonlinear equations and constraints based on mass conservation and phase equilibrium relations. Previous research on the PT or VT flash calculation typically involved introducing intermediate variables, such as the fugacity coefficient or the volume function coefficient. However, the dynamic model circumvents these intermediate variables by transforming the nonlinear system into a dynamic system, where the energy is minimized through an evolutionary process over time. Nevertheless, the dynamic model is still unable to avoid the high degree of nonlinearity.

Our study is based on the described dynamic model. The progress we make is to develop an unconditionally energy-stable method that employs a component-wise framework and results in a fully explicit algorithm. This method not only guarantees numerical convergence and efficiency but also makes implementation more convenient. To design an energy-stable linear scheme, we have utilized several techniques, including the convex-concave splitting approach [2], modified energy, and energy factorization method [5]. As a result, the molar fraction of each component and the phase volume can be easily calculated using simple arithmetic formulas, one by one.

The subsequent sections of this paper are organized in the following manner: Sect. 2 details the physical problem and defines the Helmholtz free energy. It also presents a physical and mathematical description of phase equilibrium and the dynamic model. Section 3 introduces a component-wise and unconditionally energy-stable method, along with some essential elements of its proof for energy stability. In Sect. 4, the stability and accuracy of this method are illustrated through a numerical example.

2 Mathematical Model

2.1 Physical Problem

Consider a mixture comprising q components, with a fixed temperature T and overall volume V^{tol}, the moles of each component are represented by

$\mathbf{N}^{tol}$ and their density by $\mathbf{n}$. The Helmholtz free energy is denoted by F and $f(\mathbf{n})$ is the energy density. The gas and liquid phases are denoted by G and L, with volumes V^G and V^L and moles $\mathbf{N}^G = \left[N_1^G, N_2^G, \ldots, N_q^G\right]^T$ and $\mathbf{N}^L = \left[N_1^L, N_2^L, \ldots, N_q^L\right]^T$. The pressure is p and the chemical potential of a component is μ_i. The volume and mole constraints are

$$\mathbf{N}^G + \mathbf{N}^L = \mathbf{N}^{tol}; \quad V^G + V^L = V^{tol}. \tag{1}$$

The general mathematical definition of Helmholtz free energy is

$$F = -pV + \sum_{i=1}^{q} \mu_i N_i. \tag{2}$$

Thus, for a two-phase system, if use $\mathbf{N}^G$ and V^G as primary variables, we obtain:

$$F = f\left(\mathbf{n}^G\right)V^G + f\left(\mathbf{n}^L\right)V^L = f(\frac{\mathbf{N}^G}{V^G})V^G + f(\frac{\mathbf{N}^{tol} - \mathbf{N}^G}{V^{tol} - V^G})(V^{tol} - V^G), \tag{3}$$

where $\mathbf{n}^G = \frac{\mathbf{N}^G}{V^G}, \mathbf{n}^L = \frac{\mathbf{N}^L}{V^L}$. The Helmholtz free energy density of real fluid $f(\mathbf{n})$ is composed by three parts

$$f(\mathbf{n}) = f^{\text{id}}(\mathbf{n}) + f^{\text{rep}}(\mathbf{n}) + f^{\text{att}}(\mathbf{n}). \tag{4}$$

The mathematical formulas of the three components comprising the Helmholtz free energy density of a uniform PR fluid are listed:

$$f^{\text{id}}(\mathbf{n}) = RT \sum_{i=1}^{q} n_i \left(\ln n_i - 1\right); \tag{5}$$

$$f^{\text{rep}}(\mathbf{n}) = -nRT \ln(1 - bn); \tag{6}$$

$$f^{\text{att}}(\mathbf{n}) = \frac{a(T)n}{2\sqrt{2}b} \ln\left(\frac{1 + (1 - \sqrt{2})bn}{1 + (1 + \sqrt{2})bn}\right), \tag{7}$$

where n_i is the molar density of each component in $\mathbf{n}$. n is the sum of n_i for each phase. a and b are EoS parameters for repulsion and attraction terms [5].

2.2 Dynamic Model

The system reaches equilibrium when Helmholtz free energy is minimized. This yields chemical potential and pressure balance equations. Time derivative of free energy can be expanded using the chain rule:

$$\frac{dF}{dt} = \frac{\partial F}{\partial V^G}\frac{\partial V^G}{\partial t} + \sum_{i=1}^{q} \frac{\partial F}{\partial N_i^G}\frac{\partial N_i^G}{\partial t}. \tag{8}$$

Onsager's reciprocal principle governs how components and phase volumes change. The second law of thermodynamics dictates the dissipation of Helmholtz free energy in a closed system over time. There is a symmetric full matrix $\boldsymbol{\Phi} = (\phi_{i,j})_{i,j=1}^{q+1}$ that is negative-definite, and satisfies:

$$\frac{\partial N_i^{\mathrm{G}}}{\partial t} = \sum_{j=1}^{q} \phi_{i,j} \left(\frac{\partial F}{\partial N_j^G} \right) + \phi_{i,q+1} \left(\frac{\partial F}{\partial V^G} \right); \tag{9}$$

$$\frac{\partial V^{\mathrm{G}}}{\partial t} = \sum_{j=1}^{q} \phi_{q+1,j} \left(\frac{\partial F}{\partial N_j^G} \right) + \phi_{q+1,q+1} \left(\frac{\partial F}{\partial V^G} \right). \tag{10}$$

To simplify, only the negative definite matrix $\boldsymbol{\Phi}$'s diagonal is used as coefficients in the non-linear system, decoupling the system. This simplification improves computation efficiency and convergence, resulting in the dynamic model:

$$\frac{\partial N_i^G}{\partial t} = -K_{\mu_i} \frac{\partial F}{\partial N_i^G}, \quad i = 1, 2, \cdots, q; \tag{11}$$

$$\frac{\partial V^G}{\partial t} = -K_p \frac{\partial F}{\partial V^G}. \tag{12}$$

Choose K_{μ_i} and K_p based on reference paper [4]. The dynamic model aims to quickly reach equilibrium, without focusing on non-equilibrium to equilibrium details. Using the definition of total Helmholtz free energy for the two-phase case and relations $\mu_i = \frac{\partial F}{\partial N_i}$ and $-p = \frac{\partial F}{\partial V}$, the dynamic model is expressed by

$$\frac{\partial N_i^G}{\partial t} = -K_{\mu_i} \left(\mu_i^G \left(\mathbf{n}^G \right) - \mu_i^L \left(\mathbf{n}^L \right) \right), \quad i = 1, 2, \cdots, q; \tag{13}$$

$$\frac{\partial V^G}{\partial t} = -K_p \left((-p^G) - (-p^L) \right). \tag{14}$$

When the system reaches a state of equilibrium, the resulting equations can be expressed as:

$$\mu_i^G \left(\mathbf{n}^G \right) - \mu_i^L \left(\mathbf{n}^L \right) = 0, \quad i = 1, 2, \cdots, q; \quad p^G - p^L = 0. \tag{15}$$

Therefore, the energy dissipation law holds at a continuous level as follows:

$$\frac{dF}{dt} = -\sum_{i=1}^{q} K_{\mu_i} \left| \frac{\partial F}{\partial N_i^G} \right|^2 - K_p \left| \frac{\partial F}{\partial V^G} \right|^2 \leq 0. \tag{16}$$

2.3 Prerequisites and Modified Energy

To use the convex-concave splitting method for a mixture of multiple components, the paper [3] introduces an energy parameter $\eta > 0$, allowing the free energy density to be rewritten as

$$f^{\text{convex}} (\mathbf{n}) = (1 + \eta) f^{\text{id}}(\mathbf{n}) + f^{\text{rep}}(\mathbf{n}); \tag{17}$$

$$f^{\text{concave}}(\mathbf{n}) = f^{\text{att}}(\mathbf{n}) - \eta f^{\text{id}}(\mathbf{n}); \tag{18}$$

$$f(\mathbf{n}) = f^{\text{convex}}(\mathbf{n}) + f^{\text{concave}}(\mathbf{n}). \tag{19}$$

Subsequently, the modified combination of the Helmholtz free energy can be subjected to the convex-concave approach to sustain the unconditionally energy-stable feature without affecting the overall energy quantity or the convergent outcomes. In the case of a component-wise framework, we have discovered that

$$\frac{\partial^2 f^{\text{id}}}{\partial n_i^2} = RT\frac{1}{n_i} > 0; \quad \frac{\partial^2 f^{\text{rep}}}{\partial n_i^2} = RT\left(\frac{2b_i}{1-bn} + \frac{b_i^2 n}{(1-bn)^2}\right) > 0. \tag{20}$$

The attraction term may not be concave (namely $\partial^2 f^{\text{att}}/\partial n_i^2 < 0$) even in the component-wise framework. See [1] for details. Component-wise expressions are introduced before numerical scheme design.

$$\begin{cases} f\left(n_i^{k+1}\right) = f\left(n_1^{k+1}, n_2^{k+1}, \cdots, n_{i-1}^{k+1}, n_i^{k+1}, n_{i+1}^{k}, \cdots n_q^{k}\right) \\ f\left(n_i^{k}\right) \quad = f\left(n_1^{k+1}, n_2^{k+1}, \cdots, n_{i-1}^{k+1}, n_i^{k}, n_{i+1}^{k}, \cdots n_q^{k}\right) \end{cases} i = 1, 2, \cdots, q. \tag{21}$$

$f(n_i^{k+1})$ is determined by using $n_j^k (j > i)$ and $n_j^{k+1} (j \leq i)$. The value of $f(n_i^k)$ is decided by $n_j^k (j \geq i)$ and $n_j^{k+1} (j < i)$. To keep mathematical expressions brief and concise, this notation is consistently utilized. Based on the modified energy, the discrete linear energy-stable scheme is designed by employing the convex-concave splitting approach and energy factorization method:

$$\frac{N_i^{G,k+1} - N_i^{G,k}}{\Delta t} = -K_{\mu_i}\left(\widetilde{\mu_i}^G - \widetilde{\mu_i}^L\right) \quad i = 1, 2, \cdots, q; \tag{22}$$

$$\frac{V^{G,k+1} - V^{G,k}}{\Delta t} = -K_p\left(-\tilde{p}\right). \tag{23}$$

This equation uses $\widetilde{\mu_i}^G$ and $\widetilde{\mu_i}^L$ as linear approximations of the gas-phase chemical potential with respect to $N_i^{G,k+1}$ and the liquid-phase chemical potential with respect to $N_i^{L,k+1}$, respectively. $\tilde{p}$ is also linear approximations of the pressure difference with respect to $V^{G,k+1}$.

3 Numerical Scheme and Proof

3.1 Ideal Term

In this part, we demonstrate an energy inequality involving the ideal term of free energy.

$$f^{\text{id}} = RT\sum_{i=1}^{q} n_i\left(\ln n_i - 1\right) = RT\sum_{i=1}^{q} n_i \ln n_i - RT\sum_{i=1}^{q} n_i. \tag{24}$$

A linear scheme denoted by $\widetilde{\mu}_i^{\text{id}}$ is proposed for the ideal term. The next steps describe the procedure for designing and validating of this linear scheme.

Lemma 1. *Suppose that the chemical potential associated with the ideal term within the semi-implicit linear scheme is defined as*

$$\widetilde{\mu}_i^{id} = RT\left(\ln n_i^k + \frac{n_i^{k+1}}{n_i^k} - 1\right). \tag{25}$$

Then, we have the inequality:

$$f^{id}\left(n_i^{k+1}\right) - f^{id}\left(n_i^k\right) \leq \widetilde{\mu}_i^{id}\left(n_i^{k+1} - n_i^k\right). \tag{26}$$

Proof. Because of $\ln n_i$ is concave with respect to n_i, we obtain

$$\ln n_i^{k+1} - \ln n_i^k \leq \frac{1}{n_i^k}\left(n_i^{k+1} - n_i^k\right). \tag{27}$$

As $n_i^k > 0$ and $n_i^{k+1} > 0$, using (27), we have

$$\begin{aligned} n_i^{k+1}\ln n_i^{k+1} - n_i^k \ln n_i^k &= \ln n_i^k\left(n_i^{k+1} - n_i^k\right) + n_i^{k+1}\left(\ln n_i^{k+1} - \ln n_i^k\right) \\ &\leq \left(\ln n_i^k + \frac{n_i^{k+1}}{n_i^k}\right)\left(n_i^{k+1} - n_i^k\right). \end{aligned} \tag{28}$$

Through equations (21), (24) and (28), the change in energy density contributed by ideal term is obtained

$$\begin{aligned} f^{\mathrm{id}}\left(n_i^{k+1}\right) - f^{\mathrm{id}}\left(n_i^k\right) &= RT\left(n_i^{k+1}\ln n_i^{k+1} - n_i^k \ln n_i^k\right) - RT\left(n_i^{k+1} - n_i^k\right) \\ &\leq RT\left(\ln\left(n_i^k\right) + \frac{n_i^{k+1}}{n_i^k} - 1\right)\left(n_i^{k+1} - n_i^k\right). \end{aligned} \tag{29}$$

If we define $\widetilde{\mu}_i^{\mathrm{id}}$ using (25), then it's obvious that the inequality (26) is satisfied

3.2 Repulsion Term

In this part, we will establish and demonstrate a comparable inequality to (26) that pertains to the repulsion component of the Helmholtz free energy density, denoted as f^{rep} and defined in (6). The formula can be rephrased as follows:

$$f^{\mathrm{rep}} = -nRT\ln(1 - bn) = -\sum_{i=1}^{q} n_i RT \ln\left(1 - \sum_{i=1}^{q} b_i n_i\right). \tag{30}$$

Before stating the lemma and its proof, these auxiliary variables W, w and ϵ are defined for the purpose of mathematical simplification.

$$w = \sum_{j\neq i}^{q} n_j; \qquad W = 1 - \sum_{j\neq i}^{q} b_j n_j; \qquad \epsilon = 1 - bn = 1 - \sum_{j\neq i}^{q} b_j n_j - b_i n_i = W - b_i n_i. \tag{31}$$

Regarding ϵ, w, and W, It is evident that they are all positive real numbers. We can rephrase the repulsion term of the energy density at the continuous level as:

$$f^{\text{rep}} = -\left(w + \frac{W-\epsilon}{b_i}\right) RT \ln \epsilon. \tag{32}$$

Moreover, we can define $\epsilon_i^k = 1 - \sum_{j<i}^q b_j n_j^{k+1} - b_i n_i^k - \sum_{j>i}^q b_j n_j^k$ and $\epsilon_i^{k+1} = 1 - \sum_{j<i}^q b_j n_j^{k+1} - b_i n_i^{k+1} - \sum_{j>i}^q b_j n_j^k$. Also, $w_i^k = \sum_{j<i}^q n_j^{k+1} + \sum_{j>i}^q n_j^k$ and $W_i^k = 1 - \sum_{j<i}^q b_j n_j^{k+1} - \sum_{j>i}^q b_j n_j^k$ can be defined based on the component-wise framework's Helmholtz free energy density in (21). Here we notice again that ϵ_i^k, ϵ_i^{k+1}, w_i^k and W_i^k are all positive real numbers.

Lemma 2. *The linear scheme of the chemical potential contributed by the repulsion term can be denoted as $\widetilde{\mu}_i^{rep}$, and its definition is provided below:*

For any $\epsilon_i^{k+1}/\epsilon_i^k > 0$, select a coefficient $h_i = \min\left\{\alpha, \epsilon_i^{k+1}/\epsilon_i^k\right\}$, so that there exists a corresponding coefficient $\beta_i = \frac{\frac{1}{h_i}-1}{(1-\alpha)^2}$. Then the linear scheme is

$$\begin{aligned}\widetilde{\mu}_i^{rep} &= RT\left(\frac{b_i}{W_i^k - b_i n_i^k} n_i^{k+1} - \frac{W_i^k}{W_i^k - b_i n_i^k} - \ln\left(W_i^k - b_i n_i^k\right)\right) \\ &\quad + RT\left(W_i^k + w_i^k b_i\right)\left(\frac{\beta_i b_i}{(W_i^k - b_i n_i^k)^2} n_i^{k+1} + \frac{W_i^k - (1+\beta_i) b_i n_i^k}{(W_i^k - b_i n_i^k)^2}\right).\end{aligned} \tag{33}$$

In particular, when the ratio of ϵ_i^{k+1} to ϵ_i^k is greater than or equal to a certain value, denoted by α (which is approximately equal to 0.31617), we can simplify the linear scheme form by setting β_i to 1. Then, the discrete energy dissipation law holds unconditionally.

$$f^{rep}\left(n_i^{k+1}\right) - f^{rep}\left(n_i^k\right) \le \widetilde{\mu}_i^{rep}\left(n_i^{k+1} - n_i^k\right). \tag{34}$$

Proof. By splitting the repulsion term into two parts:

$$f_1^{\text{rep}} = \frac{RT}{b_i} \epsilon \ln \epsilon; \quad f_2^{\text{rep}} = -\frac{RT}{b_i}\left(W + w b_i\right) \ln \epsilon. \quad (W + w b_i > 0) \tag{35}$$

Since the first part has the similar form with ideal part, it will yield that

$$\begin{aligned} f_1^{\text{rep}}\left(n_i^{k+1}\right) - f_1^{\text{rep}}\left(n_i^k\right) &= f_1^{\text{rep}}\left(\epsilon_i^{k+1}\right) - f_1^{\text{rep}}\left(\epsilon_i^k\right) \\ &\le \frac{RT}{b_i}\left(\ln \epsilon_i^k + \frac{\epsilon_i^{k+1}}{\epsilon_i^k}\right)\left(\epsilon_i^{k+1} - \epsilon_i^k\right) \\ &= RT\left(\frac{b_i}{W_i^k - b_i n_i^k} n_i^{k+1} - \frac{W_i^k}{W_i^k - b_i n_i^k} - \ln\left(W_i^k - b_i n_i^k\right)\right)\left(n_i^{k+1} - n_i^k\right). \end{aligned}$$

So, by defining the chemical potential $\widetilde{\mu}_{1,i}^{rep}$ as the right-hand side, we have the desired inequality for f_1^{rep} as

$$f_1^{\text{rep}}\left(n_i^{k+1}\right) - f_1^{\text{rep}}\left(n_i^k\right) \le \widetilde{\mu}_{1,i}^{\text{rep}}\left(n_i^{k+1} - n_i^k\right). \tag{36}$$

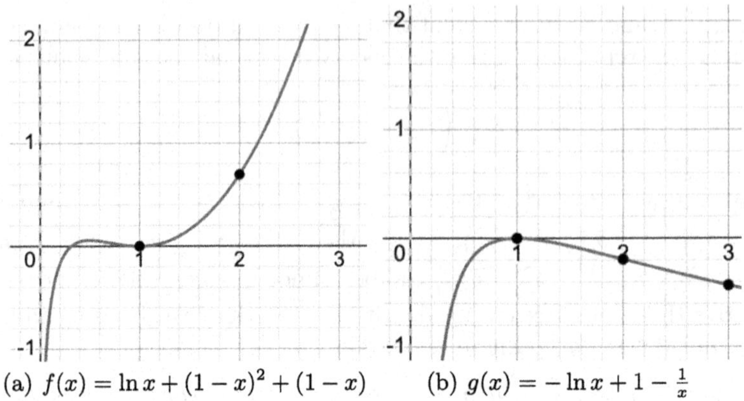

(a) $f(x) = \ln x + (1-x)^2 + (1-x)$ (b) $g(x) = -\ln x + 1 - \frac{1}{x}$

Fig. 1. The curves of auxiliary functions

Moving on to the second part, denoted as $f_2^{\text{rep}} = -\frac{RT}{b_i}(W + wb_i)\ln\epsilon$, where $W + wb_i > 0$, assume that $x = \epsilon_i^{k+1}/\epsilon_i^k$, where $\epsilon_i^{k+1} > 0$ and $\epsilon_i^k > 0$. We can make use of the properties of the following functions: (1) $g(x) = -\ln x + 1 - \frac{1}{x} \leq 0$; and (2) $f(x) = \ln x + (x-1)^2 + (1-x)$, which has two zeros at $x = 1$ and $x = \alpha \approx 0.31617$. We can conclude that $-\ln x \leqslant \frac{1}{x} - 1$ for any $x > 0$. Also, for any $x \geq \alpha$, we have:

$$-\ln x \leqslant (1-x) + (1-x)^2. \tag{37}$$

For any $h_i = x < \alpha$, we have the following inequality holds as

$$-\ln x \leq \frac{1}{x} - 1 = \frac{1}{h_i} - 1 < \frac{\left(\frac{1}{h_i} - 1\right)(1-x)^2}{(1-\alpha)^2} = \beta_i(1-x)^2 \leq (1-x) + \beta_i(1-x)^2. \tag{38}$$

Then for $\epsilon_i^{k+1}/\epsilon_i^k \geq \alpha$, the above inequality still holds because $\beta_i = 1$ obviously.

$$-\ln x \leq (1-x) + (x-1)^2 \leq (1-x) + \beta_i(1-x)^2. \tag{39}$$

The conclusion also means that for any $\epsilon_i^{k+1}/\epsilon_i^k > 0$, if choosing $h_i = \min\left\{\alpha, \frac{\epsilon_i^{k+1}}{\epsilon_i^k}\right\}$, there is always a coefficient $\beta_i = \frac{\frac{1}{h_i} - 1}{(1-\alpha)^2}$ to ensure

$$\left(-\ln\epsilon_i^{k+1}\right) - \left(-\ln\epsilon_i^k\right) \leq \left(-\frac{1}{\epsilon_i^k} + \frac{\beta_i\left(\epsilon_i^{k+1} - \epsilon_i^k\right)}{\left(\epsilon_i^k\right)^2}\right)\left(\epsilon_i^{k+1} - \epsilon_i^k\right). \tag{40}$$

By defining $\tilde{\mu}_{2,i}^{\text{rep}}$ as

$$\tilde{\mu}_{2,i}^{\text{rep}} = RT\left(W_i^k + w_i^k b_i\right)\left(\frac{\beta_i b_i}{\left(W_i^k - b_i n_i^k\right)^2} n_i^{k+1} + \frac{W_i^k - (1+\beta_i)\, b_i n_i^k}{\left(W_i^k - b_i n_i^k\right)^2}\right), \tag{41}$$

we have

$$
\begin{aligned}
& f_2^{\text{rep}}\left(n_i^{k+1}\right)-f_2^{\text{rep}}\left(n_i^k\right)=f_2^{\text{rep}}\left(\epsilon_i^{k+1}\right)-f_2^{\text{rep}}\left(\epsilon_i^k\right) \\
= & \frac{RT}{b_i}\left(W_i^k+w_i^k b_i\right)\left(-\ln\left(\epsilon_i^{k+1}\right)-\left(\ln\left(\epsilon_i^k\right)\right)\right) \\
\leq & \frac{RT}{b_i}\left(W_i^k+w_i^k b_i\right)\left(-\frac{1}{\epsilon_i^k}+\frac{\beta_i\left(\epsilon_i^{k+1}-\epsilon_i^k\right)}{\left(\epsilon_i^k\right)^2}\right)\left(\epsilon_i^{k+1}-\epsilon_i^k\right) \\
= & RT\left(W_i^k+w_i^k b_i\right)\left(\frac{\beta_i b_i}{\left(W_i^k-b_i n_i^k\right)^2} n_i^{k+1}+\frac{W_i^k-\left(1+\beta_i\right) b_i n_i^k}{\left(W_i^k-b_i n_i^k\right)^2}\right)\left(n_i^{k+1}-n_i^k\right).
\end{aligned}
$$

When $\epsilon_i^{k+1}/\epsilon_i^k \geq \alpha$, we can use $\beta_i = 1$ directly.

There are a few key points that need to be emphasized. Upon the completion of updating the mole numbers, it is necessary to verify the linear scheme expression using a reverting process because it is unclear whether $\epsilon_i^{k+1}/\epsilon_i^k < \alpha$ or $\epsilon_i^{k+1}/\epsilon_i^k \geq \alpha$. Numerical results indicate that most computations satisfy the condition $\epsilon_i^{k+1}/\epsilon_i^k \geq \alpha$, and in such cases, β_i is set to 1 for faster convergence. However, if $\epsilon_i^{k+1}/\epsilon_i^k < \alpha$, A method for adaptively selecting the value of β_i is given. This involves starting with $h_i = \frac{1}{2}\alpha$ and gradually decreasing h_i until the condition is satisfied. In our simulations, this process only happens in the states near the phase boundary, a good initial guess can ensure the reverting process is under 10 times.

3.3 Attraction Term

Despite the fact that the attraction term of free energy density is not a concave function even under the component-wise framework, by introducing the modified energy like the one shown in equations (19), (17) and (18), if η is sufficiently large, the strict convex-concave splitting for the Helmholtz free energy density of a multicomponent mixture can be attained. We view the function $f^{\text{concave}}(\mathbf{n}) = f^{\text{att}}(\mathbf{n}) - \eta f^{\text{id}}(\mathbf{n})$ as a single entity in order to ensure energy stability. By taking a derivative with respect to n_i, we can directly derive the formulation of the chemical potential μ_i^{att} from the attraction term.

$$
\begin{aligned}
\widetilde{\mu}_i^{concave} = & \frac{2\sum_{j=1}^{q} a_{ij} n_j^k b n^k - b_i a (n^k)^2}{2\sqrt{2}(bn^k)^2} \ln\left(\frac{1+(1-\sqrt{2})bn^k}{1+(1+\sqrt{2})bn^k}\right) \\
& + \frac{an^k}{2\sqrt{2}b}\left(\frac{(1-\sqrt{2})b_i}{1+(1-\sqrt{2})bn^k} - \frac{(1+\sqrt{2})b_i}{1+(1+\sqrt{2})bn^k}\right) - \eta RT \ln n_i^k.
\end{aligned} \tag{42}
$$

By integrating the schemes for all three types of terms, this linear scheme ensures the stability of the discrete free energy.

$$
f\left(n_i^{k+1}\right) - f\left(n_i^k\right) \leq \widetilde{\mu}_i \left(n_i^{k+1} - n_i^k\right). \tag{43}
$$

3.4 Linear Scheme for Updating Volume

After updating the moles of each composition in the dynamic model, A vital step of flash calculation is to utilize the provided formula to calculate the volume of each phase,

$$\frac{V^{G,k+1} - V^{G,k}}{\Delta t} = -K_p \left(-\tilde{p}^G - (-\tilde{p}^L)\right).$$

The formula for phase volume is complicated because it involves a complex non-linearity (the volume is the denominator of molar density). This makes it challenging to design a linear method. To address this issue, inspired by Kou's idea [4] on the convex-concave properties of energy with respect to volume and the relation of (2), We can conclude that the pressure terms arising from the convex part of the energy need to be handled in an implicit manner, while the concave part requires an explicit treatment to ensure energy stability. The pressure formula can be derived by taking the derivative of free energy and using the energy density of PR fluid.

$$\frac{\partial F}{\partial V} = -p = -\frac{RT}{v-b} + \frac{a(T)}{v^2 + 2bv - b^2}. \tag{44}$$

where $v = V/N$. Then, the semi-implicit scheme is

$$\begin{aligned} \frac{V^{G,k+1} - V^{G,k}}{\Delta t} = -K_p \Bigg\{ &\left[-\frac{N^G RT}{V^{G,k+1} - N^G b^G} + \frac{a(T)(N^G)^2}{(V^{G,k})^2 + 2b^G N^G V^{G,k} - (b^G)^2 (N^G)^2} \right] \\ &- \left[-\frac{N^L RT}{V^{L,k+1} - N^L b^L} + \frac{a(T)(N^L)^2}{(V^{L,k})^2 + 2b^L N^L V^{L,k} - (b^L)^2 (N^L)^2} \right] \Bigg\}. \end{aligned} \tag{45}$$

This makes the energy stability property to hold discretely. The following lemma is here to claim the existence of the linear scheme for $\tilde{p}(V^{G,k+1})$ only.

Lemma 3. *One can express the pressure term using a linear semi-implicit scheme as shown by*

$$-\tilde{p} = \bar{A} V^{G,k+1} + \bar{B}, \tag{46}$$

where the coefficient $\bar{A}\left(\mathbf{N}^{G,k+1}, V^{G,k}\right)$ *and* $\bar{B}\left(\mathbf{N}^{G,k+1}, V^{G,k}\right)$ *only involve explicit variables. It is possible to maintain the energy-stability property of the discrete system with respect to phase volume* V^G *by using the linear scheme for the pressure term* $-\tilde{p}$*, namely*

$$f\left(V^{G,k+1}\right) - f\left(V^{G,k}\right) \le -\tilde{p}\left(V^{G,k+1} - V^{G,k}\right). \tag{47}$$

Proof. When combined with the volume constraints given by equation (1), the equation used to update the volume of the gas phase can be converted to

$$\begin{aligned} \frac{V^{G,k+1} - V^{G,k}}{\Delta t} = -K_p \Bigg\{ &\left[-\frac{N^G RT}{V^{G,k+1} - N^G b^G} + \frac{a(T)\left(N^G\right)^2}{(V^{G,k})^2 + 2b^G N^G V^{G,k} - (b^G)^2 \left(N^G\right)^2} \right] \\ &- \left[-\frac{N^L RT}{V^{tol} - V^{G,k+1} - N^L b^L} + \frac{a(T)\left(N^L\right)^2}{(V^{L,k})^2 + 2b^L N^L V^{L,k} - (b^L)^2 \left(N^L\right)^2} \right] \Bigg\}. \end{aligned} \tag{48}$$

After updating the moles of all components as $N^G = \sum N_i^{G,k+1}$ and $N^L = \sum N_i^{L,k+1}$, the concave terms that are explicitly represented by notations $CA^{G,k}$ and $CA^{L,k}$ are used to simplify the mathematical expressions. Thus, the equation can be expressed as:

$$\begin{aligned}\frac{V^{G,k+1} - V^{G,k}}{\Delta t} = -K_p[(-\frac{N^G RT}{V^{G,k+1} - N^G b^G} + CA^{G,k}) - \\ (-\frac{N^L RT}{V^{tol} - V^{G,k+1} - N^L b^L} + CA^{L,k})].\end{aligned} \tag{49}$$

With some further relaxations on the time step to keep this energy dissipation property and the mathematical properties on the coefficients of linear terms. The relaxed time step will not influence the convergence of scheme. It finally reaches the following inequality:

$$\begin{aligned}f\left(V^{G,k+1}\right) - f\left(V^{G,k}\right) \leq [-N^G RT - \left(CA^{L,k} - CA^{G,k}\right)\left(V^{G,k} - N^G b^G\right) - \\ \frac{N^L RT\left(V^{G,k} - N^G b^G\right)}{V^{G,k+1} + N^L b^L - V^{tol}}]\left(V^{G,k+1} - V^{G,k}\right).\end{aligned} \tag{50}$$

It also implies that if the scheme of $-\tilde{p}$ is defined as

$$-\tilde{p} = \bar{A} V^{G,k+1} + \bar{B}, \tag{51}$$

then the discrete energy law with respect to V^G still holds as

$$f\left(V^{G,k+1}\right) - f\left(V^{G,k}\right) \leq -\tilde{p}\left(V^{G,k+1} - V^{G,k}\right).$$

Thus, the coefficients $\bar{A} = N^G RT + \left(CA^{L,k} - CA^{G,k}\right)\left(V^{G,k} - N^G b^G\right)$ and $\bar{B} = \left[N^G RT + \left(CA^{L,k} - CA^{G,k}\right)\left(V^{G,k} - N^G b^G\right)\right]\left(N^L b^L - V^{tol}\right) + ...$ $...N^L RT\left(V^{G,k} - N^G b^G\right)$, respectively. The final relaxed time step $\bar{\Delta t}$ is

$$\bar{\Delta t} = \frac{\Delta t}{\left(V^{G,k+1} - N^G b^G\right)\left(V^{tol} - V^{G,k+1} - N^L b^L\right)}. \tag{52}$$

In conclusion, the linear scheme $-\tilde{p} = \bar{A} V^{G,k+1} + \bar{B}$ is able to maintain the energy-stability property at a discrete level, as shown in Lemma (47). It is important to note that the relaxed time step will not affect the convergence or energy-stability property.

4 Numerical Experiments

This section presents several numerical experiments conducted on a mixture to show the efficiency, arrcuracy and stability of this fully explicit and unconditionally energy-stable method for solving the Peng-Robinson VT flash problem. The obtained results are then compared to the data reported in Jiri's papers [7], which are commonly used as a reference for comparison. The constant total volume of $V^{\text{total}} = 1m^3$ is established. To obtain a physically meaningful outcome, an initial guess is acquired using the Wilson correlation and the Rachford-Rice equation. For the dynamic model, the modified energy parameter is set to $\eta = 10$, and the time step is $dt = 10^8$. The values of dynamic model coefficients K_{μ_i} and K_p can typically be determined using the methods described in Kou's paper [4]. Set diffusion coefficient $D_i = 1$ for each component

$$K_{\mu_i} = \frac{D_i N_i}{RT}; \quad K_p = \frac{C_V^G C_V^L V^{tol}}{C_V^L p_G^k + C_V^G p_L^k}.$$

where the coefficients $C_V^G = C_V^L = 1$. The phase pressure p_G^k and p_L^k is the phase pressure in the last time step. The convergence criterion for the time loops is based on the relative errors in the moles of each component in the gas phase $N_i{}^G$ and the gas phase volume V^G between two adjacent time steps. These relative errors are considered as the criteria, and convergence is achieved when they are both less than 10^{-6}.

$$\left\|\Delta N_i^G\right\|_{rel} = \frac{\left\|N_i^{G,k} - N_i^{G,k+1}\right\|}{\left\|N_i^{G,k}\right\|}; \quad \left\|\Delta V^G\right\|_{rel} = \frac{\left\|V^{G,k} - V^{G,k+1}\right\|}{\left\|V^{G,k}\right\|}.$$

4.1 Binary Mixture of Methane (C_1) and *n*-penthane (nC_5)

Initially, a binary mixture of methane (C_1) and n-pentane (nC_5) is used as a case study for a two-phase flash calculation. The mixture has a total molar density of $n = 6135.3$ mol/m^3 with mole fractions of $z_{C_1} = 0.489575$ and $z_{nC_5} = 0.510425$ at a temperature of $T = 310.95$ K. Additional parameters regarding the mixture are shown in Table 1.

Table 1. Properties of the constituents in the binary $C_1 - nC_5$ mixture

Component	$\omega_i[-]$	T_{crit} [K]	P_{crit} [MPa]	M_w[g/mol]
C_1	0.011	190.56	4.599	16
nC_5	0.251	469.70	3.37	72.2

The $C_1 - nC_5$ binary interaction coefficient $\delta_{C_1 - nC_5} = 0.041$.

Table 2 shows the equilibrium results, with the resulting equilibrium pressure being $p = 10.4651$ MPa. These results show excellent agreement with the reference data from Jiri and Abbas' study [7]. The energy dissipation property can also be observed from Fig. 2, where there is an apparent decay of energy at the initial stage. To provide a better visualization of the energy trend, a zoom window of the 150*th* – 200*th* time steps is included in the plot. Convergence is reached after 224 time steps, and the resulting properties are considered to be the equilibrium properties of the mixture.

Figure 3 illustrates the variability in the accuracy of the calculation process. The general trend is a decrease in the relative error. However, there is a sag

Table 2. The resulting constitution and overall physical parameters of $C_1 - nC_5$ calculated by this novel scheme.

Property	Unit	Overall mixture	Phase Liquid	Phase Gas
molar density	mol/m^3	6135.30	10106.03	3177.74
C_1 mole fraction		0.489575	0.293459	0.954132
nC_5 mole fraction		0.510425	0.706541	0.045868
phase volume fraction			0.426881	0.573119

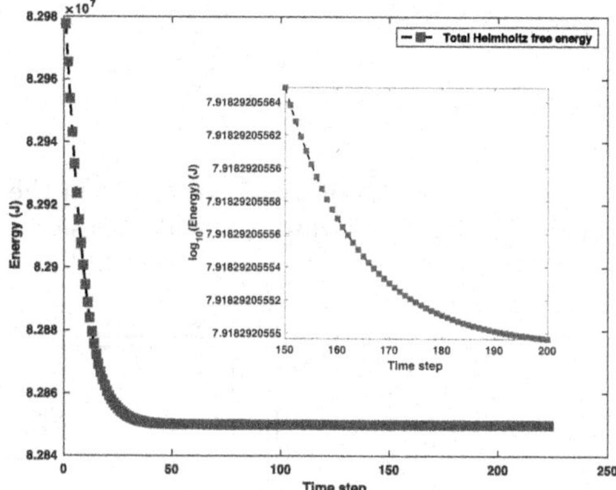

Fig. 2. Profile of total Helmholtz free energy of mixture with time step $\Delta t = 10^8$ for the binary mixture ($C_1 - nC_5$) case

in the relative error for C_1 and gas volume at the initial stage, which could be attributed to the imprecise initial estimate and the non-synchronous nature of the component-wise iteration process. As the relative error only indicates a relative difference between two consecutive time steps, possibly, a numerical adjustment process occurred to overcome the asynchrony. Fortunately, the variability in relative error for moles and phase volume in the subsequent process exhibits a similar and unanimous trend of decreasing slope.

A stem plot is depicted in Fig. 4, illustrating the energy fluctuations during the initial 10 time steps. The energy decreases as each component and then the volume are updated, which is which is in line with the proof and derivation. The energy continues to decrease within each time step, and it is observed that the energy before updating (the black stem) is higher than the energy after updating the volume of the previous time step (the magenta stem) when transitioning to a new time step. This phenomenon is caused by modifications to the PR EoS parameters a and b of the mixture at each finished time step. However, it is evident that the energy before updating (the black stem) consistently decreases throughout the entire dynamic process as expected.

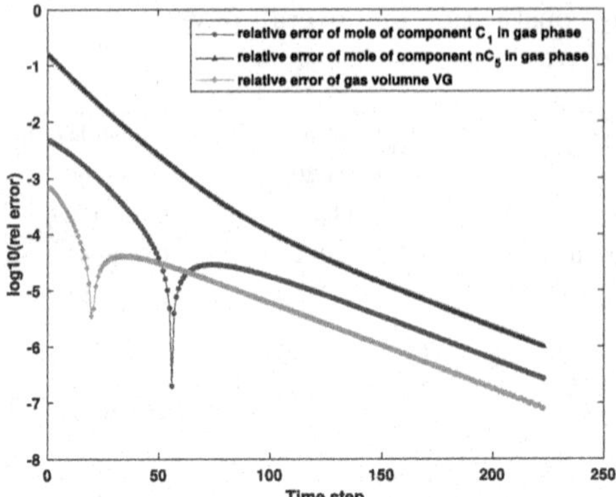

Fig. 3. Profile of relative error on moles of each component in the gas phase and gas volume with time steps for the binary mixture ($C_1 - nC_5$) case

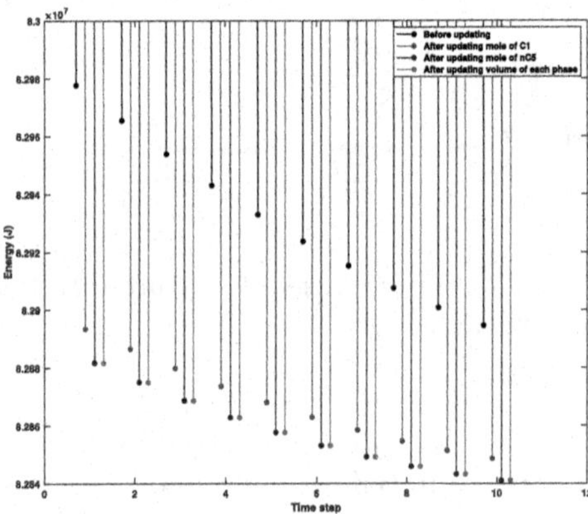

Fig. 4. Stem plot of the relative error change on moles of each component and phase volume with time steps for the binary mixture ($C_1 - nC_5$) case

5 Conclusion

This work proposes a numerical method for calculating two-phase equilibrium under constant volume, temperature, and moles, using the dynamic model for VT flash that upholds Onsager's reciprocal principle and the energy dissipation law. With a designed linear semi-implicit scheme, the moles of each component and phase volume can be updated while preserving the energy dissipation feature. The convex-concave splitting technique, energy factorization approach and component-wise iteration framework are then employed for the mixture, resulting in a fully explicit algorithm. This method avoids solving complicated linear or nonlinear systems, making it easier to implement and apply in engineering contexts. Moreover, one component-wise and unconditionally energy-stable method is proposed for the first time for the multi-component flash calculation problem.

Acknowledgments. This study is partially funded by KAUST grants BAS/1/1351-01, URF/1/4074-01, and URF/1/3769-01, as well as by the Ministry of Science and Technology, R.O.C. (No. 108-2115-M-194-004-MY2), the National Natural Science Foundation of China (No. 51874262 and No. 51936001), the Peacock Plan Foundation of Shenzhen (No. 000255) and the General Program of Natural Science Foundation of Shenzhen (No. 20200801100615003).

References

1. Fan, X., Kou, J., Qiao, Z., Sun, S.: A componentwise convex splitting scheme for diffuse interface models with van der waals and peng-robinson equations of state. SIAM J. Sci. Comput. **39**(1), B1–B28 (2017). https://doi.org/10.1137/16M1061552
2. Feng, X., Kou, J., Sun, S.: A novel energy stable numerical scheme for navier-stokes-cahn-hilliard two-phase flow model with variable densities and viscosities. In: Shi, Y., Fu, H., Tian, Y., Krzhizhanovskaya, V.V., Lees, M.H., Dongarra, J., Sloot, P.M.A. (eds.) ICCS 2018. LNCS, vol. 10862, pp. 113–128. Springer, Cham (2018). https://doi.org/10.1007/978-3-319-93713-7_9
3. Kou, J., Sun, S.: Efficient energy-stable dynamic modeling of compositional grading. Int. J. Num. Anal. Model. **14**(2) (2017). http://www.math.ualberta.ca/ijnam/Volume-14-2017/No-2-17/2017-02-04.pdf
4. Kou, J., Sun, S.: A stable algorithm for calculating phase equilibria with capillarity at specified moles, volume and temperature using a dynamic model. Fluid Phase Equilib. **456**, 7–24 (2018). https://doi.org/10.1016/j.fluid.2017.09.018
5. Kou, J., Sun, S., Wang, X.: A novel energy factorization approach for the diffuse-interface model with Peng-Robinson equation of state. SIAM J. Sci. Comput. **42**(1), B30–B56 (2020). https://doi.org/10.1137/19M1251230
6. Li, Y., Kou, J., Sun, S.: Numerical modeling of isothermal compositional grading by convex splitting methods. J. Nat. Gas Sci. Eng. **43**, 207–221 (2017). https://doi.org/10.1016/j.jngse.2017.03.019
7. Mikyška, J., Firoozabadi, A.: A new thermodynamic function for phase-splitting at constant temperature, moles, and volume. AIChE J. **57**(7), 1897–1904 (2011). https://doi.org/10.1002/aic.12387
8. Song, R., Feng, X., Wang, Y., Sun, S., Liu, J.: Dissociation and transport modeling of methane hydrate in core-scale sandy sediments: a comparative study. Energy **221**, 119890 (2021). https://doi.org/10.1016/j.energy.2021.119890
9. Song, R., Sun, S., Liu, J., Feng, X.: Numerical modeling on hydrate formation and evaluating the influencing factors of its heterogeneity in core-scale sandy sediment. J. Nat. Gas Sci. Eng. **90**, 103945 (2021). https://doi.org/10.1016/j.jngse.2021.103945
10. Zhang, T., Li, Y., Sun, S.: Phase equilibrium calculations in shale gas reservoirs. Capillarity 2(1), 8–16 (2019), https://doi.org/10.26804/capi.2019.01.02

Smart Systems: Bringing Together Computer Vision, Sensor Networks and Artificial Intelligence

Feasibility and Performance Benefits of Directional Force Fields for the Tactical Conflict Management of UAVs

Enrique Hernández-Orallo(✉), Jamie Wubben, and Carlos T. Calafate

Universitat Politècnica de València, Camino de Vera, S/N, 46022 Valencia, Spain
{ehernandez,jwubben,calafate}@disca.upv.es

Abstract. As we move towards scenarios where the adoption of unmanned aerial vehicles (UAVs) becomes massive, smart solutions are required to efficiently solve conflicts in the flight trajectories of aircraft so as to avoid potential collisions. Among the different possible approaches, adopting virtual force fields is a possible solution acknowledged for being simple, distributed, and yet effective. In this paper, we study the feasibility of a directional force field (D-FFP) approach, preliminary assessing its performance benefits compared to a standard force field protocol (FFP) using Matlab simulation. Results show that, in typical scenarios associated with aerial traffic corridors, the proposed approach can reduce the flight time overhead by 32% (on average) while maintaining the required flight safety distances between aircraft.

Keywords: UAV · Tactical conflict management · Field force · Collision avoidance

1 Introduction

The emerging Smart City paradigm is fostering developments in several research areas, including air transportation. In particular, the airspace of future cities is expected to be crowded with different aircraft performing all sorts of activities [9]. To meet this challenge, different initiatives, like U-Space [2] in Europe, attempt to regulate and standardize aerial operations so as to promote safety and efficiency.

When addressing conflicts between different aircraft in certain controlled airspace, two complementary approaches are usually considered [3]: (i) strategic conflict management, whereby possible conflicts are detected and handled before the UAVs take off by analyzing the specified flight plans, and proposing modifications to those plans if potential conflicts are detected; and (ii) tactical conflict management, which consists of managing conflicts dynamically once the UAVs are in flight. Concerning the latter, different approaches can be found in the literature [13]. Among them, force-field methods are popular for their simplicity in terms of logic and implementation, while allowing to seamlessly scale to

J. Mikyška et al. (Eds.): ICCS 2023, LNCS 14077, pp. 387–393, 2023.
https://doi.org/10.1007/978-3-031-36030-5_31

any number of obstacles/vehicles. Yet, they require precise tuning and thorough validation to address possible instability issues.

In a previous work [12] we presented FFP, a Force-Field protocol targeting UAVs of the multirotor type, that is able to achieve good performance in different realistic types of aerial conflicts while outperforming a geometric protocol [6]. Yet, simulation results have shown that there is still a margin for improvement. Hence, in this paper, we explore a novel (yet more complex) approach by introducing directional force fields. Preliminary experiments performed in the Matlab environment highlight the potential benefits of this new technique, showing performance improvements over the legacy FFP technique.

The remainder of the paper is organized as follows: in the next section, we provide a brief overview of some related works. Then, in Sect. 3, we make an overview of our proposed solution. Section 4 details how experiments were defined, and the main performance results achieved. Finally, Sect. 5 presents the main conclusions of this work, also discussing future work.

2 Related Works

The field of collision avoidance systems (CAS) has been extensively studied for all types of vehicles. Various types of solutions exist, and a good classification of these approaches can be found in [13].

An analysis of the literature shows that there are only a few works (considering UAVs) that use a force-field approach to avoid collisions. Most of them are only considering static obstacles [1,7,10,11]. In the work of [4] Choi et al., both static and moving obstacles were taken into account. Using a curl-free vector field, they were able to avoid the obstacles successfully. Their work was especially focused on solving the local minimum problem that exists for static obstacles. Furthermore, Kownacki et al. [8], also used a force-field approach to avoid collisions. In their work, they considered nonholonomic UAVs (e.g. fixed-wing planes) with several numerical simulations showing the validity of their algorithm. All the above-mentioned works were tested in simulation, and MATLAB was often used. With this type of simulation, many experiments can be performed rather quickly, which is adequate for preliminary works. However, it often omits many physical intricacies (e.g. inertia). Therefore, in a previous work (FFP) [12], we implemented a force-field approach that showed how collisions could be avoided with only a small time overhead. However, in that approach, we did not take the direction of the obstacle w.r.t the flying direction into account. That is, the magnitude of the repulsion vector had the same size independently whether the obstacle was right in front of the UAV or on the side. This made the UAVs repel some obstacles stronger than necessary. Hence, in this paper, we make a preliminary assessment of a directional force field protocol (D-FFP), as detailed below.

3 Proposed Solution

The original FFP protocol [12] uses a conventional approach whereby the flight destination is modeled as a constant attraction force, emulating a gravitational field. Instead, the repulsion between UAVs is modeled following the principles of repulsion between two similar electrical charges, having the properties of (i) being omnidirectional, and (ii) experiencing a decay of intensity with distance. In this paper, we propose a different repulsion model that is instead inspired by the repulsion properties between two magnets with the same polarity, meaning that the repulsion force depends on direction (θ). This behavior has been modeled according to the following equation:

$$R(\theta, \mu) = \begin{cases} \cos(\mu \cdot \theta) : \theta \in [-\frac{\pi}{2 \cdot \mu}, \frac{\pi}{2 \cdot \mu}] \\ C : \theta \notin [-\frac{\pi}{2 \cdot \mu}, \frac{\pi}{2 \cdot \mu}] \end{cases} \tag{1}$$

In this equation, we can modulate the repulsion behavior by adjusting parameters C and μ, whereby C is a constant value in the range between 0 and 1, which determines how large the omnidirectional component of the force field is, while μ allows regulating how narrow is the main lobe.

Figure 1 shows the resulting repulsion pattern generated when setting C to 0.25, and μ to 2, values which we will use for the experiments that will follow. We can see that, for angles close to zero (target along the line defined by the speed vector), repulsion values are very high, being reduced as we move away from that reference direction until reaching a minimum constant repulsion value, which applies for most other directions. Finally, as in the FFP protocol, the resulting direction vector is obtained using the attraction force, and the repulsion force (the latter being determined by Eq. 1).

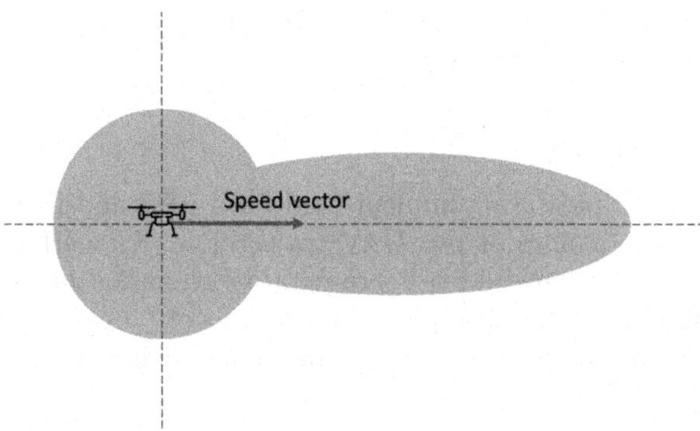

Fig. 1. Generated repulsion pattern ($C = 0.25; \mu = 2$).

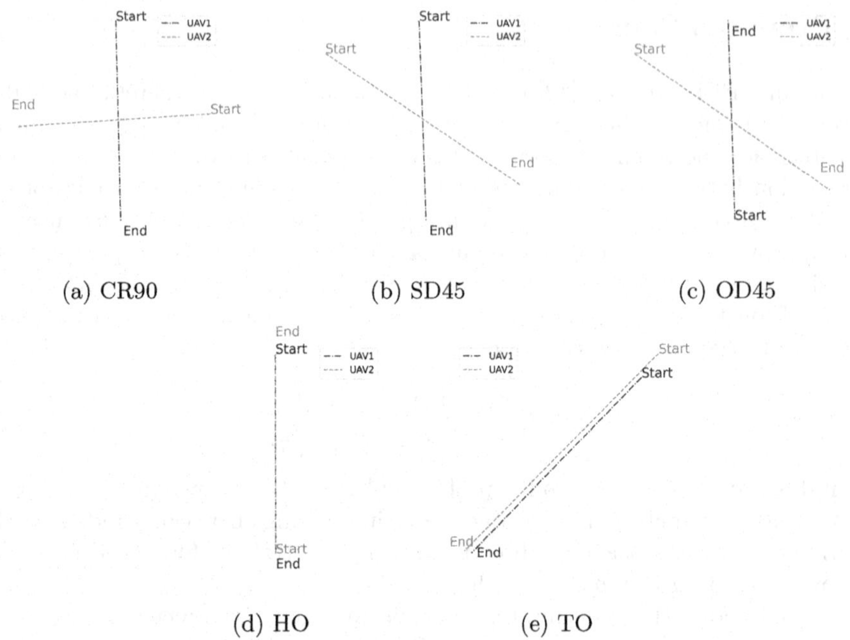

Fig. 2. Overview of the five test scenarios.

4 Simulation Results

In this section, we perform some experiments to validate the D-FFP proposal, while evaluating to which extent the proposed solution is able to improve upon the conventional FFP solution. To achieve this, we first detail how the reference simulation experiments were defined, and afterwards, we present some preliminary performance results, with discussion.

4.1 Simulation Setup

To perform our experiments, we implemented both FFP and D-FFP in Matlab and simulated the dynamics of two UAV trajectories. For performance assessment purposes, we have devised five representative scenarios whereby two UAVs have intersecting trajectories; the purpose is to provoke a conflict that must be addressed by the collision avoidance protocols. These five scenarios are shown in Fig. 2.

In terms of performance metrics, what is sought is an optimal trade-off between the flight time overhead introduced and the safety distance between UAVs. Hence, we will measure the UAVs' total flight time (TFT) to compare the time differences between both protocols. In addition, we will also measure the minimum distance between UAVs that was registered in the experiment. We should keep in mind that a minimum of 10 m of separation between them should be maintained to account for GPS error (±5 m for each).

4.2 Performance Analysis

We now present the performance results obtained in the five scenarios described above. These results are summarized in Table 1. We measured the performance using two different metrics: (i) the minimal distance between the UAVs and (ii) the time overhead (TO) introduced by our algorithm. For the time overhead, we compare the executing time with the minimal time required to finish the mission, i.e. when no collision avoidance algorithm is applied. As we can observe from Table 1, our new D-FFP algorithm does decrease the time overhead in all cases. On average, the time overhead is reduced by 8 s, which is a 32% reduction compared to the previous FFP algorithm. Notice that these gains are achieved while respecting the safety distance of 10 m between aircraft, as desired.

Table 1. Performance results for the 5 different scenarios under test when comparing the FFP and D-FFP protocols in terms of the minimum distance between UAVs and time overhead (TO) w.r.t the minimal time.

Scenario	Min. T. [s]	FFP		D-FFP	
		TO [s]	Min. distance [m]	TO[s]	Min. distance [m]
CR90	84.47	34.90	16.10	26.46	10.86
SD45	81.91	42.15	10.87	30.99	10.00
OD45	80.70	24.53	20.62	16.50	14.56
HO	83.04	16.82	10.81	8.38	11.81
TO	233.31	5.57	11.70	1.57	10.26
Average	***112.69***	***24.80***	***14.02***	***16.78***	***11.50***

To gain further insight into how such improvements are achieved, in Fig. 3 we show the actual trajectories for the two conflicting UAVs in the OD45 scenario. As shown, the adoption of D-FFP achieves a more efficient and clean avoidance trajectory, which reduces the flight time of both UAVs by reducing trajectory fluctuations. On the contrary, the FFP trajectories are slightly irregular, and sometimes the drones go around in circles, waiting to pass.

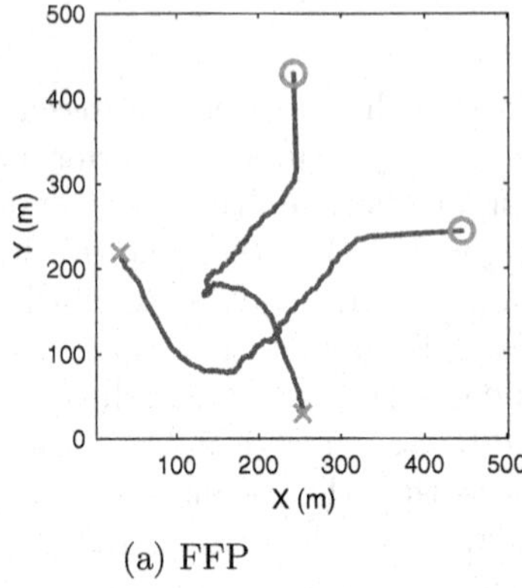

(a) FFP

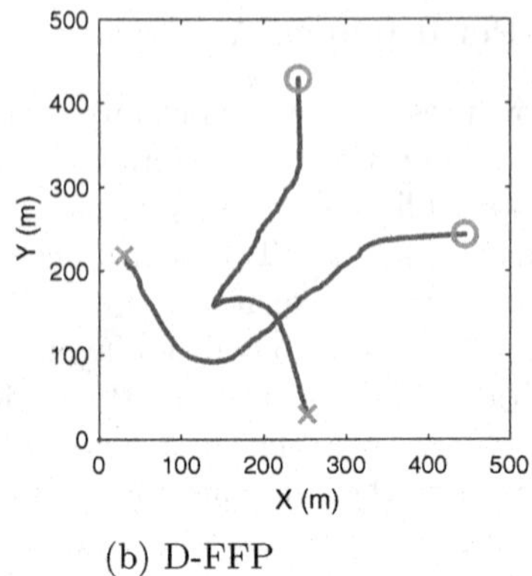

(b) D-FFP

Fig. 3. UAV flight trajectories for both protocols under test.

5 Conclusions and Future Work

In this paper, we have proposed a novel field-force approach to improve airspace management in the presence of conflicts between aircraft. In particular, we improve upon conventional field-force approaches, which assume an omnidirectional repulsion pattern around each UAV, by introducing a directional repulsion pattern.

Preliminary experiments using Matlab simulation evidence the potential of the proposed technique, which is able to improve upon the conventional approach by reducing the overall time overhead associated with collision avoidance manoeuvres, while always guaranteeing the safety distance between UAVs.

As future work, we first plan to further study the potential of this solution by analyzing the impact of the different design alternatives available while seeking the most optimal combination of parameters in terms of performance. An extension to 3D manoeuvring is also planned. We also plan to evaluate the algorithms using more real simulators such as the ArduSim [5] and, finally, the idea is to implement and test it using real drones.

Acknowledgements. This work is derived from R&D project PID2021-122580NB-I00, funded by MCIN/AEI/10.13039/501100011033 and "ERDF A way of making Europe".

References

1. Azzabi, A., Nouri, K.: Path planning for autonomous mobile robot using the potential field method. In: 2017 International Conference on Advanced Systems and Electric Technologies, pp. 389–394 (2017). https://doi.org/10.1109/ASET.2017.7983725
2. Barrado, C., et al.: U-space concept of operations: a key enabler for opening airspace to emerging low-altitude operations. Aerospace **7**(3) (2020). https://doi.org/10.3390/aerospace7030024

3. Causa, F., Franzone, A., Fasano, G.: Strategic and tactical path planning for urban air mobility: overview and application to real-world use cases. Drones **7**(1) (2023). https://doi.org/10.3390/drones7010011
4. Choi, D., Kim, D., Lee, K.: Enhanced potential field-based collision avoidance in cluttered three-dimensional urban environments. Appl. Sci. **11**(22) (2021). https://doi.org/10.3390/app112211003
5. Fabra, F., Calafate, C.T., Cano, J.C., Manzoni, P.: ArduSim: accurate and real-time multicopter simulation. Simul. Model. Pract. Theory **87**, 170–190 (2018). https://doi.org/10.1016/j.simpat.2018.06.009
6. Fabra, F., Zamora, W., Sangüesa, J., Calafate, C.T., Cano, J.C., Manzoni, P.: A distributed approach for collision avoidance between multirotor UAVs following planned missions. Sensors **19**(10) (2019). https://doi.org/10.3390/s19102404
7. Huang, C., Li, W., Xiao, C., Liang, B., Han, S.: Potential field method for persistent surveillance of multiple unmanned aerial vehicle sensors. Int. J. Distrib. Sens. Netw. **14**(1) (2018). https://doi.org/10.1177/1550147718755069
8. Kownacki, C., Ambroziak, L.: A new multidimensional repulsive potential field to avoid obstacles by nonholonomic UAVs in dynamic environments. Sensors **21**(22) (2021). https://www.mdpi.com/1424-8220/21/22/7495
9. Mohamed, N., Al-Jaroodi, J., Jawhar, I., Idries, A., Mohammed, F.: Unmanned aerial vehicles applications in future smart cities. Technol. Forecast. Soc. Chang. **153**, 119293 (2020). https://doi.org/10.1016/j.techfore.2018.05.004
10. Sun, J., Tang, J., Lao, S.: Collision avoidance for cooperative UAVs with optimized artificial potential field algorithm. IEEE Access **5**, 18382–18390 (2017). https://doi.org/10.1109/ACCESS.2017.2746752
11. Wu, E., Sun, Y., Huang, J., Zhang, C., Li, Z.: Multi UAV cluster control method based on virtual core in improved artificial potential field. IEEE Access **8**, 131647–131661 (2020). https://doi.org/10.1109/ACCESS.2020.3009972
12. Wubben, J., Calafate, C.T., Cano, J.C., Manzoni, P.: FFP: a force field protocol for the tactical management of UAV conflicts. Ad Hoc Netw. **140**, 103078 (2023). https://doi.org/10.1016/j.adhoc.2022.103078
13. Yasin, J.N., Mohamed, S.A.S., Haghbayan, M.H., Heikkonen, J., Tenhunen, H., Plosila, J.: Unmanned Aerial Vehicles (UAVs): collision avoidance systems and approaches. IEEE Access **8**, 105139–105155 (2020). https://doi.org/10.1109/ACCESS.2020.3000064

Payload Level Graph Attention Network for Web Attack Traffic Detection

Huaifeng Bao[1,2], Wenhao Li[1,2], Xingyu Wang[1,2], Zixian Tang[1,2], Qiang Wang[1,2], Wen Wang[1,2(✉)], and Feng Liu[1,2]

[1] State Key Laboratory of Information Security Institute of Information Engineering, CAS, Beijing, China
{baohuaifeng,liwenhao,wangxingyu,tangzixian,wangqiang3113, wangwen,liufeng}@iie.ac.cn

[2] School of Cyber Security, University of CAS, Beijing, China

Abstract. With the popularity of web applications, web attacks have become one of the major threats to cyberspace security. Many studies have focused on applying machine learning techniques for web attack traffic detection. However, past approaches suffer from two shortcomings: firstly, handcrafted feature extraction-based approaches are prone to introduce biases in existing perceptions, and secondly, current end-to-end deep learning approaches treat traffic payloads as non-structured string sequences ignoring their inherent structural characteristics. Therefore, we propose a graph-based web attack traffic detection model to identify the payloads in the traffic requests. Each pre-processed payload is transformed as an independent graph in which the node representations are shared through a global feature matrix. Finally, graph-level classification models are trained with graph attention networks combining global information. Experimental results on four publicly available datasets show that our approach successfully exploits local structural characteristics and global information to achieve state-of-the-art performance.

Keywords: web attack · traffic classification · graph representation learning · graph attention networks

1 Introduction

With the rapid iteration of Internet technology and the continuous improvement of computing power, the more efficient B/S architecture is becoming increasingly popular on the Internet. More and more content providers deploy their services on web pages to replace their original applications. In addition to traditional websites, various application programming interfaces (APIs) and applets have become new sources of traffic. More traffic entries and invocation methods have increased the efficiency of web application development while also bringing more complex security issues. Some malicious attackers have proposed various attack methods by studying Web technologies, such as cross-site scripting (XSS)

J. Mikyška et al. (Eds.): ICCS 2023, LNCS 14077, pp. 394–407, 2023.
https://doi.org/10.1007/978-3-031-36030-5_32

attacks, Structured Query Language (SQL) injection, command execution, directory traversal, etc. With these attack methods and the vulnerabilities in web services, web attackers can maliciously attack and infiltrate web applications to steal user data or disrupt the normal operation of the system. According to the [8], only for financial institutions, more than 736 million web attacks were recorded in 2020. It can be seen that web attacks have become one of the most mainstream cyber attacks, with a wide range of attacks and most of them posing serious privacy risks or financial losses.

In order to circumvent the above malicious attacks, researchers in related fields have conducted a lot of research. Many scholars have proposed techniques from the perspective of strengthening the security of web application , such as reverse proxy and SSL verification. However, such defenses rely heavily on web security team building, which is costly and not flexible enough to cope with changing attack methods. Therefore, more researchers focus on detecting web attacks from the application layer traffic payload. Traditional web application firewalls (WAFs) apply regular expressions to match sensitive fields from traffic requests for misuse detection. Considering that the expressive power of a single regular expression is insufficient to match the increasingly complex forms of web attack payloads, researchers have started to detect known attacks by building databases of attack signatures [3]. Such approaches require constant updating of the attack signature databases to adapt to new attack techniques and vulnerability types. The lag caused by such periodic updates poses a potential problem for system security.

With the popularity of artificial intelligence techniques, applying related techniques in web attack traffic detection has become a hot research topic. These methods can be broadly classified into two categories: handcrafted-features-based and end-to-end methods. Handcrafted-features-based methods extract features from suspicious traffic requests guided by expert experience and feed them numerically into traditional machine learning models to obtain determination results. Typical methods include extracting statistical features from the request header fields, extracting n-gram statistical features from the request payload, etc. These methods transform traffic into numerical features based on a priori knowledge of web security. So they face problems such as the inability to obtain comprehensive and in-depth information in the traffic, poor robustness against traffic pattern variations, and time-consuming pre-processing stages. In addition, the detection performance of such methods is limited by the classification capability of traditional machine learning algorithms. Recently, as deep learning has been widely used in web attack detection research, end-to-end methods have shown more robust detection capability and generalization performance with no need to design new features for novel attack methods. Researchers have started treating key fields such as uniform resource identifiers (URIs) in requests as string text and introducing natural language processing (NLP) techniques to mine the semantic information within them. Models such as TextCNN [19], BiLSTM [16], and Attention [9] have been used successively for such problems. However, such approaches treat the URI or other fields in the request as unstructured text for feature extraction and ignore the non-sequential key-value pairs dependencies inherent in web attack traffic requests.

Graph neural networks are designed to work with data embedding in non-Euclidean spaces, aggregating relevant node content information through the topology of the graph, which can achieve effective feature updates and avoid the interference of disjoint fields. In this paper, we propose Payload-Level-GAT to transform structured traffic payloads into heterogeneous graphs and update node semantic features using graph attention network (GAT) [14]. And then, node representations are aggregated by introducing global information to compute graph-level embedding and implement classification tasks.

The main contributions of this paper are as follows:

- We propose a payload-level graph representation for feature learning that abstracts traffic requests into graph structures carrying structural relationships.
- To achieve the graph-level classification task, we designed Payload-Level-GAT, combining global information and local structured contextual information to enhance the classification effect.
- We have conducted sufficient experiments on four publicly available datasets. The experimental results show that our proposed method is more efficient and generalized than baselines.

2 Related Work

2.1 Malicious Traffic Analysis Based on Machine Learning

Application layer protocol payloads play a crucial role in web attacks. Hence numerous works have focused on them. As one of the earliest studies in payload analysis, wang et al. [17] proposed PAYL to extract 1-gram frequency distribution from all bytes of the payload as features. This work achieves high detection rates and low false alarm rates but is vulnerable to mimicry attacks. After that, there are many works focused on payload analysis. Ariu et al. [1] proposed HMMPAYL, which was developed based on PAYL and used Hidden Markov Models to detect attack traffic. Oza et al. [6] extracted n-grams of HTTP traffic with various n-values. Bortolameotti et al. [2] generated features from HTTP request URIs and HTTP headers. Then, the attack traffic is detected based on the fingerprints generated by these features. In addition, end-to-end deep learning paradigms are becoming increasingly popular. Qin et al. [9] proposed an attention-based deep learning model, ATPAD, which was the first work to apply attention mechanisms in payload anomaly detection. Wang et al. [16] proposed a model to detect malicious traffic by combining CNN and LSTM.

2.2 Graph Neural Network

In recent years, there has been a growing interest in extending deep learning methods to structured data such as a graph. Researchers have proposed graph neural networks (GNN) to process graph data with neural networks.

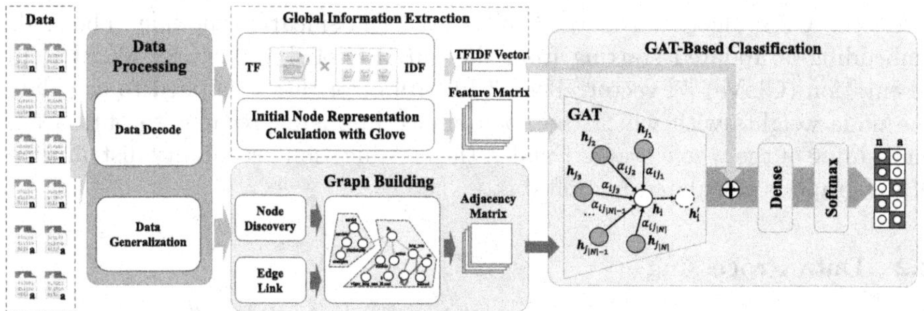

Fig. 1. Architectural view of Payload-Level-GAT.

GNN has been used in cyber security for the past few years. In [20], GNN is described to disassemble x86 instructions quickly and accurately. The key idea is to capture and propagate instruction relationships with GNN models. Another application of GNN is processing control flow graphs (CFGs) for binary code similarity detection [15]. Zhuang et al. [21] extracted control flow and data flow semantics from source code and detected vulnerabilities in smart contracts based on GNN. Recently, GNNs have also been applied to web attack detection. Liu et al. [4] proposed GraphXSS to construct graphs for the global corpus with payloads and the preprocessed items as nodes. TextGCN [18] is used to classify the payload nodes. However, this approach has two major problems: first, it uses co-occurrence relations of items in the global corpus for graph construction, ignoring local structural relations, and second, it has a large memory footprint and cannot be applied to inductive learning to determine the maliciousness of emerging traffic payloads. In contrast, our approach focuses on local structural relationships of the payloads and uses global information for their graph-level independent representation. So the method could be used for inductive learning. In addition, the memory footprint can be significantly reduced by combining mini-batch training.

3 Approach

3.1 Overall Architecture

The general architecture of our proposed method is shown in Fig. 1. First, all the data are decoded and generalized in the data preprocessing module. This is to avoid excessive dimensionality of the embedding space caused by encoding or random nonsense values and to improve the classification efficiency of the graph neural network. In the graph construction module, regular expressions are constructed to segment the payloads and discover the graph nodes from the segmented subsegments. Next, heterogeneous edges are linked, relying on contextual and structural relationships of graph nodes, forming a graph for each request. The parameters of the graph structure are learned and optimized using the GAT algorithm in the GAT-Based Classification module. The node embedding is updated starting from the initial Global Vectors for Word Representation (GloVe) [7] vector. The Tf-Idf [10] algorithm is applied to calculate the node weights with which the node embedding is aggregated to obtain the embedding of the whole graph. Finally, the classification probability distribution of the graph is calculated by SoftMax.

3.2 Data Processing

To construct a graph representation of the traffic payload, it is first necessary to extract the payload portion of the application layer traffic and perform decoding and generalization operations. Some special characters that may appear in the traffic payload could have an impact on the normal parsing process of the server, such as '?', '&' in the URI, or '⟨' in the request body of HTML format. Therefore, the server will first decode the request message after receiving it. This allows attackers to bypass defenses such as firewalls by encoding malicious fields. We attempt to decode each traffic payload to avoid the semantics of the maliciously encoded fields being segmented during subsequent graph construction. This consists of constant URI decoding, HTML entity decoding and optionally Base64 decoding, Unicode decoding, and decoding with the String.fromCharCode() method for variable or function statements in JavaScript (JS) code.

Next, we further generalize the decoded payload to prevent the graph neural network from learning particular fields and even overfitting and improve the feature learning efficiency. The above decoding step substantially reduces the vocabulary size in the whole corpus of data by restoring potentially readable fields. However, there are still some fields whose values are meaningless for determining maliciousness, although they are not encoded. We use regular expressions to identify the IP, domain, and port in the payload and generalize the fields with abstract patterns. After replacing these values, we also match the remaining numeric values in the payload and replace them with 0.

3.3 Graph Building

After data preprocessing, we propose a framework to construct a graph for each traffic payload as the input to the following graph neural network. In this subsection, we first segment the payloads into independent tokens as nodes of the graph and then link the edges of the graph from the contextual and structurally logical relationships of the tokens.

The abnormal areas of web attacks are usually found in the request path and request body of the payload (the request parameters of GET requests are treated as request bodies in this paper). For example, directory traversal attacks usually contain excessive directory hierarchies in the URI, cross-site scripting attacks, and SQL injection usually contain suspicious content in the request body, such as incomplete HTML entities and JS-sensitive functions and methods. Since these suspicious contents are mainly shown as continuous readable strings separated by some special characters after decoding, we construct a set of regular expressions to split the request URI and the request bodies into several tokens, respectively. We denote payload data containing $m + n + l$ tokens as $D = \{U_1, ..., U_i, ..., U_m, K_1, ..., K_j, ..., K_n, V_1, ..., V_k, ..., V_l\}$, U_i denotes a token of one of the URIs, and K_j and V_k denote the token separated from the keys and values in the request body, respectively. To construct a graph of the traffic payload, we treat each token appearing in it as a node of the graph.

Three classes of heterogeneous edges form the edge set of the graph. The first type of edge takes its nodes from the key-value pairs in the request body and is constructed from the logical structural relationships in the request body, i.e., such an edge starts at a key in the request body and ends at a token corresponding to the value for that key. The second type of edge is constructed with contextual adjacency relations. Each edge of this type starts at a token of the URI or request value and ends at its neighboring token. In addition, to guarantee graph connectivity, we set a 'public' node B_0 to abstractly represent the request body content. The last type of edge connects this node to the last token of the URI and to each key of the request body.

Formally, a graph of a load is defined as $G = (N, E)$, where N denotes its node set and E denotes its edge set. Concretely, the node set and the edge set can be represented as

$$\begin{aligned} N = &\{U_i, i \in [1, m]\} + \{K_j, j \in [1, n]\} \\ &+ \{V_k, k \in [1, l]\} + B_0, \end{aligned} \tag{1}$$

$$\begin{aligned} E = &\{e_{jk}, j \in [1, n], k \in [1, l]\} \\ &+ \{e_{ii'}, i \in [1, m], i^{'} \in [i - p, i + p]\} \\ &+ \{e_{kk'}, k \in [1, l], k^{'} \in [k - p, k + p]\} \\ &+ \{e_{m0}\} + \{e_{0j}, j \in [1, n]\}, \end{aligned} \tag{2}$$

where p denotes the number of neighboring tokens considered in the second class of the edge construction process. During graph building, not only the proximity dependencies of contextual relations are considered, but also the potential long-distance dependencies between keys could be iterated in short steps through the abstract node B_0.

3.4 GAT-Based Classification

In this subsection, we describe how to initialize the node representations and update them with graph attention networks to obtain a representation of the graph for classification.

The GloVe model is used to construct a globally shared initialized node representation. GloVe is an unsupervised word representation tool based on global term frequency statistics. It combines term co-occurrence information on the global corpus while capturing contextual semantic links. We pre-train the GloVe model on the global traffic payload database and obtain the initialized global shared feature matrix of graph nodes. The initial set of node embedding for graph G is defined as $N_{Emd} = \{\boldsymbol{h}_n, n \in N\}$.

There are two mainstream approaches to updating node feature vectors in graph neural networks: the spectral method and the spatial method. The spectral method maps the graph onto the spectral domain, such as the space obtained from the Laplacian matrix after feature decomposition, one of the representative methods is GCN. In this work, we use GAT [14] to update node features, a spatial method that operates directly on the graph. The model extends the information of the first-order neighbor nodes into the feature representation of the current node by masked attention, which is defined as

$$a_{ij} = a([\boldsymbol{W}\boldsymbol{h}_i||\boldsymbol{W}\boldsymbol{h}_j]), \tag{3}$$

$$\alpha_{ij} = \frac{exp(LeakyReLU(a_{ij}))}{\sum_{k \in \mathcal{N}_i} exp(LeakyReLU(a_{ik}))}, \tag{4}$$

$$\boldsymbol{h}_i^{'} = \sigma(\sum_{j \in \mathcal{N}_i} \alpha_{ij}\boldsymbol{W}\boldsymbol{h}_j). \tag{5}$$

In Eq. (3), a_{ij} denotes the importance of node j to node i, which is computed by first a linear mapping with shared parameters W to dimensionalize the features of nodes. The dimensionalized features of nodes i,j are concatenated and mapped to the real number space by $a(\cdot)$. The softmax is used to normalize the correlation coefficients to get the attention coefficients in Eq. (4). Finally, based on the computed attention coefficients, the weighted sum of the features of neighboring nodes is used as the output features of the nodes. In addition, to stabilize the learning process of self-attention, Veličković et al. found that the extension of attention with Multi-head Attention is an enhancement to the model. The computational formula of the K-head attention mechanism is defined as follows

$$\boldsymbol{h}_i^{'}(K) = ||_{k=1}^{K} \sigma(\sum_{j \in \mathcal{N}_i} \alpha_{ij}^{k}\boldsymbol{W}^k\boldsymbol{h}_j). \tag{6}$$

GAT aggregates the features of neighboring nodes to the central node with attention coefficients, which means that the updated node representation contains semantic correlations between neighboring nodes. Thus, even though tokens with large semantic gaps may appear in different payloads, their exact semantics in a particular traffic payload can be obtained by weighting the information of their neighboring nodes. Moreover, since the attention parameters a and W are shared globally during the update process, it implies that the final representation of the nodes also contains global information similar to that of other graph neural network models.

Finally, the representations of all nodes in the graph are used to predict the labels of the payloads,

$$y = softmax(ReLU(W_2 \sum_{n \in N} \beta_n \boldsymbol{h}_n^{'}(K) + b)), \tag{7}$$

where β_n serves as the contribution weights of the nodes determined by the TF-IDF algorithm, $W_2 \in \mathbb{R}^{d \times c}$ is a matrix that maps the node features to the graph representation space, $b \in \mathbb{R}^c$ is the bias vector. The training objective is to minimize the cross-entropy of the ground truth label g_i and the predicted label y_i

$$l = -g_i log y_i. \tag{8}$$

4 Experimental Evaluation

4.1 Dataset and Setup Description

To verify the detection effect of the proposed model, experiments were conducted in CSIC2010 [5], FWAF[1], TBWIDD [12], and BDCI2022[2]. CSIC2010 contains HTTP traffic data generated for e-commerce web applications, including 36000 normal requests and 25065 abnormal requests. The anomalous requests include popular attacks such as sql injection, buffer overflow, cross-site scripting, etc. FWAF is a publicly available large-scale malicious request dataset published by Fsecurify, a company aiming to develop intelligent web firewalls. They combine expert knowledge with heuristics for labeling malicious traffic. TBWIDD is published by Stevanović et al. They develop and deploy web honeypots to capture in-the-wild web attack traffic by filtering normal behavior through predefined whitelists. It contains 13048 normal requests and 9249 abnormal requests. The BDCI2022 dataset comes from the public data of the Web Attack Detection and Classification Identification track of the CCF Big Data & Computing Intelligence Contest 2022. The traffic data is divided into six labels, including normal requests, SQL injection, XSS, directory traversal, command execution, and remote code execution, totaling about 35,000 pieces. We divide the above four datasets into training, validation, and test sets according to the 6:2:2 ratio to conduct experiments, respectively. It is worth noting that except for the setting of multiclassification on the BDCI2022 dataset, the objectives of the remaining three datasets are binary classification.

We set the dimensionality of the node representation to 300 and initialize the vector with GloVe as described in Sect. 3.4. Adam is used as the optimizer. The initial learning rate is set to 0.001, the batch size to 64, the number of training epochs to 100, and the number of early stop epochs to 10. We compare the performance of our method by three widely used metrics, including accuracy (Acc), missed alarm rate (MA), and false alarm rate (FA). These metrics can be calculated as follows.

$$Acc = \frac{TP + TN}{TP + FN + FP + TN} \tag{9}$$

$$MA = \frac{FN}{TP + FN} \tag{10}$$

[1] https://github.com/faizann24/Fwaf-Machine-Learning-driven-Web-Application-Firewall.

[2] https://www.datafountain.cn/competitions/596.

$$FA = \frac{FP}{FP + TN} \quad (11)$$

where TP represents the correctly classified positive samples, FN represents the misclassified negative samples, and FP represents the misclassified negative samples.

4.2 Classification Results and Discussion

RQ1. How is the detection performance of Payload-Level-GAT? We compare Payload-Level-GAT with some handcrafted feature-based methods and end-to-end methods. Handcrafted feature-based baselines include HMM-PAYL [1], logistic regression (LR) [11], support vector machine (SVM) [11], RandomForest (RF) [13]. The end-to-end methods include TextCNN [19], LSTM [16], ATPAD [9], and GraphXSS [4] based on TextGCN. For baselines, we use the parameters described in their original paper to reimplement on the above datasets (if no relevant experimental results are available). We also take the 300-dim GloVe model for the models requiring initial embedding. Table 1 shows that our method performs competitively in detection performance on all four datasets. Our method shows the optimal performance on most metrics, with only the Fa on CSIC2010 and the Ma on BDCI2022 being suboptimal and marginally less than the optimal values. It can be seen that the end-to-end methods generally outperform the handcrafted feature-based methods. And the gap is more noticeable in challenging classification scenarios, such as multi-classification on BDCI2022, and less in more straightforward scenarios, such as TBWIDD dataset with similar detection performance across models. Remarkably, our method demonstrates perfect detection capability on the TBWIDD dataset, the potential reason for which may be that the structural information in the payload ignored by the baseline is exploited by Payload-Level-GAT. In addition to our method, another graph-based method, GraphXSS, shows superior performance in baselines, with nearly half of the metrics being suboptimal. This indicates that graph neural network-based methods hold advantages over traditional deep learning-based approaches.

To verify the detection efficiency of Payload-Level-GAT, we record the total training time and the detection time for a single test instance compared to other end-to-end baseline methods. As shown in Table 2, for the total training time and detection efficiency, the difference between Payload-Level-GAT and the baseline methods is not significant. The training time on the FWAF dataset is significantly lower than the other baselines, which can be explained by the simpler payload structure of the instances in this dataset that makes graph node updates faster, which coincides with the smaller number of edge nodes of the graphs in this dataset in Table 4. Overall, Payload-Level-GAT pervasively shows stronger detection performance on four different datasets while being comparable in efficiency to the end-to-end baseline approach.

Table 1. Comparison of Payload-Level-GAT Accuracy (Acc), Missed Alarm rate (Ma), and False Alarm rate (Fa) with baselines. Highlighted values are: (**Bold Font**) overall best classifier and (†) second best, for each metric.

Model	CSIC2010			FWAF			TBWIDD			BDCI2022		
	Acc	MA	FA	Acc	MA	FA	Acc	MA	FA	Acc	MA	FA
Handcrafted-features-based Methods												
HMMPAYL	92.63	8.27	7.97	91.26	7.51	8.32	96.84	3.48	3.76	85.32	23.81	28.93
LR	94.68	7.64	4.02	93.57	4.32	3.98	97.23	2.54	1.83	79.85	31.28	35.24
SVM	95.12	5.53	7.04	96.38	4.82	5.21	98.30	1.43	1.02	93.26	17.58	23.49
RF	95.06	3.63	3.27	95.72	3.03	3.56	98.68	1.62	0.55†	94.76	28.30	32.14
End-to-end Methods												
TextCNN	98.48†	0.38†	2.74	96.67	4.66	3.79	97.79	2.05	2.13	94.57	6.80	7.16
LSTM	98.90	1.20	1.10	94.28	2.94	3.57	98.22	1.27	1.38	94.55	6.57	9.35
ATPAD	95.86	5.60	**0.30**	97.93†	3.23	3.76	97.53	2.51	3.07	93.90	8.37	7.08†
GraphXSS	97.21	3.31	2.93	97.44	2.61†	2.59†	99.01†	1.06†	1.06	97.26†	**5.56**	9.52
Ours	**99.38**	**0.32**	0.61†	**98.69**	**1.17**	**1.31**	**100.00**	**0.00**	**0.00**	**97.47**	6.01†	**2.53**

Table 2. Comparison of Payload-Level-GAT training time consumption and detection efficiency with end-to-end methods. The **Train** column is the total training time and the **Test** column is the average time to classify a single instance in the test set.

Model	CSIC2010		FWAF		TBWIDD		BDCI2022	
	Train	Test	Train	Test	Train	Test	Train	Test
TextCNN	36 m 47 s	0.61 ms	38 m 26 s	0.59 ms	3 s	0.39 ms	21 m46 s	0.54 ms
LSTM	1 h 7 m 55 s	0.7 8ms	58 m 19 s	0.70 ms	4 s	0.52 ms	31 m30 s	0.79 ms
ATPAD	52 m 49 s	0.81 ms	46 m 25 s	0.79 ms	4 s	0.75 ms	27 m35 s	0.83 ms
GraphXSS	38 m57 s	0.7 4ms	41 m18 s	0.73 ms	3 s	0.62 ms	15 m26 s	0.56 ms
Ours	49 m5 s	0.76 ms	15 m 12 s	0.63 ms	5 s	0.58 ms	21 m42 s	0.45 ms

RQ2. What factors affect the detection performance of Payload-Level-GAT?

We performed ablation experiments to verify the effect of local structure information as well as global information on detection performance. In this work, local structure information is captured in the construction of the graph for the second and third types of heterogeneous edges. We set the ablation experiments to eliminate such edges and construct the graph from token contextual relations exclusively. In addition to the shared parameters of the graph neural network, global information is introduced in the initial GloVe embedding for graph nodes and the computation of the graph embedding weights by the Tf-Idf algorithm. In the ablation experiment, the node representations are initialized in random embedding and the graph embedding is calculated by isometric summation. Table 3 shows the detection performance for the model without global information (GI) and without local structure information (LSI) on the four datasets and the performance loss compared to the standard method. From the data in the table, it

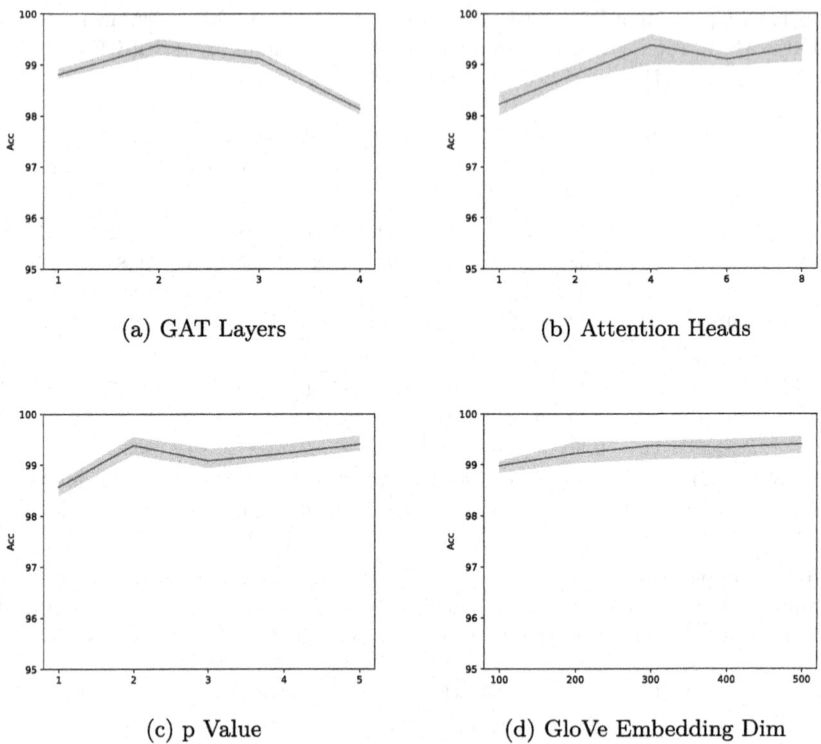

(a) GAT Layers (b) Attention Heads

(c) p Value (d) GloVe Embedding Dim

Fig. 2. Comparison results of different hyperparameters on CSIC2010.

can be concluded that both local structural information and global information bring about an improvement in detection performance.

In addition, we conducted comparative experiments on the CSIC2010 dataset for the key hyperparameters of Payload-Level-GAT to investigate the impact of hyperparameters on model detection performance. These hyperparameters include the number of GAT layers, the number of Attention headers in a single-layer GAT, the context range p considered in the construction of the first type of heterogeneous edges, and the dimensionality of GloVe encoding. Figure 2 shows the effect of different hyperparameters on the detection performance. Overall, the accuracy of the model performs stably with different parameters taken into account.

RQ3. How is the memory consumption of Payload-Level-GAT? GraphXSS, which also uses graph neural networks for payload classification, builds a graph based on the global corpus with two types of heterogeneous nodes, payload, and token in payload, achieving payload detection through node classification. As the corpus grows, the size of the global graph becomes larger, and the dependencies between nodes become more complex. The computational cost

Table 3. Comparison Results of Ablation Experiments for global information (GI) and local structural information (LSI).

Datasets		w/o GI	w/o LSI
CSIC2010	Acc %	98.96 (0.42 ↓)	98.13 (1.25 ↓)
	MA %	0.78 (0.16 ↑)	0.93 (0.31 ↑)
	FA %	0.89 (0.28 ↑)	1.36 (0.72 ↑)
FWAF	Acc %	97.82 (0.87 ↓)	97.59 (1.10 ↓)
	MA %	1.78 (0.61 ↑)	2.10 (0.93 ↑)
	FA %	1.99 (0.68 ↑)	2.03 (0.72 ↑)
TBWIDD	Acc %	99.51 (0.49 ↓)	99.22 (0.78 ↓)
	MA %	0.92 (0.92 ↑)	0.84 (0.84 ↑)
	FA %	0.77 (1.22 ↑)	1.05 (1.05 ↑)
BDCI2022	Acc %	95.37 (2.10 ↓)	96.82 (0.65 ↓)
	MA %	8.35 (2.34 ↑)	8.41 (8.40 ↑)
	FA %	6.25 (3.72 ↑)	5.42 (2.89 ↑)

and memory requirements will become unbearable. The memory footprint and the number of edges in the graph for GraphXSS and Payload-level-GAT (mean value in our method) are recorded in Table 4. It can be seen that our proposed method has a low memory footprint and is easier to deploy in practice combining with batch training strategies.

Table 4. Comparison of memory consumption and number of edge set for GraphXSS and Payload-Level-GAT.

Datasets	GraphXSS	Payload-Level-GAT
CSIC2010	84370 MB(2305449)	285 MB(92)
FWAF	76905 MB(1619610)	119 MB(68)
TBWIDD	4660 MB(69551)	28 MB(55)
BDCI2022	23964 MB(2140072)	291 MB(213)

5 Conclusion and Feature Work

In this paper, we propose Payload-Level-GAT, a graph attention network-based approach to detect web attack traffic. The core idea is to regard the payload of web attack traffic as semi-structured text and construct payload-level graphs through local contextual and structured relationships. Furthermore, a GAT-based detection framework incorporating global information to classify the payload-level graphs is proposed. The results show that Payload-Level-GAT has good detection performance and generalization ability.

In future work, there are two directions to improve Payload-Level-GAT. First, although our scheme has the potential to be deployed in decrypted traffic environments at enterprise gateways with the widespread adoption of encrypted traffic orchestration, we plan to conduct graph structure modeling of encrypted attack traffic with side-channel information to more efficiently address the trend of traffic encryption. Secondly, we plan to explore the interpretability of graph neural networks from the perspective of result analysis to support its on-the-ground application in cybersecurity.

References

1. Ariu, D., Tronci, R., Giacinto, G.: Hmmpayl: An intrusion detection system based on hidden markov models. Comput. Security **30**(4), 221–241 (2011)
2. Bortolameotti, R., et al.: Decanter: Detection of anomalous outbound http traffic by passive application fingerprinting. In: Proceedings of the 33rd Annual computer security applications Conference, pp. 373–386 (2017)
3. Clincy, V., Shahriar, H.: Web application firewall: Network security models and configuration. In: 2018 IEEE 42nd Annual Computer Software and Applications Conference (COMPSAC), vol. 1, pp. 835–836. IEEE (2018)
4. Liu, Z., Fang, Y., Huang, C., Han, J.: Graphxss: an efficient xss payload detection approach based on graph convolutional network. Comput. Secur. **114**, 102597 (2022)
5. Nguyen, H.T., Torrano-Gimenez, C., Alvarez, G., Petrović, S., Franke, K.: Application of the generic feature selection measure in detection of web attacks. In: Herrero, Á., Corchado, E. (eds.) CISIS 2011. LNCS, vol. 6694, pp. 25–32. Springer, Heidelberg (2011). https://doi.org/10.1007/978-3-642-21323-6_4
6. Oza, A., Ross, K., Low, R.M., Stamp, M.: Http attack detection using n-gram analysis. Comput. Security **45**, 242–254 (2014)
7. Pennington, J., Socher, R., Manning, C.D.: Glove: Global vectors for word representation. In: Proceedings of the 2014 Conference On Empirical Methods In Natural Language Processing (EMNLP), pp. 1532–1543 (2014)
8. Pitchkites, M.: Top cyber security statistics, facts & trends in 2022 (2022), https://www.cloudwards.net/cyber-security-statistics/
9. Qin, Z.-Q., Ma, X.-K., Wang, Y.-J.: Attentional payload anomaly detector for web applications. In: Cheng, L., Leung, A.C.S., Ozawa, S. (eds.) ICONIP 2018. LNCS, vol. 11304, pp. 588–599. Springer, Cham (2018). https://doi.org/10.1007/978-3-030-04212-7_52
10. Ramos, J., et al.: Using tf-idf to determine word relevance in document queries. In: Proceedings of The First Instructional Conference On Machine Learning. vol. 242, pp. 29–48. New Jersey, USA (2003)
11. Smitha, R., Hareesha, K.S., Kundapur, P.P.: A machine learning approach for web intrusion detection: MAMLS perspective. In: Wang, J., Reddy, G.R.M., Prasad, V.K., Reddy, V.S. (eds.) Soft Computing and Signal Processing. AISC, vol. 900, pp. 119–133. Springer, Singapore (2019). https://doi.org/10.1007/978-981-13-3600-3_12
12. Stevanović, N., Todorović, B., Todorović, V.: Web attack detection based on traps. Applied Intelligence, pp. 1–25 (2022)

13. Tama, B.A., Nkenyereye, L., Islam, S.R., Kwak, K.S.: An enhanced anomaly detection in web traffic using a stack of classifier ensemble. IEEE Access **8**, 24120–24134 (2020)
14. Veličković, P., Cucurull, G., Casanova, A., Romero, A., Liò, P., Bengio, Y.: Graph attention networks. In: International Conference on Learning Representations
15. Wang, H., et al.: jtrans: jump-aware transformer for binary code similarity detection. In: Proceedings of the 31st ACM SIGSOFT International Symposium on Software Testing and Analysis, pp. 1–13 (2022)
16. Wang, J., Zhou, Z., Chen, J.: Evaluating cnn and lstm for web attack detection. In: Proceedings of the 2018 10th International Conference on Machine Learning and Computing, pp. 283–287 (2018)
17. Wang, K., Stolfo, S.J.: Anomalous payload-based network intrusion detection. In: Jonsson, E., Valdes, A., Almgren, M. (eds.) RAID 2004. LNCS, vol. 3224, pp. 203–222. Springer, Heidelberg (2004). https://doi.org/10.1007/978-3-540-30143-1_11
18. Yao, L., Mao, C., Luo, Y.: Graph convolutional networks for text classification. In: Proceedings of the AAAI conference on artificial intelligence. vol. 33, pp. 7370–7377 (2019)
19. Yu, L., et al.: Detecting malicious web requests using an enhanced textcnn. In: 2020 IEEE 44th Annual Computers, Software, and Applications Conference (COMPSAC), pp. 768–777. IEEE (2020)
20. Yu, S., Qu, Y., Hu, X., Yin, H.: Deepdi: Learning a relational graph convolutional network model on instructions for fast and accurate disassembly. In: Proceedings of the USENIX Security Symposium (2022)
21. Zhuang, Y., Liu, Z., Qian, P., Liu, Q., Wang, X., He, Q.: Smart contract vulnerability detection using graph neural network. In: IJCAI, pp. 3283–3290 (2020)

Feature Importances as a Tool for Root Cause Analysis in Time-Series Events

Michał Kuk[1(✉)], Szymon Bobek[2], Bruno Veloso[3], Lala Rajaoarisoa[4], and Grzegorz J. Nalepa[2]

[1] AGH University of Science and Technology, Krakow, Poland
m18.kuk@gmail.com

[2] Faculty of Physics, Astronomy and Applied Computer Science, Institute of Applied Computer Science, and Jagiellonian Human-Centered AI Lab (JAHCAI), and Mark Kac Center for Complex Systems Research, Jagiellonian University, ul. prof. Stanisława Łojasiewicza 11, 30-348 Kraków, Poland

[3] Faculty of Economics - University of Porto and INESC TEC, 4200-072 Porto, Portugal

[4] CERI Digital Systems IMT Nord Europe, University of Lille, 59000 Lille, France

Abstract. In an industrial setting, predicting the remaining useful life-time of equipment and systems is crucial for ensuring efficient operation, reducing downtime, and prolonging the life of costly assets. There are state-of-the-art machine learning methods supporting this task. However, in this paper, we argue, that both efficiency and understandability can be improved by the use of explainable AI methods that analyze the importance of features used by the machine learning model. In the paper, we analyze the feature importance before a failure occurs to identify events in which an increase in importance can be observed and based on that indicate attributes with the most influence on the failure. We demonstrate how the analyses of Shap values near the occurrence of failures can help identify the specific features that led to the failure. This in turn can help in identifying the root cause of the problem and developing strategies to prevent future failures. Additionally, it can be used to identify areas where maintenance or replacement is needed to prevent failure and prolong the useful life of a system.

Keywords: explainable AI · machine learning · artificial intelligence · domain knowledge

1 Introduction

In the era of Industry 4.0, the integration of advanced technologies such as artificial intelligence (AI) and the Internet of Things (IoT) is revolutionizing the way in which industries operate. However, as these technologies become more prevalent, it is essential that they are able to provide clear and interpretable explanations for their decision-making processes. This is particularly important in the energy industry, where decisions made by AI systems can have significant consequences for the stability and sustainability of the processes.

In this paper, we aimed to demonstrate that in assets where failures are caused by component degradation the early symptoms of this degradation can be detected by the

J. Mikyška et al. (Eds.): ICCS 2023, LNCS 14077, pp. 408–416, 2023.
https://doi.org/10.1007/978-3-031-36030-5_33

classification model which furthermore allows identifying causes of the specific failures. We used an Explainable Artificial Intelligence method (XAI), specifically the SHAP (SHapley Additive exPlanations) algorithm [12] to identify these symptoms, indicating that the process of system degradation can be observed. To mitigate the presented challenges, we tested two approaches. First, we used a supervised learning problem to identify failures and XAI methods to verify the degradation process. Second, we used an unsupervised learning problem to identify anomalies in data and based on identifying this degradation also. To cope with that, we demonstrate our research with the use of the SHAP algorithm.

The paper is organized as follows: In Sect. 2 we describe the papers that cover approaches to anomaly detection and predictive maintenance. In Sect. 3 we present our method of detecting early symptoms of asset wear in the context of identifying areas where this wear occurs. We evaluate the method on two datasets in Sect. 4. Then Sect. 5 presents and discusses our results. Finally, in Sect. 6, we summarize our work.

2 Related Works and Motivation

Anomaly detection is a process of identifying unusual data points that do not conform to expected patterns. Two common methods for detecting anomalies are data-driven and model-based approaches [8, 13]. Data-driven methods can be further categorized into supervised and unsupervised, where supervised methods use labeled data to learn what is considered an anomaly, while unsupervised methods use techniques such as autoencoders or density-based clusterings to identify anomalies. In recent years, advances have been made in the field, such as using a CNN (convolutional neural network) to imitate human vision and decision-making for anomaly detection, or a multi-step approach that analyzes time series data in both the time and frequency domains [3,5].

In a study [9], the authors decided to use a variation autoencoder to calculate reconstruction error, and based on the results, they labeled this error as anomalous or not. They then built a classifier that learned which points were anomalous and used it to provide explanations with the help of a SHAP algorithm [12]. However, it is worth noting that the performance of these techniques can vary depending on the specific application and the dataset available. The generalization of these methods to other contexts remains an open problem.

In [11] authors presented a new predictive maintenance policy called Sensory-updated Degradation-based Maintenance Policy (SUDM) that utilizes real-time component degradation signals and component population data to predict residual life and schedule maintenance. The policy is evaluated using a simulation model and compared with two other benchmark policies, resulting in a lower frequency of unexpected failures and lower overall maintenance costs.

Authors in [15] use autoencoder for anomaly detection instead of traditional health index to detect bearing faults. Deep neural networks extract healthy bearing representations and decoded signal residuals for fault detection. Setting an appropriate threshold for early detection is challenging without increasing false positives. Training with healthy signals is difficult to distinguish from early degradation stages.

The study [10] predicts the degradation stages of rolling-element bearings in pharmaceuticals using high-frequency vibration data. They propose a framework that uses k-means and an autoencoder to generate a labeled dataset for training a supervised model. The results are reliable and scalable, based on experiments on the FEMTO Bearing dataset.

Considering the papers presented above, many approaches for anomaly detection do not take into account the domain of the anomalies. This means that in most cases, algorithms detect many anomalies that do not actually reflect real problems with the asset. What is more, taking into account the number of detected anomalies by the machine learning algorithm we believe that the anomalies which could have an impact on the system's working conditions should characterize by the early symptoms (some degradation process). It motivated us to develop an original method that allows for explaining which anomalies are in fact crucial for the asset.

3 Feature Understanding Method

The study consists of the following steps, which are divided into two machine-learning problems. In our research, the first case involved a supervised ML problem where failure periods were available, but in the second case we focused on finding anomalies and relying on this build classifier because such periods were not available. Finally, for both cases, we used the SHAP algorithm to analyze feature importance in time.

The Shapley value is determined by evaluating the value of the feature in all possible combinations with other features and weighting and summing the results. It is defined through the value function of the features in the model as presented in Equation:$\Phi_j = \sum_{S \subseteq 1,\ldots,p/j\}}^{s} \frac{|S|!(p-|S|-1)!}{p!}(val_x(S \cup j) - val_x(S))$

It is defined by evaluating the prediction of a subset of features (S) in the model while marginalizing the features that are not included in the subset. This is done by using a value function that takes in the vector of feature values of the instance to be explained, and the number of features (p) in the model Equation where $val_x(S) = \int f(x_1, \ldots, x_p) d\mathbb{P}_{x \notin S} - E_X(f(X))$ [14] The detailed explanation of equations is presented in [14].

Our method was evaluated on two datasets C-MAPSS [17] and real data from the steel plant which represents measurements from precess. In the first example, we used the dataset which has been already labeled [17]. So, the analyzed case was treated as a supervised problem approach. As a classifier algorithm, we used XGBoost classifier [6]. One advantage of this algorithm is its efficiency in computational time and obtained scores. The algorithm is based on decision trees, which in combination with the SHAP algorithm, is much more efficient than other classifiers that are not tree-based. In the second example, we used an unlabelled dataset where the neural network was applied to calculate the reconstruction error and obtain labels for the prediction. Then the procedure of classifier application was repeated for the labeled dataset.

3.1 Supervised Task

Both approaches use a classification algorithm to train a model which later is used by the explainer algorithm. Classification is a supervised learning task, where the goal is to predict the class or category of a given data point based on a set of input features or attributes. Mathematically, it can be represented as a mapping function $f(x)$ which maps a given input x (a feature vector) to a class label y. The function $f(x)$ is learned from a labeled training dataset $D = (x_1, y_1), (x_2, y_2), ..., (x_n, y_n)$, where x_i is the i-th input feature vector and y_i is the corresponding class label [18].

In this task, two cases can be considered. Multi-class classification is a classification task with more than two possible outcomes that should be expected as an output. In the case of anomaly detection, a binary classification type of classifier is used. In this case two possible outcomes should be expected, such as predicting whether a data point is anomalous or not. In our research, we used the XGBoost classification algorithm which is based on the tree ensemble model design to be highly scalable. XGBoost builds an additive expansion (adds new models to the existing ones) of the objective function by minimizing a loss function [4,6].

3.2 Unsupervised Task

In the unsupervised task firstly we concentrate on anomaly detection. To deal with that we used an autoencoder-based model. Autoencoders are a special type of neural network that was first introduced in [16]. They are trained to recreate their input, and their main goal is to learn an informative representation of the data in an unsupervised manner. This representation can then be used for various purposes such as clustering. The issue, as defined in [1], is to learn functions. A and B presented in the following equations $A : R^n \rightarrow R^p$ and $B : B^n \rightarrow B^p$ to solve the equation $argmin_{A,B} E[\Delta(x, B \cdot A(x))]$

The expectation of the distribution of x, represented by E, is used in conjunction with the reconstruction loss function, represented by Δ. This function measures the difference between the output produced by the decoder and the original input, typically using the L2 norm [2].

Autoencoders [7] can be used as anomaly detection methods that use dimensionality reduction to try to identify a specific subspace where the normal and abnormal data differ significantly. This is done by taking a set of normal training data, represented as d dimensional vectors $x_1, x_2, ..., x_n$, $(x_i \in R^d)$ and using a model to project them into a lower-dimensional subspace. The output of this process is a set of reproduced data $x_1^{'}, x_2^{'}, ..., x_n^{'}$. The goal is to minimize the reconstruction error, which is the difference between the original and reproduced data, in order to find the optimal subspace for anomaly detection and is defined by equation $\varepsilon(x_i, x_i^{'}) = \sum_{i=1}^{d}(x_i - x_i^{'})^2$ When the data in the test dataset is similar to the typical patterns established during training, the error in reconstructing it will be lower. However, data that deviate from these patterns will have a higher reconstruction error. By setting a threshold for the reconstruction error defined by Equation, it becomes simple to identify and classify abnormal data $c(x_i) = \{^{normal \quad \varepsilon_i < \theta}_{abnormal \varepsilon_i > \theta}$ In the next section, we present an experimental evaluation of our method on two data sets.

4 Evaluation

First, we evaluated the method using a commonly used benchmark data set with synthetic data. Then, we used a real data set obtained from our industrial partners. As it is demonstrated in the remainder of this section, both experiments resulted in promising results.

4.1 Experiment on the C-MAPSS Dataset

The C-MAPSS dataset [17] is commonly utilized for research on predicting future performance and maintenance of systems. It includes the outcomes of a simulation of a turbofan engine utilizing the C-MAPSS software, which is provided by NASA.

C-MAPSS Dataset Description. The dataset has information on hundreds of turbofan units with 3 operational parameters and 21 measurements taken during each unit's operation. The units deteriorated gradually over time, leading to the failure of the high-pressure compressor. The data is organized into cycles and split into four scenarios, reflecting the varying rates of deterioration and influencing factors.

To evaluate our research, we used the C-MAPSS dataset, which consists of 15631 rows and 29 columns. For our case, we decided to remove the columns like unit and cycle from the original dataset because data in this column were not directly connected with the sensor measurements. To train the classification model, we set an additional parameter in the model responsible for the weights of the classes. As a result, we obtained the F1-score 0.97 for class 0 and 0.83 for class 1 where class 0 means normal working condition and class 1 means failure.

In the Fig. 1, there is presented the distribution of the SHAP values takes into account the whole features available in the dataset. For this presentation, the box plots were used where the x-axis is the cycle time. The red and green lines, there are marked respectively the maximum and minimum values of the SHAP values, which were calculated based on the aggregated data. According to previous obtain results, we can see that considering the whole features in the dataset, the SHAP values increase over time.

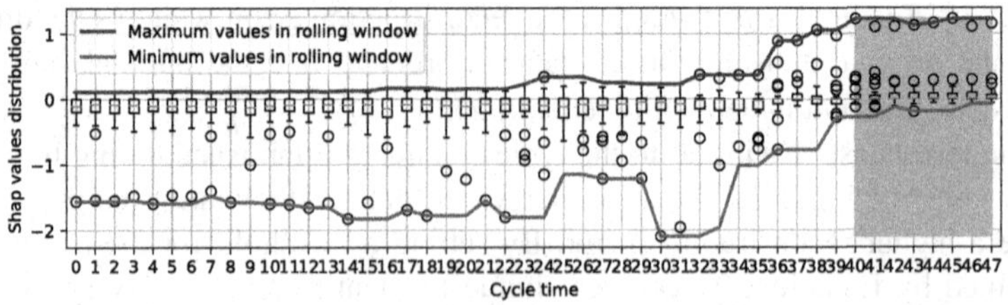

Fig. 1. SHAP values distribution during whole cycle

4.2 Experiment on Hot-Rolling Process Dataset

Our research focuses on the hot rolling process of steel at the highly automated Hot Rolling Mill (HRM) or Hot Strip Mill (HSM) at the ArcelorMittal Poland company in Krakow. The process involves heating a cold slab to around $1200°C$ in a walking beam furnace, passing it through the roughing mill (RM) to reduce thickness and width, and then through the finishing mill (FM) where the steel's thickness is further reduced. The steel is then cooled in a laminar cooler using water [9].

Hot Rolling Process Dataset Summary. The dataset consists of 14,443 instances of hot rolling process data. The measurements consist of physical values of temperatures, stresses, thickness, etc. We used 35 variables to build the model, which were generated from raw values by mapping these features to a new set of values to make the data representation more relevant and easier to process for later analysis (features transformation). We used Dense Variational Autoencoder to detect anomalies. The training dataset which where provided as input to the autoencoder. The main parameters used during the training process are follwing: Latent space shape: 4, Activation function: elu, Batch size: 32, Epochs: 300, Dropout: 0.4 After learning the model, we compared the received signal from the decoder with the signal that was treated as the input to the encoder. To use the supervised classification algorithm, we had to evaluate which data points were normal and which were abnormal (anomalies). We specify a threshold that is 0.99 percentile of the reconstruction. Based on that we labeled the dataset and perform a classification algorithm with the following scores: 1.0 F1-score for class 0 and 0.72 F1-score for class 1.

Hot Rolling Process Evaluation. As a result of using the SHAP algorithm to classify the anomalies, we obtained the results, shown in Fig. 2. Similarity to the examples presented in the Sect. 4.1 we aggregate the most important features into a single Figure, where each Figure corresponds to a different anomaly. Anomalies are marked with red rectangles or lines if an anomaly has been detected and lasted only one timestamp. In each case, we are able to observe an increase in the importance of the features before the anomaly occurred. In addition, we are able to see which features an increase in importance more and how quickly.

In the analyzed case, in three of the four graphs presented at the beginning of these anomalies, we observe an increase in the importance of temperature sensors, in two anomalies also an increase in the stress feature, and in one case an increase in torque. In order to fully evaluate these results, we asked an expert from Accerol Mittal, who was responsible for providing the data, to check the results obtained. Based on his process knowledge and experience, all temperature-related features should be reflected in the indicated anomalies because these attributes are the most important for this hot-rolling process. We choose one of the anomalies and presented it on the box plot chart in Fig. 2. The maximum values and minimum values of feature importance are marked by the red and green lines respectively. However, taking into account what happens between these two lines thanks to the analysis range of box plots we can say that the range of all features' importance increases before and are respectively high during the anomaly.

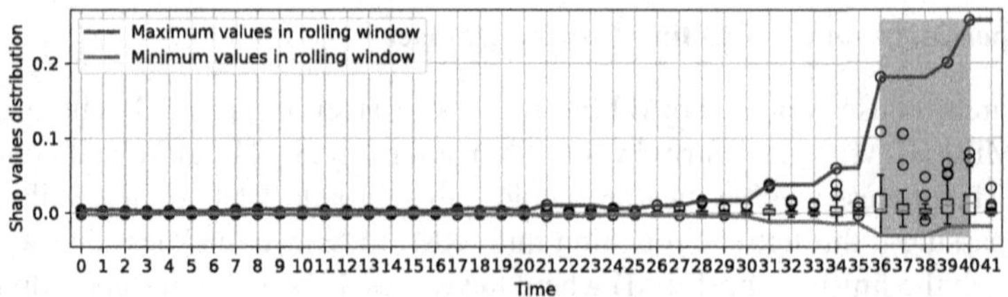

Fig. 2. SHAP values distribution before and during anomaly

5 Discussion

The work presented in this paper is based on the authors' assumption that Explainable Artificial Intelligence methods, in addition to global explanation of classification algorithms, allow to provide more information about the process and detected events in individual key events for the operation of the system.

The assumption was tested on two datasets where the events were caused by some wear processes. In both cases, we were able to identify features where the increase in the importance of features was increasing over time even before the actual event occurred. Moreover, based on this, we were able to identify which features increase the most, which allows us to better understand the behavior of the model.

The correctness of the obtained results was confirmed by an Accelor Mittal expert, who pointed out that in all the analyzed units of the hot rolling set, the temperature may be a key parameter responsible for faster wear of the rolling stage components. Given this opinion, it is reasonable to undertake further research on this topic and ultimately build an anomaly detection algorithm that takes into account the sensors which are not directly correlated with asset wear. In order to fully take into account the obtained results and expert opinions, it remains to conduct further research in order to force the model to focus on the features relevant to the wear of the element and at the same time prevent them from lowering the scores.

In addition, based on the expert's opinion and using the methods of XAI, we demonstrated that a well-prepared model built on the basis of well-prepared data is able to reproduce the actual degradation processes taking place in the plant.

6 Conclusion

In our work, we focused to analyse the causes of the detected events to validate if the anomaly characterizes early synthons. We treated these symptoms as a degradation process which led to event detection. In this work, we analyzed two examples of different cases where such behaviors can be observed. In the presented first example the events were obviously indicated by the labels. However, in the second example, these events were detected by the anomalies detection algorithm. In this example, we used an autoencoder to reproduce the original signal and find anomalies.

In both analyzed cases, we demonstrate that artificial intelligence methods are able to make predictions based on signals that are relevant to the process and can be interpreted. What's more, we were able to identify the features that are responsible for the predictions of the models, and the physical significance of these features was confirmed by an expert.

Acknowledgements. This paper is funded from the XPM (Explainable Predictive Maintenance) project funded by the National Science Center, Poland under CHIST-ERA programme Grant Agreement No. $857925(NCNUMO - 2020/02/Y/ST6/00070)$

References

1. Baldi, P.: Autoencoders, unsupervised learning, and deep architectures. In: Guyon, I., Dror, G., Lemaire, V., Taylor, G., Silver, D. (eds.) Proceedings of ICML Workshop on Unsupervised and Transfer Learning. Proceedings of Machine Learning Research, vol. 27, pp. 37–49. PMLR, Bellevue, Washington, USA (02 Jul 2012). https://proceedings.mlr.press/v27/baldi12a.html
2. Bank, D., Koenigstein, N., Giryes, R.: Autoencoders (2020). arXiv preprint arXiv:2003.05991
3. Basora, L., Olive, X., Dubot, T.: Recent advances in anomaly detection methods applied to aviation. Aerospace **6**(11) 2226–4310 (2019). https://doi.org/10.3390/aerospace6110117, https://www.mdpi.com/2226-4310/6/11/117
4. Bentéjac, C., Csörgo, A., Martínez-Muñoz, G.: A comparative analysis of xgboost. arXiv preprint arXiv:1911.01914 (2019)
5. Chalapathy, R., Chawla, S.: Deep learning for anomaly detection: A survey. arXiv preprint arXiv:1901.03407 (2019)
6. Chen, T., Guestrin, C.: Xgboost: A scalable tree boosting system. arXiv preprint arXiv:1603.02754 (2016)
7. Chen, Z., Yeo, C., Lee, B.S., Lau, C.T.: Autoencoder-based network anomaly detection. In: 2018 Wireless Telecommunications Symposium (WTS), pp. 1–5 (2018)
8. Isermann, R.: Model-based fault-detection and diagnosis - status and applications. Ann. Rev. Control **29**(1), 71–85 (2005). https://doi.org/10.1016/j.arcontrol.2004.12.002, https://www.sciencedirect.com/science/article/pii/S1367578805000052
9. Jakubowski, J., Stanisz, P., Bobek, S., Nalepa, G.J.: Anomaly detection in asset degradation process using variational autoencoder and explanations. Sensors **22**(1), 291 (2022). https://doi.org/10.3390/s22010291, https://www.mdpi.com/1424-8220/22/1/291
10. Juodelyte, D., Cheplygina, V., Graversen, T., Bonnet, P.: Predicting bearings degradation stages for predictive maintenance in the pharmaceutical industry. In: Proceedings of the 28th ACM SIGKDD Conference on Knowledge Discovery and Data Mining, pp. 3107–3115 (2022)
11. Kaiser, K., Gebraeel, N.: Predictive maintenance management using sensor-based degradation models. IEEE Trans. Syst. Man Cybern. Part A: Syst. Hum. **39**, 840–849 (2009). https://doi.org/10.1109/TSMCA.2009.2016429
12. Lundberg, S.M., Lee, S.: A unified approach to interpreting model predictions. arXiv preprint arXiv:1705.07874 (2017)
13. Mehdi, G., Naderi, D., Ceschini, G.F., Roshchin, M.: Model-based reasoning approach for automated failure analysis : An industrial gas turbine application (2015). https://doi.org/10.36001/phmconf.2015.v7i1.2719

14. Molnar, C.: Interpretable Machine Learning (2022). https://christophm.github.io/interpretable-ml-book
15. Principi, E., Rossetti, D., Squartini, S., Piazza, F.: Unsupervised electric motor fault detection by using deep autoencoders. IEEE/CAA J. Autom. Sin. **6**(2), 441–451 (2019). https://doi.org/10.1109/JAS.2019.1911393
16. Rumelhart, D.E., Hinton, G.E., Williams, R.J.: Learning Internal Representations by Error Propagation, pp. 318–362. MIT Press, Cambridge (1986)
17. Saxena, A., Goebel, K., Simon, D., Eklund, N.: Damage propagation modeling for aircraft engine run-to-failure simulation. In: 2008 International Conference on Prognostics and Health Management, pp. 1–9 (2008). https://doi.org/10.1109/PHM.2008.4711414
18. Sen, P.C., Hajra, M., Ghosh, M.: Supervised classification algorithms in machine learning: a survey and review. In: Mandal, J.K., Bhattacharya, D. (eds.) Emerging Technology in Modelling and Graphics, pp. 99–111. Springer Singapore, Singapore (2020)

Smart Head-Mount Obstacle Avoidance Wearable for the Vision Impaired

Peijie Xu(✉), Ron Van Schyndel, and Andy Song(✉)

School of Computing Technologies, RMIT University, Melbourne 3000, Australia
peijie.xu@student.rmit.edu.au, andy.song@rmit.edu.au

Abstract. Obstacles present serious risks and dangers for individuals who are blind or visually impaired (BVI), especially when they are not accompanied by a companion or assistant. In this study, we propose a head-mounted smart device to address this challenge. This study aims to establish a computationally efficient mechanism that can accurately detect the presence of obstacles on the path and provide warnings in real-time. The learned obstacle warning model needs to be reliable and small in size so that it can be embedded in the wearable device and run without consuming too much energy. Moreover, it must be able to deal with natural head turns, which can significantly impact readings from the head-mounted sensors. To determine the most appropriate model that can balance accuracy and real-time performance, we investigated more than thirty models and compared their key metrics. Our study demonstrates that a highly efficient wearable device is feasible and can help BVI individuals avoid obstacles with high accuracy. Additionally, we have collected a large data set that can serve as a benchmark for future studies in this area.

Keywords: Smart wearables · Obstacle avoidance · Vision impaired · BVI · Supervised machine learning

1 Introduction

Blind and visually impaired people (BVI), disadvantaged groups of our society, are globally estimated to be 43.3 million and 295 million, respectively, according to the 2020 stats [1]. For BVI people, navigating themselves from one place to another is a real challenge. A key part of the challenge is obstacles on their paths, posing a serious risk for a BVI person. Detecting obstacles and other hazards in real-time can greatly improve the mobility of BVI people, reducing their exposure to dangers. To help them, vision researchers and engineers have invested a large amount of effort into this area. In this study, we are to provide a wearable solution as a candidate choice for them. The solution is to be smart and fast. We introduce a low-energy consumption device that can accurately

R. Van Schyndel—Passed away prior to the submission of this paper. This is one of the last works of him.

J. Mikyška et al. (Eds.): ICCS 2023, LNCS 14077, pp. 417–432, 2023.
https://doi.org/10.1007/978-3-031-36030-5_34

warn of potential risks, e.g., obstacles, but not alarm falsely on harmless objects directly ahead. To be practical for BVI people, the detection must be fast and lightweight, running on the device itself.

The following parts of this paper are organised as such. A review of previous works on obstacle avoidance is presented in Sect. 2. The proposed head-mount smart device is described in detail in Sect. 3. Sections 4 and 5 explain our experiments including how sensor data are collected through the wearable device under a controlled environment, how they are labeled, and how learning algorithms are formulated and evaluated. The results of the experiments are in Sect. 6, while the discussions of this study are presented in Sect. 7. Section 8 concludes this study with a vision for future work.

2 Literature Review

In the past, a large number of approaches have been proposed to provide environment information to a level that can assist a BVI person to navigate around for daily activities [2]. These approaches can be grouped into three categories, which are Electronic Travel Aids (ETA), Electronic Orientation Aids (EOA) and Position Locator Devices (PLD). ETA focuses on perceiving and translating the information of the surrounding environment of the users, while EOA aims to help the BVI person maintain an accurate orientation during travel. On the other hand, PLD is to provide the position information of the person or the target in the scene [2]. In this study, we mainly review the relevant literature in the field of ETA as that is the aim of this study, although our proposed methodology could be transferred to EOA and PLD.

The effort on supplementing or replacing the white cane with ETA started around the nineteen forties [3]. After fifty years of development, electronic travel assistance for BVI people converged much more towards the navigation function [3]. Ran et al. designed a wearable assistance, Drishti, to provide dynamic interactions and adaptability to changes. This approach realised the seamless switching between indoor, outdoor environments and bus station navigation through differential GPS for the outdoor environments. The original equipment manufacturer's ultrasound sensor provides an accuracy of 22 centimetres [4]. Such low accuracy was far from meeting the needs of reliable real-time obstacle avoidance. Bousbia-Salah et al. proposed a navigation aid that relied on memory function with an integral accelerometer, computing and recording walking distance as a type of guidance [5]. Two vibrators and two ultrasonic sensors are mounted on the user's shoulders for obstacle detection. Another ultrasonic sensor was integrated into the white cane. However, this system required a sighted individual to accompany the BVI person, who was also required to carry a cane to complete the first navigation route. In addition, cumulative tracking errors were not well handled and often led to system failure after a period of operation.

Ultrasonic sensors have been popular in sensor-based solutions in the past decades. These sensors have high sensitivity and penetrative ability, hence suitable for obstacle detection tasks [6,7]. For example, "NavBelt" was capable of scanning a range of 120° by arranging eight ultrasonic sensors on the abdomen. The signals from the ultrasonic sensors were then processed by the robotic obstacle avoidance algorithms, of which the outputs were acoustically delivered to users [8]. The limitation of this system is that it cannot reliably detect obstacles above the user's head. Gao et al. created a wearable virtual cane network composed of four ultrasonic sensors on two wrists, the waist, and one ankle of the user. It is designed to detect obstacles as small as 1 cm^2 in size and located 0.7 m away [9]. A more recent study proposed a wearable assistive device with a glass frame supporting three ultrasonic sensors and a hand band equipped with a LiDAR sensor [10]. A fuzzy decision support system was integrated and achieved better obstacle avoidance performance than the conventional white cane for both outdoor and indoor environments. They reported reduced average walking time and reduced number of collisions in their experiments. However, the participants collided with some small harmless obstacles during the tests [10]. The two studies above both combined the device with the upper limb, but they did not consider the natural swing of the arm during walking.

With the proliferation of technologies like edge computing, smart sensors and AI, navigation assistance for BVI people attracts more researchers and engineers, aiming for an ultimate intelligent commericalisable wayfinding and mobile navigation mechanism. Many state-of-the-art techniques such as Radio Frequency Identification (RFID)-based model [11], Augmented Reality (AR) technology [12] or cloud system [13–15] were gradually adopted into the development of this space. But these techniques often require heavy computation power or external support (e.g., 5G, e-tags etc.). Vision sensors, especially with depth detection function, gradually become the most popular perceptive method in many systems, due to the low cost of these sensors and the fast development in deep learning-based computer vision models. Hicks and colleagues integrated a depth camera, a small digital gyroscope, and an LED display on a ski goggle to help people with poor vision utilise their functional residual vision to navigate around [16]. Yang et al. enhanced close obstacle detection by expanding the preliminary traversable area through a seeded growing region algorithm, which can break the limit of narrow depth field angle and sparse depth map [17]. A study utilised the depth and colour information captured by a consumer-grade RGB-D camera to segment the unobstructed paths in the scene [18]. They reported that the obstacle-free path segmentation algorithm could run at a rate of 2 frames per second (FPS), while the whole system including RGB image and depth data processing and user interface generation run much slower at 0.3 FPS. Lee and Medioni proposed a novel wearable navigation system based on a combination of an RGB-D camera, a laptop, a smartphone user interface, and a haptic feedback vest [19]. The system estimated a real-time ego-motion by sparse visual features, dense point clouds, and the ground plane and create a 2D probabilistic occupancy grid map for dynamic path scheme and obstacle avoidance. The heavy equipment required in this setup however compromises its usability. Other researchers also tried to adopt more accurate vision algorithms such as SSD [20], YOLO [21] on obstacle recognition scenarios with high definition camera to help

BVI people [6,22,23]. However, the problem with all these works is the large computational cost required to carry out the detection, typically a high-end laptop. Obviously, it is impractical for a BVI person to walk around with a laptop all the time. Our aim is to provide a lightweight yet real-time detection method. Hence cameras are not used in this study.

3 System Design

The system proposed in this study is a head-mount smart device. It is comprised of two main modules: the acquisition module and the processing module as shown in Fig. 1. The former is to perceive the surroundings through ultrasonic sensors and a 9-DOF orientation inertial measurement unit (IMU). These sensors are all connected to the processing module, which is responsible for data acquisition, computation, and decision-making processes. In this study, the module is a Raspberry Pi 4B, which is highly portable and versatile.

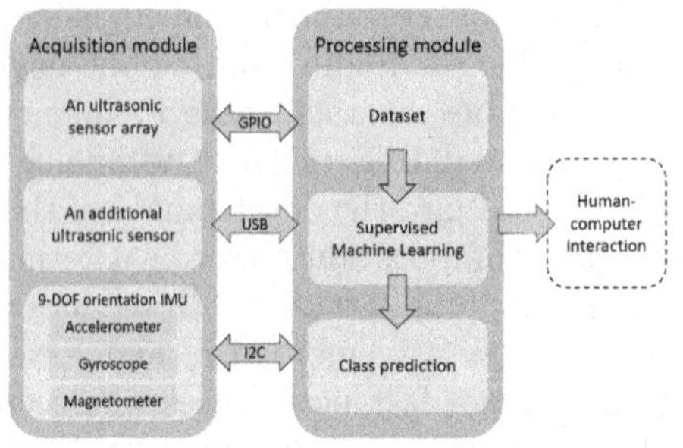

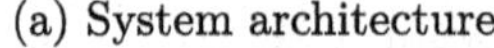

(a) System architecture

(b) The actual prototype

Fig. 1. The proposed head-mount obstacle avoidance system

The ultrasonic sensor array consists of nine sensors arranged in two rows. Four of the sensors in the top row are to detect obstacles in the upper region, while the remaining five sensors are for lower areas. These sensors are HC-SR04, with an effective detection angle of 15°, a maximum detection distance of 400 cm and a minimum of 2 cm[1]. Their accuracy is up to 3 mm. All these sensors are fully integrated with ultrasonic transmitters and receivers. The proposed system also incorporates an additional ultrasonic sensor, the MaxSonar-EZ1 from MaxBotix[2]. This sensor is placed in the middle to supplement the coverage. It communicates directly with the processing module via a USB port. This sensor has a range of 30 cm to 500 cm and is featured with compensation for target size variations, well-balanced sensitivity and specificity, built-in noise reduction, real-time background auto-calibration, real-time waveform analysis, and noise

[1] HC-SR04 Specs: https://www.handsontec.com/dataspecs/HC-SR04-Ultrasonic.pdf.

[2] HRUSB-MaxSonar®-EZ™ Series: https://www.maxbotix.com.

rejection. This sensor is expensive but helps validate readings from the sensor array. Note the sensor set position against the subject head is not guaranteed. So the learning algorithm needs to be able to handle such variations.

IMU is to measure the user's movement e.g., acceleration and rotation. Our system uses Adafruit BNO085[3], which integrates accelerometer, gyroscope and magnetometer. It communicates with the processing module via the I2C bus, transmitting three axes of linear acceleration data, three axes of gravitational acceleration data, and three axes of magnetic field sensing readings. The raw data are converted into the form of a four-point quaternion or rotation vector. That can reduce the computing burden on the processing module and minimise the impact of drift errors.

4 Data Collection

The data for the study is collected in a 15-meter-long corridor under temperature and humidity control, as depicted in Fig. 2. This study focuses on the indoor environment. Outdoor environment has a great amount of uncertainty and variations, and will be addressed in our further studies. One side of the corridor is a clean wall, while the other side is furnished with long sofas. The floor is marked 1-meter and 1.5-meter parallel to the walls, with 0.5-meter and 0.75-meter spacings between them. These marks are to guide participants walking at different paces and speeds. Additionally, angles of 75 and 60°C relative to the wall are marked to guide participants walking while looking sideways[4].

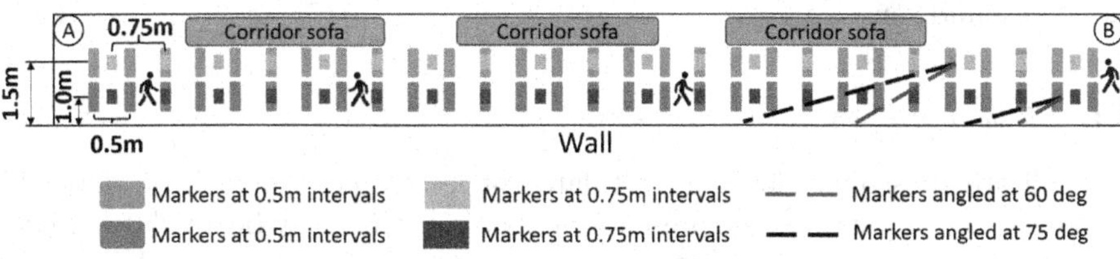

Fig. 2. The setup for data collection

Three healthy participants were invited to assist with our data collection. Firstly, they put on the head-mount device shown in Fig. 1, and adjust it to a comfortable position. We then check whether the device is functioning properly or not, and make sure the orientations of these sensors are aligned correctly. Data are collected while participants complete the following actions:

---

[3] Adafruit 9-DOF Orientation IMU BNO085 https://cdn-learn.adafruit.com.

[4] The experiments are conducted under human research ethics approval. Following the Declaration of Helsinki, participants are informed about the details of the study, including the purpose, potential risks, and obligations.

1. walking from Point A to Point B, then returning to Point A, at different speeds: 0.5 m/s; 0.75 m/s; 1.0 m/s;
2. walking from Point A to Point B, then returning to Point A, at different speeds: 0.5 m/s; 0.75 m/s; 1.0 m/s, but with other pedestrians same walking pass in the same direction or the opposite direction;
3. walking from Point A to Point B, then returning to Point A, at different speeds: 0.5 m/s; 0.75 m/s; 1.0 m/s, with the head pointing/looking to the side;
4. walking from Point A to Point B, then returning to Point A, at different speeds: 0.5 m/s; 0.75 m/s; 1.0 m/s, with the head kept turning side to side (e.g., looking around with head turns);
5. walking from Point A or Point B towards the wall at different angles: 60 deg; 75 deg.

4.1 Data Processing

All sensory data are collected with timestamps. These data are not synchronised due to the different sampling frequencies of these sensors. We manually go through timestamped data to check the validity of these data points. Redundant and duplicate data samples are removed. In total 10,234 entries of valid data are collected from three participants. That formulates our data set for the next stage of our study. Each entry contains 20 separate attributes, which are three axes of linear acceleration, three axes of gravitational acceleration, four attributes from quaternion, one distance reading from the MaxSonar ultrasonic sensor, and nine distance readings from the ultrasonic sensor array. On average, one second of movement generates 30 data points. That is about the sampling frequency of the ultrasound sensor array.

4.2 Data Labeling

To facilitate subsequent learning, all data entries are labeled with “0” and “1” to indicate the absence or the presence of obstacles that may be a risk for a BVI person. The labeling is determined according to the literature in BVI studies [8,24]. An obstacle on the pathway of the person within a distance of 1.5 m is considered positive, e.g., a risk. An obstacle that is more than 1.5 m away or is not on the trajectory of the person, e.g., on the side or way above, is labeled as zero. The processed data set contains 5,174 valid positive entries and 5,060 zero cases. That is to avoid class imbalance which may lead to bias during learning.

5 Methodology

Real-time obstacle avoidance faces three challenges, which are the delay in processing the sensory input in the ETAs system, the delay in presenting warning signals or recommendations to the BVI user, and the delay in the user’s response to the signals. This study mainly deals with the first two as the third is not in the scope of computing. In order to satisfy the requirement for real-time performance, tasks such as data acquisition, pre-processing, and decision-making need to be accomplished within a very short period of time. Hence we adopt parallelled multithread processing in our system architecture. The control of data acquisition of each sensor is separately managed as a thread. The controls of different sensors do not interfere with each other but run in a synchronised way. Due to the hardware characteristics of these sensors, they operate at different frequencies, including sampling frequency, response frequency, and feedback frequency. It would be detrimental if a certain sensor reading was blocked by the task scheduled in front of it, which may occur often if the parallel mechanism is not in place. Multithreading can effectively avoid these problems [25]. Also, multithreading facilitates resource sharing during the process. The threads share the memory and CPU time, which are scarce on the devices, e.g., Raspberry Pi that we use for this study. More specifically, the processing unit is a 64-bits Raspberry Pi OS (Raspbian) with Linux kernel 5.10.

5.1 Learning Models

To build a good obstacle detection model, we investigate a range of machine learning methods, more specifically a set of classification algorithms, that include (1) ZeroR, (2) C4.5 decision tree, (3) Naive Bayes, (4) k-nearest neighbour algorithm (kNN) and (5) Multilayer Perceptron. These methods are the most well-known and arguably the best-performing classifiers. In addition, we engage ensemble algorithms which can combine multiple classifiers together. Such approach often can improve the performance in comparison with an individual classifier. In this study we use (1) Bagging, (2) Random Forest, (3) AdaBoost, (4) Vote and (5) Stacking. These models are all evaluated through ten-fold cross-validation, a systematic way of repeated holdout that can pragmatically reduce the variance of the estimate [26]. Another factor to consider is the hyper-parameters as most of these learning methods are associated a set of hyper-parameters, for example, the size of leaf nodes in C4.5, the type of connections in multilayer perceptron. To optimise the performance of these algorithms, we conducted a set of prior empirical study on each model to find out the best combination of hyper-parameter for the model. All algorithms we used are from the Weka package, which supports a wide range of learning methods and the ten-fold cross-validation evaluation. Based on our empirical pre-study, the batch size of all the algorithms is set as 100 to ensure the learning quality.

To further compare the performance, we introduce shallow and deep learning models as well in this study, as deep learning has demonstrated its superb capability in many complex machine learning tasks such as machine vision and natural language processing. Hence this study also includes deep learning to verify its suitability for fast obstacle detection. Two binary deep neural network classifiers are therefore established. One is a 20-(10)-1 network and another is a 20-(10-10)-1 network. This means both have 20 input nodes and one output node. The former has one hidden neural layer with 10 nodes and the latter has two such hidden layers. Note, too many layers would seriously slow down the process, hence are not used in this scenario. The loss function of the network is binary cross-entropy loss, which has a strong coupling with the Sigmoid function. Stochastic gradient descent optimisation with a fixed learning rate is used to control the amount of change in weights and biases. The ten-fold cross-validation is also used here to evaluate the performance of this network model on our data set. The realisation of this model is based on PyTorch, one of the mostly used deep learning frameworks. It enables us to train our deep learning models through GPUs and CPUs in an optimised tensor library [27][5].

5.2 Model Evaluation and Analysis

The aforementioned models are compared using several important metrics, including accuracy, precision, recall, F1-Score (F-Measure) and mean absolute error (MAE). Accuracy measures the proportion of the correctly classified instances (true positives (TP) and true negatives (TN)) among the total predictions (the sum of TP, TN, false positives (FP) and false negatives (FN)) of the data set. Precision computes how close the true predictions are to the positive situation, the sum of TP and FP, e.g., $TP/(TP + FP)$. It means how many of those classified as obstacles are actual dangers ahead. On the other hand, recall evaluates the correct accuracy of prediction over the actual positive instances in the data set, which implies the ratio of the prediction classified as obstacle ahead over all the dangerous instances, $TP/(TP+FN)$. F1-Score is the harmonic mean of the precision and recall, which reflects the balance of the precision and recall, $2TP/(2TP + FP + FN)$. MAE measures the number of misclassification in the model, $\Sigma_{i=1}^{n}|error_i|/n$, where $error_i$ is the deviation from model predictions.

Another set of key indicators are efficiency related, including the model construction or training time, the model execution time, and the size of the model, as the real-time performance of obstacle avoidance is as critical as the detection accuracy. When measuring the model execution time, the trained models are applied to make predictions on a data set with 500 entries on the device, e.g., the Raspberry Pi. The average time of ten executions is calculated as the model execution time. Model sizes are also measured as large models often consume significantly more computational resources, which is not desirable due to the limited resources available on a device for data gathering and computing. Hence, a good model is considered to be high in accuracy, precision, recall and F1-Score, low in MAE, model construction time, execution time, and model size.

6 Experimental Results

6.1 Hyper-parameter Tuning

Table 1 shows different combinations of hyper-parameter settings in the Decision Tree, the minimum number of instances per leaf varying from 2, 10 to 50, and

[5] All learning is conducted using a desktop computer with an i7-12700K 3.60 GHz CPU, 16.0GB RAM, and NVIDIA GTX 1060 6GB GPU. WEKA version is 3.8.6. PyTorch version is 1.13.

the confidence factor varying from 0.25, 0.50, to 0.75. It can be seen that the performance increases as the model grows in size. With minimum instances in each leaf being 2, and a confidence factor of 0.5, the accuracy can reach 98.68%. It is worth noting that an increase in confidence factor, e.g., to 0.75, resulted in a doubled training time, especially when the minimum number of instances on each leaf is as small as 2. The model size of the generated trees is however not severely affected. From the above results we can see the optimal setting for the decision tree is 2 and 0.5.

Table 1. Example of hyper-parameter tuning - Decision Tree

Min num of objects	Confidence factor	Accuracy (%)	Construction time (s)	Model size (KB) (# Leaves/Tree size)
2	0.25	98.64	0.14	50 (127/253)
2	**0.5**	**98.68**	**0.14**	**53 (134/267)**
2	0.75	98.66	0.33	53 (134/267)
10	0.25	97.41	0.14	29 (68/135)
10	0.5	97.42	0.14	29 (69/137)
10	0.75	97.42	0.26	30 (71/141)
50	0.25	94.38	0.13	15 (29/57)
50	0.5	94.44	0.13	15 (29/57)
50	0.75	94.44	0.17	15 (29/57)

Table 2 shows the different settings for the Naive Bayes, using kernel estimator or not, using supervised discretisation parameters or not. The model reaches 91.19% accuracy when using the supervised discretisation only. That took it 0.15 s to build the model. When the supervised discretisation is disabled, the training time can be reduced quite significantly. However, the performance and the model size are not ideal. With the kernel estimator, the model size increased more than 18 times, from 38 to 715 KB. When the kernel estimator is off, the accuracy dropped below 80%.

Table 2. Example of hyper-parameter tuning - Naive Bayes

Use kernel estimator	Supervised discretisation	Accuracy (%)	Construction time (s)	Model size (KB)
False	False	79.74	0.04	7
True	False	89.25	0.05	715
False	**True**	**91.19**	**0.15**	**38**

Table 3 shows the settings of kNN, e.g., the nearest neighbour, using three distance measures, Euclidean distance, Manhattan distance and Chebyshev distance; and different k values, 1, 3 and 5. The nearest neighbour algorithm does not require training time. However the execution of the model involves all data instances. In this sense, the model size is constant and undesirably large. With Manhattan distance and $k = 1$ the accuracy can reach 99.46% validation accuracy without much overfitting.

Table 3. Example of hyper-parameter tuning - kNN

Distance function	k value	Accuracy (%)	Model size (KB)
Euclidean	1	99.11	1,974
Euclidean	3	97.40	1,974
Euclidean	5	96.98	1,974
Manhattan	**1**	**99.46**	**1,974**
Manhattan	3	98.29	1,974
Manhattan	5	98.14	1,974
Chebyshev	1	98.04	1,974
Chebyshev	3	94.16	1,974
Chebyshev	5	93.30	1,974

The result for Multilayer Perceptron is presented in Table 4. The training is done by standard backpropagation using the Sigmoid function. In the table, 'a' means the number of perceptrons in a hidden layer is half of the sum of the number of attributes and classes in the data set. Likewise, 'a, a' and 'a, a, a' mean 2 and 3 hidden layers respectively. It is obvious that training a network is time-consuming. When the network adds more layers, the training time rises quite quickly, but not necessarily leads to a better accuracy. Overall the network with one hidden layer is the best performing (97.12%) and most efficient setting.

Table 4. Example of hyper-parameter tuning - Multilayer Perceptron

Hidden layers setting	Number of hidden layers	Accuracy	Construction time (s)	Model size (KB)
'a'	**1**	**97.12**	**13.05**	**27**
'a, a'	2	96.67	22.01	36
'a, a, a'	3	96.53	27.44	44

6.2 Comparative Analysis of Learning Methods

The best hyper-parameter tuned models that are both accurate and fast are listed in Table 5, which shows accuracy, precision, recall, F1-Score, MAE, model construction time (C), model execution time (E) and model size. It can be seen that methods like kNN and some variations of Naive Bayes are less suitable for real-time detection although their accuracies are high.

6.3 Analysis of Ensemble Algorithms

Five ensemble algorithms are evaluated. Bagging, also called bootstrap aggregating, is a useful statistical estimation technique that can reduce variance and avoid overfitting. Random Forest involves a multitude of decision trees. AdaBoost

Table 5. Evaluation on learning methods

Learning Algorithm	Accuracy (%)	Precision	Recall	F1-Score	MAE	C time (s)	E time (s) (deviation)	Model size (KB)
Decision Tree	98.68	0.987	0.987	0.987	0.014	0.14	0.422 ± 0.022	53
Naive Bayes	91.19	0.915	0.912	0.912	0.091	0.15	0.880 ± 0.038	38
kNN	99.46	0.995	0.995	0.995	0.006	N/A	2.390 ± 0.189	1,974
Multilayer Perceptron	97.12	0.971	0.971	0.971	0.034	13.05	0.839 ± 0.013	27

(Adaptive Boosting) computes a weighted sum of other learning algorithms to provide a boosted classifier. Vote is a simple ensemble algorithm that trains several base estimators and predicts on the basis of aggregating each of them by weighting. Stacking is similar to Vote but with a different way of final aggregation. Four classifiers with the highest accuracy from the previous section, methods in Table 5, are used as the base classifiers in Vote and Stacking. The result of the ensemble algorithms classifiers is shown in Table 6. The Random Forest achieves the highest accuracy at 99.74%, but the size of this model went up to as large as 2,486 KB.

Table 6. Evaluation on ensemble algorithms

Algorithm	Accuracy (%)	Precision	Recall	F1-Score	MAE	C time (s)	E time (s) (deviation)	Model size (KB)
Bagging	99.12	0.991	0.991	0.991	0.042	0.56	0.501 ± 0.059	237
Random Forest	99.74	0.998	0.996	0.997	0.028	1.89	0.998 ± 0.021	2,486
AdaBoost	83.39	0.833	0.830	0.832	0.206	0.26	0.428 ± 0.031	6
Vote	99.40	0.994	0.994	0.994	0.036	11.70	2.261 ± 0.130	2,080
Stacking	99.62	0.996	0.996	0.996	0.006	118.54	2.539±0.131	2,083

6.4 Evaluation on Deep Learning Methods

Two deep networks with different numbers of hidden layers are trained for evaluation. The learning rate, batch size and max epochs are optimised in prior empirical studies. The outcomes of the ten-fold cross-validation for the optimised networks are displayed in Table 7. It shows that the neural network with one hidden layer performs better than the one with two hidden layers. That is similar to the finding in Multilayer Perceptrons. In comparison with Multilayer Perceptrons, the training cost here is much higher although the trained models are smaller and run faster.

Table 7. Evaluation on deep learning methods

Network	Accuracy (%)	Precision	Recall	F1-Score	C time (s)	E time (s) (deviation)	Model size (KB)
20-(10)-1	89.69	0.865	0.928	0.895	1241.75	0.364 ± 0.005	3
20-(10-10)-1	87.76	0.884	0.883	0.881	1342.82	0.398±0.007	4

7 Discussions

7.1 Combined Model Comparison

Following Tables 5 to 7, we further summarise the evaluation results in Fig. 3, which shows the comparison of different learning methods against the three key metrics: accuracy, execution time, and model size respectively. In total, 11 methods are included in the comparison. They are the methods listed in Table 5, Table 6, and Table 7. They are sorted according to the accuracy obtained from the cross-validation. Each method is in a different colour. The same colour means the same method. So the three figures in Fig 3 are in the same order, sharing the same x-axis. For the middle figure, execution time, the measurement is repeated more than 10 runs. The height of each bar here shows the mean, while the standard deviation can also be seen on the figure as an "I" on the top of each bar.

The leftmost six methods, Random Forest, Stacking, kNN, Vote, Bagging, and Decision Tree are all with an accuracy above 98%. However, these six

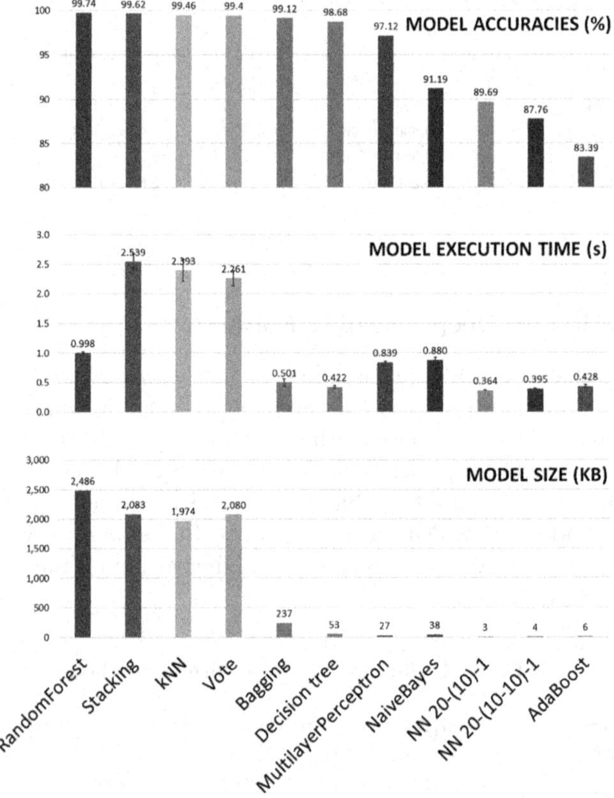

Fig. 3. Model key metrics

methods vary quite considerably in terms of execution time and model size. Three of the six, Random Forest, Bagging and Decision Tree, have execution time lower than 1 s, while the other three need more than 2.2 s to execute the trained model on 500 data entries on the test data set. Note, 2.2 s of execution time does not mean incapable of real-time performance as that is the time for processing 500 entries. When the system in operation, the number of entries coming through per second is about 30 entries, way less than 500. Nevertheless one would still prefer the models with fast execution time.

In terms of model size, Random Forest is rather disappointing in comparison. Its size is the largest, over 2 MB, not ideal for being stored on a small device. In contrast, the Bagging model is 237 KB and the Decision Tree is only 53 KB. Considering their accuracy, which is not the highest, but still amongst the best, Bagging and Decision Tree are the ideal choices. The Decision Tree is in favour, not only because of its high accuracy and low model size but also because it can be easily converted into a set of selection statements so the running time can be further reduced.

The five methods on the right side of Fig. 3, Multilayer Perceptron, Naive Bayes, two deep nets and AdaBoost, are all small in size and run faster than 1 s on the test data set. However, their accuracies are relatively low, ranging from 83.39% to 97.12%. Although fast and small are desirable characteristics for models in our obstacle detection, accuracy still takes priority. Hence in the actual operational prototype system, these models are not selected.

7.2 Real-Time Detection

To close the loop of this study, we established an operational prototype that can perform detection constantly on sensor input as long as power is on. In the system, a positive detection outcome is transformed into an audio signal. So when an obstacle is detected, the system will produce a slightly prolonged beep sound to warn the user. If the obstacle is still present 0.5 s after the beep, another beep will be triggered until there is no further positive detection. Tests on volunteers with the proposed wearable setup produced accurate and timely detection, even the person looked around while moving.

8 Conclusion

This study aims to enable a head-mount smart method to achieve efficient yet accurate obstacle avoidance to help blind and vision-impaired people (BVI) navigate around. We proposed a mechanism that integrates a set of ultrasound sensors and IMU sensors to detect obstacles in front of the person. In particular, the target is only detecting the obstacles directly on the path, excluding the objects on the side, because the head mount sensors could be seriously affected by head turns. Through the study, we show that the proposed method works well. On one hand, it can achieve high accuracy in detection, ignoring the objects directly in front of the person while he/she is looking sideways, but reporting real obstacles on his/her path even when the person is not looking straight ahead. On the other hand, the learned model is small enough to achieve real-time performance on a Raspberry Pi where the computational resource is very limited. Our study reveals that the decision tree prevails over other methods as it has the best combination of speed and accuracy, ideal for our BVI assistant task. In comparison, more sophisticated methods like ensembles and deep learning were sub-optimal due to their high computational cost.

Our future study will further improve the proposed system to extend its applicability and performance. One foreseeable improvement is to include more complex indoor settings, e.g., with furniture scattered around, with moving dynamic objects. In addition, we will investigate more types of sensors e.g., vision sensors and temperature sensors to compensate for variations. We also plan to extend to outdoor environments which are far more complex and to explore some power consumption studies to optimise the system's energy efficiency.

References

1. Bourne, R., et al.: Trends in prevalence of blindness and distance and near vision impairment over 30 years: an analysis for the Global Burden of Disease Study. Lancet Glob. Health **9**(2), e130–e143 (2021)
2. Dakopoulos, D., Bourbakis, N.-G.: Wearable obstacle avoidance electronic travel aids for blind: a survey. IEEE Trans. Syst. Man Cybernet. Part C (Appli. Rev.) **40**(1), 25–35 (2009)
3. Loomis, J.-M., Golledge, R.-G., Klatzky, R.-L.: GPS-based navigation systems for the visually impaired. Fundament. Wearable Comput. Augmen. Reality **429**, 46 (2001)
4. Ran, L., Helal, S., Moore, S.: Drishti: an integrated indoor/outdoor blind navigation system and service. In: Proceedings of the Second IEEE Annual Conference on Pervasive Computing and Communications 2004, pp. 23–30. IEEE, Orlando, FL, USA (2004)
5. Bousbia-Salah, M., Bettayeb, M., Larbi, A.: A navigation aid for blind people. J. Intell. Robotic Syst. **64**(3), 387–400 (2011)
6. Priya, T., Sravya, K.S., Umamaheswari, S.: Machine-learning-based device for visually impaired person. In: Dash, S.S., Lakshmi, C., Das, S., Panigrahi, B.K. (eds.) Artificial Intelligence and Evolutionary Computations in Engineering Systems. AISC, vol. 1056, pp. 79–88. Springer, Singapore (2020). https://doi.org/10.1007/978-981-15-0199-9_7
7. Dos, S., Aline, D.-P., Suzuki, A.-H.-G., Medola, F.-O., Vaezipour, A.: A systematic review of wearable devices for orientation and mobility of adults with visual impairment and blindness. IEEE Access **9**, 162306–162324 (2021)
8. Shoval, S., Ulrich, I., Borenstein, J.: NavBelt and the Guide-Cane [obstacle-avoidance systems for the blind and visually impaired]. IEEE Robotics Automat. Mag. **10**(1), 9–20 (2003)
9. Gao, Y., Chandrawanshi, R., Nau, A.-C., Tse, Z.-T.-H.: Wearable virtual white cane network for navigating people with visual impairment. Proc. Inst. Mech. Eng. [H] **229**(9), 681–688 (2015)

10. Bouteraa, Y.: Design and development of a wearable assistive device integrating a fuzzy decision support system for blind and visually impaired people. Micromachines **12**(9), 1082 (2021)
11. Sammouda, R., Alrjoub, A.: Mobile blind navigation system using RFID. In: 2015 Global Summit on Computer & Information Technology (GSCIT) on Proceedings, pp. 1–4. IEEE, Sousse, Tunisia (2015)
12. Joseph, S.-L., Zhang, X., Dryanovski, I., Xiao, J., Yi, C., Tian, Y.: Semantic indoor navigation with a blind-user oriented augmented reality. In: 2013 IEEE International Conference on Systems, Man, and Cybernetics on Proceedings, pp. 3585–3591. IEEE, Manchester, United Kingdom (2013)
13. Yánez, D.-V., Marcillo, D., Fernandes, H., Barroso, J.: Blind Guide: anytime, anywhere. In: Proceedings of the 7th international conference on software development and technologies for enhancing accessibility and fighting info-exclusion on Proceedings, pp. 346–352. ACM, New York (2016)
14. Li, G., Xu, J., Li, Z., Chen, C., Kan, Z.: Sensing and Navigation of Wearable Assistance Cognitive Systems for the Visually Impaired. IEEE Trans. Cognitive Developm. Syst. (2022)
15. Silva, C.-S., Wimalaratne, P.: Context-aware assistive indoor navigation of visually impaired persons. Sens. Mater **32**, 1497 (2020)
16. Hicks, S.-L., Wilson, I., Muhammed, L., Worsfold, J., Downes, S.-M., Kennard, C.: A depth-based head-mounted visual display to aid navigation in partially sighted individuals. PLoS ONE **8**(7), e67695 (2013)
17. Yang, K., Wang, K., Hu, W., Bai, J.: Expanding the detection of traversable area with RealSense for the visually impaired. Sensors **16**(11), 1954 (2016)
18. Aladren, A., López-Nicolás, G., Puig, L., Guerrero, J.-J.: Navigation assistance for the visually impaired using RGB-D sensor with range expansion. IEEE Syst. J. **10**(3), 922–932 (2014)
19. Lee, Y.-H., Medioni, G.: RGB-D camera based wearable navigation system for the visually impaired. Comput. Vis. Image Underst. **149**, 3–20 (2016)
20. Liu, W., et al.: SSD: Single Shot MultiBox Detector. In: European conference on computer vision on Proceedings, pp. 21–37. Amsterdam, The Netherlands (2016)
21. Redmon, J., Divvala, S., Girshick, R., Farhadi, A.: You only look once: Unified, real-time object detection. In: Proceedings of the IEEE Conference on Computer Vision and Pattern Recognition on Proceedings, pp. 779–788, Las Vegas, NV, US (2016)
22. Suresh, A., Arora, C., Laha, D., Gaba, D., Bhambri, S.: Intelligent smart glass for visually impaired using deep learning machine vision techniques and robot operating system (ROS). In: the 5th International Conference on Robot Intelligence Technology and Applications 5 on Proceedings, pp. 99–112, Daejeon, Korea (2019)
23. Mallikarjuna, G.-CP., Hajare, R., Pavan, P.S.S.: Cognitive IoT System for visually impaired. Mach. Learn. Approach Mater. Today: Proceed. **49**, 529–535 (2022)
24. Kassim, A.-M., et al.: Conceptual design and implementation of electronic spectacle based obstacle detection for visually impaired persons. J. Adv. Mech. Design Syst. Manufact. **10**(7), JAMDSM0094 (2016)
25. Silberschatz, A., Galvin, P.-B., Gagne, G.: Operating Syst. Concepts, 10th edn. John Wiley & Sons, New Jersey, US (2018)

26. Frank, E., Hall, M.-A., Witten I.-H.: The WEKA Workbench. Online Appendix for "Data Mining: Practical Machine Learning Tools and Techniques". 4th edn. Morgan Kaufmann, Massachusetts, US (2016)
27. Paszke, A., et al.: PyTorch: An imperative style, high-performance deep learning library. In: Annual Conference on Neural Information Processing Systems 2019 on Proceedings, pp. 8024–8035. Curran Associates Inc, Vancouver, BC, Canada (2019)

Multimodal Emotion Classification Supported in the Aggregation of Pre-trained Classification Models

Pedro J. S. Cardoso[1,2](✉), João M. F. Rodrigues[1,2], and Rui Novais[2]

[1] LARSyS – Laboratory for Robotics and Engineering Systems, ISR-Lisbon, 1049-001 Lisboa, Portugal

[2] Instituto Superior de Engenharia, Universidade do Algarve, 8005-129 Faro, Portugal
{pcardoso,jrodrig}@ualg.pt

Abstract. Human-centric artificial intelligence struggles to build automated procedures that recognize emotions which can be integrated in artificial systems, such as user interfaces or social robots. In this context, this paper researches on building an Emotion Multi-modal Aggregator (EMmA) that will rely on a collection of open-source single source emotion classification methods aggregated to produce an emotion prediction. Although extendable, tested solution takes a video clip and divides into its frames and audio. Then a collection of primary classifiers are applied to each source and their results are combined in a final classifier utilizing machine learning aggregator techniques. The aggregator techniques that have been put to the test were Random Forest and k-Nearest Neighbors which, with an accuracy of 80%, have demonstrated superior performance over primary classifiers on the selected dataset.

Keywords: Affective Computing · Multimodal Ensembles · Facial Emotions · Speech Emotions

1 Introduction

While part of the humans' communication is verbal, the truth is that a big part of our communication is nonverbal. Facial expressions, the tone of the voice (vocalization), body movements and gestures, posture, all contribute to how we communicate and understand each other. Often, humans are not even aware of that nonverbal part of what they transmit or receive, because this is inherent to them, from the day they are born. Communication is therefore achieved by using multiple sources, received by multiple "sensors", which implies that using a single source of information, such as face expressions, will provide limited information to an automated emotion detection process.

We thank the Portuguese Foundation for Science and Technology (FCT) under Project UIDB/50009/2020—LARSyS.

J. Mikyška et al. (Eds.): ICCS 2023, LNCS 14077, pp. 433–447, 2023.
https://doi.org/10.1007/978-3-031-36030-5_35

Human-centric artificial intelligence (HCAI) struggles to build automated procedures that recognize emotions which can, and in many cases should, be integrated in artificial systems, such as user interfaces or social robots [5,17,53]. For example, research on the latter subject is fundamental, as the worldwide elderly population is set to be more than double by 2050 and robots are expected to assume new roles in health and social care, to meet that higher demand [1,21]. A robot can only really interact with a person, if it achieves some degree of emotional recognition in interaction, i.e., if it understands the person's emotions and sentiments in a way to, on the fly, adjust its behavior in function of it. Emotion and sentiment analysis are therefore fundamental in the development of socially assistive robot (SAR) technologies for people care. Recent studies analyze emotional intelligence in SAR for elders [2] or which aspects may influence human-robot interaction in assistive scenarios [52]. Poria et al. [45] address the multi-modal sentiment and emotion prediction, living open to development several problems such as aspect-level sentiment analysis, sarcasm analysis, multimodal sentiment analysis, sentiment aware dialogue generation, and others. Also, Birjali et al. [10] present a study of sentiment analysis approaches, challenges, and trends, to give researchers a global survey on sentiment analysis and its related fields.

So, to implement SAR technologies, state-of-the-art results in emotions and sentiments classification are needed, being machine learning (ML) algorithms the more promising methods at the moment. Several ways are known to improve algorithms' results, being the more usual way to train them repeatedly, with available data, with different settings, until the best possible result is achieved (fine-tuning the algorithm). Training might be extremely time-consuming, as well as it implies spending a lot of energy during the training phase, also increasing the model's "carbon footprint". A solution to mitigate this is applying ensemble techniques, i.e., using the results from various algorithms previously thought and available in the community [34,35]. This use of hybridization/ensemble techniques allows empowering computation, functionality, robustness, and accuracy aspects of modelling [6], as well as it allows to reduce the referred "carbon footprint" of the models.

In this context, this paper is part of a series of studies to build a framework for emotion classification based on multiple sources (e.g., facial, speech, text, and body expression), the Emotion Multi-modal Aggregator (EMmA). The EMmA is to be supported on an ensemble of open-source code, retrieved from off-the-shelf available methods. The previous studies already presented the idea associated with single sources, namely, faces [35] and speech [34]. Here, the authors propose to integrate both sources in a single prediction, consistent with the emotions presented by the system's user. In more detail, given as input a video clip, the process starts by splitting it in its set of frames/images and its soundtrack. Then, a set of primary classifiers is applied to each source (images and speech), returning a set of probability associated to each emotion. A ML aggregator method is then fitted with those probabilities to build a final classifier. Then, to make inference over new samples, those samples will pass through the primary classifiers and the predicted probabilities are injected in the aggregator, to estimate a final prediction. Details are presented in Sect. 3.

The tested aggregator method includes two well know ML algorithm, namely: Random Forest (RF) and k-Nearest Neighbors (kNN). For faces and for speech, used independently, the proposed single source aggregators formerly proved their effectiveness over the primary classifiers, improving the individual classifiers' accuracy over the Facial Expression Recognition 2013 (FER-2013) Dataset [22], the Real-world Affective Faces Database (RAF-DB) [30], the Ryerson Audio-Visual Database of Emotional Speech and Song (RAVDESS) [32], the Toronto emotional speech set (TESS) [41], the Crowd-Sourced Emotional Multimodal Actors Dataset (CREMA-D) [15], and the Surrey Audio-Visual Expressed Emotion (SAVEE) database [23] for speech. Further, a baseline aggregator was defined using a voting methodology, i.e., each primary classifier "votes" in one emotion and the one with more votes is selected as the estimation.

The best configuration was achieved with the kNN, with an accuracy of 80.6%, 9.7% better than the best result for the individual classifiers, and 4.9% better than the voting method. Besides accuracy, the recall, the precision and the F1 scores also corroborate the attained results.

Some main contributions of the paper are the proof that the previously proposed methods in [34,35] can be extended to a multi-modal scenario, it is possible to build a high accuracy emotion detector combining off-the-shelf methods, which may allow reducing costs (e.g., computational, financial, CO_2 emissions etc.). Further, it is possible to see that the primary classifiers can be trained in different datasets than the ones on which the aggregator is to be applied, allowing a generalization of the proposed method.

The paper is structured as follows. The next section presents a brief state of the art in emotion detection from static images, speech, and video clips. The following section presents the EMmA and voting methods, along with some consideration about the used dataset. The fourth section presents the experimental results. The final section present the conclusion and future work.

2 Related Work

Recognizing expressions in order to predict interpersonal relations requires input from various sources, including sound, body language, and facial expressions, as well as factors such as age and cultural environment. For instance, Zhang et al. [56] proposed an effective multitask network that is able to learn from various auxiliary attributes, such as gender, age, and head pose, in addition to just facial expression data. Noroozi et al. [33] explored the topic of emotional body gesture classification through a comprehensive survey, concluding that, despite a plethora of research on facial expressions and speech, the recognition of the impact of body gestures remains a less explored area. The work intended to increase interest in this field by providing a new survey and highlighting the importance of emotional body gestures as a component of "body language", discussing things such as gender differences and cultural dependence, within the context of emotional body gesture recognition. A complete framework for automatic emotional body gesture recognition was also presented. Other solutions were also proposed, such as combining body posture with facial expressions for classifying affect in child-robot interaction [20]. More recent examples can be found in the literature, such as studies on mood estimation based on facial expressions and postures [14], or in other works such as the ones from Ahmed et al. [3] or Liang et al. [31].

Ekman and Friesen [19], concentrating on face expression, showed that facial displays of emotion are universal, demonstrating that the human ability to express an emotion is an evolutionary, biological fact that is independent of any particular culture. Nevertheless, even given the same input, several approaches for facial emotion classification produce varying outcomes. So, facial expression recognition utilizing a group of classifiers is not a novel concept. For instance, a pool of base classifiers developed utilizing two feature sets – Gabor filters and Local Binary Patterns – was described by Zavaschi et al. [55]. The accuracy and size of the ensemble were employed as objective functions in a multi-objective genetic algorithm that was used to find the best ensemble. Ali et al. [4] presented an ensemble method for evaluating multicultural facial expressions, proposing a set of computational strategies to manage those variations. They make use of facial photos taken from participants in the multicultural dataset, who belong to four distinct ethnic groups, namely, "Caucasians", Japan, Taiwan, and Morocco. Wang et al. [54] presented the Oriented Attention Ensemble for Accurate Facial Expression Recognition. An oriented attention pseudo-Siamese network that utilizes both global and local face information was employed by the authors. The network consists of two branches: an attention branch with a UNet-like architecture to gather local highlight information and a maintenance branch with various convolutional blocks to exploit high-level semantic features. To output the results of the classification, the two branches are combined. Benamara et al. [8] present a facial emotion recognition system that deals with automated facial detection and facial expression classification separately. The latter is carried out by a limited ensemble of only four deep convolutional neural networks, and a label smoothing technique is employed to deal with the training data that has been incorrectly labeled. The Local (Multi) Head Channel (Self-Attention) method, or LHC for short, is founded on two primary concepts [38]. First, convolution will not be replaced by attention modules like recurrent networks were in NLP (natural language processing); and second, a local approach has the potential to overcome convolutions' limitations more effectively than global attention. This is because local attention is more focused on the local region of interest than global attention, which is where the self-attention paradigm has been most extensively studied in computer vision. With LHC, the authors were able to surpass the previous state-of-the-art for the FER-2013 dataset, with a substantially reduced level of complexity and impact on the "host" architecture in terms of computational cost. An open-source Python toolbox named Py-Feat [25] supports the detection, pre-processing, analysis, and visualization of facial expression data. Py-Feat allows end users to swiftly process, analyze, and visualize face expression data while also enabling experts to share and benchmark computer vision models. For further studies in face emotion classification, please refer to, e.g., the works of Banerjee et al. [7] and Revina & Emmanuel [47] who assess multiple deep learning algorithms for effective facial expression classification and human face recognition techniques.

Popova et al. [44] described a method in which the classification of a sound fragment is reduced to an image recognition issue. The waveform and spectrogram are used by the authors to represent the sound. When they combine a Mel-spectrogram with a convolutional neural network (VGG-16), they test their method with RAVDESS and get an accuracy of 71%. The Mel-spectrogram with deltas and delta-deltas is utilized as input by Chen et al. [16] in a 3D attention-based convolutional recurrent neural network to learn discriminative features for speech emotion recognition. Experiments on the Interactive emotional dyadic motion capture database (IEMOCAP) [13] and Berlin Database of Emotional Speech (Emo-DB) corpus [11] provided cutting-edge results. By demonstrating 92.89% validation accuracy on the ESC-50 dataset and 87.42% validation accuracy on the UrbanSound8K dataset, Palanisamy et al. [37] assert that ImageNet pre-trained standard deep convolutional neural network (CNN) models can be employed as powerful baseline networks for audio categorization. De Pinto et al. [43] presented a CNN-based classification model of the emotions produced by speeches (using the RAVDESS dataset). The neutral and calm emotions, as well as those described by Ekman in 1992 [18], have also been taught to the model. They received a weighted average F1 score of 0.91. Using the RAVDESS, Emo-DB, and CaFE databases, El Seknedy and Fawzi [48]presented their findings on speech emotion classification. The key speech features are prosodic 7 features, spectral features, and energy. They employ four machine learning classifiers (Multi-Layer Perceptron, Support Vector Machine – SVM, Random Forest, and Logistic Regression). The models' accuracy was 70.56% on RADVESS, 85.97% on Emo-DB, and 70.61% on CaFE. A deep continuous recurrent neural network (C-RNN) method was presented by Kumaran et al. [29] to classify the efficiency of learning emotion changes in the classification stage. To begin with, they extract high-level spectral features using a combination of Mel-Gammatone filter in convolutional layers. The long-term temporal context is then learned from the high-level features using recurrent layers. In RADVESS, the authors had an accuracy of 80%.

Siddiqui and Javaid [50] created a framework for classifying facial and vocal emotions. Three CNNs and two detection layers make up the proposed structure. Two CNNs are trained separately utilizing visible and infrared images in the first layer, and the features they produce are then given to an SVM for classification. Another CNN was used to learn the emotions in speech using information extracted from audio spectrograms. Ankur Bhatia suggested a system that can extract sentiment and emotion from text, facial and sound [9,46]. The technique was developed over MELD (Multimodal Emotion-Lines Dataset) dataset. MELD is a dataset for spoken language emotion recognition. It is designed to be used for training and assessing models for multimodal emotion recognition and includes audio, text transcriptions, and annotations for the emotions expressed in conversations. MELD, which was taken from the Friends TV series, has more than 13,000 dialogues, more than 100,000 utterances, 7 fundamental emotions, and 2 feelings (positive and negative). On this subject, several other works can be found such as the ones from Pandeya & Lee [28], Heredia et al. [24] or Ortega el al. [36].

Despite the fact that the methodologies described have partially or globally the same overall objective as this work, they do it in different ways. The aforementioned authors concentrate on creating a single model, or on teaching a brand-new model, involving a sizable amount of data to learn from. Instead, the presented approach aims to build on the successes of earlier approaches, concentrating on employing primary classifiers that have already been built and making use of them as part of a final classification technique. I.e., the approach here presented is intended to learn from the outcomes of previously developed models, simplifying the learning phase and reducing the time needed to teach the classification model as well as the computing power that is needed for that. The proposed framework is further examined in the next sections.

3 Multi-source Aggregator and Sentiment Classifier

The proposed Emotion Multi-modal Aggregator (EMmA) is a multi-modal extension of the aggregators previously presented by Novais et al. [34,35] where face and speech were treated separately. Figure 1 illustrates the model's architecture and flow, where the primary classifiers are used in the following manner. First, Fig. 1(a), the primary classifiers models were pre-trained with some dataset proper for their objective (in the present case, face emotion dataset or speech emotion dataset). Then, Fig. 1(b), the obtained models use inference over a new dataset which includes at least faces and speech, to define the training and testing dataset that the aggregator will use to train its own model. I.e., in the present case, for each sample, each primary classifier model predicts the probability of each emotion. Since we are considering 7 emotions, following Ekman and Friesen [19] plus neutral (namely: neutral, calm, happy, sad, angry, fearful, surprised, and disgusted), and 6 primary classifiers, the output will be 42 "emotions-probabilities" values associated to a target. This 42 "emotions-probabilities" associated to a ground-truth values build the aggregator's dataset. Finally, with the primary classifiers' and aggregator's models trained, inference can be done by "passing" the new sample through the primary classifiers, which will return the emotion-probabilities, which are then fed to the aggregator, which will infer an emotion, Fig. 1 (c). Overall, one of the most cost consuming step is the first one, (a), which in this architecture can be simplified from the moment a trained model is available.

If some source (e.g., faces or speech) is not available, the aggregator's model can be retrained without requiring any change to the primary classifiers models. In this context, a solution to activate/deactivate primary classifiers models is being thought. Further, if a new primary classifier, from the same or new

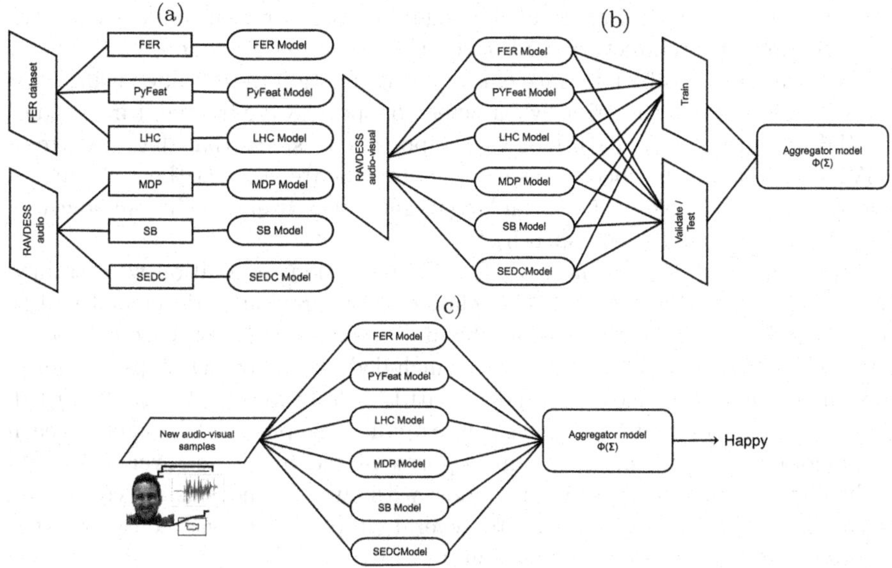

Fig. 1. Architecture and flow of the Emotion Multi-modal Aggregator (EMmA) classifier framework: (a) training of the primary classifiers' model, (b) building the dataset to be used by the aggregator for training and training of the aggregator's model, and (c) samples inference using the primary classifiers and aggregator's models.

sources (e.g., text or body posture), becomes available, this model can be added requiring, as expected, the retraining of the aggregator's model. For this latter condition, continual learning solution are being thought. Finally, as already mentioned, (i) the primary classifiers can be off-the-shelf models (see Sec. 3.1) and (ii) the aggregator method was implemented using common ML methods, namely: RF and kNN classifiers (see Sec. 3.4). Besides, complementing the threshold performance given by the individual methods, a baseline for the aggregator method was defined using a voting methodology (see Sec. 3.3).

3.1 Primary Classifiers

Three primary classifiers for faces and three for speech were taken into consideration. So, the following methods were employed in relation to face detection classifiers: (i) Local (Multi) Head Channel (Self-Attention) (LHC)[38], whose source code is accessible at [39]; (ii) Py-Feat [26], whose source code is available at [27]; and (iii) FERjs, a free implementation created by Justin Shenk, whose source code is available at [49]. It is essential to reiterate the fact that, as was already mentioned, there are other potential approaches [7,47]. On the speech side, the primary emotion classifiers were: (i) MDP [43], with its code available at [42]; (ii) SB, which is a free implementation done by Shivam Burnwal and has its code available at [12]; and SEDC (Speech Emotion Detection Classifier), which is an implementation done by the authors [34]. The six classifiers were chosen because they offer cutting-edge results, are recent implementations, represent various architecture, and have publicly accessible code.

The primary classifiers for face emotion classification were trained using FER dataset (see [35]) and the primary classifiers for speech were trained using (audio) RAVDESS dataset (see [34]). It is also important to stress that data used from RAVDESS to train the speech was not used to validate and test the multi-modal aggregator. Nevertheless, the EMmA aggregator was trained using audiovisual-RAVDESS data (image and speech).

Furthermore, since the methods for facial emotion recognition were trained for static images, the facial emotion classification previously developed had to be prepared to deal with videos, as the audiovisual-RAVDESS is composed by movie clips (see Sec. 3.2). The process included the following steps. For each clip, the (i) first 30 frames (1 s) and the (ii) last 30 frames (1 s) were discarded. Then, for the (iii) remaining frames it was applied the primary classifiers to each one, followed by a (iv) non-maximum suppression technique, i.e., over the results of the clip a sliding neighborhood window with similar emotions are considered as candidate classes, which leads to several proposals. It was considered the proposal/emotion with the highest count.

Table 1 shows the baseline results for the primary classifiers methods, as explained above[1]. The results from the face classifiers seem reasonable due to the approach that applies static faces emotion detection methods to video clips, as explained in the previous paragraph, and the use of different datasets. On the other side, the speech classifiers seem obviously overfitted, since the difference between the metric values attained with the train data set are significantly better than the ones attained with the validation and test data sets. This later fact, was somehow considered as acceptable since we are using off-the-shelf methods. So, it was decided to keep them and proceed to the following steps.

3.2 Dataset for the Aggregators

The three primary classifiers for face emotion classification were trained using FER dataset (see [35]) and the three primary classifiers for speech were trained using audio-RAVDESS (audio files, see [34]). Then, a randomly selected part of the audiovisual speech files of the Ryerson Audio-Visual Database of Emotional Speech and Song (RAVDESS) [32] (720p H.264, AAC 48kHz, MP4) were used to build a dataset for the aggregator. I.e., the primary classifiers were applied to the audiovisual-RAVDESS' files to build a set of 1437 for samples for training, 308 samples for validation and 309 samples for testing (see Table 2). The database contains 24 professional actors (12 female and 12 male), vocalizing two lexically matched statements in a neutral North American accent. Speech includes calm, happy, sad, angry, fearful, surprised, and disgusted emotions. Each expression is produced at two levels of emotional intensity (normal and strong), with an additional neutral expression. Data used from RAVDESS to train the speech were not used to validate or test the final multi-modal aggregators.

[1] See, e.g., [51] for the definitions of Accuracy, Precision, Recall, and F1-score.

Table 1. Primary classifiers' individual performance.

		FER	PyFeat	LHC	MDP	SB	SEDC
Train	Accuracy	0.555	0.460	0.531	**0.996**	0.990	0.944
	F1-score	0.552	0.461	0.516	**0.996**	0.990	0.944
	Recall	0.555	0.460	0.531	**0.996**	0.990	0.944
	Precision	0.631	0.553	0.613	**0.996**	0.990	0.945
Validation	Accuracy	0.506	0.425	0.500	0.682	**0.705**	0.636
	F1-score	0.499	0.442	0.490	0.688	**0.707**	0.637
	Recall	0.506	0.425	0.500	0.682	**0.705**	0.636
	Precision	0.569	0.575	0.588	0.706	**0.714**	0.642
Test	Accuracy	0.531	0.450	0.528	**0.709**	0.673	0.612
	F1-score	0.530	0.449	0.520	**0.713**	0.666	0.597
	Recall	0.531	0.450	0.528	**0.709**	0.673	0.612
	Precision	0.611	0.517	0.614	**0.727**	0.667	0.607

Table 2. Emotions classes' distribution on the used audio-visual RAVDESS subset.

	angry	disgust	fear	happy	neutral	sad	surprise	**total**
Train	263	134	263	263	127	252	135	**1437**
Validation	57	29	55	57	28	54	28	**308**
Test	56	29	57	56	28	54	29	**309**

3.3 Voting Aggregator's Baseline

As a very simple baseline for the aggregator method, it was decided to implement a voting method, as detailed in this section. Each of the primary classifiers return a "vote", predicting an emotion, supported on the emotion with the highest probability it produced. Then, the voting aggregator predicts an emotion as the one with most votes. In the case were two or more emotions are tie (with same amount of votes), and one of the tied emotions is the real one, for metrics purposes, the voting was accounted as correct. Table 3 shows the metrics attained with the voting aggregator, showing an increase in the accuracy over the test dataset (comparing with the best primary classifiers' results), improving the accuracy from 70.9% to 75.7%. For the test dataset, the remaining metrics (Recall, Precision and F1-score) were also improved.

Table 3. Voting aggregator metrics over the audiovisual-RAVDESS dataset.

	Accuracy	F1-score	Recall	Precision
train	0.981	0.981	0.981	0.982
val	0.740	0.748	0.740	0.786
test	**0.757**	**0.756**	**0.757**	**0.789**

3.4 ML Aggregator Methods

Two well-known machine learning methods were used to implement ML aggregator, namely: RF and kNN classifiers. The two ML methods were fitted using grid search cross validation, considering 6 training/testing cases: (i) the training data set was the dataset obtained by running the primary classifiers over the audiovisual-RADVESS samples defined for training (D); (ii) the train dataset was the previous one but standard scaled (D_{scaled}), i.e., removing the mean and scaling to unit variance, $z = (x - \mu_f)/\sigma_f$, where μ_f in the mean value and σ_f the standard deviation of the values observed for each feature f; (iii) the training dataset was set as the scaled version of the union of the train and validation datasets (DV_{scaled}); and (iv)–(vi) the training dataset was the result of applying a polynomial feature transformation of degree 2 to the previous datasets (respectively, D_{poly}, $D_{scaled,poly}$, $DV_{scaled,poly}$). In this context, since cross validation was being used, it was possible to skip validation, going directly to testing. Further, this allowed to observe if adding new samples, which were not direct part of training of the primary classifiers, and did not seem to have major liking to the primary classifiers fitting given the metrics values, could improve the aggregator's performance.

To summarize, aggregators trained with datasets (i)–(iii) receive $7 \times 6 = 42$ values, where 7 is the number of expressions (neutral, calm, happy, sad, angry, fearful, surprised, and disgusted) per primary classifier and 6 is the number of primary classifiers (3 for faces and 3 for speech emotion recognition). Similar, doing all combinations of features, 946 features are fed to the aggregator for cases (iv)–(vi).

4 Tests and Results

The experimental process was conducted using version 1.0.2 of the Scikit–Learn framework [40]. In this context, a grid-search cross validation was developed considering 5 folders and the following parameters. To fit the RF it were considered number of estimators in the set $\{50, 100, 200, 400\}$, criterion in {"gini", "entropy"}, maximum depth in $\{2, 5, 10, \infty\}$, minimum number of samples required to split an internal node in $\{2, 5, 10\}$, the minimum number of samples required to be at a leaf node in $\{1, 2, 5, 10\}$, the minimum weighted fraction of the sum total of weights (of all the input samples) required to be at a leaf node in $\{0, .25, .5\}$, and the maximum number of features to consider when looking for the best split in {"sqrt", "log2"}. For kNN case, it was considered number of neighbors in $\{3, 5, 7, 9\}$, the weights in {"uniform", "distance"}, the algorithm in {"ball_tree", "kd_tree", "brute"}, and power parameter for the Minkowski metric (p) in $\{1, 2\}$. In both cases, the remaining parameters were the default of the Scikit–Learn library.

Table 4. Accuracy, F1-score, Recall and Precision for the aggregators methods.

ML method	Training dataset	Accuracy	F1-score	Recall	Precision
RF	D	0.735	0.732	0.740	0.727
kNN	D	0.767	0.762	0.766	0.763
RF	D_{poly}	0.731	0.722	0.725	0.722
kNN	D_{poly}	0.764	0.761	0.766	0.761
RF	D_{scaled}	0.735	0.732	0.740	0.727
kNN	D_{scaled}	0.796	0.789	0.791	0.793
RF	$D_{scaled,poly}$	0.738	0.727	0.733	0.724
kNN	$D_{scale,poly}$	0.803	0.794	**0.794**	**0.805**
RF	DV_{scaled}	0.770	0.767	0.768	0.771
kNN	DV_{scaled}	0.796	0.786	0.784	0.794
RF	$DV_{scaled,poly}$	0.793	0.786	0.788	0.792
kNN	$DV_{scaled,poly}$	**0.806**	**0.796**	**0.794**	**0.805**

Table 4 summarizes the results achieved over the test set. The best results were achieved using kNN (configured with a ball_tree, 3 neighbors, $p = 1$, and uniform weights, returned from the grid search cross validation process) using scaled and polynomial features over the training and validation data, with an accuracy of 80.6%, more 9,7% than the best result for the individual classifiers (MDP achieved an accuracy of 70.9%), and 4.9% higher accuracy than the voting method (which achieved an accuracy of 75.7%). Moreover, without using validation as part of the training of the aggregator, the best result was also achieved by the kNN method, with an accuracy of 80.3%, just 0.3% less than the best case. This latter case, attained the same recall and precision than the best case, and the F1-score was just 0.002 points different (0.796 to 0.794). Relatively to the RF, it was found a big difference when considering the training and validation data (scaled and with polynomial features) to train the aggregator, passing from 73.8% accuracy ($D_{scaled,poly}$) to 79.3% accuracy.

5 Conclusion

This paper presented a framework based on an Emotion Multi-modal Aggregator (EMma), which aggregates the results extracted from the primary emotion classifications from different sources, namely facial and speech. The framework was tested using the audiovisual-RAVDESS dataset, somehow validating the initial concept: it is possible to build a state-of-the-art emotion detection system supported on methods (primary classifiers) available as open-source, trained with distinct datasets, with a minimum training of an aggregator. Another advantage of this solution is the possible speed-up in the development of an integrated solution for human emotion classification.

In conclusion, there are still a considerable number of questions that remain unanswered and represent opportunities for future research in this area. One area that could benefit from further exploration is the relationship between the number of primary classifiers used and the overall computational complexity. By delving deeper into this issue, researchers can optimize the classifier's efficiency while maintaining high levels of accuracy. Additionally, investigating the impact of incorporating primary classifiers that are significantly different in terms of accuracy compared to those already used should be addressed. Future work will also focus on the inclusion of other sources to the framework (e.g., text and body posture) and in the definition of a proper dataset that will allow testing all the strands of the model. This includes the research on the influence of other dimensions, such as gender, ethnicity, and age. By considering a broader range of factors, the model can be refined and improved to better address real-world applications and challenges. Further, if some source (e.g., faces or speech, at the moment) is not available, the aggregator model should have a solution to activate/deactivate the corresponding primary classifiers models, without retraining the former one. Also in this context, including new/removing primary classifiers, from the same or new sources (e.g., text or body posture), without retraining of the aggregator model is an objective, i.e., a procedure commonly designated as continual learning. Overall, the paper provides a foundation for further exploration and development of automated classification techniques, and future research in this area holds promises for advancing the field.

References

1. Abdi, J., Al-Hindawi, A., Ng, T., Vizcaychipi, M.P.: Scoping review on the use of socially assistive robot technology in elderly care. BMJ Open **8**(2), e018815 (2018). https://doi.org/10.1136/bmjopen-2017-018815
2. Abdollahi, H., Mahoor, M., Zandie, R., Sewierski, J., Qualls, S.: Artificial emotional intelligence in socially assistive robots for older adults: A pilot study. IEEE Trans. Affect. Comput. (2022). https://doi.org/10.1109/taffc.2022.3143803
3. Ahmed, F., Bari, A.S.M.H., Gavrilova, M.L.: Emotion recognition from body movement. IEEE Access **8**, 11761–11781 (2020). https://doi.org/10.1109/ACCESS.2019.2963113
4. Ali, G., et al.: Artificial neural network based ensemble approach for multicultural facial expressions analysis. IEEE Access **8**, 134950–134963 (2020). https://doi.org/10.1109/ACCESS.2020.3009908
5. Alonso-Martín, F., Malfaz, M., Sequeira, J., Gorostiza, J.F., Salichs, M.A.: A multimodal emotion detection system during human-robot interaction. Sensors **13**(11), 15549–15581 (2013). https://doi.org/10.3390/s131115549, https://www.mdpi.com/1424-8220/13/11/15549

6. Ardabili, S., Mosavi, A., Várkonyi-Kóczy, A.R.: Advances in machine learning modeling reviewing hybrid and ensemble methods, pp. 215–227 (2020). https://doi.org/10.1007/978-3-030-36841-8_21
7. Banerjee, R., De, S., Dey, S.: A survey on various deep learning algorithms for an efficient facial expression recognition system. Int. J. Image Graph. (2021). https://doi.org/10.1142/S0219467822400058
8. Benamara, N.K., et al.: Real-time facial expression recognition using smoothed deep neural network ensemble. Integrated Comput. Aided Eng. **28**(1), 97–111 (2020). https://doi.org/10.3233/ICA-200643
9. Bhatia, A., Rathee, A.: Multimodal emotion recognition (2020). https://github.com/ankurbhatia24/multimodal-emotion-recognition (Accessed 31 Jan 2023)
10. Birjali, M., Kasri, M., Beni-Hssane, A.: A comprehensive survey on sentiment analysis: Approaches, challenges and trends. Knowl.-Based Syst. **226**, 107134 (2021). https://doi.org/10.1016/j.knosys.2021.107134, https://www.sciencedirect.com/science/article/pii/S095070512100397X
11. Burkhardt, F., Paeschke, A., Rolfes, M., Sendlmeier, W.F., Weiss, B., et al.: A database of german emotional speech. In: Interspeech, vol. 5, pp. 1517–1520 (2005)
12. Burnwal, S.: Speech emotion recognition (2020). https://www.kaggle.com/code/shivamburnwal/speech-emotion-recognition/notebook (Accessed 31 Jan 2023)
13. Busso, C., et al.: Iemocap: Interactive emotional dyadic motion capture database. Lang. Resour. Eval. **42**, 335–359 (2008)
14. Canedo, D., Neves, A.: Mood estimation based on facial expressions and postures. In: Proceedings of the RECPAD, pp. 49–50 (2020)
15. Cao, H., Cooper, D.G., Keutmann, M.K., Gur, R.C., Nenkova, A., Verma, R.: CREMA-d: Crowd-sourced emotional multimodal actors dataset. IEEE Trans. Affective Comput. **5**(4), 377–390 (2014). https://doi.org/10.1109/TAFFC.2014.2336244
16. Chen, M., He, X., Yang, J., Zhang, H.: 3-d convolutional recurrent neural networks with attention model for speech emotion recognition. IEEE Signal Processing Lett. **25**(10), 1440–1444 (2018). https://doi.org/10.1109/LSP.2018.2860246
17. Cheng, B., Wang, Y., Shao, D., Arora, C., Hoang, T., Liu, X.: Edge4emotion: An edge computing based multi-source emotion recognition platform for human-centric software engineering. In: 2021 IEEE/ACM 21st International Symposium on Cluster, Cloud and Internet Computing (CCGrid), pp. 610–613 (2021). https://doi.org/10.1109/CCGrid51090.2021.00071
18. Ekman, P.: Facial expressions of emotion: New findings, new questions. Psychol. Sci. **3**(1), 34–38 (1992). https://doi.org/10.1111/j.1467-9280.1992.tb00253.x
19. Ekman, P., Friesen, W.V.: Constants across cultures in the face and emotion. J. Pers. Soc. Psychol. **17**(2), 124–129 (1971). https://doi.org/10.1037/h0030377
20. Filntisis, P.P., Efthymiou, N., Koutras, P., Potamianos, G., Maragos, P.: Fusing body posture with facial expressions for joint recognition of affect in child-robot interaction. IEEE Robotics Automa. Lett. **4**(4), 4011–4018 (2019). https://doi.org/10.1109/LRA.2019.2930434
21. Getson, C., Nejat, G.: Socially assistive robots helping older adults through the pandemic and life after COVID-19. Robotics **10**(3), 106 (2021). https://doi.org/10.3390/robotics10030106
22. Goodfellow, I.J., et al.: Challenges in representation learning: a report on three machine learning contests. In: Lee, M., Hirose, A., Hou, Z.-G., Kil, R.M. (eds.) ICONIP 2013. LNCS, vol. 8228, pp. 117–124. Springer, Heidelberg (2013). https://doi.org/10.1007/978-3-642-42051-1_16

23. Haq, S., Jackson, P.: Machine Audition: Principles, Algorithms and Systems, chap. Multimodal Emotion Recognition, pp. 398–423. IGI Global, Hershey PA (Aug 2010)
24. Heredia, J., et al.: Adaptive multimodal emotion detection architecture for social robots. IEEE Access **10**, 20727–20744 (2022). https://doi.org/10.1109/ACCESS.2022.3149214
25. Jolly, E., Cheong, J.H., Xie, T., Byrne, S., Kenny, M., Chang, L.J.: Py-feat: Python facial expression analysis toolbox. arXiv preprint arXiv:2104.03509 (2021)
26. Jolly, E., Cheong, J.H., Xie, T., Byrne, S., Kenny, M., Chang, L.J.: Py-feat: Python facial expression analysis toolbox (2021). https://doi.org/10.48550/arXiv.2104.03509
27. Jolly, E., Cheong, J.H., Xie, T., Byrne, S., Kenny, M., Chang, L.J.: Py-feat: Python facial expression analysis toolbox (2023). https://pythonrepo.com/repo/cosanlab-py-feat-python-deep-learning (Accessed 31 Jan 2023)
28. Kleinsmith, A., Bianchi-Berthouze, N.: Affective body expression perception and recognition: A survey. IEEE Trans. Affective Comput. **4**(1), 15–33 (2013). https://doi.org/10.1109/T-AFFC.2012.16
29. Kumaran, U., Radha Rammohan, S., Nagarajan, S.M., Prathik, A.: Fusion of mel and gammatone frequency cepstral coefficients for speech emotion recognition using deep C-RNN. Int. J. Speech Technol. **24**(2), 303–314 (2021). https://doi.org/10.1007/s10772-020-09792-x
30. Li, S., Deng, W.: Reliable crowdsourcing and deep locality-preserving learning for unconstrained facial expression recognition. IEEE Trans. Image Process. **28**(1), 356–370 (2019)
31. Liang, G., Wang, S., Wang, C.: Pose-aware adversarial domain adaptation for personalized facial expression recognition. arXiv preprint arXiv:2007.05932 (2020)
32. Livingstone, S.R., Russo, F.A.: The ryerson audio-visual database of emotional speech and song (RAVDESS): A dynamic, multimodal set of facial and vocal expressions in north american english. PLoS ONE **13**(5), e0196391 (2018). https://doi.org/10.1371/journal.pone.0196391
33. Noroozi, F., Corneanu, C.A., Kamińska, D., Sapiński, T., Escalera, S., Anbarjafari, G.: Survey on emotional body gesture recognition. IEEE Trans. Affect. Comput. **12**(2), 505–523 (2018)
34. Novais, R., Cardoso, P.J.S., Rodrigues, J.M.F.: Emotion classification from speech by an ensemble strategy. In: ACM 10th International Conference on Software Development and Technologies for Enhancing Accessibility and Fighting Info-exclusion (DSAI 2022) (2022)
35. Novais, R., Cardoso, P.J.S., Rodrigues, J.M.F.: Facial emotions classification supported in an ensemble strategy, pp. 477–488 (2022). https://doi.org/10.1007/978-3-031-05028-2_32
36. Ortega, J.D.S., Cardinal, P., Koerich, A.L.: Emotion recognition using fusion of audio and video features. In: 2019 IEEE International Conference on Systems, Man and Cybernetics (SMC), pp. 3847–3852 (2019). https://doi.org/10.1109/SMC.2019.8914655
37. Palanisamy, K., Singhania, D., Yao, A.: Rethinking cnn models for audio classification (2020). https://doi.org/10.48550/arXiv.2007.11154
38. Pecoraro, R., Basile, V., Bono, V.: Local multi-head channel self-attention for facial expression recognition. Information **13**(9), 419 (2022)
39. Pecoraro, R., Basile, V., Bono, V., Gallo, S.: Lhc-net: Local multi-head channel self-attention (code). https://github.com/bodhis4ttva/lhc_net (Accessed 29 Jan 2023)

40. Pedregosa, F., et al.: Scikit-learn: Machine learning in Python. J. Mach. Learn. Res. **12**, 2825–2830 (2011)
41. Pichora-Fuller, M.K., Dupuis, K.: Toronto emotional speech set (tess) (2020). https://doi.org/10.5683/SP2/E8H2MF
42. de Pinto, M.G.: Audio emotion classification from multiple datasets (2020). https://github.com/marcogdepinto/emotion-classification-from-audio-files (Accessed 31 Jan 2023)
43. de Pinto, M.G., Polignano, M., Lops, P., Semeraro, G.: Emotions understanding model from spoken language using deep neural networks and mel-frequency cepstral coefficients (May 2020). https://doi.org/10.1109/EAIS48028.2020.9122698
44. Popova, A.S., Rassadin, A.G., Ponomarenko, A.A.: Emotion recognition in sound, pp. 117–124 (Aug 2017). https://doi.org/10.1007/978-3-319-66604-4_18
45. Poria, S., Hazarika, D., Majumder, N., Mihalcea, R.: Beneath the tip of the iceberg: Current challenges and new directions in sentiment analysis research. IEEE Trans. Affective Comput. (2020)
46. Poria, S., Hazarika, D., Majumder, N., Naik, G., Cambria, E., Mihalcea, R.: MELD: A multimodal multi-party dataset for emotion recognition in conversations. In: Proceedings of the 57th Annual Meeting of the Association for Computational Linguistics, pp. 527–536 (2019)
47. Revina, I., Emmanuel, W.S.: A survey on human face expression recognition techniques. J. King Saud Univ. - Comput. Inform. Sci. **33**(6), 619–628 (2021). https://doi.org/10.1016/j.jksuci.2018.09.002
48. Seknedy, M.E., Fawzi, S.: Speech emotion recognition system for human interaction applications (Dec 2021). https://doi.org/10.1109/ICICIS52592.2021.9694246
49. Shenk, J., CG, A., Arriaga, O., Owlwasrowk: justinshenk/fer: Zenodo (Sep 2021). https://doi.org/10.5281/zenodo.5362356
50. Siddiqui, M.F.H., Javaid, A.Y.: A multimodal facial emotion recognition framework through the fusion of speech with visible and infrared images. Multimodal Technol. Interact. **4**(3), 46 (2020). https://doi.org/10.3390/mti4030046
51. Sokolova, M., Lapalme, G.: A systematic analysis of performance measures for classification tasks. Inform. Process. Manage. **45**(4), 427–437 (2009)
52. Sorrentino, A., Mancioppi, G., Coviello, L., Cavallo, F., Fiorini, L.: Feasibility study on the role of personality, emotion, and engagement in socially assistive robotics: A cognitive assessment scenario. Informatics **8**(2), 23 (2021). https://doi.org/10.3390/informatics8020023
53. Stock-Homburg, R.: Survey of emotions in human–robot interactions: perspectives from robotic psychology on 20 years of research. Int. J. Soc. Robot. **14**(2), 389–411 (2021). https://doi.org/10.1007/s12369-021-00778-6
54. Wang, Z., Zeng, F., Liu, S., Zeng, B.: OAENet: Oriented attention ensemble for accurate facial expression recognition. Pattern Recogn. **112**, 107694 (2021). https://doi.org/10.1016/j.patcog.2020.107694
55. Zavaschi, T.H.H., Koerich, A.L., Oliveira, L.E.S.: Facial expression recognition using ensemble of classifiers (May 2011). https://doi.org/10.1109/ICASSP.2011.5946775
56. Zhang, F., Zhang, T., Mao, Q., Xu, C.: Joint pose and expression modeling for facial expression recognition (Jun 2018). https://doi.org/10.1109/CVPR.2018.00354

Cyber-physical System Supporting the Production Technology of Steel Mill Products Based on Ladle Furnace Tracking and Sensor Networks

Piotr Hajder[1(✉)], Andrzej Opaliński[1], Monika Pernach[1], Łukasz Sztangret[1], Krzysztof Regulski[1], Krzysztof Bzowski[1], Michał Piwowarczyk[2], and Łukasz Rauch[1]

[1] AGH University of Science and Technology, Kraków, Poland
{phajder,opal,pernach,szt,regulski,kbzowski,lrauch}@agh.edu.pl
[2] CMC Poland Sp. z o.o., Zawiercie, Poland
michal.piwowarczyk@cmc.com

Abstract. The use of information technologies in industry is growing year by year. More and more advanced devices are implemented and the software needed for them becomes more complex, which increases the risk of errors. To minimize them, it is necessary to constantly monitor the condition of the system and its components. This paper presents a part of a complex production support system for steel mill, responsible for monitoring and tracking the current state on the production hall. Data on currently performed melts and their condition, collected from two sensor layers - Level1 and Level2 - combining with a camera system that allows tracking the position of the main ladle in the hall, was used to create metamodel based on linear regression and neural network for the temperature drop which is occurring during the transport of liquid steel to the casting machine. This approach enables optimization of production volume and minimizes the risk associated with a temperature drop below the optimal one for casting. Several neural network models were used: YOLOv3 for object detection, CRAFT for text detection and CRNN for text recognition. This information is published to the sensor subsystem, enabling precise determination of the state of each performed melt. The system architecture, prediction accuracies and performance analysis were presented.

Keywords: cyber-physical system · machine learning · sensors · vision processing · steelmaking automation

1 Introduction

One of the industry sectors, which is the main research area in this work, is the steel industry. Steel is a product whose use can be found in almost every other industry, even in everyday life. Depending on its purpose, various compositions

J. Mikyška et al. (Eds.): ICCS 2023, LNCS 14077, pp. 448–462, 2023.
https://doi.org/10.1007/978-3-031-36030-5_36

are used, by dosing appropriate alloy additives, and variable process conditions, such as melt temperature and casting time, or the degree of superheating. In the case of the most rigorous alloys, deviations from the process and its schedule are unacceptable, as this may result in the creation of a different steel grade, which translates into both company losses and the purpose of its use. Therefore, there is a need for continuous monitoring and control, from the very beginning of the process. An important step in the production of specific steel grades in the continuous casting process is the transfer of the main ladle (ML) from the electric arc furnace (EAF) through the ladle furnace (LF) to the continuous steel casting machine (CSC). This takes time, which causes the temperature of the molten steel to drop and may make casting difficult or impossible, possibly lowering the quality of the product.

In this paper, the development of ladle position monitoring modules, measuring systems and metamodeling of the temperature drop of liquid steel in the ML is presented. A sensory layer was created, deployed in the production hall. The measurements are aggregated with data from the database and then used in the temperature drop metamodel. Historical data were used to train this model, which is based machine learning techniques, in particular linear regression and neural network. In addition, available industrial cameras were used to track the movement (determining the position) of the ML in the production hall, enhanced with three machine learning models: one for ML detection, as well as two for detecting and recognizing text in the form of numerical labels. We present the architecture of these modules, the results of their integration and sample results, including performance and accuracy of the predictions of the models used. This work are part of a hybrid IT system designed to optimize and model continuous steel billets.

2 Preliminary Arrangements

In this work, we describe the module of the production technology support system at the steel plant, concerning the modeling of the temperature drop of liquid steel in ML. For the alloy to be properly cast, the temperature must not fall below the set value. For this purpose, we created a metamodel that evaluates the current temperature according to the time passed after ML's departure from EAF. In order to assess whether the alloy can be cast, in addition to the metamodel, we have created a subsystem for monitoring the position of the ML in the hall, to be able to assess as early as possible whether the ML will arrive on time at the CSC or whether it will need to be reheated at the LF station. This subsystem consists of two main components: sensors, measuring and collecting information about current activities in the hall, and video monitoring of the ML position, based on multiple industrial cameras. Preliminary arrangements for each of these components are described in the following sections.

2.1 Sensors

The main task of the sensory layer in the described problem is to provide the necessary data and information to the component responsible for metamodeling, based on which they will be able to model the current state of the process and control it in order to obtain its optimal result. For the proper operation of process steering component, it is necessary to provide data from the sensor layer, which will be:

- Up-to-date – delivered on an ongoing basis, preferably in real mode, immediately after obtaining them in the environment, so that the system's reaction can be immediate,
- Precise and accurate – reflecting the parameters of the real process as closely as possible in order to obtain the best representation of the metamodel,
- Complete and consistent – mapping all technical and technological parameters necessary to develop and launch a metamodel mapping the technological process carried out in the electrosteel plant hall.

In the existing hardware layer concerning sensors and measurement of the environment to which this task relates, there are already a lot of elements that can provide useful data related to sensors and parameters of such a technological process. We can distinguish three main sources among them:

1. **Level1 system**. At the lowest level – in the hardware layer – there is embedded software for devices located in the electrosteel plant hall. It is known as Level1 and stores information about the status of devices located in the hall. This data can be obtained directly from the device controllers in real mode, although in the standard mode of operation of the steel plant, the data on the process itself is not used for automatic (intelligent) process control and steering.
2. **Level2 system**. It is sensor layer available in the electrosteel plant hall, which stores part of the parameters related to the technological process carried out on main process devices. These are parameters related to the melting schedule (grade, recipient, sequence in the melting sequence), characteristics of end products related to a given ML (chemical composition, size and type of cast element) or parameters of individual stages of processing at specific positions in the steelworks plant hall (batch weight, amount of oxygen or coal used, total processing time per step). The imperfection of this layer in terms of its use in the production control process based on the metamodeling is that data are available only after the completion of the processing process on a particular device in the electrosteel plant hall (EAF, LF, CSC). Due to the characteristics of the technological process, they appear in the system every 30–50 minutes, because these are the standard processing times on individual devices included in the process.
3. **CCTV monitoring system**. The video monitoring system available in the steelworks plant hall is based on 12 cameras covering most of the hall area. The data is available in the live view system and the image is also archived,

but in the current mode of operation of the plant hall, the only image analysis mechanisms available in the system are the mechanisms of motion detection in the image, without any identification of the sources of these events. There is no distinction whether the movement in the image was generated by a person, a vehicle or a moving ladle (ML).

The main problem, common in this field of research and industry [1,2], in the sensory layer of the current infrastructure of the steelworks plant hall is that data on the steelmaking process that could enable automatic control mechanisms for such a process (based on machine learning mechanisms) are available in different hardware layers of the environment (described above), are not integrated and partly inconsistent.

The task that needs to be solved is so that they can be used as a data source for the metamodeling component based on machine learning mechanisms is the integration, aggregation and unification of data from the three sensor systems mentioned above and providing them to the metamodel component in a coherent version in real mode.

2.2 ML Position Tracking

As described in the previous subsection, there are 12 industrial cameras, providing current view of the production hall via RTSP protocol. They may be used to track position of every used ML. The main problem to be solved in the described part of the system is the detection of ML and recognition of its label. The basic solution that has been indicated for this problem is the detection of objects and text, along with its recognition. For this purpose, available Avigilon industrial cameras were used, along with machine learning models. To achieve the goal of this module, following tasks can be defined:

1. **Object detection** - detecting the ML on the image frame, which determines its position in the hall;
2. **Text detection and recognition** - detecting and recognizing the ML label, which will allow it to be identified.

Both of these tasks can be solved using machine learning methods, in particular based on convolutional neural networks [3,4]. However, since the task is not standard and publicly available training sets for the networks do not contain ML, we had to build our own dataset. For this purpose, the images from each camera were analyzed, the key ones for modeling the hall and steelmaking process were determined. Then, with the interval set per camera, the frames were uploaded to the disk. After image gathering was completed, they were selected and labeled in order to prepare them for training machine learning models. Labeling was performed for both ML and numeric labels painted on ladles. In addition, due to lot of similarities between dumped frames, we used ImgAug [5] to augment dataset with modified versions of images, in particular by 'emulation' of smoke, which can occur while during slag pouring (visible on few cameras) and may influence results obtained from models.

With an already prepared training data, the main problem to be solved is to prepare neural network models and its implementation for the indicated tasks. These models must be trained and tested for accuracy and performance. Then, they can be aggregated into one solution that receives a frame from the camera as an input and returns the position and ML label as an output.

2.3 Metamodeling for Temperature Drop Prediction

Measurements of temperature distribution of liquid steel inside a ladle is very difficult or sometimes even impossible task in industrial conditions, due to the safety reason and high cost of equipment. On the other hand precise knowledge about the overheating level is crucial to maintain optimal range of temperatures during Continuous Steel Casting process to obtain the highest quality of casted billets. Application of FEM models to predict the temperature of liquid steel in industrial conditions is impossible due to the long computation time. Therefore, in this paper, two machine learning techniques were applied for this task: linear regression (LR) and artificial neural networks (ANN) [6,7].

All computations were performed using MATLAB software. Deep Learning Toolbox was used to create and train ANNs and Optimization Toolbox was applied for gradient optimization of cost function in case of linear regression. In both cases, the aim of the model was to predict the steel temperature after specific time period based on: the type of the steel (GradeID), temperature obtained after heating (StartTemp), cooling time (TimeDiff) and the steel weight (Weight). Eventually, based on the temperature difference and cooling time, the cooling rate was computed. The collected data, after filtering, consists of 7362 records. The data concerned 81 grades of steel. Unfortunately, the number of records for individual grades of steel varied greatly. The most common type of steel was represented by over 3000 records, while for five types of steel only one record was in the dataset. Despite such differences in the number of records, preliminary training of the models revealed, that the number of steel types included in the training set did not have a major impact on the accuracy of the model. Therefore, all computations were performed for all 81 types of steel.

2.4 Related Works

There is a visible trend in the steelmaking sector to increase investment in R&D activities to optimize production [8]. The development is observable from the scientific side, which is reflected in the available literature, where many works showing the improvement of efficiency and quality of steel production can be found [9,10], including through the use of sensor networks [11]. These works are mainly focused on optimizing the steelmaking process by reducing carbon emissions into the atmosphere. An indirect impact is also observed in the form of a process whose optimization criterion is minimization of energy consumption. High emissions and energy consumption result from the high temperature that a given steel grade must reach during casting. Numerical simulations are often the basis for conducting such research, which allow for non-invasive testing of the possibilities of improving the available technology [12,13]. A situation that should be avoided at all costs in the steel production process is a drop in temperature below the casting temperature, as this will require transport of the ML for re-heating. This can be a logistically difficult task, which is why monitoring is essential.

Monitoring in the industrial sector is not a new phenomenon. This is an important element in Industry 4.0, as it also allows you to control the correct operation of support systems. One of the tasks in this paper is tracking ML. Since the appearance of the ML is almost identical for all these types of vats, this task can be treated as an Object Detection. In the industrial field, there are applications of machine learning algorithms for this class of problem [14], in particular the recognition of various workpieces on the production hall [15]. Machine learning models were used in the work, in particular convolutional networks, e.g. YOLO (You Only Look Once) and residual networks, e.g. Resnet. Their application in industrial conditions can also take place in tracking the movement of objects or detecting their absence or presence [16].

3 Proposed Methodology and Architecture

The main idea of solving the described problem was to create an IT system that implements two main functionalities:

- supporting the control of the technological process by the metamodeling component, which allows detecting potentially unfavorable situations (e.g. superheating or crystallization of the melt) and enabling the avoidance of such situations
- monitoring the state of the technological process by technologists and technical staff using mobile devices

In order to implement such a system, it was necessary to develop a component for integrating the sensory layer, which would enable the aggregation of data from all sensory systems available within the infrastructure of the steelworks plant hall. Such data would then be delivered to the component responsible for supporting the steering and monitoring of this technological process (based on machine learning mechanisms). The diagram of the main components of this system is presented in Fig. 1.

The details of the implementation of the main components of this system are presented below.

3.1 Sensors Layer

In order to implement the main functionality – supporting the control of the technological process based on the process metamodeling component – it was necessary to provide this component with the correct data necessary for its operation. For this purpose, a sensor layer data aggregation component was developed that integrates the three main sensor sources (Level1, Level2, CCTV)

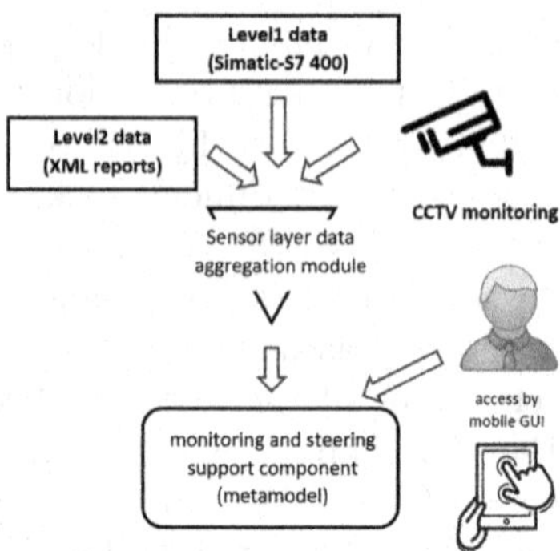

Fig. 1. Scheme of integration of the main components of the system

and unifies the data available in them, which then makes it available to the process steering component.

The requirement of the metamodeling component is to provide this data on an ongoing basis and in a coherent, unified form. The same data provided by the sensory data aggregation component is also used to implement the second basic functionality of the entire system - enabling monitoring of the process status of the steelworks hall using a mobile application for service personnel.

A component that aggregates data from the base sensor systems available at the electrosteel shop in the following way:

Level1 Integration – this hardware-level system provides data on individual devices involved in the technological process (EAF, LF, CSC). Integration with this system was carried out based on the connection with the Siemens Simatic-S7 400 controller programmed with the use of the Sharp7 library which is the C# port of Snap7 library. The data is delivered in real time and read and saved in the database with the time interval set in the configuration. Among the data obtained from this source are (among others): the current temperatures of the ML armor and the bottom of the tundish, the values of the last measurements of the liquid metal temperature, the current duration of the charge processing in the EAF or at the LF, the amount of energy consumed at individual processing stations.

Level2 Integration the mechanism of integration with the Level2 system is based on report documents generated by this system in the XML format, made available via a network drive after the completion of subsequent stages of processing in the technological process (at the EAF, LF and CSC). XML documents are automatically detected using FileWatcher, xsd and standard C# libraries used to parse XML documents. Data from parsed reports are placed in the system database in separate tables for each of the monitored devices. The data obtained by these mechanisms relate to the process stages carried out in the EAF (e.g.

charge melting time, melting process efficiency, amount of energy and oxygen used, charge weight), LF (e.g. argon mixing parameters, energy consumption in the station, current chemical composition alloy), CSC (e.g. casting mode, weight of steel, transport time to the station, number of cores).

CCTV Integration – details of the operation of the ladle detection component have been described in a separate chapter of this work, while the integration of this component has been implemented through the definition and implementation of software interfaces transferring the following information from the ladle detection component to the integrating component: camera id, detected ladle number, ladle position within the scene, detection time. The entire exchange of information is carried out by the open source message broker RabbitMQ.

Data from the sensor systems described above, operating on various hardware layers of the electrosteel plant hall, have been aggregated, unified and made available in real time as a component of sensory data aggregation. The data aggregated in this way is stored in the system database and made available to component that enable the implementation of the main functionalities of the system: the process monitoring component of the steelworks hall using an application for mobile devices and the metamodeling component, enabling support for controlling the technological process carried out in the steelworks plant hall.

3.2 ML Position Tracking

The problem of ML position monitoring has been divided into two tasks: ML detection and identification of its label. The decision was made to use machine learning to solve them. In the first approach, only the Tesseract OCR library was used for text detection and recognition, because all MLs have numerical labels, and their occurrence elsewhere is very rare on the cameras used. Unfortunately, this solution turned out to be ineffective due to the very low accuracy of detecting areas with text - in the best cases at the level of 63%.

The second approach was based on the detection of one class of objects in the ML form. Four machine learning models were used for this purpose: Mask-RCNN, MobileNet, YOLO v3 and its tiny version. All models were trained using a total pool of 7072 images, of which approximately 2000 were modified with ImgAug. Additionally, after initial testing, 700 images were added from the hall where none of the vats were located. This was due to the single-class detection - adding images without an object resulted in an increase in detection accuracy by 8%. These data were split 80-20 for training and validation, respectively. In the case of label identification, two models were used: pretrained CRAFT (Character-Region Awareness For Text) for text detection [17] and a custom convolutional recurrnet neural network as text recognizer, consisting of 28 layers (mainly Conv2D, Maxpool, BatchNormalization, and LSTM). The latter one was trained used Synth90k set enhanced with approx. 10000 cropped labels from MLs. Output from CRAFT, which are heatmaps, is postprocessed using OpenCV in order to obtain transformed boxes with text. These boxes are then processed by recognizer.

The prepared models were used to determine the position of individual ML in the hall. After ML is detected, it is cropped from original image and this crop is sent to label detection and recognition. This approach allowed to improve text detection accuracy. However, in some cases label can be detected on original image. Therefore, we decided to execute both detection models concurrently and then match boxes (ML and text) with each other using intersection. Then, ML position and its label are sent to message broker. The algorithm describing the operation of the ML tracking is shown as a flowchart in the Fig. 2.

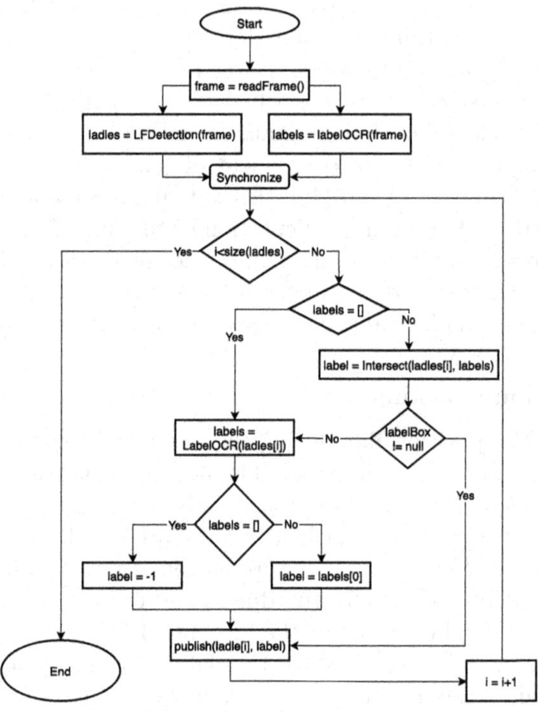

Fig. 2. ML detection and label recognition pairing algorithm

From the implementation side, the Python 3.10 with Tensorflow 2.10 was used to develop and train machine learning models. Models were built with Keras in such a way as to process image batches, e.g. text detection on multiple ML crops is possible with one inference.

3.3 Metamodeling of Temperature Drop

Linear Regression. The value predicted by linear regression model is computed using the hypothesis in the following form:

$$h_\theta(\mathbf{x}) = \boldsymbol{\theta}^T\mathbf{x} = \theta_0 + \sum_{i=1}^{n} \theta_i x_i$$

where: $\boldsymbol{\theta}$ - vector of model parameters, $\mathbf{x}$ - vector of features (input of the model with added element $x_0 = 1$), $n = 4$ - number of features

Model training consists in finding the optimal values of vector $\boldsymbol{\theta}$, which minimize the cost function:

$$J(\boldsymbol{\theta}) = \frac{1}{2m}\sum_{i=1}^{m}(\hat{y}_i - y_i)^2$$

where: m - number of training points, $\hat{y}_i$ - model prediction, y_i - training value.

The crucial for model accuracy is selection of the features (elements of the input vector). The selection of the four, listed above features was made after performing the analysis of importance. However, the small number of features makes model very simple and, as consequence, unable to learn complex relationship which may be hidden inside the training data. This situation is called underfitting or high bias problem. To avoid underfitting, new features should be added to the hypothesis. The new features may contain completely new data or, as more often, be a combination of data (features) already used. Within this paper, new feature were designated as higher power (up to the 4th) of already chosen inputs and all possible products between them. As the result, the number of features was increased to n=69, and each of the new ones was defined as:

$$x_i = x_1^{p_1} x_2^{p_2} x_3^{p_3} x_4^{p_4}$$

where: $i \geq 5$, $p_1, p_2, p_3, p_4 \in [0, 4]$, and $1 \leq p_1 + p_2 + p_3 + p_4 \leq 4$

On the other hand, introducing additional features into the hypothesis may cause the model to be too complex and it can learn the relationship hidden in the data by heart. This situation is called overfitting or high variance problem. It can be easily detected by comparing errors computed for training and testing (validation) sets. If the training error is low while the validation error if high, there is a high variance problem. To avoid the overfitting the regularization term was added to the optimized cost function:

$$J(\boldsymbol{\theta}) = \frac{1}{2m}\sum_{i=1}^{m}(\hat{y}_i - y_i)^2 + \frac{\lambda}{2m}\sum_{i=1}^{n}\theta_i^2$$

where: λ - regularization parameter, n - number of features.

Artificial Neural Network. The second model was built using artificial neural networks. Artificial neural network is an information processing system built with a given number of single elements called artificial neurons which are arranged in three layers: input, hidden and output layers. Number of neurons in the input (output) layer depends on the number of input (output) values, while the number of hidden layers is defined as the result of the user experience (usually, no more than two hidden layers for the nonlinear problems). Mostly, there is one neuron in the output layer corresponding to the predicted value. In case of ANN, adding the new features is not necessary. Therefore, only four features formed the input vector. However, the network topology (i.e. number of hidden layers and number of neurons in each one) is essential for model accuracy. Unfortunately, there are no rules that indicate the best number of hidden layers and neurons. Therefore, 49 neural networks were tested, each trained 50 times. Among them there were networks with one hidden layer and the number of neurons equal to: 6, 8, 10, 12, 14, 16, 18, 20, 22, 24, 26, 28, 30 and with two hidden layers and the number of neurons in each equal to : 5, 10, 15, 20, 25, 30.To compare trained models two Mean Absolute Errors (MAE) were calculated:

$$MAE = \frac{1}{m}\sum_{i=1}^{m} |\hat{y}_i - y_i|$$

$$R^2 = \left(\frac{\sum_{i=1}^{m} (\hat{y}_i - \bar{y})(y_i - \bar{y})}{\sqrt{\sum_{i=1}^{m} (\hat{y}_i - \bar{y})^2 \sum_{i=1}^{m} (y_i - \bar{y})^2}} \right)^2$$

4 Obtained Results

The environment used in this work is a server equipped with 2x Xeon Gold 6264 @ 3.10 GHz (36 physical cores in total), 768 GB RAM and NVidia Tesla M10, equipped with 4 graphics processors (4 CUDA visible devices). Only one GPU of this card was used for testing.

4.1 Sensor Layer Data Aggregation Module

One of the results of the component of data aggregation from sensory systems is the storage and sharing of information about the technological process carried out in the steelworks hall. The data relate to the implementation of this process at the main points of its processing - EAF, LF and CSC. Examples of processing data in the EAF are shown in the Fig. 3.

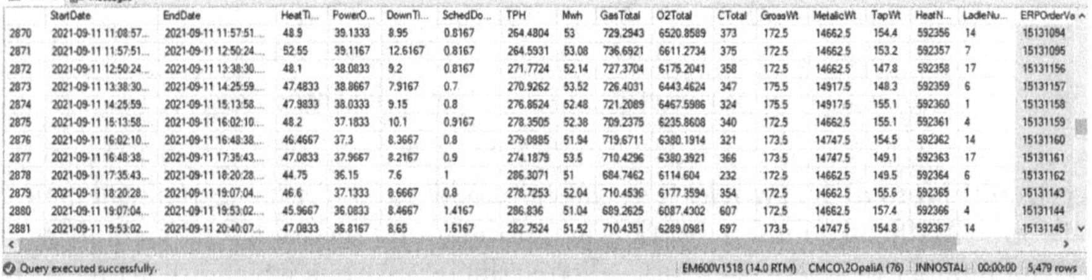

Results | Messages

	StartDate	EndDate	HeatTi...	PowerO...	DownTi...	SchedDo...	TPH	Mwh	GasTotal	O2Total	CTotal	GrossWt	MetalicWt	TapWt	HeatN...	LadleNu...	ERPOrderVa
2870	2021-09-11 11:08:57...	2021-09-11 11:57:51...	48.9	39.1333	8.95	0.8167	264.4804	53	729.2943	6520.8589	373	172.5	14662.5	154.4	592356	14	15131094
2871	2021-09-11 11:57:51...	2021-09-11 12:50:24...	52.55	39.1167	12.6167	0.8167	264.5931	53.08	736.6921	6611.2734	375	172.5	14662.5	153.2	592357	7	15131095
2872	2021-09-11 12:50:24...	2021-09-11 13:38:30...	48.1	38.0833	9.2	0.8167	271.7724	52.14	727.3704	6175.2041	358	172.5	14662.5	147.8	592358	17	15131156
2873	2021-09-11 13:38:30...	2021-09-11 14:25:59...	47.4833	38.8667	7.9167	0.7	270.9262	53.52	726.4031	6443.4624	347	175.5	14917.5	148.3	592359	6	15131157
2874	2021-09-11 14:25:59...	2021-09-11 15:13:58...	47.9833	38.0333	9.15	0.8	276.8624	52.48	721.2089	6467.5986	324	175.5	14917.5	155.1	592360	1	15131158
2875	2021-09-11 15:13:58...	2021-09-11 16:02:10...	48.2	37.1833	10.1	0.9167	278.3505	52.38	709.2375	6235.8608	340	172.5	14662.5	155.1	592361	4	15131159
2876	2021-09-11 16:02:10...	2021-09-11 16:48:38...	46.4667	37.3	8.3667	0.8	279.0885	51.94	719.6711	6380.1914	321	173.5	14747.5	154.5	592362	14	15131160
2877	2021-09-11 16:48:38...	2021-09-11 17:35:43...	47.0833	37.9667	8.2167	0.9	274.1879	53.5	710.4296	6380.3921	366	173.5	14747.5	149.1	592363	17	15131161
2878	2021-09-11 17:35:43...	2021-09-11 18:20:28...	44.75	36.15	7.6	1	286.3071	51	684.7462	6114.604	232	172.5	14662.5	149.5	592364	6	15131162
2879	2021-09-11 18:20:28...	2021-09-11 19:07:04...	46.6	37.1333	8.6667	0.8	278.7253	52.04	710.4536	6177.3594	354	172.5	14662.5	155.6	592365	1	15131143
2880	2021-09-11 19:07:04...	2021-09-11 19:53:02...	45.9667	36.0833	8.4667	1.4167	286.836	51.04	689.2625	6087.4302	607	172.5	14662.5	157.4	592366	4	15131144
2881	2021-09-11 19:53:02...	2021-09-11 20:40:07...	47.0833	36.8167	8.65	1.6167	282.7524	51.52	710.4351	6289.0981	697	173.5	14747.5	154.8	592367	14	15131145

Query executed successfully. | EM600V1518 (14.0 RTM) | CMCO\2OpaliA (76) | INNOSTAL | 00:00:00 | 5,479 rows

Fig. 3. Integrated data from EAF station.

The numerical characteristics of the data from the 18 months of collecting data from the sensory layer are as follows:

- over 16,000 reports on melts at CSC, LF and EAF stations were collected,
- number of temperature measurements taken: over 33,000 at the EAF, over 94,000 at LF stations, over 81,000 at CSCs,
- over 58,000 tests of the chemical composition of steel were carried out at LF stations,
- 75 different grades of steel were made.

More detailed information is a company secret and cannot be disclosed in this publication.

4.2 Training, Accuracy, and Performance of LF Identification

The results obtained for the indicated models are presented in the Table 1.

Table 1. Training and accuracy results of models used

Parameter	Mask R-CNN	MobileNet	YOLOv3	YOLOv3 tiny
Training time [h]	6.32	3.57	2.51	0.89
Detection time CPU [ms]	1840	1380	430	101
Detection time GPU [ms]	680	450	173	37
Accuracy (typical pos.) [%]	96	95	94	92
Accuracy (overall) [%]	88	85	82	68

Based on the obtained results, it can be seen that the accuracy of detection in typical positions of the ladle does not differ significantly - the maximum observable difference is 4%. In unusual positions, less relevant to the continuous casting process, the accuracy can vary, more depending on the specific position, less on the choice of network. However, a significant difference can be seen in the case of detection and learning times. With this criterion in mind, the YOLO network performs best, in particular its simplified version, called YOLO-tiny. The accuracy of detection slightly decreased (by approx. 13% in the general case and approx. 3% in typical cases), but the detection time decreased almost seven times (173 ms full vs 37 ms tiny). This model is therefore great for quick testing of the entire module, also on machines without GPU acceleration. Example of ML detection results from camera looking on LF is presented in Fig. 4.

Due to the high similarity of the labels on the vats, it was decided that the use of external sets would give better text detection results than the set of labels cut from ML. Final testing was performed on cut images of the ML only and achieved 89.3% prediction accuracy: 361 out of 404 images were classified correctly. We defined accuracy as percentage of cases where the network found the label on the ML at least once (some have more than one label). Obtained accuracy is

Fig. 4. ML object detection results

very high for images without any smoke (96.5%, 250 out of 259), high for small amount of smoke (87.5%, 84 out of 96). For images where manual recognition by human eye is very hard, obtained accuracy was low (55.1%, 27 out of 49 tested images were correct). Images for final tests were captured separately and were not included in any of training sets. Results from aggregated ML detection and its identification are shown in Fig. 5.

(a) ML detection on LF

(b) ML detection near EAF station

Fig. 5. ML Label detection results

4.3 Metamodeling

The smallest mean absolute error (MAE), equal 0.109, was obtained for a network with two hidden layers containing 5 and 25 neurons, respectively. The value of the coefficient of determination R2 for this network was equal to 0.764, and the training time was 7.7 s. Comparison models based of LR and ANN (Table 2) showed that neural networks allows to get more precise results.

Table 2. Assessment of models' prediction to the real values

Model	MAE	R^2
Linear regression	0.12	0.70
Artificial neural network	0.109	0.715

5 Summary and Further Works

The aim of the work was to create a multilevel system to monitor the steelworks production hall. Two modules were created to monitor the condition of the ML and the production hall, which, combined with the metamodeling module, make it possible to determine whether casting of the material is possible due to the temperature drop after leaving the EAF. The system was tested in CMC Sp. z o.o. in Zawiercie (Poland) and allows for an assessment of whether the steel can be cast after the specified time has elapsed, which may vary between each cast. The ladle monitoring module made it possible to determine the location of each ML in use, which, combined with the sensors, allowed for a precise determination of the status of each of the orders carried out by the steelworks.

At the moment, it is not used in production yet because the other components are being implemented. Among others, these include scheduling steelmaking campaigns, user interface, or aggregation with other systems in the company. The modules described in the thesis will be improved in terms of code quality, performance, and prediction accuracy, if the data collected before implementation allow it.

References

1. Myers, R.H., Montgomery, D.C., Anderson-Cook, C.M.: Response surface methodology: process and product optimization using designed experiments. John Wiley & Sons (2016)
2. Sacks, J., Welch, W.J., Mitchell, T.J., Wynn, H.P.: Design and analysis of computer experiments. Stat. Sci. **4**(4), 409–423 (1989)
3. Krzywda, M., Łukasik, S., Gandomi, A.H.: Graph neural networks in computer vision - architectures, datasets and common approaches. In: 2022 International Joint Conference on Neural Networks (IJCNN), pp. 1–10 (2022)

4. Shanthamallu, U.S., Spanias, A.: Neural Networks and Deep Learning, pp. 43–57. Springer International Publishing, Cham (2022). https://doi.org/10.1007/978-3-031-03758-0_5
5. Jung, A.B., et al.: "imgaug" (2020). https://github.com/aleju/imgaug. (Accessed 25-Jan 2023)
6. Haykin, S.: Neural Networks: A Comprehensive Foundation, 2nd ed. Prentice Hall PTR, USA (1998)
7. Joshi, A.V.: Machine Learning and Artificial Intelligence. Springer, Cham (2020). https://doi.org/10.1007/978-3-030-26622-6
8. Gajdzik, B., Wolniak, R.: Framework for r&d&i activities in the steel industry in popularizing the idea of industry 4.0. J. Open Innovation: Technol. Market Complex. **8**(3), 133 (2022)
9. Graupner, Y., Weckenborg, C., Spengler, T.S.: Designing the technological transformation toward sustainable steelmaking: A framework to provide decision support to industrial practitioners. In: Procedia CIRP, The 29th CIRP Conference on Life Cycle Engineering, April 4–6, 2022, Leuven, Belgium, vol. 105, pp. 706–711 (2022)
10. Xu, Z., Zheng, Z., Gao, X.: Energy-efficient steelmaking-continuous casting scheduling problem with temperature constraints and its solution using a multi-objective hybrid genetic algorithm with local search. Appl. Soft Comput. **95**, 106554 (2020)
11. Zhang, C.-J., Zhang, Y.-C., Han, Y.: Industrial cyber-physical system driven intelligent prediction model for converter end carbon content in steelmaking plants. J. Ind. Inf. Integr. **28**, 100356 (2022)
12. de Cassia Lima Pimenta, P.V., de Sousa Rocha, J.R., Marcondes, F.: Thermomechanical investigation of the continuous casting of ingots using the element-based finite-volume method. Euro. J. Mech. - A/Solids **96**, 104724 (2022)
13. Yang, Z., Yang, L., Guo, Y., Wei, G., Cheng, T.: Simulation of velocity field of molten steel in electric arc furnace steelmaking. In: Hwang, J.-Y., et al. (eds.) TMS 2018. TMMMS, pp. 69–79. Springer, Cham (2018). https://doi.org/10.1007/978-3-319-72138-5_8
14. Riordan, A.D.O., Toal, D., Newe, T., Dooly, G.: Object recognition within smart manufacturing. In: Procedia Manufacturing, 29th International Conference on Flexible Automation and Intelligent Manufacturing (FAIM 2019), June 24–28, 2019, Limerick, Ireland, Beyond Industry 4.0: Industrial Advances, Engineering Education and Intelligent Manufacturing, vol. 38, pp. 408–414 (2019)
15. Malburg, L., Rieder, M.-P., Seiger, R., Klein, P., Bergmann, R.: Object detection for smart factory processes by machine learning. In: Procedia Computer Science, The 12th International Conference on Ambient Systems, Networks and Technologies (ANT) / The 4th International Conference on Emerging Data and Industry 4.0 (EDI40) / Affiliated Workshops, vol. 184, pp. 581–588 (2021)
16. Ward, R., Soulatiantork, P., Finneran, S., Hughes, R., Tiwari, A.: Real-time vision-based multiple object tracking of a production process: Industrial digital twin case study. Proc. Instit. Mech. Eng. Part B: J. Eng. Manuf. **235**(11), 1861–1872 (2021)
17. Baek, Y., Lee, B., Han, D., Yun, S., Lee, H.: Character region awareness for text detection. In: Proceedings of the IEEE Conference on Computer Vision and Pattern Recognition, pp. 9365–9374 (2019)

SocHAP: A New Data Driven Explainable Prediction of Battery State of Charge

Théo Heitzmann[1,2,3(✉)], Ahmed Samet[1,2,3], Tedjani Mesbahi[1,2,3], Cyrine Soufi[1,2,3], Inès Jorge[1,2,3], and Romuald Boné[1,2,3]

[1] Université de Strasbourg, 67000 Strasbourg, France
[2] Institut National des Sciences Appliquées (INSA Strasbourg), 67000 Strasbourg, France
{theo.heitzmann,ahmed.samet,tedjani.mesbahi,cyrine.soufi, ines.jorge,romuald.bone}@insa-strasbourg.fr
[3] CNRS, ICube Laboratory UMR 7357, 67000 Strasbourg, France

Abstract. The performance and driving range of electric vehicles are largely determined by the capabilities of their battery systems. To ensure optimal operation and protection of these systems, Battery Management Systems rely on key information such as State of Charge, State of Health, and sensor readings. These critical factors directly impact the range of electric vehicles and are essential for ensuring safe and efficient operation over the long term. This paper presents the development of a battery State of Charge estimation model based on a 1-D convolutional neural network. The data used to train this model are theoretical operating data as well as driving cycles of lithium-ion batteries. An Explainable Artificial Intelligence method is then applied to this model to verify the physical behavior of the black box model. Finally, a testing platform is currently under development to assess the effectiveness of the State of Charge estimation model. Our explainable model, called SocHAP, is compared to other contemporary methods to evaluate its predictive accuracy.

Keywords: Explainable AI · Battery degradation · Deep learning · Electric vehicle · Lithium-ion battery cell · State of Charge

1 Introduction

The European Commission has approved the goal to reach a "net-zero" Greenhouse Gas (GHG) emissions level by 2050. Currently, more than 20% of the EU's GHG emissions are related to transport and almost 50% of those are caused by passenger vehicles. As opposed to other energy-intensive sectors, such as electricity generation and industry, emissions from transportation activities have been growing in the past years. Therefore, effective measures to reduce these emissions are urgently needed. The market for electric vehicles has been growing over the last few years and is becoming more and more interesting thanks to significantly reduced CO2 emissions compared to a thermal vehicle. However, if

J. Mikyška et al. (Eds.): ICCS 2023, LNCS 14077, pp. 463–475, 2023.
https://doi.org/10.1007/978-3-031-36030-5_37

the performance of electric vehicles has been greatly improved thanks to the late research advances, there are still progress to be made concerning the range and the life span of their batteries.

The range of an electric vehicle is directly linked to its energy source, which is mainly a battery composed of cells. Lithium-ion batteries are becoming overstretched on the electric vehicle market. Indeed, they have a very high energy density, a long life span and a wide range of operating temperatures [8]. The 18650 lithium-ion cell has reached a nominal capacity of 2.4 Ah which corresponds to an energy density of 200 Wh/kg [22]. In order to improve the driving range of electric vehicles, it is necessary to slow down the aging process of batteries. Indeed, as the battery ages, its storage capacity decreases, its internal resistance increases and therefore, its global performances and driving range degrade. The Battery Management System (BMS) manages the charging and discharging of the cells according to the State of Charge (SoC) and the State of Health (SoH). The advancement of battery health assessment is critical for building a clean and sustainable society, but limited access to sufficient battery aging data poses a significant challenge. This work presents an alternative solution for producing large-volume, high-quality aging datasets, which avoids the need for carrying out large-scale aging experiments in the laboratory. Although several public datasets [5] such as MIT [20] or NASA exist [19], every dataset is intended to investigate the impact of several factors on the ageing process. In this paper, we introduce a new dataset as one of our contributions. The data used in this study was obtained from a test bench installed in a research platform at INSA Strasbourg, where the cells were cycled using the WLTC (Worldwide harmonized Light vehicles Test Cycles) cycles. The data is in the form of time series measured by sensors during the cycling of the cells.

In automotive applications, SoC is considered one of the most important parameters for maintenance. Predicting an accurate value of the SoC avoids incidents such as overcharging and deep discharge of the battery. It takes values between 0 and 1 that directly indicates the amount of energy left in a battery to power an electrical device. While some studies have focused on predicting SoC with a machine learning approach, few have attempted to explain the results with an explainable model.

Therefore, our contributions in this paper are as follows: *(i)* a newly built dataset that contains time series data of battery cell meant for SoC prediction;*(ii)* a SHAP-based explainable on a top of a convolutional neural network approach, called SocHAP, to predict the SoC using multiple features; (iii) the development of a test bench implementing our estimation model for real-time battery cell SoC estimates. The paper is organized as follows, we begin by introducing the study context, then present the results of the SoC estimation training of the model. Next, we use Explainable Artificial Intelligence (XAI) to recover the physical reality of the functioning of the SoC of a battery learned by the model. Finally, we present a test bench for cell SoC estimation.

2 Battery Basics and Related Works

A lithium-ion battery is composed of several cells connected in series or in parallel. Each cell is composed of three main elements: a positive electrode (the cathode), a negative electrode (the anode) which are separated by an electrolyte [6,8].

In order to control the cells in their operation, the BMS requires several pieces of information related to the battery. While voltage, current and temperature are easily accessible via a series of sensors, SoC and SoH are cell state quantities that cannot be directly measured.

The SoC is a metric used to describe the amount of energy left in the energy storage system [21]. SoC is not a physical quantity that can be directly measured; instead, it can only be estimated by measuring strongly correlated proxy quantities such as voltage, current, and temperature [13]. Typically, SoC is expressed as a value in the range of 0 to 1, and it is defined in the literature as the ratio of the available amount of charge to the maximum amount of charge of the battery. It is computed as follows:

$$SoC = \frac{C_{rest}}{C_{nom}} \tag{1}$$

$$C_{rest} = \int i.dt \tag{2}$$

where C_{rest} is the remaining releasable capacity of the battery at a certain level of charge and C_{nom} is the nominal capacity of the battery. C_{rest} is also equal to the integral of the current i over time t. Accurate measurement of SoC is crucial for determining the appropriate charging and discharging strategies of batteries and thus avoiding any permanent damage to their internal structure [16]. The SoH of a battery is defined by the following relation:

$$SoH = \frac{C_{act}}{C_{nom}} \tag{3}$$

where C_{nom} represents the nominal capacity of the battery and C_{act} the current total capacity of the battery. This quantity gives direct information about the aging and degradation of a battery [4,12]. In the context of electric mobility, a battery is considered no longer usable and must be replaced when its SoH value reaches 80% [17].

In the literature, a few data-driven approaches and techniques have already proven to be effective in terms of accurately estimating the SoC of a lithium-ion battery cell. One of the first approaches is to use a Feed-forward Neural Network (FNN) with regression to estimate the SoC [9]. This model consists of two hidden layers with 5 neurons each, and takes voltage, current, and temperature measurements of a battery cell in the form of time series as input. Chemali et al. introduced a Long Short-Term Memory - Recurrent Neural Network (LSTM-RNN) to predict the SoC value of a cell from time series of voltage, current and temperature [3]. The model is composed of 500 stacked LSTM cells.

Machine learning and deep learning models have demonstrated good performance in classification and regression tasks across various domains. Typically, accuracy is used as an index of model quality. However, these models are often considered "black boxes" [1] because they take input features and produce output values without providing any information about the reasoning process that underlies their predictions. It is essential to provide users with explanations, and therefore, Explainable Artificial Intelligence (XAI) is employed.

Gu et al. [7] proposed an explainable model based on the SHapley Additive exPlanation (SHAP) method to determine the importance of the input parameters in a battery cell SoC estimation model. The prediction model, called SW-SHAP-LSTM, is based on a recurrent network of LSTM cells.

In this paper, we propose an explainable approach based on a Convolutional Neural Network (CNN) to estimate the SoC of a lithium-ion battery cell.

The interest in applying CNNs to time series data is that they would be able to learn filters that represent repeated patterns in the series [2,11]. The estimation model would get a better understanding of the relationships hidden in the input time-series.

We also introduce a dataset of lithium-ion cell usage cycles reproducing driving cycles of electric vehicles in urban environment. Finally, an implementation of the estimation model has been performed on an embedded system.

3 SHAP Convolutional Neural Network for SOC Prediction (SocHAP)

In this section, we introduce our SocHAPmodel that is described in Fig. 1. First, we develop a Convolution Neural Network that estimates the SoC in real-time from a window of size W = 100 time points, consisting of voltage, current and temperature measurements of the cell. Then, the SHAP method is applied to compute the importance of each feature for a particular prediction made by the model.

3.1 Data Used in the Approach and Introduction of a New Dataset

The data used for the training of the degradation models and experimentation phase are two sets of lithium-ion cell degradation data. The first set is part of our contributions, which consists of battery cell usage data in the form of driving cycles. We used lithium ferrophosphate (LFP) battery cells APR18650M1A from A123 Systems to perform our tests. These are the same cells used in the MIT dataset. The tests were generated using the Worldwide Harmonized Light Vehicle Test Procedure (WLTP). A battery cycler is programmed to wear out cells through cycles consisting of a charge phase and a discharge phase. The charging phases are carried out using the classic method of charging cells CC-CV: a constant current charge followed by constant voltage charge. The discharge phase is constituted by using the driving cycles described previously by linking several of these cycles. This phase also integrates the phenomenon of regenerative braking

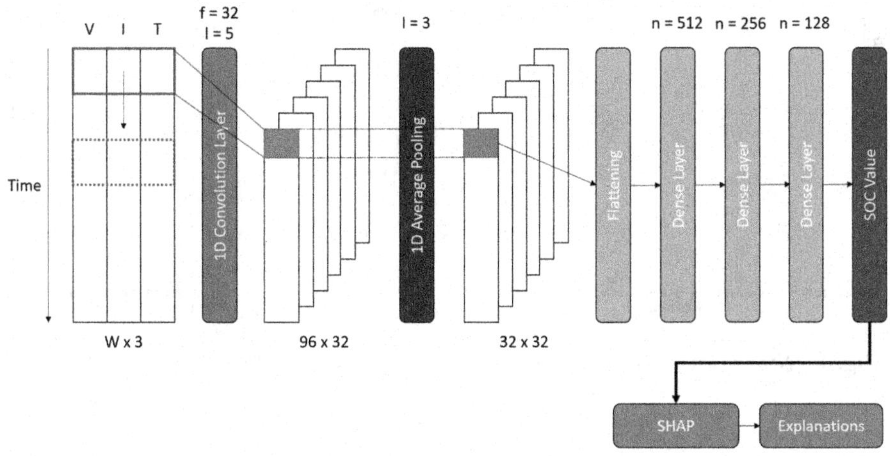

Fig. 1. Explainable Convolutional Neural Network to estimate the SOC.

which recharges the cell during the driving cycle by converting the energy dissipated during the deceleration and braking of the vehicle. This process is repeated until the cell reaches a degradation of 80%. The test system used to generate this data comes from Basytec (Basytec XCTS) and allows twelve simultaneous battery tests. This data set is a useful resource for training battery degradation models and also for data processing in the context of electric mobility. A few samples of our dataset are available in the provided link[1]. Time series of cell voltage, current, and temperature measurements for one cycle of the dataset can be seen in Figs. 2, 3, 4. The SoC time series (Fig. 5) was computed using the integral of the current.

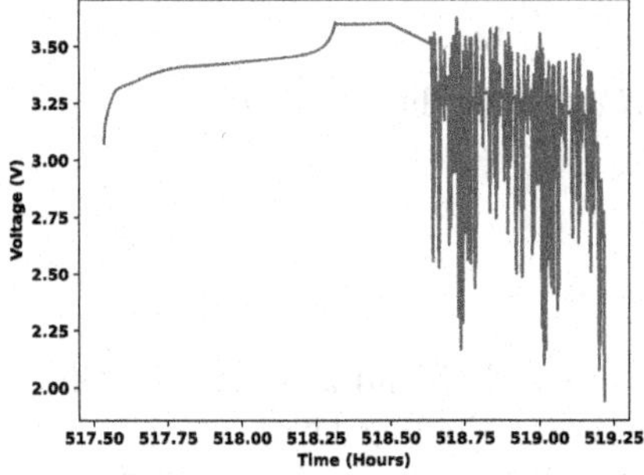

Fig. 2. INSA Cycle : Voltage

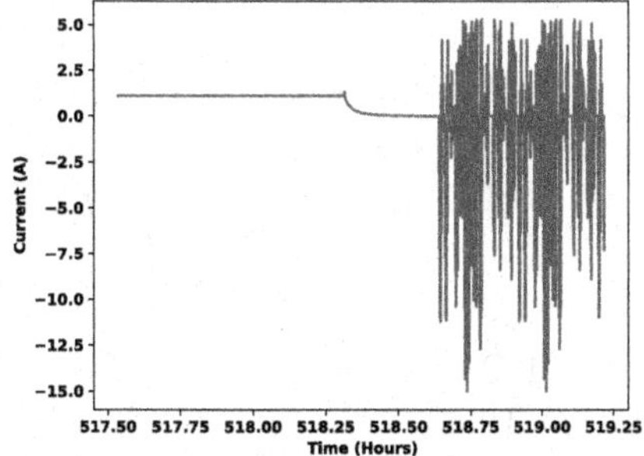

Fig. 3. INSA Cycle : Current

[1] https://github.com/thtzmn/INSA_LFP_DATASET.

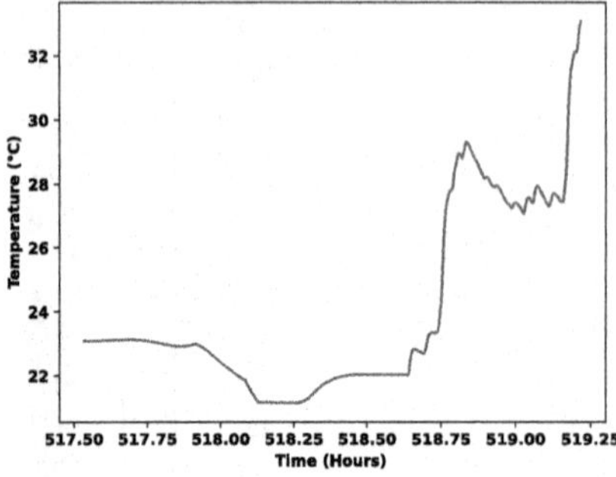

Fig. 4. INSA Cycle : Temperature

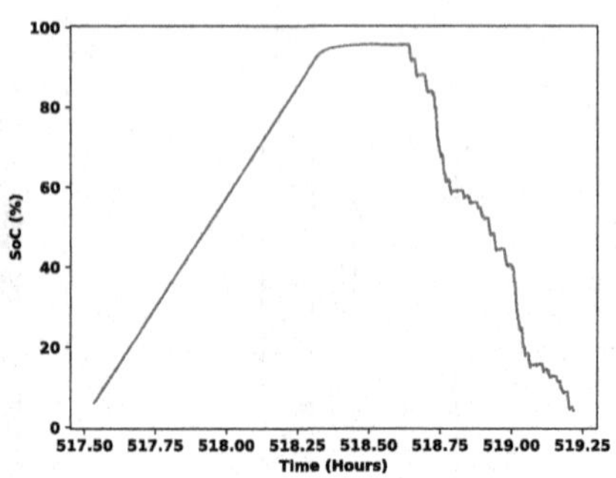

Fig. 5. INSA Cycle : SoC

The second dataset used is the one released by the Massachusetts Institute of Technology (MIT) [20], which contains data on the theoretical use of the battery with constant current discharges. To date, this is the largest public dataset containing lithium-ion cell cycling information.

3.2 State of Charge Explainable Prediction

The model used to estimate the SoC of battery cells is a convolutional neural network (CNN).

Our SocHAPmodel takes as input a sliding window of information of size W = 100 related to the battery usage, see Table 1. These data include cell voltage V, current I, and temperature T.

Table 1. Sliding window of 100 time points to perform a SoC estimation

Features		Labels
Voltage (V)	$U_0, U_1, \ldots U_{98}, U_{99}$	SoC value at t=99
Current (A)	$I_0, I_1, \ldots I_{98}, I_{99}$	
Temperature (°C)	$T_0, T_1, \ldots T_{98}, T_{99}$	

A 1-D convolution layer with f=32 filters of size l=5 and a ReLU (Rectified Linear Unit) activation followed by a 1-D Average Pooling layer of size $l = 3$ allows to perform a feature selection and data reduction operation. A flattening layer is then applied with a layer of Dropout with a coefficient of 0.05 to avoid hyper-parameter overfitting in the training process. Three fully connected layers of sizes $n = 512$, $n = 256$ and $n = 128$ respectively with ReLU activation, and a final layer of size 1 are used to estimate the SoC value.

The values of the hyperparameters have been selected after several comparisons of different network configurations, see Table 2.

Table 2. Comparison of the different configurations of the CNN estimation model

Convolution layer	layers of neurons	MAE	$RMSE$
$f = 16$	2 fully connected : $n_1 = 256, n_2 = 128$	0.045	0.0246
$f = 16$	3 fully connected : $n_1 = 512, n_2 = 256, n_3 = 128$	0.0269	0.0156
$f = 32$	2 fully connected : $n_1 = 256, n_2 = 128$	0.0424	0.0213
$f = 32$	3 fully connected : $n_1 = 512, n_2 = 256, n_3 = 128$	0.0095	0.0139

This model was trained using the Mean Absolute Error (MAE) as a loss function :

$$MAE = \frac{\sum_{i=1}^{N} |y_i - \hat{y}_i|}{N} \tag{4}$$

$$RMSE = \sqrt{\frac{\sum_{i=1}^{N} (y_i - \hat{y}_i)^2}{N}} \tag{5}$$

$$MAPE = \frac{100\%}{N} \sum_{i=1}^{N} |\frac{y_i - \hat{y}_i}{y_i}| \tag{6}$$

with N being the number of samples, $\hat{y}_i$ and y_i the estimated value of sample i and the actual value of sample i respectively. We also computed the Root Mean Square Error (RMSE) and the Mean Absolute Percentage Error (MAPE) to compare the estimation performances of our model with those of the state-of-the-art over the test samples of our training set. The AdaMax [10] gradient descent algorithm was used to train the model. It is more suitable than Adam for learning time-variant processes since it adapts the learning rate for each parameter during training.

An Explainability method is applied to determine the influence of the current as an input parameter on the SoC estimates during driving cycles. The objective is to recover the physical reality of the battery's SoC functioning learned by the model. SHAP was introduced by Lundberg and Lee in 2017 [14]. This explanation method belongs to the family of additive feature contribution methods. It assigns to each feature a coefficient or an importance value for a particular prediction $f(x)$, with f a prediction model and x a particular input of this model.

The final explanation take the form of a linear combination of the individual contributions of the input features x.

$$g(z') = \phi_0 + \sum_{i=1}^{M} \phi_i z'_i \tag{7}$$

where $g(z')$ represents a local approximation function of the original model f and ϕ_i represents the contribution of feature i to the prediction $f(x)$. $z' \in \{0,1\}^M$ equals 1 when a feature is observed or 0 otherwise and M is the number of simplified input features. The average of ϕ_i is often used as a bias value for ϕ_0. Shapley values are used to estimate the contribution ϕ_i of each feature.

$$\phi_i = \sum_{S \subseteq F \setminus \{i\}} \frac{|S|!(|F| - |S| - 1)!}{|F|!} [f(S \cup i) - f(S)] \tag{8}$$

where f is the prediction model, F the set of all input features, and S the feature subsets.

4 Experiments and Results

4.1 Evaluation of the SocHAPModel

Before applying the XAI method to explain the estimates of the SoC estimation model, we first compared its performance with that of the state-of-the-art models. All models were trained with the same samples. Slightly more than 2 500 000 samples were used for the training phase of the models. 75% as training set, 15% as validation set and 15% as test set to evaluate the performance of the models. The training was performed with Tensorflow 2 on Python 3. Their estimation performances were then compared using the MAE, the RMSE and MAPE as indicators. The training set is composed of 70% samples of our dataset with a sampling time of 0.5 s and 30% of MIT data with a sampling time of 5 s. We compare SocHAPto two other architectures. The first is an LSTM based estimation model and the second is a feed-forward neural network.

The performance of the three SoC estimation model architectures can be observed in the following Table 3:

Table 3. Comparison of performances between the different architectures.

Model	MAE	$RMSE$	$MAPE$
SocHAP	0.0096	0.0172	17.26%
He et al. FNN [9]	0.0652	0.0963	56.68%
Chemali et al. LSTM-RNN [3]	0.0106	0.0203	9.10%

Considering only the MAE and RMSE, our SocHAPapproach obtains the best estimation performances. The LSTM-based model obtains the best score for the MAPE. This means that the LSTM-based model produces more accurate estimates when the values to be estimated of SoC are close to 0 if we consider the fact that our model has a lower MAE.

Estimates of the model over the entire cycles of the MIT and INSA datasets are observable in Figs. 6, 7.

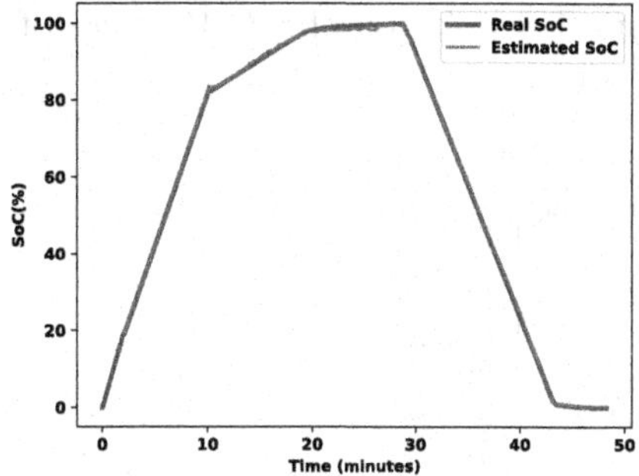

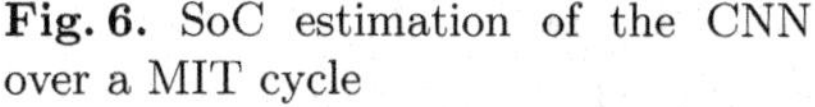

Fig. 6. SoC estimation of the CNN over a MIT cycle

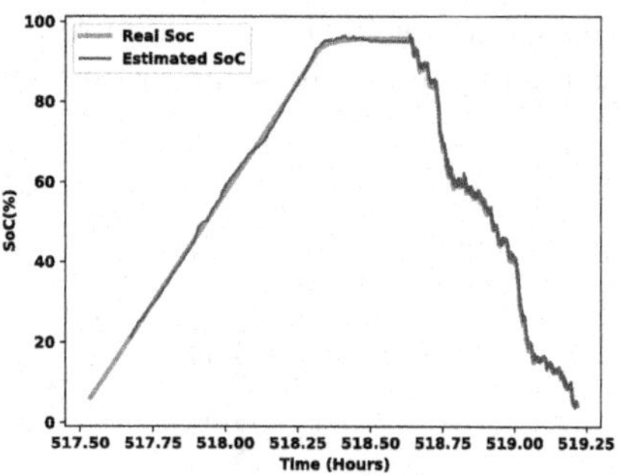

Fig. 7. SoC estimation of the CNN over a driving cycle

4.2 Explainability of the Model's Estimations

We evaluate the explanation given by SHAP over a driving cycle to understand why the model makes a certain estimate of SoC from the window of battery cell voltage, current, and temperature given as input.

By applying the SHAP and signal processing algorithms, we observed a relation between the increase of the current and the decrease of the SoC in the battery time series of a driving cycle (see Fig. 8). Based on our findings, we can conclude that the model learned the physical behavior of the battery's SoC as the SoC can be defined as the integral of the current, the amount of energy stored compared to the battery's maximum capacity during its use.

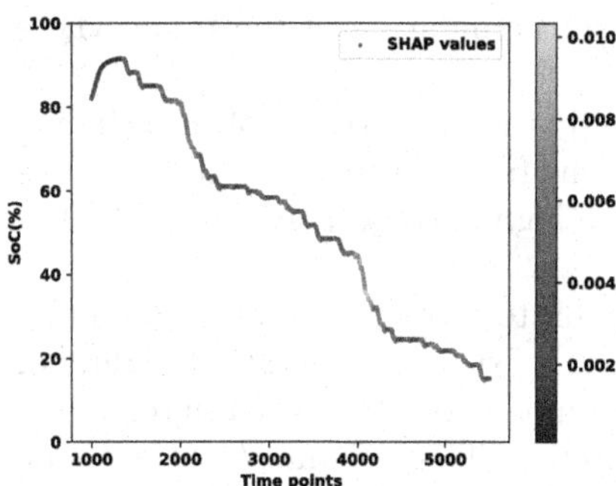

Fig. 8. Superposition of current contributions to SoC estimates during part of a driving cycle.

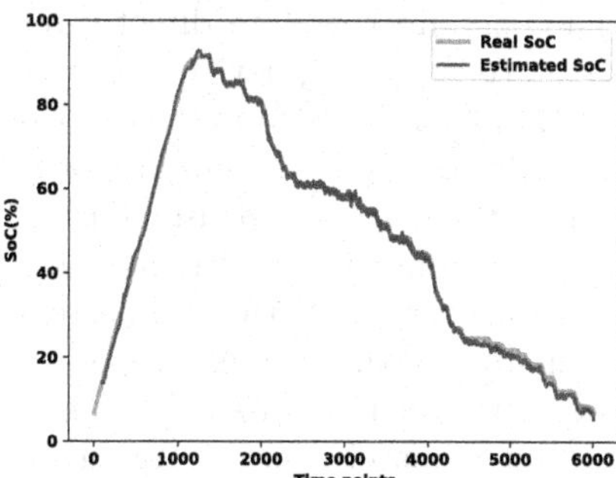

Fig. 9. Representation of the studied driving cycle.

Now that the model is validated, we can try to find the causes of the battery degradation in a driving cycle by analyzing the values of the input parameters contributions to the SoC. Figure 10 represents the evolution of SHAP values of temperature for the same driving cycle as in Fig. 9.

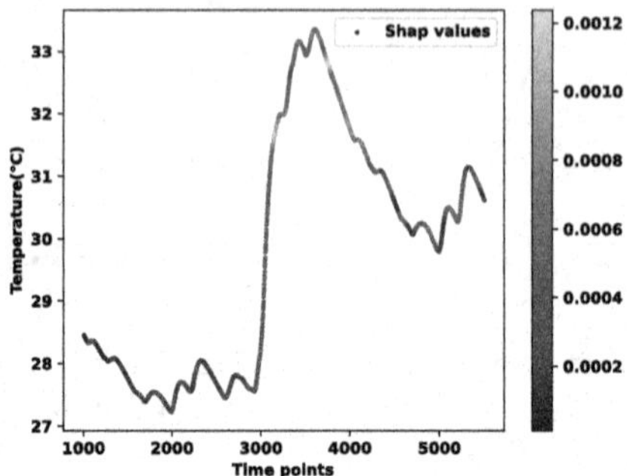

Fig. 10. Representation of the temperature evolution during the cycle with the associated SHAP values

Temperature plays a very important role in the life cycle of a lithium-ion battery. We consider as a general rule that the acceptable temperature range for the use of a lithium-ion battery is from -20 °C to 60 °C [15]. However, it has been shown that it is rather between 15 °C and 35 °C [18]. Any use of the battery outside this temperature interval can lead to an accelerated degradation of the battery. In addition, high operating temperatures will cause capacity and power losses.

The cycler used to perform battery ageing tests using driving cycles does not allow temperature control of the test chambers. The temperatures of the test chambers are therefore directly related to the temperature of the room where the battery cycling device is installed.

Based on the experiment carried out to study the temperature of the cell during a driving cycle, we initially observed that the temperature strongly increases during the discharge phase. This is due to the high current demands set by the cycler to simulate the driving cycle.

Moreover, the SHAP values related to the battery cell temperature are maximum during this strong temperature increase. By studying the SHAP values of the temperature inputs, we can isolate critical moments of battery use (e.g. high current demands, fast charging of batteries etc.). These moments of extreme battery use could then be identified and recorded. It would then be possible to compare their influence in a model predicting the SoH of the battery to quantify the degradation associated with these patterns.

4.3 Real-Time Cell SoC Test Bench Under Development

One of the perspectives of our work is to implement our SoC estimation model in a test bench. The objective is to be able to compare the estimates of our model with the real SoC of a battery cell in real conditions. Currently, we are still in the testing and adjustment phase.

This embedded estimation bench integrates a Jetson Nano which is a low-consumption microcomputer powerful enough to deploy our model. Moreover, the Jetson Nano communicates via I2C with sensors in order to retrieve measurements of voltage, current and temperature during the cycling of the cell to estimate its SoC. These data are also stored for further experimentation, such as computing the real SoC value of the battery cell using the integral of the current method.

The bench displays through an LED screen the SoC value of a lithium-ion battery cell in real time with the measurements of the voltage, current and temperature of the battery cell (Fig. 11).

The cell can be charged or discharged by connecting its power circuit either to a stabilized power supply or rheostat type resistor using banana plugs.

The operation of the bench is described by the diagram Fig. 12.

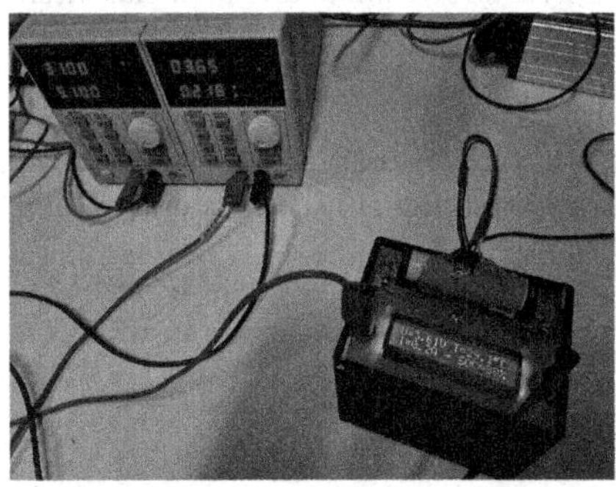

Fig. 11. Image of the test bench during the end of a charging phase

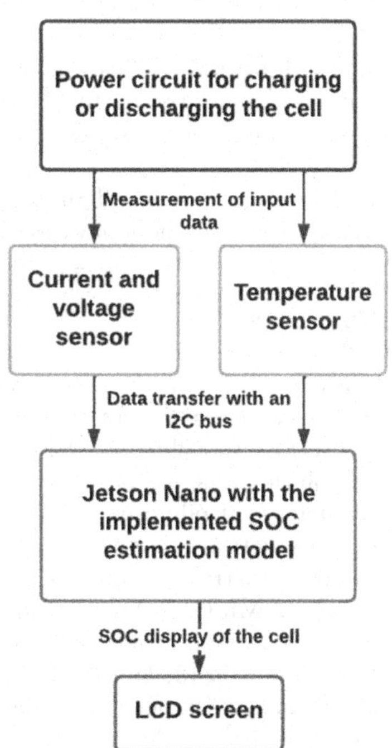

Fig. 12. Operating diagram of the SoC estimator

5 Conclusion

This paper introduces a publicly available dataset containing measurements of LFP battery cell usage during driving cycles. The dataset was used to train our proposed explainable State of Charge (SoC) estimation approach, SocHAP. Additionally, an experiment was conducted to demonstrate the explainability of the SoC estimation for a driving cycle using SHAP values. The results of this experiment suggest that patterns of battery aging acceleration can be identified by analyzing temperature data using SHAP values.

The future challenges of this work include first completing the development of the embedded SoC estimator. The second objective is to further enhance the explainability aspect of the lithium-ion battery degradation models. The aim is to make the explanations provided by the explainability methods more interpretable and to provide advice to the electric vehicle users on how to use the battery efficiently. These recommendations could be applied to both the battery charging phase and the driving phase with the goal of extending the life cycle of batteries as well as increasing the autonomy of the vehicle.

Acknowledgements. This work was carried out as part of HALFBACK and VEHICLE project, sponsored by INTERREG V A Upper Rhine Programme, FEDER and Franco-German regional funds (Bade-Wurtemberg, Rhénanie-Palatinat and Grand-Est).

This paper has received funding from the European Union under the program Horizon Europe and the innovation program under GAP-101103667.

References

1. Alicioglu, G., Sun, B.: A survey of visual analytics for explainable artificial intelligence methods. Comput. Graph. **102**, 502–520 (2022)
2. Borovykh, A., Bohte, S., Oosterlee, C.W.: Conditional time series forecasting with convolutional neural networks (2018)
3. Chemali, E., Kollmeyer, P.J., Preindl, M., Ahmed, R., Emadi, A.: Long short-term memory networks for accurate state-of-charge estimation of li-ion batteries. IEEE Trans. Industr. Electron. **65**(8), 6730–6739 (2018)
4. Chen, Z., Mi, C.C., Fu, Y., Xu, J., Gong, X.: Online battery state of health estimation based on genetic algorithm for electric and hybrid vehicle applications. J. Power Sourc. **240**, 184–192 (2013)
5. dos Reis, G., Strange, C., Yadav, M., Li, S.: Lithium-ion battery data and where to find it. Energy AI **5**, 100081 (2021)
6. Gomadam, P.M., Weidner, J.W., Dougal, R.A., White, R.E.: Mathematical modeling of lithium-ion and nickel battery systems. J. Power Sour. **110**(2), 267–284 (2002)
7. Gu, X., See, K., Wang, Y., Zhao, L., Pu, W.: The sliding window and shap theory-an improved system with a long short-term memory network model for state of charge prediction in electric vehicle application. Energies **14**(12) (2021)
8. Hannan, M., Lipu, M., Hussain, A., Mohamed, A.: A review of lithium-ion battery state of charge estimation and management system in electric vehicle applications: Challenges and recommendations. Renew. Sustain. Energy Rev. **78**, 834–854 (2017)

9. He, W., Williard, N., Chen, C., Pecht, M.: State of charge estimation for li-ion batteries using neural network modeling and unscented kalman filter-based error cancellation. Int. J. Electrical Power Energy Syst. **62**, 783–791 (2014)
10. Kingma, D.P., Ba, J.: Adam: A method for stochastic optimization (2014). https://arxiv.org/abs/1412.6980
11. Koprinska, I., Wu, D., Wang, Z.: Convolutional neural networks for energy time series forecasting. In: 2018 International Joint Conference on Neural Networks (IJCNN), pp. 1–8 (2018). https://doi.org/10.1109/IJCNN.2018.8489399
12. Lipu, M.H., et al.: A review of state of health and remaining useful life estimation methods for lithium-ion battery in electric vehicles: Challenges and recommendations. J. Clean. Prod. **205**, 115–133 (2018)
13. Lipu, M.H., Hussain, A., Saad, M., Ayob, A., Hannan, M.: Improved recurrent narx neural network model for state of charge estimation of lithium-ion battery using pso algorithm. In: 2018 IEEE Symposium on Computer Applications & Industrial Electronics (ISCAIE), pp. 354–359. IEEE (2018)
14. Lundberg, S.M., Lee, S.I.: A unified approach to interpreting model predictions (2017)
15. Ma, S., et al.: Temperature effect and thermal impact in lithium-ion batteries: A review. Prog. Nat. Sci. Mater. Internat. **28**(6), 653–666 (2018)
16. Ng, K.S., Moo, C.S., Chen, Y.P., Hsieh, Y.C.: Enhanced coulomb counting method for estimating state-of-charge and state-of-health of lithium-ion batteries. Appl. Energy **86**(9), 1506–1511 (2009)
17. Perner, A., Vetter, J.: 8 - lithium-ion batteries for hybrid electric vehicles and battery electric vehicles. In: Scrosati, B., Garche, J., Tillmetz, W. (eds.) Advances in Battery Technologies for Electric Vehicles, Woodhead Publishing Series in Energy, pp. 173–190. Woodhead Publishing (2015)
18. Pesaran, A., Santhanagopalan, S., Kim, G.H.: Addressing the impact of temperature extremes on large format li-ion batteries for vehicle applications (presentation) (May 2013)
19. Saha, B., Goebel, K.: Battery data set. NASA Ames Prognostics Data Repository (2007)
20. Severson, K., Attia, P., Jin, N., et al.: Data-driven prediction of battery cycle life before capacity degradation. Nat. Energy **4** (2019)
21. Wang, W., Wang, X., Xiang, C., Wei, C., Zhao, Y.: Unscented kalman filter-based battery soc estimation and peak power prediction method for power distribution of hybrid electric vehicles. Ieee Access **6**, 35957–35965 (2018)
22. Yoshio, M., Brodd, R.J., Noguchi, H.: Lithium-ion batteries: Science and Technologies. Springer (2009). https://doi.org/10.1007/978-0-387-34445-4

ATS: A Fully Automatic Troubleshooting System with Efficient Anomaly Detection and Localization

Lu Yuan[1,2], Yuan Meng[3], Jiyan Sun[1(✉)], Shangyuan Zhuang[1,2], Yinlong Liu[1,2], Liru Geng[1,2], and Weiqing Huang[1,2]

[1] Institute of Information Engineering, Chinese Academy of Sciences, Beijing, China
{yuanlu,sunjiyan,zhuangshangyuan,liuyinlong,gengliru, huangweiqing}@iie.ac.cn

[2] School of Cyber Security, University of Chinese Academy of Sciences, Beijing, China

[3] Xinjiang Aksu Public Security Bureau, Xinjiang, China

Abstract. As network scale expands and concurrent requests grow, unexpected network anomalies are more frequent, leading to service interruptions and degraded user experience. Real-time, accurate troubleshooting is critical for ensuring satisfactory service. Existing troubleshooting solutions adopt ensemble anomaly detection (EAD) to detect anomalies due to its robustness. However, the fixed base classifier parameters in EAD set by expert experience may reduce the efficiency of anomaly detection when faced with different data distributions. Furthermore, the binary results fed to the secondary classifier in EAD cause information loss, leading to compromised accuracy and inaccurate root cause localization. Besides, key performance indicators (KPIs) are crucial for measuring the system performance, but relying on multiple redundant KPIs to identify the root causes of anomalies is time-consuming and error-prone.

To address the above issues, we propose a fully automatic troubleshooting system, **ATS**. A new EAD method is introduced to detect anomalies, then a module is designed to trigger the root cause localization. Specifically, the EAD method updates the parameters of base classifiers to dynamically adapt to different KPI data distributions. The ensemble of soft labels generated by base classifiers is subsequently fed into the secondary classifier to achieve information-lossless anomaly detection. Then, a heuristic module is proposed to select the most appropriate KPI data based on the metric i.e., *bilayer relative difference* to trigger the efficient root cause localization. Extensive experiments demonstrate that ATS is more than twice as fast as most state-of-the-art solutions while with higher troubleshooting accuracy.

Keywords: Troubleshooting · Ensemble Anomaly Detection · Soft Label · Bilayer Relative Difference

J. Mikyška et al. (Eds.): ICCS 2023, LNCS 14077, pp. 476–491, 2023.
https://doi.org/10.1007/978-3-031-36030-5_38

1 Introduction

With the continuous expansion of network scale and rapid growth of concurrent requests, unexpected network anomalies are becoming increasingly common, such as unusual traffic patterns, data packet loss, and sudden spikes or drops in network traffic. If the anomalies are not addressed, they may result in service interruptions for users and significant deterioration in user experience. In online commercial services, such issues can even harm business profits [3,10]. Therefore, a real-time and accurate troubleshooting scheme is crucial for providing satisfactory service and ensuring a seamless user experience by rapidly identifying and resolving network anomalies.

As shown in Fig. 1, the troubleshooting process can be summarized in three steps: Key Performance Indicator (KPI) extraction, anomaly detection, and root cause localization (RCL). Firstly, operators extract various real-time KPIs by aggregating and calculating from the raw data. Secondly, Ensemble Anomaly Detection (EAD) algorithms [1,4,26,37,40] are typically employed to monitor system status by detecting variation of KPIs. In detail, the KPIs are initially input into the base classifiers, i.e. base learners [37], which generate preliminary classification results. These binary outcomes are then simultaneously fed into the secondary classifier to make the final classification decisions [30]. Finally, after detecting an anomaly, many Root Cause Localization (RCL) algorithms [2,18,22,29,35] are introduced to identify the anomaly root cause to specific dimensions such as province, city, or server ID. Considering the inevitable false positives of anomaly detection, they utilize all KPIs flagged as anomalous to locate the candidates of root cause individually. Subsequently, each candidate is scored and the one with the highest score is ultimately selected as the conclusive result.

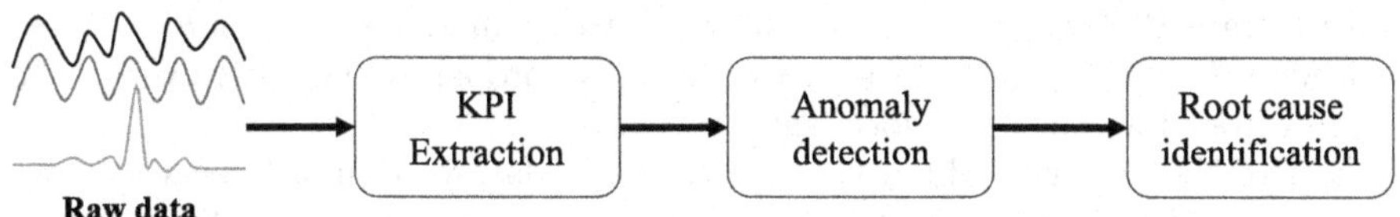

Fig. 1. Process of troubleshooting.

Despite the strong performance of these methods, there still exist several important challenges to achieving an efficient troubleshooting method in practice.

Firstly, base learners are the foundation of EAD. By combining more accurate results from base learners, a more precise and robust result can be achieved. However, existing EAD schemes mostly concentrate on optimizing the parameters of the secondary classifier, while the parameters of the base learners are set based on expert experience and remain fixed [31,34]. The fixed parameters of base learners lead to low accuracy in handling different data distributions, which ultimately affects the final accuracy of the secondary classifier.

Secondly, the final decision of EAD heavily depends on the preliminary classification results of base learners. Many EAD algorithms [1,4,37,40] utilize the secondary classifier to make the final decision only considering the binary results of base learners. However, this approach fails to capture the full classification information provided by the base learners, resulting in significant information loss. This loss has a harmful impact on the optimization of the secondary classifier and the final performance of the system.

Thirdly, current approaches [7,25,29] often utilize all impacted KPI data to identify the root cause, resulting in high computational overhead and low efficiency. Moreover, including KPI data that are irrelevant with anomaly as input can lead to a misidentification of the root cause.

To address the above challenges, we propose a *fully Automatic Troubleshooting System* (ATS) in this paper. The main contributions are summarized as follows:

- **Framework of Fully Automatic Troubleshooting System**. We propose a fully automatic troubleshooting system, namely ATS, which contains three primary components: *AutoDetect*, *Gunlock*, and *AutoRoot*. When AutoDetect identifies a system anomaly, Gunlock calculates the most relevant KPI data to trigger AutoRoot for efficient root cause localization(Section 4.1).
- **A Robust and Information Lossless Ensemble Anomaly Detection**. To improve the accuracy of anomaly detection, we propose a robust and information lossless ensemble scheme for anomaly detection (AutoDetect). AutoDetect dynamically updates base learner parameters using Bayesian Optimization to adapt to various KPI data distributions. It then combines the original probabilities of each base learner as *soft labels*, which are fed into the secondary classifier for information-lossless anomaly detection(Section 4.2).
- **Heuristic Trigger**. We propose a novel heuristic trigger called *Gunlock* to accelerate RCL (AutoRoot). Gunlock utilizes a metric known as *bilayer relative difference* (BRD) to identify the most proper KPI and transfer it to AutoRoot for root cause localization (Sect. 4.3).
- **Evaluation on real data traces**. We use a dataset from a large-scale content delivery network (CDN) to evaluate the performance of ATS. Extensive experiments demonstrate that ATS outperforms the state-of-the-art approaches. For instance, *AutoDetect* gains a high system anomaly detection performance with average 10 percent higher than state-of-art algorithms in F1-score. In addition, we note that ATS shortens troubleshooting time by half on average (Sect. 5).

2 Related Work

As the two critical components of online service system troubleshooting, anomaly detection and RCL have become popular research topics in recent years [22,23].

Ensemble learning is a machine learning technique that combines multiple base learners to achieve higher accuracy than a single model. It plays an important role in the research field of anomaly detection [1,7,37] and is widely deployed in industry [17,27,36]. To create a stronger model, Paulauskas and Auskalnis [13] propose an ensemble model consisting of four different base classifiers, which depends on the idea of combining multiple weaker learners. Vanerio et al. [37] investigate different ensemble-learning approaches and find the optimal scheme to enhance anomaly detection in network measurements. Rajagopal et al. [32] provide an ensemble paradigm based on meta-classification and stacked generalization with the goal of improving prediction accuracy. In industry, EGADS [17] in Yahoo trains multiple models for different types of KPI and assembles the appropriate models as plugins. Metis [36] and Surus [27] are proposed to achieve robust anomaly detection by aggregating different anomaly detection algorithms. These studies have demonstrated that the application of ensemble methods enhances the performance of the models, such as prediction accuracy.

However, the above solutions combine the binary results of base learners. The ensemble with binary results of base learners produces information loss which leads to inaccurate results. Moreover, they optimize the model by adjusting the parameters of the secondary classifier but do not optimize the base learners. This leads to the inefficient performance of the ensemble model.

There have been some papers proposing the advanced root cause location algorithm. HotSpot [35] develops a score metric to identify the root cause nodes. Li et al. [18] and Jing et al. [15] propose improved solutions based on clustering to address the problem of long tail distribution of KPI metrics. They locate root causes using each KPI data affected by anomalies. But some changes in KPI data are not associated with anomalies. Locating the root cause with the improper input KPI data results in wrong results and wasting of computing source.

3 Preliminaries

In this section, we illustrate some important definitions used in this paper.

To enable real-time troubleshooting, a variety of KPIs such as *visit count*, *in flow rate*, and *cache hit ratio* are collected from raw logs. KPIs that are collected with multiple dimensions are referred to as multi-dimensional KPIs (**MDKPIs**). For instance, consider the CDN data of visit count in Table 1, where the first row represents an MDKPI with three dimensions: Province, ISP (Internet Service Provider), and Website. The corresponding dimension values are $Beijing$, $ChinaMobile$, $Weibo.com$, respectively.

The KPI can be calculated for each **dimension combination**, e.g., in the first row of Table 1, the KPI value for dimension combination $\{Beijing, ChinaMobile, Weibo.com\}$ is 59 (line 1). Specifically, the character $*$ represents all the possible values of the corresponding dimension. For instance, the dimension combination $\{*, ChinaMobile, Weibo.com\}$ represents all the users that come from ISP *China Mobile* and visit the website *Weibo.com*. The *visit count* value of $\{*, ChinaMobile, Weibo.com\}$ is 59+31028+370=31457 (line 1–3).

Table 1. Example of multi-dimensional KPI.

Index	Province	ISP	Website	Value
1	Beijing	China Mobile	Weibo.com	59
2	Shanghai	China Mobile	Weibo.com	31028
3	Guangdong	China Mobile	Weibo.com	370
4	Beijing	China Mobile	QQ.com	221
5	Shanghai	China Mobile	QQ.com	10
6	Guangdong	China Mobile	QQ.com	33
7	**Beijing**	**China Unicom**	**Weibo.com**	**731**
8	**Shanghai**	**China Unicom**	**Weibo.com**	**10**
9	**Guangdong**	**China Unicom**	**Weibo.com**	**6**
10	Beijing	China Unicom	QQ.com	441
11	Shanghai	China Unicom	QQ.com	16

Especially, when there is only one dimension value and other dimensions take the value of *, the KPI is called Single-dimension KPI (**SDKPI**). For example, the KPI *visit count* with dimension *Website* is one SDKPI, which corresponds to the dimension combinations $\{*, *, Weibo.com\}$ and $\{*, *, QQ.com\}$. The visit count values of Website dimension are $(Weibo.com, 32204)$ and $(QQ.com, 721)$, which is calculated by $59+31028+370+731+10+6=32204$ (line 1–3, 7–9), $221+10+33+441+16=721$ (line 4–6, 10–11), respectively.

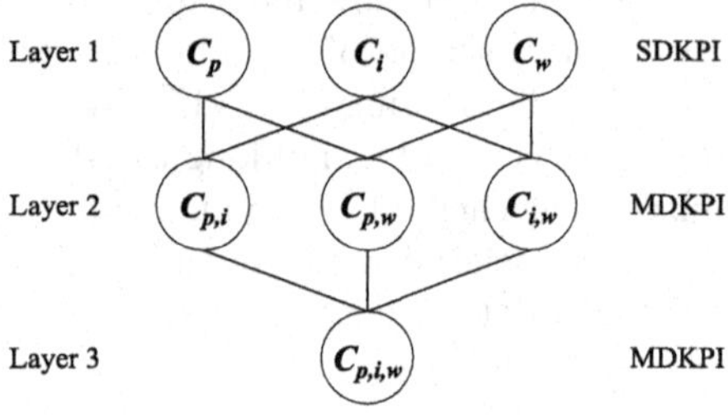

Fig. 2. Root Cause Search Tree.

As shown in Fig. 2, we assume that the search space of multi-dimensional root causes is a tree-like structure. Each node in the tree indicates a dimensional combination. A specific dimension value represents the corresponding potential fault location of it. Especially, a **leaf node** is a dimension combination without any wildcard $*$, e.g., $\{Beijing, ChinaUnicom, Weibo.com\}$. A **cabin** denotes the set of dimension combinations with the same non-wildcard dimensions. The **layer** is the number of non-wildcard dimensions of a cabin. Let the terms p, i, w denote the dimensions *Province*, *ISP*, *Website*, respectively. $\{Beijing, ChinaUnicom, *\}$ and $\{Shanghai, ChinaMobile, *\}$ belong to the cabin $C_{p,i}$, and they are in the layer 2. All leaf nodes are in the cabin $C_{p,i,w}$ of Layer 3.

4 System Design

4.1 Overview of Framework

We propose a *fully automatic troubleshooting system* (ATS) shown in Fig. 3. The ATS mainly consists of four parts: KPI extraction, *AutoDetect* for anomaly detection, *Gunlock* for trigger AutoRoot, and *AutoRoot* for RCL.

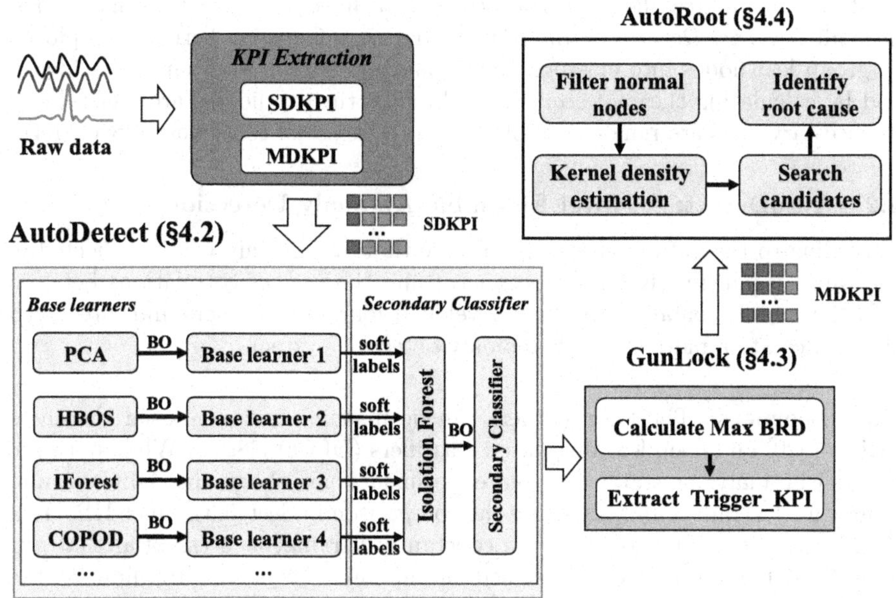

Fig. 3. The framework of ATS. The blue square denotes the KPI dimension and the green square denotes the KPI category. SDKPI consists of different types of KPI and one dimension. MDKPI includes multiple dimensions and one KPI. (Color figure online)

To **extract** KPI data, we preprocess the raw data, which includes *data filling*, *data smoothing*, and *KPI extraction*. Firstly, we fill incomplete data with the mean of the surrounding context and smooth KPI data using exponential mean average [16] to eliminate noise to some extent. Then we normalize the values to remove the influence of the order-of-magnitude differences between KPIs. We filter out KPIs with a variance close to 0 and extract SDKPI with website dimension for anomaly detection and MDKPI for RCL.

To enhance the performance of anomaly detection, **AutoDetect** is proposed which is a robust and information lossless ensemble anomaly detector (Sect. 4.2). AutoDetect's extensible first layer comprises multiple base learners elaborately selected. To accommodate diverse data distributions, we employ Bayesian optimization to automatically select the best parameters for base learners. Subsequently, the *soft labels* from the base learners are integrated and fed to the secondary classifier to avoid information loss and improve the efficiency of anomaly detection.

After detecting an anomaly, **Gunlock** (Sect. 4.3) computes the heuristic metric BRD to identify the most suitable KPI, referred to as *Trigger_KPI*. Then it extracts the MDKPI of Trigger_KPI and employs it to trigger **AutoRoot** to locate the root cause.

The multi-dimensional root cause locator AutoRoot [15] is applied to RCL. Firstly, it calculates the deviation score of each node to filter normal nodes. Secondly, *Kernel Density Estimation* (KDE) with Gaussian kernel is employed to group leaf nodes into clusters. Next, the candidate of each cluster is identified by calculating the *root score* (RS). Finally, the candidates are merged and unnecessary ones are removed to obtain a precise set of root causes (Sect. 4.4).

4.2 AutoDetect: A Novel Ensemble Anomaly Detection

According to the data analysis, we observe that an anomaly will be reflected in the value of multiple KPIs. To better capture the various correlations between KPI data and anomalies, we have developed four base learners and integrated them to create a more effective anomaly detection system.

Base Learners. The base learners can be machine learning-based classifiers (MLCs) [20] and deep learning-based classifiers (DLCs) [28,33]. While both are effective at analyzing static data, our experiments have shown that DLCs require more time to train than MLCs (see the comparison in Sect. 5.3). Since KPI data distributions in online systems are constantly varying, it is essential that our anomaly detection model be retrained and updated frequently. To minimize the time required for these processes, we choose four MLCs as our base learners due to their low training overhead. It is worth noting that this stage is scalable, so the base learners are not confined to the selected learners and other learners can be added.

To examine the KPI data from multiple angles, we choose several effective techniques as base learners. These include the proximity-based approach known as Histogram-based Outlier Score (HBOS) [8], the refactoring-based method called Principal Components Analysis (PCA) [11], the efficient classifier known as Isolation Forest (IForest) [21], and a newly developed technique called Copula-based Outlier Detection (COPOD) [19]. All of these models excel at analyzing data with high dimensions [6].

Automatic Parameter Selection. To achieve optimal performance of base learners in varying data distributions, we apply the Bayesian Optimization (BO) [9] method to optimize the base learners during training. By utilizing BO to select the most appropriate parameters, the performance of the base learners is significantly improved. It can be expressed as:

$$\lim_{p \in P} F(Clf(p); P) \tag{1}$$

where P is the set of the possible values of p. $Clf(\cdot)$ denotes the classifiers. $F(\cdot)$ is the objective function that is usually a user-defined function using different classifier metrics. In this paper, the objective function is the negative number of the F1-score of the classifier. Therefore, the objective function is represented as follows:

$$\min_{p \in P} \sum_{m=1}^{M} -F_1Score(C_m(p); P) \tag{2}$$

where C_m represents the base learners, and M represents the number of base learners which is 4 in this paper. Equation 2 represents that each base learner C_m chooses the appropriate parameters p to optimize the performance of the classifier.

Table 2. Comparison of the ensemble with soft labels and binary results.

Soft Labels					Binary Results					Label
PCA	HBOS	IForest	COPOD	Result	PCA	HBOS	IForest	COPOD	Result	
0.0061	0.0187	0.0314	0.1404	1	0	0	0	0	**0**	1
0.0079	0.0226	0.0332	0.1406	1	0	0	0	0	**0**	1
0.0091	0.0205	0.0147	0.1704	1	0	0	0	0	**0**	1
0.0080	0.0229	0.030	0.1612	1	0	0	0	0	**0**	1

Information Lossless Ensemble. Existing ensemble strategies [4,26,37] combine the binary results of base learners for comprehensive making decisions to gain better performance. The results shown in Table 2 indicate that all of the base learners generate binary results of 0 (columns 6–9). As a result, the final result of the ensemble classifier is also 0 (column 10), despite the fact that the true labels indicate a value of 1 (column 11).

It is obvious that information loss occurs during the generation of binary results by the base learners, leading to incorrect outcomes when these results are fed into the secondary classifier, ultimately compromising its accuracy.

Instead of generating binary results, the base learners of AutoDetect output the probability of data being anomalous. This probability retains all the learning information and is instrumental in identifying anomalies. We refer to this probability as the *soft label.* The soft labels then are fed into the secondary classifier to make the final decision. As shown in Table 2, the final decisions based on the soft labels of the base learners (column 5) are identical to the true labels, demonstrating the effectiveness of our approach.

Additionally, the choice of the secondary learning algorithm has a significant impact on the generalization performance of stacking integration. Therefore, after comparing with typical algorithms, we have selected Isolation Forest [21] as our secondary classifier(Section 5). To further improve the model's performance, Bayesian optimization (BO) is applied to the training of the secondary classifier.

Training of AutoDetect. As Algorithm 1 shows, the process of *AutoDetect* training is summarized as follows:

- The first step is to train individual base learners (Step 1–4). The extracted SDKPI is input into the base learners. And BO is applied to choose appropriate parameters to enhance the performance of each learner.
- The second step is to use the base learners to predict the test data and aggregation of soft labels is used as the secondary training set, which is the training set of the secondary classifier (Step 5–11).
- The final step is to train the secondary classifier (Step 12–13). To thoroughly analyze the soft labels, we adopt the stacking combination strategy and select Isolation Forest as our secondary classifier. Additionally, we utilize BO during the training process of the secondary classifier.

4.3 Gunlock: Trigger

In real-world systems, a single anomaly can impact multiple KPIs. However, analyzing each affected KPI is computationally costly, and not all changes in KPI values indicate the anomaly. This wastes computing resources. To optimize efficiency and accelerate localization, it is crucial to select the most appropriate KPI called *Trigger_KPI* as the sole input for RCL.

Existing solutions for identifying system anomalies focus on KPIs with the largest fluctuation, but this may not always be relevant or indicate anomalies. Some KPIs have a wide normal range of fluctuation without significant impact on the average value over time, leading to relative differences that appear large but are not anomalous.

To find the most proper KPI data for RCL, we propose a novel heuristic metric named *bilayer relative difference* (BRD) to evaluate the degree of change

---

Algorithm 1. AutoDetect training

Input: The SDKPIs training dataset $D = (X_1, y_1), (X_2, y_2), ..., (X_n, y_n)$, base learners $\mathcal{L}_1, \mathcal{L}_2, ...\mathcal{L}_m$, secondary learner $\mathcal{L}$
Output: Predicting result $H(X) = h'(h(X_1, X_2, ..., X_n))$
1: **for** $i = 1, 2, ..., m$ **do**
2: $\quad p = \mathit{BayesianOptimization}(\mathcal{L}_i)$.
3: $\quad h_i = \mathcal{L}_i(D, p)$
4: **end for**
5: $D' = \varnothing$
6: **for** $i = 1, 2, ..., n$ **do**
7: $\quad$ **for** $j = 1, 2, ..., m$ **do**
8: $\quad\quad z_{ij} = h_j(X_i)$
9: $\quad$ **end for**
10: $\quad D' = D' \cup ((z_{i1}, z_{i2}, ..., z_{im}), y_i)$
11: **end for**
12: $p' = \mathit{BayesianOptimization}(\mathcal{L})$.
13: $h' = \mathcal{L}(D', p')$

---

in KPIs. It considers the historical change in KPI data as follows:

$$BRD^i_\tau = \frac{k^i_\tau - k^i_{mean}}{k^i_{mean}}$$
$$k^i_{mean} = avg(med_1, med_2) \tag{3}$$

where k^i_τ represents the i-th KPI at time τ and $i = 1, 2, ..., n$. k^i_{mean} denotes the median mean of the i-th KPI in the last week. We calculate the median m_1 of the de-duplicated values of the i-th KPI data for a week. Then, the weekly data is split into two groups based on whether the values are greater than or less than m_1. med_1 and med_2 represent medians of these groups respectively. k^i_{mean} is the average of med_1 and med_2.

When an anomaly is detected, Gunlock analyzes the impact on each KPI by calculating its BRD. This helps determine which KPIs are most affected by the anomaly. Gunlock then identifies the KPI with the highest BRD as the Trigger_KPI. Finally, Gunlock extracts the MDKPI for the Trigger_KPI and sends the data to AutoRoot, which locates the root cause of the anomaly.

4.4 AutoRoot: Root Cause Localization

We implement the AutoRoot proposed in [15] to localize the root cause due to the parameter-free clustering. It is beneficial for constructing a fully automatic system.

Fliter Normal Nodes. The forecast value of the Trigger_KPI is calculated from historical data according to the ARMA (Auto-regressive moving average model) algorithm [24]. By comparing the real and forecast values, the *deviation score* for each leaf node is computed to enable differentiation between anomalous and normal nodes. The *deviation score* is defined as:

$$d(e) = \frac{f(e) - v(e)}{f(e) + v(e)} \tag{4}$$

where $f(\cdot)$ and $v(\cdot)$ are forecast value and real value functions of dimension combinations. The deviation scores of normal leaf nodes are far less than that of abnormal ones according to the GRE principle [18,35]. We use a proper range of deviation scores [-0.1,0.1] to filter out normal leaf nodes according to the observation of the real dataset [18].

Kernel Density Estimation Based Clustering. We use the KDE with the Gaussian kernel to obtain the distribution density function from the deviation score array and the relative maxima and minima of this function. The relative maxima are the centers of each cluster, while the nearby relative minima are the boundaries of the clusters. Consequently, we obtain distinct clustering intervals, and by categorizing the deviation score arrays into these intervals, we can allocate the anomalous leaf nodes inherited from the same root cause into the same clusters.

Search Candidates. The root score (RS) of each dimension combination is calculated following Eq. 5 to find the candidates for the root causes.

$$RS = avg(NPS + LF + CF) \tag{5}$$

$$NPS = 1 - \frac{avg(\frac{|v(e_{rc})-a(e_{rc})|}{v(e_{rc})}) + avg(\frac{v(e_l)-f(e_l))}{v(e_l)}}{avg(\frac{|v(e_{rc})-f(e_{rc})|}{v(e_{rc})}) + avg(\frac{v(e_l)-f(e_l))}{v(e_l)})} \tag{6}$$

$$a(e) = f(e) - f(e)\frac{f(S)-v(S)}{f(S)} = f(e)\frac{v(S)}{f(S)} \tag{7}$$

where the new potential score (NPS) is used to evaluate the probability of a node rc being the root cause. CF is denoted as the rate of descending leaf nodes in the cluster and LF is the descending rate of all its leaf nodes. The notation $avg(\cdot)$ is an average function. As shown in Eq. 6, NPS follows that if a dimension combination is a root cause, the difference between its real value and forecast value should be assigned proportionally to its leaf nodes. $v(e_{rc})$ and $a(e_{rc})$ are the real value and expected value of the leaf nodes which are inherited from the assumed root cause node rc respectively. $v(e_l)$ and $f(e_l)$ represent the real value and forecast value of all the remaining leaf nodes, respectively. Equation 7 defines the expected value $a(e)$ where S denotes a non-leaf node and leaf node e is inherited from S.

The dimension combination with the largest NPS is denoted as a candidate. The dimension combinations with the same value are also called sets in each cabin. The candidate is the most potential root cause in each set. Then we sort all the candidates by root cause score RS and extracted the candidate with the largest RS as the potential root cause in the cluster.

Identify Root Cause. We use Occam's Razor to merge all the most potential root causes to lite recommendation results. If a root cause belongs to another root cause, then the one with the wider scope is retained. For example, if {*Beijing*, *, *} and {*Beijing*, *, *Weibo.com*} are two recommended root causes from different clusters, We will eliminate {*Beijing*, *, *Weibo.com*} to make the result more concise.

5 Evaluation

In this section, we first describe the experimental dataset and performance metrics. Then, we conduct experiments on the dataset to assess the performance of ATS.

5.1 Dataset

The dataset used in this paper is collected from a real-world large-scale CDN system by ourselves. It is a synthetic dataset based on real data ranges from January

2019 to September 2020, which is collected from the CDN system. It includes five dimensions, which are *Province*, *Website*, *Operator System*, *Network Type*, and *Caching Server*. It has 30768 leaf dimension combinations. For *Website*, there are multiple system-level KPIs, such as *in_flow*, *out_flow*, *CDN_ttfb*, etc. Besides, we utilize KPIs of three websites to test the performance of *AutoDetect*. These websites are the top three video websites with massive visits and extensive traffic. The first Website (Website 1) is a live video website that lives at regular intervals.

5.2 Baseline and Evaluation Metrics

We compare the performance of anomaly detection (*AutoDetect*) with four ensemble anomaly detection (EAD) schemes as baseline. Recent research [37] mentions three ensemble schemes named *MVuniform*, *MVscore* and *MVexp*. *MVuniform* gives the same weight ($1/n$) to each learner, where n is the number of base learners and it implements simple majority voting. *MVscore* assigns weights $w_i = \frac{f_i}{\sum_{i=1}^{n} f_i}$ to the prediction of learner i, being f_i the F1-score of the learner. *MVexp* computes weights with an exponential classification F1-score, $w_i = \frac{e^{\lambda f_i}}{\sum_{i=1}^{n} e^{\lambda f_i}}$, where λ is selected to reduce the influence of low F1-score predictors. We take $\lambda = 10$ for such an effect. We also implement a baseline algorithm *Squeeze* [18] as the comparative solutions of ATS.

We apply *Precision*, *Recall* and *F1-score* to evaluate the performance of ATS. Additionally, we add *Time_cost* based on the metrics above to evaluate the performance of different MDRCL schemes. The key performance metrics *Precision* is the rate of correctly troubleshooting anomalies to the total number of anomalies troubleshooting and *Recall* is the ratio of correctly troubleshooting anomalies to all anomalies. *F1-score* is a harmonic average of *Precision* and *Recall*, which is usually used to measure the efficiency of an algorithm, denoted as $F1\_score = \frac{2*Precision*Recall}{Precision+Recall}$. We evaluate the average consumption time of each algorithm on the same Linux server.

5.3 Performance of Anomaly Detection

In this section, we describe the performance of *AutoDetect* in dataset with KPIs of three websites and compare with the performance of baseline schemes.

Table 3 reports the *Precision*, *Recall* and *F1-score* of the *AutoDetect* and the typical machine learning classifiers (MLCs) on our dataset, where the best *F1-scores* for all methods are highlighted in boldface. All of the MLCs are implemented by Pyod [39] and their parameters are optimized as same as base learners in AutoDetect. It is obvious that *AutoDetect* is the most stable and efficient. The four base learners (PCA, HBOS, IForest, COPOD) are relatively efficient among the others. HBOS, KNN [5] and LOF [14] belong to proximity-based methods, but HBOS performs much stabler and less time-consuming than others in anomaly detection. Although OCSVM [38] and CBLOF [12] have relatively good performance, we do not choose them due to time consumption.

Table 3. Performance comparison of AutoDetect with different classifiers based on Precision, Recall and F1-score.

Methods	Website1			Website2			Website3		
	Precision	Recall	F1-Score	Precision	Recall	F1-Score	Precision	Recall	F1-Score
PCA	0.924	0.956	0.940	0.668	0.226	0.337	0.863	0.524	0.652
HBOS	0.930	0.941	0.935	0.736	0.300	0.426	0.853	0.544	0.664
IsolationForest	0.936	0.922	0.929	0.751	0.229	0.351	0.883	0.530	0.663
COPOD	0.915	0.444	0.598	0.763	0.783	0.773	0.849	0.361	0.506
KNN	0.866	0.968	0.914	0.345	0.472	0.399	0.883	0.546	**0.675**
LOF	0.833	0.791	0.811	0.567	0.557	0.562	0.620	0.320	0.422
CBLOF	0.860	0.908	0.883	0.372	0.371	0.372	0.700	0.398	0.507
OCSVM	0.735	0.907	0.812	0.911	0.712	**0.799**	0.454	0.701	0.551
AutoDetect	0.929	0.976	**0.952**	0.795	0.786	**0.790**	0.883	0.546	**0.675**

As shown in Fig. 4a, we compare the *F1-score* of five anomaly detection ensemble schemes in three datasets. It is observed that *AutoDetect* performs best. We observe that *AutoDetect* outperforms Isolation Forest with the result of base learners (IFwithRe). It means that information loss impacts the performance of ensemble. Using the *soft labels* of base learners avoids the loss of information. Compared with different ensemble schemes (MVuniform, MVscore, MVexp, StackingwithKNN), *AutoDetect* achieves higher *F1-score* and more stability.

In addition, we compare the cost of time between classical MLCs and deep learning classifiers (DLCs). Figure 4b reports the result. We notice that the costs of PCA, HBOS and COPOD are the lowest, which are always lower than 5 s. The four base learners that we select are the four with the least training time. By contrast, the time cost of Autoencoder and VAE is around 3 or 4 h, respectively. Apart from OCSVM, the training time of MLCs is in seconds or minutes, while that of DLM is in hours. The cost of DLM is hundreds to thousands of times

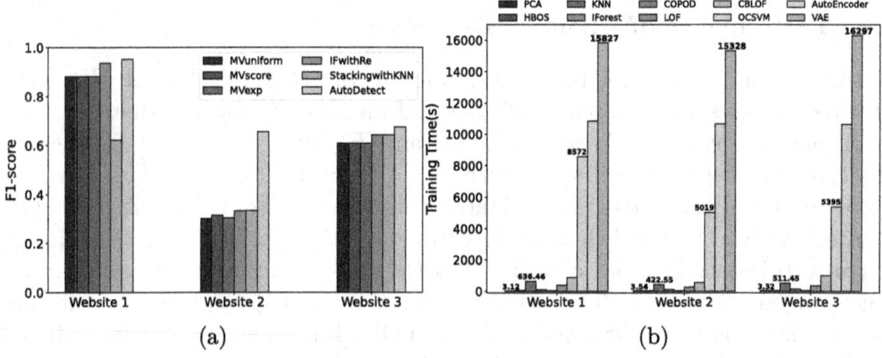

Fig. 4. (a) Performance comparison of ensemble schemes, (b) Training time comparison of learners.

than MLCs brings about failure to adapt to the time variation of the online system.

5.4 Performance of ATS

In this section, we introduce the performance of the fully automatic troubleshooting system (ATS) in terms of time consumed.

We compare the cost time of ATS, Squeeze [18], and ATS without Gunlock. As shown in Fig. 5, the average time cost by Squeeze and ATS is 46.27 and 9.75 s, respectively. It reveals ATS achieves almost 5x improvement in time cost compared to Squeeze. Since the effective parameters optimization of ATS and the reduction of search space, ATS eliminates unnecessary computations and thus achieves fast troubleshooting. Moreover, the average time cost of ATS without Gunlock (AD+AR) is more than twice as much as that of ATS. This is due to the Gunlock finding the unique *Trigger_KPI* and decreasing the computational cost.

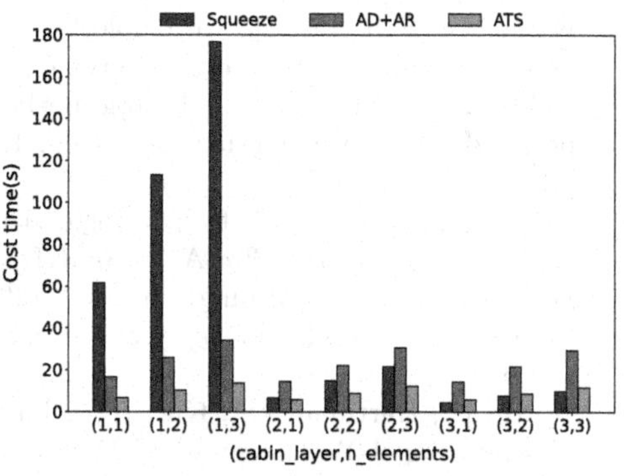

Fig. 5. Comparison of cost time.

6 Conclusion

In this paper, we propose the ATS which is a fully automatic troubleshooting system. It accomplishes troubleshooting by employing efficient ensemble soft-labels-based anomaly detection *AutoDetect*, a heuristic trigger (*Gunlock*), and fast multi-dimensional root cause localization *AutoRoot*. Furthermore, we discussed the components of the framework in detail and evaluated it in terms of anomaly detection and runtime performance. Extensive experiments on one real data trace demonstrate that ATS is more accurate and faster than traditional manual diagnostics and state-of-the-art solutions.

Acknowledgements. This work is supported by the National Key Research and Development Program of China (No. 2021YFB2910108).

References

1. Aburomman, A.A., Reaz, M.B.I.: A survey of intrusion detection systems based on ensemble and hybrid classifiers. Comput. Secur. **65**, 135–152 (2017)
2. Ahmed, F., Erman, J., et al.: Detecting and localizing end-to-end performance degradation for cellular data services based on TCP loss ratio and round trip time. IEEE/ACM Trans. Netw. **25**(6), 3709–3722 (2017)

3. Amazon: Amazon found every 100ms of latency cost them 1% in sales. http://blog.gigaspaces.com/amazon-found-every-100ms-of-latency-costthem-1-in-sales/ (Aug 2008)
4. Araya, D.B., Grolinger, K., ElYamany, H.F., Capretz, M.A., Bitsuamlak, G.: An ensemble learning framework for anomaly detection in building energy consumption. Energy Build. **144**, 191–206 (2017)
5. Chaovalitwongse, W.A., et al.: On the time series k-nearest neighbor classification of abnormal brain activity. T-SMCA **37**(6), 1005–1016 (2007)
6. Chen, Z., et al.: Combining MIC feature selection and feature-based MSPCA for network traffic anomaly detection. In: 2016 Third International Conference on Digital Information Processing, Data Mining, and Wireless Communications (DIPDMWC), pp. 176–181. IEEE (2016)
7. Folino, G., Sabatino, P.: Ensemble based collaborative and distributed intrusion detection systems: a survey. J. Netw. Comput. Appl. **66**, 1–16 (2016)
8. Goldstein, M., Dengel, A.: Histogram-based outlier score (HBOS): a fast unsupervised anomaly detection algorithm. KI-2012: Poster and Demo Track. vol. 9 (2012)
9. Golovin, D., Solnik, B., et al.: Google vizier: a service for black-box optimization. In: Proceedings of the 23rd ACM SIGKDD International Conference on Knowledge Discovery and Data Mining, pp. 1487–1495 (2017)
10. Google.http://glinden.blogspot.com/2006/11/marissa-mayer-at-web-20.html (2006)
11. Groth, D., Hartmann, S., Klie, S., Selbig, J.: Principal components analysis. In: Computational Toxicology, pp. 527–547 (2013)
12. He, Z., Xu, X., Deng, S.: Discovering cluster-based local outliers. Pattern Recogn. Lett. **24**(9–10), 1641–1650 (2003)
13. Jabbar, M.A., Aluvalu, R., Reddy, S.S.S.: Cluster based ensemble classification for intrusion detection system. In: Proceedings of the 9th International Conference on Machine Learning and Computing (ICMLC), pp. 253–257 (2017)
14. Jin, W., Tung, A.K.H., Han, J., Wang, W.: Ranking outliers using symmetric neighborhood relationship. In: Ng, W.-K., Kitsuregawa, M., Li, J., Chang, K. (eds.) PAKDD 2006. LNCS (LNAI), vol. 3918, pp. 577–593. Springer, Heidelberg (2006). https://doi.org/10.1007/11731139_68
15. Jing, P., Han, Y., Sun, J., Lin, T., Hu, Y.: AutoRoot: a novel fault localization schema of multi-dimensional root causes. In: 2021 IEEE Wireless Communications and Networking Conference (WCNC), pp. 1–7. IEEE (2021)
16. Klinker, F.: Exponential moving average versus moving exponential average. Math. Semesterberichte **58**(1), 97–107 (2011)
17. Laptev, N., Amizadeh, S., Flint, I.: Generic and scalable framework for automated time-series anomaly detection. In: Proceedings of the 21th ACM SIGKDD International Conference on Knowledge Discovery and Data Mining (SIGKDD), pp. 1939–1947 (2015)
18. Li, Z., Luo, C., et al.: Generic and robust localization of multi-dimensional root causes. In: 2019 IEEE 30th International Symposium on Software Reliability Engineering (ISSRE), pp. 47–57. IEEE (2019)
19. Li, Z., Zhao, Y., et al.: COPOD: copula-based outlier detection. In: 2020 IEEE International Conference on Data Mining (ICDM), pp. 1118–1123. IEEE (2020)
20. Liu, D., Zhao, Y., et al.: Opprentice: towards practical and automatic anomaly detection through machine learning. In: Proceedings of the 2015 Internet Measurement Conference (IMC), pp. 211–224 (2015)

21. Liu, F.T., Ting, K.M., Zhou, Z.H.: Isolation forest. In: 2008 Eighth IEEE International Conference on Data Mining, pp. 413–422. IEEE (2008)
22. Luglio, M., Romano, S.P., Roseti, C., Zampognaro, F.: Service delivery models for converged satellite-terrestrial 5G network deployment: a satellite-assisted CDN use-case. IEEE Netw. **33**(1), 142–150 (2019)
23. Ma, M., et al.: Diagnosing root causes of intermittent slow queries in cloud databases. Proc. VLDB Endowment **13**(8), 1176–1189 (2020)
24. McLeod, A.I., Li, W.K.: Diagnostic checking arma time series models using squared-residual autocorrelations. J. Time Ser. Anal. **4**(4), 269–273 (1983)
25. Meng, Y., Zhang, S., et al.: Localizing failure root causes in a microservice through causality inference. In: 2020 IEEE/ACM 28th International Symposium on Quality of Service (IWQoS), pp. 1–10. IEEE (2020)
26. Mirza, A.H.: Computer network intrusion detection using various classifiers and ensemble learning. In: 2018 26th Signal Processing and Communications Applications Conference (SIU), pp. 1–4. IEEE (2018)
27. Netflix. https://github.com/netflix/surus (2019)
28. Pang, G., Shen, C., Cao, L., Hengel, A.V.D.: Deep learning for anomaly detection: a review. ACM Comput. Surv. (CSUR) **54**(2), 1–38 (2021)
29. Persson, M., Rudenius, L.: Anomaly detection and fault localization an automated process for advertising systems. Master's thesis (2018)
30. Pham, N.T., Foo, E., et al.: Improving performance of intrusion detection system using ensemble methods and feature selection. In: The Australasian Computer Science Week Multiconference (ACSW), pp. 1–6 (2018)
31. Rahman, M.A., Shoaib, S., et al.: A bayesian optimization framework for the prediction of diabetes mellitus. In: 2019 5th International Conference on Advances in Electrical Engineering (ICAEE), pp. 357–362. IEEE (2019)
32. Rajagopal, S., Kundapur, P.P., Hareesha, K.S.: A stacking ensemble for network intrusion detection using heterogeneous datasets. Secur. Commun. Netw. **2020**, 1–9 (2020)
33. Su, Y., Zhao, Y., et al.: Robust anomaly detection for multivariate time series through stochastic recurrent neural network. In: Proceedings of the 25th ACM SIGKDD International Conference on Knowledge Discovery & Data Mining, pp. 2828–2837 (2019)
34. Sun, S., Jin, F., et al.: A new hybrid optimization ensemble learning approach for carbon price forecasting. Appl. Math. Model. **97**, 182–205 (2021)
35. Sun, Y., Zhao, Y., et al.: HotSpot: Anomaly localization for additive KPIs with multi-dimensional attributes. IEEE Access **6**, 10909–10923 (2018)
36. Tencent. https://github.com/tencent/metis (2019)
37. Vanerio, J., Casas, P.: Ensemble-learning approaches for network security and anomaly detection. In: Big-DAMA@SIGCOMM, pp. 1–6 (2017)
38. Wang, Z., Fu, Y., Song, C., Zeng, P., Qiao, L.: Power system anomaly detection based on OCSVM optimized by improved particle swarm optimization. IEEE Access **7**, 181580–181588 (2019)
39. Zhao, Y., Nasrullah, Z., Li, Z.: PyOD: a python toolbox for scalable outlier detection. J. Mach. Learn. Res. **20**(96), 1–7 (2019). http://jmlr.org/papers/v20/19-011.html
40. Zhong, Y., Chen, W., et al.: HELAD: a novel network anomaly detection model based on heterogeneous ensemble learning. Comput. Netw. **169**, 107049 (2020)

Resource Consumption of Federated Learning Approach Applied on Edge IoT Devices in the AGV Environment

Bohdan Shubyn[1,2(✉)], Piotr Grzesik[1], Taras Maksymyuk[2], Daniel Kostrzewa[1], Paweł Benecki[1], Jia-Hao Syu[4], Jerry Chun-Wei Lin[5], Vaidy Sunderam[3], and Dariusz Mrozek[1(✉)]

[1] Department of Applied Informatics, Silesian University of Technology, Gliwice, Poland
{Bohdan.Shubyn,dariusz.mrozek}@polsl.pl

[2] Department of Telecommunications, Lviv Polytechnic National University, Lviv, Ukraine

[3] Department of Computer Science, Emory University, Atlanta, GA 30322, USA
vss@emory.edu

[4] Department of Computer Science and Information Engineering, National Taiwan University, Taipei, Taiwan

[5] Department of Computer Science, Electrical Engineering and Mathematical Sciences, Western Norway University of Applied Sciences, Bergen, Norway

Abstract. Federated learning is a distributed machine learning method that is well-suited for the Industrial Internet of Things (IIoT) as it enables the training of machine learning models on distributed datasets. One of the most important advantages of using Federated Learning for Automated Guided Vehicles (AGVs) is its capability to optimize resource consumption. AGVs are typically resource-constrained systems and must operate within tight power and computational limits. By using Federated Learning, AGVs can perform model training and updating on-board, which reduces the amount of data that needs to be transmitted. This paper presents experiments to assess the consumption of resources of the Jetson Nano edge IoT device while training the Federated Learning model, and compares it with referential machine learning approaches.

Keywords: Federated Learning · predictive maintenance · smart production · Artificial Intelligence · resource consumption · recurrent neural networks

1 Introduction

Federated Learning (FL) is a cooperative Machine Learning (ML) technique that relies on the idea of gaining experience in local environments by many distributed, locally built ML models and sharing the experience to build the global

J. Mikyška et al. (Eds.): ICCS 2023, LNCS 14077, pp. 492–504, 2023.
https://doi.org/10.1007/978-3-031-36030-5_39

inference model [16]. Local models can be trained on edge IoT devices that usually have limited storage and computing capabilities compared to typical workstations. This approach reduces communications needs and energy consumptions, since after training local models on locally available data, only the local experience (e.g., ML model parameters) is shared, not the data itself. Therefore, Federated Learning is well suited to Industrial Internet of Things (IIoT) systems and IIoT-based monitoring of AGV-enabled production lines thus contributing to smart and agile manufacturing.

Automated Guided Vehicles (AGVs) and collaborative robots (cobots) play an important role in computer-integrated smart manufacturing as they support assembly tasks in an autonomous manner [23]. They can independently analyze signals from a variety of sensors to complete their operational tasks without human intervention. This requires accumulating a wealth of local data and making decisions in the context of the current environmental conditions, monitored by sensors and transportation commands acquired from external systems, such as the Transportation Management System (TMS) [17].

The inherent properties of FL make it also a suitable approach for monitoring the health of the intelligent machines working in smart production lines, like the AGVs or cobots (Fig. 1). Since smart factories employ many such autonomous units, they all can gather their experience based on operational cycles implemented in the factory. All these devices are equipped with local, edge IoT devices capable of collecting data and building local ML models for various purposes, e.g., predicting the inability to complete specific operational cycles due to low battery levels or predicting necessary maintenance tasks by finding anomalies in analyzed signals [22].

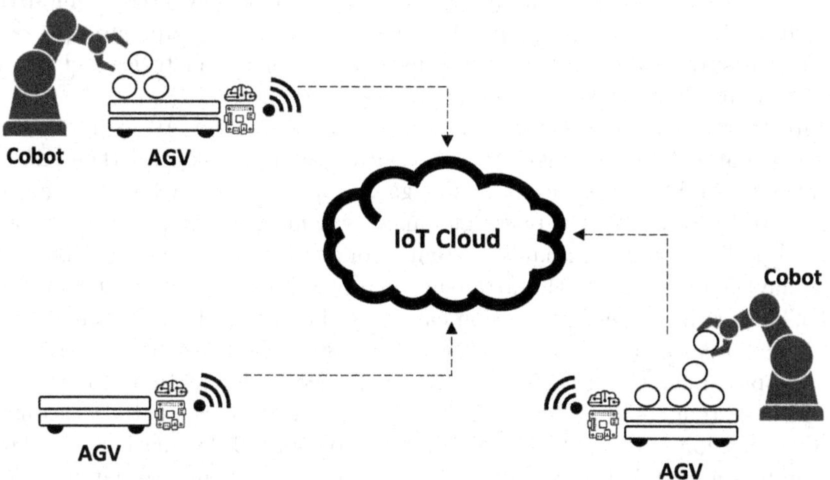

Fig. 1. AGVs performing edge IoT-based local inference and sending local models to the IoT cloud to build a global inference model.

These predictive tasks can be performed locally, but blending the experience of other AGVs with the local knowledge may bring additional improvements in the quality of inference, and opens the vehicles for the broader spectrum of operational conditions and abnormalities. However, a perfect inference model would require ideal datasets and an excellent algorithm for predictive tasks. This is not easy to achieve. The first factor depends on the existing working conditions and can evolve over time. The second factor is constrained by the capabilities of the IoT device hosting the local ML model. In this paper, we focus on the second factor. We investigate the resource and power cost of performing the predictive tasks with FL/ML algorithms on the NVidia Jetson Nano edge device mounted on the AGV. By performing predictive tasks for momentary energy consumption (MEC) of the AGV as a whole, we compare the energy cost of FL and referential ML.

The rest of the paper is organized as follows. In Sect. 2, we review the works related to Autonomous Guided Vehicles and Federated Learning. Section 3 explains the concept of the Federated Learning approach we rely on and the inference algorithm we implemented on edge IoT devices mounted on AGVs. Section 4 shows the results of experiments concerning the resource costs while performing the inference with Federated Learning. Finally, Sect. 5 discusses the obtained results and summarizes the paper.

2 Related Works

2.1 Automated Guided Vehicles

Since Automated Guided Vehicles (AGVs) operate as portable robots that transport objects with autonomous navigation, they are widely used in industrial facilities such as manufacturing plants, assembly lines, and warehouses [6]. Some AGVs consist of trailers or plates to transport materials in factories and support the development of smart factories [3,10].

Literature shows that navigation is an essential task for AGVs and aims to reduce manual work and increase onboard autonomy to control the AGV [7]. Traditional AGVs are guided and navigated by cables, and the next generation of AGVs consists of wireless guidance systems without physical guidance paths [18]. The wireless guidance systems require environmental information for AGV navigation, such as the arrangement of AGVs, the destination, and the paths. The environment information can either be provided by the central management system or acquired by AGV sensors such as radar, lidar, ultrasound, and an optical camera [15]. Furthermore, with the basic function of navigation, the AGV system could perform route planning, which makes path decisions for AGV navigation with multiple constraints or optimized destinations [1]. Apart from individual AGV routes, the AGV management system also handles traffic management and load transfer to avoid collisions and optimize AGV resource consumption [11].

2.2 Federated Learning

Federated learning has attracted researchers' attention since 2017 [8] with federated averaging (FedAvg) as a typical model [9] outperforming Federated Stochastic Variance Reduced Gradient (FSVRG) [5] and asynchronous Cooperative Learning (CO-OP) on most real-world datasets [13]. Federated learning is widely used in various research areas, such as healthcare [19], Internet of Things [12], and smart factory [4]. It is known for privacy and has a distributed nature [20], which contrasts with centralized learning algorithms (collecting and training based on all data at a powerful machine). In [2], Cho et al. studied the convergence analysis of FL for biased scheduling methods and investigated the convergence speed of FL. In [21], the authors reported a framework for federated edge learning that can adaptively schedule users to reduce total energy consumption. However, none of these works considered the power costs of the FL approach. Meanwhile, Nishio et al. noted the importance of resource consumption and designed a new FL protocol to perform the scheduling process according to computational resources channel conditions [14]. In [16], we also analyzed various training scenarios to improve the performance of the FL-based prediction model. However, in [16], we didn't focus on the consumption of resources and power efficiency. In this paper, we investigate these fields for the NVidia Jetson Nano edge device that we use in the AGV environment.

3 Federated Learning in the Distributed AGV Environment

In the AGV environment, Federated Learning (FL) is accomplished by training local models on edge IoT devices residing on the AGVs. The learning process is performed in so-called *rounds.*

Definition 1. *A round is a single iteration of distributed training of local models, sharing experience, building the global model, and updating local models with global wisdom.*

A single round consists of the steps that are presented in Fig. 2. This process is iterative, and we should repeat the course of the round periodically or when we get the needed amount of new data.

The first step of a round covers training many local models M_i on AGVs with locally acquired and collected data. Local models may rely on various algorithms, e.g., different architectures of recurrent neural networks (RNNs), but once chosen, the architecture is homogeneous for all AGVs. Next, local models are sent to the data center in the IoT cloud (step 2). In the next step (step 3), we build the global model M by averaging the weights of local models (local RNNs):

$$M = \sum_{i=1}^{N} M_i * \mathbb{I}_{M_i} \tag{1}$$

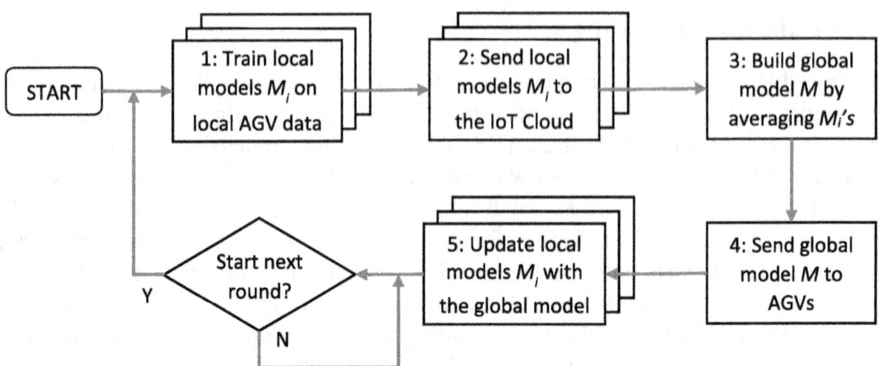

Fig. 2. Federated Learning performed in the distributed AGV environment.

where M_i is the local model built on the i-th AGV, $\mathbb{I}_{M_i}$ is the *influence* of a specific local model M_i when creating the global model, N is the number of AGVs, and $*$ is point-wise multiplication.

Definition 2. *The influence of the local model is the average relative complement of the contribution of the partial loss generated by the local model to the total loss produced by all local models.*

The influence $\mathbb{I}_{M_i}$ of a local model M_i can be calculated according to the following formula:

$$\mathbb{I}_{M_i} = \frac{1}{N-1} \frac{\sum_{j=1}^{N} L_{M_j} - L_{M_i}}{\sum_{j=1}^{N} L_{M_j}}, \tag{2}$$

where L_{M_i} is the loss calculated for the i-th AGV (MSE or Validation loss).

The global model is built specifically for the inference task performed, and its architecture should be appropriate. Within this work, we focus on predicting momentary energy consumption (MEC). We decided to predict MEC, as this parameter is important in the operation of AGVs. Any deviation of this value may indicate a particular malfunction of the vehicle or its parts. We plan to use this value to detect anomalies reflected in the significant difference between the value received by the device and the one predicted by our neural network. Our previous work [16] proved that LSTM networks for building the inference model and Mean Square Error (MSE) as a metric for verifying the influence of local models perform well when predicting MEC.

Once the global model is built centrally, it is sent back to the AGVs (step 4) and overwrites the local models M_i on each AGV (step 5). This way, global wisdom updates local knowledge, and the round completes. After acquiring another portion of data by each AGV, training local models can start in the next round.

4 Determining Jetson Nano's Resource Consumption in Federated Learning

For the experiments, we used NVidia Jetson Nano acting as an edge device on the AGVs. The edge device mounted on the AGV collects data by itself. Then, it uses the data to train a prediction model for the *Momentary energy consumption* (MEC) value over time, which is one of the indicators for faults or improper use of the AGV. During this process, we observed the resource consumption on the edge device with and without Federated Learning.

As the prediction model was implemented on edge Jetson Nano device, all the data collected by the device were used to train and build the prediction model locally without directly sending the data anywhere. Since we predicted the value of the MEC in time, we decided to use a recurrent neural network (RNN) based on the LSTM cell since this type of architecture copes quite well with time series.

In order to assess the effectiveness of Federated Learning in predicting AGV energy consumption, we collected data from nine test runs of Formica-1, one of the AGVs developed by AIUT Ltd., for which our group provides AI-based solutions. The entire data set contained roughly 12,500 samples collected with a frequency 1 Hz.

During the test drives, we executed various scenarios, including repeated circular and counterclockwise paths, driving forward and backward at a speed of 0.2 m/s, repeated emergency braking, fast acceleration in both directions, and moving the lifting plate up and down. We performed each scenario with an empty vehicle and a half-loaded payload compartment (weighing 425 kg).

In order to see how Federated Learning affects the resource consumption of edge devices at the time of training, we decided to test the following three situations:

1. Using the LSTM-based MEC prediction model without FL model trained on the entire data set (12,000 training samples).
2. Using the LSTM-based MEC prediction model trained in 4-round FL (4-R FL) on three Jetson Nano edge devices. We used three edge devices, and each of them collected 4,000 samples in 4 rounds of training (1,000 training samples per round).
3. Using the LSTM-based MEC prediction model trained in 8-round FL (8-R FL) on three Jetson Nano. We used three edge devices, and each of them collected 4,000 samples in 8 rounds of training (500 training samples per round).

In summary, we analyzed how many Jetson Nano resources are required for training the prediction model without FL, for one round of 4-R FL and one round of 8-R FL. To see how many resources are used when training the recurrent neural network, we used the *jetson stats* Python library. For all considered experiments, we were using Jetson Nano in the highest performance mode.

4.1 Using LSTM Without Federated Learning on the Entire Data Set

In the first series of experiments, we decided to check how many resources the edge device consumed without the use of Federated Learning. In this case, we trained our neural network on the entire training data set (12,000 samples). Figure 3 shows average resource consumption during training on this data set.

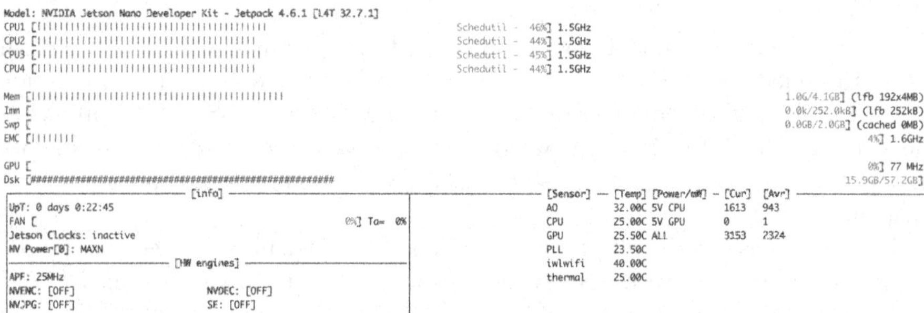

Fig. 3. Resource consumption on the edge device while training the MEC prediction model on the whole data set.

The training took 1,472 s, and from Fig. 3, we can observe that all four processors of the Jetson Nano were 45% loaded. Additionally, 1.0 GB of RAM and observed momentary power consumption of 3,153 mW during the experiment, equate to 4,641 J of total energy consumption on this device while training.

The results of the MEC prediction with the LSTM model without FL are presented in Fig. 4. We tested our model on 500 samples that were not part of the training data, resulting in an MSE value of 924.23. This figure shows two signal values, real (represented in blue) and predicted by our neural network (represented in orange). The greater the difference between these signals, the worse our neural network model performs. In this case, we see that the signal values are similar, indicating that our model performs well.

4.2 Using LSTM with 4-Round Federated Learning

In this series of experiments, we launched Federated Learning on three devices. Moreover, we used 4-R FL on each device to execute four learning rounds, build a global MEC prediction model, and share weights between these devices. This means each edge device collected 1,000 samples per round (12,000 samples in total for four training rounds). In this case, we recorded how many resources were consumed by the edge device when we trained the MEC prediction model with 1,000 samples (in one round). The average resource consumption during training on this dataset is shown in Fig. 5.

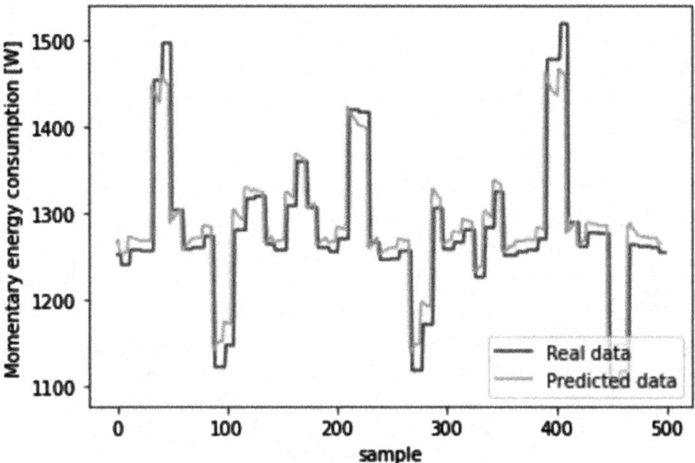

Fig. 4. Prediction performance of the LSTM model trained on the entire training data set. (Color figure online)

The training took 133 s per round, and from Fig. 5, we can observe that all four processors of the Jetson Nano were 46% loaded. Additionally, 1 GB of RAM and observed momentary power consumption of 3,232 mW during the experiment, equates to 5,158 J of total energy consumption while training. Generally, we consumed almost the same resources as in the case without Federated Learning. However, we split this resource consumption between three devices.

The MEC prediction results obtained with the LSTM model trained in 4-R FL are presented in Fig. 6. We tested our model on 500 samples that were not part of the training data, which resulted in an MSE value of 948.28.

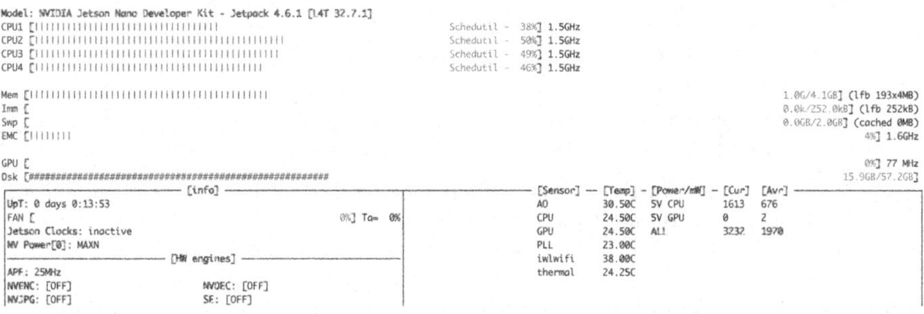

Fig. 5. Resource consumption on the edge device while training the MEC prediction model in 4 rounds of Federated Learning.

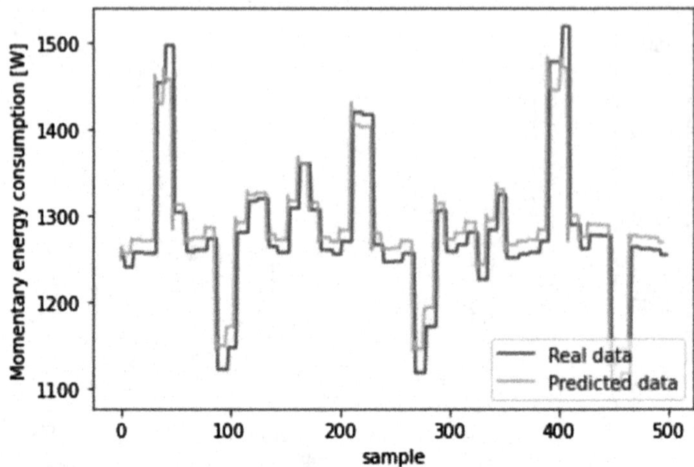

Fig. 6. Prediction performance of the LSTM model built in 4-round Federated Learning.

4.3 Using LSTM with 8-Round Federated Learning

In this series of experiments, three edge devices collected 500 samples per round (12,000 samples for 8 training rounds). The average resource consumption during the training round on this data set is shown in Fig. 7.

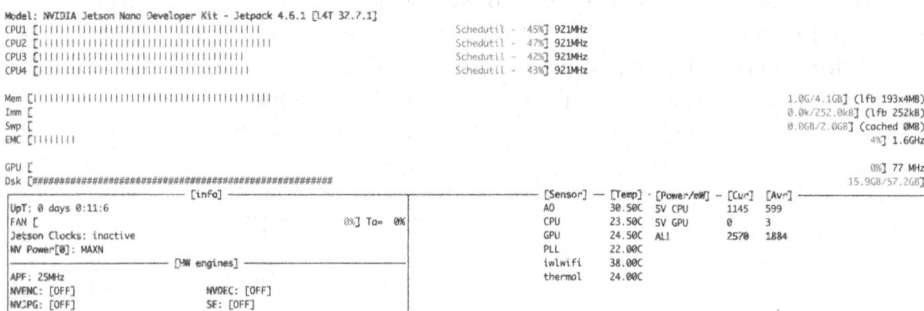

Fig. 7. Resource consumption on the edge device while training the MEC prediction model in 8 rounds of Federated Learning.

The training time took 66 s, and from Fig. 7, we can observe that all four Jetson Nano processors are, on average, 44% loaded. Additionally, 1 GB of RAM and observed momentary power consumption of 2,570 mW during the experiment, equates to 4,070 J of total energy consumption while training.

The MEC prediction results obtained with the LSTM model trained in 8-R FL are presented in Fig. 8. We tested our model on 500 samples that were not part of the training data, resulting in an MSE value of 849.39.

In this training scenario, we observed lower power consumption and noticed that the processor works at lower frequencies, which, in our opinion, is related to the lower consumption of power. Such results indicate that the 8-round Federated Learning option consumed the least edge device resources. Moreover, this approach allowed us to obtain the best prediction accuracy, which converges with the observations presented in [16].

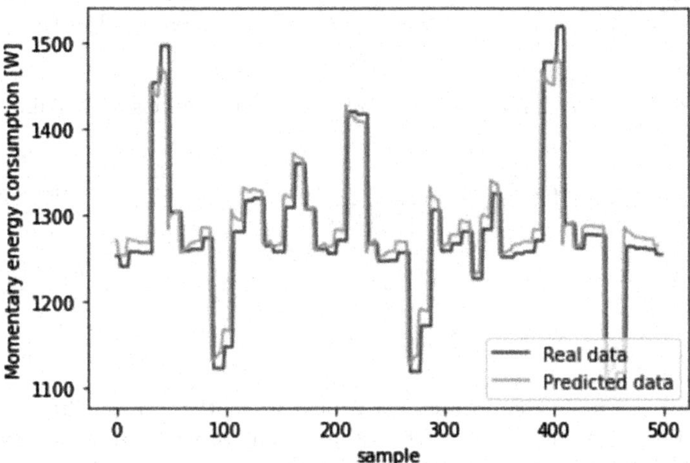

Fig. 8. Prediction performance of the LSTM model built in 8-round Federated Learning.

5 Discussion and Conclusions

Time, CPU utilization, RAM usage, and IoT device power consumption are all critical performance indicators for building the prediction models working onboard Autonomous Guided Vehicles. They all directly impact the consumption of energy (MEC) from the batteries of the AGVs, and, ultimately, their capabilities to function effectively and efficiently. For these reasons, we investigated how these resources were consumed during the training of a neural network locally on them.

This study demonstrates that the most effective and resource-saving way to train neural networks locally on Autonomous Guided Vehicles is by using a larger number of Federated Learning (FL) rounds. Table 1 summarizes the results of the different stages of the experiment. As we can observe, using 8-round training allowed for significant resource savings, reduced the overall load on the device, and improved the MEC signal prediction effectiveness, which is crucial for detecting anomalies in AGVs. Also, obtained results of Total Energy Consumption show that by using 8 rounds of Federated Training, we consume

12% less energy, not taking into account that this consumption will be additionally distributed between edge IoT devices. Furthermore, using FL provides better security for industrial data by processing data locally on the devices, reducing communication needs and the amount of data transferred. These findings can help develop more efficient and secure neural network training methods for AGVs. In future work, we plan to compare the prediction effectiveness and resource consumption of several different types of neural networks using Federated Learning. Since any neural network can be applied in FL, we have a large selection of models as an alternative to LSTM, which we used in this work.

Table 1. Resource consumption for each of the tested scenarios for training the MEC prediction model.

Approach	Time per round	CPU	RAM	Momentary power consumption	Total energy consumption
LSTM without FL	1472 s	45%	1 GB	3,153 mW	4,641 J
LSTM with 4-R FL	133 s	46%	1 GB	3,232 mW	5,158 J
LSTM with 8-R FL	66 s	44%	1 GB	2,570 mW	4,070 J

Acknowledgements. The research was supported by the Norway Grants 2014-2021 operated by the National Centre for Research and Development under the project "Automated Guided Vehicles integrated with Collaborative Robots for Smart Industry Perspective" (Project Contract no.: NOR/POL-NOR/CoBotAGV /0027/2019-00), the Polish Ministry of Science and Higher Education as a part of the CyPhiS program at the Silesian University of Technology, Gliwice, Poland (Contract No.POWR.03.02.00-00-I007/17-00), by Statutory Research funds of the Department of Applied Informatics, Silesian University of Technology, Gliwice, Poland (02/100/BK_23/0027), research funds for young scientists (grants no. PB and BS: BKM/RAu7/2023), and by the ReActive Too project that has received funding from the European Union's Horizon 2020 Research, Innovation and Staff Exchange Programme under the Marie Skłodowska-Curie Action (Grant Agreement No 871163).

References

1. Bae, J., Chung, W.: A heuristic for a heterogeneous automated guided vehicle routing problem. Int. J. Precis. Eng. Manuf. **18**(6), 795–801 (2017)
2. Cho, Y.J., Wang, J., Joshi, G.: Client selection in federated learning: convergence analysis and power-of-choice selection strategies (2020). https://doi.org/10.48550/ARXIV.2010.01243, https://arxiv.org/abs/2010.01243
3. Cupek, R., Lin, J.C.W., Syu, J.H.: Automated guided vehicles challenges for artificial intelligence. In: 2022 IEEE International Conference on Big Data (Big Data) (2022)
4. Khan, L.U., Alsenwi, M., Yaqoob, I., Imran, M., Han, Z., Hong, C.S.: Resource optimized federated learning-enabled cognitive internet of things for smart industries. IEEE Access **8**, 168854–168864 (2020)

5. Konecnỳ, J., McMahan, H.B., Ramage, D., Richtárik, P.: Federated optimization: distributed machine learning for on-device intelligence (2016). arXiv preprint arXiv:1610.02527
6. Le-Anh, T., De Koster, M.: A review of design and control of automated guided vehicle systems. Eur. J. Oper. Res. **171**(1), 1–23 (2006)
7. Martínez-Barberá, H., Herrero-Pérez, D.: Autonomous navigation of an automated guided vehicle in industrial environments. Rob. Comput.-Integr. Manuf. **26**(4), 296–311 (2010)
8. McMahan, B., Moore, E., Ramage, D., Hampson, S., Arcas, B.A.: Communication-efficient learning of deep networks from decentralized data. In: Artificial Intelligence and Statistics, pp. 1273–1282 (2017)
9. McMahan, B., Moore, E., Ramage, D., Hampson, S., Arcas, B.A.Y.: Communication-Efficient Learning of Deep Networks from Decentralized Data. In: Singh, A., Zhu, J. (eds.) Proceedings of the 20th International Conference on Artificial Intelligence and Statistics. Proceedings of Machine Learning Research, vol. 54, pp. 1273–1282. PMLR (2017)
10. Mehami, J., Nawi, M., Zhong, R.Y.: Smart automated guided vehicles for manufacturing in the context of industry 4.0. Procedia Manuf. **26**, 1077–1086 (2018)
11. Mugarza, I., Mugarza, J.C.: A coloured petri net-and d* lite-based traffic controller for automated guided vehicles. Electronics **10**(18), 2235 (2021)
12. Nguyen, D.C., Ding, M., Pathirana, P.N., Seneviratne, A., Li, J., Poor, H.V.: Federated learning for internet of things: a comprehensive survey. IEEE Commun. Surv. Tutor. **23**(3), 1622–1658 (2021)
13. Nilsson, A., Smith, S., Ulm, G., Gustavsson, E., Jirstrand, M.: A performance evaluation of federated learning algorithms. In: Proceedings of the Second Workshop on Distributed Infrastructures for Deep Learning, DIDL 2018, pp. 1–8. Association for Computing Machinery, New York (2018)
14. Nishio, T., Yonetani, R.: Client selection for federated learning with heterogeneous resources in mobile edge. In: ICC 2019–2019 IEEE International Conference on Communications (ICC). IEEE (2019). https://doi.org/10.1109/icc.2019.8761315. https://doi.org/10.1109
15. dos Reis, W.P.N., Morandin Junior, O.: Sensors applied to automated guided vehicle position control: a systematic literature review. Int. J. Adv. Manuf. Technol. **113**(1), 21–34 (2021)
16. Shubyn, B., et al.: Federated learning for improved prediction of failures in autonomous guided vehicles. J. Comput. Sci., 101956 (2023). https://doi.org/10.1016/j.jocs.2023.101956
17. Steclik, T., Cupek, R., Drewniak, M.: Automatic grouping of production data in industry 4.0: the use case of internal logistics systems based on automated guided vehicles. J. Comput. Sci. **62**, 101693 (2022)
18. Vis, I.F.: Survey of research in the design and control of automated guided vehicle systems. Eur. J. Oper. Res. **170**(3), 677–709 (2006)
19. Xu, J., Glicksberg, B.S., Su, C., Walker, P., Bian, J., Wang, F.: Federated learning for healthcare informatics. J. Healthcare Inf. Res. **5**(1), 1–19 (2021)
20. Yang, Q., Liu, Y., Cheng, Y., Kang, Y., Chen, T., Yu, H.: Federated learning. Synth. Lect. Artif. Intell. Mach. Learn. **13**(3), 1–207 (2019)
21. Zeng, Q., Du, Y., Huang, K., Leung, K.K.: Energy-efficient radio resource allocation for federated edge learning. In: 2020 IEEE International Conference on Communications Workshops (ICC Workshops), pp. 1–6 (2020). https://doi.org/10.1109/ICCWorkshops49005.2020.9145118

22. Ziebinski, A., Bregulla, M., Fojcik, M., Kłak, S.: Monitoring and controlling speed for an autonomous mobile platform based on the hall sensor. In: Nguyen, N.T., Papadopoulos, G.A., Jedrzejowicz, P., Trawiński, B., Vossen, G. (eds.) ICCCI 2017. LNCS (LNAI), vol. 10449, pp. 249–259. Springer, Cham (2017). https://doi.org/10.1007/978-3-319-67077-5_24
23. Ziebinski, A., et al.: Challenges associated with sensors and data fusion for AGV-driven smart manufacturing. In: Paszynski, M., Kranzlmüller, D., Krzhizhanovskaya, V.V., Dongarra, J.J., Sloot, P.M.A. (eds.) ICCS 2021. LNCS, vol. 12745, pp. 595–608. Springer, Cham (2021). https://doi.org/10.1007/978-3-030-77970-2_45

Quantifying Parking Difficulty with Transport and Prediction Models for Travel Mode Choice Modelling

Maciej Grzenda(✉), Marcin Luckner, and Łukasz Brzozowski

Warsaw University of Technology, Faculty of Mathematics and Information Science, ul. Koszykowa 75, 00-662 Warszawa, Poland
{Maciej.Grzenda,Marcin.Luckner,Lukasz.Brzozowski}@pw.edu.pl

Abstract. Promoting sustainable transportation necessitates understanding what makes people select individual travel modes. Hence, classifiers are trained to predict travel modes, such as the use of private cars vs bikes for individual journeys in the cities. In this work, we focus on parking-related factors to propose how survey data, including spatial data and origin-destination matrices of the transport model, can be transformed into features. Next, we propose how the impact of the newly proposed features on classifiers trained with different machine learning methods can be evaluated. Results of the extensive evaluation show that the features proposed in this study can significantly increase the accuracy of travel mode choice predictions.

Keywords: Machine learning · travel mode choices · survey data

1 Introduction

Accurately modelling travel mode choices (TMCs) is an important part of transportation planning [2,7]. It makes it possible to predict, based on multiple features such as a person's age, journey distance, and whether the person owns a car, whether the journey is likely to be made by e.g. walking, or by private car. An overview of factors impacting which mode of transportation is selected by individuals to move around can be found *inter alia* in [7]. However, the data used so far to predict travel modes do not include parking-related features even in comparative studies of different classifiers [2], or include a very limited set of features such as the parking permit feature used in [7]. Importantly, mobility is expected to be influenced, among other things, by the financial costs of mobility alternatives, which include but are not limited to parking costs [1]. Another factor related to comfort and total travel time is the time it takes to find a parking space. However, how to estimate parking difficulty for mode choice modelling is an open issue.

Hence, this work focuses on how the features capturing parking difficulty can be calculated. The proposed methods rely on the demand matrices of a

J. Mikyška et al. (Eds.): ICCS 2023, LNCS 14077, pp. 505–513, 2023.
https://doi.org/10.1007/978-3-031-36030-5_40

traffic model and predicted time of finding a parking space and getting from it to the ultimate journey destination. All these methods provide features used by machine learning (ML) models. To evaluate these features, we investigate their impact on mode choice models trained with different classification methods. The results obtained with three datasets documenting real journeys show that the features proposed in this study can help predict transport modes used for the analysed journeys.

The remainder of this work is organised as follows: in Sect. 2 the use of ML for TMC modelling is summarised. Novel parking features are proposed in Sect. 3, and evaluated in Sect. 4. Conclusions are made in Sect. 5.

2 Related Works

The travel mode prediction problem is frequently stated as a classification problem [2,4]. A multitude of methods has been used for the task, including but not limited to Support Vector Machine (SVM), Random Forest (RF), and XGBoost (XGB). Systematic reviews of ML methods for modelling passenger mode choice can be found in [2,4,7]. An important aspect of studies on TMC modelling is the data used to evaluate the models. Salas et al. in their recent work on TMC modelling with ML methods [7] observe that most of the previous studies focus on a single empirical dataset. Moreover, datasets typically used in the works, such as Dutch National Travel Survey data [8], include features such as distance travelled, age, education, and land use index, but not parking features related to journey destination. Salas et al. in [7] consider four datasets, out of which only one refers to parking issues by including the parking permit feature. An extensive review of datasets made in [4] shows that most datasets rely on trip diaries. Still, the issue of parking-related factors and their impact on mode choices is gradually being addressed. In [1], the impact of parking fees on TMC in urban environments is analysed. In [9], a fixed average parking time for the entire city is assumed. Many other works do not consider parking difficulties in TMC prediction [3,6,10] or consider features possibly related to parking difficulties such as population density [10], global traffic congestion [5] or trip density [8].

The impact of individual features on TMC models is frequently analysed. Trip distance, travellers' age, number of cars/bicycles owned, and trip density were among the predictors influencing the predictions of the models in [8]. We aim to extend extant TMC research by proposing and evaluating parking features.

3 Estimating Parking Difficulty and Costs

The goal of this work is to provide a proposal for parking-related features that would help explain some TMCs. Let $X = \{(\mathbf{x}_1, M_1), \ldots, (\mathbf{x}_N, M_N)\}$ denote the set of journeys $\mathbf{x}_i$ for which travel mode M_i actually used in the past is known. Our objective is to extend the $\mathbf{x}$ vectors by appending parking-related features i.e. use for the training and evaluation of travel mode prediction models vectors $[\mathbf{x}_i, f_1(\mathbf{x}_i), \ldots, f_F(\mathbf{x}_i)]$ including both original and F parking-related features.

The Estimation of Parking Time. For some cities, data documenting the *parking time* t_p, which we define as the approximate time needed to park a car and get from its location to the actual journey destination P is available. The higher the parking time at location P is, the more reluctant travellers may be to use private cars to travel to this destination. Hence, we propose building a regression model predicting the parking time based on records documenting t_p for individual locations (x, y). Let T be the set of parking time tuples (x, y, h, t_p), where x and y stand for the geocoordinates of the journey destination, h denotes the hour of the day at which the destination was reached and t_p denotes the parking time reported by a person reaching the destination by car. Next, a regression model $\mathcal{M}_{\text{REG}}(x, y, h)$ can be developed to predict parking time t_p and provide the value of the `PRED_PTIME` feature based on input data $(x, y, h(t))$, where t denotes the approximate time of reaching the destination present in journey record $\mathbf{x}$ and $h(t)$ denotes the hour of the day.

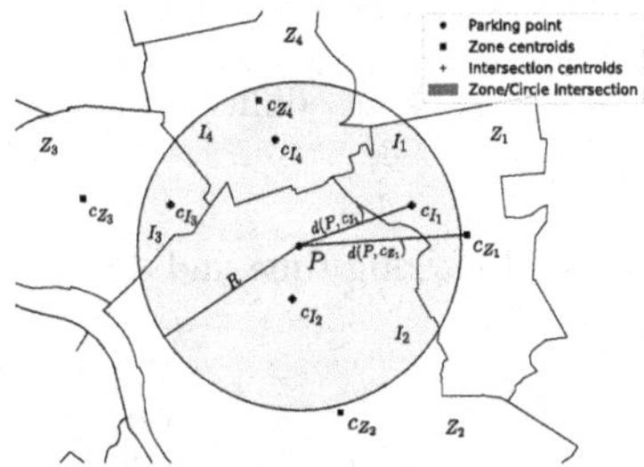

Fig. 1. Points, areas, and distances considered for parking difficulty estimation.

The Estimation of Parking Difficulty and Cost. We propose three methods to estimate parking difficulty, which we will explain using Fig. 1. Let $P = (x, y)$ be a point at which we wish to estimate the parking difficulty parameter, i.e. the journey destination in which a car ideally would be parked if used for the journey. Let $B(P, R)$ be a circle around the point P with radius R, which defines the approximate area likely to be considered by a driver to park a car. Let $Z_1, Z_2, \ldots, Z_n$ be the transport zones (polygons) that have non-empty intersections with the ball $B(P, R)$. For all zones, we denote the average number of arrivals to the zone Z at hour $h \in \{0, 1, \ldots, 23\}$ by $A_h(Z)$. Next, let $I_1, I_2, \ldots, I_n$ denote the area of the intersection of zone Z_i with the ball $B(P, R)$. Moreover, let $c_{Z_1}, c_{Z_2}, \ldots, c_{Z_n}$ denote centroids of the corresponding zones, and $c_{I_1}, c_{I_2}, \ldots, c_{I_n}$ centroids of the intersections with the corresponding zones. Let $d(\circ, \diamond)$ denote the distance between two points. Using the above notation, we define the parking difficulty feature $f_m()$ for the point $P = (x, y)$ at time t using the method m as $f_m(P, t) = \frac{1}{n} \sum_{i=1}^{n} W_m(Z_i, B(P, R)) \cdot A_h(t)(Z_i)$.

We propose three methods m to calculate the weight $W_m(Z_i, B(P, R))$. The first provides the `PDIFF_IS_AREA` feature in which the weight is proportional to the area of the intersection I_i, i.e., $W_{IA}(Z_i, B(P, R)) = I_i$. The second feature is PDIFF_IS_CENTR, where the weight is proportional to the area of the intersection I_i and inversely proportional to the distance between the parking point and the intersection centroid, i.e., $W_{IC}(Z_i, B(P, R)) = \frac{I_i}{d(P, c_{I_i})}$. Finally, in PDIFF_ZONE_CENTR, the weight is proportional to the area of the intersection I_i and inversely proportional to the distance between the parking point and the zone centroid, i.e., $W_{ZC}(Z_i, B(P, R)) = \frac{I_i}{d(P, c_{Z_i})}$.

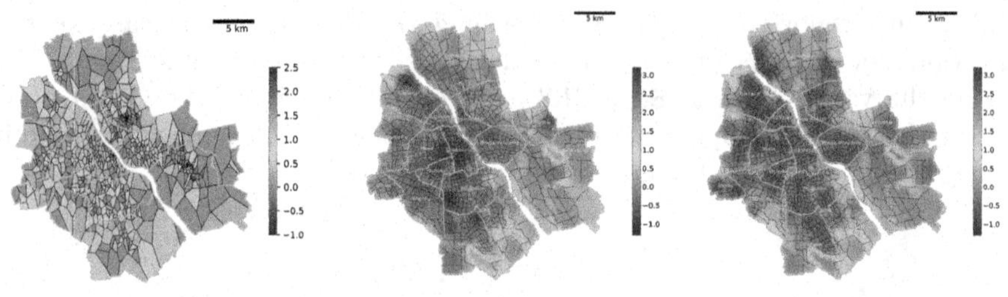

(a) Voronoi diagram of reported parking time. (b) Parking difficulty at hour $h = 8$. (c) Parking difficulty at hour $h = 17$.

Fig. 2. Spatial distribution of parking time and difficulty values. $\log_{10}$ scale

These features can be calculated for any point P in the entire city area, covered by the transport model. The features differ in how they quantify the influence of different zones. Most likely the decisions of travellers will be more affected by expected parking difficulties in the area closer to the journey destination. This is why features giving different weights to data from different zones are proposed. To illustrate both the data used to calculate the parking time features and selected PDIFF_IS_CENTR values, sample values for the City of Warsaw are provided in Fig. 2.

The feature estimating parking cost takes as an input arrival date and time s_{a}, departure date and time s_{d}, and the journey destination (x, y) i.e. the location at which a car could ideally be parked, all coming from a journey record $\mathbf{x}$. Based on the parking pricing policies of the city and these input data, the value of the PCOST feature is calculated. This yields only estimated parking costs, as factors such as the use of private parking spaces could influence actual parking costs.

4 Results

Reference Data. We selected three journey datasets to evaluate the methods proposed in this study. The datasets were collected in the City of Warsaw in 2022 and document journeys made by a representative sample of the parents of primary school children (PAR_W1 and PAR_W2 datasets) and a representative sample of all citizens of Warsaw (CIT_W1 dataset). The PAR_W1 includes journeys made by parents of children from three reference schools, while

PAR_W2 includes journeys of a representative sample of parents of children from all schools in the city. All the datasets were prepared based on the travel diaries of respondents and included all journeys of respondents within the City of Warsaw irrespective of origin and destination i.e. also including non-school journeys.

Table 1. Summary of datasets used for training and evaluation of TMC models.

Dataset	Respond.	Journeys	CAR	PT	BIKE	WALK
PAR_W1	523	1861	895 (48.54%)	344 (18.66%)	89 (4.83%)	516 (27.98%)
PAR_W2	316	798	323 (41.15%)	232 (29.55%)	18 (2.29%)	212 (27.01%)
CIT_W1	1170	2961	1044 (36.43%)	1181 (41.21%)	109 (3.80%)	532 (18.56%)

Table 1 presents the number of instances and the share of transport mode classes in the datasets. There are four modes considered in this work including public transport (PT). The survey answers were not limited to these modes, but other modes were rarely reported, which resulted in an insufficient number of examples. The raw journey records describe a respondent, i.e. education, gender, and year of birth; and information about reported journeys, such as origin, destination, departure time, and aim. Additional features were calculated using OpenTripPlanner[1] and include, separately calculated for each of the analysed travel modes, features such as distance, duration, waiting time (for PT), or estimated travel duration considering street congestion (for CAR). In this way, the journey records $\mathbf{x}_i$ used in the remainder of this work were obtained.

The Calculation of Parking Features. During the surveys, respondents were asked how long it took them to find a parking space. Answers to this question were given only if someone travelled by car. The direct usage of this data would create data leakage and bias the results. To avoid this, the missing parking time for the remaining journeys $\mathbf{x}_i$ was imputed using estimation from the k nearest neighbors (kNN) model and multivariate imputation made by the MICE algorithm overwriting the original values. Moreover, as a reported parking time equal to zero may mean not providing true data or not leaving a car at the destination at all, two attempts were used to treat zeros. While the first one used $t_p = 0$ in parking time tuples as correct values, the second one considered zeros as missing values to be imputed with proper values. The kNN-based regression model was used to predict parking time based on the set T. During optimisation, $k = 9$ was selected as an optimal value for all datasets. The average mean absolute error (MAE), calculated on the known values, reached 5.07 and 5.39 min for the estimation with and without zero parking times, respectively.

The MICE algorithm cannot be used directly to create an estimation model. The algorithm calculates only new values to replace missing data. This is done

[1] https://www.opentripplanner.org/.

using multiple imputations, in our case, based on the parking attempt's longitude, latitude and hour. Therefore, the algorithm had to be modified. In the first step, the predictive mean matching was used 5 times to calculate new values. Next, all original values were removed, and the imputation was applied again (with the same parameters) to overwrite the original data. The errors obtained by the MICE algorithm are higher and more diverse than for kNN. For the dataset with zeros considered as proper values, the MAE reached 6.46 min; after removing such instances, i.e. once zero parking time values were removed and imputed with MICE, the error exceeded 8.15 min. Thus, four parking time features `PRED_PTIME` were provided, i.e. two by kNN and two by MICE. Finally, radius R was set to 1000 meters to calculate parking difficulty features. This value reflects the distance for which walking is used most of the time.

Table 2. Summary of parking feature sets

Feature set	Features used
BASELINE	$\mathbf{x}$
C_TIME	$[\mathbf{x}$, PCOST(), PRED_PTIME()]
C_DIFF	$[\mathbf{x}$, PCOST(), PDIFF_IS_CENTR(), PDIFF_IS_AREA(), PDIFF_ZONE_CENTR()]
C_TIME_DIFF	BASELINE ∪ C_TIME ∪ C_DIFF

Algorithm 1: The evaluation of the importance of parking features

Input: D - matrix of n feature vectors, $P \in \mathbb{R}^n$ - vector of corresponding n transport modes, K - the number of CV folders, r - the number of runs

begin
  for $i = 1, \ldots r$ **do**
    $\{D_j, P_j\}_{j=1,\ldots,K}$ = DivideSetUsingStratifiedCrossValidation(D,P,K);
    for $k = 1 \ldots K$ **do**
      $D_{\mathrm{T}} = D_k$; $D_{\mathrm{V}} = D_{(k+1) \bmod K}$; $D_{\mathrm{L}} = D \setminus D_{\mathrm{T}} \setminus D_{\mathrm{V}}$;
      $\mathbf{h} = FindBestHyperParameterValues(D_{\mathrm{L}}, P(D_{\mathrm{L}}), D_{\mathrm{V}}, P(D_{\mathrm{V}}))$;
      $M = TrainWithBestHyperParams(i, D_{\mathrm{L}}, P(D_{\mathrm{L}}), D_{\mathrm{V}}, P(D_{\mathrm{V}}), \mathbf{h})$;
      $E_{\mathrm{T}}((i-1) * K + k) = E(M(D_{\mathrm{T}}), P(D_{\mathrm{T}}))$;
  $E_{\mathrm{T}} = [mean(E_{\mathrm{T}}()), median(E_{\mathrm{T}}())]$;

The Evaluation of Features. Algorithm 1 was applied separately for each dataset described in Sect. 4 and each feature set listed in Table 2. It was executed with $r = 10$ and $k = 10$. For higher diversity, a different ML technique was used for each $i = 1, \ldots, r$. Moreover, the best hyperparameter values were determined first for each of the following methods: kNN, multi-layer perception, SVMs with linear and radial kernels, XGBoost and XGBDart , ranger , naive Bayes, decision tree, and RF.

Table 3 shows the mean and median accuracy (ACC) E_T obtained for all ML methods considered together, i.e. based on 100 tests per one feature set-dataset pair. Next, for each dataset, the ML method yielding the highest median accuracy for C_TIME_DIFF was determined.

Table 3. The accuracy (ACC) of mode choice predictions on testing subsets

		(a) All methods		(b) Best method		
Dataset	Feature set	mean [%]	median [%]	method	mean [%]	median [%]
PAR_W1	BASELINE	63.95	66.00	XGBoost	67.42	67.67
	C_DIFF	64.02	66.67	XGBoost	70.43	70.33
	C_TIME	66.80	69.00	XGBoost	77.26	77.33
	C_TIME_DIFF	66.82	68.67	XGBoost	75.82	76.33
PAR_W2	BASELINE	57.62	58.93	XGBDart	64.18	63.39
	C_DIFF	57.78	60.38	XGBDart	63.93	63.39
	C_TIME	57.74	59.65	XGBDart	61.45	60.96
	C_TIME_DIFF	57.96	58.93	XGBDart	65.02	67.25
CIT_W1	BASELINE	57.03	58.56	XGBDart	62.14	62.67
	C_DIFF	57.12	58.46	XGBDart	60.55	60.00
	C_TIME	56.41	57.92	XGBDart	60.10	60.04
	C_TIME_DIFF	56.91	58.78	XGBDart	61.77	61.71

For the PAR_W1 dataset, extending the features to C_TIME_DIFF increases the mean and median ACC by about 3%, similarly to C_TIME. For the PAR_W2 dataset, C_TIME_DIFF provides the highest mean ACC, but the median ACC is better for C_DIFF. The ACC changes for CIT_W1 are only minor ones.

However, the best classifier for each dataset is sought in practice. Applying the best classifier – XGBoost – and the C_TIME set increases the mean ACC for PAR_W1 to 77.26% i.e. by nearly 10 per cent points. For PAR_W2, while the mean ACC can be slightly improved using XGBDart and all parking time features, the median ACC was improved to 67.25% when all features were included. The results for the PAR_W2 dataset show that the best classification method may yield a much higher ACC benefit arising from exploiting C_TIME_DIFF features, than suggested by mean ACC. The reason for the difference in the ACC gains between PAR_W1 and PAR_W2 may be higher MAE for parking time predictions for PAR_W2 than for PAR_W1.

The best predictor, i.e. XGBDart, cannot improve the average ACC on the CIT_W1 dataset. The method overfits training data and adding more features reduces the ACC of the models. These results show that whether additional features are helpful should be decided separately for each dataset.

5 Conclusions

In this work, we propose two novel categories of parking-related features to be used for travel mode choice modelling. The first category necessitates the use of data showing how much time it took drivers to find a parking space in different areas of the city of interest. The second group of features transforms data describing travel demands into parking difficulty features. Both feature groups contribute to the development of travel choice models. Which of them should be used depends inter alia on the available data sources. The case of the city-wide data for primary school parents (the PAR_W2 dataset) shows that the use of both feature types together may be needed to help model development.

In the case of one of the datasets, the introduction of parking-related features negatively affected some ML methods. Still, the case of the two remaining datasets shows that significant accuracy gains can be expected once these features are used together with survey-based features. This suggests that the features proposed in this work should be considered in future travel mode choice studies.

Acknowledgements. This research has been supported by the CoMobility project. The CoMobility benefits from a 2.05 million€ grant from Iceland, Liechtenstein and Norway through the EEA Grants. The aim of the project is to provide a package of tools and methods for the co-creation of sustainable mobility in urban spaces.

References

1. Ding, L., Yang, X.: The response of urban travel mode choice to parking fees considering travel time variability. Adv. Civ. Eng. **2020**, 1–9 (2020). https://doi.org/10.1155/2020/8969202
2. Hagenauer, J., Helbich, M.: A comparative study of machine learning classifiers for modeling travel mode choice. Expert Syst. Appl. **78**, 273–282 (2017). https://doi.org/10.1016/j.eswa.2017.01.057
3. Hasnine, M.S., Habib, K.N.: What about the dynamics in daily travel mode choices? A dynamic discrete choice approach for tour-based mode choice modelling. Transport Policy **71**, 70–80 (2018). https://doi.org/10.1016/j.tranpol.2018.07.011
4. Hillel, T., Bierlaire, M., Elshafie, M.Z., Jin, Y.: A systematic review of machine learning classification methodologies for modelling passenger mode choice. J. Choice Model. **38**, 100221 (2021). https://doi.org/10.1016/j.jocm.2020.100221
5. Li, M., Zou, M., Li, H.: Urban travel behavior study based on data fusion model. Elsevier Inc. (2018). https://doi.org/10.1016/B978-0-12-817026-7.00005-9
6. Lu, Y., Kawamura, K.: Data-mining approach to work trip mode choice analysis in Chicago, Illinois, area. Transp. Res. Rec. **2156**(1), 73–80 (2010). https://doi.org/10.3141/2156-09

7. Salas, P., la Fuente, R.D., Astroza, S., Carrasco, J.A.: A systematic comparative evaluation of machine learning classifiers and discrete choice models for travel mode choice in the presence of response heterogeneity. Expert Syst. Appl. **193**, 116253 (2022). https://doi.org/10.1016/j.eswa.2021.116253
8. Tamim Kashifi, M., Jamal, A., Samim Kashefi, M., Almoshaogeh, M., Masiur Rahman, S.: Predicting the travel mode choice with interpretable machine learning techniques: a comparative study. Travel Behav. Soc. **29**, 279–296 (2022). https://doi.org/10.1016/j.tbs.2022.07.003
9. Tenkanen, H., Toivonen, T.: Longitudinal spatial dataset on travel times and distances by different travel modes in Helsinki Region. Sci. Data **7**(1), 77 (2020). https://doi.org/10.1038/s41597-020-0413-y
10. Xie, C., Lu, J., Parkany, E.: Work travel mode choice modeling with data mining: decision trees and neural networks. Transp. Res. Rec. **1854**(1), 50–61 (2003). https://doi.org/10.3141/1854-06

Cloud Native Approach to the Implementation of an Environmental Monitoring System for Smart City Based on IoT Devices

Mirosław Hajder[1], Lucyna Hajder[2(✉)], Piotr Hajder[2], and Janusz Kolbusz[1]

[1] University of Information Technology and Management, Rzeszów, Poland
{mhajder,jkolbusz}@wsiz.edu.pl
[2] AGH University of Science and Technology, Kraków, Poland
{lhajder,phajder}@agh.edu.pl

Abstract. In this paper, we present the architecture and implementation of the environmental monitoring system, which is one of the main elements of the Smart City system, deployed in a small town in Poland – Boguchwała, Podkarpacie. The system is based on the Internet of Things devices and Cloud Native techniques, which allow for measuring several environmental parameters like pollution, EMF pollution, and acoustic threats. In addition to these parameters, characteristic of environmental monitoring, the system has been enhanced with video monitoring techniques, such as evaluating the traffic intensity on the main roads and crowd detection. In particular, a front-end application was implemented to visualize the results on a city map. The system is deployed on Raspberry Pi and NVidia Jetson using Kubernetes as resources orchestrator. We managed to design, implement, and deploy a system that makes measurements and predicts the parameters indicated. The proposed solution has no significant impact on the energy consumption of the measuring stations while increasing the scalability and extensibility of the system.

Keywords: Smart City · Environmental monitoring · Cloud Native · Internet of Things

1 Introduction

One of the effects of systematic improvement in the quality of life is the alarming deterioration of the environment around us. In many places around the world, the concentration of air, water, or soil pollution threatens the health of the people who live and work there. Therefore, in the last decade, much attention has been devoted to counteracting the negative effects of urbanization and industrialization of further areas of our planet. One of the methods used for this purpose is monitoring the state of the environment, which allows us to track harmful phenomena occurring in our environment and take appropriate action at the right

J. Mikyška et al. (Eds.): ICCS 2023, LNCS 14077, pp. 514–521, 2023.
https://doi.org/10.1007/978-3-031-36030-5_41

time. Monitoring can be carried out at three different levels: impact, regional, and background [1,2].

In this paper, we present the architecture and implementation of the environmental monitoring system, which is one of the main elements of the Smart City system, deployed in a small town in Poland – Boguchwała, Podkarpacie. In addition to the traditional parameters characteristic of environmental monitoring, the system has been enhanced with video monitoring techniques that allow for assessing the traffic intensity on the city's main roads and detecting clusters of people. For this purpose, machine learning methods were used, in particular object detection based on the YOLOv5 model with two setups: small and large. The system was built from IoT-based measurement stations, located in urban and suburban areas. The platforms used are Raspberry Pi for standard measurements and NVidia Jetson for video monitoring components. The system was implemented using the Kubernetes orchestrator, which enabled high scalability and expansion of the system with additional nodes. A front-end application was implemented to visualize the results on a city map.

2 Related Works and Contribution

In the available literature, a number of works of a similar nature to ours can be found. The essence of pollution measurement is the use of appropriate sensors. These devices can be an integral part of a larger platform or a separate software and hardware system. In their work [3], the authors describe the construction of an IoT class device that has the ability to measure pollutants from various ranges, together with the ability to send measurements to a remote server due to carefully planned communication. In addition to the set of many parameters and their presentation to interested parties, an important process nowadays is the prediction of pollutants, most often based on historical data. In [4], the authors present a model of a neural network based on a genetic algorithm, with which they predict the level of pollution depending on the time of day, with two defined time intervals: 1 h and 24 h. Thanks to a large set of historical data, it is possible to obtain a high accuracy, defined by a prediction error of less than 0.5%.

In the context of Smart City, in addition to monitoring the environment understood as natural phenomena and factors, it is important to monitor domains related to human activity, such as car traffic. In [5], the authors presented a model that limits the problems associated with both categories, i.e., reducing vehicle traffic and the resulting pollution. The use of historical pollution data to predict car traffic can also be found in [6]. The most important work related to the one described in this article is [7], because the authors set themselves the goal of collecting and presenting information on air pollution and street traffic. The system used the same classes of devices planned for our research, that is, Raspberry Pi and NVidia Jetson platforms.

In our work, in addition to the indicated information, forecasts of short- and long-term propagation of the indicated pollutants are made, the collection of which is much wider, because it goes beyond the aspect of air. The system also has the ability to measure electromagnetic radiation, called electromagnetic-magnetic (EMF) pollution. The big difference is also the implementation approach – in our work we focused on using Cloud Native techniques when designing the application, right from the very beginning. We use many methods of artificial intelligence to analyze the data collected by the system in order to reduce the volume of data sent to servers.

3 Methodology, System Architecture and Implementation

3.1 Data Gathering

All data were collected in the city of Boguchwała and its vicinity. Measurement points were selected for this purpose, where the measuring stations described in the next part of the work were located. Their arrangement is shown in Fig. 1. Points 1–3 are not present because, in the current version of the system, they are located outside of the city and are used for other purposes.

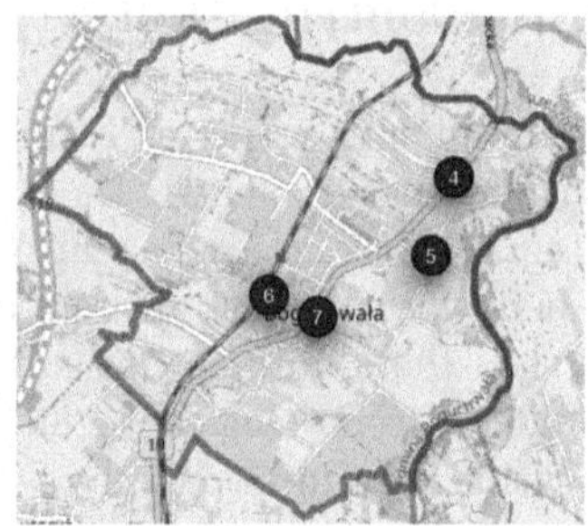

Fig. 1. Area visualization with measurement points

3.2 Data Processing

Acoustic Measurements. The task solved in the system significantly exceeds the scope covered by modeling the propagation of acoustic waves. For the needs of the project, a new task was defined to determine acoustic disturbances along the actual route of movement of motor vehicles, which were initially decomposed into two subtasks. The first of the subtasks consists in building a spatio-temporal model of the displacement of disturbances along a given route and the known acoustic parameters of a set of their sources. The second subtask includes determining the distribution of sound pressure around the emitter based on a measuring device located at any point in the area analyzed. Contrary to the currently solved problems, it is not a set of stationary point perturbations that is considered, but a set of moving sources.

We assumed that communication streams as linear sources emit cylindrical sound waves, with the emission taking place into half-space. If these sources are located on an embankment or viaduct, radiation into space is assumed if the height of the embankment (viaduct) exceeds 3 m. The distance R for which a cylindrical wave becomes spherical at the length of the sound source l is given by condition $R \geq l/\pi$. If the calculation point (the point at which the noise intensity will be determined) is the surface of a residential building, the equivalent sound pressure level L_{po} can be calculated using the expression:

$$L_{po} = L_{7.5} + 10\lg arctg\frac{l}{2R} - 10lg\,(1 - \alpha_{tlp}) - 10\lg\frac{R}{r_0}$$

where: L_{po} - sound intensity at the calculation point in dB; $L_{7.5}$ - sound pressure level at a distance of 7.5 m from the road axis; $r_0 = 7.5\,\mathrm{m}$; l - linear length of the sound source; α_{tlp} - surface attenuation coefficient.

Air Pollution Measurements. Air pollution is one of the most problematic problems for Polish residents. For this reason, state and local authorities have built measurement networks that are independent of each other. The network built by the responsible state authorities is a high-budget, low-coverage metering network. The monitoring system provides real-time data and also offers modeling of air condition, which, however, is not made available en masse. On the other hand, systems made by local governments are based on low-budget measuring equipment and are limited to sharing the results at the measurement site, or possibly historical data. Spatial pollutant modeling services are not performed in them.

The maximum value c_m (mg/m^3) of the near-ground concentration of a harmful substance in the event of an unintentional release into the environment of a gas-air mixture from a single source with a circular outlet, obtained under unfavorable meteorological conditions at a distance of x_m (m) from the source, is calculated according to the expression:

$$c_m = A \cdot M \frac{F \cdot m \cdot n \cdot \eta}{H^2 \cdot \sqrt[3]{V_1 \cdot \Delta T}}$$

where: A - atmospheric stratification coefficient, for Podkarpacie $A = 180$; M [g/s] - mass of harmful substances introduced into the atmosphere per unit of time; F - dimensionless coefficient that describes the rate of deposition of harmful substances in the atmospheric air; H [m] - height of the pollution source above ground level; η - dimensionless coefficient taking into account the influence of the terrain, in the case of flat terrain or a slight slope not exceeding 50 m per 1 km, $\eta = 1$; ΔT [°C] - difference between the temperature of the gas-air mixture T_g and the temperature of ambient air T_p; m and n - coefficients taking into account the conditions of the gas-air mixture outflow from the outlet of the emission source; V_1 [m^3/s] - gas-air mixture outflow velocity.

Electromagnetic Radiation Measurements. Information on environmental EMF pollution is currently one of the most sought-after data on the state of the environment. In the first place, EMF pollution comes from cell phone transmitters. Wireless networks at home can also be important. At present, there are no known serial field density sensors intended for use in environmental monitoring systems. The solution proposed in the system is to cooperate with the communication platform using Arduino. For spatiotemporal modeling of electromagnetic pollution, the system uses a methodology similar to that used in the case of air pollution. In this case, due to the rectilinear nature of electromagnetic wave propagation, the task of locating the source of the pollution is simplified.

Factors affecting the value of electromagnetic signal power density are most often: signal loss in free space, reflections and multipath effect, diffraction or shading effect, the need to penetrate rooms and cars by the signal, signal propagation over water and through vegetation, and various types of interference. Semi-empirical models are used to calculate the losses of radio signals during its propagation, the most famous of which are the Okamura-Hata, Walfisch-Ikegami, Ksya-Bertoni, Alsbrook and Parson models [8–10]. Statistical methods are also used to predict radio propagation losses.

Crowding and Car Traffic Detection. Gathering detection and traffic assessment are included in the same category of tasks. In both cases, the basis is a machine learning model that detects objects on frames sent by cameras. For this purpose, the YOLOv5 model was used. The model was trained to recognize two classes of objects: people and cars. It was taught using images of cars and people available on the Kaggle platform. Furthermore, the training and validation sets were enriched with 10% from the real work environment. Visualization of car traffic intensity is presented using sections of the road and a color gradient, indicating the intensity. An example is presented in Fig. 2. The figure also shows circles with a color gradient to indicate potential icing. The operation of this module is complex and is still in the testing phase, so it is not described in this paper.

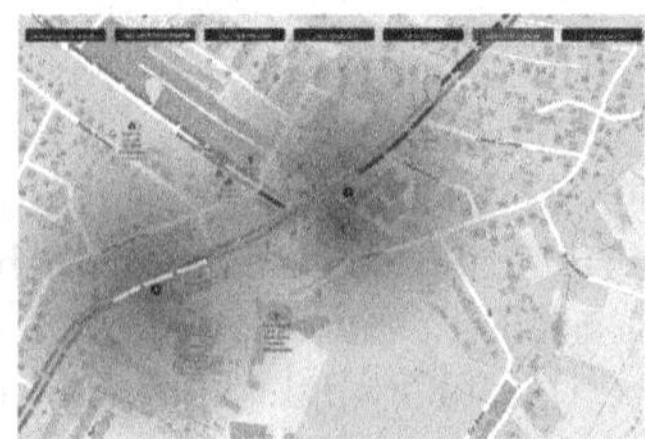

Fig. 2. Visualization of traffic and crowding in the system

3.3 Cloud Native Approach to the System

When designing the system, we had in mind that Smart City projects evolve and require the addition of new nodes; therefore, the system must be easy and highly scalable. For this reason, the system uses various Cloud Native techniques, in particular containerization and services as resources (e.g. queue, database server, FTP for updating microcontrollers). In addition, due to the nature of edge computing, we decided to use devices with high power consumption, such as Raspberry Pi for basic parameters and NVidia Jetson platforms for GPU calculations (YOLOv5 model). All edge devices do not have access to a permanent power source; they work with a battery charged with a solar panel.

The Raspberry Pi nodes are mainly used to collect information and send it to the aggregation node. In turn, Jetson nodes use the presence of CUDA cores to speed up machine learning calculations. We chose two models for testing:

- Nano, equipped with a 4-core ARM Cortex-A72 processor, 2 GB of shared RAM, and 128 CUDA cores, consumes up to 7.7 W of energy.
- AGX Orin, equipped with a 12-core ARM Cortex-A78AE processor, 64 GB of shared RAM, 2048 CUDA cores, and 64 Tensor cores, consumes a maximum of 60 W (also has 15 W and 45 W modes).

On the software side, for resource orchestration, we used the Kubernetes platform in the k3s version, dedicated to IoT class devices. The choice was dictated not only by Cloud Native issues, but also by the extensibility of the system. A broker based on Akri, which exposes sensors and cameras as resources in Kubernetes cluster, was launched on each of the nodes, detecting newly connected devices (including CSI and USB cameras). In order to increase the reliability of the system in the context of transferring information from nodes and to become independent of the frequency of their transfer, a Kafka queue was added to the system, to which measurement nodes send the results of measurements and/or analyzes.

4 Reliability and Performance Results

The last stage was to test the system in terms of performance, reliability, and accuracy of the models used. The verification procedure was divided into four stages: assessment of the possibility of adding new sensors and nodes to the cluster; checking the accuracy of the predictions of the models used; efficiency of the proposed software and hardware configuration in the form of detection times, transmission times, and numerical calculations and impact of applied solutions on energy consumption due to the power supply of the station battery.

As part of the first verification stage, efforts were made to add new sensors to measure the level of the river as the next stage of system development. For this purpose, a broker was written that reads and processes the values read from the sensor. The Akri broker automatically detected a device compatible with its implementation in the system, and Kubernetes launched the appropriate pods to support brokers. The next test involved adding a new node to the cluster. Thus, the first test was considered to have been met.

The second verification stage involves testing both specific measurement values and forecasts. The verification of the currently measured values was carried out manually with the use of specialized and calibrated instruments, i.e., a decibel meter, an air quality meter, and electromagnetic fields.

After the accuracy of the measurements made by the stations was confirmed, the forecasts of the mentioned parameters were verified. For this purpose, a simple program in Nodejs was prepared which was launched as a Job in Kubernetes. Its task was to regularly verify the measured values against the forecasts in order to continuously verify whether the accuracy did not change over time, which could suggest the omission of a factor in the models.

For numerically measured parameters, i.e. electromagnetic radiation, pollutants, and acoustics, the mean squared error (MSE) method was used. For machine learning models, standard tests were performed on a previously prepared test data set, composed of 80% (272 of 340 images) of data collected in Boguchwała. The accuracy obtained is presented in Table 1.

Table 1. Accuracy of the predictions

Parameter	Verification method	Value [%]
Acoustic threats	MSE	78
Air pollution	MSE	94
Electromagnetic radiation	MSE	89
Crowding and car traffic	Test dataset	92

Due to the nature of the parameters calculated from the images (crowding and car traffic), accuracy has a greater tolerance for errors because to mark their occurrence, a given threshold value must be exceeded. Incorrect detection in the number of two examples out of 10 does not change the classification to the indicated group. In the case of numerical parameters, it can be seen that the prediction accuracy obtained is high, although in the case of acoustic hazards, it is recommended to improve. During the implementation, it was noticed that the positioning of the microphones in each of the three dimensions has a great impact and the position opposite the measured space was not always the best position.

The third stage of verification assumed a list of performance and reliability parameters of the system. The use of the Kubernetes platform did not cause any performance loss – the CPU and memory consumption resulting from running k3s agents is less than 3% at idle. The most interesting comparison in the system was a comparison of YOLO network single inference times on Jetson Nano vs AGX Orin devices. In the case of Jetson nano, the average single frame analysis time was 147 ms for YOLOv5 small, while for Orin it was 43 ms for YOLOv5 large. It can be seen that at the moment the use of Jetson Nano is quite sufficient.

The last phase of verification required several tests at different times of the year, as it concerned the energy consumption of the solar-charged battery. In the case of Raspberry Pi and Jetson Nano devices, no impact was observed, both from the point of view of the calculations performed on them and the season. In the case of Jetson AGX Orin, the problem is to use its full power in the autumn and winter months, because the power from the solar panel is not enough to charge it. It can be used in 15 W mode or by limiting the frame rate.

5 Summary and Further Works

As part of this work, a complex Smart City system was presented to monitor the environment based on IoT devices and cloud native techniques. The project used both analytical and non-deterministic models, based mainly on machine learning methods. We managed to design, implement, and implement a system that performs measurements and makes predictions of the indicated parameters. The introduced solutions in the field of Cloud Native not only enabled the scaling of the system but also not significant energy consumption, which would exclude this approach from application. The use of the Jetson platform speeds up frame processing, but in some cases it is not possible to use it (AGX Orin in winter). It is also planned to migrate the front-end part to the cloud and test the use of its other services.

References

1. Patnaik, P.: Handbook of Environmental Analysis. CRC Press, Boca Raton (2017)
2. Wiersma, G.B.: Environmental Monitoring. CRC Press, Boca Raton (2004)
3. Duangsuwan, S., Takarn, A., Nujankaew, R., Jamjareegulgarn, P.: A study of air pollution smart sensors lpwan via nb-iot for Thailand smart cities 4.0. In: 2018 10th International Conference on Knowledge and Smart Technology (KST), pp. 206–209 (2018)
4. Jiang, X., Zhang, P., Huang, J.: Prediction method of environmental pollution in smart city based on neural network technology. Sustain. Comput. Inf. Syst. **36**, 100799 (2022)
5. Singh, S., et al.: A novel framework to avoid traffic congestion and air pollution for sustainable development of smart cities. Sustain. Energy Technol. Assess. **56**, 103125 (2023)
6. Shahid, N., Shah, M.A., Khan, A., Maple, C., Jeon, G.: Towards greener smart cities and road traffic forecasting using air pollution data. Sustain. Cities Soc. **72**, 103062 (2021)
7. Fadda, M., Anedda, M., Girau, R., Pau, G., Giusto, D.D.: A social internet of things smart city solution for traffic and pollution monitoring in Cagliari. IEEE Internet Things J. **10**(3), 2373–2390 (2023)
8. Rappaport, T.: Wireless Communications: Principles and Practice, 2nd edn. Prentice Hall PTR, Upper Saddle River (2001)
9. Walfisch, J., Bertoni, H.: A theoretical model of uhf propagation in urban environments. IEEE Trans. Ant. Propagat. **36**(12), 1788–1796 (1988)
10. Ikegami, F., Takeuchi, T., Yoshida, S.: Theoretical prediction of mean field strength for urban mobile radio. IEEE Trans. Ant. Propagat. **39**(3), 299–302 (1991)

Prediction of Casting Mechanical Parameters Based on Direct Microstructure Image Analysis Using Deep Neural Network and Graphite Forms Classification

Bartlomiej Sniezynski[1(✉)], Dorota Wilk-Kolodziejczyk[1,2], Radosław Łazarz[1], and Krzysztof Jaskowiec[1,2]

[1] AGH University of Science and Technology, Krakow, Poland
{bartlomiej.sniezynski,dwilk,lazarz}@agh.edu.pl
[2] Łukasiewicz Research Network – Kraków Institute of Technology, Krakow, Poland

Abstract. This paper presents methods of prediction of casting mechanical parameters based on direct microstructure image analysis using deep neural networks and graphite forms recognition and classification. These methods are applied to predict tensile strength of iron-carbon alloys based on microstructure photos taken with the light-optical microscopy technique, but are general and can be adapted to other applications. In the first approach EfficientNet architecture is used. In the second approach graphite structures are separated, recognized using VGG19 network, counted and classified using support vector machines, decision trees, random forest, logistic regression, multi-layer perceptron and AdaBoost. Accuracy of the first approach is better. However, the second allows to create a classifier, for which the accuracy is also high, and can be easily analyzed by human expert.

Keywords: Prediction of mechanical properties · Microstructure image analysis · Deep neural network · Graphite forms classification

1 Introduction

The main goal of this work is to present two methods for prediction of the mechanical parameters of castings based on the analysis of the microstructure image.

Initially, the prediction of mechanical properties was based on the analysis of phase diagrams. However, already in the 80s it was indicated that these techniques will be replaced by mathematical models and artificial intelligence methods [18]. In early works, neural networks were applied to predict Ferrite Number (FN) based on weight percentage of 13 elements [15,28]. In our research we also started with machine learning methods applied to process parameters

J. Mikyška et al. (Eds.): ICCS 2023, LNCS 14077, pp. 522–534, 2023.
https://doi.org/10.1007/978-3-031-36030-5_42

and chemical composition [29]. However, approach described this paper is based on microstructure image classification.

In [3] Deep Neural Networks are applied for lath-bainite segmentation in complex-phase steel. Vanilla U-Net and VGG16 neural architectures are used. In our research, which is continuation of work [11], we predict tensile strength of iron-carbon alloys based on microstructure photos taken with the Light-Optical Microscopy (LOM) technique. The first approach is similar to one proposed in [3] – EfficientNet deep neural network is applied to process the images directly. In the second approach, which is our main contribution, graphite structures are extracted from the image, and their types (forms) are recognized using VGG19 network. Feature vector consists of numbers of graphite structures assigned to every type. Classification algorithm is used to predict tensile strength based on these features. We have applied support vector machines, decision trees, random forest, logistic regression, multi-layer perceptron and AdaBoost. Performance measured by F1 score of the first approach was better in evaluation. However, the second approach allows to create classifier which can be easily analyzed by human expert. This method garnered a lot of positive feedback from domain experts because it was not a black box solution.

Another important contribution of this research is creation of a dataset consisting of microstructure images of ductile iron and their labels from a number of experiments that was used in evaluation. This dataset will be made available to interested researchers upon email request.

In the following sections related research and methodology are presented. Next experiments are described. Conclusions and future works summarize the paper.

2 Related Research

Metal-characterising properties have been a consistent subject of scientific interest for many years, due to their significant industrial importance. Consequently, various historical works attempted to automatically predict their actual values [5,15,21]. One of the first worth noting was the research conducted by Babu et al. [5], who proposed a model for predicting ferrite numbers exhibiting accuracy comparable to that of WRC-1992. Notably, this model was later improved even further by Vitek et al [28].

Classical machine learning techniques yielded similarly promising results for other analogous tasks. For instance, Badmos et al. [6] investigated prediction quality for a large array of traits, such as tensile strength, durability, hardness, and deformation rate. Similarly, Javier et al. [12] examined chemical composition, casting size, cooling speed, and thermal treatment using linear classification, k-nearest neighbours, decision trees, and Bayesian networks (BN). Yuxuan et al. [30] used a simple neural network with 20 input variables (such as chemical composition, heat treatment conditions, and test temperature) to predict the tensile characteristics of stainless steel. Additionally, as showcased by Penya et al. [20], BNs were successfully applied to recognise the presence of micro-damages in the casting exhibited before or during the casting process itself.

Sachin et al. [22] incorporated data from electron backscatter diffraction (EBSD) and designed a way to effectively identify and quantify ferrite micronutrients in complex microstructures of various steel grades. However, in a report authored by Britz et al. [9], a correlation approach based on EBSD and LOM was used—instead of using those common methods independently. Finally, a work by Gola et al. [13] dealt with the classification of the components of the microstructure of low-carbon steel by employing a data mining approach.

Presently, the employment of deep neural networks has become the prevalent approach for image-driven prediction tasks, with convolutional architectures [27] or transformers [8] being particularly favoured. Research by Durmaz et al. [3] applied U-Net networks for structure segmentation in complex phase steel, and similar methods have been effective in predicting the properties of castings [4]. In this research, transfer learning through the VGG19 [24] architecture was utilized for microstructure-based classification, while the simpler direct classification was carried out using EfficientNet [25].

3 Methodology

3.1 Direct Image Classification Using Deep Neural Networks

The primary aim of this approach is to establish a baseline for the casting quality prediction task (a binary classification problem, where one has to distinguish low tensile strength samples from those characterised by high R_m values) without over-engineering the pipeline for this specific problem. To achieve this objective, we treated the target classifier as a single-step black box and ignored the domain-specific knowledge used by the other presented approaches. Due to the limited size of the utilised dataset, we determined that employing a dedicated architecture would offer minimal advantages and may even lead to indirect overfitting. Similarly, interpretability concerns are postponed to the future works, as they were not the critical focus. Thus, we concentrated on well-established pre-trained techniques with a proven record of successful deployments.

Our initial search for a suitable foundational model showed that vanilla VGG-like arrangement [24] is enough for single graphite structure classification (see below) but together with the deep residual network ResNet50 [10] is severely underperforming in direct microstructure image classification when compared with the third candidate—a group of convolutional networks jointly known as the EfficientNets [25].

EfficientNets represent a family of convolutional networks specifically designed to balance model depth, width, and resolution. They leverage a scalable architecture that can be tuned to various sizes while maintaining a consistent level of accuracy. The main building block of EfficientNets is the mobile inverted bottleneck, first introduced as a part of the MobileNetV2 network [23]. More recently, the machine vision community proposed numerous other architectures as their successors, including EfficientNetV2 [26] (with its potentially faster fused residual blocks) and ConvNeXt [16] (proven to be a viable alternative even to the modern vision transformer models). However, we remain convinced that the original EfficientNets are still the safest choice for establishing a realistic and reproducible baseline result, mainly due to the sheer number of successful applicative studies profiting from them.

We employed the largest member of the family - the B7 model consisting of 66 million weights, pre-trained on the ImageNet dataset. The task of cast iron assessment significantly deviates from the standard object recognition problems. Specifically, the input images are near-monochromatic and uniformly filled with content. Consequently, fine-tuning the entire network (instead of adjusting only the top fully-connected layers) was deemed advantageous.

To pick the hyperparameters for the fine-tuning process, we utilised the AutoKeras meta-optimisation system [14]. The complete configuration consisted of the elements listed in Table 1. The automatic tuner itself is a hybrid oracle that performs a chosen number of trials (50 in our case). First, it aggregates parameters into conceptual categories (e.g. "augmentation" or "architecture"). Then, it generates new values for one category at a time with a greedy strategy—while using the best result obtained so far for the rest.

Table 1. EfficientNet fine-tuning parameters

Hyperparameter	The set of considered values
Adam learning rate (α)	$\{0.001, 0.0001, 0.00002\}$
top layers spatial reduction (t_s)	{GlobalAverage, GlobalMax}
classification head dropout rate (p_d)	$\{0.0, 0.25, 0.5\}$
translation augmentation factor (g_t)	$\{0.0, 0.2\}$
zoom augmentation factor (g_z)	$\{0.0, 0.2\}$
contrast augmentation factor (g_c)	$\{0.0, 0.2\}$
rotation augmentation factor (g_r)	$\{0.0, 0.2, 0.5\}$
horizontal flip (g_h)	$\{0, 1\}$
vertical flip (g_v)	$\{0, 1\}$

Both the parameter selection and the actual fine-tuning operated on a collection of images that were initially of varying sizes and aspect ratios. To ensure consistency, we first applied a pre-processing step in which each image was cropped to a square shape and rescaled to a resolution of 224×224—resulting in 3525 positive (high R_m) and 1455 negative (low R_m) samples.

The reported scores were obtained as a result of a standard 10-fold cross-validation procedure. In the case of the first analysed fold, the training set was additionally subdivided into two smaller subsets (in a ratio of 8 to 2). Those subsets were utilised to conduct an initial AutoKeras hyperparameter search. The winning arrangement of parameters obtained that way was then saved and used unchanged throughout all the remaining folds of the experiment.

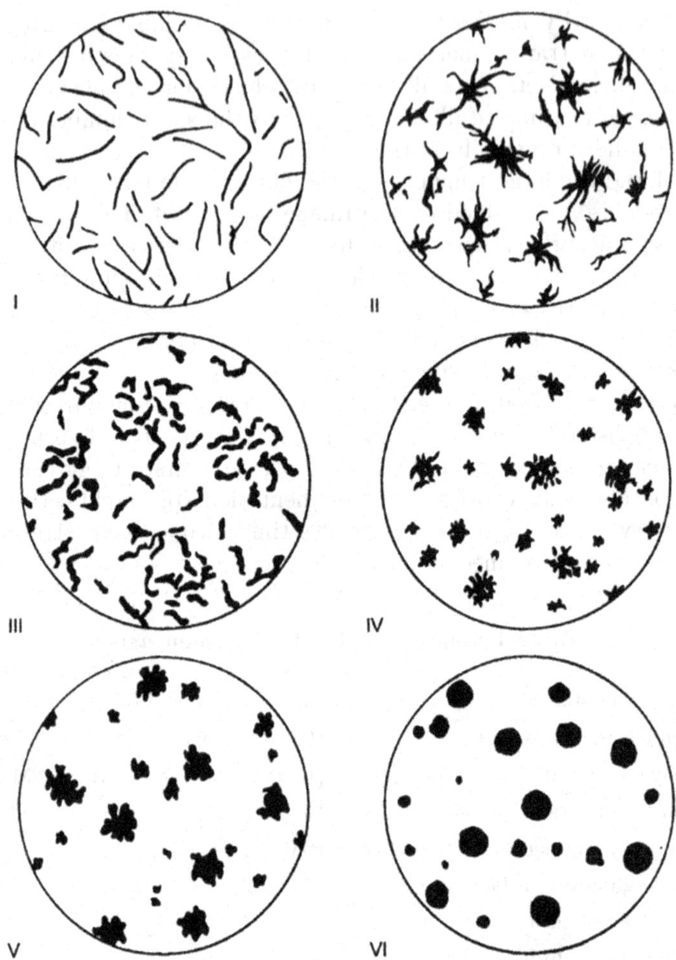

Fig. 1. Graphit forms in cast irons [1]

3.2 Microstructure-Based Classification

Microstructure-based classification idea is based on analysing graphite structures appearing in the microstrucutre LOM picture. According to the norms [1], the structures can be classified into six principal forms, which are presented in Fig. 1.

The algorithm gets LOM picture and mechanical property prediction model as input and returns predicted property, see Algorithm 1. At the beginning graphite structure type counters are initialized with 0 (lines 1–3), next separate graphite structures are distinguished in the picture P (line 4) and stored in a list S. These structures are classified to one of the structure types t and a number of structures for recognized type is updated (lines 5–8). These numbers form a feature vector X describing the picture. Next, mechanical property can be predicted using model M (line 9).

```
   Input: P – LOM picture, M – mechanical property prediction model
   Output: Predicted mechanical property
 1 foreach structure type t do
 2 |  X[t] ← 0
 3 end
 4 S ← Separate structures in picture P;
 5 foreach structure picture s ∈ S do
 6 |  t ← type of structure s;
 7 |  X[t] ← X[t] + 1;
 8 end
 9 c ← prediction of mechanical property M(X);
10 return c;
```

Algorithm 1: Microstructure-based classification

This algorithm depends on three main procedures that should be specified: structure separation (line 4), structure type classification (line 6) and mechanical properties prediction (line 9). They are described below.

Structure Separation. To separate the structures, their edges are discovered using Canny's edge detection [7] approach. For every separate edge region, part of the image surrounded by the edge is cut off and put in the middle of a white rectangle (dimensions 335×251 pixels were used in experiments). It happens that several graphite structures are in contact with each other and recognized as one, big structure. Frequency of such error is very low, below 1%.

Structure Type Classification. To determine type of the separated structures, VGG19 convolutional neural network [24] and transfer learning were applied. The last three layers of VGG19 were replaced by three fully connected layers consisting of 512 neurons each and softmax output layer. These last layers were updated during training on examples of forms of graphite structures that were manually assigned to one of six forms (Fig. 1), to which additional type 0 form, containing merged structures described above, was created.

To improve quality of classification three transformations were tried. Data augmentation was used to balance the data. Gaussian blur was applied to merge cracked structures. Image thresholding was also applied to decrease color depth to 1-bit and eliminate gray levels.

Mechanical Properties Prediction. In our research several machine learning algorithms were applied to build the model M predicting mechanical properties of casting represented by the feature vector X. Mechanical property was categorical and represented low and high tensile strength. However, presented approach is general and can be also used for other properties.

The following algorithms were applied for classification: Support vector machines, decision trees (CART algorithm), random forest, logistic regression, multi-layer perceptron and AdaBoost. Training data were only lightly unbalanced (3358 examples with high tensile strength versus 1337 with low). Training was also applied for balanced data (using random elimination of examples from over-represented category), but results were worse than on the original training data. Data augmentation techniques could be applied here but this topic is left for future research.

4 Experiments

4.1 Source Data

Photos of microstructures subjected to classification were created as a result of experiments (research projects) carried out in the former Foundry Research Institute in Krakow (currently Łukasiewicz Research Network - Krakow Institute of Technology). The basis for the classification was the information contained in the standards regarding the shape of graphite and qualifying them to the appropriate groups/classes. Another issue that should be taken into account when classifying the microstructure is the possibility of defects related to the arrangement of the graphite precipitate. The size of the precipitates and significant differences in the size of the precipitates may be an important factor. The best mechanical properties can be obtained with spheroidal graphite compared to cast iron with flake precipitates. When observing the microstructure of cast iron with flake graphite, defects related to graphite degeneration may also occur, which also deteriorate the mechanical properties.

The set consists of 223 images in two magnification levels: a hundredfold and five hundredfold magnification. Images with lower magnification have resolution 2080×1540 and with higher 1388×1040. Samples were manually classified by domain expert according to tensile strength R_m. Examples of these images are presented in Fig. 2. All images are in RGB format. Images with a hundredfold magnification are cut into 25 smaller pictures to achieve five hundredfold magnification.

Based on CRISP-DM methodology pictures with low quality and large amount of noise were removed, together with pictures done other than LOM technique and representing other types of castings, see Fig. 2-(c) for example.

4.2 Direct Classification

The initial hyperparameter search concluded by finding the following arrangement: $\alpha = 0.00002$, $t_s =$ GlobalAverage, $p_d = 0.0$, $g_t = 0.2$, $g_z = 0.0$, $g_c = 0.0$, $g_r = 0.5$, $g_h = 1$, $g_v = 1$. The gathered values are in line with our prior knowledge and expectations about the problem domain. The images of cast iron microstructures do not have a natural orientation, i.e., they have no inherent concept of top, bottom, left, right, or centre. This means that they can be freely

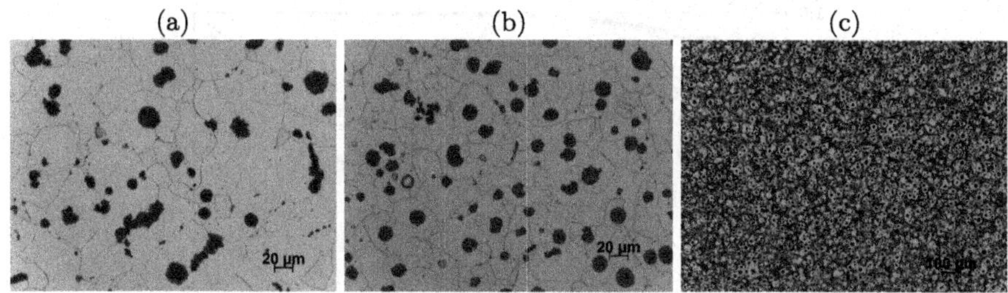

Fig. 2. Examples of LOM images representing microstructure of iron-carbon alloys with low tensile strength (a), high tensile strength (b) and outlier image removed from the training data set (c)

flipped, rotated, and shifted without affecting the representation of the underlying structures. Hence, we can easily augment our training dataset by applying those transformations to the images, without risking a negative impact on the accuracy of the final model.

The preference for low learning rates supports our decision to start with a pre-trained model, implying that a slight correction of the initial weights was enough to achieve satisfactory results for the studied problem. Finally, we acknowledge that the observed leaning towards less regularization may be a byproduct of the assumed limited epoch budget, where training was halted after 30 epochs. This is due to the fact that high-dropout networks tend to converge more slowly than their no-dropout counterparts.

Table 2 outlines the aggregated experiment results obtained for the aforementioned optimal parameter set. The small number of the utilised samples resulted in a significant between-fold variance and score deviation. The same phenomenon can be spotted when scrutinising Fig. 3: there are two clear outliers among the ROC trends, corresponding to splits that are harder to classify. On the other hand, the healthy shape of the discussed curves and the good balance between the calculated precision and recall metrics show that, fortunately, the skewed class distribution in the training set had a negligible effect on the ultimately attained performance.

4.3 Microstructure-based Classification

Structure Type Classification. Training data for structure form classification consisted of 1572 examples. Numbers of examples in every category are presented in Table 3. The network was trained in 100 epochs using Adam optimizer. Accuracy of the trained network was equal to 82% on test data (10% of the training data).

Table 2. Baseline results of a direct classification with a fine-tuned EfficientNet

Accuracy	Precision	Recall	F1 Score	AUC
$90.9\% \pm 1.4\%$	$92.9\% \pm 1.7\%$	$94.3\% \pm 2.1\%$	$93.6\% \pm 1.1\%$	$96.44\% \pm 0.93\%$

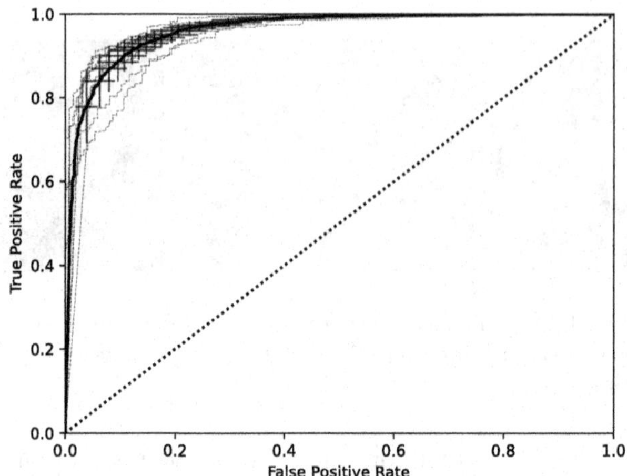

Fig. 3. Receiver Operating Characteristic (ROC) curves for direct classification. Dark red line denotes the mean ROC curve, black crosses—its standard deviation in certain points, dotted light red lines—the curves corresponding to cross-validation splits. (Color figure online)

F1 metrics was equal to 0.79. As it may be noticed in confusion matrix generated for types I-VI, which is presented in Fig. 4, the main problem is distinguishing form IV, which is mistaken with III and V. Also form II is mistaken with form III and I.

Accuracy of structure form classification using VGG19 convolutional neural network trained on data with various transformations applied are presented in Table 4. Augmentation resulted in drop of accuracy. Adding Gaussian blur allowed to improve results for classes V and VI. However, other classes were miss-classified more often (especially I and II). Therefore the overall improvement was minimal. Image thresholding resulted in accuracy decrease.

Mechanical Properties Prediction. Results obtained in 10-fold cross-validation for chosen algorithms are presented in Table 5. Hyper-parameters of these algo-

Table 3. Number of examples in every form type

Form	Number of examples
0	14
I	278
II	122
III	292
IV	76
V	289
VI	501

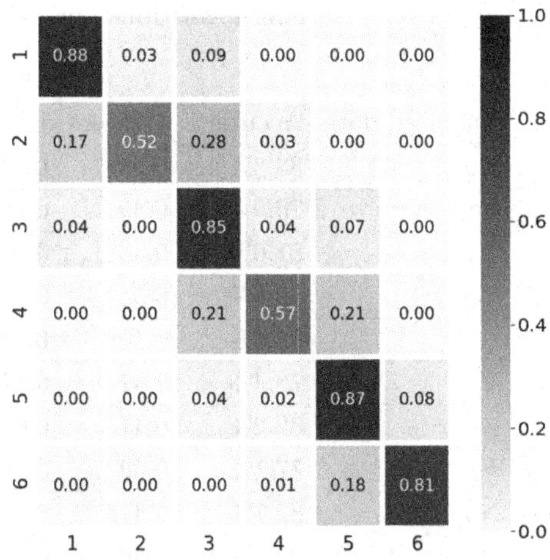

Fig. 4. Confusion matrix for VGG19 convolutional neural network for structure form classification

Table 4. Accuracy of structure form classification using VGG19 convolutional neural network trained on data with various transformations applied

Augmentation	Gaussian blur	Thresholding	Accuracy
No	No	No	82.2%
Yes	No	No	78.0%
Yes	Yes	No	79.7%
Yes	Yes	Yes	77.1%

rithms were tuned using Optuna framework [2]. Models were trained on original and balanced data. As we can see, the best Accuracy and Log loss achieved AdaBoost (83,4%, 0.67). The highest F1 values were achieved by Support vector machine and Random forest (0.90). The best Average precision achieved Random forest and AdaBoost. Accuracy of the reset of algorithms is similar. Other metrics differ more.

What can be also observed, results for balanced data are worse than for unbalanced. Better balancing techniques should be applied in the future to check their influence.

Because Decision tree algorithm achieved results close to the best models, decision trees learned were analyzed and consulted with domain expert. Example of such decision tree is presented in Fig. 5. Its accuracy is 81.2%. The expert confirmed that the decision tree is consistent with the domain knowledge.

Table 5. Results of chosen machine learning algorithms applied to mechanical properties prediction

Model	Balanced data	Accuracy	Log loss	F1	Average precision
Support vector machine	No	83.3%	0.39	**0.90**	0.90
Support vector machine	Yes	76.9%	0.49	0.78	0.81
Decision tree	No	81.6%	0.46	0.89	0.81
Decision tree	Yes	69.5%	0.58	0.74	0.65
Random forest	No	83.3%	0.40	**0.90**	**0.92**
Random forest	Yes	78.4%	0.47	0.79	0.83
Logistic regression	No	82.2%	0.41	0.89	0.91
Logistic regression	Yes	77.1%	0.52	0.78	0.80
Multilayer perceptron	No	82.7%	0.41	0.88	0.91
Multilayer perceptron	Yes	80.6%	0.46	0.83	0.84
AdaBoost	No	**83.4%**	**0.67**	0.89	**0.92**
AdaBoost	Yes	72.0%	0.62	0.76	0.77

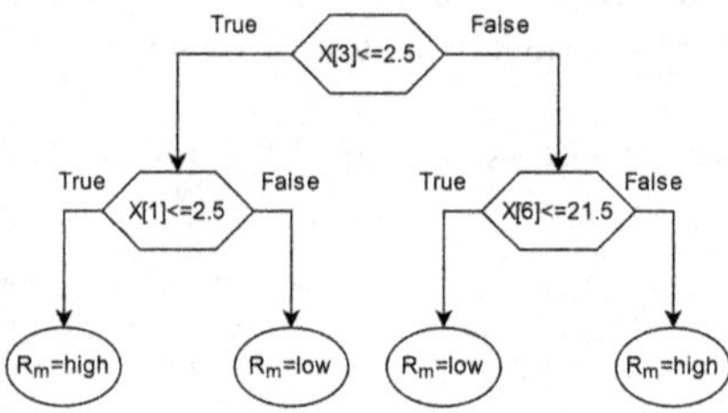

Fig. 5. Sample decision tree for mechanical properties prediction

5 Conclusions

In this paper we have shown that it is possible to predict casting mechanical parameters based on direct microstructure image analysis and recognition and classification of graphite forms. These methods were applied to predict tensile strength of iron-carbon alloys, but they can be used to predict other parameters too.

Direct image classification has better accuracy (90.1%) and F1 metrics (93.6%) than recognition and classification of graphite forms. However the latter approach allows to create models with high interpretability and still high accuracy. Decision tree allows to achieve accuracy equal to 83.3% and F1 equal to 90%, and the model learned is very simple. It was consulted with domain experts and they confirmed that it is consistent with their knowledge.

Interpretability of the model is especially important in decision support systems. Application of such a model would allow to add explanation functionality, which is very important in engineering.

In the future research we are planning to apply Grad-CAM for producing visual explanations for direct microstructure image analysis. We would like to apply other methods for data augmentation in recognition of graphite forms and other architectures than VGG19 for structure classification. We would also like to use other methods with symbolic knowledge representation for classification, e.g. scoring systems [19] that we have applied in medical domains or rule based systems, which correspond to human way of thinking well [17]. Last but not least, we would like to apply this methodology for prediction of other mechanical properties, like elastic limit (R_p) of iron-carbon alloys.

Acknowledgements. This paper received partial support from the funds assigned by the Polish Ministry of Education and Science to AGH University. Authors would like to thank Wiktor Reczek for help with Microstructure-based classification programming and experiments.

References

1. EN ISO 945–1:2018 Microstructure of cast irons - Part 1: Graphite classification by visual analysis (ISO 945–1:2017) (2018)
2. Akiba, T., Sano, S., Yanase, T., Ohta, T., Koyama, M.: Optuna: a next-generation hyperparameter optimization framework. In: Proceedings of the 25th ACM SIGKDD International Conference on Knowledge Discovery & Data Mining, pp. 2623–2631 (2019)
3. Ali, D., et al.: A deep learning approach for complex microstructure inference. Nat. Commun. **21**, 1–15 (2021)
4. Anke, S., Peter, B.: Machine learning for material characterization with an application for predicting mechanical properties. Spec. Issue: Sci. Mach. Learn. - Part I **44**(1), 1–21 (2021)
5. Babu, S.S., Vitek, J.M., Iskander, Y.S., David, S.A.: New model for prediction of ferrite number of stainless steel welds. Sci. Technol. Weld. J., 279–285 (1997)
6. Badmos, A., Bhadeshia, H.: Tensile properties of mechanically alloyed oxide dispersion strengthened iron alloys part 2 - physical interpretation of yield strength. Mater. Sci. Technol. **14**, 1221–1226 (1998)
7. Canny, J.: A computational approach to edge detection. IEEE Trans. Pattern Anal. Mach. Intell. PAMI **8**(6), 679–698 (1986)
8. Chen, C.F.R., Fan, Q., Panda, R.: Crossvit: cross-attention multi-scale vision transformer for image classification. In: Proceedings of the IEEE/CVF International Conference on Computer Vision, pp. 357–366 (2021)
9. Dominik, B., Johannes, W., Andreas, S., Frank, M.: Identifying and quantifying microstructures in low-alloyed steels: a correlative approach. In: Metallurgia Italiana, pp. 5–10 (2017)
10. He, K., Zhang, X., Ren, S., Sun, J.: Deep residual learning for image recognition. In: Proceedings of the IEEE Conference on Computer Vision and Pattern Recognition, pp. 770–778 (2016)
11. Jaśkowiec, K., et al.: Assessment of the quality and mechanical parameters of castings using machine learning methods. Materials **15**(8), 2884 (2022)
12. Javier, N., Igor, S., Yoseba, P., Sendoa, R., Mikel, S., Pablo, B.: Mechanical properties prediction in high-precision foundry production. Conferences (2009)

13. Jessica, G., et al.: Advanced microstructure classification using data mining methods. Comput. Mater. Sci. **148**, 324–335 (2018)
14. Jin, H., Chollet, F., Song, Q., Hu, X.: AutoKeras: an AutoML library for deep learning. J. Mach. Learn. Res. **24**(6), 1–6 (2023)
15. John, V., Stan, D.: Improved ferrite number prediction model that accounts for cooling rate effects part 1: model development details of a prediction model based on a neural network system of analysis are described. Weld. J. **83** (2003)
16. Liu, Z., Mao, H., Wu, C.Y., Feichtenhofer, C., Darrell, T., Xie, S.: A ConvNet for the 2020s. In: Proceedings of the IEEE/CVF Conference on Computer Vision and Pattern Recognition, pp. 11976–11986 (2022)
17. Michalski, R.S.: Attributional calculus: A logic and representation language for natural induction. Technical Report. MLI 04–2, George Mason University (2004)
18. Olson, D.L.: Prediction of austenitic weld metal microstructure and properties. Weld. J. **64**(10), 281s–295s (1985)
19. Pajor, A., Żołnierek, J., Sniezynski, B., Sitek, A.: Effect of feature discretization on classification performance of explainable scoring-based machine learning model. In: Computational Science-ICCS 2022: 22nd International Conference, London, UK, 21–23 June 2022, Proceedings, Part III. pp. 92–105. Springer, Heidelberg (2022). https://doi.org/10.1007/978-3-031-08757-8_9
20. Penya, Y.K., Bringas, P.G., Zabala, A.: Advanced fault prediction in high-precision foundry production. In: 2008 6th IEEE International Conference on Industrial Informatics, pp. 1672–1677 (2008). https://doi.org/10.1109/INDIN.2008.4618372
21. Rati, S., Khwaja, M.: Formation, quantification and significance of delta ferrite for 300 series stainless steel weldments. Int. J. Eng. Technol. Manag. Appl. Sci. **3**, 23–36 (2015)
22. Sachin, S., Andrew, B., Trimby, G.P., Simon, R.P., Julie, C.: An automated method of quantifying ferrite microstructures using electron backscatter diffraction (ebsd) data. Ultramicroscopy **137**, 40–47 (2013)
23. Sandler, M., Howard, A., Zhu, M., Zhmoginov, A., Chen, L.C.: Mobilenetv 2: inverted residuals and linear bottlenecks. In: Proceedings of the IEEE Conference on Computer Vision and Pattern Recognition, pp. 4510–4520 (2018)
24. Simonyan, K., Zisserman, A.: Very deep convolutional networks for large-scale image recognition. arXiv preprint arXiv:1409.1556 (2014)
25. Tan, M., Le, Q.: Efficientnet: rethinking model scaling for convolutional neural networks. In: International Conference on Machine Learning, pp. 6105–6114. PMLR (2019)
26. Tan, M., Le, Q.: Efficientnetv2: smaller models and faster training. In: International Conference on Machine Learning, pp. 10096–10106. PMLR (2021)
27. Tripathi, M.: Analysis of convolutional neural network based image classification techniques. J. Innov. Image Process. (JIIP) **3**(02), 100–117 (2021)
28. Vitek, J., David, S., Hinman, C., et al.: Improved ferrite number prediction model that accounts for cooling rate effects-part 2: model results. Weld. J. **82**(2), 43S (2003)
29. Wilk-Kołodziejczyk, D., et al.: Selection of casting production parameters with the use of machine learning and data supplementation methods in order to obtain products with the assumed parameters (2023)
30. Wang, Y., et al.: Prediction and analysis of tensile properties of austenitic stainless steel using artificial neural network. Metals **10**, 234 (2020)

Breaking the Anti-malware: EvoAAttack Based on Genetic Algorithm Against Android Malware Detection Systems

Hemant Rathore[1(✉)], Praneeth B[1], Sundaraja Sitharama Iyengar[2], and Sanjay K. Sahay[1]

[1] Department of CS and IS, BITS Pilani, Goa Campus, Goa, India
{hemantr,f20200096,ssahay}@goa.bits-pilani.ac.in
[2] Florida International University, Miami, USA
iyengar@cs.fiu.edu

Abstract. Today, android devices like smartphones, tablets, etc., have penetrated very deep into our modern society and have become an integral part of our daily lives. The widespread adoption of these devices has also garnered the immense attention of malware designers. Many recent reports suggest that existing malware detection systems cannot cope with current malware challenges and thus threaten the android ecosystem's stability and security. Therefore, researchers are now turning towards android malware detection systems based on machine and deep learning algorithms, which have shown promising results. Despite their superior performance, these systems are not immune to adversarial attacks, highlighting a research gap in this field. Therefore, we design and develop *EvoAAttack* based on a genetic algorithm to expose vulnerabilities in state-of-the-art malware detection systems. The EvoAAttack is a *targeted false-negative evasion attack* strategy for the *grey-box scenario*. The EvoAAttack aims to convert malicious android applications (by adding perturbations) into adversarial applications that can deceive detection systems. The EvoAAttack agent is designed to convert *maximum* malware into adversarial applications with *minimum* perturbations while maintaining syntactic, semantic, and behavioral integrity. We tested EvoAAttack against thirteen distinct malware detection systems based on machine and deep learning algorithms from four different categories. The EvoAAttack was able to convert an average of 97.48% of malware applications (with a maximum of five perturbations) into adversarial applications (malware variants). These adversarial applications force misclassifications and reduce the average accuracy of thirteen malware detection systems from 94.87% to 50.31%. Later we also designed a defense strategy (*defPCA*) to counter the adversarial attacks. The defPCA defense reduces the average forced misclassification rate from 97.48% to 59.98% against the same thirteen malware detection systems. Finally, we conclude that threat modeling improves both detection performance and adversarial robustness of malware detection systems.

Keywords: Adversarial Learning · Android Malware · Evasion Attack · Genetic Algorithm · Malware Detection

J. Mikyška et al. (Eds.): ICCS 2023, LNCS 14077, pp. 535–550, 2023.
https://doi.org/10.1007/978-3-031-36030-5_43

1 Introduction

The 21st century has witnessed a massive adoption of computing and communication devices (like smartphones, tablets, etc.) in every aspect of our daily life. Smartphones are currently used by more than two-thirds of the world's population for their professional as well as personal needs [7]. All these computing devices need an operating system for their resource management. Here, android is a dominant player and holds a market share of 71% and 48% in the smartphone and tablet ecosystem [1]. These android devices store vast amounts of users' personal as well as business data. Therefore, they have become an attractive target for malware designers who want to steal data (like contacts, SMS, pictures, videos, etc.), disrupt devices, cripple services, etc. Several recent reports indicate that numerous malware and potentially unwanted applications exist in the android ecosystem [2,12]. The latest AV-ATLAS report suggests that the total number of malicious android applications and potentially unwanted applications have reached 60 million as of January 2023 [2]. It also notes that 77.8% of newly discovered malware file types were packaged as an APK, making it the most commonly used entry point for attackers. These malware attacks can execute security breaches and result in severe consequences for users' data and privacy in the android ecosystem.

Anti-virus and anti-malware software offer the first line of defense against malware attacks. However, many recent reports suggest that the existing malware detection techniques (like signature, heuristics, etc.) are ineffective against new, complex, sophisticated malware attacks [5,12]. Researchers are now turning towards android malware detection systems based on machine and deep learning algorithms, which have shown promising results [5]. Arora et al. (2019) proposed *PermPair* to identify android permission pairs for malware detection and achieved 95.44% accuracy [3]. Wozniak et al. (2020) designed a recurrent neural network model for malware threat detection and achieved more than 99% accuracy [19]. Wang et al. (2019) designed *LSCDroid* based on locally sensitive APIs [18]. They developed a random forest model that achieved more than 96% accuracy. As more and more machine and deep learning solutions are getting integrated into real-world malware detection systems, the potential for adversarial attacks on them has gained attention.

Malware designers can develop adversarial attacks to target malware detection systems based on machine and deep learning with the aim of decreasing their performance. An adversarial attack involves introducing small perturbations to a sample to force misclassification in a malware detection system. Many researchers have demonstrated vulnerabilities in various machine and deep learning models to such adversarial attacks [15,16]. Cara et al. (2020) proposed injecting system API calls in order to force misclassification in a multi-layer perceptron model [6]. Li et al. (2020) designed a generative adversarial network that forces misclassifications in detection models [11]. These studies highlight vulnerabilities of the machine and deep learning detection systems that malware designers or adversaries can potentially exploit. On the other hand, the anti-malware community can use this information to improve the adversarial robustness of malware detection systems. The threat modeling of adversarial attacks on malware detection systems depends on the attacker's intel about the target system. It includes information about the training dataset, features, classification algorithms, and the architecture of the detection system. If the attacker possesses comprehensive knowledge about the target system, it is called a white-box scenario. Conversely, if the attacker has zero or partial knowledge about the target system, it is called a black or grey box scenario, respectively.

In this work, we propose a novel *targeted false-negative evasion attack* strategy *(EvoAAttack)* to generate malware variants that can force misclassifications in state-of-the-art malware detection systems. The EvoAAttack is designed based on a process of natural evolution (genetic algorithm) that aims to break the *Classification Robustness* of a machine and deep learning based target systems in the *grey-box scenario*. The EvoAAttack agent is designed to add minimum perturbations in malicious android applications and converts them into adversarial applications (malware variants) that can deceive detection systems. It also aims to convert maximum malware into adversarial applications by adding perturbations while ensuring their syntactic, semantic, and functional integrity. Therefore, the EvoAAttack ensures *Perturbation Measurability* to quantify the deviation of an application after perturbations, and the *Perturbation Invertibility* to enable the reconstruction of the original application after applying perturbations. The EvoAAttack investigated the adversarial robustness of thirteen distinct malware detection systems. These systems were constructed using thirteen machine and deep learning algorithms from four categories: machine learning, bagging, boosting, and neural networks. Later, we also designed and developed an adversarial defense strategy *(defPCA)* to counter the evasion attacks. We make the following contributions with this work:

1. We designed *EvoAAttack*, a novel *targeted false-negative evasion attack* strategy based on an optimized genetic algorithm to generate adversarial applications (malware variants) that can force misclassifications in next-generation malware detection systems.
2. The *EvoAAttack* exposes adversarial vulnerabilities in thirteen different malware detection systems based on four categories of classification algorithms (*machine learning*, *bagging*, *boosting*, and *neural networks*).
3. The *EvoAAttack* converted an average 97.48% of malware applications (with maximum of five perturbations) into adversarial applications (malware variants) that can force misclassifications in thirteen malware detection systems. The EvoAAttack reduced the average accuracy of thirteen malware detection systems from 94.87% to 50.31%.
4. We also designed *defPCA* defense strategy to counter the adversarial attack. The *defPCA* defense reduces the average forced misclassification rate from 97.48% to 59.98% against the same thirteen malware detection systems.

The paper is structured as follows. Section 2 explains the proposed framework for adversarial robustness, and Sect. 3 describes the experimental setup. Section 4 discusses the experimental results, and Sect.5 explains the related work. Finally, Sect. 6 concludes the paper.

2 Proposed Framework for Adversarial Robustness

In this section, we first describe the proposed framework that improves the adversarial robustness of android malware detection systems. It is followed by a detailed discussion on the *EvoAAttack* and *defPCA* defense strategy.

2.1 Framework Overview

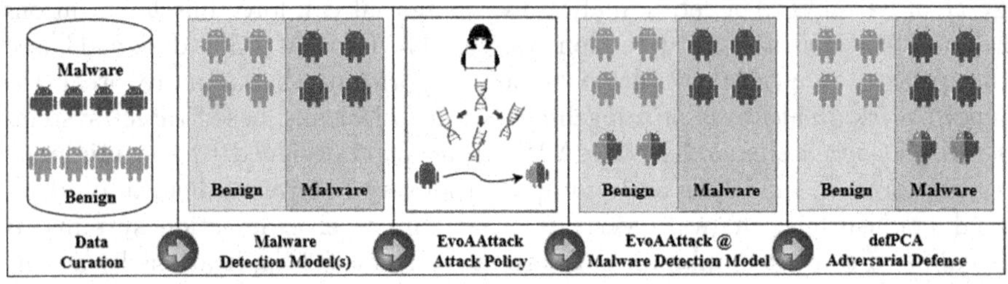

Fig. 1. Framework to improve adversarial robustness of malware detection systems.

Figure 1 illustrates the proposed five-stage framework pipeline to develop adversarially robust malware detection systems. *Stage-1* involves data curation of malware and benign android applications from authentic sources. *Stage-2* entails feature extraction from collected android applications and the construction of malware detection systems based on machine and deep learning algorithms. In *Stage-3*, we will put ourselves in the adversary's shoes and proactively design an adversarial attack strategy *(EvoAAttack)* against malware detection systems. *Stage-4* involves performing an actual *EvoAAttack* on malware detection systems developed in the second stage. The EvoAAttack aims to reduce the performance of detection systems. Then in *Stage-5*, we will develop the defense strategy *(defPCA)* to counter adversarial attacks. We employed evaluation metrics like accuracy, ROC, forced misclassification rate, etc., at various stages of the proposed pipeline. Finally, the proposed framework pipeline is expected to improve detection accuracy as well as the adversarial robustness of android malware detection systems.

2.2 Evolutionary Adversarial Attack (EvoAAttack)

Threat modeling is used to investigate the adversarial robustness of systems. Therefore we perform threat modeling of malware detection systems by stepping into adversaries' shoes and designing adversarial attacks. We propose Evolutionary Adversarial Attack (*EvoAAttack*), a *targeted false-negative evasion attack* strategy for the *grey-box scenario* against malware detection systems. The EvoAAttack aims to convert malware android applications (by adding perturbations) into adversarial malware applications that can force misclassifications in malware detection systems. The evasion attack is designed for the *grey-box scenario* that assumes the attacker possesses knowledge about the dataset and features but has no information about the classification algorithm and system architecture. The EvoAAttack agent aims to convert maximum malware applications into adversarial applications with minimum perturbations in each application. The proposed EvoAAttack agent performs perturbations while maintaining modified android applications' syntactic, semantic, and functional integrity.

The *EvoAAttack* strategy is an optimized variant of the *Genetic Algorithm (GA)*. The GA employs probability-based methods such as random selection and mutation to generate new solutions and examine the solution space. These algorithms are capable of solving problems represented as linear systems or graphs through the use of linear algebraic techniques. These necessitate the application of calculus, including gradients and partial derivatives, in the computation of the fitness function and identification of optimal candidates. The efficacy of GA is evaluated using statistical metrics like mean and standard deviation.

2.2.1 Optimization Function

The genetic algorithm optimizes the objectives using two approaches: *Pareto Optimization* and *Weighted Sum* of fitness function based optimization. The *Pareto set* contains *Pareto-optimal* solutions, representing the best trade-offs between objectives. We used *Weighted Sum* method that involves normalizing individual objectives for comparison and using a fitness function to calculate overall fitness as a weighted sum.

$$f_{raw} = \sum_{i=1}^{O} o_i * w_i \qquad \text{with} \qquad \sum_{i=1}^{O} w_i = 1 \tag{1}$$

where f_{raw} is the raw fitness value and o_i represents the objective functions and their corresponding weight w_i. Our objective is to maximize the probability of an adversarial application to be *forcefully misclassified* as *benign*. Therefore o, w_1 are set to 1 and $o_1 = 0.5 - q_c$ where q_c is the probability of the application being *malicious*. Here, q_c is 0 when the label is *benign* and 1 when the label is *malicious*.

2.2.2 Mutation and Crossover Function

A mutation event can be modeled as a *Bernoulli trial* with a random probability p of transferring a genetic marker from an exogenous genome b to a representative specimen in the genetic repository P. The objective is to imitate the benign profiles in b so that the deception of the malware detection system C is optimized. In the domain of genetic recombination (*crossing over*), two specimens in P, p_i and p_j, undergo a partial exchange of their genetic markers (*perturbations*), X_i and X_j where the exchange mechanism is defined by the operator: $E(X_i, X_j) = X_i \cup X_j$ and $X_i \cap X_j = \phi$. This exchange is subject to a fixed restriction, such that if $||X_s|| > \theta$, where $||X_s||$ represents the magnitude of genetic markers in a specimen, a process of genetic rejuvenation R must be initiated (also known as *hard restart*).

The strength of each application in the gene pool is evaluated using the *fitness function*. The fitness of each application is calculated after every generation, and the algorithm terminates upon reaching the first sample with *positive fitness*. The trace of the evading malware application is finally stored in the *trace file*. The application is marked as not evaded if the generation limit is reached before finding an evading/adversarial application. The EvoAAttack operates as follows:

1. The *EvoAAttack* commences by setting up the environment, where the attack parameters are registered, and benign dataset B along with adversarial perturbations from the past *Trace* are stored in memory. Given a malicious application m, the attack iterates through B to find the closest matching benign application b based on the *Euclidean* $L2$ norm. This sample is used as the external genome to transfer perturbations to the target application.

$$||x|| = \sqrt{x_1^2 + x_2^2 + ... + x_n^2} \tag{2}$$

   Here x is the difference between b and m and x_i represent the dimensions.
2. The next step involves populating the gene pool where a significant portion of the maximum pool capacity is filled with copies of the target sample m to which perturbations from the *trace file* have been added. Duplicates of the target sample take up the remaining slots. With every successive generation, EvoAAttack performs *crossover* and *mutation* operations on the gene pool.

Algorithm 1 shows the *EvoAAttack* policy to convert malware applications into adversarial applications. The algorithm's input is the set of benign B and malicious M applications and other hyperparameters. A target malware application m is selected from the set malware dataset M to convert it into an adversarial application. The population P is initialized with copies modified with respect to the data present in trace histories T or plain repetitions of m. Crossing over and mutations are executed with each evolving generation. The fitness value of each population member p is calculated, and the process stops with the first positive value after recording its perturbations in the trace T. The target is labeled as not evaded if the generation limit is reached before finding an adversarial application.

2.3 Adversarial Defense (defPCA)

We designed a novel defense strategy (*defPCA*) to improve the robustness of the malware detection systems. The approach involves reducing the dimensionality of the feature space using a technique known as *principal component analysis*. Feature reduction has been shown to help increase the generalization in various machine and deep learning systems.

The *defPCA* strategy begins by centering the data, where the mean of each android permission (refer Sect. 3.1) is subtracted from each android application. Let D be mean-centered dataset matrix. The covariance matrix of centered data is calculated to measure the linear relationship between each pair of permissions $VarCov = \frac{D.D^T}{n-1}$, where n is the number of instances in the dataset.

Algorithm 1. Evolutionary Adversarial Attack (EvoAAttack) Strategy.

Input:
B: set of benign applications
M: set of target malware applications
C: target android malware detection model
MAX_POP_SIZE: maximum size of population
MAX_ITER: maximum number of generations
MUT_PROB: probability of mutation
Output:
M′: set of adversarial malware applications and perturbations

1 : $T = []$ *//Traces to store perturbations*
2 : Select a sample m from M without repetition, if empty then GOTO Step 7
3 : $P = []$ *//Population*
4 : Initialize POPULATION
4.1 : If T is not empty
4.1.1 : Generate new sample t for each trace and add t to P
4.1.2 : While $len(P) < MAX\_POP\_SIZE$
4.1.2.1 : Add m to P
5 : while $iteration < MAX\_ITER$
5.1 : MUTATION
5.1.1 : Each sample p is selected with probability MUT_PROB
5.1.2 : The closest benign sample b (from B) to p is selected
5.1.3 : A random feature is transplanted to b from p
5.2 : CROSSING-OVER
5.2.1 : Select a pair of samples p_1 and p_2 from P
5.2.2 : Exchange a random set of features between them
5.3 : FITNESS-TEST
5.3.1 : For each sample p in P
5.3.1.1 : Query label from C and calculate malicious probability q_c
5.3.1.2 : If $0.5 - q_c > 0$
5.3.1.2 : Add p to M'
5.3.1.2 : Store trace of p in T
5.3.1.2 : GOTO step 2
6 : Mark m as NOT_EVADED and GOTO Step 2
7 : return M' and T

Let A be a square matrix of size $m \times n$. The eigenvector v of A is a non-zero vector that satisfies the following equation: $Av = \lambda v$, where λ is the eigenvalue corresponding to the eigenvector v. The characteristic equation of the matrix A is given by $det(A - \lambda I) = 0$, where I is the identity matrix and det is the determinant operator. The eigenvalue decomposition of the matrix A is then given by: $A = Q\Lambda Q^{-1}$, where Λ is a diagonal matrix containing the eigenvalues of A and Q is the eigenvector matrix of A. Eigen-decomposition is performed on $VarCov$ to diagonalize it as $VarCov = V(\lambda)V^T$, where V is the orthonormal matrix containing the eigen-permissions of $VarCov$ in decreasing order of importance.

A lower dimensional representation of the data can be obtained by retaining only the top k eigen-permissions. The optimal value for k was set as 8 based on experiments. It was determined by evaluating the average accuracy of three malware detection systems ($RF + LR + HGB$) before and after subjecting them to the *EvoAAttack*. Finally, the centered data is transformed to the lower dimensional space using the principal component matrix, resulting in a reduced representation of the data: $D' = D \cdot V'$, where V' is the eigen-permission matrix with k leading columns and T is the transpose of the matrix. The malware detection systems are then trained on the reduced feature space.

3 Experimental Setup

This section presents the experimental setup details, including data curation (malware and benign applications), the feature extraction process, classification algorithms, and evaluation metrics used in this work.

3.1 Data Curation and Feature Extraction

We conducted all the experiments using a large dataset collected from authentic sources. The malware applications were collected from benchmarked *Drebin Dataset* proposed by Arp et al. [4], and benign applications were downloaded from *Google Play Store*. The dataset contains 11,281 android applications of which 5,560 were malware, and 5,721 were benign. The malware dataset contains malicious applications from over twenty families, including *DroidKungFu*, *Opfake*, *BaseBridge*, etc. On the other hand, all the benign applications were validated using *VirusTotal*. Later, we disassemble each android application using *Apktool* to perform static analysis on them. We developed a parser that scans through each disassembled android application to extract its *android permission* usage and develop the feature vector. The *rows* in the *feature vector* represent *android applications* and the *columns* represent their *permission usage*.

3.2 Classification Algorithms

We examine the adversarial robustness of thirteen distinct malware detection systems based on various classification algorithms from four different categories. Table 1 lists thirteen distinct classification algorithms and their corresponding category. The detailed design and implementation of these malware detection systems are well explained in another paper [14].

Table 1. Classification algorithms and their categories for constructing malware detection models.

Category	Classification Algorithm(s)	Abbreviation
Machine Learning	Decision Trees	DT
	Support Vector Machine	SVM
	Logistic Regression	LR
Bagging based Learning	Random Forest	RF
	Extra Trees	ET
	Bagged Decision Trees	BagDT
Boosting based Learning	Adaptive Boosting	AB
	Gradient Boosting	GB
	Histogram-based Gradient Boosting	HGB
Neural Network (NN) based Learning	NN with 1 hidden Layer	NN1L
	NN with 3 hidden Layers	NN3L
	NN with 5 hidden Layers	NN5L
	Long Short Term Memory	LSTM

3.3 Platform

All the experiments were conducted in the *Google Colaboratory* environment. The runtime had a storage limit of 70 GB and a RAM limit of 12 GB. The CPU was an Intel Xeon with 1 core and 2 threads running at 2.2 GHz. *Python* was used as the main programming language. The principal libraries used include *scikit-learn*, *TensorFlow*, *Keras*, and *NumPy*. *Pandas* was used to handle CSV files, and all graphs were generated using *Matplotlib* and *Google Sheets*.

3.4 Other Parameters

The *population size* during *EvoAAttack* is the number of samples actively participating in the genetic evolution and was set to 32. The *generation cap* was determined to be 100, based on empirical verification that the chances of finding an evasive sample (adversarial application) were slim after many generations. The *trace size* or the number of trace histories stored in memory was set to 24. The *mutation probability* was set to 0.8. The *crossover rate* was set to 0.2. The *maximum number of perturbations* allowed for each sample in the gene pool was set to 5.

3.5 Evaluation Metrics

The following evaluation metrics were used to measure the detection performance and adversarial robustness of malware detection systems in our work:

- **True Positive (TP)** is the number of malware applications correctly classified as malicious by the malware detection system.
- **True Negative (TN)** is the number of benign applications correctly classified as benign/good by the malware detection system.
- **False Positive (FP)** is the number of benign applications incorrectly or wrongly classified as malicious by the malware detection system.
- **False Negative (FN)** is the number of malware applications wrongly classified as benign by the malware detection system.
- **Accuracy** is the percentage ratio of correctly classified applications to the total number of classifications performed by the malware detection system.

$$\text{Accuracy} = \frac{TP + TN}{TP + FP + TN + FN} \times 100 \tag{3}$$

- **Receiver Operating Characteristic (ROC)** is the degree of separability between malware and benign classes predicted by malware detection system.
- **Forced Misclassification Rate (FMR)** is the percentage of malware applications successfully converted into adversarial malware applications by the adversarial attack agent.

$$\text{Forced Misclassification Rate} = \frac{\#\text{ of Adversarial Apps} - FN_{baseline}}{\#\text{ of Malware Apps} - FN_{baseline}} \times 100 \tag{4}$$

4 Experimental Results

This section explains the detailed experimental results achieved by *EvoAAttack* and *defPCA* for thirteen distinct malware detection systems.

4.1 Baseline Android Malware Detection Systems

The ***Stage-1*** of the proposed framework is data curation and feature extraction as discussed in Sect. 3.1. The ***Stage-2*** focuses on the construction of thirteen distinct android malware detection systems based on data gathered in *Stage-1*. These systems were based on thirteen distinct classification algorithms from four different categories (machine learning, bagging, boosting, and neural network) (refer Table 1). The idea is to investigate the adversarial robustness of malware detection systems based on different classification algorithms from various categories. The performance evaluation of these detection systems was evaluated using accuracy, ROC, F1 score, precision, recall, etc.

Table 2 shows the performance of thirteen baseline android malware detection systems. The highest accuracy was achieved by the RF model based malware detection system (97.48%) followed by the ET model (97.47%). On the other hand, the lowest accuracy was attained by the AB model (90.70%). The thirteen detection systems achieved an average accuracy of 94.87%. Similarly, the highest and lowest ROC was achieved by the RF model (0.977) and AB model (0.91). The thirteen systems achieved an average ROC, precision and recall of 0.95, 0.97 and 0.93, respectively. The bagging based learning models performed best with an average accuracy of 97.41%, while machine learning models performed the worst with an average accuracy of 91.91%.

Table 2. Performance of thirteen distinct android malware detection systems based on various classifications algorithms from four different categories.

Classification Algorithm(s)	Accuracy (%)	ROC Score	F1 Score	Precision	Recall
DT	97.46%	0.97	0.97	0.99	0.96
SVM	92.03%	0.92	0.92	0.95	0.89
LR	91.79%	0.92	0.91	0.94	0.89
RF	97.48%	0.97	0.97	0.99	0.96
ET	97.47%	0.97	0.97	0.99	0.96
BagDT	97.29%	0.97	0.97	0.99	0.96
AB	90.70%	0.91	0.90	0.93	0.88
GB	91.92%	0.92	0.92	0.94	0.89
HGB	95.83%	0.96	0.96	0.98	0.94
NN1L	92.03%	0.92	0.92	0.94	0.90
NN3L	96.14%	0.96	0.96	0.98	0.94
NN5L	96.04%	0.96	0.96	0.98	0.94
LSTM	97.14%	0.97	0.97	0.99	0.95

4.2 EvoAAttack Against Malware Detection Systems

The ***Stage-3*** in the proposed framework is to investigate the adversarial robustness of all the above thirteen baseline malware detection systems. We designed *EvoAAttack*, a genetic algorithm based *targeted false-negative evasion attack* for the *grey-box scenario*. The fundamental idea is to emulate the process of natural evolution into the creation of new malware variants that are much deadlier and can easily deceive state-of-the-art malware detection systems. The EvoAAttack agent aims to convert maximum malware applications (by adding perturbations) into adversarial applications that force massive misclassifications in detection systems. The performance of adversarial attacks against malware detection systems was measured using forced misclassification rate, accuracy drop, etc.

4.2.1 Forced Misclassification Rate @EvoAAttack

The ***Stage-4*** of the proposed framework is to perform actual *EvoAAttack* against thirteen baseline malware detection systems constructed in Sect. 4.1. Figure 2 shows the performance of EvoAAttack, where the *y-axis* represents the forced misclassification rate, and the *x-axis* represents the maximum number of perturbations performed by EvoAAttack. The EvoAAttack agent initially adds only one perturbation and gradually increases it to a maximum of five. The EvoAAttack with just one perturbation achieved an average forced misclassification rate of 66.81% against thirteen baseline android malware detection systems. The highest forced misclassification rate (with a maximum of one perturbation) was achieved against the AB model based malware detection system (99.94%), while the lowest was attained against the RF model (16.02%). The EvoAAttack with a maximum of five perturbations achieved an average forced misclassification rate of 97.48% against thirteen detection systems with a 100% forced misclassification rate against six malware detection systems. Furthermore, RF and ET models exhibit the highest resistance to the evasion attack with FMR of 82.92% and 86.46%, respectively.

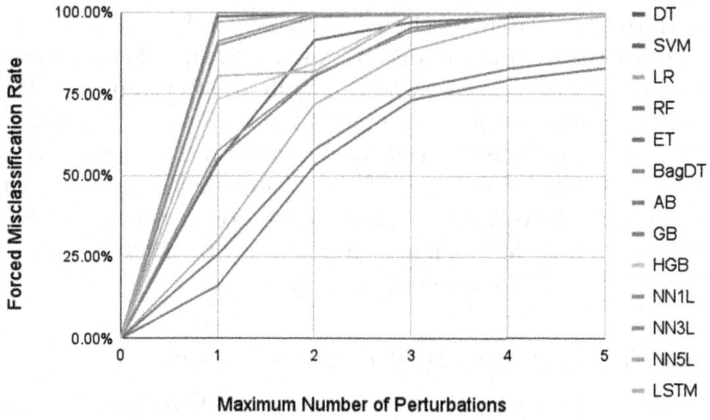

Fig. 2. EvoAAttack against different android malware detection systems.

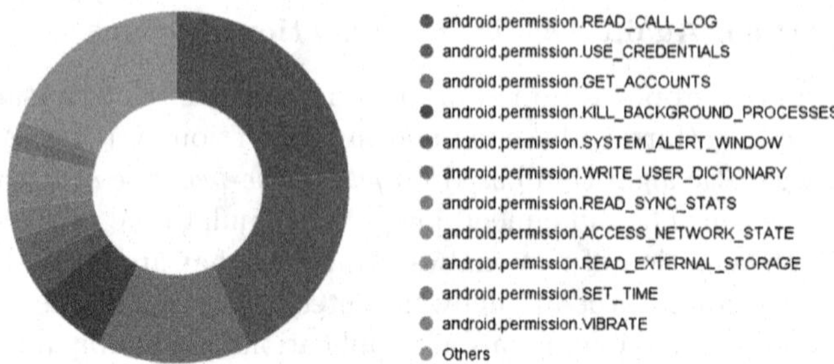

Fig. 3. Top 10 most frequently perturbed android permissions by EvoAAttack agent.

4.2.2 Vulnerable Android Permissions

Figure 3 presents data distribution of the most frequently perturbed android permissions by the EvoAAttack agent against thirteen malware detection systems. The pie chart clearly shows that few android permissions were perturbed majority of the time by EvoAAttack agent to force misclassification in malware detection systems. The top three perturbed android permissions were android.permission.READ_CALL_LOG, android.permission.USE_CREDEN TIALS, and android.permission.GET_ACCOU-NTS, which together accounted for over 55% of all perturbations. This finding suggests that malicious actors could also potentially use these android permissions to evade state-of-the-art malware detection systems.

4.3 defPCA Defense Strategy

The ***Stage-5*** of the proposed framework is to develop a defense strategy for malware detection systems to counter adversarial attacks. We propose *defPCA* based on principal component analysis that improves the generalizability and robustness of machine and deep learning systems.

Figure 4 shows the overall performance of various malware detection systems at different stages of the proposed framework. The *blue bars* represent the accuracy of thirteen different baseline malware detection systems, whereas the *red bars* represent the accuracy of detection systems after *EvoAAttack*. The *yellow bars* represent the accuracy of detection systems after implementing *defPCA* defense strategy, and the *green bars* represent *EvoAAttack on defPCA* based malware detection systems. The thirteen baseline malware detection systems (*blue bars*) achieved accuracy between 90.70% to 97.46%. Then, we performed threat modeling and designed EvoAAttack against detection systems. The EvoAAttack forced a massive number of misclassifications that drastically reduced accuracy (*red bars*) in all thirteen malware detection systems.

4.3.1 Detection Performance @ defPCA Systems

The *yellow bars* in Fig. 4 represent the accuracy of different *defPCA* based malware detection systems. The thirteen defPCA systems achieved an average accuracy of 92.31%. Here, the RF model based detection system accomplished the

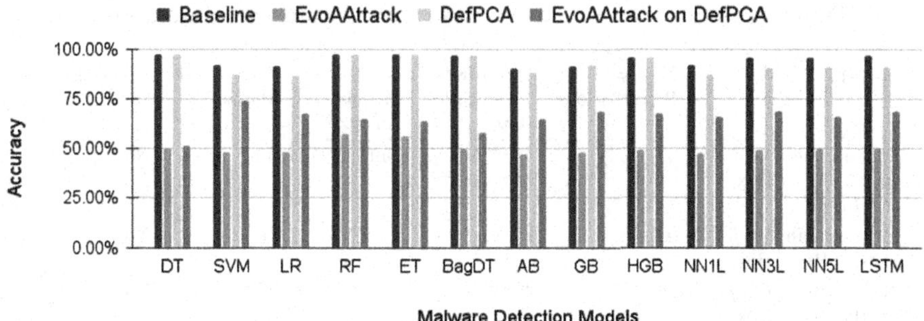

Fig. 4. Overall detection performance and robustness of malware detection systems. (Color figure online)

highest accuracy of 97.46%, whereas the LR model attained the lowest accuracy of 87.05%. On the other hand, the thirteen detection systems achieved an average ROC of 0.92.

4.3.2 Adversarial Robustness @ defPCA Systems

Now, we again performed EvoAAttack on the defPCA malware detection systems to investigate their adversarial robustness. The *green bars* in Fig. 4 show the accuracy of defPCA systems after EvoAAttack. The thirteen defPCA systems after EvoAAttack achieved an average accuracy of 65.65%, which is a 15.34% improvement without any defense strategy. The LR and SVM models with defPCA showed the biggest improvement at 26.09% and 19.74%, respectively. On the other hand, DT models with defPCA showed the least improvement of 1.25%.

5 Related Work

Table 3 lists the existing literature on adversarial attacks on malware detection models and compares them with our proposed approach. Grosse et al. (2016) used the computation of forward derivatives to create adversarial samples against neural network models and attained a misclassification rate of 85% [9]. Though it provides a good stepping stone, the lack of generalization is quite a vulnerability in the work. Hu and Tan (2017) developed a GAN-based attack model with a substitute detector to mimic the target classifier, making it a white box attack [10]. Taheri et al. (2020) [17], and Rathore et al. (2021) [13] also proposed white box strategies to fool different machine and deep learning models. However, this falls short when it comes to real-world applications, as the working mechanisms of most malware detection systems are hidden from the adversary. Fang et al. (2019) proposed grey-box approaches based on deep reinforcement learning that use limited knowledge about the target system, obtaining a success rate of 19.13% and 75%, with a maximum of 100 perturbations [8]. However, it failed to achieve a significant drop in accuracy after the attack. Taheri et al. (2020)

Table 3. Comparison of our proposed work with existing literature.

Paper	Attack Scenario	Max # of Perturbations	Forced Misclassification Rate	# of Models Analyzed	Vulnerability Analysis	Robustness Analysis
Grosse et al. (2016) [9]	White Box	20	40.97%-84.05%	Only DNNs	No	No
Hu and Tan (2017) [10]	White Box	All	99%	6	No	No
Fang et al. (2019) [8]	Grey Box	80	46.56%	1	No	No
Cara et al. (2020) [6]	Grey Box	100	80%	1	No	No
Li et al. (2020) [11]	Grey Box	100	85%	9	No	No
Rathore et al. (2021) [13]	White Box	5	44.28%	8	No	Yes
Proposed (EvoAAttack & defPCA)	**Grey Box**	**5**	**97.49%**	**13**	**Yes**	**Yes**

performed random perturbations on the ranked feature space of the dataset to achieve an accuracy drop of just 12% [17]. In contrast, Rathore et al. (2021) attained a forced misclassification rate of 44.28% using a single policy attack based on q-learning [13]. They failed to perform in-depth feature vulnerabilities analysis or suggest suitable countermeasures to enforce robustness.

There are many limitations in the existing literature that we have addressed in this work. Authors have assumed a white box scenario for threat modeling, which is impractical since most details about the target system are generally unknown. Also, authors have achieved very low forced misclassification rates even with a very high number of perturbations. The perturbations should be minimized to decrease the overall cost of generating adversarial samples. Detailed analysis of perturbations is also not discussed in the existing literature. The defense strategy is incomplete without investigating its adversarial robustness, and it is hardly discussed in the literature.

6 Conclusion

Android malware have exploded in the last few years and have become a real threat to smartphones and tablets. Researchers propose next-gen malware detection systems based on machine and deep learning. These systems have demonstrated encouraging results but might be at risk against adversarial attacks.

In this work, we performed threat modeling of malware detection systems to explore their vulnerabilities. We designed and developed *EvoAAttack*, a novel *targeted false-negative evasion attack* against state-of-the-art malware detection systems for *grey-box scenario*. The EvoAAttack agent converted an average of

97.49% of malware applications into adversarial applications (malware variants) that forced massive misclassifications in thirteen malware detection systems. The EvoAAttack also reduced the average accuracy of thirteen malware detection systems from 94.87% to 50.31%. We also published the list of ten android permissions that adversaries might use to generate more malware variants. Later, we develop a defense strategy (*defPCA*) to counter the attack on malware detection systems. The defPCA reduced the average forced misclassification rate from 97.48% to 59.98% against the same thirteen malware detection systems. Finally, we conclude that threat modeling improves both detection performance and adversarial robustness of malware detection systems.

References

1. Android - Statistics & Facts (2023). https://www.statista.com/topics/876/android/
2. AV-TEST (2023). https://portal.av-atlas.org/malware/statistics
3. Arora, A., Peddoju, S.K., Conti, M.: Permpair: android malware detection using permission pairs. IEEE Trans. Inf. Forensics Secur. **15**, 1968–1982 (2019)
4. Arp, D., Spreitzenbarth, M., Hubner, M., Gascon, H., Rieck, K.: Drebin: effective and explainable detection of android malware in your pocket. In: Network and Distributed System Security Symposium (NDSS), vol. 14, pp. 23–26 (2014)
5. Bhat, P., Dutta, K.: A survey on various threats and current state of security in android platform. ACM Comput. Surv. (CSUR) **52**(1), 1–35 (2019)
6. Cara, F., Scalas, M., Giacinto, G., Maiorca, D.: On the feasibility of adversarial sample creation using the android system API. Information **11**(9), 433 (2020)
7. Digital 2023: Global Overview Report: Simon Kemp (Hootsuite) (2023). https://datareportal.com/reports/digital-2023-global-overview-report
8. Fang, Z., Wang, J., Li, B., Wu, S., Zhou, Y., Huang, H.: Evading anti-malware engines with deep reinforcement learning. IEEE Access **7**, 48867–48879 (2019)
9. Grosse, K., Papernot, N., Manoharan, P., Backes, M., McDaniel, P.: Adversarial perturbations against deep neural networks for malware classification. arXiv preprint arXiv:1606.04435 (2016)
10. Hu, W., Tan, Y.: Generating adversarial malware examples for black-box attacks based on GAN. In: DMBDA, pp. 409–423. Springer, Heidelberg (2023). https://doi.org/10.1007/978-981-19-8991-9_29
11. Li, H., Zhou, S., Yuan, W., Li, J., Leung, H.: Adversarial-example attacks toward android malware detection system. IEEE Syst. J. **14**(1), 653–656 (2019)
12. McAfee 2023 Consumer Mobile Threat Report (2023). https://www.mcafee.com/blogs/internet-security/mcafee-2023-consumer-mobile-threat-report/
13. Rathore, H., Sahay, S.K., Nikam, P., Sewak, M.: Robust android malware detection system against adversarial attacks using q-learning. Inf. Syst. Front. **23**, 867–882 (2021)
14. Rathore, H., Sahay, S.K., Rajvanshi, R., Sewak, M.: Identification of significant permissions for efficient android malware detection. In: Gao, H., J. Durán Barroso, R., Shanchen, P., Li, R. (eds.) BROADNETS 2020. LNICST, vol. 355, pp. 33–52. Springer, Cham (2021). https://doi.org/10.1007/978-3-030-68737-3_3
15. Rathore, H., Samavedhi, A., Sahay, S.K., Sewak, M.: Robust malware detection models: learning from adversarial attacks and defenses. Forensic Sci. Int. Dig. Invest. **37**, 301183 (2021)

16. Sewak, M., Sahay, S.K., Rathore, H.: DRLDO: a novel DRL based de-obfuscation system for defence against metamorphic malware. Def. Sci. J. **71**(1), 55–65 (2021)
17. Taheri, R., Javidan, R., Shojafar, M., Vinod, P., Conti, M.: Can machine learning model with static features be fooled: an adversarial machine learning approach. Cluster Comput. **23**(4), 3233–3253 (2020). https://doi.org/10.1007/s10586-020-03083-5
18. Wang, W., Wei, J., Zhang, S., Luo, X.: LSCDroid: malware detection based on local sensitive api invocation sequences. IEEE Trans. Reliabil. **69**(1), 174–187 (2019)
19. Woźniak, M., Siłka, J., Wieczorek, M., Alrashoud, M.: Recurrent neural network model for iot and networking malware threat detection. IEEE Trans. Ind. Inf. **17**(8), 5583–5594 (2020)

Smart Control System for Sustainable Swimming Pools

Cristiano Cabrita[1(✉)], Jailson Carvalho[1], Armando Inverno[1], Jânio Monteiro[2], and Miguel J. Oliveira[1]

[1] ISE, Universidade do Algarve, Faro, Portugal
ccabrita@ualg.pt
[2] INESC-ID & ISE, Universidade do Algarve, Faro, Portugal

Abstract. Specific research programs, legislation and funding intend to protect, conserve and enhance the EU's natural capital, transforming the EU into a green, competitive, low-carbon and resource-efficient economy. These guidelines aim at protecting European citizens from health and environmental risks. Indeed, there is an increasing interest on decarbonization of the electricity generation, with a special focus on the introduction of Renewable Energy Resources (RES).

This paper is a preliminary insight into a new control approach from where smart decision is made based on predictions returned by models of sustainable thermal systems (local renewable sources generation devices) and on information gathered from an array of sensors in order to regulate swimming pool's water temperature. The information (ambient variables and sub-systems internal transfer function modelling) is then combined with an optimization framework which goal is to ultimately, reduce the requirements for human intervention in the swimming pool maintenance and provide resources savings for the final user in terms of financial and natural resources, contributing to a sustainable environment. The research work is developed within the scope of the Ecopool+++ project: Innovative heated pools with reduced thermal losses.

Keywords: Renewable Energy · Model Predictive Control · ARMAX · Outdoor Swimming Pools

1 Introduction

The European Union and its national governments have set clear objectives to guide European environmental policy up to 2020 and a longer-term vision (for the next 30 years). Specific research programs, legislation and funding intend to protect, conserve and enhance the European Union's (EU) natural capital, to transform it into a green, competitive, low-carbon and resource-efficient economy. These guidelines aim at protecting european citizens from pressures and risks to health and well-being related to the environment.

In this context, a significant portion of research is dedicated to exploring the usage of wind power, biomass and solar power for replacing traditional energy sources as they

J. Mikyška et al. (Eds.): ICCS 2023, LNCS 14077, pp. 551–565, 2023.
https://doi.org/10.1007/978-3-031-36030-5_44

pose many advantages such as little environmental risk and are envisaged as a means to comply to European and global environmental norms [1].

Consulting the power consumption data supplied by Portuguese statistics portal Pordata [2], and for the case of Portugal, in the civil year 2020 it was observed a total of 15215 thousand tep (tons equivalent petrol) due to energy consumption. The figures also show that the energy sources like petrol corresponded to 6260 thousand tep, gas to 1740 thousand tep, renewable energy mix of wind, hydro and solar a total of 2905 thousand tep, and the remaining consumption was due to electrical energy.

As such, the previous figures identify Portugal as one of the European countries where the effective penetration of renewable energy (power usage) is significant (as it is responsible for 19% of the overall energy consumption) with a higher impact when compared with the average value of 11,7% in the EU. Therefore, demand for new sources of flexibility and growing recognition of the multi-energy nature of districts are increasing interest in the interaction between energy sectors, like electricity, heating/cooling, gas and in the significant amount of flexibility available from heating [2, 3].

Solar energy, a form of renewable energy not only is abundant in our environment but can reduce the harmful environmental gas emissions resulting from the burning of fossil fuels. The most common way of using solar power is to convert sunlight into heat energy to produce hot water, through the usage of solar thermal collectors. In such case, the basic mechanism uses the incident solar radiation for generating heat as it converts the irradiated energy into thermal energy. There are many different applications where solar heat energy can be used, such as domestic water-heating systems, pool-heaters, and space-heating systems [4, 5].

In this scope, this paper proposes a new scheme to increase and control a swimming pool's water temperature using sustainable thermal systems (local renewable sources generation devices). The aim of the control system is to coordinate the functioning of a set of thermal energy sources and thermal storage, to adjust the water temperature of outdoor swimming pools according with user requirements. It relies on information gathered from an array of sensors and on weather variables forecast which are then combined with an optimization framework.

The system goes way beyond traditional systems [6, 7] where typically only efficiency is addressed and the number of systems is small (solar and gas based thermal systems). Further, we address the development of methods for system identification and future time-based forecasting, using machine learning methods, and employ optimization based in simulated data, to control the water temperature in the swimming pool's tank. The control of the water temperature is set according to a setpoint as specified by the user. The idea employed explores the ability of the control system to make a proper decision on the required control signal.

The remainder of this work has the following structure. We start by addressing other works related to this work in Sect. 2. Next, a description on the problem is made in Sect. 3. Section 4 specifies the mathematical formulation of the problem. Section 5 presents the results of the three scenarios considered. Finally, Sect. 5 draws results and presents perspectives on future work.

2 State of the Art

The work in [8] proposes a new scheme for monitoring and controlling the swimming pool's quality (pH, chlorine, water level, temperature, water pressure) through a low-cost system based on wireless sensor networks. Despite having economic benefits by consuming less natural and material resources, it does not consider thermal comfort and does not include sustainable thermal systems.

In a similar manner, a web-based swimming pool information system was presented by Marais et al. [9]. Numerous elements of pool maintenance, for instance pH, chlorine levels and water level can be remotely monitored and further configured according to user-defined schedule.

One of the most popular heating technologies for outdoor swimming pools which is also environment friendly is based on PCM (Phase Change Material) storage tanks. In [10], Y Li and G Huang discussed their application to outdoor swimming pools and showed that they can bring out economic benefits by simply shifting electricity consumption from on-peak to off-peak periods. The water temperature regulation is made by ON-OFF control of a pump using a model of the swimming pool where the assumption that the temperature of the pool water was equal to the outlet temperature of the pool. The numerical analysis and simulations were performed within a platform that combined Matlab and TRNSYS.

In a more recent work [11], Y Li and G Huang show that performance of the PCM storage tank can further be improved by proposing a new approach where thermal comfort is regarded instead of using the outlet water temperature in the pool. As such, they have incorporated solar irradiation recorders, data logger, ultrasonic anemometer, temperature sensors, and other sensors to collect field data.

Other authors [6], analyzed the energy efficiency using a combined hot water system composed of solar thermal collectors and natural gas thermal power plant. They conclude that thermal energy sources using natural gas and solar energy remain the best solutions in terms of energy efficiency, low pollution and operating costs.

In [12], the authors use predictive control to show that indoor swimming pools water temperature's regulation can become more energy efficient when a hybridized solar + boiler system (possibly powered by biomass) is used as a thermal supply. Their results indicate that regulation based on predictive control can maintain pool thermal conditions while reducing energy demand. Moreover, they also show that this approach consumes less fuel when compared to traditional Proportional Integral Derivative (PID) control. This system is yet to be applied to outdoor swimming pools.

In [13], Dong et al., propose an integrated control system composed of a fuzzy automatic optimization algorithm and the Smith predictor compensator to adjust the temperature of pool water. The simulations show it can achieve good control effects for serious delay and serious inertia pool temperature control system.

With our proposal, the system not only is able to use more than the usual renewable systems but it also has the ability to address future time-based forecasting in conjunction with machine learning methods.

3 Problem Formulation

Typically, a dynamical system is affected by external stimuli. On one hand, there are the *inputs*, which are commonly associated with the external signals and can be manipulated by the observer. There are also *Disturbances*, which correspond to signals which can or cannot be measured. Of interest to the observer there are also the outputs. Figure 1 illustrates the relations between the *inputs*, disturbances and *outputs* of a typical dynamical system.

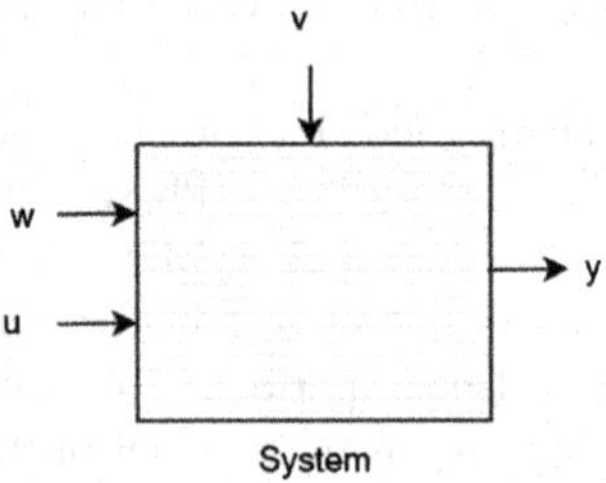

Fig. 1. System with output y, input(s) u, measured disturbance w and unmeasured disturbance v (adapted from [14]).

Generally speaking, a dynamical system is one that the future output value is related to the past inputs and disturbances according to some nonlinear function $f(y,u)$:

$$y(t) = f(\mathrm{y(t-1),\ y(t-2), \ldots y(t-n), u(t-1),\ u(t-2), \ldots} \\ \mathrm{\ldots u(t-m)}, w, v) \tag{1}$$

where n refers to the time instants or lags into the past for the output signal y and m the time instants into the past for the control signal u.

In its simplest description, the system considered in this work (see Fig. 2) is composed of several thermal sub-systems, which include: (1) the water tank (POOL), (2) Phase Change Material (PCM) accumulator, (3) solar collectors (flat-plate type SC), (4) Terrace Heat Exchanger (PC), and (5) Geothermal accumulator (PCGeo).

The various thermal sub-systems are connected by one or several ducts (one or more *inputs*), and monitoring valves which under activation promote water flow at a constant rate and preestablished direction regarding the retention valves setup. The complexity of the system is high because weather variables (seen as disturbances) condition the system response. Inclusion of weather variables as *inputs* is recommended as long as they can in some way be foreseen or predicted. In this work these data samples are considered available through WebApi requests to an online weather data server.

Due to multitude of sub-systems, several scenarios of operation (or *setups*) can be considered during the system operation. For example, to make water circulate through the solar collectors and the Pool, requires the operation of pump B_1 and the opening of valves V_2 and V_9. Another scenario includes water circulation from the solar collector into the duct system of the PCGeo sub-system.

Ultimately, the goal is to regulate the water temperature at the Pool. Readings the sensors T_i ($i = 1...12$) they allow the monitoring of the temperature in each of the subsystems and provide additional information to allow the control mechanism to decide which setup is the most effective at a given moment.

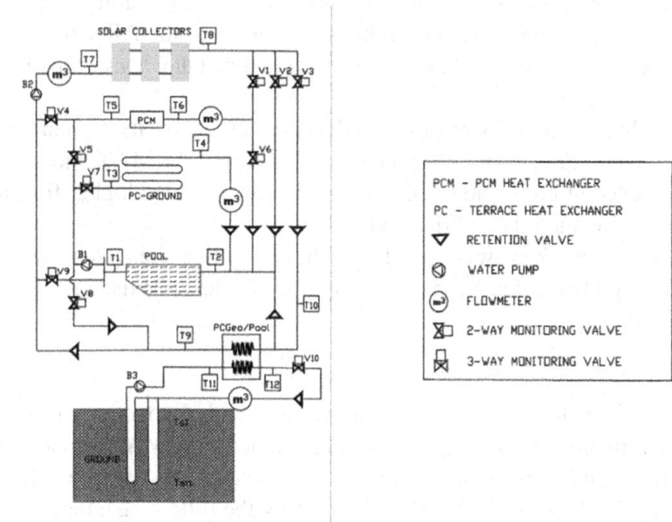

Fig. 2. Model of the Swimming Pool's water heating system using Renewable energy sources

4 Systems Identification

System identification concerns finding the appropriate model for a "real" system. This includes finding the appropriate inputs and estimating the appropriate parameters values for the model. When seen as a black-box, the utilization of the model becomes simple and provides alternative control techniques in order to identify the best control signal to apply at the input of the system.

Systems identification can be performed through the analysis of time-series records for each of the scenarios or setups previously mentioned. Appropriate algorithms help define a ARMA (Auto-Regressive-Moving-Average), ARIMA (Auto-Regressive-Integrated - Moving-Average), or SARIMA (Seasonal ARIMA with exogenous inputs) models.

The choice of an autoregressive model depends on the compromise between the simplicity of the model and the properties of a time series. If a time series is said to be stationary (i.e., if its properties are not affected by a change in the time source) the models chosen are normally the Autoregressive with Exogenous Input (ARX) and/or the Autoregressive Moving Average with Exogenous Input (ARMAX). ARMAX is more complex than ARX due to the fact that it has the ability to deal with stationary time series whose error regression is a linear combination. If a time series is non-stationary, the Autoregressive Integrated Moving Average with exogenous input (ARIMAX) and/or the

Seasonal Autoregressive Integrated Moving Average with exogenous input (SARIMAX) model can and should be used. Both models, ARIMAX and SARIMAX, are capable of handling both stationary and non-stationary series. However, if the time series has seasonal elements, the best option would be SARIMAX.

In [15], an analysis was made to forecast load demand in the context of smart grids, using ARX, Artificial Neural Networks (ANN) and Artificial Neural Networks optimized by Genetic Algorithm (ANN-GA). In this same analysis, the ARX presented a higher mean absolute percentage error but lower execution time but when compared to the ANN and ANN-GA solutions.

In [16], a hybrid model was developed to predict electricity demand as a function of outdoor temperature. It then was compared with the ARMAX model. Despite the good performance of both, and having the ARMAX presented higher forecast errors, it is simpler to define than the hybrid model.

In the current work, it was found that the modelling the system using ARMAX produced acceptable results. To this end, next we will describe the equations that define the ARMAX models.

ARMAX

Like ARX [16], Autoregressive Moving Average with Exogenous Input (ARMAX) includes additionally the moving average component. ARMAX modelling is, again, applied when a time series has regression characteristics, and the error is a linear combination [17]. The ARMAX model is ruled by the following equations:

$$\boldsymbol{\phi}(L)y(t) = \sum_{i=1}^{n} \boldsymbol{\beta}_i(L)\boldsymbol{u}_i(t) + \boldsymbol{\psi}(L)e(t) \tag{2}$$

with

$$\boldsymbol{\phi}(L) = 1 + \phi_1 L^{-1} + \cdots + \phi_p L^{-p} \tag{3}$$

$$\boldsymbol{\beta}_i(L) = \beta_{i1} + \beta_{i2} L^{-1} + \cdots + \beta_{ip} L^{-p+1} \tag{4}$$

$$\boldsymbol{\psi}(L) = 1 + \psi_1 L^{-1} + \cdots + \psi_q L^{-q} \tag{5}$$

where

- $\boldsymbol{\psi}(L)$ is the *moving average* component;
- q is the order for the *moving average* component;

and,

$$\boldsymbol{\phi}_t = [\phi_1, \phi_2, \ldots, \phi_p],^T \tag{6}$$

$$\boldsymbol{y}_t = [y(t-1), y(t-2), \ldots, y(t-p)], \tag{7}$$

$$\boldsymbol{\beta}_\mathrm{t} = \left[\beta_{11}, \beta_{12}, \ldots, \beta_{1p}, \beta_{21}, \beta_{22}, \ldots, \beta_{2p}, \beta_{\mathrm{n}1}, \beta_{\mathrm{n}2}, \ldots, \beta_{\mathrm{np}}\right],^T \tag{8}$$

$$\begin{aligned}&\left[u_{11}(t), u_{12}(t-1), \ldots, u_{1p}(t-p+1), u_{21}(t), u_{22}(t-1), \ldots\right.\\&\left.\ldots, u_{2p}(t-p+1), u_{n1}(t), u_{n2}(t-1), \ldots, u_{np}(t-p+1)\right]\end{aligned} \tag{9}$$

$$\boldsymbol{\psi} = \left[1, \psi_1, \psi_2, \ldots, \psi_q\right], \tag{10}$$

$$\boldsymbol{e}_t = \left[e(t), e(t-1), e(t-2), \ldots, e(t-q)\right]. \tag{11}$$

The coefficients of the $\boldsymbol{\phi}$, $\boldsymbol{\beta}$ and $\boldsymbol{\psi}$ polynomials are defined by Eqs. 9–11, and estimated using the CSS-MLE method [18].

4.1 Modelling the System

One reason for using a SMart system is to provide a forecast into a specified time step into the future depending on the actuation of the thermal system in action. Here, we assume that the variables influencing the model parameters are known (predictable and available) for the prediction horizon of the controlling scheme.

As the system is based on simulation, the TRNSYS 18.0 [19] simulation tool is applied for developing the mathematical models for the sub-systems that comprise the global system.

Since at a later stage every sub-system will be employed at a local testbed, the modelling approach requires using sampled data to perform model identification. In this sense, the simulated data is retrieved from the TRNSYS tool using a specified sampling period.

The model identification procedure carried on this work is given by algorithm 1. Its goal is to find an ARMAX model for each one of the sub-thermal systems.

Algorithm 1: System modelling

Input: *a matrix of the past sampled data for the n_inp input (includes three weather variables, wind velocity, ambient temperature, relative humidity and Pool input water temperature) and the one output sampled data, size of the training data (Train_size), size of the testing data (Test_size)*
Output: *the best representation of the system according to its forecasting performance on forecasting the testing data (Best_model)*

```
Generate a set of M ARMAX models with randomly chosen polynomials order
For each ARMAX model
    Estimate the polynomial parameters using the Maximum-LikeLihood-Estimator (MLE)
    Calculate the Sum-of-Squared-Error (SSE) and R² on the training data.
    Calculate the prediction values over the testing data
    Calculate SSE value for the predicted values (SSE_predict)
Best_model = the model with the lowest SSE_predict
return Best_model
```

Model Predictive Control

The typical control approach to swimming pool water temperature regulation is based on classic control methodologies where heat exchangers such as pumps or boilers are either switched ON or OFF.

In scenarios where the right conditions are met, the application of predictive control makes it possible to anticipate the control action in advance, regarding the correct identification of the system and, consequently, the model's ability to predict the system's operation for future instants. Typically, the methodology employed uses the concept of optimization for deciding on the effective control input value (u) to be applied to the system (see Fig. 3).

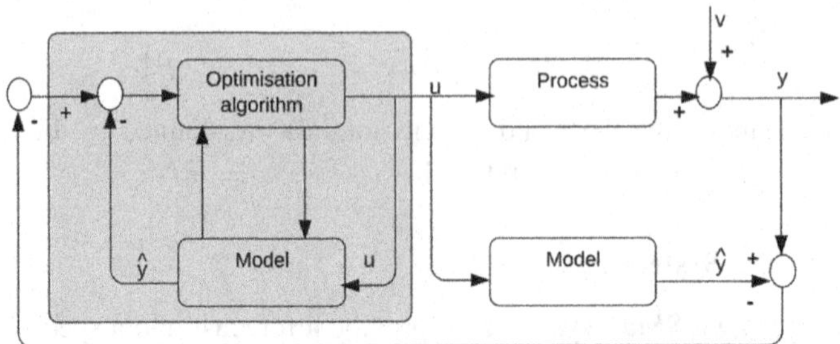

Fig. 3. Flow diagram for Model Predictive Control.

As shown in Fig. 4, using the past N_e sampled data an estimation of the best model is carried out. The model is then used to provide a sequence of control signals to the system for the prediction horizon, or *Np* future steps.

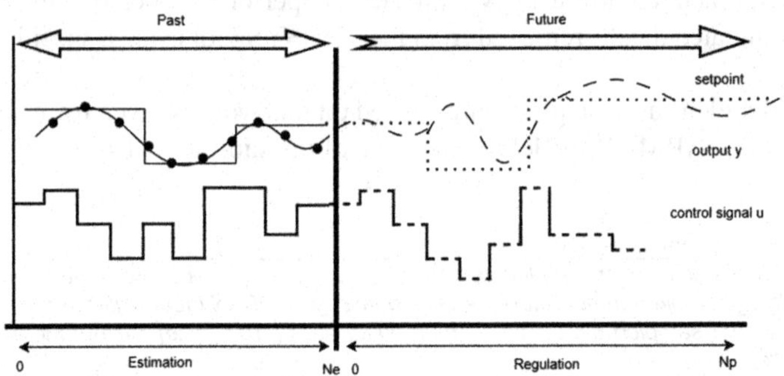

Fig. 4. System response based on generalized model predictive control.

Using an optimizer (based on a Genetic Algorithm) an optimal solution can be found, evaluating solutions that best satisfy the restrictions imposed and, in the end, selecting the best one. The optimizer considers as objective the minimization of Eq. 12:

$$\begin{cases} \sum_{i=1}^{Np} \left(setpoint - \mathrm{y}_i(u_i)\right)^2 \\ st \quad \lfloor u \rfloor \leq u_i \leq \lfloor u \rfloor,\, u \in \mathfrak{R} \end{cases} \tag{12}$$

where *u* translates the control signal, and *y* the output of the model, as shown in Fig. 1.

Under the aforementioned conditions, the control strategy to adopt must consider modelling each of the sub-systems that constitute the overall system, for each and every control scenario.

It follows that the renewable sources integrated in the system do not allow the immediate availability of the required control signals so an alternative approach must be adopted. So, from the standpoint of optimal control the system poses a drawback: not only each of the sub-systems have serious inertia (i.e., slow *input-output* responses), i.e. they cannot provide the required inlet water temperature in the Pool immediately (water flow regulation is possible but water temperature regulation is not) because they are highly dependent on weather conditions.

One of the most common approaches for water heating in swimming pools [6] uses a gas boiler. This type of action can be classified as Assured Effort. The boiler's output water temperature (Bout) is regulated and considered to be attainable instantly (within a pre-defined duration control step). This scenario is depicted by Fig. 5, left.

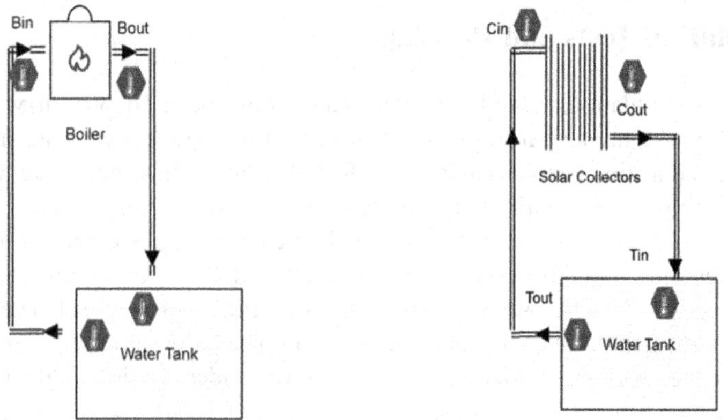

Fig. 5. The scenario considered for regulating the water temperature. Left: the boiler is used as the heating element. Right: a solar collector is used as heating element.

The type of service supported by the boiler is compared with the one supported by a solar collector of 50 m^2, represented in Fig. 5 right. The type of service provided by it could be considered as Best Effort, as it will not assure that the target temperature could be met, since it depends on weather conditions that are not controllable.

The core of the control system is based in an Energy Management System (EMS), with a Model Predictive Controller, which is responsible for finding the optimal value for the pool inlet water temperature (control signal) and according with the predicted values on the weather variables. In this sense, the system accounts for user preferences as it estimates the required set of water temperatures of the output of the boiler that will drive the swimming pool's water temperature to the desired setpoint. This scenario is accomplished with the help of an optimizer, using the Genetic Algorithm (GA).

In the following sub-section the genetic algorithm basics principles are explained.

4.2 The Genetic Algorithm for Controlling the Boiler System

The Genetic Algorithm was used to find the most appropriate set of future water temperatures for the boiler output, in a pre-defined prediction horizon. The fitness of the candidates was computed according with Eq. 12.

In this case, the encoding of the chromosome in the genetic algorithm consists of a list of real-valued genes where the i_{th} position element encodes the expected temperature value of the boiler output at the i_{th} future time step, for a pre-defined range of possible values. For instance, for a prediction horizon of 4 steps, and possible values in the range [4.0, 33.0], the i_{th} individual chromosome could be translated by:

4.0	15.0	23.4	33.0

As crossover a uniform operator using single point was used. mutation probability was based on a gaussian distribution. In summary, the default crossover and mutation operators were used. The termination criteria was the number of generations.

5 Simulation Tests and Results

To execute the simulations, sampled data (with a sampling period of 30 min) was obtained from the TRNSYS numerical simulation of the swimming pool. It was then divided into the training and testing parts. Weather records such as ambient temperature, wind speed and relative humidity were also collected and added to the training data.

The training data used a set of past values. In the following, *N* translates a parameter that indicates the number of past time steps used to define the training data. Initially different values of *N* were used to compare the variations of the model. The best value was then set constant. The testing data includes sampled values for the future time steps according to the prediction horizon considered, and considered being available at present time.

The performance of the models were evaluated using several metrics. The metrics were the Sum-of-Squared Errors (SSE), and Mean Relative Error percentage (MRE%). The first metric was also used by the search algorithms during model estimation, while the second metric is a key figure for comparing performance between solutions.

5.1 Results with the ARMAX Model

As it was pointed out before, the weather variables were used as exogenous inputs to the ARMAX model, and so, was the water temperature at the output of the collector (Cout in the right image of Fig. 5).

The period of the year used for testing was in the month of April (specifically in the second half of April) when the weather conditions traditionally are not adequate to raise the pool's water temperature above 20 ºC without introducing complementary equipment. Figure 6 represents the water temperature of the swimming pool, without solar collector versus using a solar collector.

Data measures (from the TRNSYS simulation) before 8 h of April the 28th, were used as training data. The following interval of 12 h was used as testing data (which correspond to 24 time steps of 30 min), i.e., predicting the system output from 8 h till 20 h on that day.

Table 1 shows the metrics obtained from a list of ARMAX models retrieved using the procedure described by algorithm 1 (Sect. 4.1) where M = 10, for several sizes of

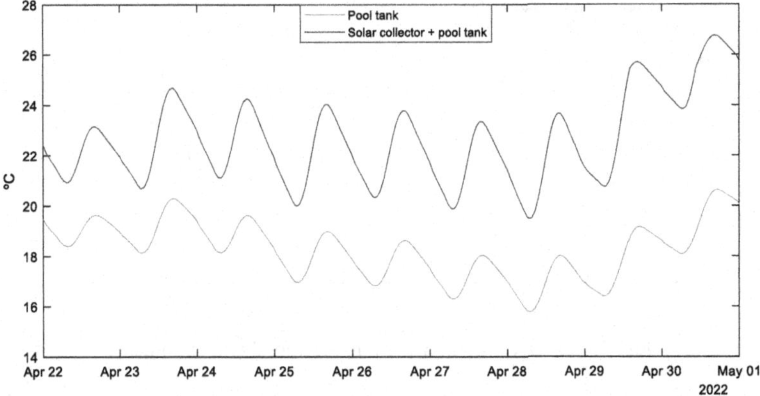

Fig. 6. Comparison of the water temperature of the swimming pool, without solar collector versus using a solar collector, between the 22nd and the 1st of May.

training data sets. As it can be observed the size of the training data plays a decisive role in the model's ability to give adequate predictions in the horizon considered. This trend is observed not only for the SSE value for the training data, but also for the SSE value on the testing data. It follows that the value for SSE is size dependent which is not the case for the MSE.

The best model is the one which presents a SSE value for the testing data of 9.86×10^{-2} which used a total training data set of $N = 320$. The prediction for 24 steps into the future (in steps of 30 min) is given in Fig. 7.

Then, to assess whether this model (the one with lowest SSE for testing data) is also capable of achieving accurate predictions, the model was used to make predictions for 24 future time steps beginning at other specific hours, according to Table 2. The results show that the MRE prediction error ranges from 0.086% to 0.24%.

Table 1. Models performance metrics for different training data sizes

	Polynomial orders $\{\phi, \beta_i, \psi\}$	SSE (training data)	SSE (testing data)	MRE % (testing data)	MSE
80	{5, [4 2 2 5],1}	5.97	5.98	1.96	1.23×10^{-4}
80	{10, [8 6 4 2],7}	7.45×10^{-1}	1.26	8.62×10^{-1}	6.41×10^{-5}
160	{9, [10 1 6 1],9}	1.19×10^{-1}	9.52×10^{-1}	8.03×10^{-1}	1.22×10^{-4}
160	{8 [2 9 10 6],9}	6.0×10^{-2}	1.48	9.06×10^{-1}	9.40×10^{-5}
320	**{6, [5 9 3 5],10}**	$\mathbf{1.72 \times 10^{-1}}$	$\mathbf{9.86 \times 10^{-2}}$	$\mathbf{2.44 \times 10^{-1}}$	$\mathbf{1.18 \times 10^{-5}}$
320	{5, [7 1 9 6],9}	1.86×10^{-2}	3.26×10^{-1}	4.11×10^{-1}	1.26×10^{-5}
1920	{5, [2 4 10 6],10}	2.82×10^{-2}	1.03	7.94×10^{-1}	2.38×10^{-4}
1920	{7, [8 5 7 2],10}	2.26×10^{-1}	9.14×10^{-1}	7.34×10^{-1}	2.41×10^{-4}

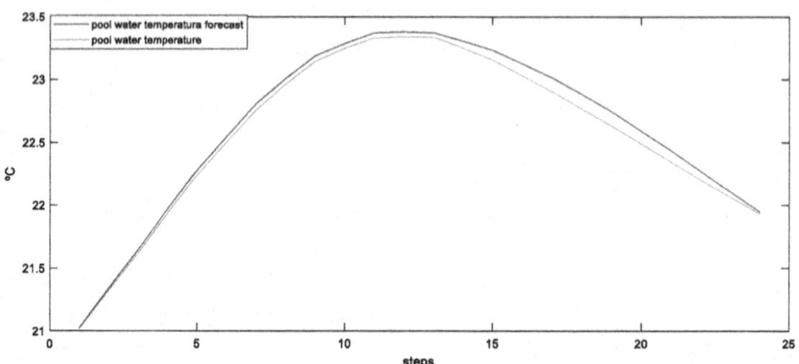

Fig. 7. ARMAX Pool Water temperature 24 steps forecast

Table 2. The best model SSE metric when used to predict 24 steps into the future, beginning at 8 h, 12 h, 15 h and 19 h into the 28th of April

Starting hours	MRE%
8 h (day and hour used for finding the model)	2.4×10^{-1}
12 h	8.6×10^{-2}
15 h	1.2×10^{-1}
19 h	6.4×10^{-2}

The three scenarios were then compared on April the 28th. In general, the system needs to decide at a given time instant, if it fully relies in the forecasted generation given by the solar collectors, or if it requires using the gas heater. In particular, and based on the difference between the highest predicted pool water temperature and the target setpoint, the system was made to decide that at 8 h it needed to regulate the water temperature at the boiler output to achieve the target temperature of 25 °C. To this aim it runs the genetic algorithm, and together with the model's predictions combined with the forecast weather data it estimates the required water temperature at the boiler output (Bout) minimizing Eq. 12, as discussed in Sect. 4.1 (case of assured effort).

Figure 8 shows how the three scenarios compare for a temperature setpoint of 25 °C for that particular day. A closer response to the setpoint temperature is expected in the assured effort case, as this is driven by the usage of the genetic algorithm capacity to find a suitable sequence of the boiler's output water temperature. Nevertheless, if sustainability is a priority, the best effort guarantees a good profile for the water temperature, although not always reaching the desired temperature, in these specific weather conditions.

The estimated temperature values of Bout (the required control signal of the system as illustrated in Fig. 4) defined by the GA optimizer for the 12 h prediction horizon (24 steps, starting at 8 h) are given in Fig. 9.

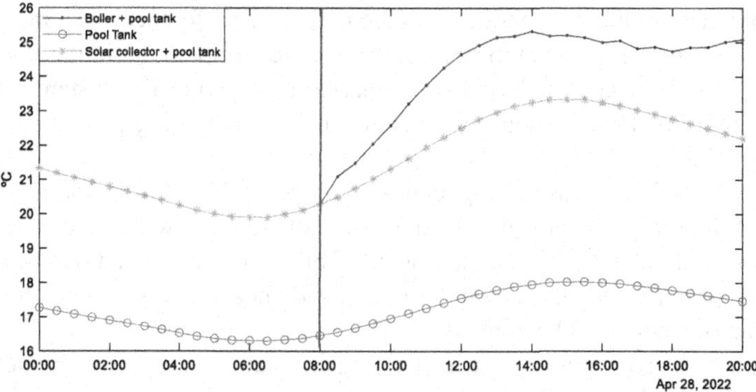

Fig. 8. Comparison between control based on assured effort (Boiler) with a MRE% value of 11.38%, best effort (Collector water temperature) with a MRE% value of 16.05% and no control (Pool water temperature) with a MRE% value of 45.65%.

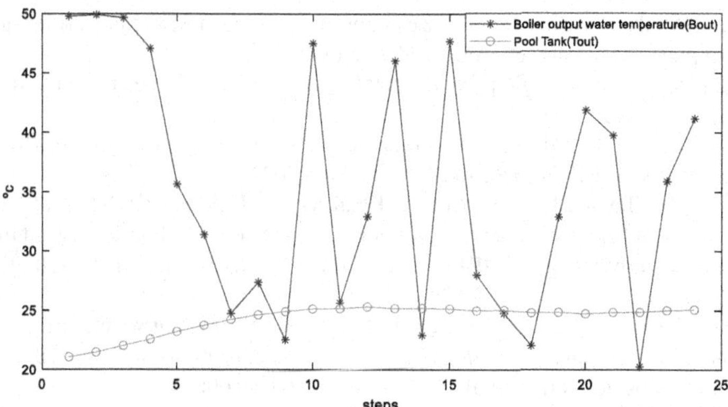

Fig. 9. Comparison between: (*) the estimated values of boiler output water temperature as a result of applying control based on assured effort and (o) the solar collectors output water temperature as a result of best effort.

6 Conclusions

This paper gives insight into a new scheme for setting up the swimming pool's water temperature through the usage of sustainable thermal systems (local renewable sources generation devices). Because of the complexity of the system, a list of several setups are considered beforehand and their application over time is set automatically according to their performance at a particular time range.

A simple setup case is explored where only the solar collectors are applied to control the pool's water temperature. The performance of this control system was compared to a setup where, traditionally, heating systems such as boiler are applied.

The results have shown that not always is the system able to satisfy the specifications set by the user but the inclusion of the renewable based system (solar collector) can make the system more eco-sustainable while ensuring the desired water temperature under very specific environmental conditions. When the user preferences become very strict the usage of the boiler can lead to an acceptable water temperature in the Pool.

This approach of control based on scenarios and having the system a degree of autonomy, requires lower human intervention in the swimming pool maintenance is gradually attained.

Future work will address the application of the SMART control approach where the selection of the setup scenario will be automatically defined within a pre-defined prediction horizon. In that case the model forecast ability here addressed will be prominent for defining which of the setups will be prioritized ahead as to guarantee that the best combination of setups will be applied.

It is the authors conviction that incorporating this methodology will provide resources savings for the final user in terms of financial and natural resources, contributing to a sustainable environment.

References

1. Official site of the European Union. https://european-union.europa.eu/priorities-and_actions/actions-topic/environment_en, Accessed 24 Nov 2022
2. Pordata. Consumo de energia primária: total e por tipo de fonte de energia. https://shorturl.at/jCE57. Accessed 24 Nov 2022
3. Solar Heating and & cooling programme, international energy agency. https://www.iea-shc.org/solar-heat-worldwide-2020. Accessed 03 Nov 2022
4. Abdunnabi, M., Tawil, I.H., Benabeid, M., Elhaj, M.A., Mohamed, F.: Design of solar powered space heating and domestic hot water system for libyan common house. In: 12th International Renewable Energy Congress (IREC), pp. 1–6 (2021). https://doi.org/10.1109/IREC52758.2021.9624800
5. Tyler, G., Harrison, T., Hulverson, R., Hristovski, K.: Estimating water, energy, and carbon footprints of residential swimming polls. In: Water Reclamation and Sustainability, pp. 343–359 (2014). https://doi.org/10.1016/b978-0-12-411645-0.00014-6
6. Popa, C.-D., Ungureanu, C.: Analysis of a hybrid water heating system for a swimming pool. In: 2021 International Conference on Electromechanical and Energy Systems (SIELMEN), pp. 521–526 (2021). https://doi.org/10.1109/SIELMEN53755.2021.9600361
7. Fahmy, F.H., Nafeh, A.A., Ahamed, N.M., Farghally, H.M.: A simulation model for predicting the performance of PV powered space heating system in Egypt, In: 2010 International Conference on Chemistry and Chemical Engineering, pp. 173–177 (2010). https://doi.org/10.1109/ICCCENG.2010.5560389
8. Simões, G., Dionísio, C., Glória, A., Sebastião, P., Souto, N.: Smart system for monitoring and control of swimming pools. In: 2019 IEEE 5th World Forum on Internet of Things (WF-IoT), Limerick, Ireland, pp. 829–832 (2019). https://doi.org/10.1109/WF-IoT.2019.8767240
9. Marais, J.M., Bhatt, D.V., Hancke, G.P., Ramotsoela, T.D.: A web-based swimming pool information and management system. In: 2016 IEEE 14th International Conference on Industrial Informatics (INDIN), Poitiers, France, pp. 980–985 (2016). https://doi.org/10.1109/INDIN.2016.7819304
10. Li, Y., Huang, G., Xu, T., Liu, X., Wu, H.: Optimal design of PCM thermal storage tank and its application for winter available open-air swimming pool. Appl. Energy **209**, 224–235 (2018)
11. Li, Y., Huang, G.: Temperature control of a PCM integrated open-air swimming pool in cold season: a numerical and experimental study. In: 2019 IOP Conference Series: Earth Environment Science, vol. 238, p. 012001 (2019). https://doi.org/10.1088/1755-1315/238/1/012001

12. Delgado Marín, J.P., Vera García, F., García Cascales, J.R.: Use of a predictive control to improve the energy efficiency in indoor swimming pools using solar thermal energy. Solar Energy **179**, 380–390 (2019). ISSN 0038–092X. https://doi.org/10.1016/j.solener.2019.01.004
13. Dong, Y., Yonghong, H., Gaohong, X.: Design of indoor swimming pool water temperature control system based on fuzzy controller and Smith predictor. In: Proceedings of 2011 International Conference on Electronic & Mechanical Engineering and Information Technology, Harbin, China, pp. 4678–4681 (2011). https://doi.org/10.1109/EMEIT.2011.6024079
14. Ljung, L.: System Identification, theory for the user. University of LinKöping Sweden (1987). ISBN 0-13-881640-9
15. Campos, B.P., da Silva, M.R.: Demand forecasting in residential distribution feeders in the context of smart grids. In: 12th IEEE International Conference on Industry Applications (INDUSCON), Curitiba, pp. 1–6 (2016). https://doi.org/10.1109/INDUSCON.2016.7874464
16. Awaludin, R.I., Rao, K.S.R.: Conventional ARX and artificial neural networks ARX models for prediction of oil consumption in Malaysia. In: 2009 IEEE Symposium on Industrial Electronics & Applications, Kuala Lumpur, pp. 23–28 (2009). https://doi.org/10.1109/ISIEA.2009.5356496[
17. Ruslan, F.A., Haron, K., Samad, A.M., Adnan, R.: Multiple input single output (MISO) ARX and ARMAX model of flood prediction system: case study Pahang. In: 2017 IEEE 13th International Colloquium on Signal Processing & its Applications (CSPA), Batu Ferringhi, pp. 179–184 (2017). https://doi.org/10.1109/CSPA.2017.8064947
18. Statsmodels.tsa.arima_model.ARMA.fit, Statsmodels. https://shorturl.at/CMSY1. Accessed 20 Jan 2019
19. Trnsys 18.0, Transient Simulation Tool Doc04-MathematicalReference; Doc07-Programmers Guide. Solar Energy Laboratory, University of Wisconsin-Madison

Solving Problems with Uncertainties

Distributionally-Robust Optimization for Sustainable Exploitation of the Infinite-Dimensional Superposition of Affine Processes with an Application to Fish Migration

Hidekazu Yoshioka[1](✉), Motoh Tsujimura[2], and Yumi Yoshioka[3]

[1] Japan Advanced Institute of Science and Technology, 1-1 Asahidai, Nomi, Japan
yoshih@jaist.ac.jp

[2] Doshisha University, Karasuma-Higashi-Iru, Imadegawa-dori, Kamigyo-Ku, Kyoto, Japan
mtsujimu@mail.doshisha.ac.jp

[3] Shimane University, Nishikawatsu-Cho 1060, Matsue, Japan
yyoshioka@life.shimane-u.ac.jp

Abstract. We consider a novel modeling and computational framework of the distributionally-robust optimization problem of jump-driven general affine process focusing on its application to inland fisheries as a case study. In particular, we consider an exploitation problem of migrating fish population as a major fishery issue. Our target process is the superposition of Ornstein–Uhlenbeck processes (supOU process) serving as a fundamental model of mean-reverting phenomena with (sub-)exponential memory. It is an affine process and admits a closed-form characteristic function being useful in applications. A theoretical novelty here is the assumption that the supOU process is allowed to have uncertain model parameter values as often encountered in engineering applications. Another novelty is the formulation of a long-term exploitation problem of the supOU process where the uncertainty is penalized through a generalized divergence between benchmark and distorted models. We present a strictly convex discretization of the optimization problem based on the model identified using the existing data of migrating fish population of a river in Japan. Further, the statistical analysis results in this paper are new by themselves. The computational results suggest the optimal harvesting policy of the fish population.

Keywords: Computational Optimization · supOU Processes · Generalized Divergence

1 Introduction

Computational optimization plays a vital role in the planning of resource and environmental management because the problems of interest are not always analytically solvable [1, 2]. Furthermore, in such problems, target dynamics (e.g., resource, environmental,

J. Mikyška et al. (Eds.): ICCS 2023, LNCS 14077, pp. 569–582, 2023.
https://doi.org/10.1007/978-3-031-36030-5_45

and biological dynamics) are often stochastic [3, 4] and hence the objective function to be minimized is given as some expectation. The expectation can then be evaluated numerically [5] although it may be computationally prohibitive if the underlying dynamics are complicated and high-dimensional.

Another potential issue encountered in the modelling and optimization related to resource and environmental management is that the target dynamics need to be identified using only a limited amount of data that is not necessarily accurate. The identified model therefore usually contains modelling errors emerging as incorrect parameter values and/or functional forms of coefficients [6, 7]. The model uncertainty should then be taken into account when considering an optimization problem using the identified model in a computationally feasible way.

In addition to the above-mentioned issues, processes of interest in resource and environmental dynamics have sub-exponential memory [8], postulating the use of a mathematical model that can capture this property. Volterra and mixed moving average processes are such candidates from both theoretical and engineering viewpoints [9, 10], while their computational optimization under model uncertainty has not been studied well to the best of the authors' knowledge.

Motivated by the issues reviewed above, the objectives of this paper are the formulation and application of a computational optimization approach for resource and environmental management under model uncertainty. We focus on the problem of exploiting a part of migrating fish population at a fixed point in a river reach of an anthropogenically-disturbed river system and then distribute them to other parts of the same river system to sustain the regional fish population [11]. This is a major problem in inland fisheries in Japan whose bottleneck is the optimization of the harvesting rate of the population. We show that the dynamics can be identified as the superposition of Ornstein–Uhlenbeck processes (supOU process) [12] whose characteristic function is available in a closed form. This useful property allows us to compute the probability density function (PDF) by a discrete Fourier transform.

Our optimization problem is to harvest the migrating fish population so that it is not excessively harvested because completely exploiting the population triggers a local extinction of the fish. The risk of the overexploitation is evaluated by a Conditional-Value-at-Risk (CVaR) measure [13], while the model uncertainty is penalized by a generalized (Tsallis) divergence [14]. The latter is a generalization of the Kullback–Leibler divergence serving as a more flexible mathematical tool.

The optimization problem is a feasible convex one once the PDF is given, which is father made strictly convex by regularizing the non-smoothness of the CVaR measure [15]; the strict convexity guarantees the unique solvability. We present computational examples using the real data [16] and analyze the impacts of model uncertainty on the distortion of the PDF and the optimal harvesting rate. Limitation and extendibility of the proposed computational approach are finally discussed. Our contribution thus covers both theory and application of a new computational optimization approach under uncertainty.

2 SupOU Process

2.1 Formulation

We formulate and review the supOU process following the literature [10, 12]. The supOU process is a superposition (i.e., integration) of infinitely many continuous-time OU processes using the jump-type random fields called Lévy bases [12]. These are tractable models of infinite-dimensional white noises having independent increments. It is known that Lévy bases associate equivalent representations based on Poisson random measures that are physically easier to understand. We exploit this fact and set the supOU process X_t at time t as

$$X_t = \int_0^{+\infty} \int_0^{+\infty} \int_{-\infty}^{t} \exp(-\rho(t-s)) z N(\mathrm{d}z, \mathrm{d}\rho, \mathrm{d}s), \quad t \in \mathbb{R}. \tag{1}$$

Here, N is a Poisson random measure on $(0, +\infty) \times (0, +\infty) \times \mathbb{R}$ with the compensator $\nu(\mathrm{d}z)\pi(\mathrm{d}\rho)\mathrm{d}s$, where π is a probability measure of a positive random variable, and ν is a Lévy measure of pure-jump Lévy processes having bounded-variations and positive jump sizes. The representation (1) is intuitive as the right-hand side can be formally understood as a limit of the superposition:

$$X_t \approx \sum_{i=1} \int_0^{+\infty} \int_{-\infty}^{t} \exp(-\rho_i(t-s)) z_i N_i(\mathrm{d}z_i, \mathrm{d}s) \tag{2}$$

with a suitable positive and strictly increasing sequence $\{\rho_i\}_{i=1,2,3,\ldots}$ and a family of mutually-independent Poisson random measures $\{N_i\}_{i=1,2,3,\ldots}$ on $(0, +\infty) \times \mathbb{R}$. Indeed, this approximation is theoretically justified in the sense of distribution [12]. This kind of discretization method is called Markovian lifts.

The supOU process (1) is assumed to be stationary that is why we do not explicitly specify an initial condition. The parameter $\rho > 0$ represents the reversion speed distributed according to the probability measure π. To guarantee the existence of the stochastic integral of (1), we need to assume that the reversion speed is not too much accumulated around the origin $\rho = 0$:

$$R \equiv \int_0^{+\infty} \rho^{-1} \pi(\mathrm{d}\rho) < +\infty. \tag{3}$$

Physically, R represents the dominant time-scale of the supOU process, and hence it is a key quantity for better understanding its dynamics. The condition (3) guarantees the existence of the right-hand side of (1) as well as the stationarity and almost sure positivity of X_t [12, 17]. We will show that this condition is harmless in practice.

2.2 Statistical Properties

An advantage of the supOU process is that its characteristic function is obtained analytically. The characteristic function $c(\xi)$ is given by [10]

$$c(\xi) = \mathbb{E}\left[e^{\mathrm{i}\xi X_t}\right] = \exp\left(R \int_0^{+\infty} \int_0^{+\infty} \left(\exp(\mathrm{i}\xi z e^{-t}) - 1\right) \nu(\mathrm{d}z)\mathrm{d}t\right), \quad \xi \in \mathbb{R} \tag{4}$$

with the imaginary unit i ($i^2 = -1$) and a suitable expectation $\mathbb{E}$. The moment generating function, if necessary, is formally obtained as $m(\xi) = c(-i\xi)$. Hereafter, we assume the Lévy measure of the form $\nu(dz) = Az^{-(\alpha_\nu+1)}\exp(-\beta_\nu z)dz$ ($A > 0$, $\alpha_\nu < 1$, $\beta_\nu > 0$) that covers the cases considered in our application. More general ν can also be utilized when preferred.

We close this section by presenting two remarks. Firstly, the closed-form availability of the characteristic function (4) allows for computing the corresponding PDF via a discrete Fourier transform [18]. We can therefore avoid computing expectations of the supOU process without resorting to a time-consuming statistical method like a Monte-Carlo method. Indeed, the infinite-dimensionality implied in -(2) suggests that we need to design a Monte-Carlo method to generate a sufficiently large number of paths of a high-dimensional system for evaluating an expectation, which would be computationally expensive and maybe infeasible. At the same time, key statistics, such as mean, variance, and skewness can also be obtained from (4). Secondly, the autocorrelation function that can handle both exponential and sub-exponentially decaying cases is obtained in a closed-form, which becomes fully-explicit if we assume Gamma-type π complying with the condition (3). We effectively use these properties in the application.

3 Optimization Problem

3.1 Generalized Divergence

We define a generalized divergence before going to the formulation of the optimization problem. Given two equivalent positive PDFs p, r of a positive random variable Z, the generalized (Tsallis) divergence $D_q(r|p)$ from p to r is set as [14]:

$$D(r|p) \equiv \int_0^{+\infty} \frac{1}{1-q}\left(-(\omega(Z))^q + q\omega(Z) + 1 - q\right)p(Z)dZ, \tag{5}$$

where $q \in (0, 1]$ with the Radon–Nikodým derivative $\omega = r/p$. If $q = 1$, then (5) is understood as the classical Kullback–Leibler divergence

$$D(r|p) \equiv \int_0^{+\infty} (\omega(Z)\ln\omega(Z) - \omega(Z) + 1)p(Z)dZ. \tag{6}$$

In both cases, we have $D(r|p) \geq 0$ and $D(r|p) = 0$ if and only if $p = r$, and that $D(p|r) \neq D(r|p)$ in general due to the asymmetry implied in (6). The divergence is therefore an index that can measure the probabilistic difference (i.e., model uncertainty) between two equivalent PDFs but is not a metric. The uncertainty is then evaluated by considering p as the benchmark model and r as a distorted model. A remark between the two cases (5)- is that the integrand as a function of the Radon–Nikodým derivative $\omega(Z)$ is more strongly convex as well as becomes larger (i.e., has a sharper profile) for larger q. It implies that the model uncertainty in our context is evaluated to be larger for larger q. Hence, the Kullback–Leibler divergence is an extreme case of the generalized divergence. There is no model uncertainty if $\omega = 1$ almost everywhere.

For later use, define the q-exponential and q-logarithm functions:

$$\exp_q(x) = (1 + (1-q)x)^{\frac{1}{1-q}} \text{ and } \ln_q(x) = \frac{x^{1-q} - x}{1-q},\ x \geq 0. \tag{7}$$

We conventionally set $\exp_1(x) = \exp(x)$ and $\ln_1(x) = \ln x$.

3.2 Problem Formulation

The optimization problem here is to determine the harvesting rate $h = h(X_t)$ $(0 \leq h \leq 1)$ as a bounded measurable function of the observed process X_t now considered as a unit-time population of migrating fish species. The set of such functions is denoted as $\mathfrak{H}$. The fish should be harvested as large as possible by the decision-maker, a fishery cooperative, while fully exploiting them should be avoided to prevent a local extinction. The optimization problem without model uncertainty is then set as

$$\text{Find } \inf_{h \in \mathfrak{H}} J(h) \text{ with } J(h) = \mathbb{E}\big[g(1-h)\big] - \lambda \text{CVaR}_{\alpha}((1-h)X). \tag{8}$$

Here, the expectation is based on the PDF p of the stationary supOU process. The first term of J with a non-negative, uniformly bounded, and strictly convex function g represents the averaged harvesting rate. The second term represents the risk of harvesting the migrating fish population when it becomes small. Here, we use the upper CVaR measure with the weight $\lambda > 0$ and the quantile $\alpha \in (0, 1)$ for generic non-negative random variable Z

$$\text{CVaR}_{\alpha}(Z) = \frac{\mathbb{E}[Z\mathbb{I}(Z \leq Z_{\alpha})]}{\mathbb{E}[\mathbb{I}(Z \leq Z_{\alpha})]} = -\inf_{u \geq 0}\left\{-u + \frac{1}{\alpha}\mathbb{E}[\max\{u - Z, 0\}]\right\} \tag{9}$$

with the indicator function $\mathbb{I}(Z \leq Z_{\alpha})$ of the set $\{Z \leq Z_{\alpha}\}$ ($\mathbb{I}(Z \leq Z_{\alpha}) = 1$ if $Z \leq Z_{\alpha}$ and is 0 otherwise) and the quantile value Z_{α} such that $\mathbb{E}[\mathbb{I}(Z \leq Z_{\alpha})] = \alpha$. The right-most side of (9) is the optimization-oriented dual formula [13]. The risk-aversion of the decision-maker becomes stronger as the penalization (λ) increases or the quantile level (α) decreases.

The problem (8) is a convex optimization problem with the objective being lower-semicontinuous in a Hilbert space of bounded functions with the norm $\|\cdot\|$ given for generic $f : (0, +\infty) \to \mathbb{R}$ by

$$\|f\|^2 = \int_0^{+\infty} f^2(X)p(X)\mathrm{d}X. \tag{10}$$

Owing to the convexity, the problem is feasible.

We extend the problem (8) so that the model uncertainty can be evaluated in the context of distributionally-robust optimization [e.g., 19]. The distributionally-robust problem assumes that the expectation $\mathbb{E}$ is distorted and evaluated by r equivalent to p but not necessarily by p. Given a PDF p, the admissible set of positive measurable functions ω such that $\int_0^{+\infty} \omega(Z)p(Z)\mathrm{d}Z = 1$ is denoted as $\mathfrak{W}$.

We formulate the distributionally-robust optimization problem as follows:

$$\text{Find } W = \inf_{h \in \mathfrak{H}} \sup_{\omega \in \mathfrak{W}} J(h, \omega), \tag{11}$$

$$J(h, \omega) = \mathbb{E}_q\big[g(1-h)\big] - \lambda \text{CVaR}_{\alpha}((1-h)X) - \mu D(\omega p|p), \tag{12}$$

where the expectation $\mathbb{E}_q$ is such that $\mathbb{E}_q[\cdot] = \mathbb{E}\big[\omega^q \cdot\big]$ to consistently reformulate the problem under uncertainty in the sense of Tsallis [20]. The last term of (12) penalizes the model ambiguity with another weight $\mu > 0$ in a way that larger model uncertainty and hence larger divergence $D(\omega p|p)$ is allowed for smaller μ. The case $\mu \to +\infty$ thus corresponds to the benchmark case (8).

The problem (11) is a min-max problem but can be rewritten as a minimization problem as shown below. Considering the dual representation formula, given $h \in \mathfrak{H}$, the inner maximization is achieved by

$$\omega^*(X) = \exp_q\left(\frac{F(X,u,h)}{\mu}\right)\left(\int \exp_q\left(\frac{F(X,u,h)}{\mu}\right)p(X)\mathrm{d}X\right)^{-1} \in \mathfrak{W} \tag{13}$$

with F given by

$$(0 \le) F(X,u,h) = g(1-h) + \frac{\lambda}{\alpha}\max\{u-(1-h(X))X, 0\} (< +\infty). \tag{14}$$

Plugging (14) into (11) yields

$$W = \inf_{u \ge 0, h \in \mathfrak{H}} \left\{-\lambda u + \mu \ln_q\left(\int \exp_q\left(\frac{F(X,u,h(X))}{\mu}\right)p(X)\mathrm{d}X\right)\right\}. \tag{15}$$

This is a convex optimization problem because F of (14) is convex with respect to all the arguments and the second term of the right-hand side of (15) defines a convex certainty equivalent of the expectation $\int F(X,u,h)p(X)\mathrm{d}X$ [Theorem 5.1 of 21]. Further, we can replace the range $[0,+\infty)$ of the auxiliary decision variable u by a sufficiently large compact set $[0,U]$ with some constant $U > 0$. This is proven by a simple contradiction argument [Proposition 4 of 19]. Therefore, this distributionally-robust optimization problem is feasible.

3.3 Regularization of CVaR

The problem (15) is convex but not necessarily strictly convex, which is due to the non-smoothness of the "max" function in the CVaR measure. We therefore regularize "max" function as the following strictly convex one with a small $\tau > 0$ [15]:

$$\max_\tau(x) = \tfrac{x+\sqrt{x^2+4\tau^2}}{2} \text{ for } x \ge 0, \tag{16}$$

satisfying

$$\max\{0,x\} \le \max_\tau(x) \le \max\{0,x\} + \tau \text{ for } x \ge 0. \tag{17}$$

The difference between max and $\max_\tau$ functions is at most τ, and can be made arbitrary small by choosing a suitably small τ.

Now, the regularized distributionally-robust optimization problem reads

$$\text{Find } W_\tau = \inf_{u \ge 0, h \in \mathfrak{H}} \left\{-\lambda u + \mu \ln_q\left(\int \exp_q\left(\frac{F_\tau(X,u,h(X))}{\mu}\right)p(X)\mathrm{d}X\right)\right\} \tag{18}$$

with F_τ given by

$$F_\tau(X,u,h) = g(1-h) + \frac{\lambda}{\alpha}\max_\tau\{u-(1-h(X))X, 0\}. \tag{19}$$

It is straightforward to show that the problem (18) is strictly convex, and that the range of the auxiliary decision variable u can be made compact without changing the value of W. Consequently, the problem (18) admits a unique minimizing pair $u \geq 0$, $h \in \mathfrak{H}$. The corresponding worst-case Radon–Nikodým derivative ω^* is then found as (13) with F replaced by F_τ. The regularized problem is therefore feasible and uniquely solvable. We numerically compute it with a small $\tau > 0$ as explained below.

3.4 Numerical Discretization

The regularized problem (18) is discretized through a replacement of the PDF p by its empirical version p_N ($N \in \mathbb{N}$) on a uniform grid, which is formally given by

$$p_N(X)\mathrm{d}X \approx \sum_{i=1}^{N} p(i)\delta(X - X_i),\ \sum_{i=1}^{N} p(i) = 1\,, \tag{20}$$

with the Dirac's delta $\delta(\cdot)$, a positive sequence $\{p(i)\}_{1 \leq i \leq N}$ serving as discrete probabilistic weights, and a non-negative and strictly increasing sequence $\{X_i\}_{1 \leq i \leq N}$ as representative points at which the process is evaluated. The harvesting rate h and the Radon–Nikodým derivative ω are accordingly discretized on the same grid. The discretization of the regularized problem is then set as

$$\text{Find } W_{\tau,N} = \inf_{u \geq 0, h \in \mathfrak{H}} \left\{ -\lambda u + \mu \ln_q \left(\int \exp_q \left(\frac{F_\tau(X, u, h(X))}{\mu} \right) p_N(X)\mathrm{d}X \right) \right\}, \tag{21}$$

with the worst-case Radon–Nikodým derivative ω^* given by (13) with F replaced by F_τ and p by p_N.

The computational procedure of the problem (21) in practice is as follows. Firstly, we identify the supOU process, namely the measures π and v, from some time series data. Then, the characteristic function (4) is Fourier inverted on a sufficiently fine uniform grid to compute p_N following Hainaut [18]. The size and spacing of the grid are chosen sufficiently large and fine so that the computed PDF is entirely positive (i.e., no or at most small Gibbs oscillations). The minimizing pair u, h is computed where the former is a real constant and the latter is distributed at each grid point. This process is carried out by using a gradient descent method with an inertia [22], but other optimization methods can be utilized if preferred. Finally, the worst-case Radon–Nikodým derivative is accordingly obtained by the optimized pair u, h.

4 Application

4.1 Study Site

The study site of this paper is the weir called Meiji-yosui irrigation head works located at the midstream of Yahagi River pouring to Pacific Ocean, Aichi Prefecture, Japan (35° 02′ 52″ E 137° 10′ 43″). This weir is an observation point of the spring upstream migration of the fish ayu *Plecoglossus altivelis altivelis* from Pacific Ocean to the river [16]. *P. altivelis* is one of the most important inland fishery resources in Japan from ecological, fisheries, and cultural standpoints [23]. Recent river development critically degraded habitat and food availability for the fish, thereby their population as well as catches have been decreasing. In some case, a part of the migrating fish population is harvested at a point in a river reach and distributed to the other reaches to prevent them from extinctions [11]. Such a project is usually planned and executed by a local inland fishery cooperative. When and how much of the fish should be harvested for this purpose has been a crucial issue for inland fishery cooperatives in Japan.

The supOU process is fitted against the data of each year based on the verified least-squares approach [10, 17] that firstly fits the probability measure π by comparing theoretical and empirical autocorrelation function and then the Lévy measure v by using mean, standard deviation, and skewness. We assume the Gamma-type $\pi(\mathrm{d}\rho) \propto \rho^{\alpha_\rho - 1}\exp(-\rho/\beta_\rho)\mathrm{d}\rho$ ($\alpha_\rho > 1$, $\beta_\rho > 0$). Then, the autocorrelation of the supOU process with lag $s \geq 0$ is sub-exponential given by $\left(1+\beta_\rho s\right)^{1-\alpha_\rho}$, which approaches to the exponential case as $\alpha_\rho \to +\infty$ with fixed $\beta_\rho\left(\alpha_\rho - 1\right)$. We need $\alpha_\rho > 1$ by (3). We set $\alpha_v = -2$ as it has preliminary been found to effectively work.

The collected data of daily migrating populations of the *P. altivelis* from 2010 to 2020 is available in the recent report [16]. This kind of fine and large amount of data is rarely available, which is why we have chosen the present study site. Figure 1 plots the reported daily migrating population in each year. Table 1 shows the total, mean, standard deviation, skewness, and kurtosis of the time series data in each year. Hereafter, we normalize X_t in each year by the total migrated population without significant loss of generality. Table 2 shows the fitted parameter values for the data and the time-scale R of each year for self-centeredness of the paper. Indeed, this kind of statistical results against the time series data, namely Tables 1 and 2, are unique contributions by themselves.

Tables 1 and 2 show that the time series data is probabilistically positively skewed and has sharper PDFs than Gaussian ones. In addition, the yearly difference of the memory structure is found to be significantly different; α ranges from $O(10^0)$ (polynomial) to $O(10^4)$ (almost exponential). The relative errors of the statistics are on average 0.014 (mean), 0.034 (standard deviation), 0.011 (skewness), and 0.0084 (kurtosis). Hence, the supOU process can capture the wide range of statistical behavior of the migrating fish population dynamics. The time scale R ranges from $O(10^{-1})$ (day) to $O(10^1)$ (day), suggesting that the decaying speed of peaks are broadly different among different years. Interestingly, the total migrated population is significantly different among different years. Biological understanding of this large difference is beyond the scope of this paper, but will be an interesting research topic to be resolved in the future.

4.2 Computational Results and Discussion

We present demonstrative computational examples of the regularized distributionally-robust optimization problem for years 2018 and 2019. The year 2018 has a relatively longer memory with α_ρ closer to 2 than the other years and the PDF in 2018 is more skewed as well as sharper than 2019 (See, Tables 1 and 2). The PDF was computed using the discrete Fourier transform with the space increment of 1/2,500 and the degree-of-freedom $N = 2,500$. We needed a post-processing using an exponential spectral filter [23] to obtain oscillation-free PDF profiles. This procedure can be considered as a part of

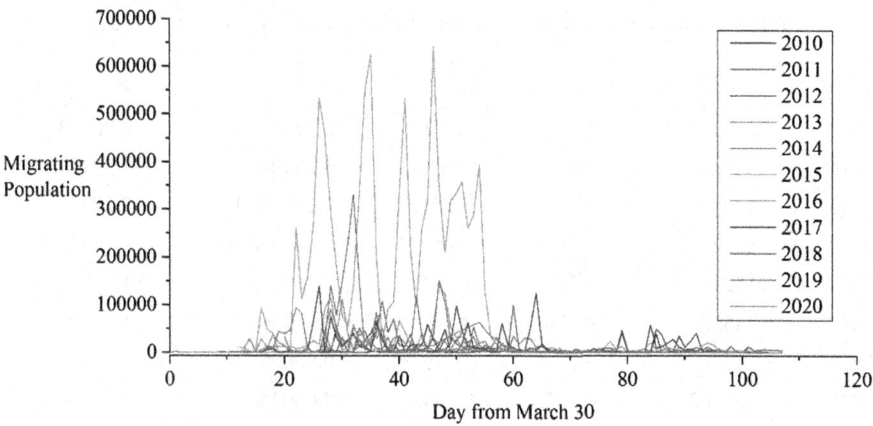

Fig. 1. Reported time series data of the daily migrating population of *P. altivelis* in each year.

Table 1. Total, mean, standard deviation, skewness, and kurtosis of the migrating population. The statistics are computed for the supOU process X normalized (divided) by the total.

Year	Total	Mean	Standard deviation	Skewness	Kurtosis
2010	487,951	0.0118	0.0205	3.39	12.5
2011	985,637	0.0106	0.0184	3.32	12.8
2012	761,990	0.0126	0.0141	2.30	4.94
2013	839,587	0.00741	0.0187	4.62	30.4
2014	601,147	0.0114	0.0215	3.47	15.8
2015	1,276,048	0.0124	0.0185	2.91	9.12
2016	10,030,840	0.0133	0.00890	1.57	1.69
2017	1,440,609	0.0133	0.0176	2.55	7.37
2018	2,307,520	0.0134	0.0223	3.09	11.7
2019	447,134	0.0185	0.0230	2.35	6.67
2020	1,103,486	0.0138	0.0128	1.85	3.44

model uncertainty although its influences would be small in our case. The regularization parameter is $\tau = 0.00000125$ that has been found to be sufficiently small. We report that if $\tau = 0$ then the gradient descent fails to converge due to discontinuous variational derivatives. We set the nominal parameter values $q = 0.5$, $\mu = 10$, and $\alpha = 0.30$ unless otherwise specified. All the decision variables below are obtained from (21). All the solutions have been obtained without numerical instabilities.

Firstly, we analyze the harvesting rate. Figure 2 compares the computed harvesting rates in 2018 and 2019 as a function of the (normalized) migrating population X and the weight λ. The optimal harvesting rate is 1 at $X = 0$, while it does not play a role in practice because no fish can be harvested in such as case. For $X > 0$, the harvesting

Table 2. Fitted parameter values of the supOU process. The time unit is day.

Year	A	β_V	α_ρ	β_ρ	R
2010	1553.5	42.037	2.3720	2.5778	0.28273
2011	344.85	47.009	3.4618	0.25484	1.5940
2012	2375.1	95.032	3,754.2	0.00011700	2.2774
2013	80.293	31.808	25,577	0.000026310	1.4861
2014	162.09	36.756	12.838	0.048555	1.7397
2015	545.64	54.224	3.5410	0.217560	1.8089
2016	13,781	251.33	2,270.0	0.000057746	7.6321
2017	2,424.2	64.165	5,334.1	0.00025953	0.72249
2018	265.73	40.360	2.8859	0.32082	1.6528
2019	5,288.0	52.455	3.2520	1.7553	0.25298
2020	4,005.6	125.28	63,896	0.0000046289	3.3810

rate is increasing in the migrating population. No fish should be harvested for relatively small positive X, which appears due to the use of the CVaR term effectively penalizing the local extinction. This non-harvesting area in the figure contracts as the weight λ and hence the extinction risk increases. This observation suggests that the decision-maker should not be too much afraid of the risk of local extinction for the global minimization of the objective (12). The impacts of the risk-aversion are concentrated more on the small population for the sharper PDF in 2018. The analysis below focuses on the year 2019 as the impacts of the uncertainty are more visible for a wider range of the population.

Secondly, we analyze the uncertainty and its influences on the distortion of the PDF. Figure 3 compares the computed worst-case uncertainties, the Radon–Nikodým derivatives, for the nominal $\mu = 10$ and smaller $\mu = 1$ potentially allowing for larger uncertainty. Figure 4 then compares the corresponding computed PDFs. Larger model uncertainty leads to the worst-case Radon–Nikodým derivatives and PDFs be more concentrated on small X, thereby underestimating the mean population.

Finally, we analyze impacts of the parameter q in the generalized divergence. Figure 5 shows the computed harvesting rates for $q = 0.25$ and $q = 1$. It is shown that the use of the sharper divergence (larger q) results in less conservative harvesting rate having smaller range of no-harvesting. This is due to that the Radon–Nikodým derivative is restricted to be smaller for the sharper divergence.

Consequently, the decision-maker can design his/her harvesting strategy flexibly as demonstrated in this paper.

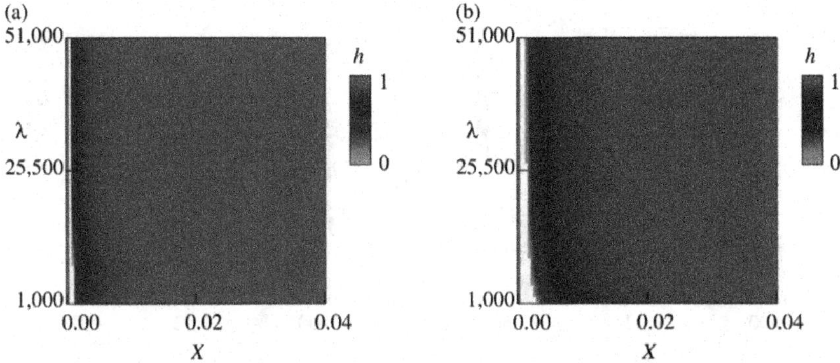

Fig. 2. Harvesting rate in (a) 2018 and (b) 2019. $h = 0$ is optimal in the white area.

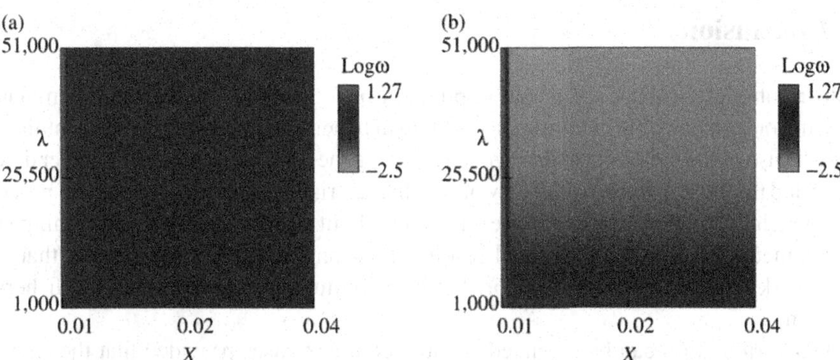

Fig. 3. Worst-case ambiguity in 2019 with (a) nominal and (b) larger uncertainty.

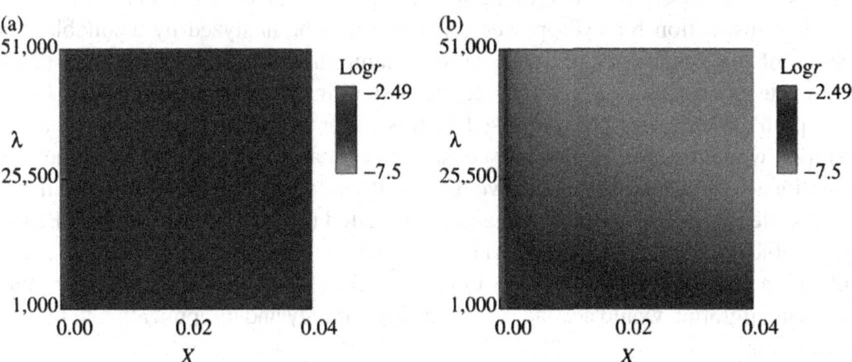

Fig. 4. Worst-case PDF in 2019 with (a) nominal and (b) larger uncertainty.

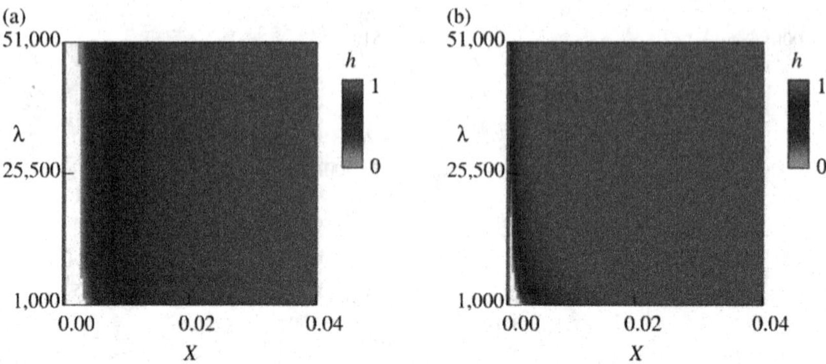

Fig. 5. Harvesting rate in 2019 with (a) $q = 0.25$ and (b) $q = 1$. Compare with Fig. 2.

5 Conclusion

A distributionally-robust optimization problem of the supOU process was formulated focusing on a sustainable exploitation problem in fisheries. Key points in our formulation were the use of the generalized divergence covering the classical Kullback–Leibler divergence and the closed-form availability of the characteristic function of the supOU process with which the expectations can be evaluated without resorting to time-consuming statistical methods. The computational results based on the real data suggested that our framework can be utilized to support the decision-making for resolving the fisheries problem.

Our formulation can be extended to more complex cases provided that the characteristic function of the target process is accessible in a closed form. Affine stochastic Volterra processes [25] and self-exciting affine processes [10] would be such examples. In this view, a variety of modern engineering issues related to sustainability such as the water abstraction for hydropower generation can be analyzed by a suitable modification of our framework. We focused on a jump-driven process, while considering jump-diffusion processes would not encounter significant technical difficulties.

A problem that was not addressed in this paper is multi-stage problems as fish migration would be more reasonably considered as seasonal population dynamics. The (conditional) characteristic function would be still available in a closed-form even in such cases, while our preliminary investigations suggested that computing transient PDFs is highly subject to Gibbs oscillations. The oscillations will be mitigating by a more careful application of a spectral filter [24]. From a biological standpoint, better understanding of the fish migration would advance the modeling strategy and its application to fisheries optimization.

Acknowledgements. This work was supported by Japan Society for the Promotion of Science (22K14441, 22H02456) and Grant from MLIT Japan (B4R202002).

References

1. Ullah, G.W., Nehring, M.: A multi-objective mathematical model of a water management problem with environmental impacts: an application in an irrigation project. PLoS ONE **16**(8), e0255441 (2021)
2. Boddiford, A.N., Kaufman, D.E., Skipper, D.E., Uhan, N.A.: Approximating a linear multiplicative objective in watershed management optimization. Eur. J. Oper. Res. **305**(2), 547–561 (2023)
3. Buser-Young, J.Z., Peck, E.K., Chace, P., Lapham, L.L., Vizza, C., Colwell, F.S.: Biogeochemical dynamics of a glaciated high-latitude wetland. J. Geophys. Res. Biogeosci. **127**(6), e2021JG006584 (2022)
4. Huang, Y., Insley, M.: The impact of water conservation regulations on mining firms: a stochastic control approach. Water Res. Econ. **36**, 100185 (2021)
5. Pichler, A., Xu, H.: Quantitative stability analysis for minimax distributionally robust risk optimization. Math. Program. **1191**, 47–77 (2018)
6. Doser, J.W., Leuenberger, W., Sillett, T.S., Hallworth, M.T., Zipkin, E.F.: Integrated community occupancy models: a framework to assess occurrence and biodiversity dynamics using multiple data sources. Methods Ecol. Evol. **13**(4), 919–932 (2022)
7. Wu, X., Marshall, L., Sharma, A.: Incorporating multiple observational uncertainties in water quality model calibration. Hydrol. Process. **36**(1), e14452 (2022)
8. Beran, J., Feng, Y., Ghosh, S., Kulik, R.: Long-Memory Processes, pp. 1–106. Springer, Heidelberg (2013). https://doi.org/10.1007/978-3-642-35512-7
9. Hamaguchi, Y.: Variation of constants formulae for forward and backward stochastic Volterra integral equations. J. Differ. Equ. **343**, 332–389 (2023)
10. Yoshioka, H., Tanaka, T., Yoshioka, Y., Hashiguchi, A.: Stochastic optimization of a mixed moving average process for controlling non-Markovian streamflow environments. Appl. Math. Model. **116**, 490–509 (2023)
11. Yoshioka, H., Tanaka, T., Aranishi, F., Tsujimura, M., Yoshioka, Y.: Impulsive fishery resource transporting strategies based on an open-ended stochastic growth model having a latent variable. Math. Methods Appl. Sci. (2021). https://doi.org/10.1002/mma.7982
12. Barndorff-Nielsen, O.E., Stelzer, R.: Multivariate supOU processes. Ann. Appl. Probab. **21**(1), 140–182 (2011)
13. Rockafellar, R.T., Uryasev, S.: Conditional value-at-risk for general loss distributions. J. Bank. Finan. **26**(7), 1443–1471 (2002)
14. Tsallis, C.: Possible generalization of Boltzmann-Gibbs statistics. J. Stat. Phys. **52**(1), 479–487 (1988)
15. Luna, J.P., Sagastizábal, C., Solodov, M.: An approximation scheme for a class of risk-averse stochastic equilibrium problems. Math. Program. **157**(2), 451–481 (2016). https://doi.org/10.1007/s10107-016-0988-4
16. Yamamoto, T., Yamamoto, D., Nagatomo, M.: Data on numbers of ayu Plecoglossus altivelis ascending the Yahagi River at Meiji-yousui irrigation head works (2010–2020). Yahagi River Res. **26**, 13–18 (2021). in Japanese
17. Yoshioka, H.: Fitting a superposition of Ornstein-Uhlenbeck process to time series of discharge in a perennial river environment. ANZIAM J. **63**, C84–C96 (2021)
18. Hainaut, D.: Pricing of spread and exchange options in a rough jump–diffusion market. J. Comput. Appl. Math. **419**, 114752 (2023)
19. Liu, W., Yang, L., Yu, B.: Kernel density estimation based distributionally robust mean-CVaR portfolio optimization. J. Global Optim. **84**(4), 1053–1077 (2022)
20. Ma, H., Tian, D.: Generalized entropic risk measures and related BSDEs. Statist. Probab. Lett. **174**, 109110 (2021)

21. Ben-Tal, A., Teboulle, M.: An old-new concept of convex risk measures: the optimized certainty equivalent. Math. Financ. **17**(3), 449–476 (2007)
22. Calder, J., Yezzi, A.: PDE acceleration: a convergence rate analysis and applications to obstacle problems. Res. Math. Sci. **6**(4), 1–30 (2019). https://doi.org/10.1007/s40687-019-0197-x
23. Khatun, D., Tanaka, T., Aranishi, F. (2022). Stock assessment of landlocked ayu Plecoglossus altivelis altivelis in Japan through length-based models. Environ. Sci. Pollut. Res. **30**, 2649–2664 (2023)
24. Ruijter, M., Versteegh, M., Oosterlee, C.W.: On the application of spectral filters in a Fourier option pricing technique. J. Comput. Finan. **19**(1), 75–106 (2015)
25. Abi Jaber, E.: The characteristic function of Gaussian stochastic volatility models: an analytic expression. Finan. Stochast. **26**(4), 733–769 (2022)

Global Sensitivity Analysis Using Polynomial Chaos Expansion on the Grassmann Manifold

Valentina Bazyleva(✉), Victoria M. Garibay, and Debraj Roy

Faculty of Science, Informatics Institute, University of Amsterdam, Science Park 904, 1098 XH Amsterdam, The Netherlands
valya.bazyleva@student.uva.nl, {v.m.garibay,d.roy}@uva.nl

Abstract. Traditional global sensitivity analysis (GSA) techniques, such as variance- and density-based approaches, are limited in cases where a comprehensive understanding of temporal dynamics is critical, especially for models with diverse timescales and structural complexity, such as system dynamics and agent-based models (ABMs). To address this, we propose a novel manifold learning-based method for GSA in systems exhibiting complex spatiotemporal processes. Our method employs Grassmannian diffusion maps to reduce the dimensionality of the data and polynomial chaos expansion (PCE) to map stochastic input parameters to diffusion coordinates of the reduced space. We calculate sensitivity indices from PCE coefficients, aggregating multiple outputs and their entire trajectories for a more general estimation of parameter sensitivities. We demonstrate the capabilities of the proposed approach by applying it to the Lotka-Volterra model and an epidemic dynamics ABM and capturing diverse temporal dynamics. We establish that the new methodology meets all "good" properties of a global sensitivity measure, making it a valuable alternative to traditional GSA techniques. We anticipate that it will potentially expand the application of manifold-based approaches and deepen the understanding of complex spatiotemporal processes.

Keywords: Global sensitivity analysis · Sobol' indices · Agent-based modelling · Grassmannian diffusion maps · Polynomial chaos expansion

1 Introduction

Parametric global sensitivity analysis (GSA) is an essential tool for enhancing model efficiency. The determination of which parameters and combinations thereof contribute the most to model uncertainty can allow for the development of simplified models. This directs the focus of experiments toward the parameters of greatest influence. Eliminating or fixing parameters at a certain value can also provide a substantial computational advantage. When developing a computationally intensive agent-based model (ABM), reducing model complexity is of great interest. Frequently in ABMs, heterogeneous, simulated populations interact, make decisions, and take action at every time step, so increasing the speed of these calculations has a cumulative advantage.

J. Mikyška et al. (Eds.): ICCS 2023, LNCS 14077, pp. 583–597, 2023.
https://doi.org/10.1007/978-3-031-36030-5_46

The current standard for GSA relies on the use of either Sobol' or density-based methods. A critical disadvantage of using these methods is the inability to aggregate the results of calculations for each individual time step into a single index or, alternatively, the loss of information when only considering the final outcome of the model in the analysis, ignoring the progression at individual time steps. ABMs are characteristically stochastic and often subject to non-linear interaction effects. Due to the frequent appearance of multimodal and fat-tailed distributions in ABM results, conducting a GSA presents a unique challenge [21]; sensitivity is not well captured with traditional Sobol' analyses as the technique is variance-based. While it is possible to employ density-based methods in cases with poorly defined variance, the described issues in aggregation still apply.

To address these limitations, the proposed analysis method inclusively considers the outputs of interest at each time step. Then, the resulting high-dimensional data is organised into a tensor and projected onto a Grasmann manifold, with the goal of detecting that the data can be mapped to a subspace of lower dimension. The reduced-dimension data is used in conducting a sensitivity analysis with polynomial chaos expansion (PCE) methods. PCE, used in mapping numerical model input to output, becomes difficult to apply to cases with high dimensionality. However, assuming sparse effects makes it possible to create representative surrogate models from fewer samples, lowering the computational expense of PCE. The next sections begin with the context under which the methods in this study were developed, including a guided review of related work. Then, an overview of the proposed GSA method is provided, followed by demonstrative applications to a classic Lotka-Volterra dynamical system and a large-scale ABM of disease dynamics. These applications culminate in an assessment and discussion of the performance of the novel methodology.

2 Related Work

Parametric variance-based GSA, specifically with Sobol'/Saltelli ANOVA techniques, has become widely adopted for ABMs across various fields due to its robust sensitivity estimates for non-linear models with parameter interactions [2,23,29,33]. However, a crucial limitation of these methods is that variance in model outputs is not always attributed to uncertainty [21], and sensitivity assessment in ABMs for verification and validation is often insufficiently explored [22]. This could be attributed to the lack of tools and methodologies focusing on a comprehensive analysis of ABM dynamics. One such inefficient use of Sobol'/Saltelli variance-based GSA is estimating sensitivity based solely on the final time step, disregarding the preceding trajectory. This issue has been addressed by time-dependent GSA, which calculates Sobol' sensitivity indices at multiple time steps throughout the simulation [16].

The computational cost of the Sobol'/Saltelli method based on estimating high-dimensional integrals via crude Monte-Carlo (MC) simulation is a practical concern for researchers due to slow error convergence [29]. More efficient sampling schemes have been proposed, such as Latin hypercube sampling and low-discrepancy or Sobol' sequences [19,28], along with direct formulas for evaluating Sobol' indices with fewer model evaluations [15,20,24,27,32]. However, calculating Sobol' indices for complex and large-scale spatial ABMs remains challenging due to computational costs, particularly in models with systemic variability from aleatory uncertainty [1]. Averaging over multiple ABM repetitions has been suggested to address this issue [18], with some researchers leveraging multi-GPU parallel computing and high-performance computing resources for GSA of large-scale ABMs [31].

Besides MC and quasi-Monte-Carlo (QMC) methods, stochastic polynomial methods such as PCE can be used for constructing surrogate models to approximate ANOVA decomposition [5,30]. The calculation of Sobol' indices for PCE emulators is analytical, reducing the computational cost to that of computing PCE coefficients [30]. Despite PCE's advantages and the growing research on sparse methods tackling the issue of exponential increase in the number of polynomial basis functions with increasing input dimension—resulting in an excessive computational cost for models with high-dimensional input, also known as "curse of dimensionality" in uncertainty quantification (UQ) literature [8,13,17]—PCE-based ANOVA decomposition is not commonly applied for GSA in ABMs with only few examples of successful applications [4,8,10].

Sparse PCE methods help reduce the size of basis functions and experimental design required for GSA but can become computationally intractable for high-dimensional output [13]. Dimensionality reduction techniques, including linear and non-linear methods, can address the challenge of model fidelity for UQ tasks when, instead of a scalar or low-dimensional vector, high-dimensional responses are of interest [6,25]. While linear methods can extract the dominant modes of data, they are ineffective in capturing the non-linear geometries of a dataset. Transcending the limitation of linear techniques, non-linear methods for dimensionality reduction posit that high-dimensional data resides on a manifold, which is a low-dimensional and more informative space [6,25]. Thus, kernel-based diffusion maps (DMaps) can discover low-dimensional manifolds embedded in Euclidean space and be exploited to construct accurate, lower-cost surrogate models [11,12]. However, exceedingly high-dimensional data, such as numerical simulations with many degrees of freedom, may not be well-described in Euclidean space and may inherently reside on a submanifold of a Riemannian manifold [6].

A proposed extension from and complement to DMaps is Grassmannian diffusion maps (GDMaps) addressing limitations when dealing with data exhibiting geometric structures on a Riemannian submanifold [6,25]. This is achieved by combining pointwise linear dimensionality reduction with a multipoint, non-linear dimensionality reduction step using DMaps with a suitable Grassmannian kernel. GDMaps are particularly fitting for high-dimensional data represented by vectors or matrices, where Euclidean metrics cannot meaningfully describe distances between objects, thus capturing the geometric structures spanning the data [6]. Dos Santos et al. present a simple example illustrating GDMaps' capability to capture intrinsic geometric structures in data [6]. This example demonstrates that while conventional DMaps can find a low-dimensional representation, the resulting manifold may not accurately represent the underlying subspace structure of the data (see SI Section A).

Leveraging the suitability of GDMaps for latent representation of very high-dimensional data on a lower-dimensional manifold, Kontolati et al. [14] proposed a surrogate model construction method capable of generating out-of-sample predictions from a limited number of observations. The "encoder" path of the technique combining GDMaps and PCE for dimensionality reduction and mapping between input parameters and diffusion coordinates provides a framework for statistical moment estimation from PCE coefficients in the latent space [14], potentially enabling sensitivity index calculations from PCE coefficients. However, to the authors' knowledge, GSA methods employing GDMaps and PCE have not yet been suggested.

3 Methods

3.1 Variance-Based Global Sensitivity Analysis: Sobol' Indices

We consider a set of d independent random variables (RVs) $\boldsymbol{X} = \{X_i\}_{i=1}^d$, serving as an input into a model $Y = f(\cdot)$. For simplicity, we assume that the RVs X_i are uniformly distributed on $[0,1]$: $Q_i \sim \mathcal{U}(0,1)$, $\Gamma = [0,1]^d$ and write the Sobol' decomposition of the response $f(\mathbf{X})$ as the finite, hierarchical expansion:

$$\begin{aligned} f(\boldsymbol{X}) &= f_0 + \sum_{i=1}^{d} f_i(X_i) + \sum_{i,j\neq i}^{d} f_{ij}(X_i, X_j) + \cdots + f_{12\ldots d}(\boldsymbol{X}) \\ &= f_0 + \sum_{\boldsymbol{u}\subset\{1,\ldots,d\}} f_{\boldsymbol{u}}(\boldsymbol{X}_{\boldsymbol{u}}), \end{aligned} \tag{1}$$

where $\boldsymbol{X}_{\boldsymbol{u}} := \{X_{i_1}, \ldots, X_{i_s}\}$ and the summands satisfy the orthogonality condition: $\int_\Gamma f_{\boldsymbol{u}}(\boldsymbol{X}_{\boldsymbol{u}}) f_{\boldsymbol{v}}(\boldsymbol{X}_{\boldsymbol{v}}) d\boldsymbol{X} = 0 \quad \forall \boldsymbol{u} \neq \boldsymbol{v}$. In Eq. (1), f_0 is the mean response of f, the univariate functions $f_i(X_i)$ quantify independent contribution given the individual parameters, the bivariate functions $f_{ij}(X_i, X_j)$ represent the interactions of X_i and X_j on the response with similar interpretations for higher-order interaction effects [26].

From the total variance theorem, the total variance $\mathbb{V}[Y] = D$ can be decomposed as $D = \sum_{i=1}^d D_i + \sum_{1\leq i<j\leq d} D_{ij} + \cdots + D_{12\ldots d}$, which we use to define first- and total-order Sobol' indices as

$$S_i = \frac{D_i}{D} = \frac{\mathbb{V}[\mathbb{E}(Y|X_i)]}{\mathbb{V}[Y]}, \quad S_{T_i} = 1 - \frac{D_{\sim i}}{D} = 1 - \frac{\mathbb{V}[\mathbb{E}(Y|X_{\sim i})]}{\mathbb{V}[Y]} = \frac{\mathbb{E}[\mathbb{V}(Y|X_{\sim i})]}{\mathbb{V}[Y]}. \tag{2}$$

Calculation of Sobol' Indices with Conventional Methods Using MC, we obtain the following estimators for mean as $\hat{f}_0 = 1/N \sum_{n=1}^N f(X^{(n)})$, and for total variance as $\hat{D} = 1/N \sum_{n=1}^N f^2(X^{(n)}) - \hat{f}_0^2$. To obtain the estimates for D_i

and $D_{\sim i}$, we use Saltelli's algorithm (explained in [24,26]) to reduce the number of evaluations from N^2 for crude MC to $N(d+2)$ by constructing three types of X samples: $X = (X_1, \dots, X_d)^\top$, its complete resample $X' = (X'_1, \dots, X'_d)^\top$, and $(X_i, X'_{\sim i}) = (X'_1, \dots, X'_{i-1}, X_i, X'_{i+1}, \dots, X'_d)^\top$, with $i = 1, \dots, d$, where all factors except for X_i are resampled. Thus, the estimates for D_i and $D_{\sim i}$ become $\hat{D}_i = 1/N \sum_{n=1}^{N} f(X_i^{(n)} X_{\sim i}^{(n)}) f(X_i^{(n)} X'^{(n)}_{\sim i}) - \hat{f}_0^2$ and $\hat{D}_{\sim i} = 1/N \sum_{n=1}^{N} f(X_i^{(n)} X'^{(n)}_{\sim i}) f(X'^{(n)}_i X'^{(n)}_{\sim i}) - \hat{f}_0^2$, respectively. The derived estimators $\hat{D}$, $\hat{D}_i$ and $\hat{D}_{\sim i}$ are then used to calculate the first- and total-order Sobol' indices in Eq. (2).

Sobol' Indices Using PCE PCE describes the input-output relationship using polynomials orthogonal with respect to the probability density function (PDF) of the input RVs. Sobol' decomposition of a PCE results from reordering terms of the truncated PCE approximating $f(\boldsymbol{X})$, written as $\widetilde{\mathcal{E}}(\boldsymbol{X}) = \sum_{\boldsymbol{\alpha} \in \mathcal{A}} \eta_{\boldsymbol{\alpha}} \Phi_{\boldsymbol{\alpha}}(\boldsymbol{X})$, where $\mathcal{A}$ is a total-degree multi-index set, $\eta_{\boldsymbol{\alpha}}$ are corresponding PCE coefficients, and $\Phi_{\boldsymbol{\alpha}}(\mathbf{X})$ are multivariate orthonormal polynomials with respect to $f_{\mathbf{X}}$ such that $\langle \Phi_{\boldsymbol{\alpha}}(\boldsymbol{X}) \Phi_{\boldsymbol{\beta}}(\boldsymbol{X}) \rangle = \int_Z \Phi_{\boldsymbol{\alpha}}(\boldsymbol{X}) \Phi_{\boldsymbol{\beta}}(\boldsymbol{X}) f_{\boldsymbol{X}}(\boldsymbol{X}) d\boldsymbol{X} = \gamma_{\boldsymbol{\alpha}} \delta_{\boldsymbol{\alpha}\boldsymbol{\beta}}$. We can obtain interaction sets as $\mathcal{A}_{\boldsymbol{u}} = \{\boldsymbol{\alpha} \in \mathcal{A} : t \in \boldsymbol{u} \Leftrightarrow \alpha_t \neq 0\}$ for a given $\boldsymbol{u} := \{i_1, \dots, i_s\}$, leading to the following decomposition: $\widetilde{\mathcal{E}}(\boldsymbol{X}) = \mathcal{E}_0 + \sum_{\boldsymbol{u} \subset \{1,\dots,d\}} \mathcal{E}_{\boldsymbol{u}}(\boldsymbol{X}_{\boldsymbol{u}})$, with $\mathcal{E}_{\boldsymbol{u}}(\boldsymbol{X}_{\boldsymbol{u}}) := \sum_{\boldsymbol{\alpha} \in \mathcal{A}_{\boldsymbol{u}}} \eta_{\boldsymbol{\alpha}} \Phi_{\boldsymbol{\alpha}}(\boldsymbol{X})$, resulting in the following general expression for PCE-based Sobol' indices, which can be derived analytically at any order from the PCE coefficients [30]:

$$S_{\boldsymbol{u}} = D_{\boldsymbol{u}}/D = \sum_{\boldsymbol{\alpha} \in \mathcal{A}_{\boldsymbol{u}}} \eta_{\boldsymbol{\alpha}}^2 / \sum_{\boldsymbol{\alpha} \in \mathcal{A} \setminus \mathbf{0}} \eta_{\boldsymbol{\alpha}}^2. \tag{3}$$

3.2 GSA Using Grassmannian Diffusion Maps and PCE

The methodology presented in Algorithm 1 largely draws from GDMaps technique proposed by Dos Santos et al. [6] and manifold learning-based PCE developed by Kontolati et al. [14]. Refer to the corresponding papers for a thorough description of the Grassmann manifold principles and other elements of differential geometry essential for developing GDMaps. See SI Section B.1 for more details on the proposed methodology.

The GDMaps method is an extension of the conventional DMaps that consists of two stages: a linear pointwise dimension reduction and a non-linear multipoint dimension reduction. The implementation of GDMaps is outlined in Lines 1–5 of Algorithm 1. This first step projects each element of the data on the Grassmann manifold, or Grassmannian, denoted as $\mathcal{G}(p, n)$ and defined as a p-dimensional subspaces embedded in n-dimensional Euclidean space $\mathbb{R}^n$. The parameter p, which relates p-dimensional subspaces embedding, is closely tied to the notion of matrix rank or the number of linearly independent matrix columns. The non-linear dimensionality reduction step consists of building a valid kernel (Lines 3–4 in Algorithm 1) and running the DMaps algorithm (Line 5 in Algorithm 1). The projection kernel (see SI Section B.1) is adopted throughout this research.

Algorithm 1: GSA using PCE on the Grassmann manifold

Input: Experimental design $\mathscr{X} = \{\boldsymbol{X}_i \in \mathbb{R}^k\}_{i=1}^{\mathcal{N}}$ and model response concatenated into a single vector and reshaped into $n \times m$ matrix $\mathscr{Y} = \{\mathcal{M}(\boldsymbol{X}_i)\}_{i=1}^{\mathcal{N}} = \{\boldsymbol{Y}_i \in \mathbb{R}^{n\times m}\}_{i=1}^{\mathcal{N}}$; Grassmannian dimension p; retained diffusion coordinates g; maximal polynomial degree $s_{\max}$.

Output: First-order Sobol' indices $\{\boldsymbol{S}_i \in \mathbb{R}^g\}_{i=1}^{k}$; total-order Sobol' indices $\{\boldsymbol{S}_{T_i} \in \mathbb{R}^g\}_{i=1}^{k}$; approximated PCE error ϵ_{val}.

1 **for** $i \leftarrow 1$ **to** $\mathcal{N}$ **do**

2 Perform singular value decomposition (SVD): $\boldsymbol{Y}_i = \boldsymbol{U}_i \boldsymbol{\Sigma}_i \boldsymbol{V}_i^T$, where $\boldsymbol{\Sigma}_i \in \mathbb{R}^{p\times p}$ is a diagonal matrix containing singular values. $\boldsymbol{U}_i \in \mathbb{R}^{n\times p}$ and $\boldsymbol{V}_i \in \mathbb{R}^{m\times p}$ are orthonormal matrices.

3 For each pair $[\boldsymbol{U}_i, \boldsymbol{U}_j]$ and $[\boldsymbol{V}_i, \boldsymbol{V}_j]$ compute the entries $k_{i,j}$ of the kernel matrices $k_{ij}(\boldsymbol{U})$ and $k_{ij}(\boldsymbol{V})$ using e.g., projection kernel as $k_{pr}(\boldsymbol{X}_i, \boldsymbol{X}_j) = \left\|\boldsymbol{X}_i^T \boldsymbol{X}_j\right\|_F^2$.

4 (Optionally) construct a composed Grassmannian diffusion kernel $K(\boldsymbol{U}, \boldsymbol{V})$ by taking the Hadamard product of the corresponding kernels: $k(\boldsymbol{U}, \boldsymbol{V}) = k_{i,j}(\boldsymbol{U}) \circ k_{i,j}(\boldsymbol{V})$, or by summing $k_{i,j}(\boldsymbol{U}) + k_{i,j}(\boldsymbol{V})$.

5 Run the DMaps with a Grassmannian kernel to obtain first g non-trivial diffusion coordinates $\{\boldsymbol{\Theta}_i \in \mathbb{R}^g\}_{i=1}^{N}$, and their respective eigenvectors $\{\psi_k\}_{k=1}^{g}$ with $\psi_k \in \mathbb{R}^{\mathcal{N}}$ and eigenvalues $\{\lambda_k\}_{k=1}^{g}$.

6 Construct a total-degree multi-index set Υ (with cardinality $\#\Upsilon = S$) that satisfy $\|\mathrm{s}\|_1 \leq s_{\max}, s_{\max} \in \mathbb{Z}_{\geq 0}$, leading to a PCE basis of size $\frac{(k+s_{\max})!}{k!s_{\max}!}$.

7 Construct PCE approximation $\widetilde{\mathcal{E}}(\mathbf{X}) = \sum_{s\in\Upsilon} \eta_s \Phi_s(\boldsymbol{X})$ where $\eta_s \in \mathbb{R}^g$ is computed by solving the least square problem $\underset{\zeta\in\mathbb{R}^{\#\Lambda}}{\arg\min} \frac{1}{\mathcal{N}} \sum_{i=1}^{\mathcal{N}} \left\{\mathcal{E}(\boldsymbol{X}_i) - \sum_{s\in\Upsilon} \eta_s \Phi_s(\boldsymbol{X}_i)\right\}^2$.

8 Approximate PCE generalisation error by calculating the validation error as $\epsilon_{\text{val}} = \frac{\sum_{i=1}^{\mathcal{N}_*}(\boldsymbol{\Theta}_i^* - \tilde{\mathcal{E}}(\boldsymbol{X}_i^*))^2}{\sum_{i=1}^{\mathcal{N}_*}(\boldsymbol{\Theta}_i^* - \bar{\boldsymbol{\Theta}}^*)^2}$, where $\{\boldsymbol{X}_i^* \in \mathbb{R}^k\}_{i=1}^{\mathcal{N}_*}$ and $\{\boldsymbol{\Theta}_i^* \in \mathbb{R}^g\}_{i=1}^{\mathcal{N}_*}$ comprise a test set, chosen to be of size $\mathcal{N}_* = \frac{1}{3}\mathcal{N}$; $\bar{\boldsymbol{\Theta}}^* = \frac{1}{\mathcal{N}_*}\sum_{i=1}^{\mathcal{N}_*} \boldsymbol{\Theta}_i^*$ is the mean response of the test set on the latent space.

9 Obtain first-order Sobol' indices $\{\mathbf{S}_i \in \mathbb{R}^g\}_{i=1}^{k}$ and total-order Sobol' indices $\{\mathbf{S}_{T_i} \in \mathbb{R}^g\}_{i=1}^{k}$ using Eqs. (8) and (9) from SI Section B.1, respectively.

Using GDMaps, we acquire g diffusion coordinates $\boldsymbol{\Theta}_i$ for the top g non-trivial eigenvalues (non-parsimonuous implementation). To address the issue of "repeated eigendirections" in complex data, we also utilise parsimonious representation employing local linear regression to identify unique eigendirections [7].

Next, we calculate Sobol' indices on the manifold using PCE following Kontolati et al.'s approach [14]. PCE is used to approximate mapping between input parameters and corresponding model responses projected on the latent space (i.e., coordinates on the diffusion manifold) $\mathcal{E} : \boldsymbol{X} \rightarrow \boldsymbol{\Theta}$ as $\widetilde{\mathcal{E}}(\boldsymbol{X}) = \sum_{s\in\Upsilon} \eta_s \Phi_s(\boldsymbol{X})$. The implementation of the approach employing the least square method to obtain vector-valued PCE coefficients $\eta_s \in \mathbb{R}^g$ is outlined in Lines 6–8 of Algorithm 1, with Line 8 used to calculate validation error to evaluate the surrogate's accuracy. From Sect. 3.1, Sobol' indices can be acquired without extra calculations by collecting multi-indices related to partial variance caused by individual random inputs (first-order effects) or combined with other random inputs (total-order effects) into two multi-index sets. As the PCE coefficients have a vector-valued dimension equal to the retained diffusion coefficients g, we estimate the first- and total-order Sobol' indices for each of the g diffusion coordinates.

3.3 Applications

A classic dynamical system (Lotka-Volterra [14]) and an ABM (DeepABM-COVID [3]) were selected as a sample demonstration of the range of model types for which the framework is applicable. These were utilised to compare, albeit indirectly, the performance of the proposed GSA framework and GSA employing conventional Sobol' index calculation methods over multiple time steps. The two models used in illustrating the application of the proposed framework and the setup used for the evaluation are described in SI Sections B.2 and B.3.

4 Results

Application 1: Lotka-Volterra Dynamical System. First- and total-order sensitivity indices, along with 95% bootstrap confidence intervals, were computed for the Lotka-Volterra dynamical system with two uncertain parameters α and β at fifteen evenly spaced time steps (Figs. 1a to 1d). Oscillatory behaviour in both main and total-effect indices corresponds to the behaviour of the model outputs for the defined parameter ranges for α and β (see SI Fig. C.1). The difference between the resulting first- and total-order indices for both outputs is small, hence the variance in model output is predominantly due to main effects rather than interactions.

Mean and variance of first- and total-order Sobol' indices were obtained using the GSA framework with GDMaps PCE from fifty resampled input matrices (Figs. 1e and 1f). Three first- and total-order indices were derived from the PCE coefficients for each output. Three non-trivial, parsimoniously selected diffusion coordinates were used, converging to $\boldsymbol{\Theta}_i = \{\theta_1, \theta_2, \theta_5\}$ for one solution[1], which can be found as 2D plots in SI Fig. C.2. In Figs. 1e and 1f, both main and interaction effects of α and β have close values, with β slightly higher for the first two diffusion coordinates. Interaction effects are significantly larger for the second and third coordinates. While direct comparison with Figs. 1a to 1d is inappropriate due to different data representations, the proposed framework arguably better highlights parameter differences in terms of their influence on output variance. The new GSA approach reveals a more apparent distinction between main and total-effect indices compared to the conventional time-dependent GSA methods. A similar comparison for the Lotka-Volterra model with four uncertain parameters is presented in SI Fig. C.3.

[1] Different sets of diffusion coordinates are possible for each resampled solution.

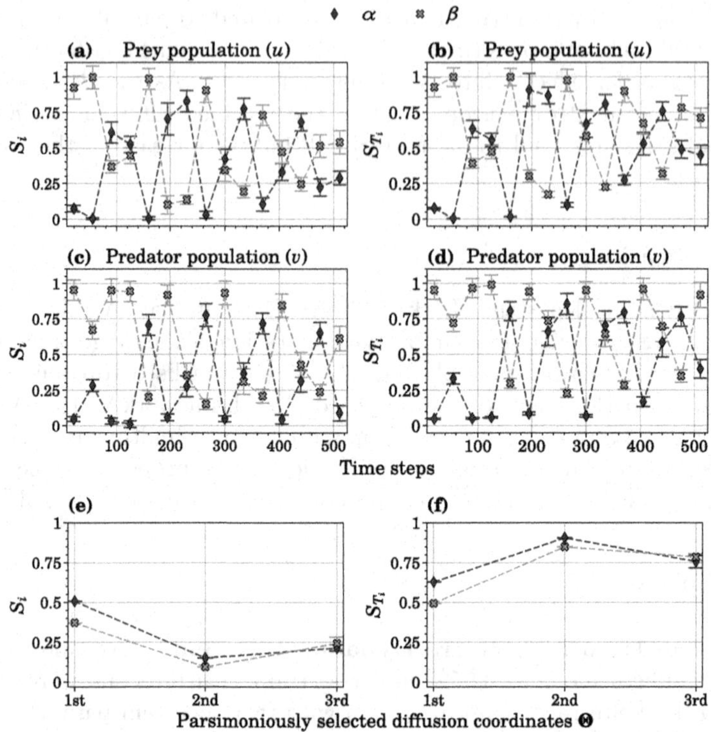

Fig. 1. Estimates of first- and total-order sensitivity indices, S_i and S_{T_i}, respectively, of α, and β (see SI Section B.2 Tab. 1) for two model output measures: number of prey per time step u, and number of predators per time step v, using (a-d) conventional Sobol' index calculation methods and (e, f) the GSA framework employing GDMaps PCE. Error bars indicate 95%-bootstrap confidence intervals in (a-d), and variance from fifty resamples in (e, f). For GDMaps PCE,, Grassmannian dimension $p = 10$ and maximal polynomial degree $s_{\max} = 6$ were used.

Application 2: DeepABM-COVID. We applied the proposed framework to estimate Sobol' indices on the Grassmann manifold using the DeepABM-COVID model outputs. Data generation and the general procedure for GSA framework are outlined in SI Section B.3. For GDMaps, we considered six dimensions, $p = \{3, 13, 17, 22, 23, 24\}$, corresponding to most frequently occurring ranks in the entire dataset (twenty runs). The frequency of occurrence and selection threshold are shown in Fig. 2a. Non-trivial eigenvalues[2] for chosen dimensions are presented in Fig. 2b. Given the physical interpretation of DMaps based on Markov Chain timescales, the regions around unstable equilibria (slow modes) with the largest eigenvalues correspond to the slowest possible ergodic dynamics in a system. From Fig. 2b, $p = 3$ is attributed to the slowest dynamics on the

[2] The first zero-indexed eigenvalue, λ_0, is a trivial eigenvalue, which is always $\lambda_0 = 1$.

manifold compared to other dimensions. Smaller p values correspond to larger λ_i values, particularly for $i = 1$ and $i = 2$, due to lower p allowing less detailed data representation on the Grassmannian, resulting in a more coarse-grained subspace structure revealed by DMaps performed on the manifold.

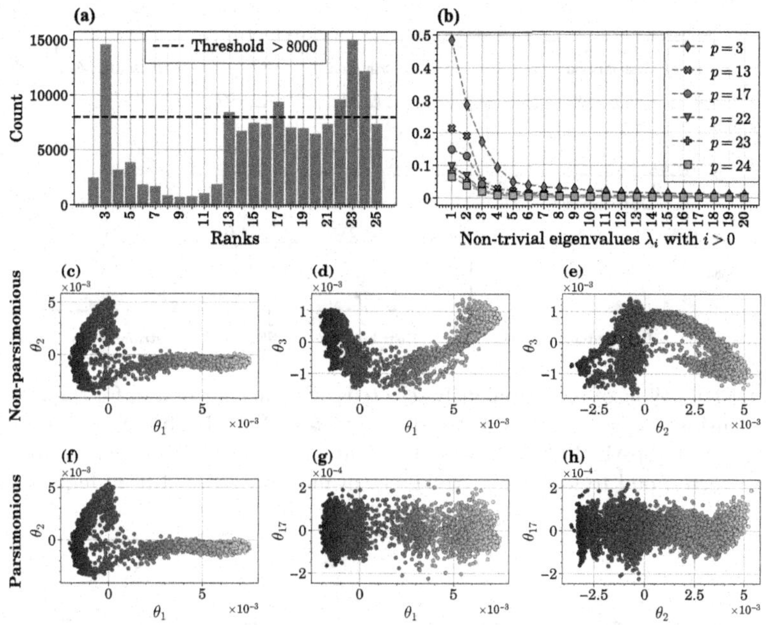

Fig. 2. (a) The frequency of rank occurrence in the entire data (twenty runs) with a threshold used for selection of the Grassmannian dimension, p. (b) Scree-plot of eigenvalues from GDMaps on DeepABM-COVID model output matrix $\mathscr{Y} \in \mathbb{R}^{7168 \times 900}$ for run 3 and six Grassmannian dimensions p. (c-h) 2D plots of three diffusion coordinates from GDMaps on DeepABM-COVID model output for run 16, using $p = 13$. Diffusion coordinates converged to $\boldsymbol{\Theta}_i = \{\theta_1, \theta_2, \theta_{17}\}$ for parsimonious representation (f-h).

We examined both parsimonious and non-parsimonious implementations to retain diffusion coordinates. Figures 2c to 2h present 2D plots of retained diffusion coordinates ($g = 3$) for both implementations. The example demonstrates the case when the first two coordinates coincide, but the third parsimoniously selected one corresponds to shorter timescale dynamics indicated by the scale of the y-axis. The remaining 2D plots for other Grassmannian dimensions are in SI Figs. C.5 (non-parsimonious) and C.6 (parsimonious). Notably, for larger p, parsimonious representation selected diffusion coordinates with lower corresponding eigenvalues more frequently than for smaller p. This relates to Fig. 2b, where larger p values are attributed to lower initial eigenvalues and exhibit lower decay rates. Figure 3 shows three examples of distributions of parsimoniously selected diffusion coordinates. Multimodal distributions correspond to higher eigenvalues, while unimodal and Gaussian-like distributions are attributed to lower eigenvalues, as in Fig. 3f. Additional examples can be found in SI Figs. C.7 and C.8.

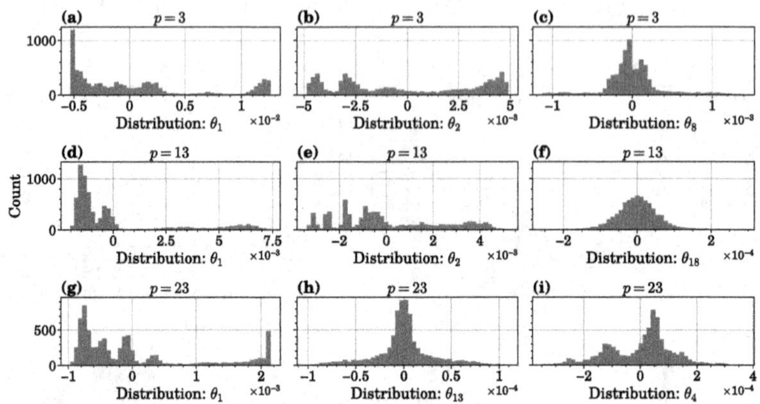

Fig. 3. Distributions of parsimoniously selected diffusion coordinates for the Grassmannian dimensions: $p = 3$ (a-c), $p = 13$ (d-f), and $p = 23$ (g-i) obtained from GDMaps run 17 of the DeepABM-COVID model output $\mathscr{Y} \in \mathbb{R}^{7168\times900}$. Higher values in the subscript correspond to lower non-trivial eigenvalues λ_i, ordered from high to low.

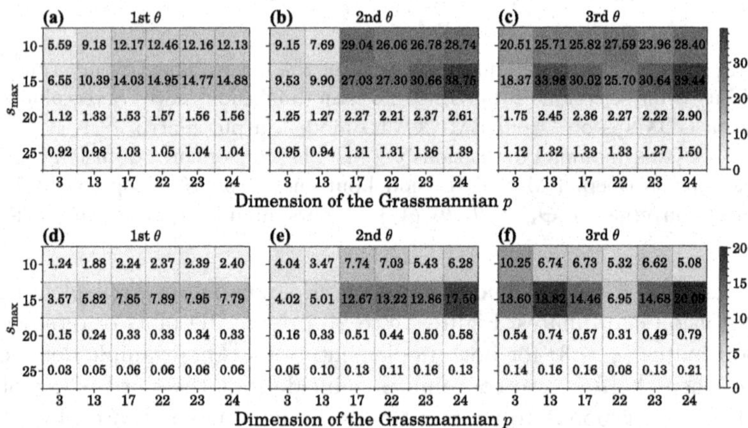

Fig. 4. Heatmaps of the mean (a-c) and standard deviation (d-f) of validation error from GDMaps PCE averaged over twenty runs for Grassmannian dimensions $p = \{3, 13, 17, 22, 23, 24\}$ and maximal polynomial degrees $s_{\max} = \{10, 15, 20, 25\}$. The colour maps are between zero and the maximal values rounded to the second decimal place.

To construct PCE surrogates with the total-degree PCE basis, we evaluated different maximum polynomial degrees $s_{\max} = \{10, 15, 20, 25\}$. Validation errors averaged over twenty runs, and standard deviations are presented in Figs. 4a to 4c and Figs. 4d to 4f, respectively, for each Grassmannian dimension p, and maximum polynomial degree $s_{\max}$ with parsimonious representation used for retaining three diffusion coordinates. While smaller $s_{\max}$ values generally yielded less accurate surrogates, $s_{\max} = 15$ exhibited the largest validation errors for all p. Larger manifold dimensions led to increased mean errors and variability. Based on this analysis, we used $s_{\max} = 25$ to obtain Sobol' indices on the manifold.

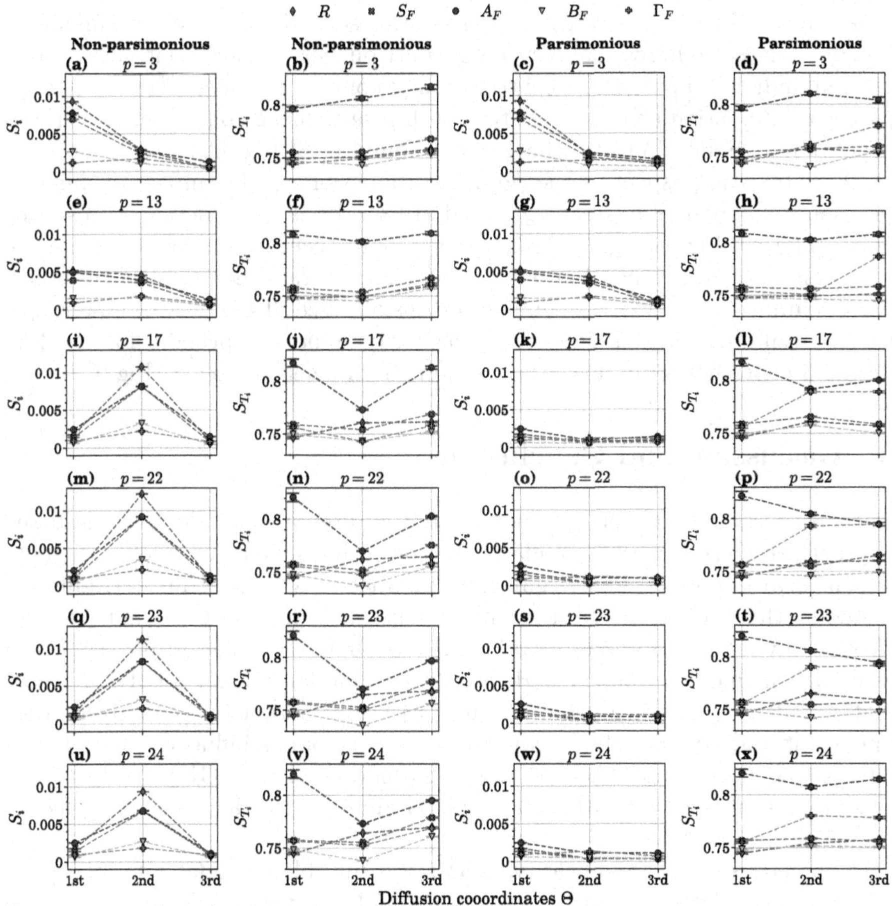

Fig. 5. Estimates of first- and total-order sensitivity indices, S_i and S_{T_i}, respectively, of five uncertain input parameters (see SI Section B.3 Tab. 2) for five model output measures, obtained from the GSA framework using GDMaps PCE. Error bars indicate variance across 20 runs.

Figure 5 presents first- and total-order sensitivity indices, averaged over twenty runs, from PCE using GDMaps with six Grassmannian dimensions, $p = \{3, 13, 17, 22, 23, 24\}$, and both non-parsimonious and parsimonious representations to keep three diffusion coordinates. As observed in Figs. 5a to 5d and Figs. 5e to 5h, S_i and S_{T_i} are comparable across representations for $p = 3$ and $p = 13$. This is due to GDMaps with lower p values resulting in larger initial eigenvalues, leading to parsimonious representation selecting diffusion coordinates that correspond to longer timescale dynamics more frequently. The main difference between the two implementations can be seen for $p = \{17, 22, 23, 24\}$, especially in the main effect indices, S_i. Larger p values provide more detailed data representation on the manifold, while non-parsimonious selection results in diffusion coordinates capturing the longest timescale dynamics on the manifold, possibly related to individual parameter contributions to model output variance. Overall, individual parameter contributions to output variance are smaller compared to interaction effects, consistent with results using conventional methods for obtaining Sobol' indices, which can be found in SI Fig. C.4.

An interesting aspect of the proposed framework is the impact of different maximum polynomial degrees $s_{\max}$ used in PCE basis construction on resulting Sobol' indices. Increasing $s_{\max}$ allowed for more accurate model output reconstruction (reducing validation error) and higher total-order sensitivity indices for all parameters and Grassmannian dimensions (see SI Fig. C.10). Lower $s_{\max}$ values resulted in higher first-order sensitivity indices, especially for the first diffusion coordinates and lower Grassmannian dimensions (see SI Fig. C.9).

5 Discussion and Conclusions

The proposed method for parametric GSA utilises a manifold learning-based approach to construct PCE emulators on lower-dimensional manifolds for high-dimensional problems with significant interaction effects. Unlike traditional methods, this technique enables a more general estimation of parametric sensitivities by aggregating entire trajectories of multiple model responses. Using an oscillating model example, we demonstrated that traditional Sobol' index estimation failed to provide a definitive answer to what parameter is relatively more important and whether the variance in model output is influenced by main or interaction effects. Conversely, the GSA method employing GDMaps PCE successfully revealed clear relations between parameters and their relative influence on the output variance.

Characterised by non-linearity, ABMs are suitable candidates for the GSA approach using PCE on the Grassmann manifold. In this paper, we applied the method to a large-scale spatial ABM of epidemic dynamics and captured strong interaction effects of uncertain parameters on the variance of multiple aggregated model responses. We also investigated the influence of hyper-parameters, such as the dimension of the Grassmann manifold and maximal polynomial degree and two approaches for retaining the desired number of diffusion coordinates, parsimonious and non-parsimonious, on sensitivity measures. Lower Grassmannian dimensions yielded higher main effect indices due to a more coarse-grained data representation. Non-parsimonious implementation produced larger first-order indices, resulting from longer timescales of diffusion coordinates corresponding to larger eigenvalues. Additionally, a higher maximal polynomial degree was attributed to a smaller validation error, as expected, which was mainly caused by the resolution of the interactions between parameters, leading to larger total-effect indices.

A simulation outcome or trajectory is a time series of state variables with dimension kN, where N is the number of agents and k is each agent's state variables. As typical agent-based simulations may involve a considerable number of agents, these time series can become exceedingly high-dimensional. As part of our future work, we aim to leverage the power of GDMaps to reduce the complexity of simulation trajectories at the micro level. By doing so, we can explore the impact of parameter sensitivity on the system's dynamic modes, key long-lasting states, transitional pathways, and essential degrees of freedom. Furthermore, we plan to address a limitation in the current implementation of not following general advice to perform over-sampling for the regression used in calculating PCE coefficients [9]. Model selection, like Least Angle Regression, should be added to the implementation to circumvent this issue.

In conclusion, the capabilities of the GSA framework utilising GDMaps PCE to aggregate entire trajectories of multiple model outputs and capture different timescales and degrees of structural complexity satisfy all the "good" properties of a global sensitivity measure extensively discussed in [21], providing a more comprehensive estimation of parameter sensitivities. This methodology is expected to open new avenues for ABM practitioners and Complexity Science scholars to deepen their understanding of systems exhibiting complex spatiotemporal dynamics.

Supporting Information. The Supporting Information can be found at Valentina Bazyleva, Victoria Garibay, & Debraj Roy. (2023). Supporting Information: Global Sensitivity Analysis using Polynomial Chaos Expansion on the Grassmann Manifold. Zenodo. https://doi.org/10.5281/zenodo.8050579.

Data and Code Availability. The code used in generating data for testing the methodology proposed in this study can be found at https://github.com/bazvalya/GSA_using_GDMaps_PCE. The output data of the DeepABM-COVID simulations is at https://figshare.com/articles/dataset/output_data_zip/22216921.

Acknowledgements. This research was conducted with support from the Dutch Research Council (NWO) under contract 27020G08, titled "Computing societal dynamics of climate change adaptation in cities".

References

1. Baustert, P., Benetto, E.: Uncertainty analysis in agent-based modelling and consequential life cycle assessment coupled models: A critical review. J. Clean. Prod. **156**, 378–394 (2017)
2. Cacuci, D.G.: Sensitivity & Uncertainty Analysis, vol. 1. Chapman and Hall/CRC (2003). https://doi.org/10.1201/9780203498798
3. Chopra, A., et al.: Deepabm: Scalable, efficient and differentiable agent-based simulations via graph neural networks (2021). https://doi.org/10.48550/ARXIV.2110.04421
4. Colas, F., Gauchi, J.P., Villerd, J., Colbach, N.: Simplifying a complex computer model: sensitivity analysis and metamodelling of an 3d individual-based crop-weed canopy model. Ecol. Model. **454**, 109607 (2021)
5. Crestaux, T., Le Maıtre, O., Martinez, J.M.: Polynomial chaos expansion for sensitivity analysis. Reliab. Eng. Syst. Safety **94**(7), 1161–1172 (2009)
6. Dos Santos, K.R., Giovanis, D.G., Shields, M.D.: Grassmannian diffusion maps-based dimension reduction and classification for high-dimensional data. SIAM J. Sci. Comput. **44**(2), B250–B274 (2022)
7. Dsilva, C.J., Talmon, R., Coifman, R.R., Kevrekidis, I.G.: Parsimonious representation of nonlinear dynamical systems through manifold learning: A chemotaxis case study. Appl. Comput. Harmon. Anal. **44**(3), 759–773 (2018)
8. Edeling, W., Arabnejad, H., Sinclair, R., Suleimenova, D., Gopalakrishnan, K., Bosak, B., Groen, D., Mahmood, I., Crommelin, D., Coveney, P.V.: The impact of uncertainty on predictions of the covidsim epidemiological code. Nat. Comput. Sci. **1**(2), 128–135 (2021)
9. Hosder, S., Walters, R., Balch, M.: Efficient sampling for non-intrusive polynomial chaos applications with multiple uncertain input variables. In: 48th AIAA/ASME/ASCE/AHS/ASC Structures, Structural Dynamics, and Materials Conference, p. 1939 (2007)
10. Hu, Y.: Agent-based models to couple natural and human systems for watershed management analysis. Ph.D. thesis, University of Illinois at Urbana-Champaign (2016)
11. Kalogeris, I., Papadopoulos, V.: Diffusion maps-based surrogate modeling: An alternative machine learning approach. Int. J. Numer. Methods Eng. **121**(4), 602–620 (2020)
12. Kalogeris, I., Papadopoulos, V.: Diffusion maps-aided neural networks for the solution of parametrized PDES. Comput. Methods Appl. Mech. Eng. **376**, 113568 (2021)
13. Konakli, K., Sudret, B.: Polynomial meta-models with canonical low-rank approximations: Numerical insights and comparison to sparse polynomial chaos expansions. J. Comput. Phys. **321**, 1144–1169 (2016)
14. Kontolati, K., Loukrezis, D., dos Santos, K.R., Giovanis, D.G., Shields, M.D.: Manifold learning-based polynomial chaos expansions for high-dimensional surrogate models. Int. J. Uncertain. Quant. **12**(4) (2022)
15. Kucherenko, S., Song, S.: Different numerical estimators for main effect global sensitivity indices. Reliab. Eng. Syst. Safety **165**, 222–238 (2017)
16. Ligmann-Zielinska, A., Sun, L.: Applying time-dependent variance-based global sensitivity analysis to represent the dynamics of an agent-based model of land use change. Int. J. Geograph. Inf. Sci. **24**(12), 1829–1850 (2010). https://doi.org/10.1080/13658816.2010.490533

17. Lüthen, N., Marelli, S., Sudret, B.: Sparse polynomial chaos expansions: Literature survey and benchmark. SIAM/ASA J. Uncertain. Quantif. **9**(2), 593–649 (2021)
18. Marino, S., Hogue, I.B., Ray, C.J., Kirschner, D.E.: A methodology for performing global uncertainty and sensitivity analysis in systems biology. J. Theoret. Biol. **254**(1), 178–196 (2008)
19. McKay, M.D., Beckman, R.J., Conover, W.J.: A comparison of three methods for selecting values of input variables in the analysis of output from a computer code. Technometrics **42**(1), 55–61 (2000)
20. Owen, A.B.: Better estimation of small sobol' sensitivity indices. ACM Trans. Model. Comput. Simulat. (TOMACS) **23**(2), 1–17 (2013)
21. Pianosi, F., Wagener, T.: A simple and efficient method for global sensitivity analysis based on cumulative distribution functions. Environ. Model. Softw. **67**, 1–11 (2015). https://doi.org/10.1016/j.envsoft.2015.01.004
22. Richiardi, M.G., Leombruni, R., Saam, N.J., Sonnessa, M.: A common protocol for agent-based social simulation. J. Artif. Soc. Soc. Simulat. **9** (2006)
23. Saltelli, A. (ed.): Global Sensitivity Analysis: The Primer. John Wiley, Chichester; Hoboken (2008). oCLC: ocn180852094
24. Saltelli, A.: Making best use of model evaluations to compute sensitivity indices. Comput. Phys. Commun. **145**(2), 280–297 (2002)
25. dos Santos, K.R., Giovanis, D.G., Kontolati, K., Loukrezis, D., Shields, M.D.: Grassmannian diffusion maps based surrogate modeling via geometric harmonics. Int. J. Numer. Methods Eng. **123**(15), 3507–3529 (2022)
26. Smith, R.C.: Uncertainty Quantification: Theory, Implementation, and Applications, vol. 12. Siam (2013)
27. Sobol', I.M., Myshetskaya, E.: Monte Carlo estimators for small sensitivity indices **13**(5–6), 455–465 (2008). https://doi.org/10.1515/mcma.2007.023
28. Sobol', I.M.: On the distribution of points in a cube and the approximate evaluation of integrals. Zhurnal Vychislitel'noi Matematiki i Matematicheskoi Fiziki **7**(4), 784–802 (1967)
29. Sobol, I.: Global sensitivity indices for nonlinear mathematical models and their Monte Carlo estimates. Math. Comput. Simulat. **55**(1–3), 271–280 (2001)
30. Sudret, B.: Global sensitivity analysis using polynomial chaos expansions. Reliab. Eng. Syst. Safety **93**(7), 964–979 (2008)
31. Tang, W., Jia, M.: Global sensitivity analysis of a large agent-based model of spatial opinion exchange: A heterogeneous multi-GPU acceleration approach. Annal. Assoc. Am. Geograph. **104**(3), 485–509 (2014)
32. Tarantola, S., Gatelli, D., Kucherenko, S., Mauntz, W., et al.: Estimating the approximation error when fixing unessential factors in global sensitivity analysis. Reliab. Eng. Syst. Safety **92**(7), 957–960 (2007)
33. Wainwright, H.M., Finsterle, S., Jung, Y., Zhou, Q., Birkholzer, J.T.: Making sense of global sensitivity analyses. Comput. Geosci. **65**, 84–94 (2014). https://doi.org/10.1016/j.cageo.2013.06.006

Fuzzy Solutions of Boundary Problems Using Interval Parametric Integral Equations System

Eugeniusz Zieniuk, Marta Czupryna(✉), and Andrzej Kużelewski

Institute of Computer Science, University of Bialystok, Ciołkowskiego 1M, 15-245 Białystok, Poland
{e.zieniuk,m.czupryna,a.kuzelewski}@uwb.edu.pl

Abstract. This paper investigated the possibility of obtaining fuzzy solutions to boundary problems using the interval parametric integral equations system (IPIES) method. It focused on the IPIES method because, thanks to the analytical modification of the boundary integral equations (BIE), it does not require classical discretization. In this method, an original modification of directed interval arithmetic was also proposed. Solutions obtained using classical and directed interval arithmetic (known from the literature) were also presented for comparison. The extension of the IPIES method (to obtain fuzzy solutions) was to divide the fuzzy number into α-cuts (depending on the assumed confidence level). Then, such α-cuts were represented as interval numbers. Preliminary tests were carried out in which the influence of boundary condition uncertainty on fuzzy solutions (obtained using IPIES) was investigated. The analysis of solutions was presented on examples described by Laplace's equation. The accuracy verification of the fuzzy PIES solutions required a modification of known, exactly defined analytical solutions. They were defined using intervals and calculated using appropriate interval arithmetic in α-cuts to obtain fuzzy analytical solutions finally. The research showed the high accuracy of fuzzy solutions obtained using IPIES and confirmed the high potential of the method in obtaining such solutions.

Keywords: boundary problems · uncertainty · fuzzy solutions · interval arithmetic

1 Introduction

The wide application of computer simulation in practice shows that defining input data exactly (by real numbers) significantly limits and idealizes reality. Practically, these data are always given with some uncertainty resulting from experimental data or measurement errors. One of the more intuitive ways to model uncertainty is to use interval numbers. In the boundary problems, they have been used in the interval finite element method (IFEM) [1] and interval boundary element method (IBEM) [2]. However, most of the IFEM or IMEB

J. Mikyška et al. (Eds.): ICCS 2023, LNCS 14077, pp. 598–605, 2023.
https://doi.org/10.1007/978-3-031-36030-5_47

research focuses on boundary conditions or various parameters defined uncertainly. Researchers often omit the problem of modelling the uncertainty of the boundary shape by interval coordinates of points. Such modelling in IFEM and IBEM is troublesome due to the necessity of interval discretization. As a result of a significant increase in the number of interval input data, solutions are overestimated and useless in practice.

To solve the problem (by significantly reducing the amount of interval input data), the method of interval parametric integral equations system (IPIES) was proposed [3,4]. The method's main advantage is the unnecessity of classical discretization [5,6]. The functions that model the boundary's shape are included directly in the mathematical formalism of PIES. As a result, a small amount of input data is required to model the shape of the boundary and boundary conditions. This significantly reduces the number of equations in the system to be solved, which reduces the number of calculations, shortens the time and reduces the required computer resources. The previous studies [3,4] proved insignificant overestimations and high accuracy of the interval solutions obtained by IPIES.

The paper presents the IPIES application to obtain fuzzy solutions to the boundary problem. The fuzzy set theory [7,8] is another way that can be used to define the uncertainty of boundary problems. The advantage (in comparison with intervals) is obtaining additional information about the behaviour of the solutions inside the interval bounds. As in the previous methods of uncertainty modelling, the fuzzy finite element method (FFEM) [9] and the fuzzy boundary element method (FBEM) [10] can also be found in the literature.

2 Interval Parametric Integral Equations System

Including uncertainly defined boundary conditions in the PIES requires defining interval boundary functions. Therefore, the solution on the boundary (of the problem modelled by the Laplace equation) can be obtained by solving the interval PIES defined as follows:

$$0.5\boldsymbol{u}_l(s_1) = \sum_{j=1}^{n} \int_{\widehat{s}_{j-1}}^{\widehat{s}_j} \left\{ U_{lj}^*(s_1,s)\boldsymbol{p}_j(s) - P_{lj}^*(s_1,s)\boldsymbol{u}_j(s) \right\} J_j(s)ds, \tag{1}$$

where $l = 1, 2, ..., n$, and $\widehat{s}_{l-1} \leq s_1 \leq \widehat{s}_l$, $\widehat{s}_{j-1} \leq s \leq \widehat{s}_j$. The $\widehat{s}_{l-1}, \widehat{s}_{j-1}$ are the beginnings, and the $\widehat{s}_l, \widehat{s}_j$ are endings of the boundary segments exactly defined in a parametric coordinate system. The function $J_j(s)$ is the Jacobian to the segment of the curve S_m (where $m = j, l$).

The functions $\boldsymbol{p}_j(s)$, $\boldsymbol{u}_j(s)$ are interval parametric boundary functions on individual segments S_j of the boundary. One will be given as uncertainly defined (interval) boundary conditions, while the other will be searched for in the numerical solution of the interval PIES. The paper assumes an exactly defined boundary shape, so the kernels are defined classically:

$$U_{lj}^*(s_1,s) = \frac{1}{2\pi} \ln \frac{1}{[\eta_1^2 + \eta_2^2]^{0.5}}, \; P_{lj}^*(s_1,s) = \frac{1}{2\pi} \frac{\eta_1 n_j^{(1)}(s) + \eta_2 n_j^{(2)}(s)}{\eta_1^2 + \eta_2^2}, \tag{2}$$

where $n_j^{(1)}(s), n_j^{(2)}(s)$ are components of the normal vector $\boldsymbol{n}$ to the boundary segment S_j. Kernels allow for the analytical inclusion in its mathematical formalism of the boundary shape by appropriate relations between segments $\eta_1 = S_l^{(1)}(s_1) - S_j^{(1)}(s), \eta_2 = S_l^{(2)}(s_1) - S_j^{(2)}(s)$.

Since the direct application of interval arithmetic known from the literature caused significant overestimations, it was decided to propose a modification of directed interval arithmetic for calculations in the above method [3,4]. It consists of shifting the operations to the positive semi-axis in multiplication.

3 Fuzzy Solutions to Boundary Problems

The interval numbers give only the values of a certain set's lower and upper bounds. So, researchers became interested in ways that also define the interior of such a set. In the fuzzy set theory [7,8], it was found that the human ability to make the right decisions decreases due to the appearance of uncertainty in more complex systems. Such uncertainty can be easily expressed in words, i.e. using a linguistic variable, which can be fuzzified and defined by a fuzzy set. It is a set without clearly defined boundaries. The values inside have an additional function determining their degree of belonging to the set. Therefore, the fuzzy set A is represented by the function $\mu_A(x)$ called the membership function:

$$A = [(x, \mu_A(x))|x \in R, \mu_A(x) \in [0, 1]]. \tag{3}$$

Corespondingly, $\mu_A(x) = 1$ it means that x is in the set A, while $\mu_A(x) = 0$ means that x does not belong to this set. The shape of this function depends on the fuzzification method used. The best-known types of membership functions are the triangular and Gaussian functions. One special kind of fuzzy set is a fuzzy number. This term denotes a special case of a convex, normalized fuzzy set with a continuous membership function. An example is a triangular fuzzy number (TFN), a fuzzy set with a triangular membership function. In a simplified form, it is reduced to the L-R representation [8], where the fuzzy number is represented as $x = (m, a, b)$ (Fig. 1).

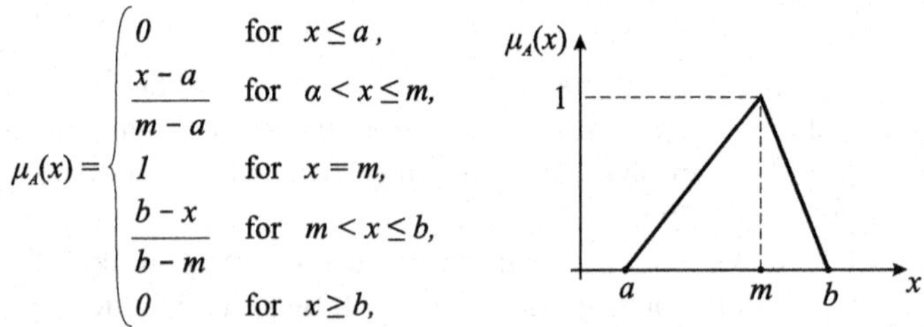

$\mu_A(x) = \begin{cases} 0 & \text{for } x \le a, \\ \frac{x-a}{m-a} & \text{for } a < x \le m, \\ 1 & \text{for } x = m, \\ \frac{b-x}{b-m} & \text{for } m < x \le b, \\ 0 & \text{for } x \ge b, \end{cases}$
$\mu_A(x)$
1
a
m
b
x

Fig. 1. A triangular fuzzy number

Obtaining fuzzy solutions with the interval PIES was to use a simplified notation of a fuzzy number using α-cuts. This method divides the membership function into certain levels called α-cuts. Each defines an interval in which the degree of values membership is greater than the given value α. So, the membership function can be defined using interval values: $\boldsymbol{x}_\alpha = [\underline{x}, \overline{x}]_\alpha$ easily. Respectively $\underline{x}$ is the smallest, and $\overline{x}$ is the largest value whose degree of membership is greater than or equal to α. A general schema of the application of fuzzy logic in modelling and simulating boundary problems is shown in Fig. 2.

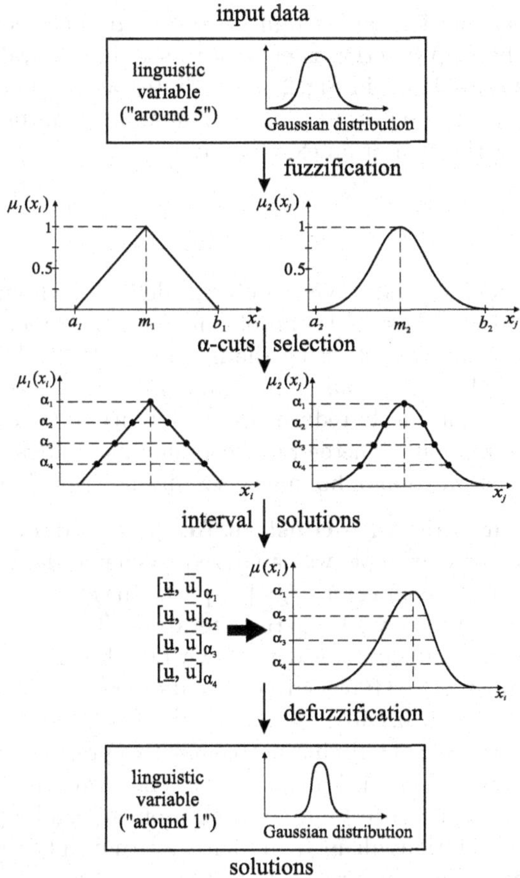

Fig. 2. A general schema of the application of fuzzy logic in modelling and simulating boundary problems.

The first stage is to define the input data with uncertainty. It can be a linguistic value, a real number, or a probability distribution. However, the most important thing is determining the appropriate way of fuzzification for a given variable. Fuzzification methods include intuition, reasoning, genetic algorithms, neural networks, induction or statistical distribution. After the fuzzification, the obtained data is defined as fuzzy numbers with the corresponding membership functions.

The next step is to divide the defined membership functions into the appropriate number of α-cuts for all uncertainly defined input data. The amount of information transferred to the simulation will depend on their number, which affects the amount of information obtained as a solution. Each α-cut is defined as an interval number. Interval solutions are calculated for α-cuts using appropriate interval arithmetic.

Finally, interval solutions are obtained for each α-cut. Therefore, obtaining the solution membership function is enough to set the interval solutions in the appropriate α-cuts (as shown in Fig. 2 at the "interval solutions" step). Finally, the defuzzification process on the membership function allows obtaining solutions as a probability distribution or an exact value.

4 Tests

The strategy of modelling and solving problems defined with uncertainty will be tested and analyzed on the example of problems described by Laplace's equation. The mentioned strategy (Fig. 2) using the method of IPIES [3,4] from formula (3) was implemented as a computer program. Defining the uncertainty of all input data at once significantly reduces the possibility of identifying the causes of possible overestimations. Therefore, the preliminary research in this paper was limited to the verification of boundary conditions defined by fuzzy numbers.

Example 1. Square domain - constant fuzzy boundary conditions
The first elementary example was analyzed to enable the comparison of the proposed uncertainty modelling strategy [3,4] with the classical [11] and directed [12] interval arithmetic known from the literature. The problem is shown in Fig. 3. One of the four segments was defined with the boundary condition by triangular fuzzy number $u_1 = (100, 95, 105)$. α-cuts were considered for the membership function $\mu(x) = 0, 0.1, 0.2, \ldots 1$.

The analytical solution of a similar problem with a precisely defined boundary condition $u = 100$ can be defined as $u = 100y$. Therefore, the modified interval analytical solution for the uncertainly defined problem can be presented as $u = [95, 105]y$. Interval solutions defined in this way (multiplication only by real number) reach the same values regardless of the interval arithmetic used. The interval solutions obtained on the α-cuts were saved as membership functions to obtain the final fuzzy analytical solution. Apart from the u_1 value, the boundary condition has been precisely defined (Fig. 3) to enable direct comparison between the IPIES and the interval analytical solutions.

Fuzzy PIES solutions (obtained using IPIES) and fuzzy analytical solutions (obtained using interval analytical solutions) are shown in Fig. 4. The solutions obtained using the proposed strategy (Fig. 4a) are almost equal to the proposed fuzzy analytical solutions (Fig. 4b). Additionally, solutions obtained directly applying for calculations in IPIES classical or directed interval arithmetic (known from the literature) are presented. Even such an elementary example causes significant overestimations using classical interval arithmetic (Fig. 4d).

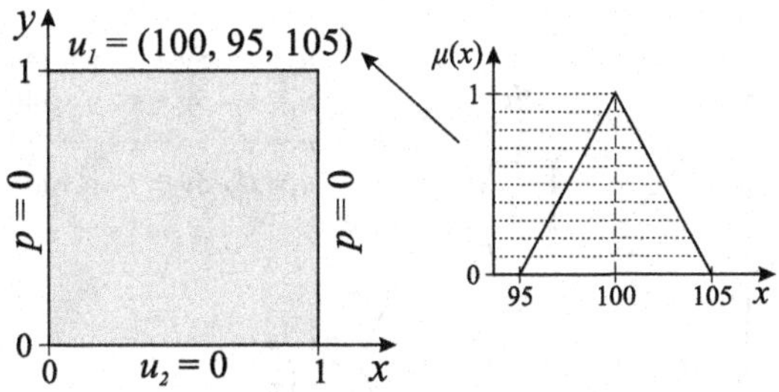

Fig. 3. Modelling example with fuzzy boundary conditions

In comparison, the directed interval arithmetic narrows the interval solutions' radii (Fig. 4c).

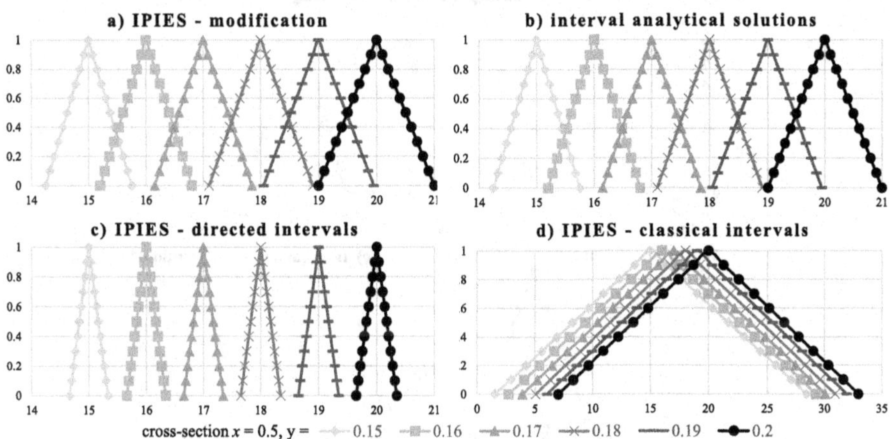

Fig. 4. Comparison of obtained fuzzy solutions in the domain.

Example 2. Complex area - a fuzzy function of boundary conditions
Another example is shown in Fig. 5. The analytical solution defined in the general form is $u = x^2 - 5y + x - y^2 + k$. Fuzzy solutions of the problem defined in this way (in the cross-section marked with black dots in Fig. 5), in analogous α-cuts as in the previous example, are shown in Fig. 6a. In addition, an exact analytical solutions were presented (for the middle value of the fuzzy number, i.e. $k = 50$). Additionally, an example with a constant fuzzy value of $k = (40, 50, 60)$ was assumed to enable direct comparison between fuzzy analytical solutions (Fig. 6c) and fuzzy PIES (Fig. 6b). The solutions obtained using fuzzy PIES are almost equal to those obtained using fuzzy analytical solutions.

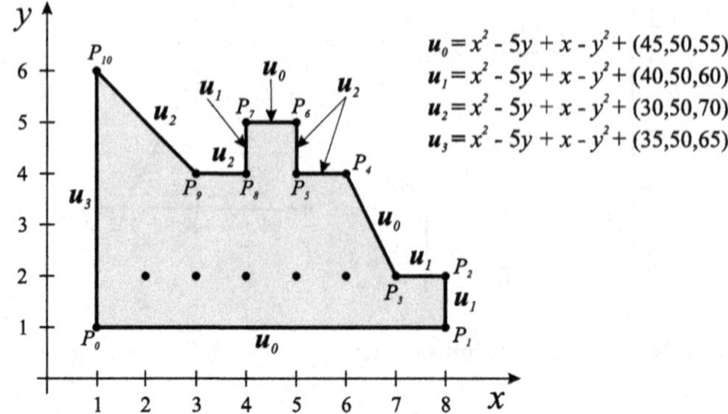

Fig. 5. Modelling example with fuzzy boundary conditions

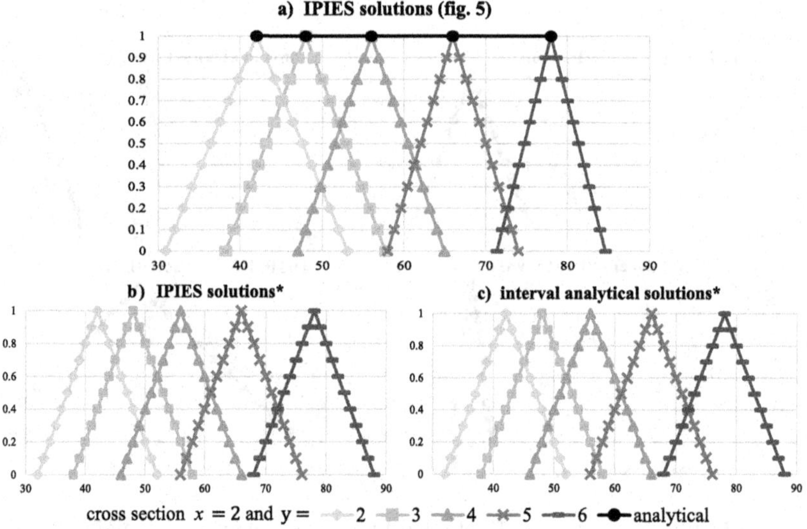

Fig. 6. Comparison of obtained fuzzy solutions in domain

5 Conclusions

The paper proposes the application of interval PIES to obtain fuzzy solutions to boundary problems. It was decided to focus only on the fuzzy boundary conditions to draw unambiguous conclusions. Such data were considered in subsequent α-cuts to allow IPIES application. After implementing the proposed algorithm, tests were carried out on the example of problems described by the Laplace equation. The disadvantages of the classical and directed interval numbers known from the literature and the advantages of the modification of directed interval arithmetic proposed in IPIES are highlighted in the first elementary example. Another more complex example also confirms the correctness of the solutions.

Ultimately, as a result of the tests, the high potential of the IPIES method in obtaining fuzzy solutions was presented. In the subsequent research, more comprehensive tests are also planned for the uncertainty of the boundary shape and modifications of the algorithm to the other differential equations.

References

1. Doan, Q.H., Luu, A.T., Lee, D., Lee, J., Kang, J.: Non-stochastic uncertainty response assessment method of beam and laminated plate using interval finite element analysis. Smart Struct. Syst. **26**(3), 311–318 (2020)
2. Piasecka-Belkhayat, A.: Interval boundary element method for transient diffusion problem in two layered domain. J. Theor. Appl. Mech. **9**(1), 265–276 (2011)
3. Kapturczak, M., Zieniuk, E.: IPIES for uncertainly defined shape of boundary, boundary conditions and other parameters in elasticity problems. In: Rodrigues, J.M.F., et al. (eds.) ICCS 2019. LNCS, vol. 11540, pp. 261–268. Springer, Cham (2019). https://doi.org/10.1007/978-3-030-22750-0_20
4. Zieniuk, E., Czupryna, M.: The strategy of modeling and solving the problems described by Laplace's equation with uncertainly defined boundary shape and boundary conditions. Inf. Sci. **582**, 439–461 (2022)
5. Kapturczak, M., Zieniuk, E.: PIES Modeling the boundary shape of the problems described by Navier-Lame equations using NURBS curves in parametric integral equations system method. J. Comput. Sci. **53**, 101367 (2021)
6. Zieniuk, E., Szerszen, K., Kapturczak, M.: A numerical approach to the determination of 3D stokes flow in polygonal domains using PIES. In: Wyrzykowski, R., et al. (eds.) Parallel Processing and Applied Mathematics 2011, LNCS 7203, pp. 112–121. Springer, Heidelberg (2012). https://doi.org/10.1007/978-3-642-31464-3_12
7. Zadeh, L.A.: Fuzzy Sets. Inf. Control **8**, 338–353 (1965)
8. Dubois, D., Prade, H.: Fuzzy Set and Systems - Theory and Applications. Academic Press, New York (1980)
9. Zhao, B., Song, H.: Fuzzy Shannon wavelet finite element methodology of coupled heat transfer analysis for clearance leakage ow of single screw compressor. Eng. Comput. **37**, 2493–2503 (2021)
10. Zalewski, B.: Fuzzy boundary element method for material uncertainty in steady state heat conduction. SAE Int. J. Mater. Manufact. **3**(1), 372–379 (2010)
11. Moore, R.E.: Interval Analysis. Prentice-Hall, Englewood Cliffs, New York (1966)
12. Markov, S.: Extended Interval Arithmetic Involving Infinite Intervals, MathematicaBalkanica, New Series, vol. 6, pp. 269–304 (1992)

Allocation of Distributed Resources with Group Dependencies and Availability Uncertainties

Victor Toporkov(✉), Dmitry Yemelyanov, and Alexey Tselishchev

National Research University "MPEI", Moscow, Russia
{ToporkovVV,YemelyanovDM}@mpei.ru

Abstract. In this work, we introduce and study a set of tree-based algorithms for resources allocation considering group dependencies between their parameters. Real world distributed and high-performance computing systems often operate under conditions of the resources availability uncertainty caused by uncertainties of jobs execution, inaccuracies in runtime predictions and other global and local utilization events. In this way we can observe an availability over time function for each resource and use it as a scheduling parameter. As a single parallel job usually occupies a set of resources, they shape groups with common probabilities of usage and release events. The novelty of the proposed approach is an efficient algorithm considering groupings of resources by the common availability probability for the resources' co-allocation. The proposed algorithm combines dynamic programming and greedy methods for the probability-based multiplicative knapsack problem with a tree-based branch and bounds approach. Simulation results and analysis are provided to compare different approaches, including greedy and brute force solution.

Keywords: Distributed Computing · Resource · Uncertainty · Availability · Probability · Job · Group · Knapsack · Branch and Bounds

1 Introduction and Related Works

High-performance distributed computing systems, such as Grids, cloud, and hybrid infrastructures, provide access to substantial amounts of resources. These resources are typically required to execute parallel jobs submitted by users and include computing nodes, data storages, network channels, software, etc. The actual requirements for resources amount and types needed to execute a job are defined in resource requests and specifications provided by users [1–5]. Distributed computing systems organization and support bring certain economical expenses: purchase and installation of machinery equipment, power supplies, user support, etc. As a rule, users and service providers interact in economic terms and the resources are provided for a certain payment. Economic models [3–5] are used to efficiently solve resource management and job-flow scheduling problems in distributed environments such as cloud computing and utility Grids. Majority of scheduling solutions for distributed environments implement scheduling strategies on a basis of efficiency criteria [1–5].

J. Mikyška et al. (Eds.): ICCS 2023, LNCS 14077, pp. 606–620, 2023.
https://doi.org/10.1007/978-3-031-36030-5_48

Traditional models consider scheduling problem in a deterministic way. Such an approach is sometimes justified by the strict market rules for resources acquisition and utilization during the purchased period of time. Commercial Grids and cloud service providers usually own full control over the resources and may reliably consider their local schedules for some scheduling horizon time [1, 3]. Besides, market-based interactions and QoS constraints compliance require deterministic model for the resources utilization profile. Thus, it is convenient to represent available resources as a set of slots: time intervals when particular nodes are idle and may be used for user jobs execution [4–8]. However general distributed computing systems with non-dedicated resources usually cannot rely on deterministic utilization schedules and instead make predictions based on the utilization predictions and probabilities [9–12]. The probabilities of the resources' availability and utilization at any given time may originate from jobs execution and completion time uncertainties, local activities of the resource provider, maintenance, or numerous failure events. Particular utilization characteristics and patterns usually strongly depend on the resource types. However, according to [9] about 20% of Grid computational nodes exhibit truly random availability intervals.

The scheduling problem in Grid is *NP*-hard due to its combinatorial nature and many heuristic solutions have been proposed. When scheduling under uncertainties, proactive and reactive approaches are usually distinguished [12]. Proactive algorithms concentrate on the resources' utilization predictions and heuristic-based advanced resources allocations and reservations. Reactive algorithms analyze current state of the computing environment and make decisions for jobs migration and rescheduling. Both types of algorithms may be used in a single system to achieve even greater resource usage efficiency. The resources availability predictions for the considered scheduling interval may be obtained based on the historical data processing, linear regression models or with help of expert and machine learning systems [9–11]. In [10], a set of availability states is defined to model resource behavior and probabilities state transitions. On the other hand, sometimes it is possible to identify distributions of resources utilization and availability intervals [9]. Economic scheduling models are implemented in modern distributed and cloud computing simulators GridSim and CloudSim [13]. They provide reliable tools for resources co-allocation but consider price constraints on individual nodes and not on a total window allocation cost. However, as we showed in [6], algorithms with a total cost constraint can perform the search among a wider set of resources and increase the overall scheduling efficiency. Algorithms [14–16] implement knapsack- based slot selection optimization according to a probability-based criterion with a total job execution cost constraint.

This paper extends scheduling algorithms and model presented in [14–16]. We propose proactive algorithms for resources selection and co-allocation computing environments with non-dedicated resources and corresponding availability uncertainties. The uncertainties are modeled as resources availability events and probabilities: a natural way of machine learning and statistical predictions representation [16]. Common resources' allocation and release times are modeled with interdependent resource groups.

The novelty of the proposed approach consists in a dynamic programming scheme performing resources selection with a total availability criterion maximization. The paper is organized as follows. Section 2 presents availability-based scheduling problem and several greedy, knapsack and branch and bounds-based approaches for its solution. Section 3 contains an experiment setup and simulation results obtained for the considered algorithms. Section 4 summarizes the paper and highlights further research topics.

2 Resource Selection Algorithm

2.1 Probabilistic Model for Resource Utilization

In our model we consider a set R of heterogeneous computing nodes with price c_i characteristics under utilization uncertainties. The probabilities (predictions) $p_i(t)$ of the resources' availability and utilization for the whole scheduling interval L are provided as input data. We model a resource utilization schedule as an ordered list of utilization events, such as resource's *allocation, occupation (execution)* and *release* events. An individual job execution on a single resource is modeled as a sequence of *allocation, occupation* (actual execution) and *release* events (see Fig. 1). Additionally, global resources utilization uncertainties, such as maintenance works or network failures, are modeled as a continuous *occupation* event with $P_o << 1$ during the whole considered scheduling interval.

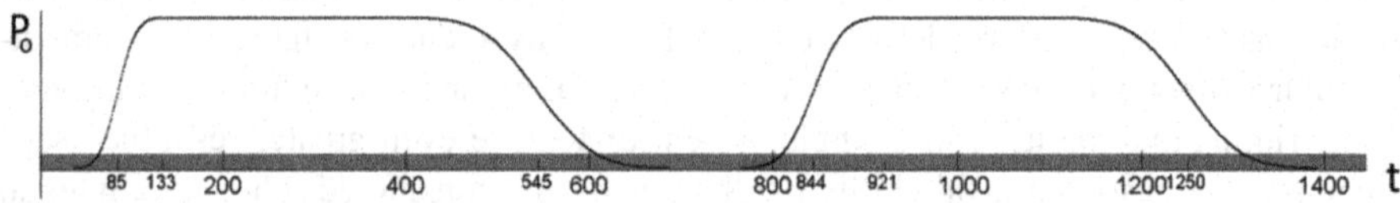

Fig. 1. Example of a single resource occupation probability schedule.

Figure 1 shows an example of a single resource occupation probability P_o schedule. With two jobs already assigned to the resource, there are two resources allocation events (with expected times of allocation at 85- and 844-time units), two resources occupation events (starting at 133- and 921-time units) and two resources release events (expected release times are 545- and 1250-time units respectively). Gray translucent bar at the bottom of the Fig. 1 represents a sum of global utilization events with a total resource occupation probability $P_o = 0.06$. During the whole *execution* interval, the resource's occupation (utilization) probability is assumed as $P_o = 1$. Utilization probability for *allocation* events is modeled by random variable with a normal distribution, and for *release* events - with a *lognormal distribution* to consider the long tails [15]. Expected allocation and release times are derived from the job's replication and execution time estimations. Corresponding standard deviations depend on the job's features and may be predicted based on user estimations or historical data [9–11, 15]. Hence, in Fig. 1 the resource occupation probability at expected times of allocation and release events are: $P_o(85) = P_o(545) = P_o(844) = P_o(1250) = 0.5$.

However, to execute a job, a resource should be allocated for a specified time period T. Based on the model above, we propose the following procedure to calculate a total availability probability P_a of a resource r during time interval T. P_a describes probability, that the resource r will be fully available and will not be interrupted during T.

1. Retrieve a set of independent utilization events e_i active for the resource r during the time interval T. When a subset of dependent events is active during the interval, then only a single event providing the maximum occupation probability P_o is retrieved. For example, from the *allocation-occupation (execution)-release* events chain only the *execution* event is retrieved with $P_o = 1$.
2. For each independent event e_i a maximum occupation probability during the interval l is calculated: $P_o^{max}(e_i) = \max_{t \epsilon T} P_o(e_i, t)$. Corresponding partial availability probability $P_a(e_i)$ is calculated for each event e_i as a probability that the resource will not be occupied by the event during the interval T: $P_a(e_i) = 1 - P_o^{max}(e_i)$.
3. The resource will be available during the whole-time interval T only in case it will not be occupied by any of the active utilization events. Thus, the total availability probability for the resource r is a product of all partial availability probabilities calculated for independent events e_i:

$$P_a^r = \prod_i P_a(e_i). \tag{1}$$

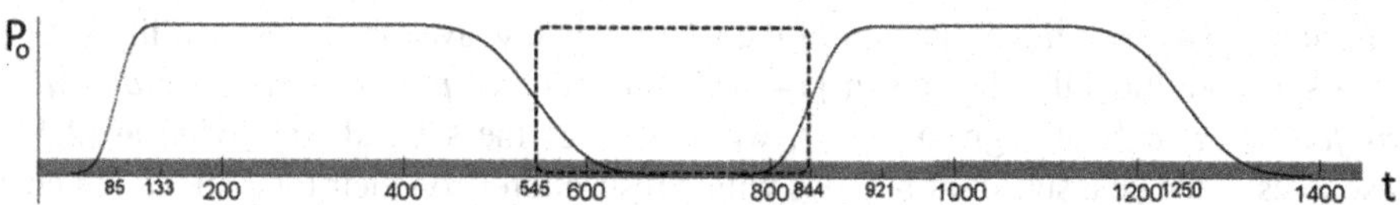

Fig. 2. Example of a resource occupation probability schedule.

For example, consider a resource availability probability for an interval T : [545; 844] presented as a dotted rectangle in Fig. 2. Three independent events are active during the interval: 1) resource release event e_1 with the expected release time at 545 time units, 2) resource allocation event e_2 with the expected allocation time at 844 time units, and 3) a global utilization event e_3 with a constant occupation probability $P_o = 0.06$ (related details were provided with a Fig. 1 example). Corresponding partial occupation and availability probabilities are: $P_o^{max}(e_1) = 0.5$, $P_o^{max}(e_2) = 0.5$, $P_o^{max}(e_3) = 0.06$, while $P_a(e_1) = 0.5$, $P_a(e_2) = 0.5$, $P_a(e_3) = 0.94$. So, the total probability of the resource availability during the whole interval T is $P_a^r = 0.235$.

2.2 Parallel Job Scheduling and Group Dependencies

To execute a parallel job, a set of simultaneously available nodes (a *window*) should be allocated ensuring user requirements from the resource request. The resource request usually specifies number n of nodes required simultaneously for a time period T and a maximum available resources allocation budget C. The total cost of a window allocation is calculated as $C_W = \sum_{i=1}^{n} T * c_i$, where c_i is resource i price for a single time unit.

These parameters constitute a formal generalization for resource requests common among distributed computing systems and simulators [13–16]. Period T of the resources acquisition is usually the same for all resources selected for a parallel job. Common allocation and release times ensure the possibility of inter-node communications during the whole job execution. In this way, the *total window availability* is a function of availability probabilities of all the selected resources during the considered time interval T. More formally, when a set of n resources is selected for a job, the total window availability P_a^{w} during the expected job execution interval can be estimated as a product of availability probabilities $P_a^{\mathrm{r_i}}$ of each *independent* window nodes:

$$P_a^{w} = \prod_{i}^{n} P_a^{r_i}. \quad (2)$$

Here $P_a^{\mathrm{r_i}}$ can be calculated for each resource by the algorithm described in Sect. 2.1. If any of the window nodes will be occupied during the expected job execution interval (i.e., $P_a^{r_i} = 0$), the whole parallel job will be postponed or even aborted. Therefore, in general, the window allocation procedure should consider *maximization of the total probability of availability* $P_a^{\mathrm{w}} \rightarrow$ max. Based on the model above the general statement of the window allocation problem is as follows: during a scheduling interval L allocate a subset of n nodes with performance $p_i \geq p$ for a time T, with common allocation and release times and a restriction C on the total allocation cost. As a target optimization criterion, we assume maximization of the whole window availability probability (2).

As we additionally showed in [14, 15], this *general problem can be reduced to the following task*: at a given time t, which defines the set and state of m available resources, allocate a subset of n nodes with a restriction C on their total allocation cost while performing maximization of their total availability probability (2). In [14, 15] we proposed several approaches to solve the problems above. However, their statement and solution assume *independence* of individual resources as well as their utilization events. That is why in (2) we calculate the total window availability as a product of the availability probabilities of its elements.

In a more general and realistic model, the resources and their utilization events *are not independent*. On the contrary, there are group dependencies between the resources' parameters. The most typical example of such a dependency is a result of a parallel job execution. When a parallel job is scheduled, a set of selected resources is allocated for a common period T. That is, all the selected resources will share allocation, occupation, and release times. So, they should be modeled with a common chain of *allocation-occupation-release* events. In another words, these resources have a *group dependency.*

Figure 3 shows example of utilization events modeled for a parallel job, which requested three nodes. Red areas present resources' utilization probability for allocation and release events. As the exact allocation and release times are unknown, the corresponding occupation probabilities $P_o(t) < 1$. Green areas show execution event with the occupation probability $P_o = 1$. The main issue is that criterion (2) becomes inaccurate when applied to a resources' set with many internal group dependencies. For example, in Fig. 3 if we consider total availability probability of resources 1, 4 and 5 at time $t = 400$, criterion (2) will calculate it as a product $P_a^{\mathrm{w}} = P_a^1 * P_a^3 * P_a^4$. However, as these resources are used by the same parallel job (and have a common group dependency), their actual total availability probability $P_a^{\mathrm{w}} = P_a^1 = P_a^3 = P_a^4 \geq P_a^1 * P_a^3 * P_a^4$.

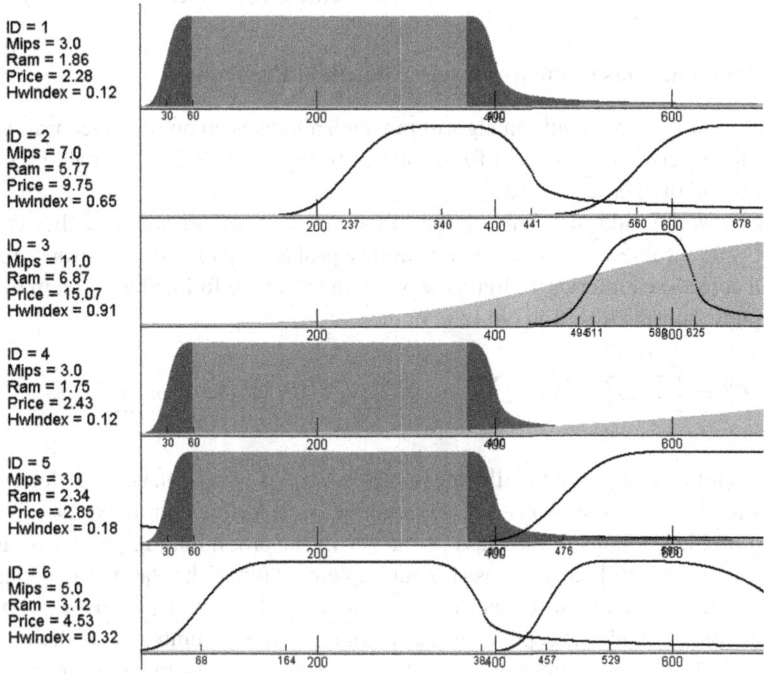

Fig. 3. Example of a parallel job execution schedule.

To describe it more formally we consider a set of groups G over the set R of the available resources. Each component group $G_i \in G$ represents a subset of resources $r_j \in R$ with a common group dependency. For example, one scheduled job, like in the example above, forms a single group G_i which includes all the resources selected for the job. So, for example, if one resource $r_j \in G_i$ is selected for a window W, the common group availability $P_a^{\mathrm{G_i}}$ should be used for calculation of a total W availability probability P_a^{w}. However, additionally selecting any other resources from G_i will not affect P_a^{w}, as their group probability component $P_a^{\mathrm{G_i}}$ is already considered.

So, the total window W availability probability can be calculated as follows:

$$P_a^{w} = \prod_{i}^{n^*} P_a^{G_i}, \tag{3}$$

where n^* is a number of diverse groups used for the window W, and $P_a^{G_i}$ is availability probability for each different group G_i used for the window. Group G_i is added to (3) if at least one of its resources is selected for the window. It is worth noting, that in the extreme case each group G_i can contain only one resource, and thus (3) will converge to

(2). In this paper we propose and study resources allocation algorithm which performs (3) $P_a^w \rightarrow$ max optimization considering economic constraint on the total window cost and group dependencies G. However firstly we should introduce helper algorithms performing (2) $P_a^w \rightarrow$ max optimization without the group dependencies configuration.

2.3 Direct Solutions of the Resources Allocation Problem

Let us discuss in more details an algorithm which allocates an optimal (according to the probability criterion P_a^w) subset of n resources from the set R of m available resources with a limit C on their total cost.

Firstly, we consider maximizing the following total resources availability criterion $P_a^w = \prod_j^n p_a^{r_j}$, where $p_a^{r_j} = p_j$ is an availability probability of a single resource $r_j \in R$ during a considered interval T. In this way we can state the following problem of an n -size window subset allocation out of m nodes:

$$P_a^w = \prod_j^m x_j p_a^{r_j} \rightarrow \max, \sum_j^m x_j c_j \leq C, x_j \in \{0, 1\}, j = 1..m, \sum_j^m x_j = n, \quad (4)$$

where c_j is total cost required to allocate resource r_j, x_j - is a decision variable determining whether to allocate resource r_j ($x_j = 1$) or not ($x_j = 0$) for the current window.

This problem relates to the class of integer linear programming problems, and we used 0–1 knapsack problem as a base for our implementation. The classical 0–1 knapsack problem with a total weight C and items-resources with weights c_j and values p_j have a similar formal model except for extra restriction on the number of items required: $x_1+x_2+\cdots+x_m = n$. Therefore, we implemented the following dynamic programming recurrent scheme:

$$f_j(c, v) = \max\{f_{j-1}(c, v), f_{j-1}(c - c_j, v - 1) * p_j\}, \quad (5)$$

$$j = 1, .., m, c = 1, .., C, v = 1, .., n,$$

where $f_j(c, v)$ defines the maximum availability probability value for a v-size window allocated from the first j resources of m for a budget c. After the forward induction procedure (4) is finished the maximum availability value $P_{a\,max}^w = f_m(C, n)$. x_j values are then obtained by a backward induction procedure.

An estimated computational complexity of the presented knapsack-based algorithm *KnapsackP* is $O(m * n * C)$.

Another approach for n-size window allocation is to use a more computationally efficient greedy approach. We outline four main greedy algorithms to solve the problem (3). The task is to select n out of m resources providing maximum total availability probability P_a^w with a constraint on their total allocation cost n.

1. *MaxP* selects first n nodes providing maximum availability probability p_j values. This algorithm does not consider total usage cost limit and may provide infeasible solutions. Nevertheless, *MaxP* can be used to determine the best possible availability options and estimate a budget required to obtain them.

2. An opposite approach *MinC* selects first n nodes providing minimum usage cost c_j or an empty list in case of exceeding a total cost limit C. In this way, *MinC* does not perform any availability optimization, but always provides feasible solutions when it is possible. Besides, *MinC* outlines a lower bound on a budget required to obtain a feasible solution.
3. Third option is to use a weight function to regularize nodes in an appropriate manner. *MaxP/C* uses $w_j = p_j/c_j$ as a weight function and selects first n nodes providing maximum w_j values. Such an approach does not guarantee feasible solution, but nonetheless performs some availability optimization by implementing a compromise solution between *MaxP* and *MaxC*.
4. Finally, we consider a composite approach *GreedyUnited* for an efficient greedy-based resources allocation. The algorithm consists of three stages.

   a.Obtain MaxP solution and return it if the constraint on a total usage cost is met.
   b.Else, obtain MaxP/C solution and return it if the constraint on a total usage cost is met.
   c.Else, obtain MinC solution and return it if the constraint on a total usage cost is met.

This combined algorithm *GreedyUnited* is designed to perform the best possible greedy optimization considering a restriction on a total resources' allocation cost C.

Estimated computational complexity for the greedy resources' allocation step is $O(m * \log m)$. More details regarding the algorithms above are provided in [14–16].

2.4 Resources Allocation Algorithms with Group Dependencies

Based on *KnapsackP* and *GreedyUnited* implementations above we propose the following algorithm for a general resource allocation problem considering group dependencies between the available resources. It takes as input set R of the available resources (each resource is characterized with costc_i) and set G of groups over R (each group G_i has a common availability probabilityp_i). The algorithm then allocates a subset of n resources with a restriction C on their total cost while performing maximization of their total availability probability (3). The problem is solved by branch and bounds method by maintaining max-heap data structure H containing interim candidate solutions S_j. The higher the achieved availability probability P_a^w (3) or its upper bound, the closer the solution S to the top of the heap$H$. For each solution S we maintain two subsets of groups that should (G^+) and should not (G^-) be used in the current solution. Both G^+ and G^- are initialized as empty sets. Additionally, we consider subset G^0 as all groups from G not included in G^+ orG^-, so G^0 is initialized as$G$.

Initial candidate solution S^0 with empty $G^0 = G$ and empty sets G^+ and G^-, is placed into H with $P_a^w = -\textit{infinity}$. Next, we perform the following steps.

1. Retrieve next solution candidate S from H. If S is marked as valid solution, then return S as a result, end of the algorithm.
2. Prepare list of resources R_s to calculate P_a^w for S.
   a. Init R_s as empty set.
   b. For each group G_j from G^+ add the cheapest resource to the solution window W_s with the $p_i = P_a^{G_j}$; add other resources from this group $r_i \in G_j$ to R_s with $p_i = 1$.

   c. For each group G_j from G^0 add all resources $r_i \in G_j$ to R_s with $p_i = \sqrt[k]{P_a^{G_j}}$, where k is number of resources in G_j.
3. Use algorithm *KnapsackP* or *GreedyUnited* to perform direct solution of S to allocate resources into W_s (it can be partially filled during step 2.b) from set R_s of prepared resources with (2) $P_a^w \rightarrow \max$ optimization without group dependencies.
4. Check if the resulting solution is valid.
   a. If all resources from W_s are included in groups from G^+, then put this solution S into H with key P_a^w and mark it as a *valid solution.*
   b. If at least one resource r_s from W_s is included in some group G_s from G^0, then we need to split this solution S into two candidates: S^+ and S^-. For S^+ remove group G_s from G^0 and add into G^+. For S^- remove group G_s from G^0 and add into G^-. Put both solution candidates S^+ and S^- into H with key P_a^w as an upper estimate.
5. Go to step 1.

The algorithm above performs branch and bounds approach by splitting candidate solutions by sets of resources groups G^+ and G^- required to use or skip correspondingly. A special resource set R_s preparation in step 2 allows us to use (2) optimization algorithms and obtain either a final valid solution or a candidate solution with pretty accurate upper estimate. The algorithm finishes when the next solution obtained from the max-heap data structure is a valid solution composed of resources from G^+ groups and, thus, its P_a^w calculated with (2) satisfies rules for group dependencies availability calculations (3).

3 Simulation Study

3.1 Considered Algorithm Implementation

For the simulation study we consider and compare the following algorithm implementations.

1. Firstly, we implemented *brute-force* algorithm to solve the resources allocation problem with (3) $P_a^w \rightarrow \max$ optimization. We used this algorithm for a preliminary analysis in small experiments with up to 21 resources to compare its optimization efficiency with other approaches.
2. Next, we prepared three implementations of a general branch and bounds algorithm described in Sect. 2.4. First implementation *KnapsackGroup* uses *KnapsackP* for all interim allocations during the algorithm step 3. *Greedy* performs interim optimizations at step 3 with *GreedyUnited* algorithm. Finally, *Greedy+* runs *GreedyUnited* for all interim optimizations, but once the solution is found, the final solution optimization is performed again using more accurate *KnapsackP* approach.
3. Finally, we consider *KnapsackP (KnapsackSingle)* as standalone algorithms for the comparison. This algorithm does not support group dependencies and performs (2) $P_a^w \rightarrow \max$ optimization. The obtained solution is then recalculated accordingly to (3) to compare it to the algorithms above.

For the simulation study we execute and collect resulting data for all the considered algorithms (*BruteForce, KnapsackGroup, Greedy, Greedy* + and *KnapsackSingle)* in different resource environments with randomized characteristics c_i, p_i and group dependencies. An experiment was prepared using a custom distributed environment simulator [6, 14, 15]. For our purpose, it implements a heterogeneous resource domain model: nodes have different usage costs and availability probabilities. Each node supports a list of active global and local job utilization events. Figure 3 shows an example of such an environment with many resources and a Gantt chart of the utilization events.

Additionally, we generate random uniformly distributed group dependencies between the resources. So, the resources allocation problem can be defined with the following parameters: N – number of available resources (each characterized with cost c_i and availability probability p_i), G – number of different groups (containing random non-intersecting subsets of resources), n – number of resources required for allocation and C – available budget, i.e. constraint on the total cost of the selected resources.

3.2 Proof of Optimization Efficiency

The first experiment series studies algorithms optimization and computational efficiency in comparison with *BruteForce* approach. Brute force is usually inapplicable in real-world tasks due to its exponential computational complexity. However, it guarantees exact optimization solution, and can be used to evaluate optimization characteristics of other considered algorithms. During each simulation experiment, the resources allocation was independently performed by algorithms *BruteForce, KnapsackGroup, Greedy, Greedy+* and *KnapsackSingle.* The comparison is obtained with different values of G, n, C of the allocation problem. As *BruteForce* applicability is limited, firstly we performed resources allocation simulation with only $N = 21$ available resources.

Figure 4 shows resulting availability probability P_a^w depending on number $n \in [1; 21]$ of requested resources in environment with $N = 21$ available resources, $G = 8$ different groups and without the total cost restriction ($C = \sum_i^N c_i$). The main result is that proposed algorithms *KnapsackGroup, Greedy* and *Greedy+* provided the same P_a^w value as *BruteForce* (that is why they are not presented in Fig. 4). *KnapsackGroup* theoretically guarantees exact problem solution in integers and is expected to provide results identical to *BruteForce.* Greedy algorithms provided optimal solution due to the lack of the total cost limit (see *GreedyUnited* and *MaxP* descriptions in Sect. 2.3). However, *KnapsackSingle* in most cases failed to provide optimal solution with up to 5% lower availability probability compared to *BruteForce.* The equality is achieved only in two simplified scenarios with $n = 1$ and $n = 21$, when group dependencies are not relevant for the problem.

Figure 5 shows actual algorithms' execution time required to achieve allocation results from Fig. 4. As can be seen, *BruteForce* calculation time dramatically increases for $n \in [7; 15]$ and exceeds half a second for $n = 11$. This is explained by the combinatorial nature of selecting subset of n from N available resources. Even the most computationally complex *KnapsackGroup* algorithm, which combines pseudo polynomial 0–1 knapsack implementation with branch and bounds approach is presented in Fig. 5 as a straight line 100 times lower compared to the *BruteForce* maximum. *Greedy* approaches were up to 1000 times faster than *BruteForce.* So, according to the trend in Fig. 5, in environments with $N > 25$ *BruteForce* becomes practically inapplicable and other exact algorithms and approximations should be considered. The accuracy of such approximations in general

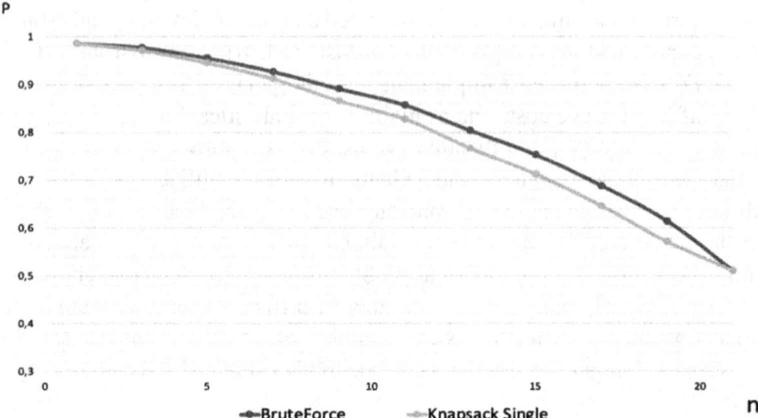

Fig. 4. Simulation results: resulting availability probability P_a^w depending on number n of requested resources.

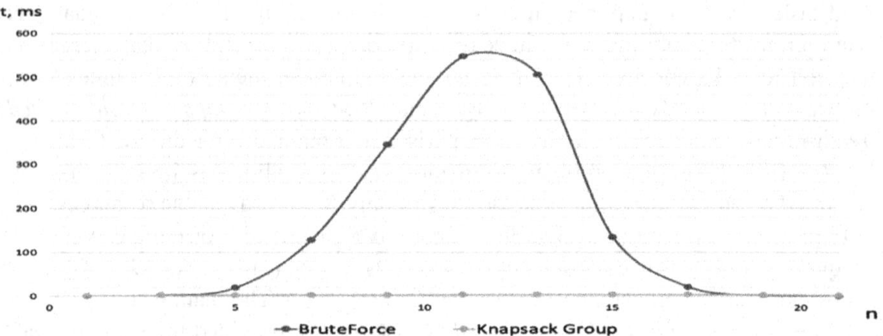

Fig. 5. Simulation results: average calculation time depending on number n of requested resources.

should be estimated with the economical restriction C on the total window allocation cost.

Figure 6 shows how window availability probability depends on the allocation budget $C \in [30; 120]$ in problem setup with $n = 8$, $G = 8$ and $N = 21$. In this environment only *KnapsackGroup* was able to obtain exact solutions (identical to *BruteForce*) for all C values. Additionally, *KnapsackGroup* provides almost constant 5% advantage over *KnapsackSingle*. The results of Greedy algorithms are also within 5% of the exact solution and reaches *BruteForce* for $C > 90$. In general, the obtained simulation result confirms accuracy of *KnapsackGroup* algorithm and gives an approximate estimate of the accuracy of the more computationally simple algorithms.

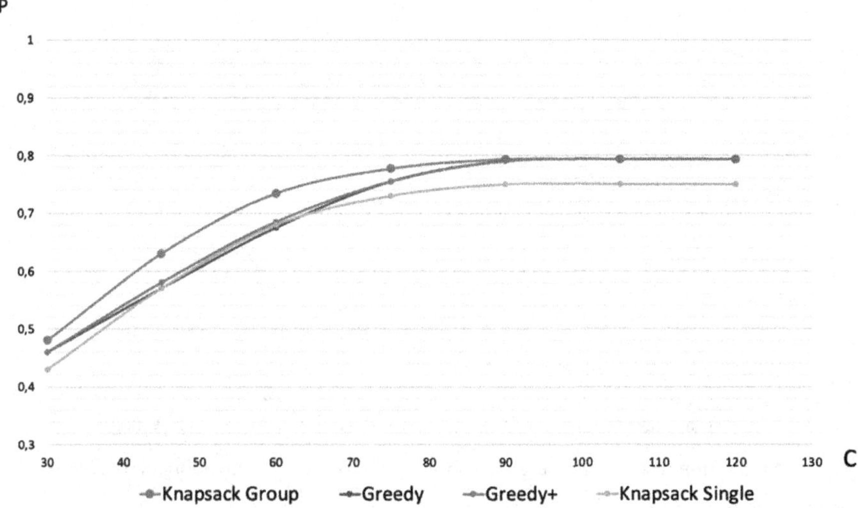

Fig. 6. Simulation results: resulting availability probability P_a^w depending on the budget C.

3.3 Practical Optimization Efficiency Study

Next experiment series studies proposed algorithms in more complex problem settings with $N = 200$, $G = 40$ and $n = 20$. As brute force becomes impractical for such figures, we use *KnapsackGroup* as a reference and accurate solution of (3).

Firstly, Fig. 7 shows availability probability as a function of $C \in [40; 220]$. Lower bound was selected so that it was almost impossible just to allocate any 20 resources with budget $C < 40$, without any optimization. So, the resulting P_a^w generally increase with increasing C. Upper bound $C > 200$ allows to select almost any resources without checking for the total cost limit. In this experiment setup with more resources and optimization variability, Greedy algorithms are already seriously losing the accuracy of the solution. The advantage of *KnapsackGroup* exceeds 20% for some values of C. This result generally correlates with works [14, 15]. At the same time, *KnapsackSingle* provides availability probability only 10% lower than the exact solution. In this way, the absence of group dependencies information turns out advantageous compared to the accuracy of greedy approximations of the multiplicative knapsack problem. Only in scenarios with $C > 200$, i.e., without the cost restriction, *Greedy* can outperform *Knasack Single* in environment with group dependencies between the resources.

Another crucial factor for the practical applicability is the algorithms' calculation time presented in Fig. 8 for the same environment settings. *Greedy* and *Greedy+* have almost identical calculation times and so their graphs fully overlap. The obvious trend is that tighter restrictions on the budget C cause a strong increase in working time for branch and bounds - based algorithms (*KnapsackGroup, Greedy, Greedy+*). This is explained by the necessity to select resources with respect to the C constraint, rather than by the target criterion. And this strategy requires consideration of more diverse groups and splitting in branch and bounds approach. For example, with$C = 40$, an average size of the solution tree for *KnapsackGroup* was almost 5000 elements causing nearly 8 s of

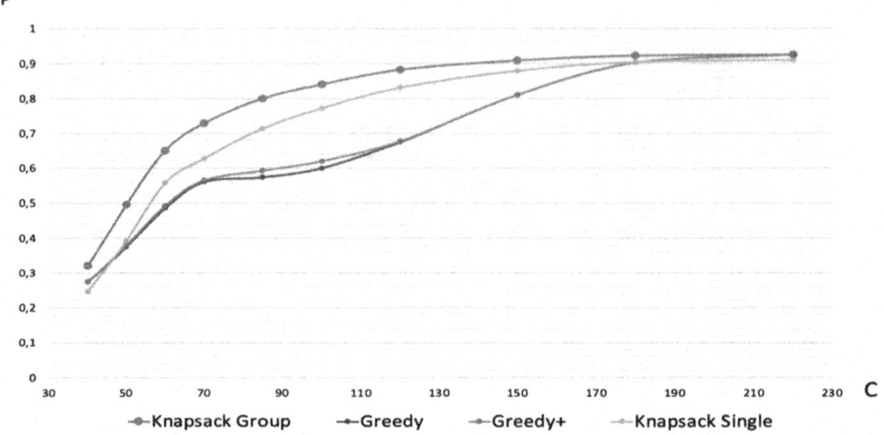

Fig. 7. Simulation results: resulting availability probability P_a^w depending on the budget C.

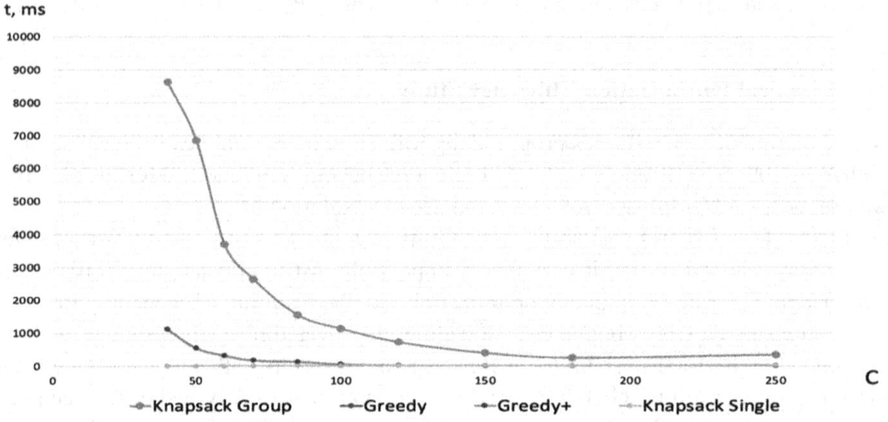

Fig. 8. Simulation results: average calculation time depending on the budget C.

the execution time. And with $C = 120$ the tree size decreased to nearly 100 elements leading to a sub second execution time. Similar calculation time trend applies to Greedy tree algorithms as well.

Thus, based on Figs. 7 and 8 we conclude, that with tight economical budget restrictions the most practically adequate option is a simple multiplicative 0–1 knapsack algorithm [14], as such problem setup requires greater emphasis on the cost optimization and less on the groups' combinations. With a looser cost restriction ($C \in [100; 200]$ in our experiment) tree-based *KnapsackGroup* becomes a preferred option as it provides exact optimization solution for an adequate calculation time. Finally, when there is no cost restriction, a tree-based *Greedy* algorithm can provide exact optimization result in the least amount of calculation time.

4 Conclusion and Future Work

In this work, we address the problem of dependable resources co-allocation for parallel jobs in distributed computing with group dependencies over the resources. Such group dependencies usually define utilization events common for subsets of resources, such as simultaneous allocation or release events. To handle this problem, we designed several branches and bounds algorithms based on a multiplicative 0–1 knapsack problem. In a simulation study we proved accuracy of the proposed algorithms in comparison with a brute force approach, estimated their calculation time and practical applicability in a more complex scheduling problems with up to 200 available computing nodes.

Future work will concern additional optimization in the algorithms' complexity and calculation time. In addition, we plan to consider similar allocation task based on an additive 0–1 knapsack problem.

Acknowledgments. This work was supported by the Russian Science Foundation (project No. 22-21-00372, https://rscf.ru/en/project/22-21-00372/).

References

1. Lee, Y.C., Wang, C., Zomaya, A.Y., Zhou, B.B.: Profit-driven scheduling for cloud services with data access awareness. J. Parall. Distrib. Comput. **72**(4), 591–602 (2012)
2. Garg, S.K., Konugurthi, P., Buyya, R.: A linear programming-driven genetic algorithm for meta-scheduling on utility grids. Int. J. Parall. Emerg. Distrib. Syst. **26**, 493–517 (2011)
3. Buyya, R., Abramson, D., Giddy, J.: Economic models for resource management and scheduling in grid computing. J. Concurren. Comput.: Pract. Exp. **5**(14), 1507–1542 (2002)
4. Ernemann, C., Hamscher, V., Yahyapour, R.: Economic scheduling in grid computing. In: Feitelson, D.G., Rudolph, L., Schwiegelshohn, U. (eds.) JSSPP 2002. LNCS, vol. 2537, pp. 128–152. Springer, Heidelberg (2002). https://doi.org/10.1007/3-540-36180-4_8
5. Kurowski, K., Nabrzyski, J., Oleksiak, A., Weglarz, J.: Multicriteria aspects of grid re-source management. In: Nabrzyski, J., Schopf, J.M., Weglarz, J. (eds.) Grid Resource Management, pp. 271–293. Kluwer Academic Publishers, State of the Art and Future Trends (2003)
6. Toporkov, V., Toporkova, A., Bobchenkov, A., Yemelyanov, D.: Resource selection algorithms for economic scheduling in distributed systems. In: ICCS 2011, June 1–3, 2011, Singapore, Procedia Computer Science, vol. 4. pp. 2267–2276. Elsevier (2011)
7. Netto, M.A.S., Buyya, R.: A flexible resource co-allocation model based on advance reservations with rescheduling support. In: Technical Report, GRIDSTR-2007-17, Grid Computing and Distributed Systems Laboratory, The University of Melbourne, Australia, October 9 (2007)
8. Jackson, D., Snell, Q., Clement, M.: Core algorithms of the Maui scheduler. In: Feitelson, D.G., Rudolph, L. (eds.) JSSPP 2001. LNCS, vol. 2221, pp. 87–102. Springer, Heidelberg (2001). https://doi.org/10.1007/3-540-45540-X_6
9. Javadi, B., Kondo, D., Vincent, J., Anderson, D.: Discovering statistical models of availability in large distributed systems: An empirical study of SETI@home. IEEE Trans. Parall. Distrib. Syst. **22**(11), 1896–1903 (2011)
10. Rood, B., Lewis, M.J.: Grid resource availability prediction-based scheduling and task replication. J. Grid Comput. **7**, 479 (2009)

11. Tchernykh, A., Schwiegelsohn, U., El-ghazali, T., Babenko, M.: Towards understanding uncertainty in cloud computing with risks of confidentiality, integrity, and availability. J. Comput. Sci. **36** (2016)
12. Chaari, T., Chaabane, S., Aissani, N., Trentesaux, D.: Scheduling under uncertainty: Survey and research directions. In: 2014 International Conference on Advanced Logistics and Transport (ICALT), pp. 229–234 (2014)
13. Calheiros, R.N., Ranjan, R., Beloglazov, A., De Rose, C.A.F., Buyya, R.: CloudSim: A toolkit for modeling and simulation of cloud computing environments and evaluation of resource provisioning algorithms. J. Softw. Pract. Exp. **41**(1), 23–50 (2011)
14. Toporkov, V., Yemelyanov, D.: Availability-based resources allocation algorithms in distributed computing. In: Voevodin, V., Sobolev, S. (eds.) RuSCDays 2020. CCIS, vol. 1331, pp. 551–562. Springer, Cham (2020). https://doi.org/10.1007/978-3-030-64616-5_47
15. Toporkov, V., Yemelyanov, D., Grigorenko, M.: Optimization of resources allocation in high performance computing under utilization uncertainty. In: Paszynski, M., Kranzlmüller, D., Krzhizhanovskaya, V.V., Dongarra, J.J., Sloot, P.M.A. (eds.) ICCS 2021. LNCS, vol. 12747, pp. 540–553. Springer, Cham (2021). https://doi.org/10.1007/978-3-030-77980-1_41
16. Toporkov, V., Yemelyanov, D., Bulkhak, A.: Machine learning-based scheduling and resources allocation in distributed computing. In: Groen, D., et al. (eds.) Computational Science – ICCS 2022. ICCS 2022. Lecture Notes in Computer Science, vol. 13353, pp. 3–16. Springer, Cham (2022). https://doi.org/10.1007/978-3-031-08760-8_1

A Bayesian Optimization Through Sequential Monte Carlo and Statistical Physics-Inspired Techniques

Anton Lebedev[1(✉)], M. Emre Şahin[1], and Thomas Warford[2]

[1] The Hartree Centre, Keckwick Ln, Warrington, UK
{anton.lebedev,emre.sahin}@stfc.ac.uk

[2] University of Manchester, M13 9PL Manchester, UK
thomas.warford@student.machester.ac.uk

Abstract. In this paper, we propose an approach for an application of Bayesian optimization using Sequential Monte Carlo (SMC) and concepts from the statistical physics of classical systems. Our method leverages the power of modern machine learning libraries such as NumPyro and JAX, allowing us to perform Bayesian optimization on multiple platforms, including CPUs, GPUs, TPUs, and in parallel. Our approach enables a low entry level for exploration of the methods while maintaining high performance. We present a promising direction for developing more efficient and effective techniques for a wide range of optimization problems in diverse fields.

Keywords: Stochastic Methods · High-Performance Computing · Bayesian Inference

1 Introduction

Bayesian optimization of ever-growing models has become increasingly important in recent years and significant effort has been invested in achieving a reasonable runtime-to-solution. Unfortunately, most optimization tasks are implemented and optimized within a specific framework, resulting in a single optimized model. The proliferation of such implementations is difficult, as it requires both domain expertise and knowledge of the specific framework and programming language.

Probabilistic programming frameworks such as Stan [6] and (Num)Pyro [4], provide such support for efficient optimisation methods such as Hamiltonian Monte Carlo (HMC) algorithm which can explore complex high-dimensional probability distributions. These frameworks are powerful tools for statistical analysis and inference.

Stan excels at handling complex hierarchical models with ease, which is often challenging in other probabilistic programming frameworks. Its user-friendly interface makes it accessible to those with little experience in Bayesian inference and statistical modelling, making it a popular choice for the accurate and

J. Mikyška et al. (Eds.): ICCS 2023, LNCS 14077, pp. 621–628, 2023.
https://doi.org/10.1007/978-3-031-36030-5_49

efficient analysis of complex models. It provides domain experts with a performant tool to perform the said task with little to no programming knowledge. It achieves this by defining a "scripting language" for models and translation of these into C++ code. It suffers, however, from a lack of inherent parallelism and a formulation of its methods in heavily-templated C++. NumPyro, built on JAX [5], enables efficient exploration of high-dimensional probabilistic models using different methods, while JAX itself combines the flexibility and ease-of-use of NumPy with the power and speed of hardware accelerators for efficient numerical computing. Additionally, JAX offers compatibility and portability by supporting code execution on a variety of hardware.

Inspired by HMC and SMC descriptions in [7,8], we implemented an HMC algorithm in Python using the NumPyro and JAX frameworks and with DeepPPL [1] utilized to translate existing Stan models into their Python equivalents. In this paper we present preliminary findings from our implementation of SMC for Bayesian parameter searches developed with its physical origins intact.

2 Method Description

Upon review of the SMC algorithm [7,8] and its HMC kernel, it has become apparent that the approach resembles the cooling process of an ideal gas in a potential field with unknown minima, and that the algorithm's development would benefit from an understanding based on physical systems. To facilitate this understanding, we have undertaken a reformulation of the SMC and HMC algorithms to more accurately reflect the particle ensembles of statistical physics. A simple first step was the reintroduction of a temperature into the expressions for probability, seeing as the probabilities used in these methods are the maximum-entropy probabilities of an ensemble (collection) of particles at a fixed energy:

$$p_i = \frac{e^{-E_i/(k_B T)}}{\sum_{j=1}^{N} e^{-E_j/(k_B T)}} . \tag{1}$$

Here k_B is the Boltzmann constant (carrying the dimensions of energy per degree of temperature), T the temperature and $E_i = \mathcal{H}(q_i, p_i)$ is the energy of the particle i at position q_i with a momentum p_i given a Hamiltonian:

$$\mathcal{H}(q,p) = \frac{p^2}{2m} + V(q) . \tag{2}$$

As is common, the Hamiltonian encodes the dynamics of the system. Since we seek to determine the parameters that maximise the log-likelihood of the model the potential can be defined [7] as:

$$V(q) := -\ln(P(q|X)) , \tag{3}$$

where $P(q|X)$ is the posterior probability density of the model, where X is the vector containing the observations and q is the (position) vector in parameter space - the parameters of the model.

The formulation of (1) commonly used in mathematics implies $T = \frac{1}{k_B}$ or an arbitrary definition of $k_B := 1$. The former fixes a degree of freedom of the method, whilst the latter allows for a variation, but removes the dimensionality of the constant which, in conjunction with T ensures that the argument of the exponential remains a dimensionless number or quantity. Whilst *functionally* of no consequence retaining the physical dimensions of the respective quantities allows for sanity checks of formulae during development.

The distribution of the "auxiliary momentum" described in [7] is naturally dependent on T, with a higher temperature T resulting in a broader distribution of the momenta (faster particles) and a wider area of the initial sample space covered.

The HMC process is rather simple and described in [7] in sufficient detail. In contrast, sequential Monte-Carlo is provided in a less legible form in [8]. Hence we privde here the simple version we turn our attention to:

1. Initialise the position samples to a normal distribution on $\mathbb{R}^n$ and the momentum samples to a normal distribution with temperature T.
2. Iterate for a given number of SMC steps:
   (a) Propagate each particle in the ensemble using HMC.
   (b) Determine the lowest energy for all particles, subtract it from every energy (renormalisation) and store it for future processing.
   (c) Compute and store the average parameter value.
   (d) Determine the effective size $N_{\mathit{effective}}$ of the ensemble according with [8].
   (e) If resampling is necessary select $N_{\mathit{effective}}$ particles with largest weights (lowest energies) and duplicate them according to their probability until the original ensemble size is reached. Then reset the momenta to a thermal distribution and the weights of the particles to $\frac{1}{N}$.
3. Compute the moving average of the stored averages, using the stored energies as weights in accordance with (1).

Here we must note the rescaling of the weights by subtraction of the lowest energy, which is physically motivated by the freedom to choose the origin of the energy scale. Similarly, the weighted moving average in the last step results in a smoothing of the excursions of the mean after a resampling step by weighting means with large associated energies exponentially smaller (the higher the energy the more unlikely a configuration is to occur).

3 Numerical Experiments

3.1 Models

To demonstrate the effectiveness of the implementation we have selected two simple models:

1. A sequence of M independent tosses of two coins - the Coin Toss (CT) model.
2. Item Response Model with Two-Parameter Logistic (IRT 2PL).

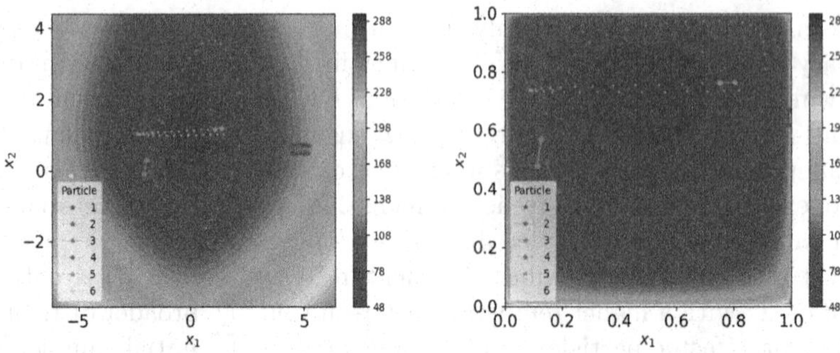

(a) HMC trajectories on the unconstrained R^2 for the CT model.

(b) HMC trajectories and the potential of the CT model on the unit square support of the model.

Fig. 1. Trajectories of our HMC implementation in the potential defined by (3) for the CT model. Note the dense zig-zag trajectories that result if the momentum inversion proposed in [7] is implemented.

CT Model. The CT model assumes complete ignorance of the a-priori coin bias p and its maximum a-posteriori probability estimator can be determined formally to be:

$$\hat{P}_{MAP} = \frac{K}{N} . \tag{4}$$

Here K is the number of observed heads and $N = 40$ is the total number of observations. This allows us to check the numerical approximation of (3) as well as its gradients, ensuring proper functioning of the implementation. The true parameters of the coin bias for each of the two coins are

$$p_1 = \frac{1}{2} , \; p_2 = \frac{3}{4} . \tag{5}$$

The potential of the CT model, along with a few sample trajectories of the particles propagating therein are shown in Fig. 1a prior to the constraining to the support of the model, and in Fig. 1b after. Here we note that we chose *not* to invert the momentum in case the new phase-space point (q, p) passes the Metropolis-Hastings acceptance test, contrary to alg. 1 of [7]. As can be seen in the figures, such an inversion results in a rather slow sampling of the potential in the case of a smooth potential. It is, however, beneficial for a rough potential and hence likely better in most practical applications.

IRT 2PL. Item Response Theory (IRT) [2] is a statistical model that is widely used in a variety of research fields to analyze item responses in assessments or surveys. The two-parameter logistic (2PL) model is a specific type of IRT model that assumes each item has two parameters: the difficulty parameter and the discrimination parameter. The model assumes that the item responses y_i for

$i = 1, \ldots, I$ are Bernoulli distributed with a logit link function. The probability of responding correctly to item i is given by:

$$P(y_i = 1|\theta, a_i, b_i) = \frac{1}{1 + \exp(-a_i(\theta - b_i))} \tag{6}$$

where θ is the person's latent ability, a_i is the discrimination parameter for item i, and b_i is the difficulty parameter for item i.

3.2 Experimental Results

Given the physical interpretation of the SMC iteration and the weighted average derived therefrom we expected - a-priori - a rather rapid and smooth convergence towards the true parameter values. Given a smooth potential, as in the case of the CT model, it is furthermore expected that convergence to the estimated parameters is faster for lower temperatures T.

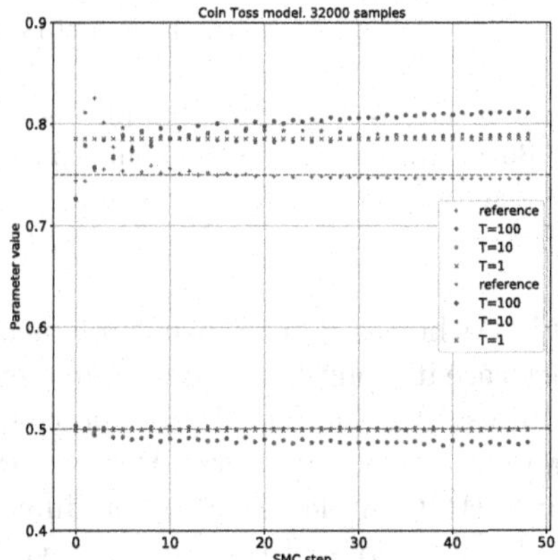

Fig. 2. Parameter estimates obtained with SMC with physics-motivated moving averages. The dashed lines represent the true parameter values. The crosses refer to a reference implementation not available to the public.

Quality of Estimation. As can be seen in Fig. 2 , all estimates of the bias of a fair coin converge within 5 iterations to the true value $\frac{1}{2}$ (indicated by the dashed red line). In case of a rather high temperature (more precisely: thermal energy) of $T = 10\frac{1}{k_B}$ the apparent limit value deviates noticeably from the true value, given the remarks above this is not unexpected and will be remedied in a future iteration of the method.

In the case of the biased coin, with $p_2 = \frac{3}{4}$ it becomes obvious that the current development stage of the method may suffer from a freeze-in (c.f. $T =$

1 case) resp. a bias in the implementation. For a comparison reference, non-public, implementation results are marked as '+', showing a similar convergence behaviour but without the offsets.

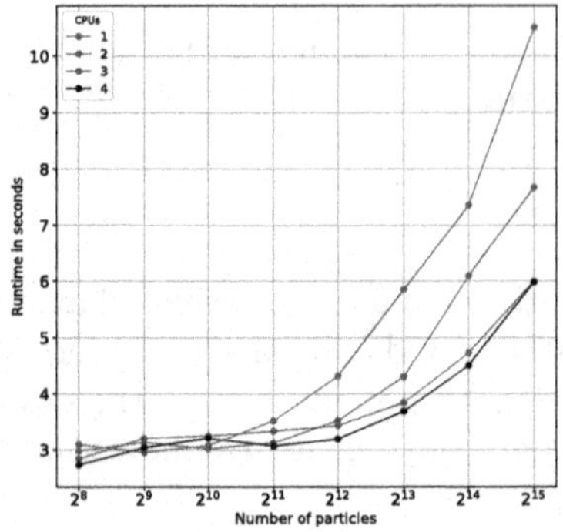

(a) Execution time of HMC for the Coin Toss model.

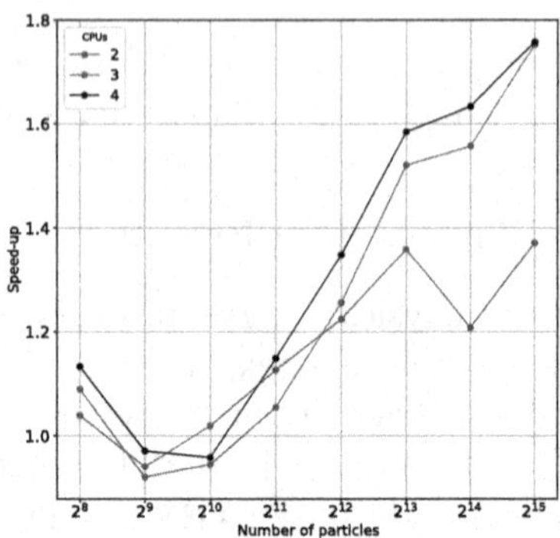

(b) Speed-up in comparison to 1 CPU core for the Coin Toss model.

Fig. 3. Execution time and speed-up of our HMC implemetation when using mutliple CPU cores of a Ryzen 5 3600X to determine the coin biases of the CT model using HMC with 1000 Leap-Frog steps.

Performance - HMC. Although a reference implementation is available for comparison, its drawbacks are its complexity and limited execution architectures.

Our development utilises JAX and NumPyro, allowing us to run the method for variable models on a variety of architectures. Here we present the performance data obtained for the pure HMC implementation on different architectures. One can observe, in Fig. 3a the behaviour of the run time when running our version of HMC on multiple CPUs, parallelised via MPI. For the CT model run times for up to 2048 particles are dominated by communication and data management overhead. This is a surprisingly small number, given the simplicity of the model and the limited amount of observed data fed into it. Overall the run time growth is sub-linear up until 65538 particles for 1 CPU. The resulting speed-up of using more than one CPU is depicted in Fig. 3b, demonstrating a sub-linear increase even with two CPUs. This observation confirms that the model is too small to scale effectively, at least for fewer than 128k particles (c.f. the IRT model below).

In contrast to the CT model is the IRT model, whose runtimes are displayed in Fig. 4a. One can immediately see, that the run time is dominated by overhead only below 256 particles *per device* (here: CPU). It is also important to note, that the run time using 4 CPU cores is larger than when only 3 are utilised. We attribute this to the apparent and unintentional spawning of multiple Python

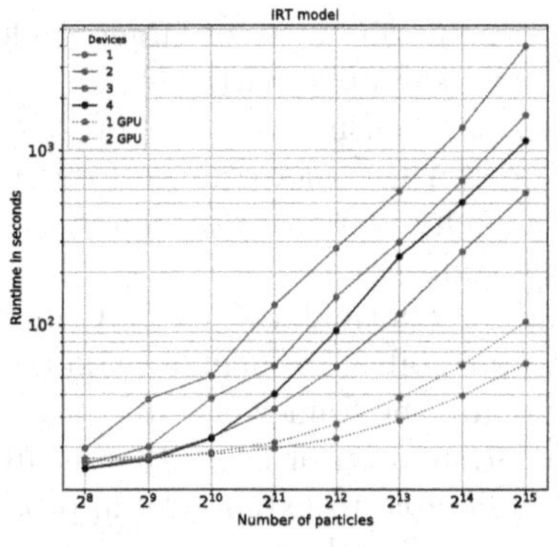

(a) Execution time for the IRT model.

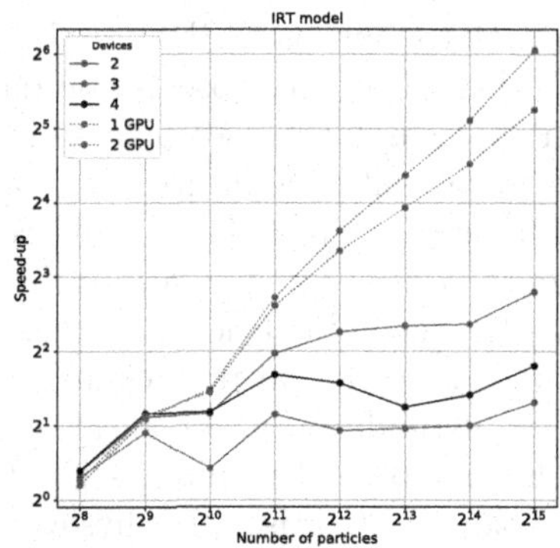

(b) Speed-up in comparison to 1 CPU core.

Fig. 4. Execution time and speed-up of our HMC implementation when using mutliple CPU cores of a Ryzen 5 3600X as well as a RTX 3060 and GTX 1080 Ti to determine the coin biases of the IRT model for HMC with 1000 Leap-Frog steps.

processes per MPI process. Our observations show that each MPI process spawns roughly 2 Python processes. This will be a point of future investigation. This observation explains the apparent super-linear speed-up for the CPUs displayed in Fig. 4b. One can also observe the order of magnitude reduction in the overall run time in the case of 65538 particles, when one GPU is used. The sub-linear speed-up from 1 GPU to 2 GPUs can here be explained by the fact, that the devices were asymmetrically bound to the system (PCIE-x16 vs. PCIE-x8) as well as that two devices of different generations were used: a GTX 1080 Ti and an RTX 3060.

The latter point shows the flexibility of JAX and NumPyro, which allowed us to run the same method on multiple CPU cores and multiple, heterogeneous, GPUs using MPI without having to modify the code!

4 Conclusions and Future Work

In conclusion, we have demonstrated that using the NumPyro and JAX frameworks along with intuition from classical mechanics and statistical physics, it is possible to re-create an SMC Bayesian optimisation process, whilst enriching it with an intuitive understanding of the respective steps.

The selected framework enabled us to create a simple, easy to maintain, implementation of HMC and SMC that can be parallelised to multiple computing devices of varying architectures on demand. Our implementation outperforms a similar implementation in Stan by a factor of ~ 2 on a single CPU (core) and scales well to multiple CPUs/GPUs, with larger gains obtained for larger models (or models with more observation data) and more sampling particles.

In the near future, we plan to extend MPI parallelism from the core HMC stage to the entire SMC iteration, as well as perform a thorough performance analysis and optimisation of our code, since preliminary checks indicate the existence of, e.g., unnecessary host-device data transfers.

On the theoretical side we plan to continue the reformulation of SMC in the language of statistical physics and expect the possibility to include maximum-entropy methods and thermalisation/annealing into the framework, utilising the long history of MC in physics [3]. Having a formulation of SMC that utilises terminology of (statistical) physics we hope to be able to extend the approach from the classical onto the quantum domain. This holds the potential of including existing quantum resources into Bayesian optimisation processes without requiring the user to know the intricacies of the new architecture.

Acknowledgements. During PRACE Summer of High-Performance Computing 2022, TW was able to implement parallel HMC using the discussed formulations and included correctness checks based on physical systems and models with known closed-form solutions in the implementation.

References

1. Baudart, G., Burroni, J., Hirzel, M., Mandel, L., Shinnar, A.: Compiling stan to generative probabilistic languages and extension to deep probabilistic programming. In: Proceedings of the 42nd ACM SIGPLAN International Conference on Programming Language Design and Implementation, pp. 497–510 (2021)
2. Béguin, A.A., Glas, C.A.W.: MCMC estimation and some model-fit analysis of multidimensional IRT models. Psychometrika **66**, 541–561 (2001)
3. Binder, K., Heermann, D.W.: Monte Carlo Simulation in Statistical Physics an Introduction, 6 edn. Springer (2019)
4. Bingham, E., et al.: Pyro: Deep universal probabilistic programming. J. Mach. Learn. Res. **20**(1), 973–978 (2019)
5. Bradbury, J., et al.: JAX: composable transformations of Python+NumPy programs (2018). http://github.com/google/jax
6. Carpenter, B., et al.: Stan: A probabilistic programming language. J. Statist. Softw. **76**(1) (2017)
7. Hoffman, M.D., Gelman, A.: The no-u-turn sampler: Adaptively setting path lengths in Hamiltonian Monte Carlo. J. Mach. Learn. Res. **15**, 1593–1623 (2014)
8. Moral, P.D., Doucet, A., Jasra, A.: Sequential Monte Carlo samplers. J. R. Statist. Soc. B **68**, 411–436 (2006)

Creating Models for Predictive Maintenance of Field Equipment in the Oil Industry Using Simulation Based Uncertainty Modelling

Raul Ramirez-Velarde(✉), Laura Hervert-Escobar, and Neil Hernandez-Gress

Tecnologico de Monterrey, Eugenio Garza Sada 2501 Sur, Col. Tecnológico, 64849 Monterrey, NL, Mexico

rramirez@tec.mx

Abstract. Determining what causes field equipment malfunction and predicting when those malfunctions will occur can save large amounts of money for corporations that are capital-intensive. To avert equipment downtime, field equipment maintenance departments must be adequately resourced. Herein, we demonstrate the efficacy of machine learning to determine time between failure, repair time (equipment downtime) and repair cost. Additionally, a mean value analysis is carried out to determine the maintenance department capacity. Uncertainty is modelled using statistical analysis and simulation.

Keywords: predictive maintenance · machine learning · heavy-tail simulation

1 Introduction

Predictive equipment maintenance is one of the most important areas in industries that are heavily reliant on capital, as all other processes depend on the correct performance of its "clients" (equipment). It is well-known that over time, the performance of field equipment decreases, and their failure rate increases, so it is logical and a good strategy to diagnose areas of opportunity for minimizing repair costs, equipment down-time, production delays, and equipment failure frequency. In this paper, we will analyse maintenance work orders data for field equipment maintenance provided by an oil extraction corporation corresponding to upstream gas extraction operations to create predictive models that can allow companies to reduce costs and improve efficiency. However, data and the models developed with it bring inherent uncertainty both epistemically and ontologically. Uncertainty modelling techniques include stochastic simulation, chance-constrained models, Markov processes, stochastic optimization, Bayesian models, evidence theory, fuzzy theory, information-gap theory, and statistics and probability. In this paper, we present and study how uncertainty is modelled using statistics and probability, and stochastic simulation. We carry out a thorough analysis of the company's maintenance area by applying unsupervised and supervised data science analysis to its work orders database, which consists of more than 55 variables and over 500K records.

J. Mikyška et al. (Eds.): ICCS 2023, LNCS 14077, pp. 629–643, 2023.
https://doi.org/10.1007/978-3-031-36030-5_50

Specifically, the research goals are:

1. Determine how Time between Failures (TBF), repair costs, and repair duration can be characterized and identify which factors influence these variables. Create statistical and mathematical models to make predictions.
2. Develop models to predict how repair wait time and the number of repair reports waiting to be serviced can change with changes in the organization, characteristics, and structure of the company's maintenance areas.
3. Provide a framework for uncertainty modelling through simulation.

2 Related Work

The research presented in this paper falls under the predictive maintenance field, which is an area that has gained increased attention in the context of equipment maintenance systems. Predictive maintenance is defined by [1] as "regular monitoring of the actual mechanical condition, operating efficiency, and other indicators of the operating condition of machine-trains to ensure the maximum interval between repairs as well as to minimize the number and cost of unscheduled outages caused by machine-train failures". Due to its importance for the efficiency of field operations, many studies have performed predictive maintenance. In [2], a comprehensive review of the state of the art of equipment maintenance systems is presented, and it is concluded that predictive maintenance technology is a growing research field with fast-increasing contributions from different application areas, such as mechanical, chemical, energy, automation, etc. Predictive maintenance proceeds by creating machine behaviour models. One example is shown in [3], where a bi-level optimization model is presented, which addresses the need to balance an optimization procedure locally (level one) and globally (level two) while minimizing system average interruption frequency. In [4], an application of predictive maintenance using a model-based analysis is presented, to enhance the accuracy of performance diagnostics. Different models under different operating conditions are created and then performance is recalculated and compared for efficiency. Another model-based approach is shown in [5].

Recently, the body of knowledge related to machine learning, big data, artificial intelligence, and cloud computing has been combined with predictive maintenance, resulting in the emergence of Cyber-Physical Systems (CPS). An extensive literature review of CPS as of 2020 is presented in [6], which concludes that the main keywords are internet of things, industry 4.0, predictive maintenance, machine learning, artificial intelligence, cloud computing, and big data. In [7], a comprehensive review of machine learning algorithms applied to tool wear selection is provided, with adversarial neural networks and random forests being identified as the best models for tool wear prediction. A comparison of machine learning algorithms used to reliably estimate performance and detect anomalies in a representative combined cycle power plant is found in [8]. Here, unlike in [7], the prediction is carried out using multivariable models with features such as temperature, humidity, and pressure. Early-stage malfunctions are detected using anomaly estimation via unsupervised machine learning algorithms and principal component analysis for dimension reduction. A definition of Cyber-Physical Systems and its relation to Industry 4.0 is presented in [9], and the concepts of big data, cloud computing, and machine learning and their applications to industry 4.0 are reviewed. Big data and cloud computing are mentioned in [10] in which a systematic architecture is proposed. An application using neural networks and deep learning to improve maintenance support and wear prediction of field equipment is shown in [11].

An important part of the research presented in this paper is modelling uncertainty. There is extensive bibliography regarding uncertainty management. Uncertainty is usually modelled using statistical analysis, probability theory, and Bayesian methods, stochastic optimization, constrained models, fuzzy theory, information-gap theory, Monte Carlo simulation, and discrete event simulation based on statistical characterization of critical variables. An important reference for modelling uncertainty using probability theory is found in [12] which gives the mathematical background for uncertainty diffusion and draws heavily on inferential statistics to model systematic errors. Another important reference is [13], which contains a wide range of applications of uncertainty modelling, from probability and statistics principles to applications in economics, art, psychology, and sciences. Another important reference is [14], which is a compendium of mathematical tools for engineering, ranging from decision making to optimization, to probability applications and curve fitting, and certainly, modelling uncertainty. In chapters 7, uses of probability theory, Monte Carlo simulation, chance constrained models, Markov processes, and stochastic optimization are reviewed. [15] is a specific survey of modern techniques for uncertainty modelling. In this paper, uses of probability theory and probability theory-derived techniques such as Monte Carlo methods, Bayesian methods, and evidence theory are reviewed. Also, fuzzy theory and Information-Gap theory are reviewed. The mathematical background for uncertainty propagation found in [12] is applied in [16] for modelling uncertainty in linear regression, which shows how difficult modelling uncertainty can be even for the simplest of machine learning algorithms. An important discussion of the differences between uncertainty and risk can be found in [17]. In this paper, it is argued that a change in risk is a change in the spread of the probability spectrum conserving the mean, whereas uncertainty is a change in the probability curves that can change the mean. It is also argued that some agents might be fond of risk but averse to uncertainty, whereas some agents might accept high uncertainty if the risk is low. High-risk almost always indicates high gains but high uncertainty might indicate the contrary. An example of the use of standard statistics such as standard deviation to model uncertainty can be found in [18], whereas in [19] bootstrapping is used to run several simulations taking traditional standard deviation statistics to create confidence intervals for biomanufacturing processes. An interesting approach is carried out in [20], where uncertainty is managed by using a genetic algorithm to improve a simulation carried out with a mixed integer linear programming model. Simulation is again utilized, and uncertainty is modelled using Bayesian averaging for modelling watersheds and stream flows in [21], whereas in [22], Monte Carlo simulation is thoroughly studied and compared against conventional uncertainty estimation methods such as uncertainty diffusion mathematical models. The paper indicates when it is appropriate to use Monte Carlo simulation.

3 Methodology

The following methodology for analytics and model creation, called **Predictive Variance Association**, was followed for three types of analysis: Time between failures (TBF), repair cost, and repair duration. We are given a matrix X with data in which the rows are samples (m samples) with numerical values, and the columns are variables (n variables). The explanatory variables are separated into matrix E (o variables), and the dependent variables (p variables) or variables of interest, are separated into matrix G. Naturally, $n = o + p$. Additionally, from PCA [23], samples scores and variables loadings, dimension reduction and clustering can also be carried out. See Fig. 1, which shows the process.

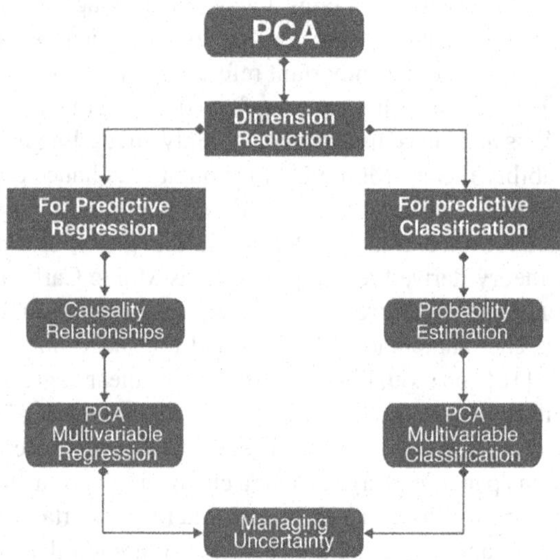

Fig. 1. Predictive Factor Association Steps.

3.1 Predictive Analytics for Continuous Variables by Regression

This process is accomplished when the objective is to predict the future value of a continuous variable of interest:

1. **First PCA**. Carry out PCA for matrix X and find matrices $X = PDQ^t$ [23, 24].
2. Let $F = XQ_{1...k}$ be the **principal component scores**, where $k < o$. Let $L = Q\sqrt{D}$ be the loadings matrix. The square cosine table is generated by squaring L.
3. Use square cosines and a clustering algorithm over the columns of F to determine:
   a. **Collinearity and dimension reduction**. Explanatory variables that are grouped together, that is, have a high cosine in the same component column or are clustered in the circle correlation plot are collinear and thus can eliminate those columns from matrix X. Repeat PCA, and then go to step 1.

   b. **Explain components**. Explanatory variables with the highest squared cosine in a particular column of L can be used to associate a component to a particular variable of interest, thus elucidating the meaning of the component through the variable of interest. Explanatory variables with high squared cosine in the same column can also be associated with that variable of interest.
   c. **Causality relationships**. If explanatory variables are grouped with variables of interest, then that is evidence of a causality relationship.
   d. The **square cosine** is the square of the cosine of the angle α between the vectors of variable loadings, i.e., the correlation. If variables are close, the cosine will tend to one. If variables are separated, the cosine will approach zero. If $\rho^2(x, y) = cos^2(\alpha) > 0.5$ then $\cos(\alpha) > \mp 0.7071$, thus associate variables with squared cosine greater than 0.5.
4. **Generate PCA based synthetic attributes.** Eliminate variables of interest leaving only matrix E. Carry out PCA over E.
5. **Curve Fitting**. Carry our curve fitting between PCA derived explanatory variables and variables of interest clustered together in step 3.b using the new sample scores from step 4.
6. **Evaluate Curve Fitting**. Carry out curve fitting. Discard all $R^2 < 0.5$. This indicates which explanatory variables influence the most the variable of interest. Then $\widehat{G_i} = f_{ij}(F_j) = f_{ij}(q_1X_1, q_2X_2, \ldots, q_oX_o)$ where the coefficients q_i are determined by matrix Q.
7. **New samples** called X' can give predictions on variables, for $F' = X'Q$ and $\widehat{G_i'} = f_{ij}\left(F_j'\right)$.

3.2 Categorical Variable Prediction by Multi-label Classification

This process is carried out when the objective is to predict categorical variables. The procedure is similar to the one for continuous variables, except that the prediction is accomplished with a classification algorithm instead of a curve fitting algorithm:

1. Same as 3.1
2. Same as 3.1
3. Same as 3.1
4. Same as 3.1
5. **Apply classification algorithm**. Since the relationship between variables of interest and components is not always evident in classification problems, a search algorithm is carried out to determine the best components to use to predict a particular categorical variable of interest.
   a. **Match** F_i row by row with variable of interest j (assumed to be categorical with 0 or 1 as value), where $i = 1 \ldots o$ and $j = 1 \ldots p$.
   b. **Estimate class probabilities**. This can be carried out using different procedures, such as window-based probability estimation shown in [25]. Other machine learning classification algorithms are also used. Once this is completed, calculate performance measures. Each Fi is a linear combination of explanatory variables that potentially has enough information to give good predictions for the variables of

interest. Then $\widehat{G_i} = f_{ij}(F_j) = f_{ij}(q_1X_1, q_2X_2, \ldots, q_oX_o)$ where the coefficients q_i are determined by matrix Q. If $\widehat{G_i} > 0.5$ then assume a result of 1, and 0 otherwise.

6. **New samples** called X' can give predictions on variables, for $F' = X'Q$ and $\widehat{G_i'} = f_{ij}\left(F_j'\right)$. If $\widehat{G_i} > 0.5$ then assume a result of 1, and 0 otherwise.

4 Results and Discussion

Throughout this document, an analysis of the company's maintenance area is done by applying data analysis to its Work Orders database, which uses the standard **ISO 14224:2016** and consists of 55 variables and 1,429,919 observations. After cleaning up of invalid values and some critical missing entries, and keeping records from 2015 to 2018, the database was reduced to 590,604 entries. Additionally, when talking about corrective jobs, we refer to jobs of JobType labeled "corrective", consisting, after value clean-up, of 89,480 elements (rows).

4.1 Dataset Description

New variables were created by establishing thresholds for the variables of interest which created classes. For Duration, the threshold was **7 days** for all jobs, which was 95% percentile and **6 days for corrective jobs**, which was also 95% percentile. A Duration of less than the threshold was labelled as 0. A Duration above the threshold was labelled as 1. For **Total cost**, the threshold was established at **$6,520.00** which is 99% percentile (above is 1, below is 0) and **$6,700.00 for corrective jobs** (99% percentile). For **time between reports** for the same equipment, the threshold was set to **40 days** for both general and corrective reports (63% percentile general, 42% corrective, above is 1, below is 0).

Many of the variables had to be recoded as they were originally coded as labels or text. Statistical Analysis was conducted. Table 1 summarizes statistics of all reports and Table 2 summarizes statistics of only corrective reports. The "Time between Reports same Equipment" and "Time between Failures same Equipment" were calculated by searching the unique equipment ID and subtracting the report dates, giving a time span in days. The statistics given are for data corresponding to years 2015 through 2018.

Table 1. Statistical summary for all reports

Variable	Mean	Std Dev	80% per	90% per	95% per	99%per
Duration	2.7929	13.1885	0	1	7.3	26.4300
Total Cost	397.67	8,675.53	0	0	0	6,520.00
Time bet Reports	0.00214	0.046213	0	0	0	0.0010
TBR Same Equip	91.7292	165.6745	133.94	322.59	443.33	799.88

Table 2. Statistical summary for corrective reports

Variable	Mean	Std Dev	80% per	90% per	95% per	99%per
Duration	1.9790	12.6575	0	1	6	16.65
Total Cost	724.52	17,677.36	0	0	0	16,700.00
Time bet Reports	0.0141	0.1213	0	0	0	1.0000
TBR Same Equip	124.0433	168.1949	213.72	343.63	473.54	800.41

Time Between Failures

We analysed time between reports of any kind for any equipment. That is, the time between global consecutive reports. Data shows that there is always more than one report per day of any kind. The information given for time between reports does not have enough granularity to determine probability distribution as data was given in days, but actual arrival rate is minutes.

Analysis for time between reports of any kind (corrective, verification, update, scheduled maintenance, etc.) for a particular equipment indicates that reports for any given individual equipment follow an exponential distribution which will allows to establish probability bands on the time between reports for the same equipment.

As for corrective reports, analysis of time between any failures of any equipment indicates that corrective reports arrive at a frequency of several reports per day. The event of less than one report a day is very rare.

Analysis pf time between corrective reports, that is, time between failures for a particular equipment indicates that reports for any given individual equipment follow an exponential distribution which will allows to establish probability bands on the time between failures for the same equipment.

Duration

A variable called "Duration" was also analyzed. Analysis of report duration for all reports showed a very long tail, indicating that most reports take a day or less to be completed, but a significant number take more than 6 days. Of those, around 14,000 take 7 days, more than 16,000 take 26 days, and a very small fraction take more than 1,000 days.

For corrective reports, repair duration has the same characteristics. 91% of reports are resolved within a day, while a small percentage can take up to 1,000 days.

Total Cost

Analysis of Total Cost for all reports shows a very long tail. 57% of all report have zero cost, whereas 99.59% of all reports have a cost of $6,524.00 or less. As for corrective reports, 41.51% report a total cost of zero and 99.76% report a total cost of $16,758.00 or less.

4.2 Principal Component Analysis

The next step is to analyse the relationship between our main variables of interest, Duration, Time between Reports/Failures and Cost and the other variables by carrying out principal component analysis.

4.2.1 Duration

Principal component analysis was also carried out. Correlation plots and square cosine lead us to conclude that **Equipment Type**, **Equipment Criticality**, **Area**, **Material Cost**, **Labor Cost**, **IsAffectingProduction** and **WordOrder** are the variables that correlate the most to duration, and these can be used to predict repair duration. Duration is associated mostly to the second principal component.

4.2.2 Repair Cost

Principal component analysis was also carried out to determine the factors affecting repair cost. According to the square cosines table, most of the variation of the total cost is captured by the first two principal components. We determined that the most important relations for the cost are the material cost and labour cost, which we know are directly related (total cost being the sum of labour and material). Thus, it was inferred that other factors which affect the cost are the level of **equipment criticality, WOType, TradeGroup, IsAffectingProduction and Job Type**.

4.2.3 Time Between Reports/Failures

The results of PCA for time between reports indicate that there are no variables that influence the time between reports, and that there are very few variables that influence the time between reports for the same equipment, mainly: **IsAffectingProduction, Equipmentcode, EquipmentRollupCode, ActualDuration and LabortCost**. But the influence is limited.

Also, PCA results show that there are no variables that influence the time between failures, that is, corrective reports, when considering all reports, and that there are very few variables that influence the time between failures for the same equipment. These variables are: **IsAffectingProduction, Equipmentcode, ActualDuration, CauseC and SafetyC**. But the influence is limited.

4.3 Machine Learning Prediction

Prediction models for our variables of interest were developed. These are classification type models, meaning that given a set of features represented as columns in a data matrix, the output is 1 or 0 for each sample; 1 indicates that the sample corresponds to a class and 0 indicates that it does not. In our case, the classes were determined as follows (Table 3):

Table 3. Thresholds use to create the classification classes.

Variable	Threshold (equals to 1)
Repair Duration	≥1 *day*
Total Repair Cost	≥$250.00
Time Between Failures	≥40 *days*

Repair Duration

For repair duration machine learning prediction, we used the class "Repair Time > 1 Day". The machine learning algorithm used for repair duration was a multi-layer perceptron classifier (MLP, an artificial neural network). The performance parameters for repair duration prediction using MLP are shown in Table 4. The global accuracy is 0.984, and the F1 obtained for class 1 is 0.99, which are considered excellent results. The likelihood ratio obtained for class 1 is 200, which is considered very good.

Table 4. Repair duration prediction performance parameters using deep learning.

Class	Precision	Recall	F1/score
0	0.86	0.74	0.80
1	0.99	1.00	0.99

Repair Cost

A MLP model was also used to create a model to predict whether the repair cost would be $250.00 dollars or higher. As predictors, the variables determined by PCA to be more closely related to repair cost were used. The results are shown in Table 5. The global accuracy achieved was 0.90, but the likelihood ratio for class 1 was only 1.25, which is not considered good. Additionally, the recall for class 1 was too low at 0.56.

Table 5. Total cost prediction performance parameters using deep learning. Accuracy of 0.90 is considered good, but F1 of 0.643 is considered low. Likelihood ratio of 1.25 in not good.

Class	Precision	Recall	F1-score
0	0.92	0.97	0.94
1	0.76	0.56	0.64

Time Between Failures

To approximate a prediction model for time between failures, we used the somewhat loose category of time between failures greater than or equal to 40 days. We used MLP and Logistic Regression classifiers and obtained similar results. The results for the MLP classifier are shown in Table 6. This variable proved to be the hardest to predict. The likelihood ratio of 3.823 for class 1 is considered adequate, but the recall for class 0 is too low at 0.38. The global accuracy achieved is 0.615.

4.4 Performance Analysis: Mean Value Analysis

What resources are necessary to service all reports with the correct quality level? Firstly, we will address this question using mean-value analysis; the notation for this follows in Table 7. Secondly, we will use simulation to model variability and uncertainty.

Table 6. Repair duration prediction performance parameters using deep learning. Accuracy obtained is 0.615, considered regular, but F1 of 0.701 not so much. Likelihood ratio obtained is 3.823 which is adequate.

Class	Precision	Recall	F1-score
0	0.58	0.38	0.46
1	0.63	0.79	0.70

Table 7. Notation for mean value analysis

Variable	Meaning	Units
X	Throughput. Reports carried out per unit time	$\frac{Job}{day}$
S	Service time. Time required to complete a report	$\frac{day}{job}$
U	Utilization. Fraction of the time a worker is busy	None
P	Parallel workers. Workers working concurrently at any given moment	Workers
R	Residence Time. Total time required to finish a single job from report to conclusion	Days
N	Jobs in System. Jobs waiting to be finished per worker including the active one	$\frac{Job}{report}$
D	Delay. The time a report will wait before been worked upon	Days

As mentioned in Sect. 5.2, on average there are $X = 467.239$ reports per day, which represents both the report arrival rate λ and the service throughput X, as $\lambda = X$. Thus, the average inter-arrival time per report is 1/467.239 = 0.00214 days per report, or 0.0514 h per report (a report every 3.0816 min). In that same section, we learned that the average duration time per report is $S = 2.7929$ days. This can only be possible if many personnel are working in parallel in different problems.

Thus, by the utilization law:

$$U = \frac{SX}{P}$$

where P is the number of parallel maintenance workers. If we assume $U = 0.7, 0.8$ *or* 0.9 the following numbers of simultaneous maintenance workers P must be employed at project operation time (Table 8):

Let us assume a utilization of $U = 0.8$. By queueing theory:

$$N = \frac{U}{1-U} = \frac{0.8}{1-0.8} = 4 \text{ reports}$$

Table 8. Number of parallel maintenance workers by utilization factor

U	P
0.7	1,864
0.8	1,631
0.9	1,450

Indicating that each worker has $N = 4$ jobs pending, including the one being serviced now. And by Little's law:

$$R = \frac{N}{\left(\frac{X}{P}\right)} = \frac{4}{\frac{467.239}{1,631}} = 13.963 \; days$$

Indicating that each report will wait $R = 13.963$ days from creation to conclusion or 5 times the service time 2.7929 days. The wait time, the time a report will remain unattended, would be $D = 13.963 - 2.7929 = 11.1701$ days.

Additionally, we can repeat the procedure to determine the number of workers carrying out corrective repairs. For corrective repairs, we have a duration of 1.9790 days, an actual duration of 1.4961 days, and a time between failures of 0.0141 days per job, giving a demand of 70.922 jobs per day. Again assuming $U = 0.8$, we find:

Using Duration, $X = 70.922$ jobs/day and $S = 1.9790$ days/job (Duration). Assuming again $U = 0.8$ we find that $P = \frac{1.9790(70.922)}{0.8} = 175.4433$ workers doing repairs in parallel, total time $R = \frac{4}{\frac{70.922}{175.4433}} = 9.90$ days (5 times the service time), and wait time $D = 9.90 - 1.9790 = 7.92$ days.

Results are summarized in Table 9.

Table 9. Summary of mean value analysis results. Always U = 0.8 and thus N = 4 jobs in system per worker including the current one.

S is	S (Service Time)	X (Throughput)	R (Total time)	P (Parallel Workers)	D (Wait to service)
Duration	2.7929	467.239	13.963	1,631	11.170
Dur Corrective	1.9790	70.922	9.900	175	7.920

4.5 Modelling Uncertainty Using Simulation

The mean value analysis carried out in Sect. 5.5 gives us an idea of the capacity required to provide quality service in field equipment maintenance. To be able to establish different scenarios in which both workloads for repair personnel and available resources can vary substantially, more information is needed. Now, we turn to simulation to establish probability distributions on R, or total system time, and N, the number of jobs pending.

To carry out this simulation, we take the data from Sect. 5.2 and determine probability distributions for service times (called Duration) and time between failures (TBF). TBF will determine the job arrival rate. We find that service time has a long-tailed probability distribution, which we model using the Pareto probability, since it is the best fit (see Figs. 2 and 3), with a shape parameter,α, of 1.4525 and a scale parameter, A, of 1.2344. The information given for time between arrivals does not have enough granularity to determine a probability distribution, so it will be assumed to be exponential, as is the normal in discrete-time simulations (data was given in days, but actual arrival rate is in minutes).

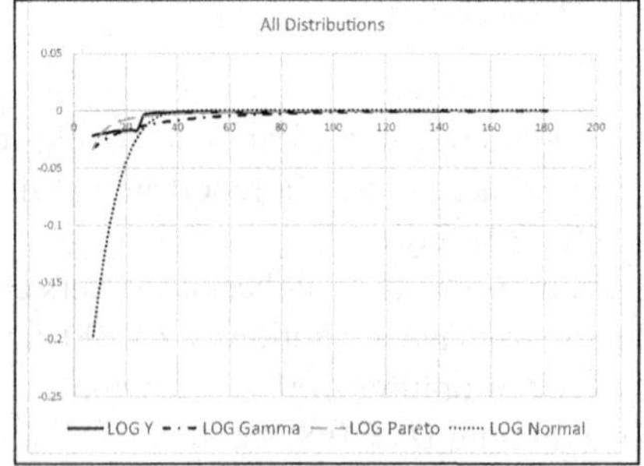

Fig. 2. Fitting different probability distributions to Duration data

Fig. 3. QQ plot Pareto distribution vs Duration Data

In this simulation, a simplified model was used. The workload was divided between workers. Each worker receives a new job every 5 days, and on average, a worker finishes the job in 3.9623 days. The ratio between arrival rate and service rate is $\rho = 0.7925$. Average simulation results are given in Table 10.

Table 10. Average simulation results

Parameter	Average
U (Utilization)	0.8111
W (R, total wait time)	2,243.58
L (N, pending Jobs on queue)	171.01
Max L (Max N, maximum queue)	1,001

Max L in Table 10 shows that even with a ratio between arrival rate and service rate of $\rho = 0.7925$, which is usually not considered a heavy load, the extreme variability of the service time can create severe bottlenecks. As a result, the probability distribution of the number of jobs in the system has a rather heavy tail, for which the best fit is a Pareto distribution with shape parameter $\alpha = 0.2859$ and scale parameter $A = 1.5$ (see Figs. 4 and 5).

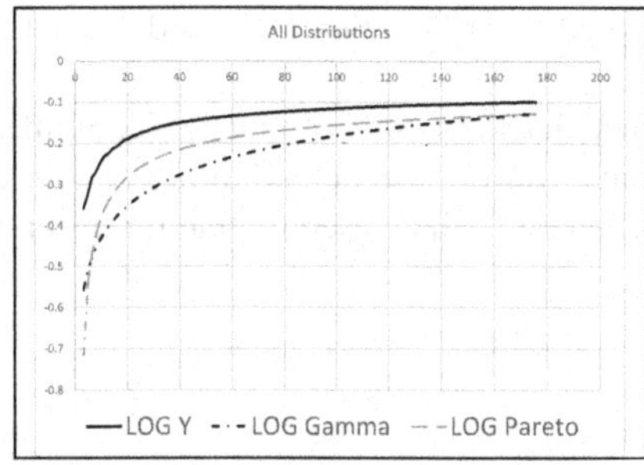

Fig. 4. Fitting different probability distributions to jobs on queue simulation result

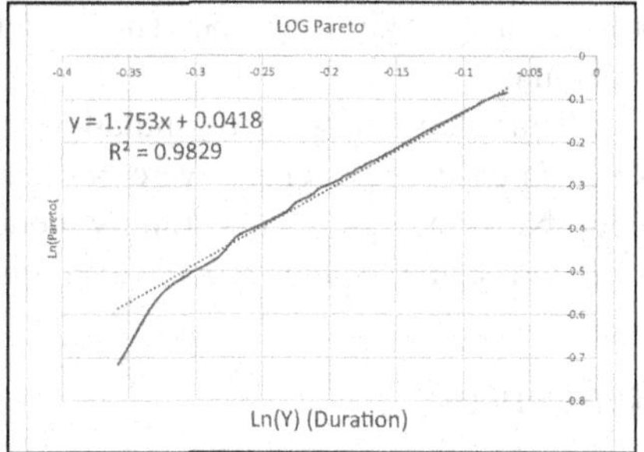

Fig. 5. QQ plot Pareto distribution vs jobs on queue simulation result

Surprisingly, the wait time is not long tailed. The best fit for the total time in the system (wait time R or W) is a Gamma probability distribution (see Figs. 6 and 7).

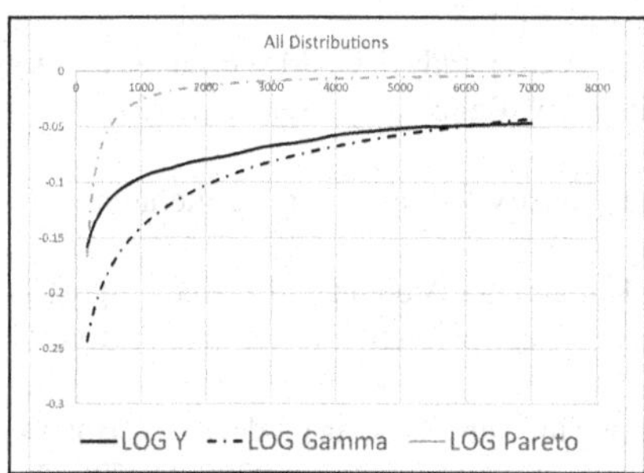

Fig. 6. Fitting different probability distributions to total wait time

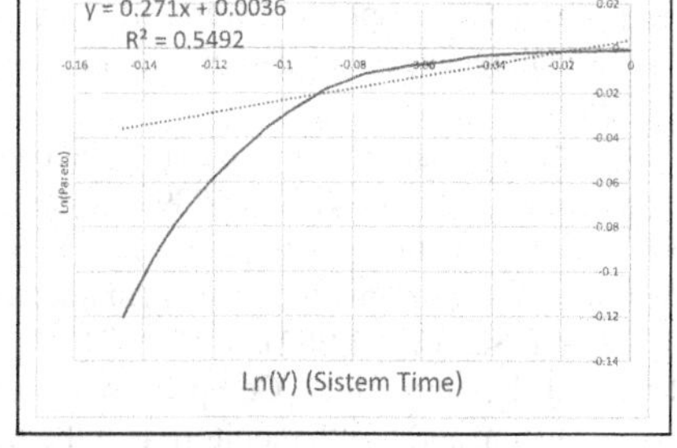

Fig. 7. QQ plot Pareto distribution vs total wait time simulation result

5 Conclusions

Statistical analysis showed that, even though the time between reports or failures for the same equipment follows the well-known exponential decay, the totality of reports duration (all types of reports) and repair cost exhibits a long-tail probability distribution. This is true for both general reports and specific corrective reports, interpreted as equipment failures. This is caused by the fact that the most common value in the three variables is zero.

Principal component analysis showed that the correlations between the independent variables (50 of them) and the variables of interest, mainly Duration, Total Cost and Time Between Reports/Failures, are not very strong, although not so small that the information is totally random. Principal component analysis provides a roadmap for constructing predictive models.

Following PCA's pointers, predictive models were constructed. The predictions of Total Cost and Time Between Reports/Failures had lower confidence results, but pointed in the right direction. The model for predicting Duration had good predictive performance.

Mean value analysis and simulation confirmed that the job of keeping field equipment is not an easy one, facing severe bottlenecks due to the extreme variability of service times. Nevertheless, the uncertainty models developed can be used to plan for different workload scenarios and resource availability.

References

1. Mobley, R.K.: An Introduction to Predictive Maintenance. Elsevier Science New York City, NY, USA (2002)
2. Hoppenstedt, B., et al.: Techniques and emerging trends for state of the art equipment maintenance systems—a bibliometric analysis. Appl. Sci. **8**(6), 916 (2018)
3. Matin, S.A.A., Mansouri, S.A., Bayat, M., Jordehi, A.R., Radmehr, P.: A multi-objective bi-level optimization framework for dynamic maintenance planning of active distribution networks in the presence of energy storage systems. J. Energy Storage **52**, 104762 (2022)
4. Kim, T.S.: Model-based performance diagnostics of heavy-duty gas turbines using compressor map adaptation. Appl. Energy **212**, 1345–1359 (2018)
5. Barberá, L., Crespo, A., Viveros, P., Stegmaier, R.: A case study of GAMM (graphical analysis for maintenance management) in the mining industry. Reliab. Eng. Syst. Saf. **121**, 113–120 (2014)
6. Zonta, T., et al.: Predictive maintenance in the Industry 4.0: a systematic literature review. Comput. Ind. Eng. **150**, 106889 (2020)
7. Wu, D., Jennings, C., Terpenny, J., Gao, R.X., Kumara, S.: A comparative study on machine learning algorithms for smart manufacturing: tool wear prediction using random forests. J. Manuf. Sci. Eng. **139**(7), 071018 (2017)
8. Hundi, P., Rouzbeh, S.: Comparative studies among machine learning models for performance estimation and health monitoring of thermal power plants. Appl. Energy **265**, 114775 (2020)
9. Javaid, M., Haleem, A., Singh, R.P., Suman, R.: An integrated outlook of cyber–physical systems for Industry 4.0: topical practices, architecture, and applications. Green Technol. Sustain. **1**(1), 100001 (2023)
10. Lee, J., Ardakani, H.D., Yang, S., Bagheri, B.: Industrial big data analytics and cyber-physical systems for future maintenance & service innovation. Procedia CIRP **38**, 3–7 (2015)
11. Li, Z., Hao, J., Gao, C.: Equipment maintenance support effectiveness evaluation based on improved generative adversarial network and radial basis function network. Complexity **2021**, 1–11 (2021)
12. Kirkup, L., Frenkel, R.B.: An Introduction to Uncertainty in Measurement: Using the GUM (Guide to the Expression of Uncertainty in Measurement). Cambridge University Press (2006)
13. Bammer, G., Smithson, M.: Uncertainty and Risk: Multidisciplinary Perspectives. Routledge (2012)
14. Loucks, D.P., van Beek, E.: Water Resource Systems Planning and Management. Springer, Cham (2017). https://doi.org/10.1007/978-3-319-44234-1
15. Li, Y., Chen, J., Feng, L.: Dealing with uncertainty: a survey of theories and practices. Trans. Knowl. Data Eng. **25**(11), 2463–2482 (2012)
16. Klauenberg, K., Martens, S., Bošnjaković, A., Cox, M.G., van der Veen, A.M.H., Elster, C.: The GUM perspective on straight-line errors-in-variables regression. Measurement **187**, 110340 (2022)
17. Ilut, C.L., Martin, S.: Modeling uncertainty as ambiguity: a review. National Bureau of Economic Research, Cambridge, MA, USA (2022)

18. Papadakos, G., Marinakis, V., Konstas, C., Doukas, H., Papadopoulos, A.: Managing the uncertainty of the U-value measurement using an auxiliary set along with a thermal camera. Energy Build. **242**, 110984 (2021)
19. Xie, W., Barton, R.R., Nelson, B.L., Wang, K.: Stochastic simulation uncertainty analysis to accelerate flexible biomanufacturing process development. Eur. J. Oper. Res. **310**(1), 238–248 (2023)
20. Badakhshan, E., Ball, P.: A simulation-optimization approach for integrating physical and financial flows in a supply chain under economic uncertainty. Oper. Res. Perspect. **10**, 100270 (2023)
21. Moknatian, M., Mukundan, R.: Uncertainty analysis of streamflow simulations using multiple objective functions and Bayesian model averaging. J. Hydrol. **617**, 128961 (2023)
22. Papadopoulos, C.E., Yeung, H.: Uncertainty estimation and Monte Carlo simulation method. Flow Meas. Instrum. **12**(4), 291–298 (2001)
23. Jolliffe, I.: Principal Component Analysis. Springer, New York (2002). https://doi.org/10.1007/b98835
24. Golub, G.H., Van Loan, C.F.: Matrix Computations. Johns Hopkins University Press, Baltimore (2012)
25. Ramirez-Velarde, R., Hervert-Escobar, L., Hernandez-Gress, N.: Predictive analytics with factor variance association. In: Rodrigues, J.M.F., et al. (eds.) ICCS 2019. LNCS, vol. 11540, pp. 346–359. Springer, Cham (2019). https://doi.org/10.1007/978-3-030-22750-0_28

On the Resolution of Approximation Errors on an Ensemble of Numerical Solutions

A. K. Alekseev and A. E. Bondarev(✉)

Keldysh Institute of Applied Mathematics RAS, Moscow, Russia
bond@keldysh.ru

Abstract. Estimation of approximation errors on an ensemble of numerical solutions obtained by independent algorithms is addressed in the linear and nonlinear cases. In linear case the influence of the irremovable uncertainty on error estimates is considered. In nonlinear case, the nonuniform improvement of estimates' accuracy is demonstrated that enables to overperform the quality of linear estimates. An ensemble of numerical results, obtained by four OpenFOAM solvers for the inviscid compressible flow with an oblique shock wave, is used as the input data. A comparison of approximation errors, obtained by these methods, and the exact error, computed as the difference of numerical solutions and the analytical solution, is presented. The numerical tests demonstrated feasibility to obtain the reliable error estimates (in the linear case) and to improve the accuracy of certain approximation error in the nonlinear case.

Keywords: approximation error · ensemble of numerical solutions · Irremovable uncertainty · Euler equations · OpenFOAM

1 Introduction

The development of the verification methods (*a posteriori* error estimation) is complicated by the formation of discontinuities for CFD problems, governed by equations of hyperbolic or mixed types. By this reason the corresponding progress is less considerable if compare with the finite element domain (elliptic equations) [1–3]. Nevertheless, modern CFD standards [9, 10] require a verification of numerical solution. At present, the set of methods, such as defect correction [4] and Richardson extrapolation [5–8], are used for the approximation error estimation in CFD. Their recommendations are mainly based on the grid convergence in the form of the Richardson extrapolation or the Runge approach (which considers the solution on a fine grid as true one). However, the common methods of *a posteriori error* estimation in CFD have significant troubles. The defect correction is an intrusive method (it requires the modification of the code) and is based on some linearization and assumption of the smallness of error, which may violate at the strong shock waves. The Richardson extrapolation is not intrusive (based on postprocessing), but, unfortunately, requires several consequent refinements of the grid that cause high computational expenses.

By these reasons, papers [8, 11–13] consider nonintrusive methods of the pointwise error estimation, which do not require the mesh refinement. The need for at least

J. Mikyška et al. (Eds.): ICCS 2023, LNCS 14077, pp. 644–655, 2023.
https://doi.org/10.1007/978-3-031-36030-5_51

three independent solvers and a relatively small accuracy of error estimates are their drawbacks.

The present paper considers the conditions that improve the accuracy of error estimates for the linear method [8, 11, 12]. Some properties of the nonlinear method [13] are discussed, which increase the accuracy for certain solution at the expense of the accuracy of others. The considered nonlinear version of the algorithm enables to estimate the approximation error with higher accuracy, if compare with the linear case.

2 The Linear Problem for the Approximation Error Estimation Using the Differences of Numerical Solutions

An ensemble of n numerical solutions $u_m^{(i)}$ $(i = 1...n)$, computed on the same grid by different numerical algorithms contains some information regarding their approximation errors. Herein u is the gas-dynamics variables, i is the number of algorithm, m is the index $(m = 1, ..., L)$, marking the grid node in a vectorized form. Papers [8, 11, 12] address the estimation of the approximation errors on the ensemble of numerical solutions. The Inverse Problem, posed in the variational statement with the zero order Tikhonov regularization, is used to treat the differences between the solutions in these papers. The paper [13] applies the Inverse Problem for the nonlinear statement of approximation error estimation in the context of regularization.

Herein, we address estimation of the approximation errors on the ensemble of numerical solutions using several approaches. Some important features of the both linear and nonlinear formulations are considered, including non-uniform resolution of errors.

We start from the linear statement, based on differences of solutions. We denote the projection of the exact solution $\tilde{u}$ onto the grid as $\tilde{u}_m$ and the approximation error for $i-th$ solution as $\Delta u_m^{(i)} = u_m^{(i)} - \tilde{u}_m$. The point-wise differences of the numerical solutions $d_{ij,m} = u_m^{(i)} - u_m^{(j)} = \tilde{u}_{h,m} + \Delta u_m^{(i)} - \tilde{u}_{h,m} - \Delta u_m^{(j)} = \Delta u_m^{(i)} - \Delta u_m^{(j)}$ are computable and depend on approximation errors. The following relation can be stated:

$$D_{kj}\Delta u_m^{(j)} = f_{k,m}. \tag{1}$$

Herein, $f_{k,m}$ is a vectorized form of the differences of solutions, D_{kj} is a rectangular $N \times n$ matrix, the summation over a repeating index is implied (no summation over m). Since the relation $N = n \cdot (n-1)/2$ (N is the number of equations and RHS terms) is valid, the simplest case when the number of equations is equal the number of unknowns corresponds to the ensemble of three numerical solutions:

$$\begin{pmatrix} 1 & -1 & 0 \\ 1 & 0 & -1 \\ 0 & 1 & -1 \end{pmatrix} \begin{pmatrix} \Delta u_m^{(1)} \\ \Delta u_m^{(2)} \\ \Delta u_m^{(3)} \end{pmatrix} = \begin{pmatrix} u_m^{(1)} - u_m^{(2)} \\ u_m^{(1)} - u_m^{(3)} \\ u_m^{(2)} - u_m^{(3)} \end{pmatrix}. \tag{2}$$

At $n = 3$, the number of unknowns is equal to the number of data elements, at $n > 3$ it is greater. Unfortunately, the matrix inversion is infeasible at any n. It is caused by the invariance of the difference of solutions to a shift transformation: $u_m^{(j)} = \tilde{u}_m^{(j)} + b + \Delta u_m^{(j)}$

and $\Delta u_m^{(j)} = \Delta \tilde{u}_m^{(j)} + b$ for any $b \in (-\infty, \infty)$, where $\Delta \tilde{u}_m^{(j)}$ is the true error. By this reason, the problem (1) is underdetermined (non-unique).

An addition of linear terms containing solutions (without differences) to RHS can not remove the degeneracy, since the relation $u_m^{(j)} = \tilde{u}_m + b + \Delta \tilde{u}_m^{(j)} - b$ holds. Fortunately, the shift invariance is a purely linear effect. So, a transition to the nonlinear statement may cure this degeneration.

3 The Nonlinear Statements for the Estimation of the Approximation Error

The paper [13] consider the opportunities, presented by the quasilinear (regarding the parameters of interest) equation $A(\vec{x})\vec{x} = f$, $\vec{x}_m = \{\Delta u_m^{(1)}, \Delta u_m^{(2)}, \Delta u_m^{(3)}, \tilde{u}_m)$ having the form:

$$\begin{pmatrix} 2u_m^{(1)} - \Delta u_m^{(1)} & 0 & 0 & u_m^{(1)} - \Delta u_m^{(1)} \\ 1 & 0 & -1 & 0 \\ 0 & 1 & -1 & 0 \\ 1/3 & 1/3 & 1/3 & 1 \end{pmatrix} \cdot \begin{pmatrix} \Delta u_m^{(1)} \\ \Delta u_m^{(2)} \\ \Delta u_m^{(3)} \\ \tilde{u}_m \end{pmatrix} = \begin{pmatrix} (u_m^{(1)})^2 \\ u_m^{(1)} - u_m^{(3)} \\ u_m^{(2)} - u_m^{(3)} \\ (u_m^{(1)} + u_m^{(2)} + u_m^{(3)})/3 \end{pmatrix}. \tag{3}$$

The Eq. (3) contains the nonlinear term $(u_m^{(1)})^2$ at right hand side, which prohibit the shift invariance $u_m^{(j)} = \tilde{u}_m + \Delta u_m^{(j)} + b$.

The determinant of matrix A of Eq. (3) is equal $u_m^{(1)}$. So, this equation may be solved by the simple matrix inversion (for non-zero solutions). In order to account for the nonlinearity, the linearization of the following form $A(\vec{x}^q)\vec{x}^{q+1} = f$ can be used at the iterative solution of Eq. (3) (q is the number of iteration).

Another nonlinear option (without differences of solutions) has the appearance:

$$\begin{pmatrix} 1 & 0 & 0 & 1 \\ 0 & 1 & 0 & 1 \\ 0 & 0 & 1 & 1 \\ 2u_m^{(1)} - \Delta u_m^{(1)} & 0 & 0 & u_m^{(1)} - \Delta u_m^{(1)} \end{pmatrix} \cdot \begin{pmatrix} \Delta u_m^{(1)} \\ \Delta u_m^{(2)} \\ \Delta u_m^{(3)} \\ \tilde{u}_m \end{pmatrix} = \begin{pmatrix} u_m^{(1)} \\ u_m^{(2)} \\ u_m^{(3)} \\ (u_m^{(1)})^2 \end{pmatrix}. \tag{4}$$

It may be directly checked that $\det A(x) = -u_m^{(1)}$ and this matrix also can be inverted, and the result of the inversion has the form:

$$A^{-1}(x) = \begin{pmatrix} -1 + \Delta u_m^{(1)}/u_m^{(1)} & 0 & 0 & 1/u_m^{(1)} \\ -2 + \Delta u_m^{(1)}/u_m^{(1)} & 1 & 0 & 1/u_m^{(1)} \\ -2 + \Delta u_m^{(1)}/u_m^{(1)} & 0 & 1 & 1/u_m^{(1)} \\ 2 - \Delta u_m^{(1)}/u_m^{(1)} & 0 & 0 & -1/u_m^{(1)} \end{pmatrix}. \tag{5}$$

The matrix $A(x)$ may be decomposed into a sum the known matrix and a disturbance, which depends on the unknown approximation error $A(x) = A + \Delta A$.

$$A(x) = \begin{pmatrix} 1 & 0\,0 & 1 \\ 0 & 1\,0 & 1 \\ 0 & 0\,1 & 1 \\ 2u_m^{(1)} & 0\,0 & u_m^{(1)} \end{pmatrix} + \begin{pmatrix} 0 & 0\,0 & 0 \\ 0 & 0\,0 & 0 \\ 0 & 0\,0 & 0 \\ -\Delta u_m^{(1)} & 0\,0 & -\Delta u_m^{(1)} \end{pmatrix}. \tag{6}$$

The matrix A may be applied to solve approximately the problem.

4 The Numerical Algorithms

In order to solve Eqs. (2), (3), (4) we apply three different numerical methods. The coincidence of results ensures the reliability of analysis. In order to solve Eqs. (3), (4) we apply the matrix inversion (Gauss-Jordan elimination) [14]. The matrix inversion may be implemented for non degenerated square matrices only. In order to solve degenerate Eq. (2) we apply the variational statement and the Moore-Penrose pseudoinverse [15, 16]. The Moore-Penrose pseudoinverse for the degenerated matrices applies the regularization and has an appearance $A^+ = (A^*A + \alpha E)^{-1}A^*$. Herein A_{ij} may be rectangular, $i \geq j$ (j- number of unknowns, i- number of equations and RHS terms). The following index form was used for the Moore-Penrose pseudoinverse

$$u_k = (A_{nj} \cdot A_{nk})^{-1} A_{ij} f_{i,m} = A_{ki}^+ f_i, \quad A_{ki}^+ = (A_{nj} \cdot A_{nk})^{-1} A_{ij}. \tag{7}$$

In order to obtain the steady and bounded solution of Eq. (2) we also use the variational statement of the Inverse Problem that includes the zero order Tikhonov regularization [17, 18]. In this statement the minimization of the functional is applied:

$$\varepsilon_m(\Delta\vec{u}) = 1/2(D_{ij}\Delta u_m^{(j)} - f_{i,m}) \cdot (D_{ik}\Delta u_m^{(k)} - f_{i,m}) + \alpha/2(\Delta u_m^{(j)} \Delta u_m^{(j)}), \tag{8}$$

where α is the regularization parameter. The steepest gradient descent is applied for the search of minimum of the functional.

The minimum of the regularizing term enables to obtain solutions with the minimum shift error, since:

$$\min_{b_m} \sum_j^n (\Delta u_m^{(j)})^2/2 = \min_{b_m}(\delta(b_m)) = \min_{b_m} \sum_j^n (\Delta\tilde{u}_m^{(j)} + b_m)^2/2. \tag{9}$$

The variation of the regularizing term over shift change may be stated as

$$\Delta\delta(b_m) = \sum_j^n (\Delta\tilde{u}_m^{(j)} + b_m)\Delta b_m. \tag{10}$$

So, its extremum occurs at:

$$b_m = -\frac{1}{n}\sum_j^n \Delta\tilde{u}_m^{(j)} = -\Delta\overline{u}_m. \tag{11}$$

Thus, the shift b_m is equal to the mean true error (with the opposite sign). By this reason, in linear approach (2), the error estimate $\left|\Delta u_m^{(j)}\right|$ cannot be less than $|b_m|$ and contains some irremovable uncertainty.

Formally, the presence of an irremovable uncertainty b_m is the main drawback of linear approach. Nevertheless, if errors magnitudes differ significantly (one is much greater others), then

$$b_m = -\frac{1}{n}\sum_j^n \Delta\tilde{u}_m^{(j)} \sim \Delta\tilde{u}_m^{(\max)}/n \tag{12}$$

and the irremovable uncertainty of greatest error is moderate, while the irremovable fault of small errors may be great. So, some errors (great) may be estimated enough accurately. Really, the numerical tests demonstrate the good resolution of great errors and the poor resolution of small errors.

The variational statement of problems (3) and (4) corresponds to the minimum of the following functional:

$$\varepsilon_m(\vec{x}_m) = 1/2(A_{ij}^m x_m^{(j)} - f_{i,m}) \cdot (A_{ik}^m x_m^{(k)} - f_{i,m}) + \alpha/2(x_m^{(j)} x_m^{(j)}), \tag{13}$$

where $\vec{x}_m = \{\Delta u_m^{(1)}, \Delta u_m^{(2)}, \Delta u_m^{(3)}, \tilde{u}_m)$. The dependence of the matrix A_{ij}^m on flow parameters is specific for this statement. The component of the solution $\tilde{u}_m$ does not suffer from the shift invariance, so the irremovable uncertainty is the same as in the linear case.

5 The Test Problem

The flowfield engendered by the oblique shock waves governed by two dimensional compressible Euler equations is used in the tests due to the availability of analytic solutions and the high level of the approximation errors. The flowfield is engendered by a plate at the angle of attack $\alpha = 20°$ in the uniform supersonic flow (M = 4). The analytic solution is engendered by the Rankine-Hugoniot relations. The projection of the analytic solution on the computational grid is considered as a true solution and used for estimation of the true error.

The conditions at the left boundary ("*inlet*") and at the upper boundary *("top")* are specified by the inflow parameters. The conditions at the right boundary ("*outlet*") are specified by the zero gradient condition. The conditions at the down boundary, which ensure the non-penetration on the plate surface, are posed by the zero normal gradient for the pressure and the temperature and the "*slip*" condition for the speed.

The following solvers are used that belong to the OpenFOAM software package [19]:

- *rhoCentralFoam* (rCF), based on the central-upwind scheme [20, 21].
- *sonicFoam* (sF), based on the PISO algorithm [22].
- *pisoCentralFoam* (pCF) [23], which combines the Kurganov-Tadmor scheme [20] and the PISO algorithm [22].
- *QGDFoam (QGDF),* which implements the quasi-gas dynamic algorithm [24].

6 Numerical Results

The numerical results for error analysis are provided in Figs. 1–8 for $\alpha = 20^o, M = 4$. The index along the abscissa axis $i = N_y(k_x - 1) + m_y$ is defined by indexes along X (k_x) and Y (m_y). The jump of variables corresponds to the shock wave. The magnitudes of the numerical and analytical density are provided. The Figures present also the true density error $\Delta\rho_m^{(i)} = \rho_m^{(i)} - \tilde{\rho}_m,$, the error estimates using linear (2) and nonlinear (3), (4) approaches (practically coinciding in tests), and the irremovable uncertainty $b_m = -\frac{1}{n}\sum_j^n \Delta\tilde{\rho}_m^{(j)}$.

First, the ensemble of rCF, pCF, sF solvers is used. The Fig. 1 presents the vectorized density and estimates of error for the solution computed by rCF method. The numerical solution, the analytical solution, true error $\Delta u_m^{(i)} = u_m^{(i)} - \tilde{u}_m$, the irremovable uncertainty, the linear (Eq. (2)), and nonlinear estimates of error (Eq. (4)) are presented.

The Fig. 2 presents the analogical information for rCF based solution, Fig. 3 demonstrates the information regarding sF based solution.

The Fig. 1 and Fig. 2 present the cases when the irremovable uncertainty is close to the true error. One may see the relative poor quality of the linear error estimation for these cases. The Fig. 3 presents the case when the irremovable uncertainty is lesser the true error and it is unable to spoilt the estimates.

The nonlinear estimates for the case of Fig. 1 are zero, so they are not presented. In Fig. 2 they are close to zero and nonrealistic. The nonlinear estimate (Eq. (4)) presented in Fig. 3 is close to the true error and overperforms the linear estimate (Eq. (2)). One can see that the nonlinear estimates (Eq. (4)) have significant anisotropy of sensitivity.

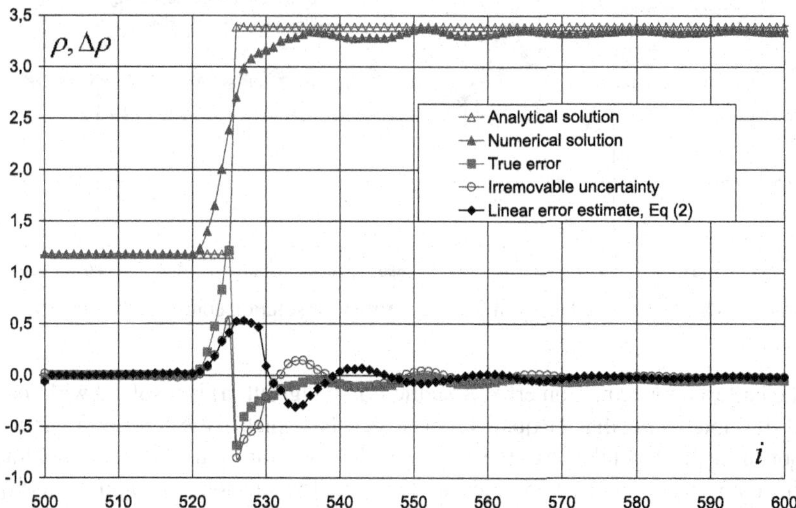

Fig. 1. The vectorized density and density errors for solution computed by rCF solver.

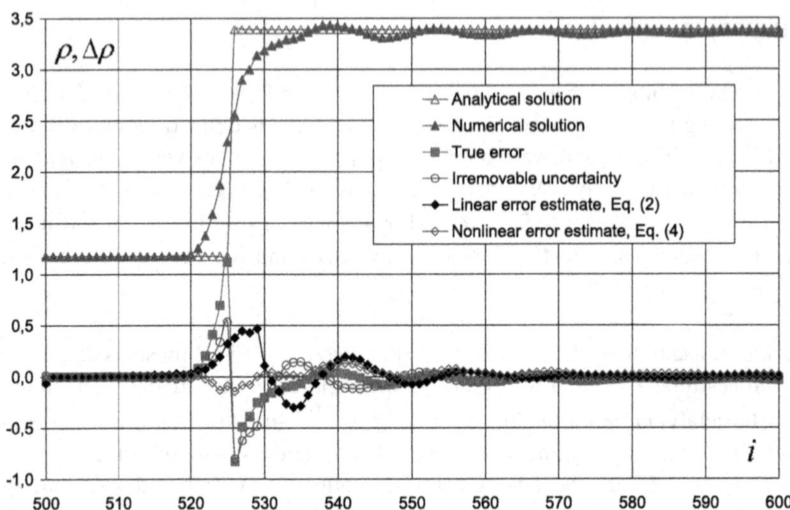

Fig. 2. The vectorized density and density errors for solution computed by pCF solver.

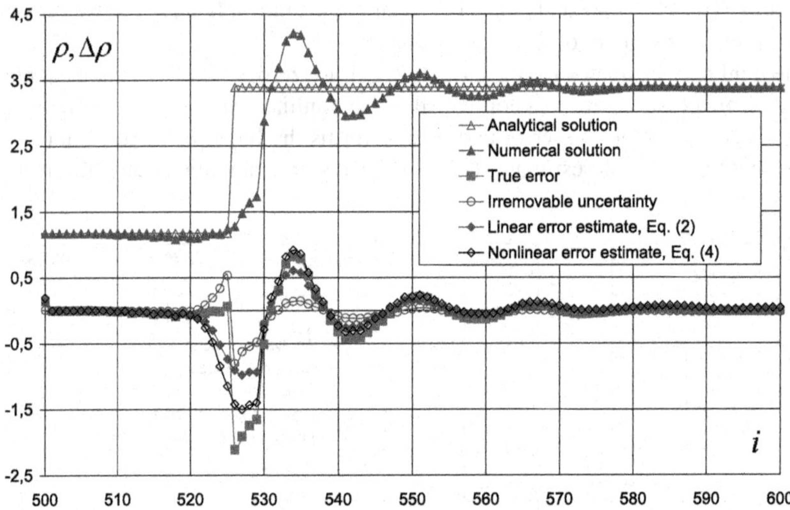

Fig. 3. The vectorized density and density errors for solution computed by sF solver.

The largest approximation error (specific for sF algorithm) is resolved with the best quality. It is interesting that the quality depends on the choice of the algorithm. Figures 1–3 demonstrate the dependence of the quality on the algorithm in use (linear or nonlinear). The linear approach (Eq. (2)) provides moderate quality, the nonlinear approach (Eq. (3) and (4)) provides high resolution for the estimate of the largest error.

The error structure is presented by Figs. 4,5, which provide the true error, irremovable uncertainty, linear error estimate, and improved estimate (estimate + irremovable uncertainty). One may see that the irremovable uncertainty $b_m = -\frac{1}{n}\sum_{j}^{n} \Delta\tilde{\rho}_m^{(j)}$ is the main source of the uncertainty at both the linear and nonlinear estimation of approximation error. Fortunately, it spoils significantly the small errors and leaves safe the great errors. The improved estimate is close to the true error for both linear and nonlinear approaches.

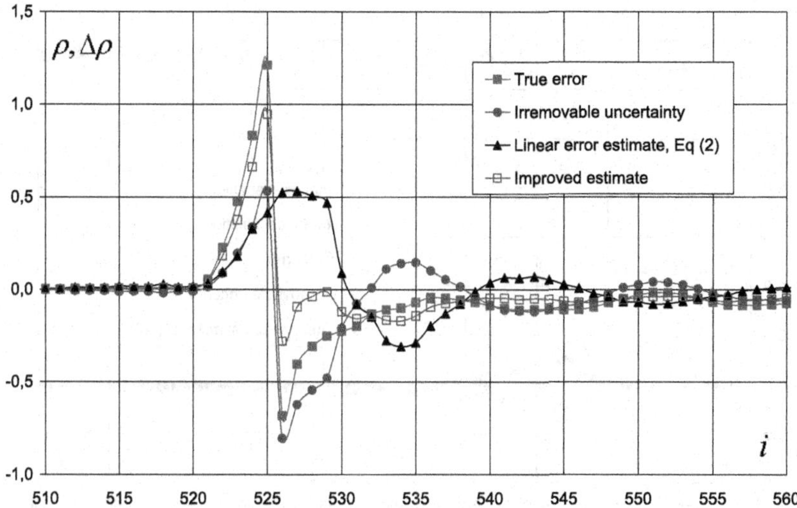

Fig. 4. The errors' structure for solution computed by rCF solver.

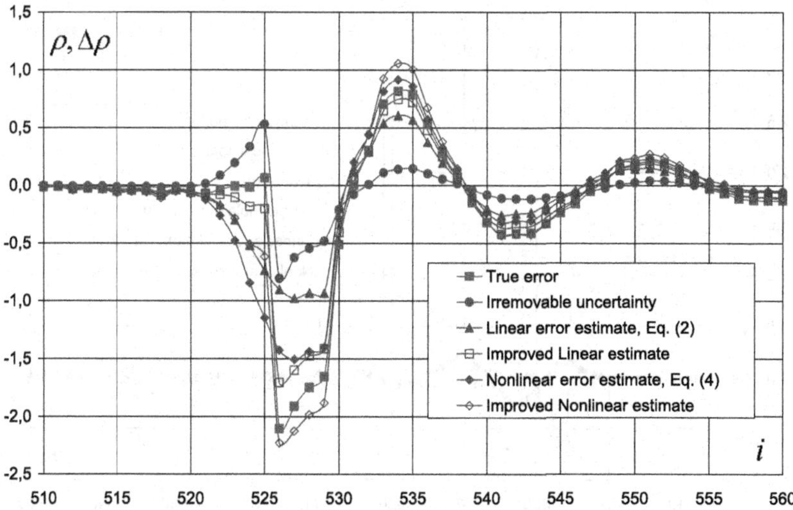

Fig. 5. The errors' structure for solution computed by sF solver

Second, the ensemble of rCF, pCF, QGDF solvers is used. The Fig. 6 presents the vectorized density and estimates of errors for rCF based solution. The Fig. 7 presents the analogical information for pCF based solution, Fig. 8 demonstrates QGDF based solution. The results are qualitatively close to previous (Figs. 1–5). The magnitude of the approximation error for solution computed by QGDF is less if compare with solution by sF. By this reason, the error of QGDF based solution is estimated with less accuracy if compare with sF based solution.

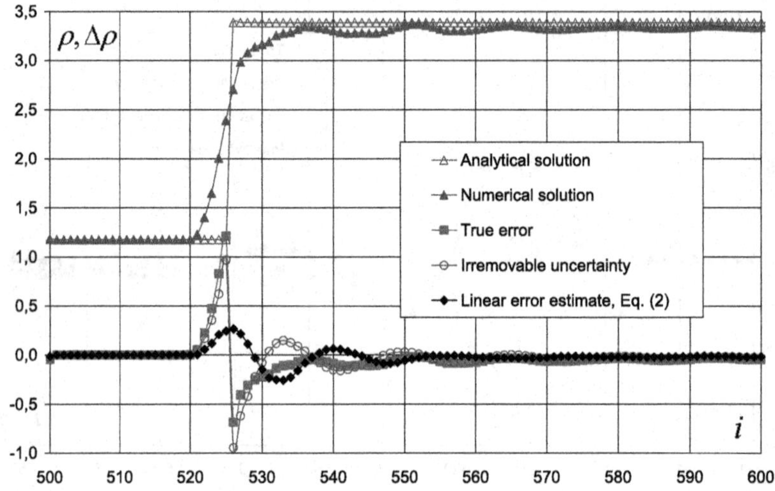

Fig. 6. The vectorized density and density errors for solution computed by rCF solver

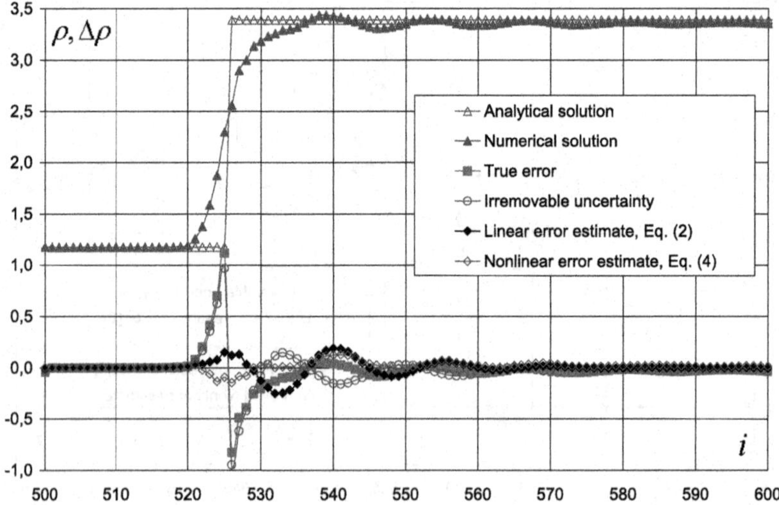

Fig. 7. The vectorized density and density errors for solution computed by pCF solver.

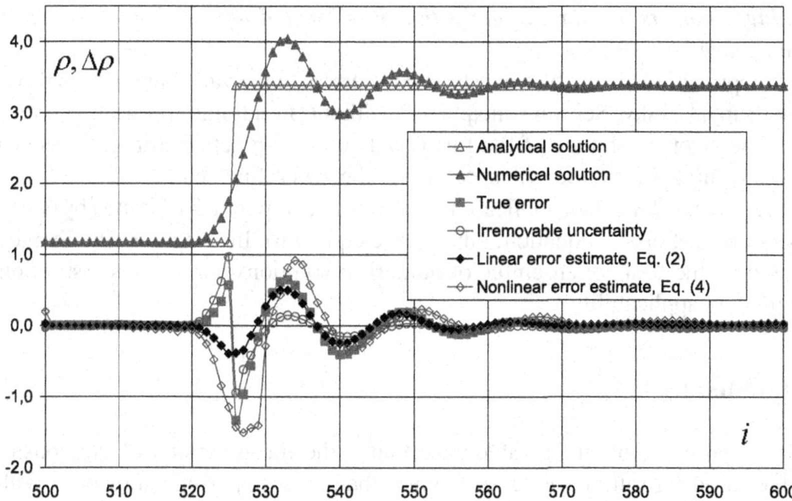

Fig. 8. The vectorized density and density errors for solution computed by QGDF solver.

7 Discussion

The present results demonstrate the feasibility of the precise estimation of the approximation error for the most inaccurate solution from the ensemble of numerical solutions, obtained by different algorithms.

The numerical tests demonstrated drastic difference of the error resolution provided by the linear and nonlinear approaches. The linear approach resolves all errors with the same quality (if abstract from the unremovable uncertainty). The sensitivity of the solution component in the nonlinear case may be estimated from expression $\delta x_i = A_{ij}^{-1}(x)\delta f_j + \delta A_{ij}^{-1}(x)f_j = B_{ij}(x)\delta f_j$ as $|\delta x_i| \le C_i \cdot \|\delta f_j\|$, where $C_i = \|B_{ij}(x)\|$ is the norm of matrix row. So, the sensitivity of different components of the solution may be different and some components of the solution may belong to the nullspace of matrix [26]. The numerical tests demonstrated that the nonlinear approach practically does not resolve small errors, while the great error is resolved with the high accuracy.

At first glance, the numerical solution having a great approximation error can not be valuable. However, in the frame of the ensemble based error estimation [25] such solution enables to generate some hypersphere (with the centre at some precise solution) that contain the true solution. So, the precise estimation of the approximation error is very important, since it justifies the application of the ensemble based method [25] for capture of true solution.

In general, it enables to construct the numerical solution in the sense of Synge [1, 2]. At present, the numerical solution is commonly considered to be an element of the sequence converging to the exact solution as the step of discretization decrease. By this reason, the mesh refinement is the key element of the modern CFD practice. Contrary to this approach, Synge stated ([1], p. 97): ***"In general, a limiting process is not used, and we do not actually find the solution.... But although we do not find it, we learn***

something about its position, namely, that it is located on a certain hypercircle in function space".

This approach easily enables to estimate the errors of the valuable functionals (using Cauchy–Bunyakovsky–Schwarz inequality) without an adjoint approach. The acceptance of the error magnitude of the functionals used in practices (drag, lift etc.) may serve as a natural criterion for the necessity of the grid refinement.

At present, the domain of applicability of the Singe's approach is limited by equations of special form (Poisson equation, biharmonic equation). In this work, the Synge idea is realized on the basis of ensemble of numerical solutions that relaxes restrictions of Synge method applicability.

8 Conclusion

Due to the presence of irremovable uncertainty, the linear version of the considered ensemble based algorithm enables to estimate the greatest approximation error with the acceptable accuracy, while the errors of more precise solutions are resolved with the lesser quality.

The nonlinear version of the ensemble based algorithm enables the estimation of the greatest approximation error with the higher accuracy (if compare with the linear case), while the errors of more precise solutions are practically not resolved. The precisely resolved component of the error may be detected from the analysis of the distances between numerical solutions and corresponds to the most distant solution.

References

1. Synge, J.L.: The Hypercircle in Mathematical Physics. CUP, London (1957)
2. Synge, J.L.: The Hypercircle method. In: Studies in Numerical Analysis, pp. 201–217. Academic Press, London (1974)
3. Repin, S.I.: A posteriori estimates for partial differential equations, vol. 4. Walter de Gruyter (2008). https://doi.org/10.1515/9783110203042
4. Roy, C., Raju, A.: Estimation of discretization errors using the method of nearby problems. AIAA J. **45**(6), 1232–1243 (2007)
5. Richardson, L.F.: The approximate arithmetical solution by finite differences of physical problems involving differential equations with an application to the stresses in a masonry dam. Trans. Roy. Soc. London Ser. A **2**(10), 307–357 (1908)
6. Banks, J.W., Aslam, T.D.: Richardson extrapolation for linearly degenerate discontinuities. J. Sci. Comput. **57**, 1–15 (2012)
7. Roy, C.: Grid convergence error analysis for mixed-order numerical schemes. AIAA J. **41**(4), 595–604 (2003)
8. Alekseev, A.K., Bondarev, A.E., Kuvshinnikov, A.E.: A comparison of the Richardson extrapolation and the approximation error estimation on the ensemble of numerical solutions. In: Paszynski, M., Kranzlmüller, D., Krzhizhanovskaya, V.V., Dongarra, J.J., Sloot, P.M.A. (eds.) ICCS 2021. LNCS, vol. 12747, pp. 554–566. Springer, Cham (2021). https://doi.org/10.1007/978-3-030-77980-1_42
9. Standard for Verification and Validation in Computational Fluid Dynamics and Heat Transfer, ASME V&V 20-2009 (2009)

10. Guide for the Verification and Validation of Computational Fluid Dynamics Simulations, American Institute of Aeronautics and Astronautics, AIAA-G-077-1998 (1998)
11. Alekseev, A.K., Bondarev, A.E., Kuvshinnikov, A.E.: A posteriori error estimation via differences of numerical solutions. In: Krzhizhanovskaya, V.V., et al. (eds.) ICCS 2020. LNCS, vol. 12143, pp. 508–519. Springer, Cham (2020). https://doi.org/10.1007/978-3-030-50436-6_37
12. Alekseev, A.K., Bondarev, A.E.: The Estimation of approximation error using inverse problem and a set of numerical solutions. Inverse Prob. Sci. Eng. **29**(13), 3360–3376 (2021). https://doi.org/10.1080/17415977.2021.2000604
13. Alekseev, A.K., Bondarev, A.E., Kuvshinnikov, A.E.: On a nonlinear approach to uncertainty quantification on the ensemble of numerical solutions. In: Groen, D., de Mulatier, C., Paszynski, M., Krzhizhanovskaya, V.V., Dongarra, J.J., Sloot, P.M.A. (eds) ICCS 2022, LNCS 13353, pp. 637–645. Springer, Cham (2022). https://doi.org/10.1007/978-3-031-08760-8_52
14. Press, W.H., Flannery, B.P., Teukolsky, S.A., Vetterling, W.T.: Numerical Recipes in Fortran 77: The Art of Scientific Computing. Cambridge University Press (1992)
15. Moore, E.H.: On the reciprocal of the general algebraic matrix. Bull. Am. Math. Soc. **26**(9), 394–395 (1920)
16. Penrose, R.: A generalized inverse for matrices. Proc. Camb. Philos. Soc. **51**(3), 406–413 (1955)
17. Tikhonov, A.N., Arsenin, V.Y.: Solutions of Ill-Posed Problems. Winston and Sons, Washington DC (1977)
18. Alifanov, O.M., Artyukhin, E.A., Rumyantsev S.V.: Extreme Methods for Solving Ill-Posed Problems with Applications to Inverse Heat Transfer Problems. Begell House (1995)
19. OpenFOAM. http://www.openfoam.org
20. Kurganov, A., Tadmor, E.: New high-resolution central schemes for nonlinear conservation laws and convection-diffusion equations. J. Comput. Phys. **160**(1), 241–282 (2000). https://doi.org/10.1006/jcph.2000.6459
21. Greenshields, C., Wellerr, H., Gasparini, L., Reese, J.: Implementation of semi-discrete, non-staggered central schemes in a colocated, polyhedral, finite volume framework, for high-speed viscous flows. Int. J. Numer. Meth. Fluids **63**(1), 1–21 (2010). https://doi.org/10.1002/fld.2069
22. Issa, R.: Solution of the implicit discretized fluid flow equations by operator splitting. J. Comput. Phys. **62**(1), 40–65 (1986). https://doi.org/10.1016/0021-9991(86)90099-9
23. Kraposhin, M., Bovtrikova, A., Strijhak, S.: Adaptation of Kurganov-Tadmor numerical scheme for applying in combination with the PISO method in numerical simulation of flows in a wide range of Mach numbers. Procedia Comput. Sci. **66**, 43–52 (2015). https://doi.org/10.1016/j.procs.2015.11.007
24. Kraposhin, M.V., Smirnova, E.V., Elizarova, T.G., Istomina, M.A.: Development of a new OpenFOAM solver using regularized gas dynamic equations. Comput. Fluids **166**, 163–175 (2018). https://doi.org/10.1016/j.compfluid.2018.02.010
25. Alekseev, A.K., Bondarev, A.E., Kuvshinnikov, A.E.: On uncertainty quantification via **the ensemble of independent numerical solutions**. J. Comput. Sci. **42**, 10114 (2020)
26. Strang, G.: The fundamental theorem of linear algebra. Amer. Math. Monthly **100**(9), 848–859 (1993)

Semantic Hashing to Remedy Uncertainties in Ontology-Driven Edge Computing

Konstantin Ryabinin[1,2(✉)] and Svetlana Chuprina[2]

[1] Saint Petersburg State University, 7/9 Universitetskaya Emb., Saint Petersburg 199034, Russia
kostya.ryabinin@gmail.com
[2] Perm State University, 15 Bukireva Str., Perm 614068, Russia

Abstract. This paper discusses the specific kind of uncertainties, which appear in ontology-driven software development. We focus on the development of IoT applications whose source code is generated automatically by an ontology-driven framework. So-called "compatibility uncertainties" pop up when the ontology is being changed while the corresponding generated application is in operation. This specific kind of uncertainties can be treated as a variant of implementation uncertainties. The algorithm of its automated handling is presented. The proposed algorithm is implemented within the SciVi platform and tested in the real-world project devoted to the development of custom IoT-based hardware user interfaces for virtual reality. We use the SciVi platform as a toolset for the automatic generation of IoT devices firmware for ontology-driven Edge Computing but the problem discussed is common for any tools which are used for the generation of ontology-driven software.

Keywords: Ontology-Driven Edge Computing · IoT · Firmware Generation · Semantic Hashing · Implementation Uncertainties

1 Introduction

The spread of Ubiquitous Computing faces the challenges of configurability, interoperability, and context awareness of programmable microelectronic devices, which build up the Internet of Things (IoT). These devices are expected to interconnect into sustainable computing networks, maintain the data flow, and support user-friendly human-machine interaction to fulfill the Ubiquitous Computing paradigm [1,3,11]. One of the strategies to improve networking stability and performance within the Ubiquitous Computing environment is to push the

This study is supported by the research grant No. ID92566385 from Saint Petersburg State University.

J. Mikyška et al. (Eds.): ICCS 2023, LNCS 14077, pp. 656–671, 2023.
https://doi.org/10.1007/978-3-031-36030-5_52

capabilities of artificial intelligence to the networks' edge. This is all about providing the end-point lightweight devices (so-called Edge Nodes) with the ability to track the context of their work and make decisions on their own without relying on more powerful hub devices (so-called Fog Nodes) or remote services (so-called Cloud Nodes).

The promising approach to enhance the intelligent features of smart devices is the bridging of the IoT with the Semantic Web technologies that is called the Semantic Web of Things (SWoT) [9]. SWoT assumes leveraging ontologies to build formal models of knowledge about the devices, interfaces, networks, and processed data to overcome the interoperability, configurability, and accessibility limitations of the traditional IoT.

The logical step of SWoT development is the ontology-driven Edge Computing (ODEC) vision [1,25,27,28] that enables ontologies to act not just as descriptive models of IoT artifacts, but also as full-fledged drivers of computation and communication processes on the network's edge.

In our previous work [25], we contributed to the development of the ODEC by the following. First, we created the semantic compression algorithm to fit the task ontologies [31] to the memory of resource-constrained edge devices. Second, we implemented a tiny reasoner that is capable of running on edge devices, interpreting the compressed task ontologies, and performing the computation and communication tasks according to ontological specifications. The reasoner consists of an immutable core and a dynamic set of functions (so-called operators), which can be referenced from within task ontologies. Actual firmware for a particular edge device is generated automatically by means of services provided by our platform SciVi[1] [24], and namely, its special toolset called EdgeSciVi Workbench. The firmware generation is driven by an application ontology that combines domain and task ontologies and offers a specified system of concepts for a particular application [31]. In our case, the application ontology describes a set of supported hardware components, data processing mechanisms, communication protocols, and interoperability techniques (examples of such kind of application ontologies can be found in [26]). Once generated, the firmware can be installed on the desired device. After that, the actual programs this device should execute can be encoded as task ontologies and pushed to the device without reflashing. To compose the task ontologies, we have a special high-level visual programming tool based on the data flow paradigm. The building blocks for the data flow diagrams are operators, which are derived from the application ontology and correspond to the functions incorporated in the embedded reasoner. The entire workflow is based on the so-called low-code concept [16] allowing users with no programming hard skills to declaratively describe the data processing pipelines by means of high-level building blocks. All the low-level details of computation, communication, and configuration are automatically covered by the SciVi platform, which extracts all the relevant knowledge from the underlying ontological knowledge base.

[1] https://scivi.tools/.

Although this approach provides a fast and easy way to organize the Edge Computing process, it still has limitations. The one we address in the present work is the uncertainty that can be called "compatibility uncertainty" of a random edge device discovered in the network.

The problem is as follows. Let us have an edge device, whose firmware is an embedded reasoner that has been generated by SciVi according to the application ontology $\mathcal{D}$. This reasoner incorporates a set of functions (operators) Φ that is a subset of all the operators described by $\mathcal{D}$ (the actual subset of operators for a particular firmware is being chosen by the user). After the firmware installation, let us turn off this device for a while. Let us change $\mathcal{D}$ while the device is offline. After this device reappears in the network, it will be discovered as capable of ODEC. However, as $\mathcal{D}$ has been changed since the firmware generation, attempts to use $\mathcal{D}$ as a source for task ontologies for this device will lead to undefined behavior because the operators described by $\mathcal{D}$ can be incompatible anymore with the functions in the device firmware.

The obvious solution for this problem would be version numbering control. A version number can be assigned to ontology $\mathcal{D}$ and stored in the embedded reasoner by its generation. With each change, the version number of $\mathcal{D}$ should be increased. When used in ODEC context, the device should indicate the version of $\mathcal{D}$ it contains and this version should be matched with the actual version of $\mathcal{D}$. If these versions differ, the firmware has to be regenerated and reinstalled. However, this naive solution has two major drawbacks. First, ontology $\mathcal{D}$ is in fact a merge of small ontologies, which describe individual data processing operators, data types, communication protocols, hardware elements, etc. The merging of these ontologies is performed by SciVi automatically at runtime [4], while the knowledge engineers work with the small initial ontologies, which are much more readable and handy. So, version numbering of $\mathcal{D}$ cannot be implemented directly. The second drawback is the uncertainty of whether the changed $\mathcal{D}$ is really incompatible with the current version of the device firmware or not. If any single change of any part of $\mathcal{D}$ requires a firmware update, all the ODEC profit in device reconfiguration speed and easiness is nullified.

In fact, only those changes of $\mathcal{D}$ matter, which affect the description of operators incorporated in the particular reasoner. To track these changes, we propose using special semantic hashing of the application ontology. This hashing allows us to build unique fingerprints of individual operators, store them in the generated firmware, and use them to check the compatibility of firmware with the current version of $\mathcal{D}$.

The proposed hashing algorithm is implemented in SciVi along with the new improved version of embedded task ontologies representation format (so-called EON, stands for Embedded ONtology [25]) to drive the Edge Computing process. The new version of EON (EON 2.0) allows efficient and straightforward encoding of operators' instances with their fingerprints. We tested the proposed improvements of our ODEC implementation by creating a custom configurable hardware control panel to automate the debugging of experimental scenes in virtual reality (VR).

In this paper, we present an improvement of our ontology-driven Edge Computing implementation. The following key results can be highlighted:

1. The semantic hashing algorithm for application ontologies to build unique 16-bit fingerprints of data processing operators described by these ontologies.
2. The improved representation format to store task ontologies in the memory of resource-constrained edge devices.
3. The approach to tackling the problem of edge devices' "compatibility uncertainty" within the ontology-driven Edge Computing paradigm.

2 Related Work

Reducing the different types of uncertainties plays a pivotal role not only in optimization and decision-making processes but also in solving a variety of real-world problems of software development/validation, including the context of programmable microelectronic devices within the IoT ecosystem [14]. In this paper, we focus on some kind of uncertainties that occurs when the ontology is being changed while the corresponding generated ontology-driven application is in operation. We consider this specific kind of uncertainties as a variation of implementation uncertainties [21].

F. Scioscia and M. Ruta are among the pioneers, who proposed blending Semantic Web and IoT technologies to build so-called SWoT [22,29]. This vision was first introduced to tackle the issues of "data management in pervasive environments" [29] and further extended to address the "challenges associated with standardization, interoperability, discovery, security, and description of IoT resources and their corresponding data" [18]. The most valuable contributions to SWoT are systematically reviewed by A. Rhayem et al. [18]. Along with this review, F. Qaswar et al. fulfilled an analysis of the most interesting cases when ontologies were been applied in IoT [17]. This analysis shows up, that ontologies are nowadays intensively used in IoT design and development to tackle interoperability, integration, and privacy issues.

One of the possible directions of SWoT evolution is ODEC [25], the vision of using ontologies to describe the entire functioning process of IoT devices and corresponding data flows, which allows declaratively specified management of Edge Computing. The pioneering works in ODEC are related to describing edge device capabilities by ontologies [27]; ontology-driven edge device virtualization [28]; using ontologies to retrieve devices, infer their interoperability, and select their operation mode [5]; development of embedded ontology-based expert systems to detect anomalies within IoT and robotics [30]; model-driven middleware generation for semantic integration of devices within Ubiquitous Computing environment [1]. Our contribution to ODEC is the development of a full-fledged ontology reasoner that is embedded into edge devices and capable of executing functions described by task ontologies [25].

One of the vast problems of ODEC is adequate ontology representation suitable for edge devices [32]. The most popular standard formats for storing ontologies are OWL[2] and RDF[3], but both of them are too verbose and thereby unacceptable for most resource-constrained edge devices. To tackle this problem, several compression algorithms are proposed, which aim to reduce syntactic, semantic, and structural redundancies in the representation of ontologies [8,12]. As shown by M. A. Hernández-Illera et al., exploiting structural redundancies is the most promising direction in terms of compression ratio [8]. The results of different state-of-the-art approaches to reducing structural redundancies are reported independently by M. Röder et al. and T. Sultana et al. M. Röder et al. state that the best compression can be achieved by so-called k^2-trees [20], while T. Sultana et al. propose a grammar-based knowledge graph compression algorithm that outperforms all the most popular ontology compression techniques [33]. However, all the above-mentioned approaches deal with regular desktop-based ontology reasoners and cannot be efficiently ported to edge devices due to limitations of RAM capacity and CPU frequency.

X. Su et al. [32] and K. Sahlmann et al. [27] propose promising approaches to bridge the gap between knowledge graph representation requirements and edge devices capabilities, but none of them ensures the fitting of complete task ontologies into the edge device RAM for the full-fledged reasoning. Our previous contribution was a mix of reducing structural and semantic redundancies to achieve an extreme compression ratio of task ontologies specifically for ODEC use cases [25]. We proposed a special format called EON to represent embedded ontologies. But applying this format to practical ODEC use cases revealed some limitations, which we overcome in its next version, EON 2.0, described in more detail in Sect. 4.

To make ODEC accessible not only for programmers and knowledge engineers but also for casual users without special IT hard skills, high-level management tools should be provided. In this regard, an emerging trend is the spread of so-called low-code development platforms, which allow composing software in visual editors without writing source code [16]. This approach can be efficiently combined with ontology-driven software development [13]. The toolset for the visual editor can be automatically generated according to the application ontology that describes available software building blocks. Then, the visual software model composed by the user can be converted to task ontology. Finally, the result software product can be automatically generated with a help of a semantic reasoner that processes this task ontology [4].

To remedy the "compatibility uncertainty" between the application ontology and the task ontologies, so-called semantic hashing can be utilized to efficiently track the relevant changes in the corresponding ontologies. Semantic hashing is an approach of encoding specific documents (for example, ontologies) in compact binary vectors (hash codes) to allow efficient and effective similarity search [7]. According to the specification of EON, codes to operate with should be only 16 bits long to fit in the RAM of target edge devices [25]. So, we decided to use Pearson hashing [15] that can be easily adapted to the hash values of any length starting with 8 bits. With a predefined lookup table [15], this hashing algorithm

[2] https://www.w3.org/OWL/.

[3] https://www.w3.org/RDF/.

gives good results in terms of collision avoidance. To fully prevent collisions, additional MD5 hashing [19] is proposed (see Sect. 3).

3 Semantic Hashing to Reduce Uncertainties

As mentioned above, in this paper we discuss the so-called "compatibility uncertainty".

3.1 Formal Model of Operator

The SciVi platform is based on a microservice architecture, where each microservice corresponds to a data processing operator. We formalize the operator as

$$\Delta : \{I, S\} \rightarrow O, \qquad (1)$$

where $I = \{I_k | k = \overline{1, |I|}\}$ is a set of typed inputs, $S = \{S_l | l = \overline{1, |S|}\}$ is a set of typed parameters (also denoted as settings), $O = \{O_t | t = \overline{1, |O|}\}$ is a set of typed outputs of the Δ operator [4]. Let the set of available types be denoted as Q.

Theorem 1. *Any operator that adheres to (1) can be described by the* $\mathcal{D}_\Delta$ *lightweight application ontology.*

Proof. According to [6], $\mathcal{D}_\Delta$ ontology can be expressed as $\mathcal{D}_\Delta = \{T, R, A\}$, where T is a thesaurus of concepts, R is a set of relationships between concepts from T, and A is a set of axioms related to the elements of T and R.

By the assumption, $\mathcal{D}_\Delta$ is a lightweight ontology, which means $A = \emptyset$. Let us identify the basic categories of elements of T:

1. *Input* (element of I) – independent variable within an operator.
2. *Setting* (element of S) – constant within an operator.
3. *Output* (element of O) – dependent variable within an operator.
4. *Operator* (element Δ of formula (1)) – transformation that maps the values of independent variables (inputs) and constants (settings) to the values of dependent variables (outputs).
5. *Type* (element of Q) – a concept that defines the set of values that an operator's input, setting, and output can take, as well as the allowed operations on these values.

For any particular Δ operator, the T set can be composed as

$$\begin{aligned} T = \{&Operator, Input, Setting, Output, Type, \Delta, \\ &I_1, I_2, \ldots, I_{|I|}, S_1, S_2, \ldots, S_{|S|}, \\ &O_1, O_2, \ldots, O_{|O|}, Q_1, Q_2, \ldots, Q_{|I|+|S|+|O|}\}. \end{aligned}$$

Let us use two types of relationships to build R:

1. *is_a* - paradigmatic relationship "subclass-class".
2. *has* - paradigmatic relationship "class-property".

Let us use description logic attributive language with complement, so-called $\mathcal{ALC}$ [2], to formulate a terminology component (so-called TBox):

$$\begin{aligned}
&I_k \sqsubseteq Input, k = \overline{1, |I|},\\
&S_l \sqsubseteq Setting, l = \overline{1, |S|},\\
&O_t \sqsubseteq Output, t = \overline{1, |O|},\\
&Q_p \sqsubseteq Type, p = \overline{1, |I| + |S| + |O|},\\
&I_k \sqsubseteq Q_{p(k)},\\
&S_l \sqsubseteq Q_{p(|I|+l)},\\
&O_t \sqsubseteq Q_{p(|I|+|S|+t)},\\
&\Delta \equiv Operator \sqcap \exists has.I_1 \sqcap \ldots \sqcap \exists has.I_{|I|}\\
&\qquad \sqcap \exists has.S_1 \sqcap \ldots \sqcap \exists has.S_{|S|}\\
&\qquad \sqcap \exists has.O_1 \sqcap \ldots \sqcap \exists has.O_{|O|}.
\end{aligned}$$

Here $p(x)$ is a function that maps indices of elements of the I, S, and O sets to the indices of corresponding types from the Q set. This TBox corresponds to the knowledge graph shown in Fig. 1.

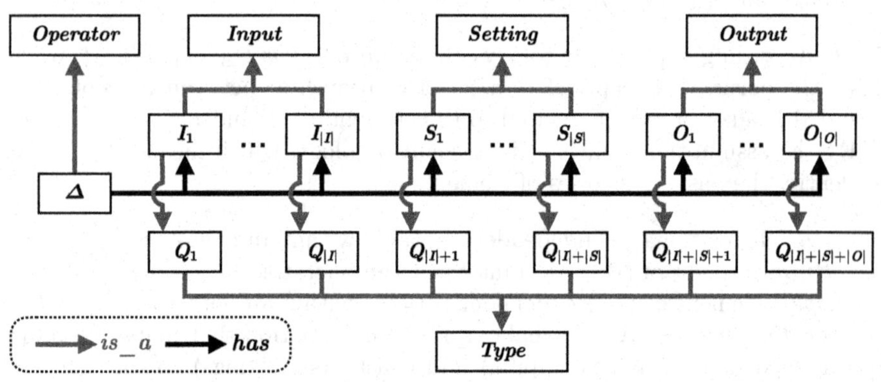

Fig. 1. Generalized description of Δ operator in the form of knowledge graph

From the above, $\mathcal{D}_\Delta = \{T, R\}$ is the lightweight ontology that precisely describes the Δ operator. This is an application ontology because it was been engineered for specific use within an ontology-driven data processing platform. □

To describe complex types, for example, arrays, the *base_type* relationship is added to R to represent type specifications.

Let us assume that we have n operators which relate to some specific class of data processing problems. Then, according to Theorem 1, this set of operators can be described by the following application ontology:

$$\mathcal{D} = \bigcup_{i=1}^{n} \mathcal{D}_{\Delta_i}. \tag{2}$$

3.2 Unique Identifier of Operator

According to the EON format specification, operators should have unique 16-bit identifiers to be encoded into the concise semantically compressed ontology representation [25]. Previously, we used just a regular numbering of operators according to the order of their appearance in the $\mathcal{D}$ ontology (i.e. the i index in Eq. 2). However, this approach leads to "compatibility uncertainty" as mentioned above because having a particular identifier of the operator we cannot check whether it corresponds to the current version of $\mathcal{D}$. To remedy this uncertainty, we propose calculating the operator's identifier as a hash of the operator's structure to be able to detect possible ontology changes. This hash should preserve the operator's semantics within the context of the operator's execution.

Only those changes of $\mathcal{D}$ break the operator's compatibility, which affect the operator's execution process, so the hash function should be invariant to any irrelevant changes. In the compilers theory, the function is primarily identified by its name and type signature [10]. Similarly, an operator adhering to (1) can be identified by its name (denoted as name(Δ)) and the names of types of its inputs, settings, and outputs (denoted as name(Q_p), $p = \overline{1, |I| + |S| + |O|}$). It must be noted, that in case of complex types (for example, enumerations or structures) or hierarchical types, name(Q_p) should produce a concatenation of all the names of types in hierarchical order, with the ">" sign as a delimiter. Assuming that operator and all the types have string names, the following equation can be used to build the operator's string identifier:

$$\begin{aligned}\sigma(\Delta) = \text{name}(\Delta) + \text{"@I"} + \sum_{i=1}^{|I|} \text{name}(Q_i) + \text{"@S"} + \sum_{i=|I|+1}^{|I|+|S|} \text{name}(Q_i) \\ + \text{"@O"} + \sum_{i=|I|+|S|+1}^{|I|+|S|+|O|} \text{name}(Q_i),\end{aligned} \tag{3}$$

where sums denote string concatenation of operands sorted in lexicographical order with the ":" sign as a delimiter.

To get the operator's unique identifier, we propose calculating a 16-bit Pearson hash [15] of $\sigma(\Delta)$:

$$\pi(\Delta) = \text{Pearson}(\sigma(\Delta)). \tag{4}$$

Usually, the operators are described as a taxonomy. In this case, Q_p is assembled by the reasoner taking into account the operators' inheritance. Figure 2 demonstrates an example of application ontology that describes two operators handling general input-output pins of a microcontroller within an edge device.

Operators *Input Pin* and *Output Pin* inherit from their parent operator *GPIO* (General-Purpose Input-Output) the *Pin Number* that is the setting of

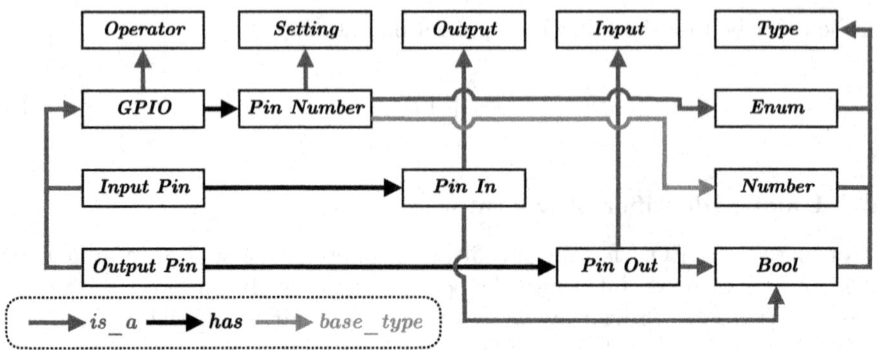

Fig. 2. Fragment of application ontology describing ODEC operators

complex type (enumeration of numbers). In addition, the *Input Pin* operator introduces a Boolean output *Pin In* and the *Output Pin* operator introduces a Boolean input *Pin Out*. The instances of these operators provide the ability to read and write logical values to the microcontroller's pins and thereby control different peripheral electronic components of custom edge devices. For these operators, the following signatures are calculated according to the described algorithm:

$$\sigma(Input\ Pin) = \text{“}\texttt{Input Pin@SEnum>Number@OBool}\text{”};$$
$$\sigma(Output\ Pin) = \text{“}\texttt{Output Pin@IBool@SEnum>Number}\text{”}.$$

The actual hash values depend on the particular Pearson's lookup table. In our case:

$$\pi(Input\ Pin) = 19218;$$
$$\pi(Output\ Pin) = 57372.$$

The described hashing algorithm is invariant to changes of the names and the order of inputs, settings, and outputs, as well as possible taxonomy changes of the operator as long as the taxonomy changes do not lead to changing the set of operators' inputs, outputs, or settings due to inheritance. Of course, the changes of other operators' descriptions do not affect the hash of the current one. Only the change of the current operator's signature that directly affects this operator's execution process, will alternate the hash result.

The obvious problem of short Pearson hashing is the relatively high probability of collisions, but the total number of operators used in a single class of data processing problems is pretty low. On average, the SciVi knowledge base currently contains 40 operators per problem class (although this number may grow with further SciVi development), which is ca. 1600 times less than the capacity of a 16-bit Pearson hash. In this case, the collision probability is about 1.2%. To get the collision probability higher than 50%, more than 300 operators are needed. However, to detect possible collisions, we use an additional check that is described in Sect. 3.3.

3.3 Embedded Reasoner Compatibility Check

As described in [25], the ODEC implies the following working principle. Let us assume, $\mathcal{D}$ is an ontology that describes a set of n operators as in Eq. (2). First, the user chooses a subset $\Phi = \{\Delta_1, \Delta_2, \ldots, \Delta_m\}$ of operators from $\mathcal{D}$ to be included in the embedded reasoner, $m \leq n$. Usually, $m \ll n$ due to the program memory restrictions of edge devices. Next, the firmware for the edge device is generated. This firmware contains the embedded reasoner with a special module of functions (FM) that incorporates implementations of operators from Φ. After this firmware is installed on the target edge device, the user can compose task ontologies, describing the particular data collection, data processing, and communication tasks the edge device should execute. For this, SciVi provides high-level visual programming tools based on data flow diagrams. These data flow diagrams are automatically converted to task ontologies referring to the operators from Φ [4,24,25]. These task ontologies are uploaded to the edge device and processed by the reasoner that triggers the necessary functions of FM, which correspond to the operators. The triggering order, input parameters, and related settings of functions are inferred from the task ontology.

To call the appropriate function for the corresponding operator, a special lookup table is generated that maps the operators' identifiers to the addresses of corresponding functions in program memory. To check if the particular reasoner installed on the particular edge device is compatible with the current ontology $\mathcal{D}$, the set of lookup table identifiers $\Pi = \{\pi(\Delta_1), \pi(\Delta_2), \ldots, \pi(\Delta_m)\}$ (see Eqs. (3) and (4)) is sent from the reasoner to the SciVi server and matched with the identifiers of all the operators described by $\mathcal{D}$. This requires $m \cdot n$ comparisons of 16-bit integer values. The total number of comparisons is fairly small (as mentioned above, the average value for n is 40, and $m \leq n$), moreover, the comparisons are performed on the server side, not on the edge device, so this operation is very fast despite its quadratic complexity.

To defeat potential collisions, an additional check is performed. Along with 16-bit Pearson hashes of operators' signatures $\sigma(\Delta_1), \sigma(\Delta_2), \ldots, \sigma(\Delta_m)$ (see Eq. (3)), a 128-bit MD5 hash [19] of the following value is calculated:

$$\xi = \sum_{i=1}^{m} \sigma(\Delta_i), \quad \mu = \mathrm{MD5}(\xi), \tag{5}$$

where sum denotes string concatenation of operands sorted in lexicographical order with the ";" sign as a delimiter.

This MD5 value is stored in the reasoner's source code along with the lookup table and sent to the SciVi server together with the lookup table identifiers.

SciVi server first searches the elements of Π among the operators' identifiers from $\mathcal{D}$. If at least one $\pi(\Delta_i), i = \overline{1, m}$ has no corresponding operator described by $\mathcal{D}$, the reasoner is treated as incompatible. Otherwise, the Φ set is reconstructed containing operators with the identifiers from Π. Then, to make sure there were no collisions, the received MD5 value μ is compared with the MD5 calculated for the signatures of operators from Φ. If these hashes are not equal, the reasoner is treated as incompatible (collision of Pearson hashes took place). Otherwise, the reasoner is compatible with $\mathcal{D}$.

The incompatible reasoner should be updated, while the compatible one can be used as is for subsequent ODEC process.

4 EON 2.0 Ontology Representation Format

In the EON 1.0 format [25], it was not efficient enough to represent different instances of the same operator. However, the edge devices like custom control panels, which are supposed to handle many similar buttons, often require executing the same operator many times during a single data processing iteration. Herewith, each call of the operator has its own settings and each operator's execution result is transmitted to its own branch of a data processing pipeline.

To improve the handling of operators' instances, we have upgraded EON to version 2.0 by slightly altering the memory layout of compressed task ontology. The EON 2.0 blob structure is as follows:

`| DFChunkLen | DFChunk | SChunkLen | SChunk | IChunk |`

`DFChunkLen` (stands for Data Flow Chunk Length, 1 byte) is a number of 3-byte elements in `DFChunk`. `DFChunk` (stands for Data Flow Chunk, 3 times `DFChunkLen` bytes) is a chunk containing the sequence of data transmission links of the task ontology formed like this:

`| OpInstA | Output | Input | OpInstB |`

`OpInstA` (stands for Operator Instance A, 1 byte) and `OpInstB` (stands for Operator Instance B, 1 byte) are task ontology identifiers of the operators' instances, whereby the output of `OpInstA` is linked to the input of `OpInstB`. `Output` (4 bits) is an index of output of `OpInstA`. `Input` (4 bits) is an index of input of `OpInstB`.

`SChunkLen` (stands for Settings Chunk Length, 2 bytes) is a length in bytes of `SChunk`. `SChunk` (stands for Settings Chunk, `SChunkLen` bytes) is a chunk containing the sequence of settings formed like this:

`| OpInst | Setting | Type | Value |`

`OpInst` (stands for Operator Instance, 1 byte) is a task ontology identifier of the operator's instance that has the corresponding setting. `Setting` (4 bits) is an index of setting of `OpInst`. `Type` (4 bits) is an internal type identifier of setting (currently, the following types are supported: signed and unsigned 8-bit, 16-bit, 32-bit integers, 32-bit float, and string). `Value` is an encoded setting's value (length depends on the type, strings are null-terminated).

`IChunk` (stands for Instances Chunk, length is not stored) is a chunk containing the sequence of operators' instances formed like this:

`| UIDOp | OpInst1 | OpInst2 | ... | 0x0 |`

`UIDOp` (stands for Unique Identifier of Operator, 2 bytes) is a unique identifier of the operator (calculated as described in Sect. 3.2), whose prototype is described in the application ontology and instances are described in the task ontology. `OpInst1, OpInst2, ...` (stand for Operator Instance, each is 1 byte) are task ontology identifiers of corresponding operator's instances. The `0x0` (1 byte) is a zero-byte terminating the list of instances.

The described format allows a very concise representation of task ontologies, including the instancing of operators. As evaluated in [25], EON-formatted ontologies require about 800 times less storage space than OWL-formatted ones. The new EON memory layout introduced in this paper does not decrease the EON efficiency and allows for explicit encoding of operators' instances. Thereby, it increases the area of EON applications within ODEC.

5 Discussion and Conclusion

WWe introduce the above-mentioned approach to remedy ODEC uncertainty and improvements of EON format within the research project "Text processing in L1 and L2: Experimental study with eye-tracking, visual analytics and virtual reality technologies"[4] (supported by the research grant No. ID92566385 from Saint Petersburg State University). One of the goals of this project is to study the peculiarities of the reading process of humans within a VR environment using eye tracking to estimate the differences in information perception in virtual and physical reality. For this, VR scenes with different texts are sequentially shown to informants. Their eye gaze tracks are sampled with a tracker embedded into the VR head-mounted display (HMD) and transmitted to SciVi for subsequent analysis and storage [23].

Under normal circumstances, the experiment requires a director who uses the SciVi Web interface to switch the scenes, start/stop eye gaze recording, and tune the analytics pipeline settings if needed. However, during the debugging of the analytics pipeline, it is more efficient when the pipeline developer plays both the role of an informant and a director simultaneously, without engaging additional people and wasting their time. However, the problem is that the informant has their eyes covered with the HMD, so it is nearly impossible to use a SciVi Web interface at that time. Taking the HMD on and off is not an option, because for the correct eye tracking, calibration is needed after taking the HMD on, and continuous re-calibrations would be very tedious. To solve this problem, custom reconfigurable edge-device-based controllers can be used, whose hardware control elements (buttons, potentiometers, etc.) are mapped to the particular pipeline settings, which are debugged. In this use case, ODEC is very efficient, because it allows both hardware and software reconfiguration and remapping without reprogramming and reflashing. Utilizing the built-in VR controllers instead of custom edge devices will not be as efficient, because it would require rebuilding the VR scene each time when the controllers' role should change. Moreover, the limited set of buttons on the built-in controller does not provide as much reconfiguration freedom as a custom device.

We successfully adopted ODEC in our VR-based research project. The proposed method to remedy the ODEC "compatibility uncertainty" through the semantic hashing of application ontologies allowed the efficient reuse of edge

[4] L1 and L2 stand for the native and foreign languages respectively.

devices powered by ODEC in a situation of continuous enrichment of the SciVi repository with new operators and ontologies. The implementation of the proposed compatibility checking reduces the amount of firmware generation and device reflashing cycles. It works fast even though formally it has quadratic complexity. Calculating the semantic hash of a particular operator described by an application ontology with 328 nodes and 845 relationships takes 2.15 ms on average (on a MacBook Pro 2.3 GHz 8-Core Intel Core i9 CPU, 16 Gb RAM). This enables the real-time processing of compatibility requests. The memory footprint of the semantic hash is fairly small: it requires appending just 16 bytes of MD5 hash sum to the device firmware, while the hashes of individual operators are stored in the identifiers of a functions module lookup table and do not require extra storage space.

The average time of updating our ODEC-powered edge device (based on the ESP8266 microcontroller) in different development cases is shown in Table 1. The operators used in the ODEC device are denoted as "related operators", the unused ones are denoted as "unrelated operators". In each development case, the behavior of the ODEC device is changed by uploading a new task ontology. If the task ontology is treated as compatible with the reasoner installed on this device, the only action needed to update the device behavior is transmitting this ontology from the computer, which takes 16 ms on average (just a single tick of ODEC device processing loop, which is set up to the rate 60 Hz in our implementation). Otherwise, if the task ontology is treated as incompatible, its uploading should be preceded by a firmware regeneration, device reflashing, reboot, and WiFi reconnection, which takes about 30 s on average in total.

Table 1. Performance comparison of ODEC device updating

Development case: type of changes in $\mathcal{D}$	No changes	Changes of related operators' structure	Changes of related operators' parameters naming	Changes of unrelated operators	Average
Conventional versioning	16 ms	30000 ms	30000 ms	30000 ms	22504 ms
Semantic hashing	16 ms	30000 ms	16 ms	16 ms	7512 ms

As seen in the table, the introduced semantic hashing increases the performance of updating the ODEC device 3 times on average. The software implementation of operators' descriptions semantic hasher and EON 2.0 format encoder are available in SciVi open source repository: https://github.com/scivi-tools/scivi.web/blob/master/onto/hasher.py.

For the next step of ODEC development, we plan to design an ontology-driven bus for joining hardware components of edge devices on plug-and-play principles. This will further increase the efficiency of the device reconfiguration process, which is crucial in the cases of creating custom hardware user interfaces within the IoT ecosystem.

References

1. Abdulrab, H., Babkin, E., Kozyrev, O.: Semantically enriched integration framework for ubiquitous computing environment. In: Babkin, E. (ed.) Ubiquitous Computing, chap. 9, pp. 177–196. IntechOpen (2011). https://doi.org/10.5772/15262
2. Baader, F., Calvanese, D., McGuinness, D.L., Nardi, D., Patel-Schneider, P.F. (eds.): The Description Logic Handbook: Theory, Implementation and Applications. Cambridge University Press (2003)
3. Calderon, M., Delgadillo, S., Garcia-Macias, A.: A more human-centric internet of things with temporal and spatial context. Procedia Comput. Sci. **83**, 553–559 (2016). https://doi.org/10.1016/j.procs.2016.04.263
4. Chuprina, S., Ryabinin, K., Koznov, D., Matkin, K.: Ontology-driven visual analytics software development. Program. Comput. Softw. **48**(3), 208–214 (2022). https://doi.org/10.1134/S0361768822030033
5. Dibowski, H., Kabitzsch, K.: Ontology-based device descriptions and device repository for building automation devices. EURASIP J. Embed. Syst. **2011**(1), 1–17 (2011). https://doi.org/10.1155/2011/623461
6. Golitsyna, O.L., Maksimov, N.V., Okropishina, O.V., Strogonov, V.I.: The ontological approach to the identification of information in tasks of document retrieval. Autom. Documentation Math. Linguist. **46**, 125–132 (2012). https://doi.org/10.3103/S0005105512030028
7. Hansen, C., Hansen, C., Simonsen, J.G., Alstrup, S., Lioma, C.: Unsupervised multi-index semantic hashing. In: Proceedings of the Web Conference 2021, pp. 2879–2889 (2021). https://doi.org/10.1145/3442381.3450014
8. Hernández-Illera, A., Martínez-Prieto, M.A., Fernández, J.D.: RDF-TR: exploiting structural redundancies to boost RDF compression. Inf. Sci. **508**, 234–259 (2020). https://doi.org/10.1016/j.ins.2019.08.081
9. Jara, A.J., Olivieri, A.C., Bocchi, Y., Jung, M., Kastner, W., Skarmeta, A.F.: Semantic web of things: an analysis of the application semantics for the IoT moving towards the IoT convergence. Int. J. Web Grid Serv. **10**(2/3), 244–272 (2014). https://doi.org/10.1504/IJWGS.2014.060260
10. Kernighan, B.W., Ritchie, D.M.: C Programming Language. Prentice-Hall (1988)
11. Mao, S., et al.: Chapter 14 - Ubiquitous Computing. In: Godfrey, A., Stuart, S. (eds.) Digital Health, pp. 211–230. Academic Press (2021). https://doi.org/10.1016/B978-0-12-818914-6.00002-8
12. Martínez-Prieto, M.A., Fernández, J.D., Hernández-Illera, A., Gutiérrez, C.: RDF compression. In: Sakr, S., Zomaya, A. (eds.) Encyclopedia of Big Data Technologies, pp. 1–11. Springer, Cham (2018). https://doi.org/10.1007/978-3-319-63962-8_62-1
13. Pan, J.Z., Staab, S., Aßmann, U., Ebert, J., Zhao, Y. (eds.): Ontology-Driven Software Development. Springer, Heidelberg (2013). https://doi.org/10.1007/978-3-642-31226-7

14. Patel, A., Debnath, N.C., Bhushan, B. (eds.): Semantic Web Technologies: Research and Applications, 1st edn. CRC Press (2022). https://doi.org/10.1201/9781003309420
15. Pearson, P.K.: Fast hashing of variable-length text strings. Commun. ACM **33**(6), 677–680 (1990). https://doi.org/10.1145/78973.78978
16. Pinho, D., Aguiar, A., Amaral, V.: What about the usability in low-code platforms? A systematic literature review. J. Comput. Lang. **74** (2023). https://doi.org/10.1016/j.cola.2022.101185
17. Qaswar, F., et al.: Applications of ontology in the internet of things: a systematic analysis. Electronics **12**(1) (2023). https://doi.org/10.3390/electronics12010111
18. Rhayem, A., Mhiri, M.B.A., Gargouri, F.: Semantic web technologies for the internet of things: systematic literature review. Internet Things **11** (2020). https://doi.org/10.1016/j.iot.2020.100206
19. Rivest, R.: The MD5 Message-Digest Algorithm. RFC 1321, RFC Editor (1992). https://doi.org/10.17487/RFC1321
20. Röder, M., Frerk, P., Conrads, F., Ngomo, A.-C.N.: Applying grammar-based compression to RDF. In: Verborgh, R., et al. (eds.) ESWC 2021. LNCS, vol. 12731, pp. 93–108. Springer, Cham (2021). https://doi.org/10.1007/978-3-030-77385-4_6
21. Roza, M.: Verification, Validation and Uncertainty Quantification Methods and Techniques (An Overview and their Application within the GM-VV Technical Framework). Science and Technology Organization, NATO (2014)
22. Ruta, M., Scioscia, F., Di Sciascio, E.: Enabling the semantic web of things: framework and architecture. In: 2012 IEEE Sixth International Conference on Semantic Computing, pp. 345–347 (2012). https://doi.org/10.1109/ICSC.2012.42
23. Ryabinin, K., Belousov, K.: Visual analytics of gaze tracks in virtual reality environment. Sci. Vis. **13**(2), 50–66 (2021). https://doi.org/10.26583/sv.13.2.04
24. Ryabinin, K., Chumakov, R., Belousov, K., Kolesnik, M.: Ontology-driven visual analytics platform for semantic data mining and fuzzy classification. Frontiers Artif. Intell. Appl. **358**, 1–7 (2022). https://doi.org/10.3233/FAIA220363
25. Ryabinin, K., Chuprina, S.: Ontology-driven edge computing. In: Krzhizhanovskaya, V.V., et al. (eds.) ICCS 2020. LNCS, vol. 12143, pp. 312–325. Springer, Cham (2020). https://doi.org/10.1007/978-3-030-50436-6_23
26. Ryabinin, K., Chuprina, S., Labutin, I.: Tackling IoT interoperability problems with ontology-driven smart approach. In: Rocha, A., Isaeva, E. (eds.) Perm Forum 2021. LNNS, vol. 342, pp. 77–91. Springer, Cham (2022). https://doi.org/10.1007/978-3-030-89477-1_9
27. Sahlmann, K., Scheffler, T., Schnor, B.: Ontology-driven device descriptions for IoT network management. In: 2018 Global Internet of Things Summit (GIoTS) (2018). https://doi.org/10.1109/GIOTS.2018.8534569
28. Sahlmann, K., Schwotzer, T.: Ontology-based virtual IoT devices for edge computing. In: Proceedings of the 8th International Conference on the Internet of Things (2018). https://doi.org/10.1145/3277593.3277597
29. Scioscia, F., Ruta, M.: Building a semantic web of things: issues and perspectives in information compression. In: 2009 IEEE International Conference on Semantic Computing, pp. 589–594 (2009). https://doi.org/10.1109/ICSC.2009.75
30. Seitz, C., Schönfelder, R.: Rule-based OWL reasoning for specific embedded devices. In: Aroyo, L., et al. (eds.) ISWC 2011. LNCS, vol. 7032, pp. 237–252. Springer, Heidelberg (2011). https://doi.org/10.1007/978-3-642-25093-4_16
31. Slimani, T.: Ontology development: a comparing study on tools, languages and formalisms. Indian J. Sci. Technol. **8**(24), 1–12 (2015). https://doi.org/10.17485/ijst/2015/v8i34/54249

32. Su, X., Riekki, J., Haverinen, J.: Entity notation: enabling knowledge representations for resource-constrained sensors. Pers. Ubiquit. Comput. **16**, 819–834 (2012). https://doi.org/10.1007/s00779-011-0453-6
33. Sultana, T., Lee, Y.K.: gRDF: an efficient compressor with reduced structural regularities that utilizes gRePair. Sensors **22**(7) (2022). https://doi.org/10.3390/s22072545

Ontological Modelling and Social Networks: From Expert Validation to Consolidated Domains

Salvatore Flavio Pileggi(✉)

School of Computer Science, Faculty of Engineering and IT, University of Technology Sydney, Ultimo, Australia
SalvatoreFlavio.Pileggi@uts.edu.au

Abstract. Data from Social Networks is a valuable asset within both a scientific and a business world. In the context of this work, ontological modelling from Social Networks is understood as a knowledge building process to generate a shared domain model. Such a technique relies on a balanced co-existence of human intuition/creativity and technological support, referred to as Hybrid Intelligence. Additionally, it assumes collaborative modelling and collective/social intelligence. The method implies a certain degree of uncertainty that is, in principle, inversely proportional to the achieved consensus. There are two clear different convergence points between the proposed process and collective/social intelligence: (*i*) at a data level, because of the nature of the input which is generated by different individuals, communities, stakeholders and actors; and (*ii*) at a modelling level, where human and automatically generated inputs, design decisions and validations are expected to involve several contributors, experts, modellers or analysts. Although looking holistically at the modelling process, this paper concisely focuses mostly on the ontological structure and the associated uncertainty, while resulting systems and studies are object of future work.

Keywords: Ontology · Social Networks · Ontological Modelling · Knowledge Engineering · Uncertainty · Collaborative Modelling · Hybrid Intelligence

1 Introduction

Social Networks are a massive source of data and, potentially, of information and knowledge. Indeed, it is largely assumed that, if properly exploited, such data is a valuable asset within both a scientific [1] and a business [5] world. This huge amount of information may become extremely effective in a context of data-intensive modelling, where patterns across massive diverse data may be identified by applying sophisticated approaches and methods.

In general terms, Social Networks present an intrinsic complexity [7] that reflects somehow the corresponding complexity of real world systems and human

J. Mikyška et al. (Eds.): ICCS 2023, LNCS 14077, pp. 672–687, 2023.
https://doi.org/10.1007/978-3-031-36030-5_53

behaviour/interaction. In fact, data from Social Networks present a significant level of entropy and noise. Additionally, human or bot-generated malicious activity is a well-known issue within online platforms, with serious social implications (e.g. on democratic elections [29]).

The great availability of complex data is well integrated with the capability of conceptualization in a machine-processable context [16], as well as AI-based technology is providing more and more support for automated classification and analysis, contributing to further increase the scalability of systematic processes.

Ontological modelling assumes the outcome of the knowledge building process to be expressed as a formal ontology [16]. In the context of this work, the mainstream process relies on Social Network data and on Hybrid Intelligence [13], which is understood as a balanced co-existence of human and artificial intelligence.

In this specific case, we assume a broad definition for human intelligence, which is a combination of analytical capabilities and intuition/creativity. Moreover, as the ultimate goal of the aimed process is a shared domain model, a collaborative approach [50] based on collective/social intelligence [25] becomes a driver factor. There are two clear different convergence points between the proposed knowledge-building process and collective/social intelligence: (*i*) at a data level, because of the nature of the input which is generated by different individuals, communities, stakeholders and actors; and (*ii*) at a modelling level, where human inputs, design decisions and validations are expected to involve several contributors, experts, modellers or analysts.

The target process implies a certain degree of uncertainty, which is a key modelling component, critical to establish trust and transparency. In principle, the resulting uncertainty is considered to be inversely proportional to the achieved consensus.

Another relevant aspect is the actual role of AI in collaborative modelling. Pragmatically and at a very theoretical level, AI could be a kind of "special" collaborator, with more or less relevance in a given context. Ideally, it should be in charge to perform the "dirty work", while humans can focus on the most creative and critical aspects. On the other side, an improper use of AI could de facto nullify the hybrid approach by establishing a kind of AI-driven dominant factor. It intrinsically sets up a major challenge to establish effective hybrid environments resulting from the analysis of the actual implications for the cognitive process.

Although looking holistically at the modelling process, this paper concisely focuses on the Ontology and related uncertainty, while resulting systems and studies, as well as the exhaustive discussion of Hybrid Intelligence are object of future work.

Structure of the Paper. The introductory part follows with a concise overview of the background concepts and their state of the art (Sect. 2), while methodological aspects are addressed in Sect. 3. The core part of the paper includes a conceptual description of the proposed ontology (Sect. 4), followed by an overview of the current implementation (Sect. 5) and by a discussion of potential applications (Sect. 6).

2 Research Background

An exhaustive review of ontology, its current application within computer systems and its role within hybrid systems is out of the scope of this paper. This section rather aims at providing a concise overview of the relationship between Ontology and data and of the application of Ontology related to Social Networks. Moreover, some considerations on Hybrid Intelligence in the context of the current technological momentum are provided.

2.1 Ontology and Data

Within computer systems, ontologies are commonly understood as resources that represent agreed domain semantics [48]. They are rich data models [16], normally characterised by a relative independence from particular applications or tasks [48].

Ontology-based data management [31] is a well-established approach to access and use data in a underlying information system by means of an ontology which provides a conceptual representation of the domain of interest [31]. Additionally, many applications implicitly need to access multiple heterogeneous data sources from internal and/or external databases as an integrated data space. Ontology enables such an abstraction via interoperability [6]. In general terms, linking data to ontologies increases the capabilities of complex systems within a Semantic Web context [4]. Linked Data assumes structured data from different sources linked by semantics [37].

By enabling a semantic knowledge space, ontologies become valuable also at a functional level to enhance and support complex tasks and processes, such as data mining [14] or enrichment [43].

Ontologies are broadly adopted in the different application domains [42], including, among the very many, Medicine [20], Biology [3] and Software Engineering [18]. Sophisticated applications may be developed over the provided semantic infrastructure. For instance, gene Ontology [3] is an ontology-based tool that ambitiously aims to the unification of biology; AmiGO [9] supports the retrieval of gene product data and associated semantics, while [52] converts high-throughput data to clinical relevance.

2.2 Ontology and Social Networks

Looking holistically at the intersection between social and semantics in the Web [23, 39], ontologies are largely adopted in different contexts, at a different level of abstraction and as part of different kinds of system.

At a very general level, Social Networks are a huge source of information and of course such data may be used to populate ontologies [17], as well as, according to an opposite perspective, Social Networks may be described or conceptualised by ontologies [10]. Knowledge/information extraction from Social Network is assumed to be a common practice [14] and the role of ontology is often explicit and relevant.

Within Social Networks, ontologies are normally used to achieve specific goals, such as, for example, collaborative filtering [8], recommendation system [24], access control [35], enhanced user profiling [49] and privacy preservation [11].

Applications that deserve particular attention in perspective because of their relevance, popularity and potential are related to analysis (e.g. [32]) and prediction (e.g. [38]).

2.3 Hybrid Intelligence

Although it is not probably possible to provide a simple, formal, and universally accepted definition of intelligence, "natural intelligence", proper of humans and animals, is commonly associated with the ability to abstract a given reality to generate some kind of mental model. Such models, which may be very subjective, allow the mental simulations that underlay our normal understanding of thinking and reasoning [33]. Our intelligence enables our capability of analysis, of problem solving and decision making in everyday life [27].

Such a definition is often integrated with "Emotional Intelligence" [36], that puts emphasis on emotions (e.g. empathy), with intuition [26], and with creativity. Additionally, the social context affects the way in which a given reality is perceived and, therefore, the associated mental models. Interactions with other individuals contribute to progressively establish a higher level of intelligence, often referred to as "collective intelligence" [30]. Collective Intelligence is an evolving concept that becomes critical in the era of online Social Networks.

On the other side, computers have naturally pointed out the concept of "thinking machines" [34] and so, indirectly, of AI. AI is currently object of an intense discussion given the last advances in the field, which have generated a mixed of excitement and concern both with a generic advice for reflection.

In this specific technological momentum, Hybrid Intelligence is gaining popularity as, in its most modern definition, it focuses on expanding the human intellect with AI, instead of replacing it [19].

3 Methodology and Approach

Because of the intrinsic complexity of conceptualization, especially within a computational context, Ontology Engineering is, in general terms, object of research interest. Methodological aspects on ontology design and maintenance with a focus on both knowledge processes and meta- processes have been addressed with a certain degree of generality [51]. While it is largely assumed that ontology design is a creative activity which extensively relies on human intuition from experts and practitioners [2], an engineering approach is required in order to establish an effective and efficient process to generate exploitable outcomes with a focus on usability/re-usability [22]. Moreover, the quality of ontologies may affect, in general, the quality of semantic datasets and structures [21].

Despite a significant number of contributions [44] and different possible approaches (e.g. human-centred [28], collaborative [46]), as far as the author knows, there is no consensus on a reference methodological framework as well as on a relatively systematic process. However, a number of principles have progressively emerged and are commonly accepted [12]. OBO Foundry principles for ontology engineering [47] are commonly adopted within the biomedical domain and may be reasonable applied more in general [45].

In line with this set of principles, the development of the proposed ontology re-iterates the importance of:

- open and public-ally available source
- implementation in a standard formal language
- global identifiers
- annotations and meta-data as part of the implementation
- unambiguous definition of concepts and relationships
- orthogonality and link to other vocabularies.

A conceptual description of the ontology is provided in the next section, while some details of major interest on its implementation are proposed later on in the paper.

4 Ontology in Concept

A simplified view of the aimed knowledge building process is proposed in Fig. 1a. Such a conceptualization doesn't reflect a fine-grained process but rather a theoretical holistic architecture.

Ideally, three different phases (and associated architectural components) can be identified as follows:

- *Integration.* It aims to systematically establish a knowledge space from input datasets and knowledge/information sources. Such a purpose overcomes data integration as it requires a semantic approach.
- *Analysis.* The knowledge-space is processed as a whole to generate an additional level of knowledge, eventually also including human inputs. Automated capabilities become determinant to approach scalable analysis and modelling.
- *Synthesis.* Final outputs, including analytics and models, are generated from the previous step.

A conceptual representation of *SS-Dom* ontology is proposed in Fig. 1b. Despite the ontology presents a seamless structure at an implementation level (see Sect. 5), from a logical perspective it is possible to ideally distinguish among at least three different modules (or sub-ontologies): *Data*, *CORE* and *Application.* Such a logical structure results from the underlining process and reflects

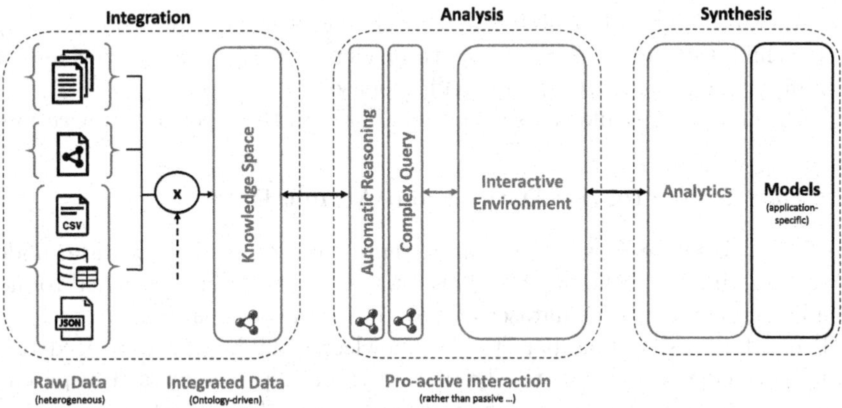

(a) Conceptualization of the aimed knowledge building process.

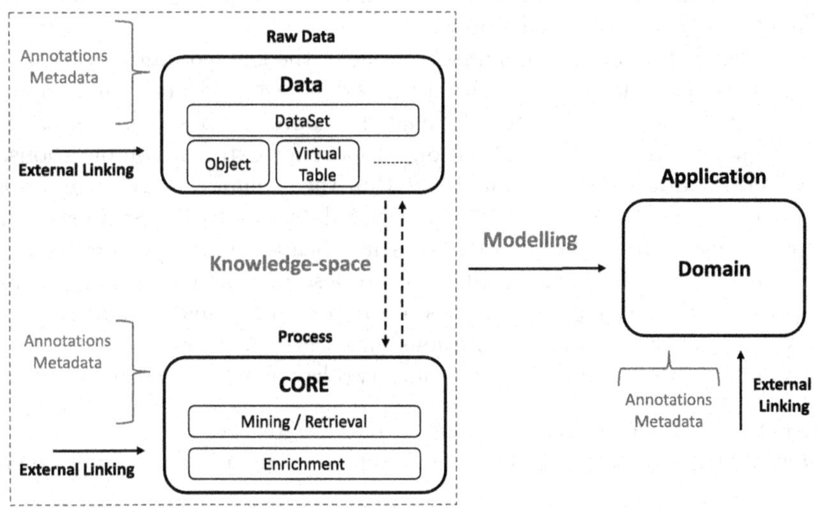

(b) Ontology overview from a conceptual perspective.

Fig. 1. The knowledge engineering process and its ontological view.

a focus on usability and flexibility along different possible applications and systems. While *Data* and *CORE* support in close synergy the knowledge-space enabling, *Domain* supports the modelling phase.

Each semantic module is discussed separately in the following sub-sections.

4.1 Data Sub-ontology: Dataset and Atomic Data

The *Data* sub-set deals with a semantic representation of the raw data, understood as the input for the process. This lower semantic level is designed to meet some key requirements for dataset and atomic data specification.

The datasets that take part to the knowledge building process need to be formally specified and semantically characterised [40]. It assures transparency, traceability, as well as enhanced analysis capabilities. The definition of a dataset can vary from case to case and it is not necessarily the traditional one. For instance, a very common case working with data sourced from Social Networks is to consider a data endpoint isolated by a specific retrieval query as a dataset. It allows a further degree of analysis as results may be produced or interpreted as a function of the retrieval method.

The holistic specification of datasets is one of the key requirements at a data level but it is not a sufficient condition by itself to assure a functional semantic environment. Indeed, in order to fully enable an ontology-based process, including internal inference and co-ordination/support for external computations, in general terms data has to be available within the semantic space at an atomic level or a proper access method to external data has to be established. For example, a given dataset from Twitter is semantically specified but also the single objects - i.e. posts and or authors - are described at an ontological level. At an atomic level, typical approaches are based on the mapping of relational models or object-oriented representations into semantic formats.

In summary, the sub-ontology provides two key assets to higher layers:

- Semantically-enriched holistic specification of datasets
- Integrated specification of raw data to support semantic data management.

4.2 CORE Sub-ontology: Retrieving and Processing Raw Data

The sub-ontology previously described provides a relatively static view of raw data and eventual semantic enrichments. The CORE sub-ontology may be considered its natural complement as it focuses on relevant related processes.

Typical examples, where applicable, are the data retrieval process and unstructured text analysis. In the former case, it is important to formally specify the logic adopted to isolate a given dataset, for instance the keywords adopted within a given query. The latter case normally relies on NLP techniques and may typically include, among others, text, topic and sentiment analysis.

At a more semantic level, semantic linking, equivalence and other relationships may be established to generate an enhanced knowledge space. Many recent studies and developments shows how the capability to effectively adopt cutting-edge technology may affect in a determinant way the quality of the final outcome. Therefore this phase may be considered as critical to establish Hybrid Intelligence. While boundaries are assumed to be blurred, an effective approach should take into account the complexity of a cognitive process that largely relies on AI technology.

4.3 Domain Sub-ontology: From Expert Validation to Consolidated Domains

This is the most abstracted and, in a way, most application-specific ontological subset. It includes the characterising aspects of the target domain resulting from the knowledge space.

Such a synthesis effort may include different aspects and aim to different interrelated components. A relatively common understanding includes (*i*) the definition of the domain taxonomy, (*ii*) the classification of domain elements according to that taxonomy, (*iii*) relationships among the different domain elements and (*iv*) semantic enrichments, annotations and linking.

As previously introduced, in the context of this work we are assuming a human effort supported by advanced technology to fully exploit the potential of hybrid approach. In order to effectively model shared domains, a collaborative approach reflecting expert and collaborative intelligence becomes critical. While it is reasonable to assume that such a philosophy ultimately increases the quality of the outcome, it also generates in fact a certain degree of uncertainty, which is, in principle, inversely proportional to the achieved consensus. Additionally, collaboration may be one of the key issue to establish a real balanced coexistence of human and artificial intelligence.

5 Current Implementation

A detailed description of the Ontology implementation is out of the paper scope. This section rather aims to overview the evolving implementation, which is currently adopted in the different experiments and studies. Such empirical experimentation is providing valuable feedback that is contributing to a progressive consolidation and generalization of the resulting system.

Additionally, the second part of the section addresses the uncertainty model, with a focus on concept classification. Such uncertainty is addressed at an holistic level to reflect an extended collaboration model that involve humans and machines.

5.1 Ontological Modelling

The current prototype is based on a formal specification of datasets [40]. The original model has been adapted to import social content from Twitter in a JSON format. Most semantic descriptions and enrichments adopt the PERSWADE-CORE vocabulary [41], which also provides a generic relationship model among ontological elements to be particularised as a part of the modelling process.

A simplified example of ontological model for a given domain is represented in Fig. 2. Such a conceptualization shows an example of class hierarchy at a very small scale, for instance resulting from a preliminary phase. It is composed of two parts, one of which is fully inferred from the other by automatic rules. The taxonomy overall represents the domain model as perceived by final users. However, outputs are generated from the inferred part of the ontology.

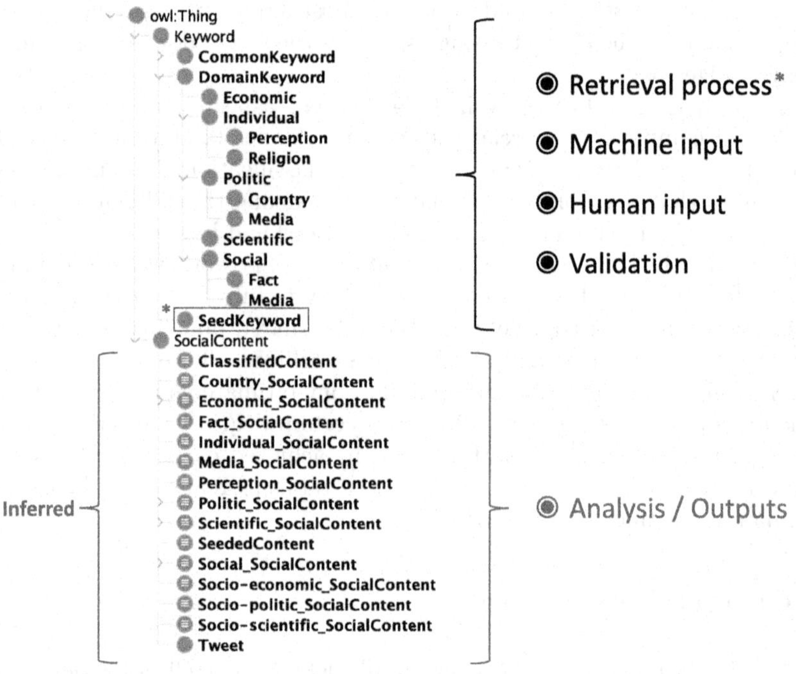

Fig. 2. Example of class hierarchy in Protege [15] and its interpretation.

The ontological model results from a combination of machine inputs, which are expected to be predominant given the amount of data, and human inputs, typically validations or structural modifications. In addition, the retrieval strategy plays a key role as it determines the data endpoints, that are the actual input for the process. There are a number of mechanisms in the system to explicit the retrieval process and to make it part of the final ontological model. For instance, as shown in Fig. 2, the keywords used in the retrieval process are classified under the category *SeedKeyword*; it may facilitate the design of specific filters, if requested at an application level.

From the experience achieved so far, the benefits of ontology and ontological modelling can be informally summarised as follows:

- reduce the gap between humans and machine within hybrid environments
- help to address the complexity of the process
- foster transparency (potentially)

5.2 Modelling Uncertainty

As previously addressed, the hybrid approach and the need for automated processing of massive data, intrinsically generate an uncertainty. In general terms, because of the collaborative approach, we distinguish between individual (e.g. one user or expert) and shared or collective view of the ontological model.

Given an individual view of a classification for the keyword k from the contributor m, $k|_{u_m}$, a collective view associated with M participants ($K|_M$) may be generated from individual views by union or intersection as in Eq. 1 and Eq. 2 respectively.

$$k|_M = \cup\, k|_{u_m} \qquad u_m \in M \tag{1}$$

$$k|_M = \cap\, k|_{u_m} \qquad u_m \in M \tag{2}$$

As an example, we assume to aim at modelling a generic COVID-19 domain from Social Networks data. Given the genericness of the topic, it is expected to have a very diversified taxonomy that may include concepts at a different level of abstraction.

Let's assume a collaborative approach for keyword classification involving 3 independent participants. Looking at two specific keywords, *Sweden* and *Italy*, two contributors provide the same individual view ($Sweden/Italy|_{u_1} = \{Country\}$ and $Sweden/Italy|_{u_2} = \{Country\}$) according to a generic-purpose classification, while the third participant assumes a more contextual interpretation by providing an additional association ($Sweden/Italy|_{u_3} = \{Country, Politics\}$). Depending on the intent and extent of the model, a collective view $Sweden/Italy|_{u_1,u_2,u_3}$ may be generated by union (Eq. 3) or intersection (Eq. 4).

$$\begin{aligned} &Sweden/Italy|_{u_1,u_2,u_3} = \\ &\quad = Sweden/Italy|_{u_1} \cup Sweden/Italy|_{u_2} \cup Sweden/Italy|_{u_3} = \\ &\quad = \{Country\} \cup \{Country\} \cup \{Country, Politics\} = \\ &\quad = \{Country, Politics\} \end{aligned} \tag{3}$$

$$\begin{aligned} &Sweden/Italy|_{u_1,u_2,u_3} = \\ &\quad = Sweden/Italy|_{u_1} \cap Sweden/Italy|_{u_2} \cap Sweden/Italy|_{u_3} = \\ &\quad = \{Country\} \cap \{Country\} \cap \{Country, Politics\} = \\ &\quad = \{Country\} \end{aligned} \tag{4}$$

The uncertainty p associated with the collective view (k_M^p) is a function of the dis-alignment among individual views and, therefore, may be computed by estimating the distance between the collective view and the individual views. The size of a view is the number of associations that it includes.

Assuming the maximum size of the collective view S computed according to Eq. 5 and s_m to be the size of the difference between S and the individual view from the contributor m (Eq. 6), uncertainty can be computed as in Eq. 7.

$$S = sizeOf\ \{\cup\ k|_{u_m}\} \tag{5}$$

$$s_m = sizeOf\ \{S - k|_{u_m}\} \tag{6}$$

$$p\ =\ \sum^{m}(S - s_m)\ \ /\ \ M \tag{7}$$

Regardless of the method adopted to generate the collective view - i.e. Eq. 3 or 4 - from a uncertainty point of view, the partial dis-alignment may be computed for the example provided as per Eq. 7, where $S = 2$ (maximum size of the collective view), $M = 3$ (number of contributors), while the size of individual views is $s_{u_1} = 1$, $s_{u_2} = 1$ and $s_{u_3} = 0$. It results in $p = 0.67$.

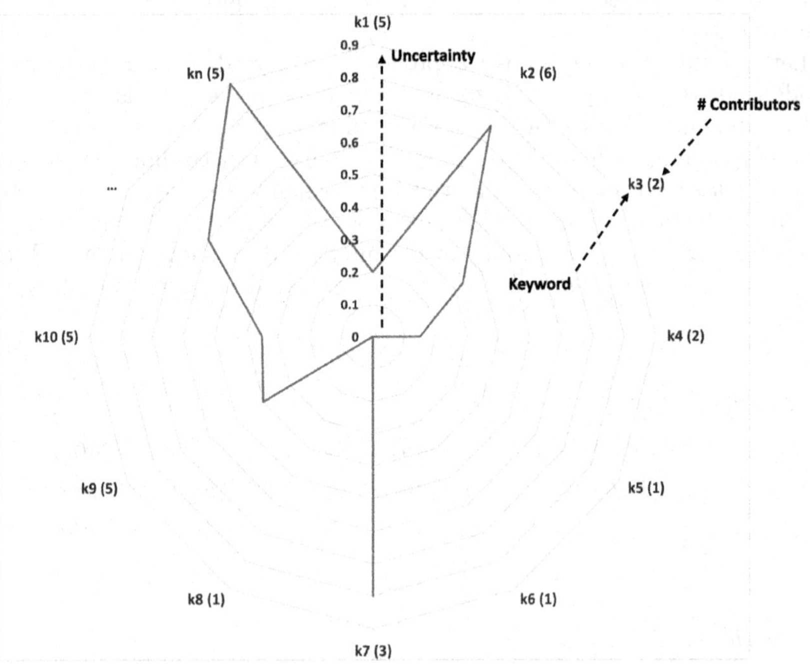

Fig. 3. Holistic visualization of the uncertainty associated with a given classification. This is a mock-up generated from fake data as an example.

This simple model intuitively reflects the idea that a higher consensus - i.e. averagely smaller s_m - is synonymous with a lower uncertainty. There is no clear understanding of the relationship between uncertainty and the number of contributors in a given context. We prefer, therefore, to do not consider the number of participants explicitly in the uncertainty quantification.

In general terms, the same model can be adopted to express (*i*) the uncertainty associated with a given concept classification (*ii*) a holistic understanding of the uncertainty (example in Fig. 3).

6 Applications

The concept of ontological model from Social Networks implicitly presents a certain generality. Its effective exploitation depends mostly on the quality of the achieved outcome against the intent/scope of the aimed model.

Ontological modelling becomes valuable where conceptualization plays (or may play) a key role, especially when the conceptual model is expected to be machine-processable and adopted within sophisticated computer systems. The benefits emerged from empirical experimentations have been briefly discussed in the previous section.

As previously discussed, Social Networks are a massive source of data, although it is not always easy to convert such a huge amount of data into information or knowledge. It applies also to ontological modelling that requires human intuition, expertise and capability to be integrated with sophisticated computer-based techniques to assure an effective and scalable process along the different stages.

Because of the nature of data from Social Networks, which is mostly human-generated, applications with a social focus may be extremely relevant in this context, as well as those in which collaborative modelling or collective intelligence provide an effective value. Additionally, ontological modelling may contribute to address entropic data at a significant scale.

Ongoing studies adopting the techniques and models addressed in the paper focuses on hybrid technology, situation awareness, controversial (e.g. NO-Vax) and socio-scientific (e.g. Climate Change) issues.

7 Conclusions and Future Work

This paper has discussed ontological modelling from Social Networks in context by concisely summarizing the related body of knowledge. The approach proposed aims to generate a consolidated shared conceptualization of a target domain in an ontological format by applying hybrid techniques resulting from a balanced integration of human intuition/creativity with cutting-edge technology. The collaborative approach considerably increases the quality of the outcome but also introduces a potential uncertainty that is inversely proportional to the achieved consensus.

Although the paper presents a holistic foci on the process, it puts emphasis on ontological structures. Such a semantic infrastructure is expected to be applied within different systems and studies, mostly in the social science domain.

The current implementation as presented in the paper is supporting the initial design of system, the analysis technique, and the integration of the different functional components of the associated research prototype. The ontology is expected to evolve accordingly, in line to an agile philosophy which prioritises effective application, usability and re-usability.

The empirical experimentation conducted so far has allowed to informally identify possible benefits of ontological models within hybrid environments to reduce the gap between humans and machine, help to address the complexity of the process, and potentially foster transparency.

Acknowledgements. I would like to thank the conference organisers, the workshop chair as well as the anonymous reviewers, who have provided valuable constructive feedback.

References

1. Ackland, R.: Social network services as data sources and platforms for e-researching social networks. Soc. Sci. Comput. Rev. **27**(4), 481–492 (2009)
2. Alterovitz, G., et al.: Ontology engineering. Nat. Biotechnol. **28**(2), 128–130 (2010)
3. Ashburner, M., et al.: Gene ontology: tool for the unification of biology. Nat. Genet. **25**(1), 25–29 (2000)
4. Berners-Lee, T., Hendler, J., Lassila, O.: The semantic web. Sci. Am. **284**(5), 34–43 (2001)
5. Bonchi, F., Castillo, C., Gionis, A., Jaimes, A.: Social network analysis and mining for business applications. ACM Trans. Intell. Syst. Technol. (TIST) **2**(3), 1–37 (2011)
6. Buccella, A., Cechich, A., Rodríguez Brisaboa, N.: An ontology approach to data integration. J. Comput. Sci. Technol. **3** (2003)
7. Butts, C.T.: The complexity of social networks: theoretical and empirical findings. Soc. Netw. **23**(1), 31–72 (2001)
8. Cantador, I., Castells, P.: Multilayered semantic social network modeling by ontology-based user profiles clustering: application to collaborative filtering. In: Staab, S., Svátek, V. (eds.) EKAW 2006. LNCS (LNAI), vol. 4248, pp. 334–349. Springer, Heidelberg (2006). https://doi.org/10.1007/11891451_30
9. Carbon, S., et al.: AmiGO: online access to ontology and annotation data. Bioinformatics **25**(2), 288–289 (2009)
10. Chen, L., Wei, S., Qingpu, Z.: Semantic description of social network based on ontology. In: 2010 International Conference on E-Business and E-Government, pp. 1936–1939. IEEE (2010)

11. Chou, J.K., Bryan, C., Ma, K.L.: Privacy preserving visualization for social network data with ontology information. In: 2017 IEEE Pacific Visualization Symposium (PacificVis), pp. 11–20. IEEE (2017)
12. Corcho, O., Fernández-López, M., Gómez-Pérez, A.: Ontological engineering: principles, methods, tools and languages. In: Ontologies for Software Engineering and Software Technology, pp. 1–48. Springer, Heidelberg (2006). https://doi.org/10.1007/3-540-34518-3_1
13. Dellermann, D., Ebel, P., Söllner, M., Leimeister, J.M.: Hybrid intelligence. Bus. Inf. Syst. Eng. **61**(5), 637–643 (2019)
14. Dou, D., Wang, H., Liu, H.: Semantic data mining: a survey of ontology-based approaches. In: Proceedings of the 2015 IEEE 9th International Conference on Semantic Computing (IEEE ICSC 2015), pp. 244–251. IEEE (2015)
15. Gennari, J.H., et al.: The evolution of protégé: an environment for knowledge-based systems development. Int. J. Hum. Comput. Stud. **58**(1), 89–123 (2003)
16. Guarino, N., Oberle, D., Staab, S.: What is an ontology? In: Staab, S., Studer, R. (eds.) Handbook on Ontologies. IHIS, pp. 1–17. Springer, Heidelberg (2009). https://doi.org/10.1007/978-3-540-92673-3_0
17. Hamasaki, M., Matsuo, Y., Nisimura, T., Takeda, H.: Ontology extraction using social network. In: International Workshop on Semantic Web for Collaborative Knowledge Acquisition (2007)
18. Happel, H.J., Seedorf, S.: Applications of ontologies in software engineering. In: Proceedings of Workshop on Sematic Web Enabled Software Engineering (SWESE) on the ISWC, pp. 5–9. Citeseer (2006)
19. Ho, L., de Boer, V., van Riemsdijk, M.B., Schlobach, S., Tielman, M.: Argumentation for knowledge base inconsistencies in hybrid intelligence scenarios. In: KR4HI First International Workshop on Knowledge Representation for Hybrid Intelligence (2022)
20. Ivanović, M., Budimac, Z.: An overview of ontologies and data resources in medical domains. Expert Syst. Appl. **41**(11), 5158–5166 (2014)
21. Janowicz, K., Hitzler, P., Adams, B., Kolas, D., Vardeman, C., II.: Five stars of linked data vocabulary use. Semant. Web **5**(3), 173–176 (2014)
22. Jarrar, M., Meersman, R.: Ontology engineering – the DOGMA approach. In: Dillon, T.S., Chang, E., Meersman, R., Sycara, K. (eds.) Advances in Web Semantics I. LNCS, vol. 4891, pp. 7–34. Springer, Heidelberg (2008). https://doi.org/10.1007/978-3-540-89784-2_2
23. Jung, J.J., Euzenat, J.: Towards semantic social networks. In: Franconi, E., Kifer, M., May, W. (eds.) ESWC 2007. LNCS, vol. 4519, pp. 267–280. Springer, Heidelberg (2007). https://doi.org/10.1007/978-3-540-72667-8_20
24. Kadima, H., Malek, M.: Toward ontology-based personalization of a recommender system in social network. In: 2010 International Conference of Soft Computing and Pattern Recognition, pp. 119–122. IEEE (2010)
25. Kihlstrom, J.F., Cantor, N.: Social intelligence (2000)
26. Klein, G.: The power of intuition: how to use your gut feelings to make better decisions at work. Currency (2004)
27. Klein, G.A.: Sources of Power: How People Make Decisions. MIT Press (2017)
28. Kotis, K., Vouros, G.A.: Human-centered ontology engineering: the HCOME methodology. Knowl. Inf. Syst. **10**(1), 109–131 (2006)
29. Lee, T.: The global rise of "fake news" and the threat to democratic elections in the USA. Public Administration and Policy (2019)
30. Leimeister, J.M.: Collective intelligence. Bus. Inf. Syst. Eng. **2**, 245–248 (2010)

31. Lenzerini, M.: Ontology-based data management. In: Proceedings of the 20th ACM International Conference on Information and Knowledge Management, pp. 5–6 (2011)
32. Li, T., Yang, H., He, J., Ai, Y.: A social network analysis methods based on ontology. In: 2010 Third International Symposium on Knowledge Acquisition and Modeling, pp. 258–261. IEEE (2010)
33. Ma, J., Wen, J., Huang, R., Huang, B.: Cyber-individual meets brain informatics. IEEE Intell. Syst. **26**(5), 30–37 (2011)
34. Machinery, C.: Computing machinery and intelligence-am turing. Mind **59**(236), 433 (1950)
35. Masoumzadeh, A., Joshi, J.: OSNAC: an ontology-based access control model for social networking systems. In: 2010 IEEE Second International Conference on Social Computing, pp. 751–759. IEEE (2010)
36. Mayer, J.D., Roberts, R.D., Barsade, S.G.: Human abilities: emotional intelligence. Annu. Rev. Psychol. **59**, 507–536 (2008)
37. Parundekar, R., Knoblock, C.A., Ambite, J.L.: Linking and building ontologies of linked data. In: Patel-Schneider, P.F., et al. (eds.) ISWC 2010. LNCS, vol. 6496, pp. 598–614. Springer, Heidelberg (2010). https://doi.org/10.1007/978-3-642-17746-0_38
38. Phan, N., Dou, D., Wang, H., Kil, D., Piniewski, B.: Ontology-based deep learning for human behavior prediction with explanations in health social networks. Inf. Sci. **384**, 298–313 (2017)
39. Pileggi, S.F., Fernandez-Llatas, C., Traver, V.: When the social meets the semantic: social semantic web or web 2.5. Future Internet **4**(3), 852–864 (2012)
40. Pileggi, S.F., Crain, H., Yahia, S.B.: An ontological approach to knowledge building by data integration. In: Krzhizhanovskaya, V.V., et al. (eds.) ICCS 2020. LNCS, vol. 12143, pp. 479–493. Springer, Cham (2020). https://doi.org/10.1007/978-3-030-50436-6_35
41. Pileggi, S.F., Voinov, A.: PERSWADE-CORE: a core ontology for communicating socio-environmental and sustainability science. IEEE Access **7**, 127177–127188 (2019)
42. Poli, R., Healy, M., Kameas, A.: Theory and Applications of Ontology: Computer Applications. Springer, Dordrecht (2010). https://doi.org/10.1007/978-90-481-8847-5
43. Quattrini, R., Pierdicca, R., Morbidoni, C.: Knowledge-based data enrichment for HBIM: exploring high-quality models using the semantic-web. J. Cult. Herit. **28**, 129–139 (2017)
44. Roussey, C., Pinet, F., Kang, M.A., Corcho, O.: An introduction to ontologies and ontology engineering. In: Ontologies in Urban Development Projects, pp. 9–38. Springer, London (2011). https://doi.org/10.1007/978-0-85729-724-2_2
45. Schober, D., Kusnierczyk, W., Lewis, S.E., Lomax, J., et al.: Towards naming conventions for use in controlled vocabulary and ontology engineering. In: The 10th Annual Bio-Ontologies Meeting (2007)
46. Simperl, E., Luczak-Rösch, M.: Collaborative ontology engineering: a survey. Knowl. Eng. Rev. **29**(1), 101–131 (2014)
47. Smith, B., et al.: The obo foundry: coordinated evolution of ontologies to support biomedical data integration. Nat. Biotechnol. **25**(11), 1251–1255 (2007)
48. Spyns, P., Meersman, R., Jarrar, M.: Data modelling versus ontology engineering. ACM SIGMOD Rec. **31**(4), 12–17 (2002)

49. Stan, J., Egyed-Zsigmond, E., Joly, A., Maret, P.: A user profile ontology for situation-aware social networking. In: 3rd Workshop on Artificial Intelligence Techniques for Ambient Intelligence (AITAm I2008) (2008)
50. Sure, Y., Erdmann, M., Angele, J., Staab, S., Studer, R., Wenke, D.: OntoEdit: collaborative ontology development for the semantic web. In: Horrocks, I., Hendler, J. (eds.) ISWC 2002. LNCS, vol. 2342, pp. 221–235. Springer, Heidelberg (2002). https://doi.org/10.1007/3-540-48005-6_18
51. Sure, Y., Staab, S., Studer, R.: Ontology engineering methodology. In: Staab, S., Studer, R. (eds.) Handbook on Ontologies. IHIS, pp. 135–152. Springer, Heidelberg (2009). https://doi.org/10.1007/978-3-540-92673-3_6
52. Yu, G., Wang, L.G., Yan, G.R., He, Q.Y.: DOSE: an R/Bioconductor package for disease ontology semantic and enrichment analysis. Bioinformatics **31**(4), 608–609 (2015)

Teaching Computational Science

The Idea of a Student Research Project as a Method of Preparing a Student for Professional and Scientific Work

Krzysztof Nowicki, Mariusz Kaczmarek, and Pawel Czarnul(✉)

Faculty of Electronics, Telecommunications and Informatics, Gdansk University of Technology, Narutowicza 11/12, 80-233 Gdańsk, Poland
{know,markaczm}@pg.edu.pl, pczarnul@eti.pg.edu.pl

Abstract. In the paper we present the idea and implementation of a student research project course within the master's program at the Faculty of Electronics, Telecommunications and Informatics, Gdańsk Tech. It aims at preparing students for performing research and scientific tasks in future professional work. We outline the evolution from group projects into research project and the current deployment of both at bachelor's and master's levels respectively, management of projects i.e. steps, reporting and monitoring at both faculty and individual project's levels within our custom-built Research Project System (RPS). We further elaborate on adopted formal settings and agreements especially considering the possibility of external clients taking part in the projects. Methodology of conducting and several examples of awarded projects are presented along with statistics on the number of submitted/conducted projects as well as those finalized with actual submitted/published research papers/patents proving actual (inter)national impact of the course.

Keywords: PBL · research student project · university-industry cooperation · Research Project System

1 Introduction

Today's students and skilled workers must learn to function in a climate of constant technological change, innovation and social change. To meet this challenge and prepare them for their future careers, universities need to implement new, attractive and creative forms of education and training. Students are trained to be subject matter experts, highly skilled technology problem solvers. At the same time, they must be team players, be able to work in interdisciplinary teams, identify problems and react to them quickly, and take the role of team leaders. This is especially true for engineering students. Therefore, in order to prepare students for the 21st century environment, academic teachers try to create educational activities that will help students develop substantive knowledge, solve problems and collaborate to meet workplace challenges. One strategy to help students

J. Mikyška et al. (Eds.): ICCS 2023, LNCS 14077, pp. 691–706, 2023.
https://doi.org/10.1007/978-3-031-36030-5_54

achieve these skills is project-based learning (PBL), eg. [1]. *Project Base Learning* is a key term in the new learning strategy paradigm. It is a teaching strategy that prioritizes the activity of a student or a group of students whose task is to independently, with the support of a mentor (research supervisor) solve a certain engineering or research problem. The students are the subject in the process of completing the task. The main foundation of PBL is the constant interaction and active participation of all members of the project group and the mentor. Thanks to this strategy, students engage in projects by formulating research hypotheses, defining design assumptions, collecting and analyzing information from various fields of knowledge in order to develop the final product. The main focus of PBL teaching is on allowing students to interact and communicate with their peers while working on their projects and to engage in reflective and critical thinking about what they are learning and doing. Therefore, project-based learning is considered an important approach to learning, which can also support the improvement of students' communication skills during project implementation and the acquisition of so-called *Soft skills*, increasingly required from job candidates by employers [3,4] and prepare them for life long education [2]. In the article, we present the concept of a mandatory student research project aimed at transferring competences that allow you to work in research and development teams in companies.

2 Literature Review

Over the last several years, many articles have been published on the implementation of the practical concept of project-based learning, or more broadly, challenge-based learning. Most of these articles/reports come from universities and concern the education of future engineers in various fields. These are future engineers in the field of ICT, IoT, Machine Learning [5], but we also have energy, civil engineering, environmental engineering and chemical technologies [6,7]. Summaries of learning outcomes in the PBL system for periods of 10 and more years are already available [8].

An important element of the efficient implementation of the subject of the student project (PBL) is an appropriate IT system for managing and reporting progress in PBL. Many solutions are described, ranging from systems based on the moodle platform, ending with own implementations of CMS systems [9–11]. The next step in changing the education system at universities is the challenge base learning concept. Extension of the PBL idea, where we assume even greater interdisciplinarity of projects and the combination of three stakeholders: the design team, a representative of the industry and the end customer who would like to buy such a developed product in the final stages [12]. At Polish universities, student research projects are carried out as projects carried out by student research clubs (SRC) or as special student activity selected as part of special competitions announced by the authorities of individual faculties. An example of research projects carried out by student research clubs and financed by the university are projects carried out at the Wroclaw University of Environmental and Life Sciences [13]. The conditions for project settlement include: publication of the work in any form; presentation of the obtained results. After completion of the research project, the SRC is obliged to present a final report on the implementation of the research project and the use of the allocated funds. Another approach, giving an opportunity to carry out research projects in self-organizing student groups, selected as part of competitions announced by university faculties, are examples from the Jagiellonian University [14]. The amount of funding and the number of awarded projects are not strictly defined and depend on the Faculty's budget for a given year. There are also thematic competitions, e.g. in 2013/2014 for the best study programs (the so-called One Million for Biochemistry and One Million for Biotechnology). One of the requirements of the competition is that the project is submitted and (if funds are awarded) implemented by a group of students from at least two different fields of study. In this way, student cooperation and interdisciplinary projects are supported. Students prepare and settle projects both in terms of content (e.g. presentation at a conference) and financially. It is a preparation for independent raising of funds for research after graduation. Although undergraduate students can also participate in the projects, the role of initiators and managers of these mini-projects is performed primarily by MSc students. Cooperation with the industry while defining the topics of student research projects and their subsequent implementation was put to the fore at the Institute of Thermal Technology at the Silesian University of Technology [15]. As part of SRP, a student or a team of students (contractor) under the supervision of a researcher (project supervisor) performs a specific research task. The mentor undertakes to properly prepare the contractors for the task. Contractors are obliged to perform the work diligently and to prepare test reports. After completing the project, the contractors receive a certificate confirming their commitment to its implementation. The results obtained as part of SRP should be presented at the seminar of the Scientific Circle, and also be the basis for the implementation of engineering and diploma theses. Student research projects are also carried out at medical universities, for example in Pomeranian Medical University in Szczecin [16]. As part of the research and development subsidy, the Vice-Rector defines a fund for the implementation of student research projects selected in the competition. Unfortunately, the competition is limited to members of scientific circles affiliated with the Student Scientific Society. Projects can apply for funding of up to 5,000PLN, and the implementation time can be up to 12 months. Also at the ETI faculty of the Gdańsk University of Technology, we have experience with learning through the implementation of student team projects in the field of High performance computing (HPC) [17] as well as using underlying ICT infrastructure for teaching HPC [18]. All the examples listed above concern the implementation of projects selected through competitions and are most often addressed to members of scientific clubs operating at the faculty or university. This limits the number of beneficiaries, which is related to the financing provided.

3 Concept of Students Research Project

3.1 Outline

In the years 2004–2019, the ETI faculty conducted a subject called "Group project" for all fields of study. The concept was described in [19]. It was a team student project, obligatory for all students of the 1st and 2nd semester of MSc studies. The project group of students could consist of 3 to 5 people, they selected a leader/manager from among the members of the group, who supervised the implementation of the project. The topics of the projects were more engineering and application than typically research. In all editions we assume that more then 5000 students take part in the project implementation. In the last edition of 2019/2020, several research projects were allowed to be implemented to test the idea of learning through a research project with a research hypothesis.

In line with the research university profile of Gdansk University of Technology (Gdansk Tech), the Research project was designed as a course to teach and prepare students for conducting work in research teams, possibly in R&D teams of high-tech IT companies, institutes or possible careers at universities.

Starting from February 2021, the Research project has been carried out in second degree (master's) studies in a two-semester system. It starts in the first semester and continues in the second semester with a three-semester Master's degree program.

By design, a student or possibly a team of students is expected to verify a research hypothesis defined by a client – either internal (faculty member) or external (company, another university, faculty, institute). Performing required research and tests for the final verification of the hypothesis may require prior development of devices, implementing code, data preparation etc. This is typically known at the project definition phase and stated by the client.

Verification of the hypothesis should be performed in line with a proper methodology and supported by scientifically significant and meaningful experiments and/or tests. Students are expected to finalize the course with preparation of a report formatted as a scientific publication, written in English.

For projects defined by external clients, the scope of the project might alternatively focus on preparation of an application/product in which case the final outcome might be a patent application.

3.2 Steps, Reporting and Monitoring Project Execution in the RPS Electronic System

Companies are offered both an opportunity to define projects as well as to influence the topics that, in their opinion, students should be interested in, e.g. through competitions for students – see the IHS Markit competition for the "Best Research Project in the Field of Artificial Intelligence" - (https://spb.eti.pg.edu.pl/pages/contest). Projects are implemented on the basis of an agreement signed by the company, students and the faculty, which defines the rules of cooperation and the possibility of acquiring rights to the results of work performed under the project as outlined in Sect. 4. Enterprises specify the purpose of the project work and the products expected as its result. The service work such as "maintenance of the network" but also work that does not include research tasks (reproductive, copy of a product already existing on the market) are not allowed for submitting and implementation. In the general scheme of the research project implementation process for companies or the University, it is assumed that they are responsible for ensuring the conditions for project implementation by students (production environment), defining project goals and accepting plans and work results.

It is worth emphasizing that during developing the concept of the Research Project, efforts were made to reflect the typical course of the project implementation process in companies – from the moment of recruitment of employees for the project, through the creation of a project group, project implementation, to its final acceptance [19]. The first stage is collecting offers(topics) from project principals and publishing them in the Research Project System (RPS) (a student finds several hundred proposals for project topics in RPS – this corresponds to hundreds of job offers for ICT engineers and others covered by the Project, which can be found on the Internet and in press advertisements). In the completed 2021/22 edition, over 282 topics were collected, of which 66 came from external companies and additionally 14 from other GUT faculties. In the present edition of 2022/23, 301 topics were proposed to students, including 51 from external companies and additionally 19 from other GUT faculties. In previous editions of the group project also included research topics (over 20, including several from external companies). About 350 students participate in each edition. A student may apply for participation in one or more projects (this is equivalent to applying for a job with multiple employers). At RPS, the student obtains volunteer status. It often happens that course participants apply for even a dozen or so projects, becoming volunteers in each of them (this corresponds to submitting an application for employment in many companies).

The next step involves meeting and an interview. The project supervisor (mentor) organizes individual meetings with volunteers applying for the project and conducts an interview with them (this corresponds to the interview stage). Often, a joint meeting of all volunteers interested in a given project is organized. This allows both parties, the mentor and the students, to get to know each other. The mentor can find out about the skills and knowledge of potential project participants and their aspirations, e.g. regarding team leadership. The scope of work to be performed by individual persons is also determined, as well as the rules of assessment. At this stage, property rights (not only intellectual) are also determined. This is a key topic in projects commissioned by external companies. However, also in the topics offered by academic teachers, the issue of copyright and property rights plays an important role. After these matters have been clarified, final decisions are made by students about staying in or withdrawing from the project group and by the supervisor regarding the selection of group members from among those still declared (this corresponds to hiring an employee). Accepting a student's participation in a given project is possible only if he/she has not been accepted in another project (and information about this is provided by RPS). In the next stage, the leader of the group is selected from among its members. It is often proposed by students, although the decisive word belongs to the mentor, acting as the principal. Then we have, for the first time, cooperation of the entire team in decision-making and cooperation that allows the group to self-organize (this corresponds to the organization of the project team). Of course, during the implementation of the project, such cooperation of the team, also in technical matters, will occur many times, e.g. when determining the details of the schedule. Setting the schedule consists of adding to the generally imposed deadlines (end of semesters) exact dates of completing individual tasks, participants responsible for these and methods of documenting (companies also know when the project should be completed, and, depending on the method of management, more or less detailed planning of the next stages of its implementation). Of course, in certain circumstances, the schedule may be changed. Always, however, after discussion with the group, it is decided by the mentor. The decision on the choice of a project management methodology (e.g. agile or classic) is made by team members after consultation with the supervisor.

In parallel with the implementation of projects in teams, in order to reduce the gap between the knowledge acquired at university, students' visions and the actual scope and form of engineer's work in a research team, a series of meetings with representatives of enterprises and universities was conducted as a part of the Research Project course. They presented, from various points of view, problems related to the implementation of projects in companies and universities, in particular the research process, expectations that companies have towards students and young employees starting work, required soft skills, etc. Examples of meeting topics include:

- Scientific research and conducting research projects; Systematic literature review; Reporting research and scientific articles; Methods of preparing a presentation.
- Differences between a research project and a non-research project; Where to get money to implement your own ideas and how to use them well; How to prepare a project well to get financing.
- The role of documentation at Intel.
- Project challenges on the example of failures and successes of development projects.
- Management of research projects in an IT corporation.
- Agile in business projects – How not to slip on fashion?; Analyst as an interface between the client and the ICT project team; Testing – "different tests same culture, different cultures same tests".
- Innovative, international R&D projects in the military area – issues of the manufacturing process and its documentation on examples.
- Work in an open source project.

Companies participating in the presentations and offering projects include: Intel, Excento, Vector Technologies, Aiton Caldwell SA, IHS Markit, Radmor

S.A., ADVA Optical Networking, Smart4Aviation. A consultant of The World Bank and the National Center for Research and Development, winner of one of the editions of "Odyssey of Minds", as well as the organizer of the GUT project "Startup School" (https://pg.edu.pl/en/startup) shared their knowledge with students. Thanks to the variety of implemented projects and two-level support for students: by project supervisors and through a package of knowledge provided in the form of lectures, the Research Projects subject meets the requirements of project-based education and team learning.

4 Formal Setting and Agreements

In the case a research project is performed for an external entity i.e. a company, another faculty of Gdansk Tech, another university, principles of cooperation among the three parties: Students, the Faculty of ETI Gdansk Tech as well as the client is regulated by an agreement of participation in a research project. Specifically, the agreement states that the university appoints Students to carry out a research project on the specific subject and the Students undertake to implement the Project. A project supervisor is a faculty employee whose tasks include supervision designing and implementing the project, its intermediate and final evaluation. It is important that Students declare that they will perform the project personally, will not infringe the rights of other parties. The parties agree they will keep confidential information marked as such.

It an event that a Company acting as a client declares an interest in acquiring the copyrights to the results of a project it will submit a statement to Students or Students and the university, the latter when the university is a co-author of the work. A deadline is declared for such submission after which all the parties agree to conduct negotiations of conditions based on which the copyrights to the results will be acquired by the Company. Formal details of the latter will then be regulated by a separate agreement signed by the involved parties. The vice-dean for cooperation and promotion and the appropriate project supervisor, in consultation with the head of the department, are responsible for negotiating the contract for the implementation of the project. Therefore, students and supervisors have the support of the faculty authorities, and if necessary - the Attorneys' Office for Intellectual Property and Projects or Patent Attorney or Centre for Knowledge and Technology Transfer. If the agreement is not signed, the property rights to the project result remain the property of the students.

From the point of view of the university, it is guaranteed by the first agreement that relevant information may be provided for the evaluation of disciplines within the University and that results might be used by the university based on a free license, either granted by Students or the Company, for research and teaching purposes.

Companies are encouraged to supply a fund for awards granted to the best research projects conducted at the Faculty of ETI as well as organize their own thematic competitions for such.

In an event the university expresses an intent to use outcome (in particular applications and tools) of a Research project in its further research including commercial purposes (primarily the works conducted by a department hiring the supervisor), both the university and the student might agree to sign an agreement on the transfer of property rights.

In an event, the Research project scope realized by the Student(s) is conducted within a project financed by the university, copyrights are transferred to the University based on a respective project agreement.

Based on the Regulations for management and commercialization of intellectual property within the University, the University has the right to use scientific material contained within a scientific work of an author including Students of the University for research and didactic purposes.

5 Implementation of Research Projects

The way a research project is managed as an academic subject is hierarchical. The responsible person at the faculty is the faculty coordinator who is assisted by department coordinators (16 departments – 16 department coordinators) who manage the course implementation process in the departments. They help supervisors of research projects in formal matters, manage the project presentation schedule. They may also have contact with external clients and students. Project teams composed of any number (in practice up to 10 people) of students undertake the implementation of a topic selected from among the proposals submitted by external companies, academic teachers (supervisors, including those often implementing projects financed from external sources) and students (see Fig. 1).

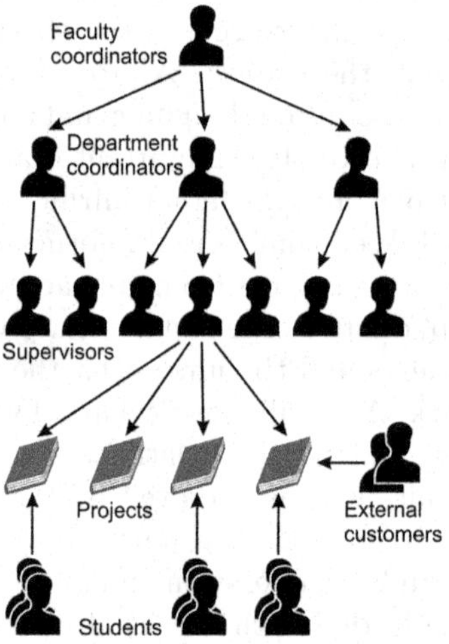

Fig. 1. The process of creating research topics

The result of a year-long work on a selected problem is a product and appropriate documentation, including a proposal for a scientific article or patent application. The course of work is supervised by mentors appointed by the departmental or faculty coordinator (see Fig. 1).

5.1 Methodology of Conducting Student Research Projects

The system supporting efficient organization of work concerning the research project is a custom IT system – the Research Project System (RPS) [https://spb.eti.pg.edu.pl/]. In its creation, experience from the implementation of the system developed in 2010–2020 for running a student group project was used.

The implementation of the student's research project is documented in the RPS informatics system by:

1. Defining by the client (university employee - mentor, external company (external client) or students) the topic and research hypothesis.
2. Defining the schedule including stages with control dates(milestones). The stages should correspond to the tasks leading to the verification of the hypothesis, e.g.: data collection, data preparation/normalization/processing, algorithm/solution design, implementation of the solution, conducting experiments, analysis of results, verification of the hypothesis based on the obtained results, publication/report.
3. Interim reports documenting the achievement of milestones in the schedule.
4. Developing a poster (in Polish and English) after the 1st and 2nd semester informing about the contractors, basic tasks and achieved results.
5. Development of a report in the form of a publication according to a template (IEEE, Elsevier) in English or a report in the form of a patent application.

Subsequent stages of the Research Project implementation are presented in Fig. 2. The whole process begins with the submission of the topic of the research project by the client, i.e. by an external company, a researcher of the ETI Faculty or a researcher of another faculty or university. The author of the project topic from the ETI Faculty automatically becomes a supervisor. On the other hand, for external topics, a faculty member who is willing to lead a given topic is assigned as a supervisor. The next stage is reporting the desire to implement the project by students and forming a project group, after talks and acceptance of group members by the supervisor. All these stages are supported by the RPS system. The members of the project group selects a project leader. Then, the project implementation schedule is set, including milestones and key points that require partial reports to be sent to the system. The schedule must be approved by the supervisor. In contrast to the previously implemented student group project [19], changes in the project implementation schedule are possible. This makes it possible to take into account the more dynamic nature of work in a research project, e.g. problems in conducting experiments with an external contractor, changes in the publishing process - reviewing publications, which are very difficult to define in time in the schedule. The schedule may be changed/updated after obtaining the supervisor's approval. Reports are sent to key stages, which must also be approved by the supervisor. At the end of the second semester of the project, a scientific publication is prepared according to the general IEEE.org template. The project is fully successful when the article is published at a scientific conference or in a scientific journal. It is also allowed to file a patent application as equivalent reporting.

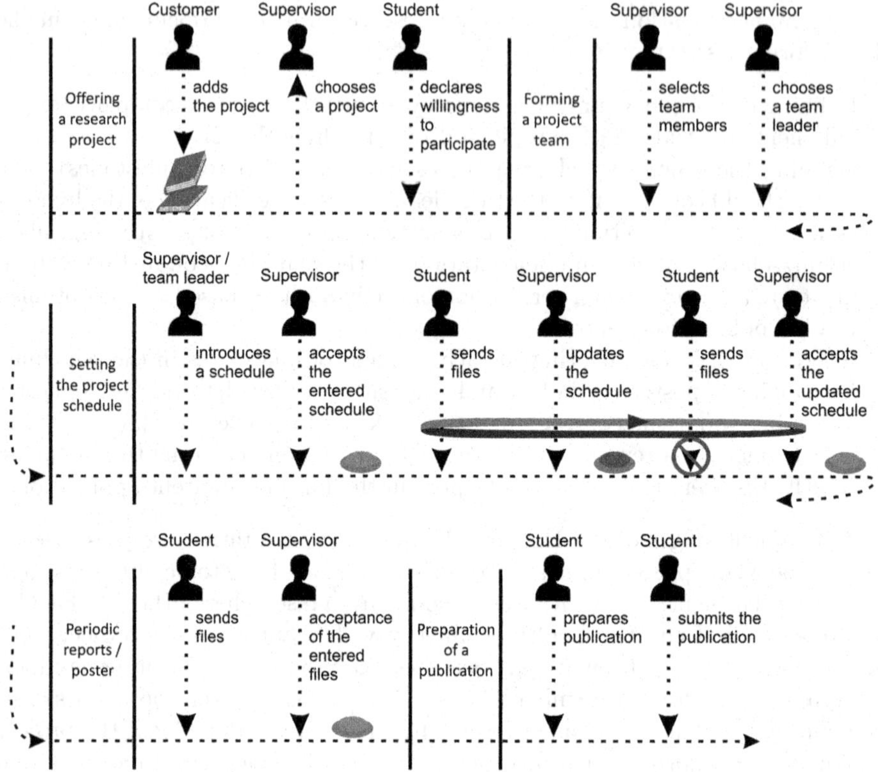

Fig. 2. Research Projects Service – project life cycle

The number of submitted research project topics in the two editions implemented so far is presented in Tables 1 and 2. Table 1 summarizes the number of projects submitted by external clients. Note: RP1-cooperation with companies (projects undertaken by students), RP2-cooperation with other GUT faculties (projects undertaken by students), RP3-cooperation with companies (projects not undertaken by students), RP4-cooperation with other GUT faculties (projects not undertaken by students), RP5-proposed by ETI faculty teachers (projects undertaken by students), PR6-proposed by ETI faculty teachers (projects not undertaken by students); "+"-AI Tech projects - according to grant government program "Digital Poland" [https://www.gov.pl/web/aitech]. The table summarizes all the projects submitted in three editions, divided into projects submitted by employees of the ETI Faculty and by external entities.

Table 1. The submitted research projects in two editions

	Research projects					
Edition	RP1	RP2	RP3	RP4	RP5	RP6
2021	23	5	43	9	13+	15+
2022	10	10	41	9	12+	30+
Summary	33	15	84	18	25+	45+

Table 2. Overview of research projects

	Research projects				
Edition	ETI Faculty	Other Faculty	AI Tech	Industry	Status
2022	189	19	42	51	submitted 301
	60	10	12	10	taken by students 92
2021	174	14	28	66	submitted 282
	43	5	13	23	taken by students 84
2020[a]	23	0	0	4	submitted 23
	5	0	0	0	taken by students 5

Note: [a] formally students' group project with partial research aspects

5.2 Awarded Research Projects

In order to encourage students to implement innovative projects and devote enough time to them, competitions are organized: dean's competition for the best project and competitions sponsored by one of the IT companies – for the best project in the field of artificial intelligence. The awarded student projects' implementations are presented during the inauguration of the first year of second-degree studies at the ETI faculty (400 people in the auditorium). In the completed 2021/22 edition, the following projects were distinguished: "Fast automatic design of planar antennas using optimization algorithms" [22–25], "Implementation of a WBAN radiolocalization system prototype with the use of deep learning" [26], "Supporting the safety of people and car intelligence with the use of automatic pedestrian detection in thermal image sequences" [28,29], "Analysis of infrared images and segmentation of facial features on thermograms using SI for the purposes of COVID-19 prevention" [30].

Additionally, the research project and awarded projects were picked up and covered by local media TV Gdańsk and Radio Gdańsk [31,32].

5.3 Published Research Projects Results

Many of the implemented projects ended with a scientific publication in an international journal, at an international conference or a patent [20–30,33–38]. Table 3 summarizes the publication results in the completed 2021/2022 edition. Out of 80 completed research projects, 11 obtained results suitable for publication in conference materials or scientific journals. There was also one patent application. In total, eight scientific project supervisors published papers as co-authors. In total, 16 papers or conference papers were published. It also happened that two or more articles were written concerning one research project, e.g. [28,29] or [22–25].

Table 3. Number of publications of research projects results

Published research papers		Under review research papers	
Conference	Journal	Journal	Patent Application
10	4	1	1

6 Conclusions

At the ETI faculty, the idea of PBL has evolved. In the first phase, a student group project was implemented. As part of this subject, engineering/application projects were implemented. Over time, the idea has evolved into the implementation of more research projects (with elements of an experiment that would allow to verify a stated research hypothesis), and ended up as a full-featured research project with a report in the form of a publication for a scientific journal (IEEE style formatted). We have observed that the subject Research Project is a supplement and a certain alternative to Student Science Clubs. There is a relatively small number of students in science clubs and often the projects carried out in clubs are not of a research but rather of an application nature. In our opinion the most valuable elements of the Research project subject are as follows:

- within the class schedule – hours common to all students enabling creation of interdisciplinary groups,
- supported by lectures – theory taught by practitioners from industry,
- students finalize their work in the form of a scientific report/article,
- RPS platform not only supporting the organization of students' work but also enabling external companies to define topics and cooperate with students.

In terms of teaching, computer science serves as a tool allowing to solve research challenges from various domains. Specifically, the following have been frequently and regularly used and applied across various projects:

- Various (programming) languages to solve a research problem e.g. C++, C#, Python, Matlab etc.
- Versioning solutions e.g. svn, git and platforms such as github, gitlab etc.
- Conducting selected projects using commercial technologies.
- The RPS system for definition of phases, milestones, schedules, storage and assignment of documentation, code and results to the phases. A need for a uniform scheme and an environment for conducting and documenting projects using either open or commercial standards has become apparent.

- Modern methods and environments for projects that required AI based solutions. Specifically, the ETI faculty offers NVIDIA DGX systems while selected departments offer advanced resources such as workstations and servers with 2 multi-core CPUs and 2–8 GPUs for time-consuming applications such as training DNN models. Additionally, students have an option to apply for research grants at the Centre of Informatics Tricity Academic Supercomputer and network (CI TASK)[1] and use e.g. the Tryton supercomputer and/or GPU enabled systems.

In terms of universal benefits of the student research project, we shall emphasize the following aspects:

- Conducting the research project requires and teaches cooperation of both IT and non-IT specialists which is an inevitable necessity in professional work – these skills are not usually subject of the regular university courses,
- Apart from very specialized topics there are those that result in solutions and tools applicable to a variety of contexts, needs and domains. Examples of such projects include: "Comparison of the effectiveness of different types of neural network learning with a small dataset", "Evaluation of performance and energy of DNN training as well as network quality for a selected application in a multi GPU environment under power capping", "Web-based application for automatic checking of exams using neural networks", "Web-based application for automatic exam checking using neural networks and handwriting recognition", "A tool for proofreading research papers", "Automation of application conversion between IT technologies", "Elder's wellbeing analysis system in domestic conditions", "Fake news detection techniques".
- The actual concept of the student research project is not tied to a specific faculty but is rather universal. Consequently, good experiences from the first two editions resulted in the rector's authorities considering extending student projects in this form to all faculties of the university. This would enable even greater interdisciplinarity of projects, as it would be possible to combine issues such as mechanical issues with aspects of electronic control and finally reporting the system status to the IT system. Such an idea fits perfectly into the mission of a research university of the 21st century. Starting with the next edition of the project, the Faculty of Management and Economics as well as the Faculty of Civil and Environmental Engineering will launch the student research project as well.

[1] https://task.gda.pl/en/.

References

1. Berglund, F., Johannesson, H., Gustafsson, G.: Multidisciplinary project-based product development learning in collaboration with industry. In: Proceedings of the 3rd International CDIO Conference, Cambridge, Massachusetts (2007)
2. Bhattacharya, S.M.: Technologically enhanced PBL environment for preparing lifelong learners. In: Sixth IEEE International Conference on Advanced Learning Technologies (ICALT 2006), pp. 1125–1126. https://doi.org/10.1109/ICALT.2006.1652655
3. Serce, F.C., Alpaslan, F.N., Swigger, K., Brazile, R., Dafoulas, G., Lopez, V.: Strategies and guidelines for building effective distributed learning teams in higher education. In: 9th International Conference on Information Technology Based Higher Education and Training (ITHET), pp. 247–253 (2010). https://doi.org/10.1109/ITHET.2010.5480079
4. Sababha, B.H., Al-Qaralleh, E., Al-Daher, N.: A new student learning outcome to strengthen entrepreneurship and business skills and mindset in engineering curricula. In: 2021 Innovation and New Trends in Engineering, Science and Technology Education Conference (IETSEC), pp. 1–4 (2021). https://doi.org/10.1109/IETSEC51476.2021.9440489
5. Khan, M., Ibrahim, M., Wu, N., Patil, R.: Interdisciplinary project based learning approach for machine learning and internet of things. In: IEEE Integrated STEM Education Conference (ISEC) 2020, pp. 1–6 (2020). https://doi.org/10.1109/ISEC49744.2020.9280619
6. Sungur, S., Tekkaya, C.: Effects of problem-based learning and traditional instruction on self-regulated learning. J. Educ. Res. **99**(5), 307–320 (2006). https://doi.org/10.3200/JOER.99.5.307-320
7. Armarego, J., Clarke, S.: Problem-based design studios for undergraduate SE education. In: 18th Conference on Software Engineering Education I& Training (CSEET 2005), pp. 249–254 (2005). https://doi.org/10.1109/CSEET.2005.24
8. Hoffman, K., Hosokawa, M., Blake, R., Headrick, L., Johnson, G.: Problem-based learning outcomes: ten years of experience at the University of Missouri-Columbia School of Medicine. Acad. Med. **81**(7), 617–625 (2006). PMID: 16799282. https://doi.org/10.1097/01.ACM.0000232411.97399.c6
9. Ravankar, A., Imai, S., Ravankar, A.: Managing the project: the essential need for project management training and education in graduate schools. In: 2019 8th International Congress on Advanced Applied Informatics (IIAI-AAI), pp. 420–425 (2019). https://doi.org/10.1109/IIAI-AAI.2019.00092
10. Qiu, L., Riesbeck, C.K.: An incremental model for developing computer-based learning environments for problem-based learning. In: IEEE International Conference on Advanced Learning Technologies, Proceedings, pp. 171–175 (2004). https://doi.org/10.1109/ICALT.2004.1357397
11. Bousmah, M., Elkamoun, N., Berraissoul, A., Aqqal, A.: Online method and environment for elaborate the project-based learning specifications in higher education. In: Sixth IEEE International Conference on Advanced Learning Technologies (ICALT 2006), pp. 769–773 (2006). https://doi.org/10.1109/ICALT.2006.1652555
12. Kohn Rådberg, K., Lundqvist, U., Malmqvist, J., Hagvall Svensson, O.: From CDIO to challenge-based learning experiences-expanding student learning as well as societal impact? Eur. J. Eng. Educ. **45**(1), 22–37 (2020)
13. https://upwr.edu.pl/studia/studencka-aktywnosc/studenckie-projekty-badawcze
14. https://wbbib.uj.edu.pl/badania-projekty/studenckie-projekty-badawcze

15. https://itc.polsl.pl/index.php?option=com_content&view=article&id=290&Itemid=166
16. https://old.pum.edu.pl/administracja/badania-naukowe/studenckie-projekty-naukowe. https://stn.pum.edu.pl/dokumenty/
17. Czarnul, P.: Teaching high performance computing using BeesyCluster and relevant usage statistics. In: International Conference on Computational Science, ICCS 2014, Cairns, Queensland, Australia, 10–12 June 2014, pp. 1458–1467 (2014). https://doi.org/10.1016/j.procs.2014.05.132
18. Czarnul, P., Matuszek, M.: Use of ICT infrastructure for teaching HPC. In: 2019 IEEE 14th International Conference on Computer Sciences and Information Technologies (CSIT), pp. xvii–xxi (2019). https://doi.org/10.1109/stc-csit.2019.8929841
19. Krawczyk-Brylka, B., Nowicki, K.: Projekty grupowe jako przygotowanie do wspolpracy w zespolach wirtualnych (Team projects as preparation for virtual team collaboration), e-mentor no 3 (85) (2020). ISSN 1731–6758
20. Cygert, S., et al.: Towards cancer patients classification using liquid biopsy. In: Rekik, I., Adeli, E., Park, S.H., Schnabel, J. (eds.) PRIME 2021. LNCS, vol. 12928, pp. 221–230. Springer, Cham (2021). https://doi.org/10.1007/978-3-030-87602-9_21
21. Wysocki, M., Nicpon, R., Trzaska, M., Czapiewska, A.: Research of accuracy of RSSI fingerprint-based indoor positioning BLE system. Przeglad Elektrotechniczny, 86–89. Krajowa Konferencja Elektroniki (2022). https://doi.org/10.15199/48.2022.09.17
22. Czyz, M., Olencki, J., Bekasiewicz, A.: Design and optimization of a compact planar radiator for UWB applications and beyond. In: EuCAP, pp. 1–3 (2022) https://doi.org/10.23919/EuCAP53622.2022.9769307
23. Czyz, M., Olencki, J., Bekasiewicz, A.: A compact spline-enhanced monopole antenna for broadband/multi-band and beyond UWB applications. AEUE - Int. J. Electron. Telecommun. **146**(3) (2022). art. Nr. 154111
24. Bekasiewicz, A., Kosiel, S., Czyz, M.: Accurate non-anechoic radiation pattern measurements of small antennas using time-gating method with automatic calibration. IEEE Antenna Wirel. Propag. Lett
25. Bekasiewicz, A., Czyz, M.: The way one defines specification matters: on the performance criteria for efficient antenna optimization in aggregated bi-objective setups. In: 2022 16th European Conference on Antennas and Propagation (EuCAP), pp. 1–5 (2022). https://doi.org/10.23919/EuCAP53622.2022.9768959
26. Urwan, S., Wysocka, D., Pietrzak, A., Cwalina, K.: Position estimation in mixed indoor-outdoor environment using signals of opportunity and deep learning approach. Int. J. Electron. Telecommun. **68**, 594–607 (2022). https://doi.org/10.24425/ijet.2022.141279
27. Cygert, S., Czyzewski, A., Wroblewski, B., Slowinski, R., Wozniak, K.: Closer look at the uncertainty estimation in semantic segmentation under distributional shift (2021). https://doi.org/10.1109/ijcnn52387.2021.9533330
28. Gorska, A., Guzal, P., Namiotko, I., Wedolowska, A., Wloszczynska, M., Ruminski, J.: AITP - AI thermal pedestrians dataset. In: 2022 15th International Conference on Human System Interaction (HSI), pp. 1–4 (2022). https://doi.org/10.1109/HSI55341.2022.9869478
29. Gorska, A., Guzal, P., Namiotko, I., Ruminski, J., Wloszczynska, M.: Pedestrian detection in low-resolution thermal images. In: 2022 15th International Conference on Human System Interaction (HSI), pp. 1–4 (2022). https://doi.org/10.1109/hsi55341.2022.9869447

30. Filipowicz, P., Kowalewski, M., Slominska, K., Kaczmarek, M.: Analysis of the influence of external conditions on temperature readings in thermograms and adaptive adjustment of the measured temperature value. In: 2022 15th International Conference on Human System Interaction (HSI), pp. 1–6 (2022)
31. Gdańsk, T.V.: Good morning, this is Gdańsk, March 2022. https://gdansk.tvp.pl/59111554/18032022
32. Gdańsk, R.: Best student research projects selected. Interesting Innovations Awarded, February 2022. https://radiogdansk.pl/wiadomosci/region/trojmiasto/2022/02/23/najlepsze-studenckie-projekty-badawcze-na-politechnice-gdanskiej-wybrane-nagrodzono-ciekawe-innowacje/
33. Wierciński, T., Rock, M., Zwierzycki, R., Zawadzka, T., Zawadzki, M.: Emotion recognition from physiological channels using graph neural network. Sensors **22**(8), 2980 (2022). https://doi.org/10.3390/s22082980
34. Boinski, T., Zawora, K., Szymanski, J.: How to sort them? A network for LEGO bricks classification. In: Groen, D., de Mulatier, C., Paszynski, M., Krzhizhanovskaya, V.V., Dongarra, J.J., Sloot, P.M.A. (eds.) Computational Science - ICCS 2022. LNCS, pp. 627–640. Springer, Cham (2022). https://doi.org/10.1007/978-3-031-08757-8_52
35. Boinski, T., Szymanski, J., Krauzewicz, A.: Active learning based on crowdsourced data. Appl. Sci. Basel **12**(1), 409 (2022)
36. Sledz, B., Zarazinski, S., Boinski, T.: Patent Application. Uklad do sortowania klockow, Application number: P. 440912 (2021)
37. Olewniczak, S., Boinski, T., Szymanski, J.: Towards extending wikipedia with bidirectional links. In: 31st ACM Conference on Hypertext and Social Media (2020)
38. Olewniczak, S., Boinski, T., Szymanski, J.: Bidirectional fragment to fragment links in Wikipedia. In: European Conference on Knowledge Management (2020)

Empirical Studies of Students Behaviour Using Scottie-Go Block Tools to Develop Problem-Solving Experience

Paweł Gepner[1(✉)], Martyna Wybraniak-Kujawa[1], Jerzy Krawiec[1], Andrzej Kamiński[1], and Kamil Halbiniak[2]

[1] Warsaw University of Technology, Warsaw, Poland
{pawel.gepner,martyna.kujawa,jerzykrawiec,andrzej.kaminski}@pw.edu.pl
[2] Czestochowa University of Technology, Czestochowa, Poland
khalbiniak@icis.pcz.pl

Abstract. This research is introducing and evaluating the new method supporting grow of programming skills, computational thinking and development of problem-solving approach that evolutionarily introduce programming good practices and paradigms through a block-based programming. The proposed approach utilizes problem-based Scottie-Go game followed by Scratch programming environment implanted to Python programming course to improve the learners' programming skills and keeps motivation for further discovery of computational problem-solving activity. To date, practically little work has been devoted to examining the relationship between beginner development environments and the development practices they stimulate in their users. This article tries to shed light on this aspect of learning programming by carefully examining the behaviour of novice programmers using the innovative block-based programming learning method.

Keywords: Computer programming · Problem solving · Computational thinking · Empirical studies · User behaviour · Innovative interaction techniques

1 Introduction

In the 21st century the problem solving, and computer programming are perceived as the most critical competencies for students to be used in the solving real-world problems and one of the most demanding area of teaching ranked by educators [2]. It is already understood that programming and coding is not only restricted to computer scientist or professional coders but is fundamental expertise for resolving most sophisticated problems of the era and constructing complicated systems [9]. There are studies describing the relationship of computer science in education that demonstrate the importance of teaching computer thinking from an early age as a fundamental skills. Nevertheless, the same class

J. Mikyška et al. (Eds.): ICCS 2023, LNCS 14077, pp. 707–721, 2023.
https://doi.org/10.1007/978-3-031-36030-5_55

of problems are important limitations also outside the educational sector and might be interesting for software development industries.

Developing an effective programming learning concept and curriculum in the right way is a serious and difficult challenge, but many analyses have been done to successfully incorporate programming and coding into the curriculum. Majority of the researches discovered that at the final stages of initial programming courses, most students had problems in decomposition and implementation of algorithms using coding techniques to solve actual problems [9,10].

Numerous studies have been concentrating on new innovative, alternative approaches and discussing the results of implementing into the teaching program visual programming environment, such as Scratch, Etoys, Alice, RoboMind [15], LighBot, Play LOGO 3D, or Karel, Jeroo, B# [5]. Those studies found positive results connected to visual programming environments and discussed the improvement for novice programmer's engagement in programming with help them grow programming skills [3]. No doubt, that use of visual programming ecosystem has proved its precise benefits to support learning for novice's programmer's [3], but how this is stimulating computational problems solving ability needs to be deeper discovered and better understood. This paper tries to address the gap of this problem and covers two fundamental issues. As the first is to propose a new learning method dedicated to novice programmers, aimed to develop computational thinking and developing a problem-solving approach using a block-based development environment. The second goal is to review the efficiency of solving computational problems by validating three teaching scenarios differing in the sequence of methods and tools used and the visual programming environments in computational problem-solving activities by novice programmers. This paper pursues to address by answering the following research questions:

- Does the new proposed method stimulates the behaviour of the problem-solving and coding thinking skills of novice programmers to solve computational problems?
- Do the novice programmers' computational solutions to the problems are different and they are related, to the way they were learned programming?
- What impact does new method on the novice programmers' perceived difficulty in programming?

It seems obvious that the outcomes of this research can become useful to organizations and individuals responsible for curriculum development for novice programmers to promote programming strategy and computational problem-solving skills.

2 Related Work

The aim of this section of the work is to present the fundamental ideas, highlighting the presented method and to present the need for the planned work on improving the problem-solving skills and computer thinking of novice programmers.

2.1 Problem Solving for Programming Learning

Computational problem-solving contains the development of computer codes and is considered as the fundamental expertise of computer science education but path to the solution mainly implicating extensive problem-solving skills, instead of basic technical coding activities [15]. According to literature study, problems in the programming process were mainly in the problem-solving phase (such as analysis and design) and the implementation phase and was mainly driven by the lack of problem-solving skills complemented less by lack of semantic knowledge, and weakness in testing the design.

This observation encouraged many researched to propose new schema of teaching; problem-solving first and then programming, they also emphasized the importance of teaching problem-solving independently from programming language. In this approach, students can concentrate on problem solving and testing strategies totally independently from syntax limitations. When the solution is created then particular programming language can be used to translate the solution into code and to test the final program. This idea can be implemented and described by following 8 stages process:

1. Presentation and definition of the problem;
2. Transition from an initial understanding of the problem to a detailed formulation;
3. Building a solution plan using the decomposition of goals into sub-goals, as well as tasks to implement each sub-goal;
4. Developing a solution design from a high-level design to a detailed design by transforming sub-goals into corresponding algorithms;
5. Identify the best type of language to implement the algorithm;
6. Learning the syntax of a programming language;
7. Execution of a detailed design in code;
8. Code testing.

Based on the method described above, there is evidence that problem-solving based learning improves student achievement from 50% to 68% [14]. The study also showed that students in the software development process are enthusiastic about the new learning environment, this significantly affects their enthusiasm and removes blockades related to learning and the use of syntax in programming.

In parallel to the research described above, many scientists also noted that when learning computer-based problem-solving through games, students are more experience pleasure in the learning process than with traditional lectures [15]. Similarly, game development involves both design and programming, and can support the learning of IT concepts by students [8]. It turned out that using educational games, students enjoy e-learning and achieve better learning results as in traditional courses [15]. The use of narrative tools, visual programming and flow modelling tools allowed students to concentrate on algorithms rather than coding [7], as well focus more on the abstract layer of design and problem-solving skills [16]. Research shows that Introduction to Programming courses using object visualization technics and attractive 3D animation environment radically improve student performance [6].

2.2 Computational Thinking in Educational Context

Jeanette Wing introduced the concept of computational thinking as the first person, she published the definition first time in 2006: "Computational thinking involves solving problems, designing systems, and understanding human behaviour, by drawing on the concepts fundamental to computer science. Computational thinking includes a range of mental tools that reflect the breadth of the field of computer science" [13]. Another definition proposed by the "International Society for Technology in Education and the Computer Science Teachers Association" describe computational thinking as: the way how the people use computers to analyse the data, representing data through abstractions and automating solutions through algorithmic thinking. Of the many different proposed definitions, one of the most cited is that of the Royal Society, proposed in 2012 "Computational thinking is the process of recognizing aspects of computation in the world that surrounds us, and applying tools and techniques from Computer Science to understand and reason about both natural and artificial systems and processes".

According to the observation made by [12], there are several basic computational concepts that programmers have to use:

- Sequence: To build a program, think about the sequence of steps.
- Iteration (looping): Repeat are used for repetition (iterating a series of instructions).
- Conditional statements: if-elif-else checks conditions.
- Threads (parallel execution): Running simultaneously threads-generate independent runs that are executed in parallel.
- Event handling: e.g., pressing a key performs some defined action.
- User interface design: For example, using clickable objects to create buttons.
- Keyboard input: to prompt users to type.

If the developer is unfamiliar with concepts such as conditional statements, data structures, loops, or objects, it will be difficult to plan a suitable solution. Therefore, to build a solution in a specific programming language, basic knowledge of the syntax seems to be necessary [11].

In fact, many researchers, including Jeanette Wing, tried to describe and systematize these processes and based on these explorations following skills have been identified which are important for computer thinking:

- Abstraction is the process of making things more understandable by reducing unnecessary detail [4].
- Algorithmic thinking is a method of getting to a solution via a clear definition of the actions.
- Automation is a process that minimizes the work done, in which a machine-computer performs a set of repetitive commands and tasks quickly and efficiently [4].
- Decomposition is a way of thinking about things in terms of their component parts. The parts can then be understood, solved, developed, and evaluated separately [4].

- Debugging is the systematic process of analysis and evaluation using competences such as testing, tracing, and logical thinking to predict and verify outcomes [4].
- Generalization is a way of quickly solving new problems based on previous solutions to problems and building on prior experience. Algorithms that solve some specific problems can be adapted to solve a whole class of similar problems [4].

In this study, the conventional Introduction to Programming curriculum is enhanced to tackle some problems of learning programming. Consequently, problem solving learning, game-based learning and computational thinking have being implemented to improve the course curriculum. The main idea of this study aims to show novice programmers the "complete view" of software development process, provide a learning environment based on game-based experience, increase student enthusiasm, and remove syntactic obstacles to coding. The method and results of the study are presented below.

3 Materials and Methods

This study explores, how the new method of teaching using Scottie-Go block tools stimulate the behaviour of the problem-solving and coding-thinking skills by novice programmers. To make the comparison three scenarios have been examined and the classic Introduction to Python programming course was used as a reference and starting point.

3.1 Classic Course

Typical novice coder, who starts to learn programming, should focus on programming concepts and problem-solving algorithm rather than on language specifics, because they are different from one to another. Python provides the highest level of programming abstraction. So, the student does not have to think about memory management, which is necessary in C++, or class hierarchy, that is unavoidable in Java, or variable types and declarations, that exist in almost each programming language.

In our study we have followed typical curriculum of teaching Python for beginners based on three fundamental development principles which novice programmers need to understand and follow. First, analysing a given problem and determining the best programming strategy to implement, second creating an algorithm to solve the problem, finally converting the algorithm into the Python code. During the course all the programming concepts that novice programmers are expected to know and practically use have been introduced and verified. In addition, numerous of the computation thinking elements and problem-solving exercises and projects have been implemented in final stage of the class to build the foundation for comparison with other analysed teaching methods.

3.2 Scratch Extension to the Classic Course

In Poland, teachers have quite a lot of freedom in the selection of tools for learning programming in contemporary K-12 computer science education and among the two most frequently chosen block programming solutions, they use Balti and Scratch. Balti is a relatively unknown environment and, apart from Poland, Czech Republic and Slovakia is practically not used. In Poland, Balti has been deployed in over 4,500 schools and used by 79,755 students.

Scratch is much more often chosen by teachers and has a greater installed base and share in the education sector. Utilising the block-based programming environments like Scratch for introduction to novice programmers basic programming principles and computational thinking is not new and has been analysed and discussed by many researchers [1]. The overall result of this research has already been reviewed and it shows that Scratch enhances learning and problem-solving skills and significantly improves the interest of aspiring programmers to learn codding and develop computer science interest. Other work has shown that Scratch programming is an appropriate way to introduce younger students of all ages to the programming world. The advantage of using Scratch is that it is visually attractive and promotes active learning. Some studies have found a decrease in the number of students who dropped out and lost interest in computer once Scratch was implemented for introduction the concept of programming [17]. Unfortunately, Scratch users may be reluctant and have a difficulty to do transition to the traditional programming environment where syntax becomes an essential and important component. There is, of course, a concern that without the appropriate tools and interfaces to facilitate the transition from block tools introduced in Scratch and methods in the classic programming approach, it can be difficult.

In our scenario, we do not use Scratch as an independent module and we do not start the education process with an introduction to Scratch and then Python, but we rather introduce a specific programming concept (loop, condition, etc.) and outline it in Scratch with nice and easy to understand graphical form, and once the idea is understood and digest the same concept is described in Python. Essentially, Scratch is a nice programming block format to illustrate the idea of a specific programming concept that is much easier for a novice programmer to learn. There is a study comparing students answering multiple-choice programming questions that showed that students cope better with questions and tasks using block representation and graphical methods compared to tasks presented in text form [18].

The same concepts and plans like in classic course were also implemented to the students who participated in the extended course with Scratch enabled, also the same exercises and tests from "BEBRAS" repository have been used to validate the level of adoption for to computational thinking and finally the same testing project has been implemented in the final stage to check computer problem-solving and computational thinking ability and skills.

3.3 Scottie Go and Scratch Extensions to the Classic Course

Scottie Go products are a series of innovative and unique game-based approaches to introduce novice programmers to the world of programming and creative technology. The game comprises a free to download application, with no ads nor in-play purchases, which features 91 quests of increasing difficulty, a board and 179 cardboard tiles. The app can be installed on all major operating systems. The game covers all basic concepts of programming, ranging from basic instructions, algorithms, and parameters to loops, conditional expressions, variables and functions. To play the game, students start the app and read the challenge of a specific quest. This typically requires the main character to move around the board and to perform specific actions. With the use of the cardboard tiles, players create simple coding instructions for the main character to follow. The instructions are scanned with the app, which then illustrates how the character performs the instructions on the board. This solution gives the players an immediate feedback on whether the instructions were accurate, where the issue may lie, and whether the result can be further optimized. The game requires students to complete a level before moving to the next one. However, teachers are provided with a digital key to let more able students skip specific, less challenging sections. The game enables students to gain basic and advanced programming skills and to boost their logical thinking skills in an interactive and enjoyable approach. At the same time, it allows non-specialist teachers to integrate learning activities in their lessons, which will deliver against the standards of the Computer Science curriculum. Often used to promote teamwork, Scottie Go allows learners to discover programming through a tactile and kinesthetic approach. This means that the computational thinking behind programming is illustrated in an enticing and appealing way. Furthermore, by manipulating the sequences of code with their hands on the actual board, as opposed to typing strings of code on a device, Scottie Go focuses the learner on the actual process of problem solving. The typical try and error approach, so common in traditional programming, is replaced by a more considered, thought-based process and computational thinking.

The purpose of the study was to propose and evaluate the new learning method for a novice programmer, which will incubate the computational thinking and develop a problem-solving ability, rather than only concentrate the efforts on Python syntactic and coding skills. Scottie Go as the block based and gaming-based toll of learning is great candidate to stimulate these abilities. Our method expanded the approach described in the Subsect. 3.2 with Scottie Go module added on the beginning of the course.

We introduce the selected quest and games from Scottie Go during first 8 weeks of the class to develop the ability of novice programmers to discover and practice planning a suitable solution. The idea is to develop a solution from a high-level design concept of building the path to the solution. This approach requires propose detailed design by transforming sub-goals into corresponding algorithms. Additionally, the system identifies the best type of implemented algorithm and rank it with the proper number of stars. More stars are the higher ranking meaning better optimized solution to the problem. Competition aspect to plan and implement best solution, stimulates a new programmer to optimize their algorithms and incorporates fun aspect to the learning process.

3.4 Participants

This study was conducted at four public schools in Poland, each school had minimum three classes in the same level of education - eighth grade and almost similar size in terms of students in each class. The research sample consists of 361 students from four different schools, from twelve classes. As for gender, 184 are girls (51%) and 177 are boys (49%). Among these 361 students, 121 have taken the classic course, 118 took Scratch extension to the classic course and 122 Scottie Go and Scratch extensions to the classic course. Polish ministry of education formalizes the curriculum and allowing introduction to programming in Python in grade eight but not clarify or define the method how the subject is going to be educate. Most students usually have no prior knowledge of Python programming concepts and skills. Accordingly, their prior knowledge and programming skills are assumed to be equivalent. Four teachers participated in the experiment, each of them had 3 class, one from each type of tested methods, each teacher specialized in teaching computer science and had over 10 years of teaching experience.

3.5 Research Method

The purpose of the study was to propose and determine the impact of a new learning method for novice programmers to stimulate the behaviour of computer thinking and problem-solving skills. To quantity the effect of the new learning method and compare it with other tested methods, two tests were developed and carried out at the end of the curriculum as a learning outcomes control mechanism. The ideal idea seems to be to test the solution using a comparative method, this approach allows to better understand the effects of learning the concept of programming developing computer thinking and stimulating problem solving behaviour.

In its classical form, the experimental method assesses the results achieved from the control group in contrast to the results obtained by the treated group. In our scenario, the treatment consisted of implementing the proposed new method compared to the classical Python curriculum. In experimental evaluation, both groups should be equivalent if participants are randomly assigned to control and treatment groups. In our case, it was difficult to randomly assign scholars to different schools, but each school had 3 different categories of tested methods, and the experiment involved 4 different teachers and each used each method. The research procedure for the classic course was embedded with typical curriculum introduction to Python. In the first 34 weeks of the course the basics and extended programming concepts have been introduced to the students according to the curriculum and enabled all the novice programmers with all the competencies from classic course. Through this period of the class, students were introduced to the basic programming idea and abstraction such as: if-then conditions, variables, constants, for and while loops, nested loops, operation on strings, arrays, two-dimensional array, functions and use of modules in Python. In week 35 s phase took a place, and the instructor clarified the importance of "BEBRAS" test and mechanics behind it and special test to validate the ability of the computational problem-solving has been implemented. We used the one of the tests from the repository of "BEBRAS"- International Challenge on Informatics and Computational Thinking customized for eight grade. The implemented 45 min online test includes 10 problems to solve and students can achieve from 0–40 points. The test is very intuitional and does not require any knowledge of any particular programming language or any environment to be installed. All the questions in the test were dedicated to verifying the computational thinking and provide clear proof if proposed method enhance novice programmers to solve computational problems. The second testing procedure started week after in week 36 and was dedicated to verifying ability of the problem-solving and computational thinking ability and skills. First, the instructor clarified the idea of test, then students got a Python script (Fig. 1) to modify it in the way to solve the problem in most optimal way. We have been ready to see various programming strategies and implementation expose by the participants could manifest, but idea was noticeably clear the most efficient algorithm is going to win. Finally, we built the ranking of the solution based on the most effective algorithms from computational problem-solving perspective not from most efficient and condensed version of Python implementation. The assignment was to "calculate the number of zero appearance in central row and column in given matrix - table". We have used the markers of programming actions for problem solving and computational thinking to judge the solution. The indicators were group in two categories:

```
test.py
from random import randint
A=[]
for i in range(9):
    m=[]
    for j in range(9):
        n=randint (0,1)
        m.append(n)
    A.append(m)
    print (m)
```

```
[0, 0, 0, 1, 1, 0, 1, 0, 0]
[0, 0, 0, 1, 0, 0, 1, 0, 1]
[0, 1, 1, 1, 1, 1, 1, 1, 0]
[1, 1, 0, 0, 0, 1, 1, 1, 0]
[1, 1, 1, 0, 0, 1, 1, 1, 0]
[1, 0, 1, 1, 0, 1, 1, 0, 0]
[1, 1, 1, 1, 1, 0, 1, 0, 1]
[0, 1, 0, 1, 0, 0, 0, 1, 1]
[1, 0, 0, 0, 1, 1, 1, 0, 0]
```

Fig. 1. Python code to be modified to solve the problem

- Computational thinking practice: simply iteration, nested or multi-conditional iteration, agile style.
- Computational solving practice design: problem decomposition, sequential approach, selective approach.

Based on the indicators and overall algorithm quality (if the problem was solved or not) we have assessed the solution. We have observed different solution to the problem which can be group in 3 categories. Categories were created based on the way the problem was decomposed and numbers of iteration in the loops. The first category was 9 steps looping and there we have observe two type of implementations two sequential loops of 9 steps or one loop of 9 steps. Second category was loping with 4 steps and there we have seen also 2 subcategories 4 independent loops or 2 loops with 4 steps each. The final most sophisticated category was we called agile or spiral approach where students in 4 steps loop completed the task. Last approach is presented in Fig. 2. In addition to the category of implementation we have also judge how the exception and replication have been handled and how the central A[4, 4] element of the matrix has been managed (some solutions counted A[4, 4] element many times).

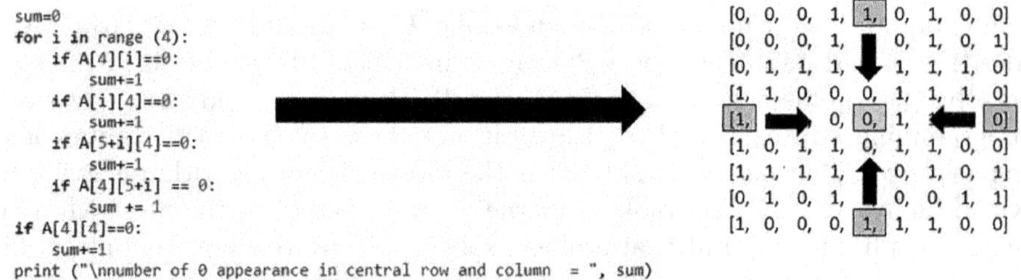

Fig. 2. One of the six type of solution: spiral approach with 4 steps loop (centripetal)

Taking into account indicators, type and quality of the proposed algorithm as well how the exceptions have been handled students were able to achieve from 0–100 points from the final test. Of course, they were fully aware that their results from the project would be considered as important component of their mark for the course. The exactly the same components have been used for verification of two other teaching methods, the only difference is that in Scratch extension to the classic course and Scottie Go and Scratch extensions to the classic course, the curriculum has been modified, but testing methodology implemented in week 35 and week 36 remains exactly the same. Figure 3 shows the timeline for all 3 type of courses.

4 Results

The data gathered in this study included information about the results of computational problem-solving activities practices, strategies, and as well results of computational thinking test. Collected indicators are representing the computational thinking activities in the action, which is a computational problem-solving task. These indicators could be classified as two dimensions: computational thinking practice and computational solving design. To discover the

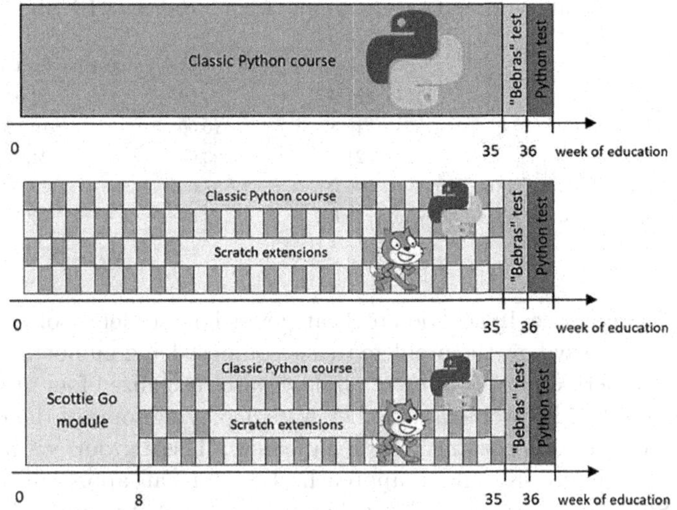

Fig. 3. Timeline of all 3 courses

student's patterns of computational thinking in the action of problem solving, we deployed "BEBRAS" test. In order to verify our postulate that students with the Scottie Go and Scratch extension to the classic course perform better than students from the classic course (our control group), the "BEBRAS" test was carried out. In addition, there was also an evaluation of the intermediate group, which was based on the adding Scratch modification to classic course. In this evaluation, the dependent variable was student performance determined after taking the "BEBRAS" test. On average, the students in the Scottie Go and Scratch extensions to the classic course performed better (M = 26.00, SD = 10.16, AVG = 24.01) than those in the classic course (M = 21.00, SD = 10.82, AVG = 21.66) and the Scratch extensions to the classic course (M = 24.00, SD = 10.60, AVG = 23.64). The percentage of scores between 20–40 points (best scores) is also important. The highest score is 70% in favor of Scottie Go and Scratch extensions to the classic course. The worst results of this parameter were achieved after completing classic Python course (55%). Score for Scratch extensions to the classic course was better than classic course (63%) (Table 1). The assessment of all courses is performed in identical procedures.

It must be also distinguished that the failure rate for all type of courses is similar. The "BEBRAS" test was not passed by those students who did not attend classes regularly and we do not see any correlation with the version of the course they were assigned.

The proposed test - "calculate the number of zero appearance in central row and column in given matrix - table" examines computational thinking models and problem-solving trials. This procedure resulted in a seven group of results based on the indicators and overall algorithm quality. This solution provided clear distinctions and meaningful explanations for the different models of

Table 1. Statistical comparison of the results from 3 courses ("BEBRAS" test).

	Classic Python	Scratch extensions	Scottie extensions
Median	21.00	24.00	26.00
Standard deviation	10.82	10.60	10.16
Average	21.66	23.64	24.01
Percentage of results between 0 and 20 points	45.00	37.00	30.00
Percentage of results between 20 and 40 points	55.00	63.00	70.00

computational practice. In addition to 3 categories how students solved the problem based on the way how the problem was decomposed and numbers of iteration in the loops (described in the Subsect. 3.5), we also recognized four group, which were not able solve the problem, but they were deploying some of the techniques which can be judged according to our indicators. These group we named and evaluated accordingly: Kamikaze approach, Try and Fail approach, Sequential approach, Selective approach. Table 2 shows 7 groups of students and how they are performing according to our indicators cross the three different teaching methods we have examined.

Table 2. The results of the Python task divided into 7 groups of solutions.

Name of group	Points	Classic Python	Scratch extensions	Scottie extensions
Kamikaze	0–5	6	5	5
Try and Fail	6–15	15	12	13
Sequential	16–34	32	28	26
Selective	35–64	41	43	44
9 Loops	65–84	13	12	13
4 Loops	85–94	11	13	15
Agile/Spiral	95–100	3	5	6

The dependent variable in this comparison was the number of points obtained after completing the test "calculate the number of zero appearance in central row and column in given matrix - table". On average, students accomplished the Scottie Go and Scratch extensions to the classic course much better (M = 44.00, SD = 28.41, AVG = 47.77) than these in classic Python course (control group) (M = 42.00, SD = 26.92, AVG = 42.83) and also better than these in Scratch extensions to the classic course (M = 41.00, SD = 27.99, AVG = 46.86). The highest percentage of scores between 20–40 points is for Scottie Go and Scratch extensions to the classic course - 44%. The worst result of this parameter was achieved after completing classic Python course (39%). Score for Scratch extensions to the classic course was better than classic course (42%) (Table 3).

Table 3. Statistical comparison of the results from 3 courses (Python exam).

	Classic Python	Scratch extensions	Scottie extensions
Median	42.00	41.00	44.00
Standard deviation	26.92	27.99	28.41
Average	42.83	46.86	47.77
Percentage of results between 0 and 20 points	61.00	58.00	56.00
Percentage of results between 20 and 40 points	39.00	42.00	44.00

Finally, after two test we have collected feedback from the classes where Scottie Go and Scratch extensions to the classic course has been implemented and students have declared that Scottie missions cause them to spend more time and to be more concentrated on the curriculum, also this extension motivated them to solve and complete the mission at home, helped to learn algorithms and programming concepts in the subsequent part of the course, improve creativity, expand problem solving skills and made programming pleasurable more. In addition to these results, this paper looked at students' progress on the Scottie Go and Scratch extensions to the classic course to understand the impact of reducing the length of the classic course by 8 weeks and adding instead a computational problem-solving module at the start of the course. On the one hand, this module enriches the development of these skills, but at the same time reduces the time needed to learn the syntax elements related to Python programming.

5 Discussions and Conclusions

The goal of the research was to validate the effect of a new learning method dedicated to novice programmers, specifically designed to develop computational thinking and problem-solving approach utilising a block-based development environment. The impact of the Scottie Go and Scratch extensions to the classic course on novice programmers was evaluated by two research tests (Sect. 4). Each of the research test was assessed by comparing to classic curriculum Introduction to Programming and also to extended version of this course with incorporated Scratch.

The study shows that problem solving through using a block-based development environment is effective way to support new inexperience programmers to absorb coding and computational design concepts. The results obtained during the study show a significant improvement in the skills of computer understanding, thinking and solving computer problems at this stage of education. This suggests the possibility of recommending to the education authorities to introduce block tools in the core curriculum on Introduction to Programming.

This article also contributes to our understanding and perception of the relationship between designing development environments and leveraging the development practices they generate. It helps us analyse the relationship between design and learning, especially with regard to programming. Nevertheless, the authors recognise a further need to explore the relationship between the ability of programmers to better model physical phenomena and the modelling of the natural world and the usage of visual programming tools and good practices.

The analysis in this study also recognized different computational practice patterns and design strategies for solving computational problems - we observed three categories of proposed solutions. Taking into account the difference in solving problems by the participants, further research into the use of block design development environment should take into account such individual differences. However, the areas of computational problems in this study are specific, which may reduce other behaviour patterns and project strategies. In such case future research should analyse different aspects of novice programmers' processes of using a block-based development environment.

It is worth analysing potential actions taken by novice programmers during the development process, such as reviewing bugs, improving plans, and tracking code, that could provide better insight into behaviour patterns and design strategies presented by novice developers. It is essential to have a comprehensive and holistic view of how these design choices affect newcomers [19].

Taking into consideration the feedback given by students and teachers at the end of the course it would also be valuable to integrate game elements into the course, to see how this advancement motivates students to increase computation problem solving capability.

In future research it is also valuable to investigate, if possible, the introduction of an admission test to better understand the level of knowledge and skills of the groups at the beginning of the course and potentially improve the adaptation of the method used.

The fundamental goal of this research is that it will help form the next generation of initial computer science learning environments and thus shape the next generation of learners, improving programming skills utilising block-based tools to develop problem solving abilities. This approach has the potential to help students achieve better results when introduced to a programming course, which in turn can have an impact on their performance in future IT projects.

Acknowledgement. Authors want to thank the anonymous reviewers whose suggestions significantly improved the quality of this manuscript.

References

1. Aivaloglou, E., Hermans, F.: How kids code and how we know: an exploratory study on the Scratch repository. In: Proceedings Of The 2016 ACM Conference On International Computing Education Research, pp. 53–61 (2016)
2. Bower, M., et al.: Improving the computational thinking pedagogical capabilities of school teachers. Aust. J. Teach. Educ. **42**, 4 (2017)
3. Chao, P.: Exploring students' computational practice, design and performance of problem-solving through a visual programming environment. Comput. Educ. **95**, 202–215 (2016)
4. Fagerlund, J., Häkkinen, P., Vesisenaho, M., Viiri, J.: Computational thinking in programming with scratch in primary schools: a systematic review. Comput. Appl. Eng. Educ. (2020)

5. Garcıa-Peñalvo, F., Mendes, A.: Exploring the computational thinking effects in pre-university education (2018)
6. Hosseini, R., et al.: Improving engagement in program construction examples for learning Python programming. Int. J. Artif. Intell. Educ. **30**, 299–336 (2020)
7. Hu, Y., Chen, C., Su, C.: Exploring the effectiveness and moderators of block-based visual programming on student learning: a meta-analysis. J. Educ. Comput. Res. **58**, 1467–1493 (2021)
8. Israel-Fishelson, R., Hershkovitz, A.: Persistence in a game-based learning environment: the case of elementary school students learning computational thinking. J. Educ. Comput. Res. **58**, 891–918 (2020)
9. LaToza, T.D., Arab, M., Loksa, D., Ko, A.J.: Explicit programming strategies. Empir. Softw. Eng. **25**(4), 2416–2449 (2020). https://doi.org/10.1007/s10664-020-09810-1
10. Mathew, R., Malik, S., Tawafak, R.: Teaching problem solving skills using an educational game in a computer programming course. Inf. Educ. **18**, 359–373 (2019)
11. Papadakis, S., Kalogiannakis, M., Orfanakis, V., Zaranis, N.: The appropriateness of scratch and app inventor as educational environments for teaching introductory programming in primary and secondary education. In: Early Childhood Development: Concepts, Methodologies, Tools, and Applications, pp. 797–819 (2019)
12. Papadakis, S.: Apps to promote computational thinking concepts and coding skills in children of preschool and pre-primary school age. In: Mobile Learning Applications In Early Childhood Education, pp. 101–121 (2020)
13. Pellet, J., Dame, A., Parriaux, G.: How beginner-friendly is a programming language? A short analysis based on Java and Python examples. University of Cyprus (2019)
14. Theobald, E., et al.: Active learning narrows achievement gaps for underrepresented students in undergraduate science, technology, engineering, and math. Proc. Nat. Acad. Sci. **117**, 6476–6483 (2020)
15. Topalli, D., Cagiltay, N.: Improving programming skills in engineering education through problem-based game projects with Scratch. Comput. Educ. **120**, 64–74 (2018)
16. Visnovitz, M.: Classical Programming Topics with Functional Programming, pp. 41–55. Central Eur. J. New Technol. Res. Educ. Pract. (2020)
17. Weintrop, D., Wilensky, U.: Between a block and a typeface: designing and evaluating hybrid programming environments. In: Proceedings of the 2017 Conference On Interaction Design and Children, pp. 183–192 (2017)
18. Weintrop, D., et al.: Defining computational thinking for mathematics and science classrooms. J. Sci. Educ. Technol. **25**, 127–147 (2016)
19. Weintrop, D., Holbert, N.: From blocks to text and back: programming patterns in a dual-modality environment. In: Proceedings of the 2017 ACM SIGCSE Technical Symposium on Computer Science Education, pp. 633–638 (2017)

Symbolic Calculation Behind Floating-Point Arithmetic: Didactic Examples and Experiment Using CAS

Włodzimierz Wojas(✉) and Jan Krupa

Department of Applied Mathematics, Warsaw University of Life Sciences (SGGW), ul. Nowoursynowska 159, 02-776 Warsaw, Poland
wlodzimierz_wojas@sggw.edu.pl

Abstract. Floating-point arithmetic (FP) is taught at universities in the framework of different academic courses, for example: Numerical Analysis, Computer Architecture or Operational Systems. In this paper we present some simple "pathological" and "non-pathological" examples for comparison with symbolic calculations which lie behind calculations in FP with double precision. The CAS programs: Mathematica and wxMaxima are used for calculation. We explain, by making calculations directly from the definition of double precision, why in the presented examples such final results were obtained. We present a didactic experiment for students of the Informatics Faculty of Warsaw University of Life Sciences within the course of Numerical Methods.

Keywords: floating-point arithmetic · symbolic calculation · mathematical didactic · CAS

1 Introduction

The FP [1–7] is ubiquitous in scientific computing. It is taught at universities in the framework of Numerical Analysis, Computer Architecture, Operational Systems and other courses. A rigorous approach to FP is seldom taught in academic courses and also not enough students, scientists and programmers study numerical analysis in details [3]. In this paper we present some didactic examples of symbolic calculations which lie behind calculations in FP (with double precision [1,2]). These examples were prepared by us for students of Informatics faculty of Warsaw University of Life Sciences and demonstrated to them in the framework of Numerical Methods course. Each operation in FP is performed according to a precise-symbolic algorithm. In spite of the fact that FP is based on symbolic operations, it gives approximate results with some exceptions, e.g.: adding, subtracting and multiplying integers; adding, subtracting, multiplying and dividing negative integer powers of 2. We present in this paper simple examples using Mathematica and wxMaxima where the result of operations depend on the interpretation of the user input data (numbers) by CAS functions (such as Solve, Limit, Det) - as symbolic or approximate. The result may also depend whether these CAS functions use more or less clever algorithms. In the second part of this paper, by making calculations

J. Mikyška et al. (Eds.): ICCS 2023, LNCS 14077, pp. 722–736, 2023.
https://doi.org/10.1007/978-3-031-36030-5_56

directly from the definition of double precision, we explain why in the examples presented such final results were obtained. The didactic approach presented in this article is multidisciplinary - it combines theoretical issues in the field of numerical analysis with the use of practical CAS computing tools.

2 Didactic Motivations

Sauer's book "Numerical Analysis" [6] presents the basics of calculations in FP with double precision with examples not found in our opinion in other literature. The examples presented in this article are, on the one hand, inspired by the examples in this book, but on the other hand, complement and extend them. Such examples are extremely important in teaching FP arithmetic with double precision. The calculations with double precision presented in this article were exercise tasks for students of Informatics faculty of Warsaw University of Life Sciences within the subject of Numerical Methods. These students were presented with floating point arithmetic, including double precision in the subjects: Computer Architecture and Operating Systems. In the framework of these subjects, students calculated with double precision, e.g.: $1+2^{-5}$, $1+2^{-52}$, $1+3\cdot 2^{-53}-1$. Such examples are important in double precision arithmetic, but the calculations presented in our examples are a significant complement to them, deepening the understanding of double precision calculations, although they require more tedious calculations (but still within the reach of students' skills).

3 "Pathological" and "Non - Pathological" Example of Linear Equations System Solution

Example 1. ("Pathological")

Consider the following system of linear equations and solve them in Mathematica and Maxima

$$\begin{cases} 3x + y = 0, \\ 0.3x + 0.1y = 1. \end{cases}$$

Example 2. ("Non - Pathological")

Consider the following system of linear equations and solve them in Mathematica and Maxima

$$\begin{cases} 2x + y = 0, \\ 0.2x + 0.1y = 1. \end{cases}$$

Let's us notice that generally in **double precision** we have:

Listing 3.1: Mathematica code:

```
 In[1]:=0.1+0.1−0.2
Out[1] =0
 In[2]:=0.1+0.1+0.1−0.3
Out[2] =5.551115123125783∗10^−17
 In[3]:=N[2^(−54),16]
Out[3] =5.551115123125783∗10^−17
```

Listing 3.2: Mathematica code:

```
 In[4]:=−0.3+0.1+0.1+0.1
Out[4] =2.775557561562891∗10^−17 (= 2^−55)
```

It follows that in double precision we have:
$-0.3 + 0.1 + 0.1 + 0.1 \neq 0.1 + 0.1 + 0.1 - 0.3$ (no additivity).

Let's solve the Example 1. First, we solve it standardly using function Solve in Mathematica and solve in Maxima. Next, we solve it again in Mathematica and Maxima using Cramer's rule.

Listing 3.3: Mathematica code for Example 1:

```
 In[1]:=eq1={3 x + y == 0, 0.3 x + 0.1 y == 1};
 In[2]:= Solve[eq1, {x, y}]
Out[2] = {}
 In[3]:= NSolve[eq1, {x, y}]
Out[3] = {}
 In[4]:=M = {{3, 1}, {0.3, 0.1}};
 In[5]:=Det[M]
Out[5] =5.551115123125783∗10^−17
 In[6]:={ M0 = Det[M],
 M1 = {{0, 1}, {0.3, 0.1}}; MatrixForm[M1], M10 = Det[M1],
 M2 = {{3, 0}, {0.3, 1}}; MatrixForm[M2], M20 = Det[M2],
 r1 = {x −> M10/M0, y −> M20/M0}}
Out[6] =r1={x −> −2.40192∗10^16, y −>7.20576∗10^16}
 In[7]:=eq1 /. r1
Out[7] ={True, True}
```

Listing 3.4: Maxima code for Example 1:

```
(%i1)  solve([3*x+1*y=0,0.3*x+0.1*y=1],[x,y]);
rat: replaced 0.3 by 3/10 = 0.3
rat: replaced 0.1 by 1/10 = 0.1
(%o1)  []
(%i2)  M:matrix([3,  1],[0.3,0.1]);
(M)    matrix(
                [3,     1],
                [0.3,   0.1]
       )
(%i3)  d:determinant (M);
(%o3)  5.551115123125783*10^-17
(%i4)  [M1:matrix ([0,1],[1,0.1]),  M2:matrix ([3,0],[0.3,1])];
(%o4)  [matrix(
                [0,     1],
                [1,     0.1]
       ), matrix(
                [3,     0],
                [0.3,   1]
       )]
(%i5)  [d1:determinant(M1),d2:determinant(M2)];
(%o5)  [-1,3]
(%i6)  [x0:d1/d,y0:d2/d];
(%o6)  [-1.801439850948199*10^16,5.404319552844595*10^16]
(%i7)  [xy:matrix([x0],[y0]), b:matrix ([0],[1]), M.xy-b];
(%o7)  [matrix(
                [-1.801439850948199*10^16],
                [5.404319552844595*10^16]
       ), matrix(
                [0],
                [1]
       ), matrix(
                [0.0],
                [0.0]
       )]
```

We get the following results. Using functions Solve in Mathematica and solve in Maxima we get that system of equations has no solution. But using Cramer's rule in double precision we get a single solution of the system. Solutions in Mathematica and Maxima are different.

Let's solve the Example 2 in Mathematica and Maxima.

Listing 3.5: Mathematica code for Example 2:

```
 In[1]:=eq1={2 x + y == 0, 0.2 x + 0.1 y == 1};
 In[2]:= Solve[eq1, {x, y}]
Out[2] = {}
 In[3]:= NSolve[eq1, {x, y}]
Out[3] = {}
 In[4]:=M = {{2, 1}, {0.2, 0.1}};
 In[5]:=Det[M]
Out[5] =0
```

Listing 3.6: Maxima code for Example 2:

```
(%i1)  solve([2*x+1*y=0,0.2*x+0.1*y=1],[x,y]);
rat:  replaced 0.2 by 2/10 = 0.2
rat:  replaced 0.1 by 1/10 = 0.1
(%o1)  []
(%i2)  M:matrix([2,    1],[0.2,0.1]);
(M)    matrix(
                [2,      1],
                [0.2,    0.1]
       )

(%i3)  determinant (M);
(%o3)  0.0
```

We get the same results in Mathematica and Maxima that system of equations has no solution. The determinant of the coefficient matrix is equal to 0 so we cannot use Cramer's rule.

4 "Pathological" and "Non - Pathological" Example of Limit of a Real Function

Example 3. ("Pathological")

Let us calculate the following limit: $\lim_{x\to 0.1^-} \frac{1}{-3x+0.3}$

Example 4. ("Non - Pathological")

Let us calculate the following limit: $\lim_{x\to 0.1^-} \frac{1}{-2x+0.2}$

Let's solve the Examples 3 and 4 in Mathematica and Maxima.

Listing 4.1: Mathematica code for Examples 3 and 4:

```
In[5]:=f[x_] = 1/(−3 x + 0.3);
In[6]:=Limit[f[x], x −> 0.1, Direction −> 1]
Out[6] =∞
In[7]:=f[0.1]
Out[7]=−1.80144∗10^16
In[8]:=g[x_] = 1/(−2 x + 0.2)'
In[9]:=Limit[g[x], x −> 0.1, Direction −> 1]
Out[9] =∞
In[10]:=g[0.1]
Power::infy: Infinite expression 1/0. encountered. >>
ComplexInfinity
```

Listing 4.2: Maxima code for Examples 3 and 4:

```
(%i1) f(x):=1/(0.3−3∗x);
(%o1) f(x):=1/(0.3−3∗x)
(%i2) f(0.1);
(%o2) \−1.801439850948199∗10^16
(%i4) limit(f(x),x,0.1);
(%o4) \−1.801439850948199∗10^16
(%i5) g(x):=1/(0.2−2∗x);
(%o5) g(x):=1/(0.2−2∗x)
(%i6) g(0.1);
(%o6) expt: undefined: 0 to a negative exponent. −− an error. To debug this
        try: debugmode(true)
(%i7) limit(g(x),x,0.1);
(%o7) infinity
```

We get in Mathematica and Maxima that limit in Example 3 of function $f(x)$ at point 0.1 is equal to infinity but the value of function $f(x)$ at point 0.1 is finite. But in Example 4 the limit of function $g(x)$ at point 0.1 is equal to infinity and the value of function $g(x)$ at point 0.1 is undefined.

5 Floating-Point Arithmetic: Some Basic Definitions

In this section, we present a model for computer arithmetic of floating point numbers see [1–7]. There are several models, but to simplify matters we will choose one particular model and describe it in detail. The model we choose is the so-called IEEE 754 Floating Point Standard.

Rounding errors are inevitable when finite-precision computer memory locations are used to represent real, infinite precision numbers.

The IEEE standard consists of a set of binary representations of real numbers. A floating point number consists of three parts: the sign (+ or −), a mantissa, which contains the string of significant bits, and an exponent. The three parts are stored together in a single computer word. There are three commonly used levels of precision for floating point numbers: single precision, double precision, and extended precision, also known as long-double precision. The number of bits allocated for each floating point number in the three formats is 32, 64, and 80, respectively. The bits are divided among the parts as follows: all three types of precision work essentially the same way. The form of

precision	sign	exponent	mantissa
single	1	8	23
double	1	11	52
long double	1	15	64

a normalized IEEE floating point number is

$$\boxed{\pm 1.bbb\ldots b \times 2^p,}$$

where each of the N b's is 0 or 1, and p is an $M = 11$ (or 8 or 15) bit binary number representing the exponent. Normalization means that, as shown in, the leading (leftmost) bit must be 1.

When a binary number is stored as a normalized floating point number, it is leftjustified, meaning that the leftmost 1 is shifted just to the left of the radix point.

The shift is compensated by a change in the exponent. For example, the decimal number 9, which is 1001 in binary, would be stored as $+1.001 \times 2^3$ because a shift of 3 bits, or multiplication by 2^3, is necessary to move the leftmost one to the correct position. The number machine epsilon, denoted $\varepsilon_{\text{mach}}$, is the distance between 1 and the smallest floating point number greater than 1. For the IEEE double precision floating point standard, $\varepsilon_{\text{mach}} = 2^{-52}$.

We must truncate the binary representation of a number in some way, and in so doing we necessarily make a small error. The most common method of the truncation is rounding. In base 10, numbers are customarily rounded up if the next digit is 5 or higher, and rounded down otherwise. In binary, this corresponds to rounding up if the bit is 1. Specifically, the important bit in the double precision format is the 53rd bit to the right of the radix point, the first one lying outside of the box.

The default rounding technique, implemented by the IEEE standard, is to add 1 to bit 52 (round up) if bit 53 is 1, and to do nothing (round down) to bit 52 if bit 53 is 0, with one exception: if the bits following bit 52 are $10000\ldots$, exactly halfway between up and down, we round up or round down according to which choice makes the final bit 52 equal to 0. (Here we are dealing with the mantissa only, since the sign does not play a role.) Why is there the strange exceptional case? Except for this case, the rule means rounding to the normalized floating point number closest to the original number–hence its name, the Rounding to Nearest Rule. The error made in rounding will be equally likely to be up or down. Therefore, the exceptional case, the case where there are two equally distant floating point numbers to round to, should be decided in a way that doesn't prefer up or down systematically. This is to try to avoid the possibility of an unwanted slow drift in long calculations due simply to a biased rounding. The choice to make the final bit 52 equal to 0 in the case of a tie is somewhat arbitrary, but at least it does not display a preference up or down.

IEEE Rounding to Nearest Rule For double precision, if the 53rd bit to the right of the binary point is 0, then round down (truncate after the 52nd bit). If the 53rd bit is 1, then round up (add 1 to the 52 bit), unless all known bits to the right of the 1 are 0's, in which case 1 is added to bit 52 if and only if bit 52 is 1.

Definition. Denote the IEEE double precision floating point number associated to x, using the Rounding to Nearest Rule, by $\mathrm{fl}(x)$. In computer arithmetic, the real number x is replaced with the string of bits $\mathrm{fl}(x)$.

$$\textbf{absolute error} \text{ (of approximation } x \text{ by } \mathrm{fl}(x)) = |\,\mathrm{fl}(x) - x|,$$

and

$$\textbf{relative error} \text{ (of approximation } x \text{ by } \mathrm{fl}(x)) = |\,\mathrm{fl}(x) - x|/x, \ (x \neq 0).$$

In the IEEE machine arithmetic model, the relative rounding error of $\mathrm{fl}(x)$ is no more than one-half machine epsilon: $|\,\mathrm{fl}(x) - x| \leq \frac{1}{2}\varepsilon_{\mathrm{mach}}$

Example 5. (**Double precision representation of 9**)

From the above description we can formulate the three steps algorithm which determines the double precision representation of real number. We will illustrate it on the example of number 9.

First step: find enough binary digits of 9: $9 = (1001.000\ldots0)_2$.

Second step: normalise the expansion from the first step:
$9 = (1.001000\ldots0)_2 \times 2^3$.

When a binary number is stored as a normalized floating point number, it is leftjustified, meaning that the leftmost 1 is shifted just to the left of the radix point.

The shift is compensated by a change in the exponent. For example, the decimal number 9, which is 1001 in binary, would be stored as $+1.001 \times 2^3$ because a shift of 3 bits, or multiplication by 2^3, is necessary to move the leftmost one to the correct position.

Third step: Apply IEEE Rounding to Nearest Rule:
$9 = (1.\underbrace{001000\ldots0}_{52\text{digits}})_2 \times 2^3$.

The number machine epsilon, denoted $\varepsilon_{\mathrm{mach}}$, is the distance between 1 and the smallest floating point number greater than 1. For the IEEE double precision floating point standard, $\varepsilon_{\mathrm{mach}} = 2^{-52}$.

In the next sections we will use the three steps algorithm illustrated in Example 5.

6 Double Precision Representation of $\frac{1}{10}$

See Figs. 1, 2 and 3.

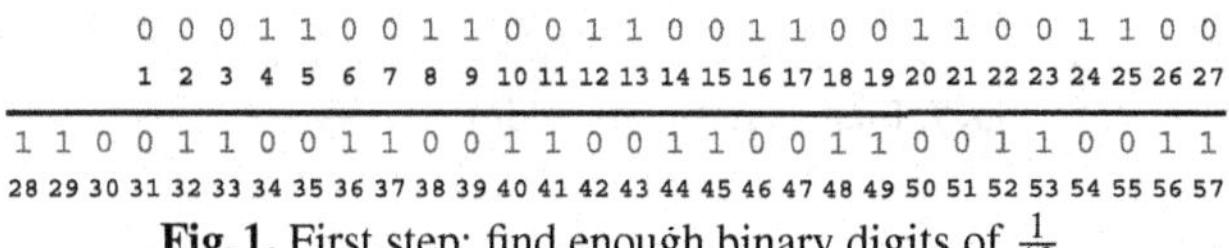

Fig. 1. First step: find enough binary digits of $\frac{1}{10}$

2^{-4} 1 1 0 0 1 1 0 0 1 1 0 0 1 1 0 0 1 1 0 0 1 1 0 0 1 1 0 0 1
1 2 3 4 5 6 7 8 9 10 11 12 13 14 15 16 17 18 19 20 21 22 23 24 25 26 27 28

1 0 0 1 1 0 0 1 1 0 0 1 1 0 0 1 1 0 0 1 1 0 0 1 1 0 0 1
29 30 31 32 33 34 35 36 37 38 39 40 41 42 43 44 45 46 47 48 49 50 51 52 53 54 55 56

Fig. 2. Second step: Normalisation of binary expansion of $\frac{1}{10}$

2^{-4} 1 1 0 0 1 1 0 0 1 1 0 0 1 1 0 0 1 1 0 0 1 1 0 0 1 1
1 2 3 4 5 6 7 8 9 10 11 12 13 14 15 16 17 18 19 20 21 22 23 24 25

0 0 1 1 0 0 1 1 0 0 1 1 0 0 1 1 0 0 1 1 0 0 1 1 0 1 0
26 27 28 29 30 31 32 33 34 35 36 37 38 39 40 41 42 43 44 45 46 47 48 49 50 51 52

Fig. 3. Third step: Rounding of binary expansion of $\frac{1}{10}$

7 Double Precision Representation of $\dfrac{2}{10}$

See Figs. 4, 5 and 6.

0 0 1 1 0 0 1 1 0 0 1 1 0 0 1 1 0 0 1 1 0 0 1 1 0 0 1
1 2 3 4 5 6 7 8 9 10 11 12 13 14 15 16 17 18 19 20 21 22 23 24 25 26 27

1 0 0 1 1 0 0 1 1 0 0 1 1 0 0 1 1 0 0 1 1 0 0 1 1 0 0 1 1
28 29 30 31 32 33 34 35 36 37 38 39 40 41 42 43 44 45 46 47 48 49 50 51 52 53 54 55 56

Fig. 4. First step: find enough binary digits of $\frac{2}{10}$

2^{-3} 1 1 0 0 1 1 0 0 1 1 0 0 1 1 0 0 1 1 0 0 1 1 0 0 1 1 0 0 1
1 2 3 4 5 6 7 8 9 10 11 12 13 14 15 16 17 18 19 20 21 22 23 24 25 26 27 28

1 0 0 1 1 0 0 1 1 0 0 1 1 0 0 1 1 0 0 1 1 0 0 1 1 0 0 1
29 30 31 32 33 34 35 36 37 38 39 40 41 42 43 44 45 46 47 48 49 50 51 52 53 54 55 56

Fig. 5. Second step: Normalisation of binary expansion of $\frac{2}{10}$

2^{-3} 1 1 0 0 1 1 0 0 1 1 0 0 1 1 0 0 1 1 0 0 1 1 0 0 1 1
1 2 3 4 5 6 7 8 9 10 11 12 13 14 15 16 17 18 19 20 21 22 23 24 25

0 0 1 1 0 0 1 1 0 0 1 1 0 0 1 1 0 0 1 1 0 0 1 1 0 1 0
26 27 28 29 30 31 32 33 34 35 36 37 38 39 40 41 42 43 44 45 46 47 48 49 50 51 52

Fig. 6. Third step: Rounding of binary expansion of $\frac{2}{10}$

8 Double Precision Representation of $\dfrac{1}{10} + \dfrac{1}{10}$

See Figs. 7, 8 and 9.

2^{-4} 1 1 0 0 1 1 0 0 1 1 0 0 1 1 0 0 1 1 0 0 1 1 0 0 1 1 0
1 2 3 4 5 6 7 8 9 10 11 12 13 14 15 16 17 18 19 20 21 22 23 24 25

0 1 1 0 0 1 1 0 0 1 1 0 0 1 1 0 0 1 1 0 0 1 1 0 1 0 0
26 27 28 29 30 31 32 33 34 35 36 37 38 39 40 41 42 43 44 45 46 47 48 49 50 51 52

Fig. 7. First step: Binary addition of $\frac{1}{10} + \frac{1}{10}$

2^{-3} 1 1 0 0 1 1 0 0 1 1 0 0 1 1 0 0 1 1 0 0 1 1 0 0 1 1
1 2 3 4 5 6 7 8 9 10 11 12 13 14 15 16 17 18 19 20 21 22 23 24 25

0 0 1 1 0 0 1 1 0 0 1 1 0 0 1 1 0 0 1 1 0 0 1 1 0 1 0 0
26 27 28 29 30 31 32 33 34 35 36 37 38 39 40 41 42 43 44 45 46 47 48 49 50 51 52 53

Fig. 8. Second step: Normalisation of binary addition of $\frac{1}{10}+\frac{1}{10}$

2^{-3} 1 1 0 0 1 1 0 0 1 1 0 0 1 1 0 0 1 1 0 0 1 1 0 0 1 1
1 2 3 4 5 6 7 8 9 10 11 12 13 14 15 16 17 18 19 20 21 22 23 24 25

0 0 1 1 0 0 1 1 0 0 1 1 0 0 1 1 0 0 1 1 0 0 1 1 0 1 0
26 27 28 29 30 31 32 33 34 35 36 37 38 39 40 41 42 43 44 45 46 47 48 49 50 51 52

Fig. 9. Third step: Rounding of binary expansion of $\frac{1}{10}+\frac{1}{10}$

Comparing double precision representations of $\frac{2}{10}$ and $\frac{1}{10}+\frac{1}{10}$ we get:

2^{-3} 1 1 0 0 1 1 0 0 1 1 0 0 1 1 0 0 1 1 0 0 1 1 0 0 1 1
1 2 3 4 5 6 7 8 9 10 11 12 13 14 15 16 17 18 19 20 21 22 23 24 25

0 0 1 1 0 0 1 1 0 0 1 1 0 0 1 1 0 0 1 1 0 0 1 1 0 1 0
26 27 28 29 30 31 32 33 34 35 36 37 38 39 40 41 42 43 44 45 46 47 48 49 50 51 52

Fig. 10. Third step: double precision representations of $\frac{2}{10}$

2^{-3} 1 1 0 0 1 1 0 0 1 1 0 0 1 1 0 0 1 1 0 0 1 1 0 0 1 1
1 2 3 4 5 6 7 8 9 10 11 12 13 14 15 16 17 18 19 20 21 22 23 24 25

0 0 1 1 0 0 1 1 0 0 1 1 0 0 1 1 0 0 1 1 0 0 1 1 0 1 0
26 27 28 29 30 31 32 33 34 35 36 37 38 39 40 41 42 43 44 45 46 47 48 49 50 51 52

Fig. 11. Third step: double precision representations of $\frac{1}{10}+\frac{1}{10}$

Finally, we can see that: $2*0.1-0.2=0$

9 Double Precision Addition of $\frac{1}{10}+\frac{1}{10}+\frac{1}{10}$

We have already double precision representation of $\frac{1}{10}+\frac{1}{10}$ and $\frac{1}{10}$:

2^{-3} 1 1 0 0 1 1 0 0 1 1 0 0 1 1 0 0 1 1 0 0 1 1 0 0 1 1
1 2 3 4 5 6 7 8 9 10 11 12 13 14 15 16 17 18 19 20 21 22 23 24 25

0 0 1 1 0 0 1 1 0 0 1 1 0 0 1 1 0 0 1 1 0 0 1 1 0 1 0
26 27 28 29 30 31 32 33 34 35 36 37 38 39 40 41 42 43 44 45 46 47 48 49 50 51 52

Fig. 12. Double precision representations of $\frac{1}{10}+\frac{1}{10}$

2^{-4} 1 1 0 0 1 1 0 0 1 1 0 0 1 1 0 0 1 1 0 0 1 1 0 0 1 1
1 2 3 4 5 6 7 8 9 10 11 12 13 14 15 16 17 18 19 20 21 22 23 24 25

0 0 1 1 0 0 1 1 0 0 1 1 0 0 1 1 0 0 1 1 0 0 1 1 0 1 0
26 27 28 29 30 31 32 33 34 35 36 37 38 39 40 41 42 43 44 45 46 47 48 49 50 51 52

Fig. 13. Double precision representations of $\frac{1}{10}$

and we determine double precision representation of $\left(\frac{1}{10}+\frac{1}{10}\right)+\frac{1}{10}$:
"Pre normalisation" of Double precision representation of $\frac{1}{10}+\frac{1}{10}$ and $\frac{1}{10}$:

2^{-3} 1 1 0 0 1 1 0 0 1 1 0 0 1 1 0 0 1 1 0 0 1 1 0 0 1 1

1 2 3 4 5 6 7 8 9 10 11 12 13 14 15 16 17 18 19 20 21 22 23 24 25

0 0 1 1 0 0 1 1 0 0 1 1 0 0 1 1 0 0 1 1 0 0 1 1 0 1 0 0

26 27 28 29 30 31 32 33 34 35 36 37 38 39 40 41 42 43 44 45 46 47 48 49 50 51 52 53

Fig. 14. "Pre normalisation" of double precision representations of $\frac{1}{10} + \frac{1}{10}$

2^{-3} 0 1 1 0 0 1 1 0 0 1 1 0 0 1 1 0 0 1 1 0 0 1 1 0 0 1

1 2 3 4 5 6 7 8 9 10 11 12 13 14 15 16 17 18 19 20 21 22 23 24 25

1 0 0 1 1 0 0 1 1 0 0 1 1 0 0 1 1 0 0 1 1 0 0 1 1 0 1 0

26 27 28 29 30 31 32 33 34 35 36 37 38 39 40 41 42 43 44 45 46 47 48 49 50 51 52 53

Fig. 15. "Pre normalisation" of double precision representations of $\frac{1}{10}$

2^{-3} 1 0 0 1 1 0 0 1 1 0 0 1 1 0 0 1 1 0 0 1 1 0 0 1 1 0 0

1 2 3 4 5 6 7 8 9 10 11 12 13 14 15 16 17 18 19 20 21 22 23 24 25

1 1 0 0 1 1 0 0 1 1 0 0 1 1 0 0 1 1 0 0 1 1 0 0 1 1 1 0

26 27 28 29 30 31 32 33 34 35 36 37 38 39 40 41 42 43 44 45 46 47 48 49 50 51 52 53

Fig. 16. Binary addition of $\left(\frac{1}{10} + \frac{1}{10}\right) + \frac{1}{10}$

2^{-2} 1 0 0 1 1 0 0 1 1 0 0 1 1 0 0 1 1 0 0 1 1 0 0 1 1 0

1 2 3 4 5 6 7 8 9 10 11 12 13 14 15 16 17 18 19 20 21 22 23 24 25

0 1 1 0 0 1 1 0 0 1 1 0 0 1 1 0 0 1 1 0 0 1 1 0 0 1 1 1 0

26 27 28 29 30 31 32 33 34 35 36 37 38 39 40 41 42 43 44 45 46 47 48 49 50 51 52 53 54

Fig. 17. Second step: Normalisation of binary addition of $\left(\frac{1}{10} + \frac{1}{10}\right) + \frac{1}{10}$

2^{-2} 1 0 0 1 1 0 0 1 1 0 0 1 1 0 0 1 1 0 0 1 1 0 0 1 1 0

1 2 3 4 5 6 7 8 9 10 11 12 13 14 15 16 17 18 19 20 21 22 23 24 25

0 1 1 0 0 1 1 0 0 1 1 0 0 1 1 0 0 1 1 0 0 1 1 0 1 0 0

26 27 28 29 30 31 32 33 34 35 36 37 38 39 40 41 42 43 44 45 46 47 48 49 50 51 52

Fig. 18. Rounding of binary expansion of $\left(\frac{1}{10} + \frac{1}{10}\right) + \frac{1}{10}$

10 Double Precision Representation of $\frac{3}{10}$

Comparison of double precision addition of $\frac{1}{10} + \frac{1}{10} + \frac{1}{10}$ and of $\frac{3}{10}$ (Figs. 19, 20 and 21).

0 1 0 0 1 1 0 0 1 1 0 0 1 1 0 0 1 1 0 0 1 1 0 0 1 1 0

1 2 3 4 5 6 7 8 9 10 11 12 13 14 15 16 17 18 19 20 21 22 23 24 25 26 27

0 1 1 0 0 1 1 0 0 1 1 0 0 1 1 0 0 1 1 0 0 1 1 0 0 1 1 0

28 29 30 31 32 33 34 35 36 37 38 39 40 41 42 43 44 45 46 47 48 49 50 51 52 53 54 55

Fig. 19. First step: Binary digits of $\frac{3}{10}$

2^{-2} 1 0 0 1 1 0 0 1 1 0 0 1 1 0 0 1 1 0 0 1 1 0 0 1 1 0 0 1 1

1 2 3 4 5 6 7 8 9 10 11 12 13 14 15 16 17 18 19 20 21 22 23 24 25 26 27 28

0 0 1 1 0 0 1 1 0 0 1 1 0 0 1 1 0 0 1 1 0 0 1 1 0 0 1 1

29 30 31 32 33 34 35 36 37 38 39 40 41 42 43 44 45 46 47 48 49 50 51 52 53 54 55 56

Fig. 20. Second step: Normalisation of binary expansion of $\frac{3}{10}$

2^{-2} 1 0 0 1 1 0 0 1 1 0 0 1 1 0 0 1 1 0 0 1 1 0 0 1 1 0
1 2 3 4 5 6 7 8 9 10 11 12 13 14 15 16 17 18 19 20 21 22 23 24 25

0 1 1 0 0 1 1 0 0 1 1 0 0 1 1 0 0 1 1 0 0 1 1 0 0 1 1
26 27 28 29 30 31 32 33 34 35 36 37 38 39 40 41 42 43 44 45 46 47 48 49 50 51 52

Fig. 21. Third step: Rounding of binary expansion of $\frac{3}{10}$

2^{-2} 1 0 0 1 1 0 0 1 1 0 0 1 1 0 0 1 1 0 0 1 1 0 0 1 1 0
1 2 3 4 5 6 7 8 9 10 11 12 13 14 15 16 17 18 19 20 21 22 23 24 25

0 1 1 0 0 1 1 0 0 1 1 0 0 1 1 0 0 1 1 0 0 1 1 0 1 0 0
26 27 28 29 30 31 32 33 34 35 36 37 38 39 40 41 42 43 44 45 46 47 48 49 50 51 52

Fig. 22. Third step: Rounding of binary expansion of $\frac{1}{10}+\frac{1}{10}+\frac{1}{10}$

2^{-2} 1 0 0 1 1 0 0 1 1 0 0 1 1 0 0 1 1 0 0 1 1 0 0 1 1 0
1 2 3 4 5 6 7 8 9 10 11 12 13 14 15 16 17 18 19 20 21 22 23 24 25

0 1 1 0 0 1 1 0 0 1 1 0 0 1 1 0 0 1 1 0 0 1 1 0 0 1 1
26 27 28 29 30 31 32 33 34 35 36 37 38 39 40 41 42 43 44 45 46 47 48 49 50 51 52

Fig. 23. Third step: Rounding of binary expansion of $\frac{3}{10}$

So, we can see that in double precision: $3*0.1-0.3=2^{-54}=5.551115123125783*10^{-17}\neq 0$.

Similarly we can see that in double precision: $-0.3+0.1+0.1+0.1=2^{-55}=2.775557561562891*10^{-17}\neq 0$

We can see also that:

1. The **absolute error** $=|0.1-\mathrm{fl}(0.1)|=|2^{-4}(2^{-52}+2^{-56}9\frac{1}{1-2^{-4}}-2^{-51})|=2^{-56}\frac{2}{5}=2^{-55}\frac{1}{5}$. Thus the **relative error** $=|0.1-\mathrm{fl}(0.1)|/0.1=2^{-54}<(1/2)\varepsilon_{\mathrm{mach}}=2^{-53}$. Of course there are other good ways to calculate the **absolute error**.
2. See Figs. 10, 11. The **absolute error** $=|0.2-\mathrm{fl}(0.2)|=2^{-3}(2^{-52}-2^{51}+2^{-56}9\frac{1}{1-2^{-4}})=2^{-55}\frac{2}{5}$. Thus the **relative error** $=|0.2-\mathrm{fl}(0.2)|/0.2=2^{-54}<(1/2)\varepsilon_{\mathrm{mach}}=2^{-53}$.
3. $|0.3-\mathrm{fl}(0.3)|=|2^{-2}2^{-56}3\frac{1}{1-2^{-4}}|=2^{-54}\frac{1}{5}$. Thus $|0.3-\mathrm{fl}(0.3)|/0.3=2^{-54}\frac{2}{3}<(1/2)\varepsilon_{\mathrm{mach}}=2^{-53}$.
4. Because $\mathrm{fl}(0.1+0.1)=\mathrm{fl}(0.2)$, thus the **relative error** $=|0.2-\mathrm{fl}(0.1+0.1)|/0.2=2^{-54}<(1/2)\varepsilon_{\mathrm{mach}}=2^{-53}$.
5. We saw that $\mathrm{fl}(0.1+0.1+0.1)-\mathrm{fl}(0.3)=2^{-54}$ (See Figs. 22, 23). Thus $|0.3-\mathrm{fl}(0.1+0.1+0.1)|=|0.3-\mathrm{fl}(0.3)+\mathrm{fl}(0.3)-\mathrm{fl}(0.1+0.1+0.1)|=|2^{-54}\frac{1}{5}-2^{-54}|=2^{-54}|1/5-1|=2^{-52}/5$. Thus $|0.3-\mathrm{fl}(0.1+0.1+0.1)|/0.3=2^{-52}2/3>2^{-52}1/2=2^{-53}=(1/2)\varepsilon_{\mathrm{mach}}=2^{-53}$. Thus the **relative error** of approximation (in double precision) 0.3 by $\mathrm{fl}(0.1+0.1+0.1)$ exceeds $(1/2)\varepsilon_{\mathrm{mach}}=2^{-53}$.
6. Similarly like in Figs. 12, 13, 14, 15, 16, 17 and 18 we can perform $0.3-0.1$ in double precision and get: $\mathrm{fl}(0.2)-\mathrm{fl}(\mathrm{fl}(0.3)-\mathrm{fl}(0.1))=2^{-55}$. Thus $|0.2-$

$\text{fl}(\text{fl}(0.3)-\text{fl}(0.1))| = |0.2-\text{fl}(0.2)+\text{fl}(0.2)-\text{fl}(\text{fl}(0.3)-\text{fl}(0.1))| = \frac{7}{5}2^{-55}$. Thus $|0.2-\text{fl}(\text{fl}(0.3)-\text{fl}(0.1))|/0.2 = 7\cdot 2^{-55} > 2^{-53} = (1/2)\varepsilon_{\text{mach}}$. Thus the **relative error** of approximation (in double precision) 0.2 by $\text{fl}(0.3)-\text{fl}(0.1)$ exceeds $(1/2)\varepsilon_{\text{mach}} = 2^{-53}$.

7. Similarly like in Figs. 12, 13, 14, 15, 16, 17 and 18 we can perform $0.3-0.1-0.1$ in double precision and get: $\text{fl}(0.1)-\text{fl}(\text{fl}(0.3)-\text{fl}(0.1)-\text{fl}(0.1)) = 2^{-55}$. Thus $|0.1-\text{fl}(\text{fl}(0.3)-\text{fl}(0.1)-\text{fl}(0.1))| = |0.1-\text{fl}(0.1)+\text{fl}(0.1)-\text{fl}(\text{fl}(0.3)-\text{fl}(0.1)-\text{fl}(0.1))| = \frac{1}{5}2^{-55}+2^{-55} = \frac{3}{5}2^{-54}$. Thus $|0.1-\text{fl}(\text{fl}(0.3)-\text{fl}(0.1)\text{fl}(0.1))|/0.1 = 3\cdot 2^{-53} > 2^{-53} = (1/2)\varepsilon_{\text{mach}}$. Thus the **relative error** of approximation (in double precision) 0.1 by $\text{fl}(0.3)-\text{fl}(0.1)-\text{fl}(0.1)$ exceeds $(1/2)\varepsilon_{\text{mach}} = 2^{-53}$.
8. Similarly like in Figs. 12, 13, 14, 15, 16, 17 and 18 we can perform $0.3-0.1-0.1-0.1$ in double precision and get: $\text{fl}(\text{fl}(0.3)-\text{fl}(0.1)-\text{fl}(0.1)-\text{fl}(0.1)) = -2^{-55}$. Thus $|0-\text{fl}(\text{fl}(0.3)-\text{fl}(0.1)-\text{fl}(0.1)-\text{fl}(0.1))| = 2^{-55}$. The relative error is not definite.

After the above examples, student can easily see that $1+2^{-53} = 1$ in double precision and is prepared to study the very important "pathological" examples in [3–5].

11 The Didactic Experiment with Examples 1–4

The experiment was carried out in two independent auditorium groups (group A and group B) of third-year students of the Informatics Faculty of Warsaw University of Life Sciences within the course of Numerical Methods. Groups A and B consisted of 28 and 31 students, respectively. Students of both groups solved a test consisting of two tasks 1 and 2 with subquestions a) and b) in each task. The time allotted for the test was 30 min in each group. But in group B, the test was preceded by a 30-minute presentation in which the double precision numerical representations presented in Sects. 6, 7, 8, 9 and 10 of this article were shown and discussed. Group A students wrote the test without such a prior presentation.

In task 1, two linear systems of equations were presented: in 1a - the system from Example 1 ("pathological") and in 1b - the system from Example 2 ("non - pathological") with the comment below. Solving the above systems of equations symbolically (exactly) we get a contradictory system in each case. But when solving them with CAS programs such as Mathematica or Maxima, that is, by calculating the determinants: $detM$, $detM_1$ and $detM_2$ in double precision and then applying Cramer's formulas, we obtain the following results in cases 1a and 1b. For 1a: $detM \neq 0$, the system has exactly one solution. For 1b: $detM = 0$, Cramer's formulas cannot be used to solve the system in this way. The question is: how, on the basis of properties of double precision floating point arithmetic, can one explain the difference in the answers obtained for linear systems in 1a and 1b?

In task 2, two limits of real function were presented: in 2a - the limit from Example 3 ("pathological") and in 2b - the limit from Example 4 ("non - pathological") with the following comment. When solving these limits symbolically we get an infinity in each case. But when solving them in double precision for example using Maxima, we obtain the following results: a finite number in case 2a and infinity in case 2b. The question is:

how, based on properties of double precision floating point arithmetic, can one explain the difference in the answers obtained in 2a and 2b?

For tasks 1 and 2, we gave a hint: consider in double precision arithmetic the representations of the following numbers: $0.3, 3 * 0.1$ and $0.2, 2 * 0.1$. Since many students provided quite extensive descriptive explanations, we adopted a two-stage grading scale: positive (if the explanation was sufficient) and negative (if the explanation was not sufficient). We got the following test results. In group A - task 1: 30% positive, 70% negative; task 2: 39% positive, 61% negative. In group B, the results were identical in both tasks: 61% positive, 39% negative. We received a higher percentage of positive answers to both test questions in group B in which the test was preceded by a presentation.

12 Conclusions

In this paper we presented some simple "pathological" and "non-pathological" examples for comparison, using Mathematica and wxMaxima with double precision calculation. Simple systems of linear equations and limits were considered. We used standard CAS functions for solve the systems of linear equations and to calculate limits and determinants. Comparison of results for analogous, "pathological" and "non-pathological" examples may seem surprising. In the second part of the paper, by analyzing double precision representations of used in examples numbers, the authors explain why the such final results were obtained. We showed that in double precision: $0.1 + 0.1 - 0.2 = 0$, $0.1 + 0.1 + 0.1 - 0.3 = 2^{-54}$ and explained briefly the way how to calculate that: $-0.3 + 0.1 + 0.1 + 0.1 = 2^{-55}$ and $0.3 - 0.1 - 0.1 - 0.1 = 2^{-55}$. The calculation could be done by hand but we think that it is good idea to implement own function in e.g. Mathematica and wxMaxima to perform this calculation. From the obtained results it follows the main reason of "pathological" character of the presented Examples 1 and 3. In our opinion these examples seems to be helpful to get the basic understanding of the nature of calculation in double precision generally an particularly in the case of used CAS programs. The examples are simple enough to do the calculations by hand but the symbolic calculations behind them are not easy for beginner students in our opinion. We did not find any analogical examples in available literature, taking into account the details we provide. Especially we could not find the calculation of **absolute error** and **relative error** of approximations: 0.1 by $\mathrm{fl}(0.1)$, 0.2 by $\mathrm{fl}(0.2)$, 0.3 by $\mathrm{fl}(0.3)$, 0.2 by $\mathrm{fl}(0.1+0.1)$, 0.3 by $\mathrm{fl}(0.1+0.1+0,1)$, 0.2 by $\mathrm{fl}(0.3) - \mathrm{fl}(0.1)$, 0.1 by $\mathrm{fl}(0.3) - \mathrm{fl}(0.1) - \mathrm{fl}(0.1)$ and comparison it to $(1/2)\varepsilon_{\mathrm{mach}}$. We would recommend discussing similar numeric examples in teaching the FP within some subjects in the framework of academic courses. The Examples 1–4 were prepared for students of Informatics Faculty of Warsaw University of Life Sciences. We presented in this article the results of a didactic experiment: students in two independent groups answered test questions related to Examples 1–4. In one of the two student groups, the test was preceded by a presentation discussing numerical representations in double precision arithmetic (Sects. 6, 7, 8, 9 and 10 of this article). We received a higher percentage of positive answers to both test questions in group in which the test was preceded by a presentation. It is worth emphasizing the multidisciplinary nature of the presented didactic approach - combining a theoretical problem in the field of numerical analysis with the use of practical CAS computing tools.

References

1. The GNU MPFR (multiple-precision floating-point computations) Library. https://www.mpfr.org
2. IEEE 754 Floating Point Standard. https://en.wikipedia.org/wiki/IEEE_754
3. Arnold, J.: An Introduction to Floating-Point Arithmetic and Computation, CERN OpenLab, 9 May 2017. https://indico.cern.ch/event/626147/attachments/1456066/2247140/FloatingPoint.Handout.pdf
4. Goldberg, D.: What every computer scientist should know about floating point arithmetic. ACM Comput. Surv. **23**, 5–48 (1991)
5. Muller, J.-M., et al.: Handbook of Floating-Point Arithmetic. Birkhäuser, Boston (2010)
6. Sauer, T.: Numerical Analysis, 3rd edn. Pearson (2017)
7. Stallings, W.: Computer Organization and Architecture, 6th edn. Prentice Hall, Upper Saddle River (2003)

Code Semantics Learning with Deep Neural Networks: An AI-Based Approach for Programming Education

Md. Mostafizer Rahman[1,2](✉), Yutaka Watanobe[2], Paweł Szmeja[3], Piotr Sowiński[3,4], Marcin Paprzycki[3], and Maria Ganzha[4]

[1] Dhaka University of Engineering & Technology, Gazipur, Bangladesh
[2] The University of Aizu, Aizuwakamatsu, Japan
mostafiz26@gmail.com,mostafiz@duet.ac.bd, yutaka@u-aizu.ac.jp
[3] Systems Research Institute, Polish Academy of Sciences, Warsaw, Poland
pawel.szmeja@ibspan.waw.pl, piotr.sowinski@ibspan.waw.pl, marcin.paprzycki@ibspan.waw.pl
[4] Warsaw University of Technology, Warsaw, Poland
maria.ganzha@ibspan.waw.pl

Abstract. Modern programming languages are very complex, diverse, and non-uniform in their structure, code composition, and syntax. Therefore, it is a difficult task for computer science students to retrieve relevant code snippets from large code repositories, according to their programming course requirements. To solve this problem, an AI-based approach is proposed, for students to better understand and learn code semantics, with solutions for real-world coding exercises. First, a large number of solutions are collected from a course titled "Algorithms and Data Structures" and preprocessed, by removing unnecessary elements. Second, the solution code is converted into a sequence of words and tokenized. Third, the sequence of tokens is used to train and validate the model, through a word embedding layer. Finally, the model is used for the relevant code retrieval and classification task, for the students. In this study, a bidirectional long short-term memory neural network (BiLSTM) is used as the core deep neural network model. For the experiment, approximately 120,000 real-world solutions from three datasets are used. The trained model achieved an average precision, recall, F1 score, and accuracy of 94.35%, 94.71%, 94.45%, and 95.97% for the code classification task, respectively. These results show that the proposed approach has potential for use in programming education.

Keywords: Code Semantics · Deep Learning · Programming Education · Software Engineering · Natural Language Processing

1 Introduction

Nowadays, the importance of programming is increasing due to the information and communication technology (ICT) needs of modern society. Thus, higher edu-

J. Mikyška et al. (Eds.): ICCS 2023, LNCS 14077, pp. 737–750, 2023.
https://doi.org/10.1007/978-3-031-36030-5_57

cational institutions are placing more emphasis on improving students' programming skills [13,15]. In addition, there is a relationship between programming and other academic courses. For instance, a data-driven analysis found that better programming skills have a positive impact on overall academic performance [15]. Therefore, it is important to provide more meaningful and effective support for the programming learning process. In this case, an AI model can provide more dynamic support, such as code searching, relevant code suggestions, algorithm identification, and classification. In particular, code classification is considered one of the main foundations of many of coding-related activities. However, classifying a large number of code fragments[1], in a non-automated manner, is a formidable task. Considering this problem, the aim of this work is to establish if machine learning models can be effectively used to understand the meaning of code and become the basis for supporting coding-related activities.

In recent years, many academic and industrial research experiments have been conducted to facilitate programming tasks. For example, identifying the location of errors in code [5,18], error detection, code refactoring [10,16], code evaluation and repair [9], etc. In addition, deep neural network (DNN) models are effectively used for coding tasks [2]. Considering the need to handle lines of programming code, recurrent neural networks (RNNs) are widely used to develop models for programming tasks. RNN models include long short-term memory (LSTM) and bi-directional long short-term memory (BiLSTM), and their variants are also commonly used in this context. In [21], structural features of the solution code were used to identify the algorithm in the code, using the convolutional neural network (CNNs) model. Also, binary code classification, code completion, code repair, and code evaluation are performed using different BiLSTM and LSTM models [8,9].

Due to its unidirectional processing of input sequences [3], the LSTM model performed worse than the BiLSTM [19] model in coding tasks [14]. On the other hand, the BiLSTM model can process input sequences in both directions (forward and backward). Since variables, functions, and classes may depend on past or future lines of code, the BiLSTM model is more effective in such cases. The aim of this contribution is to build an AI engine, using BiLSTM neural networks, to better understand the semantics of code, and to generate better code-related services. For this purpose, real-world program code fragments have been collected from an "Algorithms and Data Structures" (ADS) programming course. Next, data preprocessing, word tokenization, word embedding, model training, validation, and evaluation were performed. The experimental results show that the model achieves significant success in code classification. Moreover, the trained model can be integrated with an AI-enabled web client, in a programming learning platform, to provide students more with convenient coding-related services. Hence, this work delivers the following key contributions:

- An AI engine based on BiLSTM neural networks was trained on a large dataset of real-world solution code collected from a programming course. The engine was used to build an AI-based framework for learning code semantics.

[1] The terms code, source code, solution code, and program code are used to denote a similar meaning.

Experimental results show that the trained model obtained very good results in classifying code fragments.
- The integration scheme of the proposed approach with a programming learning platform is investigated.
- The proposed code semantics learning model can be useful for teachers to identify students' deficiencies in programming, and these problems can be discussed in class.

2 Related Work

Programming techniques and programming environments have evolved significantly over the past few decades, and intelligent coding platforms such as Crescendo [20], scratch [17], and CloudCoder [4] have been used to alleviate selected basic challenges faced by programmers while coding. However, these platforms have had limited success when programmers attempted to solve highly structured problems, using C, C++, and Python [20]. On the other hand, programming support platforms are also being developed for learning programming, with many of them focused on only one programming language [12,24,25]. Abe et al. proposed a programming environment dedicated to the C language, with basic functions such as variable declarations, expressions, and statements [1]. Similarly, Nguyen and Chua [11] proposed a web-based interface for PHP programming. In their study, rules for various expressions, such as for-loop, while-loop, and equality conditions, are developed. Our proposed approach for learning code semantics using the BiLSTM model is different from the existing methods because the proposed AI model is designed to understand the semantics of multilingual (e.g., C, C++, and Python) code. Therefore, the proposed AI model can be more effective within a programming learning platform, to assist students in coding. Moreover, the proposed AI model was trained with diverse, real-world multilingual solution code fragments, in order to provide support for many programming languages.

3 Motivation

In a recent study [15], a data-driven analysis was conducted, based on evaluation logs of solution code in a programming learning platform. The evaluation logs were collected from the works of 357 students in the ADS course. The results showed that about 37.27% of all submissions were accepted, while the remaining 62.73% were rated as incorrect. It was also found that the error verdicts were categorized into 5 main groups, i.e., Wrong Answer (WA), Compile Error (CE), Presentation Error (PE), Runtime Error (RE), and Resource Limit Error (RLE). RLE includes Time Limit Exceedance (TLE), Memory Limit Exceedance (MLE), and Output Limit Exceedance (OLE). Figure 1 shows the detailed breakdown of errors. Here, approximately 44% were WA, 21.25% were CE, 13% were RE, 10.5% were PE, and 11.25% were RLE.

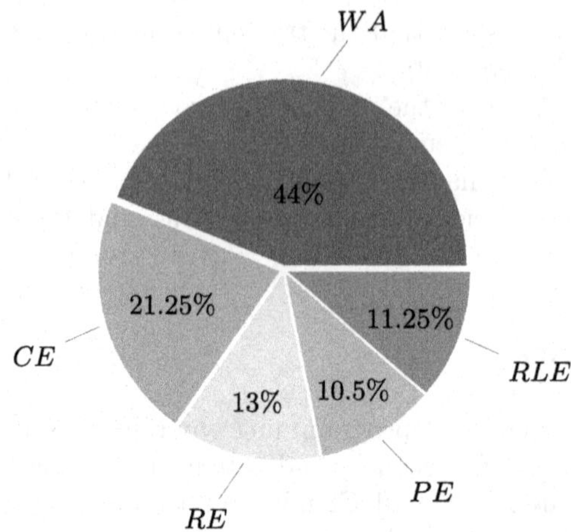

Fig. 1. Breakdown of errors, based on solution code evaluation.

In general, CE errors were caused by syntax errors in the solution code, while non-CE errors were caused by logical errors and inefficiency in the code. Figure 1 shows that a large percentage of submitted solutions received error decisions for the above reasons. To address this important problem in programming learning, and to help students improve their programming skills by providing coding-related services (e.g., code refactoring, code suggestions, code search, and classification), a machine learning model has been developed.

4 Proposed Approach

Figure 2 presents the outline of the proposed framework for AI-based code semantics learning. In the first stage of the process, the solution code fragments of the ADS course are collected from the Aizu Online Judge (AOJ) platform [22,23]. AOJ is a globally-recognized learning and competition platform for programming. In the second stage, the code is preprocessed by removing irrelevant elements such as comments, whitespaces, and tabs[2].

Next, in the tokenization step, the preprocessed code fragments are converted into a sequence of words, and each word is encoded with a unique integer number. Let $I = \{i_1, i_2, i_3, \cdots, i_l\}$ be the sequence of words of a code fragment, and $K = \{k_1, k_2, k_3, \cdots, k_l\}$ be the corresponding integer IDs for words. The overall process of code tokenization is shown in Fig. 3.

Each tokenized word sequence is converted into a vector of real numbers through the embedding layer. The embedding matrix is $\boldsymbol{S} \in \mathbb{R}^{l \times d}$, where l is the token dictionary size and d is the embedding size. For this study, the dimensions of the embedding matrix are $\boldsymbol{S} \in \mathbb{R}^{10000 \times 200}$. The vectorized data is passed to the BiLSTM neural network [19]. The bidirectional information processing

[2] Which elements are irrelevant depends on the programming language.

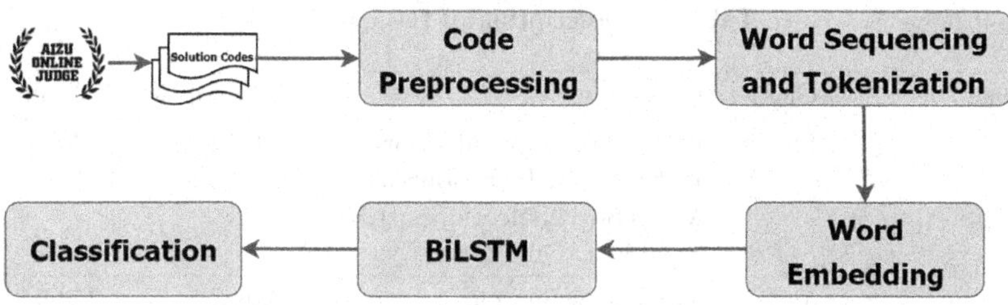

Fig. 2. AI-based code semantics learning framework

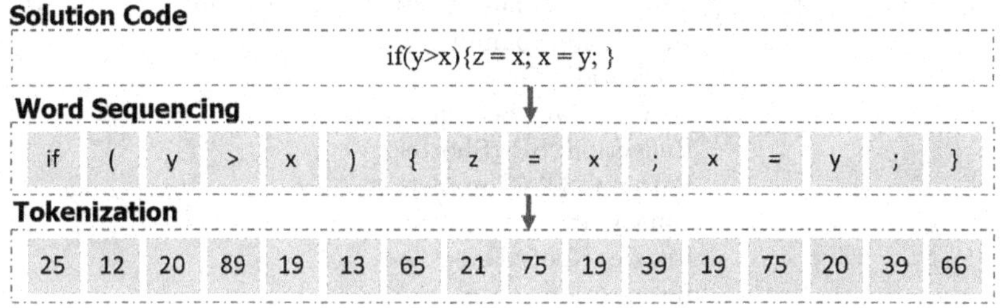

Fig. 3. Code tokenization process for the BiLSTM model training

nature of the BiLSTM model enables it to understand the complex context of the long dependent information, i.e., solution code fragments. Here, the final code classification is performed in the output layer.

Figure 4 presents the integration of the proposed AI Engine built with BiLSTM, the trained AI model, and the web client in the programming learning platform. To provide the programming services, the web client connects the AI Engine and the trained AI model.

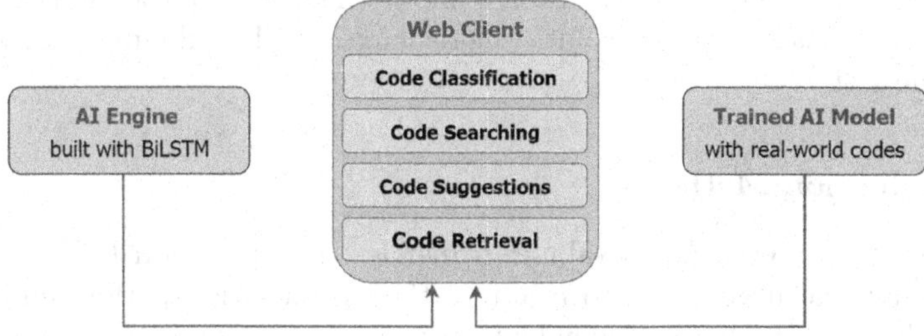

Fig. 4. AI Engine-enabled platform for programming learning

Table 1. Description of the datasets

Dataset	Classes	Class names	Avg. sol. len.	# solutions
Sorting	8	CountingSort, StableSort, BubbleSort, InsertionSort, MergeSort, SelectionSort, ShellSort, QuickSort	840.44	53,308
Searching	5	ExhaustiveSearch, StringSearch, LinearSearch, BinarySearch, PatternSearch	650.22	25,994
Graphs and trees	14	TreeWalk, BinaryTrees, MST (Minimum Spanning Tree), Projection, Graph, Area, BST (Binary Search Tree), Reflection, SP (Shortest Path), CBT (Complete Binary Tree), Convex, Puzzle, RootedTrees, Intersection	1658.46	38,761

5 Experimental Results

5.1 Datasets

For the experiments, all correct solutions from the AOJ platform, of the ADS course[3] were collected. There are 15 topics in this course and each topic involves solving 3 or 4 programming problems. The problems include various algorithms (e.g., sorting, searching, graphs, and trees), elementary data structures (e.g., stack, queue, and linked list), and numerical computations. For the experiments, three datasets were built, that group the problems thematically into: sorting, searching, and graphs and trees. Table 1 shows the details of the datasets including number of classes, average solution code length, and total number of solutions in each dataset.

5.2 Evaluation Methods

In this paper, we used four evaluation metrics, i.e., precision ($\mathcal{P}$), recall ($\mathcal{R}$), F1 score, and accuracy ($\mathcal{A}$), to evaluate the classification performance of the model. These methods are dependent on the false positive (fp), true positive (tp), false negative (fn), and true negative (tn) rates. A detailed description of these performance evaluation methods can be found in [14].

[3] https://onlinejudge.u-aizu.ac.jp/courses/lesson/1/ALDS1/1.

5.3 Implementation Details

When implementing the BiLSTM classification model, several different sets of hyperparameters were tried, by manual fine-tuning, to achieve better results. In this paper, the reported number of training epochs is 50, the batch size is 32, the maximum code sequence length is 500, the learning rate is $\alpha = \{0.01, 0.005, 0.001\}$, the number of BiLSTM nodes is 400 ($\overrightarrow{200} + \overleftarrow{200}$), the number of nodes in the dense layer is 200, the training data ratio is 80%, the validation data ratio is 13%, the test data ratio is 7%, the activation function for the dense layer is ReLU [6] and the optimization function is Adam [7]. In addition, sparse categorical cross-entropy is used as the loss function ($\mathcal{L}$) for the multiclass classification model [19]. These hyperparameters resulted in best overall performance. However, the aim of this work was *not* to exhaustively search the hyperparameter space. Hence, no claim is made that this set of hyperparameters is optimal. All experimental tasks are computed in Google Colab using the Keras+TensorFlow (2.11.x) framework and Python 3.

5.4 Results

Figure 5 shows the accuracy of BiLSTM model training and validation, on all three datasets (e.g., sorting, searching, graphs and trees) when $\alpha = 0.001$. Figure 6, on the other hand, shows the model training and validation losses for these datasets. From Figs. 5 and 6, the following observations can be made: (*i*) the model achieved the highest training and validation accuracy for the sorting dataset and the lowest for the graph and tree dataset, (*ii*) the validation accuracy curve in the searching dataset shows the inconsistency during validation, and (*iii*) the model has the highest validation losses for the searching dataset.

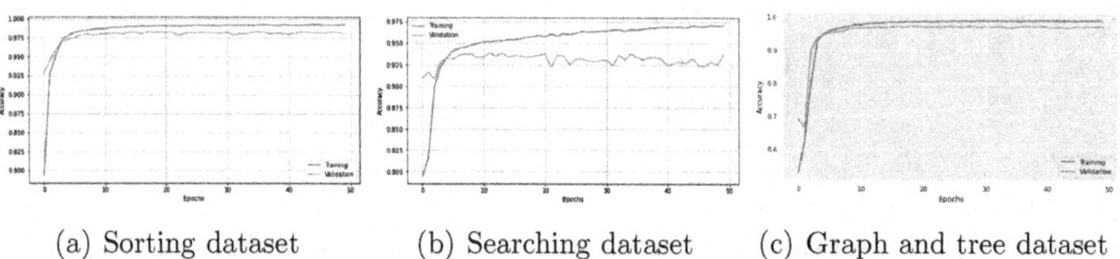

(a) Sorting dataset (b) Searching dataset (c) Graph and tree dataset

Fig. 5. Comparison of training and validation accuracy of BiLSTM models on three datasets

Tables 2, 3, and 4 show the class-wise $\mathcal{P}$, $\mathcal{R}$, and $F1$ values for three datasets. The following observations can be made from these tables: (*i*) the model achieved higher $\mathcal{P}$, $\mathcal{R}$, and $F1$ values between 96% and 99% for most classes in the sorting dataset, (*ii*) the model achieved relatively lower $\mathcal{P}$, $\mathcal{R}$, and $F1$ values for the *ExhaustiveSearch*, *BinarySearch*, and *LinearSearch* classes in the searching dataset, (*iii*) in contrast, the model obtained very low $\mathcal{P}$, $\mathcal{R}$, and $F1$ values for

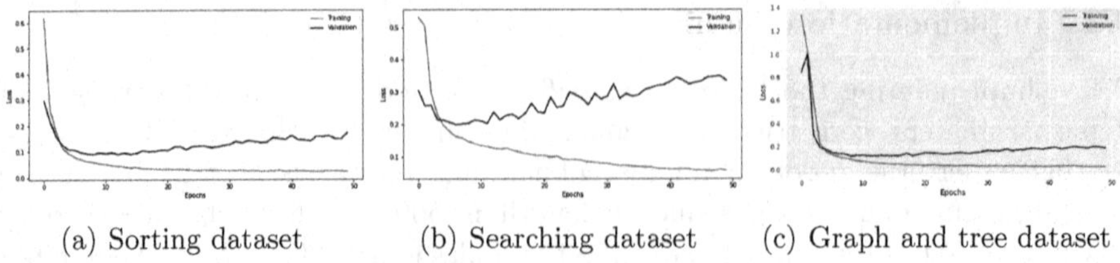

(a) Sorting dataset (b) Searching dataset (c) Graph and tree dataset

Fig. 6. Comparison of training and validation losses of BiLSTM models on three datasets

Table 2. Class-wise $\mathcal{P}$, $\mathcal{R}$, and $F1$ for the sorting dataset

Classes	$\mathcal{P}$	$\mathcal{R}$	$F1$
CountingSort	0.98	0.96	0.97
StableSort	0.99	0.98	0.98
BubbleSort	0.97	0.99	0.98
InsertionSort	1.00	0.99	0.99
MergeSort	0.98	0.99	0.99
SelectionSort	0.97	0.96	0.96
ShellSort	0.99	0.99	0.99
QuickSort	0.99	0.98	0.99

Table 3. Class-wise $\mathcal{P}$, $\mathcal{R}$, and $F1$ for the searching dataset

Classes	$\mathcal{P}$	$\mathcal{R}$	$F1$
ExhaustiveSearch	0.91	0.91	0.91
StringSearch	0.99	0.99	0.99
LinearSearch	0.90	0.91	0.90
BinarySearch	0.86	0.92	0.89
PatternSearch	0.99	0.99	0.99

many classes in the graph and tree dataset. This is due to the greater heterogeneity, complexity, classes, and code length in the graph and tree dataset.

Table 5 shows the average $\mathcal{P}$, $\mathcal{R}$, and $F1$ values for all datasets. Experimental results show that the BiLSTM model achieved $\mathcal{P}$, $\mathcal{R}$, and $F1$ of 98.37%, 98.04%, and 98.20%, respectively, for the sorting dataset. This is due to the lower diversity and higher similarity of the solutions for the sorting algorithms. In contrast, the model achieved relatively low $\mathcal{P}$, $\mathcal{R}$, and $F1$ of 91.56%, 91.77%, and 91.45% for the graph and tree datasets, respectively.

Table 4. Class-wise $\mathcal{P}$, $\mathcal{R}$, and $F1$ for graph and tree dataset

Classes	$\mathcal{P}$	$\mathcal{R}$	$F1$
TreeWalk	0.77	0.82	0.80
BinaryTrees	0.96	0.99	0.97
MST	0.98	0.98	0.98
Projection	0.99	0.99	0.99
Graph	0.93	0.95	0.94
Area	0.99	0.98	0.98
BST	0.89	0.83	0.86
Reflection	0.99	0.94	0.96
SP	0.87	0.71	0.78
CBT	1.00	0.99	1.00
Convex	0.50	0.71	0.59
Puzzle	0.98	0.98	0.98
RootedTrees	0.98	0.99	0.98
Intersection	1.00	0.98	0.99

Table 5. Average $\mathcal{P}$, $\mathcal{R}$, and $F1$ values for all datasets

Dataset	$\mathcal{P}$(%)	$\mathcal{R}$(%)	$F1$(%)
Sorting	98.37	98.04	98.20
Searching	93.13	94.32	93.70
Graph and Tree	91.56	91.77	91.45

A comparison of $F1$ and $\mathcal{A}$ values is shown in Fig. 7. It can be seen that (i) the $F1$ and $\mathcal{A}$ values for the datasets sorting and searching are mostly similar, and (ii) the difference between $F1$ and $\mathcal{A}$ value for the dataset graph and tree is comparatively high. This is due to the unbalanced instances in the classes of this dataset.

Separately, Figs. 8, 9, and 10 show the confusion matrices for these three datasets, which also demonstrate the effectiveness of the BiLSTM model for classification tasks. These higher values of $F1$ and $\mathcal{A}$ ensure that the model better understands the semantics of the code.

In addition, further experiments have been conducted with different learning rates i.e. $\alpha = \{0.01, 0.005, 0.001\}$, as shown in Table 6. It can be seen that the BiLSTM model achieves higher $\mathcal{P}$, $\mathcal{R}$, $F1$ and $\mathcal{A}$ values when $\alpha = 0.001$ is used for all datasets. It also indicates that, as could be expected, the model learns the semantics of the code better, when the learning rate becomes slower.

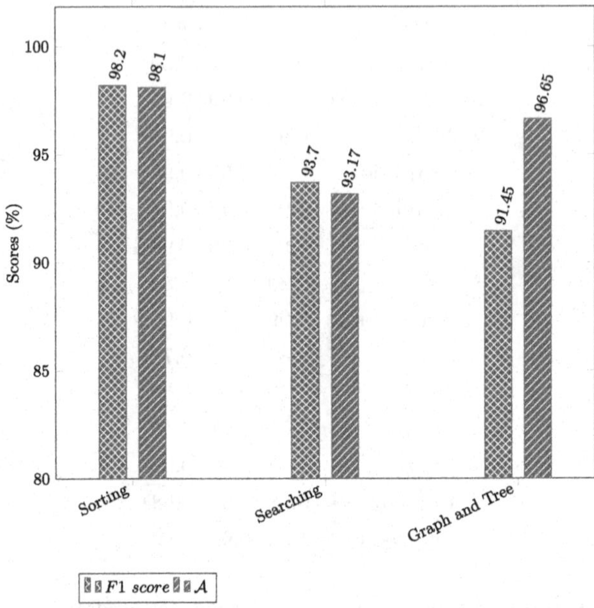

Fig. 7. Comparison of $F1$ and $\mathcal{A}$ values for all datasets

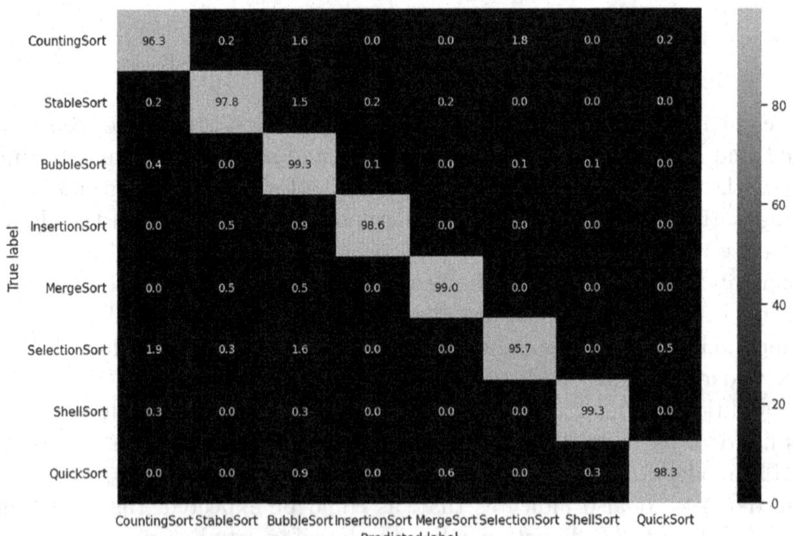

Fig. 8. Confusion matrix for Sorting dataset

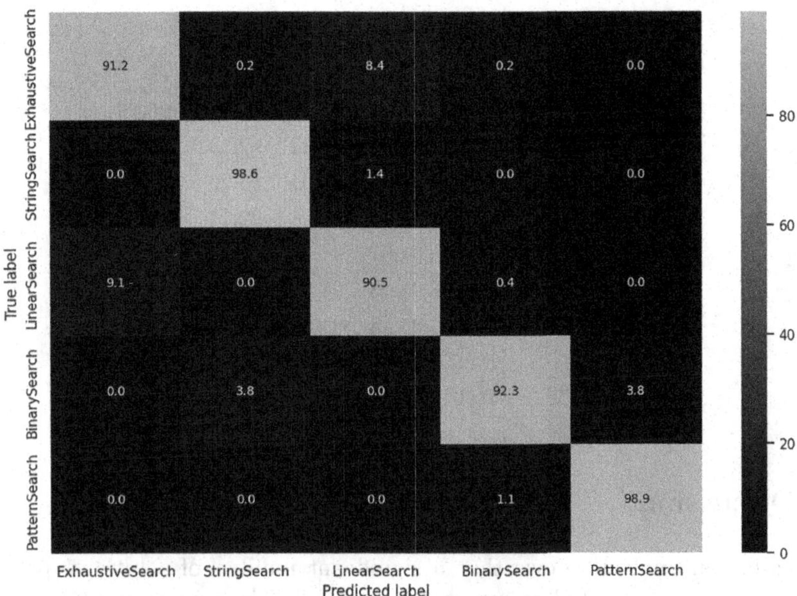

Fig. 9. Confusion matrix for Searching dataset

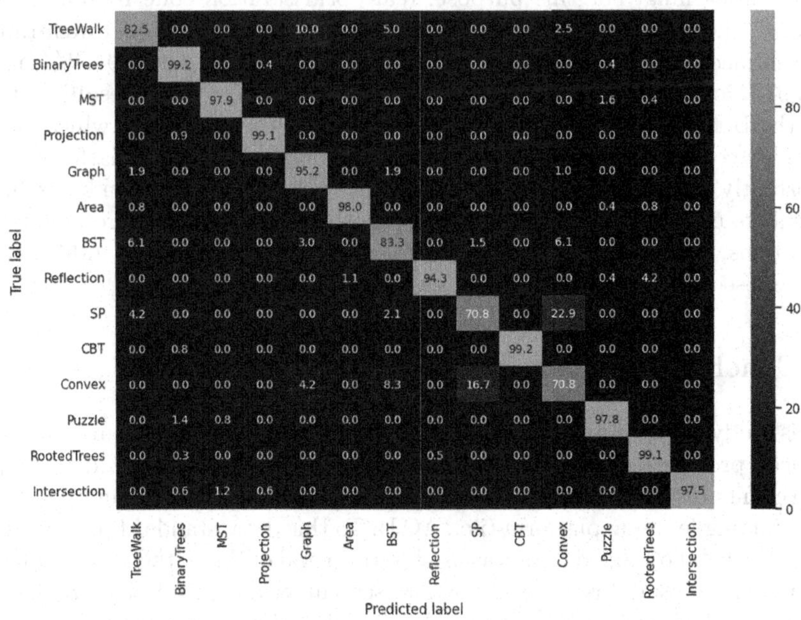

Fig. 10. Confusion matrix for Graph and Tree dataset

Table 6. Average $\mathcal{P}$, $\mathcal{R}$, $F1$, and $\mathcal{A}$ values for all datasets when $\alpha = \{0.01, 0.005, 0.001\}$

Dataset	α	$\mathcal{P}$(%)	$\mathcal{R}$(%)	$F1$(%)	$\mathcal{A}$
Sorting	0.010	96.49	96.09	96.26	95.79
	0.005	97.77	97.32	97.53	97.26
	0.001	98.37	98.04	98.20	98.10
Searching	0.010	92.54	91.98	92.14	92.08
	0.005	94.51	94.19	94.33	92.19
	0.001	93.13	94.32	93.70	93.17
Graph and Tree	0.010	85.14	83.83	83.86	93.99
	0.005	85.40	86.58	85.72	95.68
	0.001	91.56	91.77	91.45	96.65

6 Discussion

A data-driven analysis shows that a significant number of solutions, provided by students, are evaluated as incorrect, i.e., 62.73% in a programming learning platform [15]. We believe that AI-based support can be useful for students in solving problems, and can reduce the number of wrong answers. In this paper, we investigated the performance of the BiLSTM neural network model for code semantics learning. For this purpose, real-world solution code fragments were collected, the code was preprocessed and tokenized, and the model was trained. In performed experiments, the model achieved an average $F1$ of 94.45% and $\mathcal{A}$ of 95.97% for all three datasets (see Table 5 and Fig. 7). These results indicate that the BiLSTM model learned the semantics of the code well, which can be useful for coding-related services in the programming learning platform. More importantly, AI Engine enabled programming learning platform can be useful for students to understand the code errors, algorithms, and syntax, by receiving code suggestions, refactoring, and classification services during coding. Additionally, the proposed approach can support multilingual coding tasks.

7 Concluding Remarks

In this study, code semantics learning approach using BiLSTM neural networks has been proposed. A substantial number of real-world solutions, delivered by the students of the University of Aizu, have been collected from an operational programming learning platform (i.e., AOJ). To this data, standard preprocessing was applied and obtained data was used to train and evaluate the BiLSTM model for three different datasets (e.g., sorting, searching, and graph and tree), while applying various hyperparameters. The model achieved an average $\mathcal{P}$, $\mathcal{R}$, $F1$, and $\mathcal{A}$ values of 94.35%, 94.71%, 94.45%, and 95.97%, respectively, in classifying solutions. These results also indicate that the BiLSTM model understands code semantics with a high degree of accuracy. Moreover, the proposed AI model can be useful in the programming learning platform to provide students with various programming-related services such as code suggestion, code refactoring, classification, etc. Future work will consider improving the model performance with more diverse datasets.

Acknowledgment. This research was financially supported by the Japan Society for the Promotion of Science (JSPS) KAKENHI (Grant Number 19K12252).

References

1. Abe, K., Fukawa, Y., Tanaka, T.: Prototype of visual programming environment for c language novice programmer. In: 2019 8th International Congress on Advanced Applied Informatics (IIAI-AAI), pp. 140–145 (2019). https://doi.org/10.1109/IIAI-AAI.2019.00037
2. Dam, K.H., Tran, T., Pham, T.: A deep language model for software code. ArXiv abs/ arXiv: 1608.02715 (2016)
3. Hochreiter, S., Schmidhuber, J.: Long short-term memory. Neural Comput. **9**(8), 1735–1780 (1997). https://doi.org/10.1162/neco.1997.9.8.1735
4. Hovemeyer, D., Spacco, J.: Cloudcoder: A web-based programming exercise system. J. Comput. Sci. Coll. **28**(3), 30 (2013)
5. Huo, X., Li, M., Zhou, Z.H.: Learning unified features from natural and programming languages for locating buggy source code. In: Proceedings of the Twenty-Fifth International Joint Conference on Artificial Intelligence, IJCAI 2016. pp. 1606–1612. AAAI Press (2016)
6. Javid, A.M., Das, S., Skoglund, M., Chatterjee, S.: A relu dense layer to improve the performance of neural networks. In: ICASSP 2021–2021 IEEE International Conference on Acoustics, Speech and Signal Processing (ICASSP), pp. 2810–2814. IEEE (2021)
7. Kingma, D.P., Ba, J.: Adam: A method for stochastic optimization. In: 3rd International Conference on Learning Representations, ICLR 2015, San Diego, CA, USA, 7–9 May 2015, Conference Track Proceedings (2015). http://arxiv.org/abs/1412.6980
8. M. Mostafizer, R., Watanobe, Y., Nakamura, K.: A neural network based intelligent support model for program code completion. Scient. Program. **2020** (2020). https://doi.org/10.1155/2020/7426461
9. M. Mostafizer, R., Watanobe, Y., Nakamura, K.: A bidirectional lstm language model for code evaluation and repair. Symmetry **13**(2) (2021). https://doi.org/10.3390/sym13020247
10. Meng, N., Hua, L., Kim, M., McKinley, K.S.: Does automated refactoring obviate systematic editing? In: 2015 IEEE/ACM 37th IEEE International Conference on Software Engineering, vol. 1, pp. 392–402 (2015). https://doi.org/10.1109/ICSE.2015.58
11. Nguyen, T., Chua, C.: A logical error detector for novice php programmers. In: 2014 IEEE Symposium on Visual Languages and Human-Centric Computing (VL/HCC), pp. 215–216 (2014). https://doi.org/10.1109/VLHCC.2014.6883062
12. Price, T.W., Dong, Y., Lipovac, D.: Isnap: Towards intelligent tutoring in novice programming environments. In: Proceedings of the 2017 ACM SIGCSE Technical Symposium on Computer Science Education, SIGCSE 2017, pp. 483–488.

Association for Computing Machinery, New York (2017). https://doi.org/10.1145/3017680.3017762
13. Rahman, M.M.: Data analysis and code assessment using machine learning techniques for programming activities (2022). https://doi.org/10.15016/00000215
14. Rahman, M.M., Watanobe, Y., Kiran, R.U., Kabir, R.: A stacked bidirectional lstm model for classifying source codes built in mpls. In: Machine Learning and Principles and Practice of Knowledge Discovery in Databases, pp. 75–89. Springer International Publishing, Cham (2021). https://doi.org/10.1007/978-3-030-93733-1_5
15. Rahman, M.M., Watanobe, Y., Kiran, R.U., Thang, T.C., Paik, I.: Impact of practical skills on academic performance: A data-driven analysis. IEEE Access **9**, 139975–139993 (2021). https://doi.org/10.1109/ACCESS.2021.3119145
16. Raychev, V., Schäfer, M., Sridharan, M., Vechev, M.: Refactoring with synthesis. SIGPLAN Not. **48**(10), 339–354 (2013). https://doi.org/10.1145/2544173.2509544
17. Resnick, M., et al.: Scratch: Programming for all. Commun. ACM **52**(11), 60–67 (2009). https://doi.org/10.1145/1592761.1592779
18. Saha, R.K., Lawall, J., Khurshid, S., Perry, D.E.: On the effectiveness of information retrieval based bug localization for c programs. In: 2014 IEEE International Conference on Software Maintenance and Evolution, pp. 161–170 (2014). https://doi.org/10.1109/ICSME.2014.38
19. Schuster, M., Paliwal, K.: Bidirectional recurrent neural networks. IEEE Trans. Signal Process. **45**(11), 2673–2681 (1997). https://doi.org/10.1109/78.650093
20. Wang, W., Zhi, R., Milliken, A., Lytle, N., Price, T.W.: Crescendo: Engaging students to self-paced programming practices. In: Proceedings of the 51st ACM Technical Symposium on Computer Science Education, SIGCSE 2020, pp. 859–865. Association for Computing Machinery, New York (2020). https://doi.org/10.1145/3328778.3366919
21. Watanobe, Y., Rahman, M.M., Kabir, R., Amin, M.F.I.: Identifying algorithm in program code based on structural features using cnn classification model. Appli. Intell. (2022)
22. Watanobe, Y.: Aizu online judge (2018). https://onlinejudge.u-aizu.ac.jp/
23. Watanobe, Y., Rahman, M.M., Matsumoto, T., Rage, U.K., Ravikumar, P.: Online judge system: Requirements, architecture, and experiences. Int. J. Software Eng. Knowl. Eng. **32**(06), 917–946 (2022). https://doi.org/10.1142/S0218194022500346
24. Watanobe, Y., Rahman, M.M., Vazhenin, A., Suzuki, J.: Adaptive user interface for smart programming exercise. In: 2021 IEEE International Conference on Engineering, Technology & Education (TALE), pp. 01–07 (2021). https://doi.org/10.1109/TALE52509.2021.9678757
25. Wiggins, J.B., et al.: Exploring novice programmers' hint requests in an intelligent block-based coding environment. In: Proceedings of the 52nd ACM Technical Symposium on Computer Science Education, SIGCSE 2021, pp. 52–58. Association for Computing Machinery, New York (2021). https://doi.org/10.1145/3408877.3432538

A Framework for Effective Guided Mnemonic Journeys

David Iclanzan(✉) and Zoltán Kátai

Faculty of Technical and Human Sciences, Sapientia Hungarian University of Transylvania, Târgu-Mureş, Romania
{iclanzan,zoltan_katai}@ms.sapientia.ro

Abstract. The memory palace, also known as the memory journey, is a mnemonic technique where information to be remembered is encountered along a predetermined path through envisioned places, creating strong spatial and visual connections between the material and specific locations and vivid images. Constructing a memory palace for complex computer science concepts may prove challenging for students, as it requires the identification, selection, and organization of essential ideas from the material. This task is better suited to an expert on the curriculum, such as the instructor.

In this paper, a framework for designing and delivering Guided Mnemonic Journeys is proposed. Led by a teacher in person or via audio or video, the approach uses a virtual tour of the university campus at its core. The approach combines various mnemonic techniques to create a more comprehensive approach to memory enhancement. The instructor plans the story arc and places various mnemonic cues along the path and then guides the students through the narrative and imagery, providing context, clear structure, and instructions for each step.

A pilot study indicated that students perceived the delivered Guided Mnemonic Journeys positively, appreciating the rich didactic activity and the opportunity to learn new mnemonic techniques.

Keywords: Mnemonics · Memory Palace · Virtual Tour

1 Introduction

According to the revised Bloom's Taxonomy the verb associated with the first level of learning is remembering [13]. This explains why improving memory performance is so important. Numerous techniques have been suggested to enhance memorization, among which mnemonics stands out as one of the most popular [31]. Mnemonic strategies support both information encoding and retrieval. One of the most attractive ancient mnemonic devices is the method of loci (MOL),

This study was partially supported by the Sapientia Foundation Institute for Scientific Research.

J. Mikyška et al. (Eds.): ICCS 2023, LNCS 14077, pp. 751–765, 2023.
https://doi.org/10.1007/978-3-031-36030-5_58

also called the memory palace (MP) technique [33]. The effective implementation of this method implies the invention of a familiar visuo-spatial mnemonic environment (familiarity criterion) and the repeated imaginary walks through it. During the "encoding walk" objects serving as landmarks are paired with to-be-remembered items. Information is retrieved sequentially by retracing the encoding walk (from landmark to landmark).

Of course, ancient Greek orators used the method of loci based on a completely mentally created environment (traditional MOL). However, in recent decades, as computer-supported education has gained more and more space, the concept of virtual MOL (vMOL) has been introduced [6,7,14]. Researchers in computer science (CS) and psychology have begun to explore the potential of virtual worlds to provide students with a template and support for their mental representation of the memorial palace. The findings seem to establish the prediction that virtual reality (VR) solutions will play a key role in the effective implementation of MOL in the future [11,23].

Most vMOL implementations are based on computer-generated MPs. To be familiar to most students, environments such as a common room or building were usually created. Obviously, these imagined environments can only partially satisfy the "familiarity criterion". In this paper, we present a VR-solution that provides a very familiar environment by immersing students in a virtual tour of their university campus. Unlike previous research, in which students navigated the virtual environment on their own, in the present study, we used a teacher-led guided journey. The adoption of this approach enabled a teacher-level implementation of several principles suggested by previous studies on the MOL technique [7]. Furthermore, several other mnemonic techniques were utilized to augment the students' MP-experience, including but not limited to, the story method, chunking, alliteration, acrostics, spelling, visual and association mnemonics, as reported in [1]. Increased teacher involvement is also justified by survey studies, indicating that while students may be aware of mnemonic techniques, they tend to opt for alternative study strategies instead [21,28,29]. Complex methods such as the MOL are the least popular.

2 Background

2.1 Method of Loci: A Mnemonic-Specific Technique to Improve Memory

"If nothing has altered in long-term memory, nothing has been learned" [30]. This powerful statement clearly emphasizes the key role of memory in learning. Consequently, cognitive psychology defines learning as an adjustment in long-term memory [3]. Although there are multiple ways to improve memory [20], the peculiarity of mnemonic techniques lies in their focus on enhancing the recall of information.

The term "mnemonics" is rooted in the Greek word mnēmonikos, which translates to "of memory" or "relating to memory". This word is associated with Mnemosyne, the Greek goddess of memory and remembrance in mythology [12].

The concept of mnemonics involves finding better methods for encoding information for easier retrieval and recall. Furthermore, [12] emphasize that mnemonic devices can be considered both as systematic memory enhancement techniques and as learning strategies that can improve the learning process and increase recall of information. They involve using methods such as elaborative encoding, retrieval cues, and imagery to encode information in a way that makes it easier to remember. Mnemonics connect new information to something more familiar or meaningful, resulting in improved retention.

The Method of Loci is an ancient mnemonic technique that involves linking information to be learned with specific locations. Accordingly, the MOL serves as the foundation for constructing a MP. To apply the MOL or build a MP, one must mentally place the information to be learned in familiar places, referred to as "loci" (place = lat. locus, pl. loci), which serve as landmarks for navigating the memory palace. A considerable body of research confirmed that this method of association aids in the memorization and retrieval of information [33]. For example, [17] examined ten superior memorizers (eight of them placed at the highest levels in the World Memory Championships). The interviews of the participants revealed that all of them used mnemonics during the learning phase, and nine out of the ten superior memorizers used the MOL technique for some or all of the tasks. On the other hand, previous reviews on the use of mnemonics in the classroom have reported inconsistent results [5,15,19]. In addition, Putman [24] emphasizes that importance of the right circumstances [31].

Hedman and Bäckström [7] collected nine key principles to effectively implement the MOL technique: place-like environments (information is organized into location-based settings), striking imagery (visual impression is generated to anchor information within loci), association (visual representation must be linked to the information or concept to be remembered), creative use of spatiality (the constructed mental locations do not have to mirror physical locations), architectural guidance (environment should be designed with a degree of architectural diversity), positioning of elements (information elements in the environment should be easily visible), perspicuity (it should be simply to obtain a sense of the location of items), calmness (create environments that enable deep concentration), grouping and order (systematic organization of the information). These principles can also be interpreted as a fusion of the fundamental MOL with other mnemonic techniques [1].

2.2 Brief Review of Previous Work on Virtual Memory Palaces

According to [32], creating and using a MP is challenging. It requires personal and familiar locations, extensive training, and a significant amount of cognitive effort and attention. To reduce cognitive demands and make the learning process easier, researchers have begun integrating virtual MPs (vMPs) into the mnemonic process. The vMP acts as a template for the traditional mental representation of a MP and helps ease the burden. Several studies have investigated various aspects of the MOL in virtual settings. A fundamental question examined was whether vMOLs could be more effective than traditional ones. Previous work has also revealed several elements that can contribute to more effective vMP implementations. In the following, we provide a concise overview of some relevant studies in this field.

Hedman and Bäckström [7] developed a virtual museum-like platform where loci were essentially text displayed on virtual walls. Although the study did not demonstrate any superiority of vMP over traditional learning methods, the authors also reported that the distance between rooms and loci could cause navigation difficulties for students.

The study by Legge et al. [14] consisted of two experimental groups, using traditional and virtual MOL respectively, and found that both groups outperformed the control group (which had no specific learning strategy), with no significant differences between the MOL groups. The peculiarity of this study was that the vMOL approach did not include loci, which required participants to invest more mental effort to generate the item-location pairings.

Jund et al. [11] conducted a study to evaluate the impact of two different frames of reference (egocentric vs. allocentric) on the recall performance of participants when using vMOL. They found that designing the loci as images of colorful textures and patterns can help avoid semantic bias in participants' memory processes.

Huttner et al. [9,10] investigated whether participants perform better with learning content that was presented virtually through visualized loci, or if the MOL is more effective when users have to imagine the content through text-only loci. The researchers noted that using visualized loci proved to be more effective for the memorization process and concluded that greater immersion in virtual environments leads to better recall.

Recently, Yang et al. [32] conducted an exploratory study to examine the use of VR techniques and mnemonic devices to aid in the retrieval of knowledge from scholarly articles (participants were asked to read, memorize, and recall abstracts of scientific publications). Three different conditions were analyzed: without a predefined strategy (control condition), image-based memory palace, and VR-based memory palace. The results showed that the use of a vMP significantly improved the amount of knowledge retrieved and retained compared to the control condition and showed moderate improvement over the image-based memory palace. These authors conclude that their findings support the usefulness of VR for certain cognitive tasks and offer insights for enhancing future VR and visualization applications.

The prior research mentioned above demonstrates the potential of the vMP technique to enhance memorization. Most of the research investigated how the method helps to recall lists of items, which is a relevantly elementary task [32]. In the realm of CS education, a pertinent utilization of the vMP approach would undoubtedly involve more than just aiding in the recollection of a word list. Interestingly, recently Robins et al. [25] noted that they have not found any particular investigation of study skills or the use of mnemonic techniques in computing education. On the other hand, the findings of the study conducted by Yang et al.[32] are promising, as they effectively used the vMP technique for more complex tasks, also involving comprehension and retention.

Furthermore, the studies examined have revealed some possible difficulties and suggested effective implementations of vMP. We anticipate that the proposed increase in teacher involvement will help alleviate the identified deficiencies and achieve significant improvements. For example, navigation difficulties such as those indicated by [7] can be avoided by using teacher-led guided journeys. If the teacher is responsible for selecting the locations and planning the item-locus pairings, it can eliminate the additional mental effort required by the student (noticed by [14]) and prevent the occurrence of the phenomenon of semantic bias in the memory process, as brought to attention by Jund et al. [11]. Obviously, such an implication of the teacher can support higher-level immersion in virtual environments, as highlighted by Huttner et al. [9,10].

3 Guided Mnemonic Journeys

The Guided Mnemonic Journey (GMJ) is built upon and emerges from a meticulously planned [7] virtual tour of a familiar environment. The virtual environment is purposefully customized with strategically placed embedded virtual cues and actionable hotspots that serve as contextual anchors for mnemonic exercises and information item-locus pairings. An overarching narrative is woven into the journey, serving as a unifying thread that binds together the various mnemonic exercises and information items that students will experience along the way.

In the classroom setting, the instructor assumes the role of a navigator that guides students through the vMP. The instructor reveals the accompanying story in a step-by-step fashion, presenting each item and mnemonic exercise as they are encountered in the virtual environment. By doing so, the instructor creates a synchronous learning experience for the students, ensuring that they are all on the same page and working together to build their mnemonic skills and retain the presented information.

3.1 Virtual Tours as Memory Palaces

A virtual tour of a well-known and frequently visited environment, such as a house, workplace, or school, possesses the essential elements required to build an effective vMP, namely navigability and familiarity. A virtual tour allows for the exploration of the environment in a structured and controlled manner, allowing users to move through the environment at their own pace and take in details and information in a logical sequence. Additionally, using a virtual tour of an everyday environment ensures that users are already familiar with the layout and features of the space, which makes it easier to associate new information with specific locations within the environment. This familiarity also helps reduce the cognitive load, making it easier to focus on encoding and retaining information [32].

Fig. 1. Two similarly outfitted computer laboratories. Neural Style Transfer and live bidirectional morphing was used to infuse uniqueness and achieve a striking imagery.

Using readily available virtual tour editors, the virtual memory palace can be easily populated with (actionable) virtual cues that are used to anchor the information that needs to be retained. Triggering virtual cues can also prompt pre-programmed actions such as displaying detailed content, opening a web page, streaming a YouTube video, and more. These mechanisms can be used to provide additional information and context that will be tied through the method of loci to the spot within the tour.

Compared to computer-generated virtual reality environments, 360 virtual tours are more accessible and easier to create. Virtual tour editors offer sophisticated tools to achieve interactivity and customization, such as adding animations or avatars within the 360 environment. This ease of use makes it effortless for instructors to create virtual memory palaces quickly and with minimal technical skills.

While computer-generated virtual reality can deliver a higher degree of interactivity because objects and elements within the environment can be freely moved and even manipulated to create new experiences, constructing a photorealistic computer-generated VR requires advanced technical skills to program and develop. As a result, we believe that 360 virtual tours are a more practical and effective tool for creating vMPs for most instructors.

When designing and constructing the virtual tour, it is essential to carefully consider the effective implementation of the MOL technique by adhering to established key principles [7]. To illustrate, we utilized a strategy to achieve architectural guidance for comparable locations. This was achieved by gradually back-and-forth transforming the 360 scenery of similarly outfitted computer laboratories to assimilate the style of distinct renowned artworks. This approach had the added benefit of presenting remarkable and visually striking imagery, which is another key tenet of an effective MOL. The style transfer process and morphing, which involves a gradual and seamless transition between the photorealistic depiction and the painting style, is depicted in Fig. 1.

3.2 Combination of Mnemonic Techniques

The use of multiple mnemonic techniques is a sound strategy, since different methods can aid in the retention and recall of different types of information.

MP is particularly useful in remembering sequential lists, such as the steps in an algorithm or recipe, and the names of famous computer scientists throughout history in chronological order. However, it may not be the optimal choice for other types of non-sequential information, such as facts, relationships, or definitions. By combining different mnemonic techniques, a more comprehensive and adaptable memory system can be established [1].

The Story Method, for instance, involves creating a narrative or story that connects different pieces of information, thereby assisting in recalling related concepts or ideas. The chunking technique involves breaking up large amounts of information into smaller, more manageable chunks. Grouping related concepts together and placing them in specific locations within the MP helps students to recall the connections between items as well. Acronyms and acrostics techniques involve creating a word or phrase from the first letter of a series of words to be remembered and incorporating them into the corresponding locations in the MP.

3.3 Guided Learning Experience

In previous studies, the MP technique was used mainly as a self-directed learning activity. On the contrary, the third central concept of our approach is to have the instructor plan, build, and deliver the learning experience in the form of a guided tour of the MP. The reasoning behind this is two-fold. First, students often lack proficiency in the use of mnemonic techniques, and second, constructing a memory palace for CS topics that are not yet mastered can be challenging, since it requires the identification, selection, and organization of key concepts from the material.

Mnemonic techniques have been proven to be effective in aiding memory and learning, but, like any skill, they require practice and mastery to be used effectively. This can be challenging or intimidating at first for many students, particularly those who may not be familiar with mnemonic techniques or have had little exposure to them. As guided meditation can be an excellent tool for beginners to learn and establish regular practice, we believe that guided mnemonic exercises led by an instructor can be useful for students to become comfortable and adept at using mnemonic techniques. Guided exercises can provide clear instructions, structure, and support that can help students understand the techniques and practice them effectively.

Second, learning complex CS topics goes beyond memorizing and remembering, the first level of the Revised Bloom's Taxonomy [13]. To facilitate higher-level learning objectives such as understanding and analysis, the memory palace and information delivery must be constructed in a way that logically structures concepts, visualizes the relationships between them, and provides a depiction on how they fit together in a broader framework [32]. Suitable visual and spatial associations must be constructed to make abstract concepts more concrete and easier to understand. Additionally, the mental images that are established often must simulate problem-solving scenarios and have to engage and reinforce the application of concepts in different contexts. Producing such clever associations, highlighting nuances and corner cases, providing additional meaning and understanding for a given context, etc. can only be effectively achieved by someone who already mastered the material, which in a classroom setting is the instructor.

3.4 Journey Dynamics

Our proposed approach enhances the logical structure provided by the story and locations by integrating additional mnemonic techniques that facilitate retention. The instructor can design and customize the mnemonic anchors and virtual cues depending on the level of active audience participation they aim to achieve.

Passive Journeys. In this approach, audience participation is limited mainly to observing the unfolding of the mnemonic journey, and engagement and retention of information must be achieved through activation of emotional responses, primarily through the use of humor, shock, and surprise. The release of dopamine, a neurotransmitter associated with pleasure and reward, in response to startling stimuli enhances the encoding of information in memory, leading to better subsequent recall [16]. Moreover, there is a positive correlation between the level of emotional arousal evoked by the material and the extent of subsequent recall [22].

Humor can also make information more relatable and easier to remember by connecting it to something familiar or amusing. Shock and surprise can break our expectations and create a strong impression that can enhance our recall of the information.

An effective technique that we tried is assigning each room or lab in a virtual tour to host algorithms of a certain complexity time. As we pass down the "ever increasing complexity hallway", after the linear running time room, we "enter the realm of $O(n \log n)$". In a passive journey, a virtual cue displayed on the entrance door might contain some seemingly random item, such as a fish, as shown in the left pane of Fig. 2. To grab the attention of the audience, the narrator could open with a shocking and unexpected line: "Let us stop briefly to honor a H&M Quilted Jacket wearing mosquitofish". What does the clothing item and mosquitofish have to do with algorithms?

Then the narrator activates the hotspot and a Wikipedia page appears, displaying a monument constructed in Sochi honoring the western mosquitofish[1]. This hopes to surprise the audience and arouse curiosity as the narrator asks "You wonder why?". After a dramatic pause, it continues with the following, strongly emphasizing the highlighted words: "For *merging* into many habitats to *quickly* eradicate Malaria by eating *heaps* and *heaps* of mosquito larvae." This opens the process where gradually the associations are revealed and explained, connections with the Heapsort, Mergesort and Quicksort are established.

To strengthen the mnemonic anchors, the instructor can present fun trivia, such as the fact that the quilted jacket seems to be the only wearable starting

[1] https://en.wikipedia.org/wiki/Mosquitofish#/media/File:Monument_of_a_Fish.JPG.

Fig. 2. Virtual cues relying on outlandish associations in the left panel, and discovery and recognition in the right panel.

with the letter 'Q'[2] or that there are very few common words that contain the letters HMQ concomitantly, with mosquitofish being one of them.

Active Journeys. In an alternative approach, the guided journey serves as a broad framework, and the nuances and intricacies are uncovered collaboratively with the audience. The focus is on the social experience and, therefore, the design of virtual cues and exercises should be tailored accordingly. Rather than being jarring, they should act as clues and enable the identification and comprehension of the underlying meaning. In this scenario, students embark on a detective or reverse engineering journey to decipher and reveal the meaning behind the mnemonic anchors. In case of confusion or difficulty, the instructor provides additional context and support to aid in discovery and understanding.

For instance, the premise of the story could be that related software design patterns are being held captive in different rooms. To rescue them, participants must solve the "seal" on each door by interpreting the symbols and correctly identifying the patterns being held hostage. This requires active participation and critical thinking, as participants must analyze clues and metaphors and recall the names and characteristics of each pattern. When the experience is interactive and challenging, participants are more likely to remember the information and be motivated to learn more.

An example of such "seal" or ideogram is illustrated in the right panel of Fig. 2. Upon closer inspection, students may recognize that it is based on the most prominent symbol in Michelangelo's fresco painting "The *Creation* of Adam", depicting the outstretched arms of Adam and God. If not, on the left of the door there is help available: the entire fresco is displayed. Clicking on the image of the fresco opens the dedicated Wikipedia page, providing additional context and information. Additional clues, such as Adam being a *singleton* immediately after creation and the *prototype* for humanity - The Primordial Man, the step-by-step approach for creating / *building* something complex, *factory* and *builder* cap

[2] https://englishenglish.biz/a-z-things-you-can-wear/.

iconography, and other mnemonic cues, help guide students in deciphering and coming up with the right solution.

Another way to encourage active participation of the audience is to introduce mnemonic exercises, where students are challenged to devise a solution, such as creating a humorous or catchy wordplay or acrostic for a mnemonic anchor, such as the starting letters of sorting algorithms HMQ (for example, "Hunky Male Quartet", "Happy Meal Queen" or "Holy Molly Quacamole"). This type of exercise not only activates the audience but also promotes creativity and enhances the personalization of the mnemonic device. It encourages students to think beyond the surface level and come up with something unique and memorable to aid them in their retention of information.

4 Pilot Study

In a pilot study with 30 freshman CS students, we conducted a 20 min GMJ to reinforce their learning of ten basic search and sorting algorithms. Participants observed the virtual tour-based journey on a projector screen while the narrator followed a carefully planned script to provide context and deliver the learning content. The style of the journey was passive, allowing participants to focus on visuals and narration with no active involvement required.

To create a clear and logical structure, the algorithms were grouped based on their time complexities, and each group was anchored to a different computer laboratory on campus.

The overarching narrative was crafted to be engaging, unusual, and also relatable to the students. It begins with a tired and unmotivated student visiting the university campus to seek inspiration before the upcoming exam on search and sorting algorithms. As the story progresses, the student embarks on a journey characterized by surreal and abstract elements that challenge the boundaries of reality. The narrative is punctuated by humorous and absurd situations where often teachers are the protagonists, wordplay, unexpected developments, and imagery that at first appear nonsensical. However, as the journey continues, the seemingly disconnected elements are gradually linked to the algorithms studied through associations and connections, hopefully making them and their characteristics and properties easier to understand and remember. The accompanying virtual tour depiction reinforces the narrative by incorporating visually striking imagery and cues that correspond to the various mnemonic exercises and information items encountered throughout the journey.

To make the virtual environment more engaging and memorable, each laboratory was designed with specific characteristics, such as a billiard room, a bistro, a fishing colony, etc. To aid in the retention of algorithmic characteristics such as worst-case, best-case running times, and other relevant information, a variety of mnemonic exercises were presented as the tour explored the premises of each computer laboratory. By combining these elements, a clear spatial representation of the algorithms and their respective time complexities was provided.

Following the guided activity led by the instructor, the students were prompted to mentally revive and retrace the plot and recall as many specific details as possible related to the characteristics of the algorithms. This exercise not only solidified their understanding of the material but also served as a form of self-assessment, allowing them to somewhat gauge their mastery of the content.

We measured the Perceived Usefulness (PU) of the experienced learning technique by asking participants to complete a questionnaire at the end of the class. PU is a central construct in the Technology Acceptance Model (TAM) [4] that is widely used in the field of information systems to understand how users regard the value and utility of a particular technology or system [2,10,27]. The TAM proposes that users' intention to use a technology or system is influenced by their perceived usefulness of the technology, which refers to the extent to which a user believes that a technology or system will help them perform a task more effectively or efficiently. Users are more likely to adopt a new technology or system if they perceive it to be useful in achieving their goals.

We adapt the original survey elements to the educational setting, assessing the degree to which students believe that using the proposed technique would ease and enhance their learning performance and outcomes. Participants indicated one a 7 point Likert scale their level of agreement with statements regarding the guided memory journey 1) helping them learn new concepts more rapidly (time saving), 2) making it easier to remember complex things (effort saving), 3) improve knowledge, 4) improve learning performance and grades, 5) help make most out of their time while learning (effectiveness), 6) being a flexible learning technique, 7) being overall useful in learning.

At the end of the survey, students had the opportunity to express their opinion and provide written feedback on the educational experience.

5 Results and Discussion

Figure 3 displays the distribution of responses using Likert plots, also known as diverging stacked bar charts. The width of each horizontal bar represents the percentage of respondents who gave a specific response. The sum of the percentages for the disagreement, neutral, and agreement responses are shown on the left, middle, and right of each item. Additionally, the bars are color-coded to represent the degree of agreement or disagreement, with darker shades indicating stronger responses.

The figure clearly shows that the participants found the evaluated method valuable with functional utility, expressed by the predominantly positive responses to the questionnaire items. The level of agreement ranged from 100% for perceived flexibility to 60% for perceived effectiveness in supporting the learning process. The neutral and disagreement percentages were low, ranging from 0% to a maximum of 20%, again for effectiveness. Other studies have also reported that vMP implementations are perceived as useful by students. For example, [10] conducted a TAM evaluation of their vMP and found a positive trend with respect to this factor. The central tendency of the responses to all items, except for Performance and Effectiveness, is reflected in a median score of

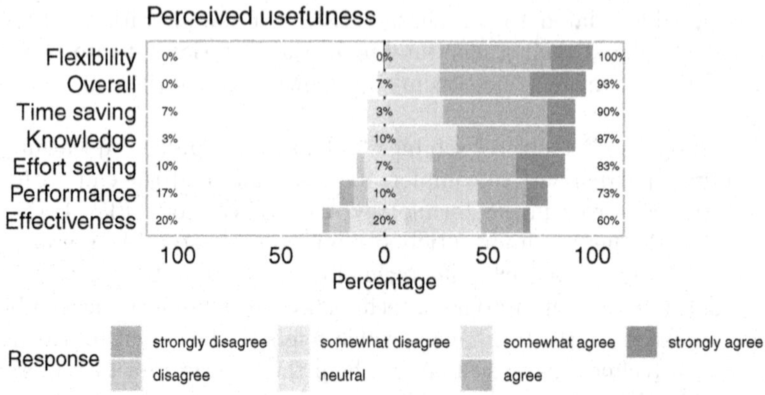

Fig. 3. Distribution of responses along the Likert scale for the various factors assessing Perceived Usefulness.

2. This finding reveals that a minimum of 50% of the participants either "agree" or "strongly agree" that the presented technique can save time and effort, enhance retention and knowledge, and possess general utility and adaptability.

The 100% level of agreement for the flexibility factor is an extremely promising result, demonstrating that students recognize the significant potential of the method in terms of its applicability to other topics as well. In addition, the high level of agreement regarding the Time saving factor (the mean score on the scale $[-3, 3]$ was $M = 1.6$, with the standard deviation $SD = 1.0034$) and Knowledge ($M = 1.5333$, $SD = 0.9732$) factors indicate that most of the participants perceived the proposed GMJ method as an effective way to improve their knowledge and accelerate their learning of new concepts. These encouraging findings are consistent with other recent studies that have predicted that the use of vMPs can significantly improve the overall learning experience [8,18,23].

The 10%-83% disagreement-agreement ratio, median score or 1 ("somewhat agree") for the Effort saving factor reveals that some students did not find the application of this technique self-evident, especially when trying to remember complex information. Furthermore, written feedback indicated that participants recognized the significant time and training required to master this method, which is a common drawback mentioned in previous studies on MOLs implemented by students [14,26]. On the other hand, these findings further emphasize the benefits of teacher-led guided journeys.

Participants rated the Performance and Effectiveness factors the lowest, which is a plausible result. It is understandable that after only one use, students may not fully recognize and evaluate how the method can contribute to their learning outcomes. However, the fact that the students disagreed the most with the statement that the method could help them make the most of their learning time suggests that this technique should be used as a complementary tool.

The students' comments also provide valuable feedback on the evaluated method. These remarks reveal that some students had a complete change of opinion, starting with negative prejudice and ending with a highly positive experience. For example, one of them commented, "Honestly, I used to think that teaching using unusual things just to aid memorization and learning was foolish, but my opinion changed after today's class. It can really help a lot in learning! In my opinion, it would be very useful to incorporate this technique into education, without exaggerating it. People need to use creativity and imagination because in the modern world, we have fewer chances to do so." Appreciation for the creativity, imagination, and invested work behind the implemented mnemonic journeys appeared in other comments as well. Another participant stated that "It can be a useful learning method, but it takes a lot of time to work well and effectively. It requires a complex technique and creativity to be truly effective and useful. It may not be suitable for everyone, but those who master it can learn more efficiently and in a shorter period of time, and the learned material will stick."

Overall, it can be stated that developing effective GMJs involves a significant initial investment in effort on the part of the instructor. Creating an amusing and captivating mnemonic script demands meticulous planning, ideation, and a wealth of experience in identifying the curriculum's intricate and challenging areas. However, the first feedback results are optimistic, indicating that the benefits derived from this one-time investment are considerable, and the created mnemonic journey can be repurposed for subsequent student cohorts annually.

Limitations. While the study provides a valuable first insight into the effectiveness of the GMJ technique in CS education, it is important to note that the results are limited to a single context and a specific set of participants. The potential of this technique in various educational contexts, such as different academic levels and subject areas, and the comparison of passive and active journey styles has yet to be explored. Additionally, the long-term effectiveness of the technique in promoting memory retention and enhancing learning outcomes requires further investigation.

Furthermore, it is important to evaluate the capacity of the foundational virtual tour for repeated use. Although it may be effective for initial use, its saturability must be gauged in terms of repeated application for different topics. As locations become associated with specific concepts and information, the tour may become less effective in anchoring new information and creating new associations. Overuse of the same loci and cues can also cause interference with existing associations and may lead to confusion and decreased retention.

6 Conclusions

This paper presented a detailed account of our methodology and first outcomes in developing and administering Guided Mnemonic Journeys covering computer science topics. The proposed approach is based on three key principles: the extension of 360 virtual tours to leverage the benefits of familiarity, navigability, and easy technological accessibility, the integration and layering of various mnemonic techniques, and the provision of guided instruction within a synchronous learning context.

Feedback provided by students highlighted the appreciation for the guided approach and its potential as a tool for enhancing memory retention and improve learning outcomes. The positive reception of the approach underscores the importance of incorporating mnemonic techniques in CS educational settings, and provides evidence for the value of utilizing Guided Mnemonic Journeys as a means of enhancing learning and memory. More research is needed to explore the full potential of this technique, including its applicability to various educational settings and its effectiveness in promoting long-term memory retention.

References

1. Akhter, N., Nawshin, N., Khatun, M.: Mnemonic techniques: Characteristics, educational suitability and research prospects in the field of anatomy. Mymensingh Med. J. **30**(4) (2021)
2. An, S., Eck, T., Yim, H.: Understanding consumers' acceptance intention to use mobile food delivery applications through an extended technology acceptance model. Sustainability **15**(1), 832 (2023)
3. Anderson, J.R.: Learning and memory: An integrated approach. John Wiley & Sons Inc. (2000)
4. Davis, F.D.: Perceived usefulness, perceived ease of use, and user acceptance of information technology. MIS Q., 319–340 (1989)
5. Dunlosky, J., Rawson, K.A., Marsh, E.J., Nathan, M.J., Willingham, D.T.: Improving students' learning with effective learning techniques: Promising directions from cognitive and educational psychology. Psychol. Sci. Public interest **14**(1), 4–58 (2013)
6. Fassbender, E., Heiden, W.: The virtual memory palace. J. Comput. Inform. Syst. **2**(1), 457–464 (2006)
7. Hedman, A., Bäckström, P.: Rediscovering the art of memory in computer based learning-an example application. In: E-Learn: World Conference on E-Learning in Corporate, Government, Healthcare, and Higher Education, pp. 1024–1031. Association for the Advancement of Computing in Education (AACE) (2003)
8. Heersmink, R., Knight, S.: Distributed learning: Educating and assessing extended cognitive systems. Philos. Psychol. **31**(6), 969–990 (2018)
9. Huttner, J.P., Pfeiffer, D., Robra-Bissantz, S.: Imaginary versus virtual loci: Evaluating the memorization accuracy in a virtual memory palace. In: Proceedings of the 51st Hawaii International Conference on System Sciences,HICSS, pp. 274–282 (2018)
10. Huttner, J.P., Robbert, K., Robra-Bissantz, S.: Immersive ars memoria: evaluating the usefulness of a virtual memory palace. In: Proceedings of the 52nd Hawaii International Conference on System Sciences, HICSS, pp. 83–92 (2019)
11. Jund, T., Capobianco, A., Larue, F.: Impact of frame of reference on memorization in virtual environments. In: 2016 IEEE 16th international conference on advanced learning technologies (ICALT), pp. 533–537. IEEE (2016)

12. Jurowski, K., Jurowska, A., Krzeczkowska, M.: Comprehensive review of mnemonic devices and their applications: State of the art. Int. E-J. Sci. Med. Educ. **9**(3) (2015)
13. Krathwohl, D.R.: A revision of bloom's taxonomy: An overview. Theory Practice **41**(4), 212–218 (2002)
14. Legge, E.L., Madan, C.R., Ng, E.T., Caplan, J.B.: Building a memory palace in minutes: Equivalent memory performance using virtual versus conventional environments with the method of loci. Acta Physiol. (Oxf) **141**(3), 380–390 (2012)
15. Levin, J.R.: Mnemonic strategies and classroom learning: A twenty-year report card. Elem. Sch. J. **94**(2), 235–244 (1993)
16. Lisman, J.E., Grace, A.A.: The hippocampal-vta loop: controlling the entry of information into long-term memory. Neuron **46**(5), 703–713 (2005)
17. Maguire, E.A., Valentine, E.R., Wilding, J.M., Kapur, N.: Routes to remembering: the brains behind superior memory. Nat. Neurosci. **6**(1), 90–95 (2003)
18. Mäkelä, M., Löytönen, T.: Rethinking materialities in higher education. Art, Design Commun. Higher Educ. **16**(2), 241–258 (2017)
19. Manalo, E.: Uses of mnemonics in educational settings: A brief review of selected research. Psychologia **45**(2), 69–79 (2002)
20. Mastropieri, M.A., Scruggs, T.E.: Enhancing school success with mnemonic strategies. Interv. Sch. Clin. **33**(4), 201–208 (1998)
21. McCabe, J.A., Osha, K.L., Roche, J.A., Susser, J.A.: Psychology students' knowledge and use of mnemonics. Teach. Psychol. **40**(3), 183–192 (2013)
22. McGaugh, J.L.: The amygdala modulates the consolidation of memories of emotionally arousing experiences. Annu. Rev. Neurosci. **27**, 1–28 (2004)
23. Peeters, A., Segundo-Ortin, M.: Misplacing memories? an enactive approach to the virtual memory palace. Conscious. Cogn. **76**, 102834 (2019)
24. Putnam, A.L.: Mnemonics in education: Current research and applications. Trans. Issues Psychol. Sci. **1**(2), 130 (2015)
25. Robins, A.V., Margulieux, L.E., Morrison, B.B.: Cognitive sciences for computing education. The Cambridge Handbook of Computing Education Research, pp. 231–275 (2019)
26. Rosello, O., Exposito, M., Maes, P.: Nevermind: Using augmented reality for memorization. In: Adjunct Proceedings of the 29th Annual ACM Symposium on User Interface Software and Technology, pp. 215–216 (2016)
27. Sengik, A.R., Lunardi, G.L., Bianchi, I.S., Wiedenhöft, G.C.: Using design science research to propose an it governance model for higher education institutions. Educ. Inf. Technol. **27**(8), 11285–11305 (2022)
28. Soler, M.J., Ruiz, J.C.: The spontaneous use of memory aids at different educational levels. Appl. Cogn. Psychol. **10**(1), 41–51 (1996)
29. Stalder, D.R.: Learning and motivational benefits of acronym use in introductory psychology. Teach. Psychol. **32**(4), 222–228 (2005)
30. Sweller, J.: Implications of cognitive load theory for multimedia learning. Cambridge Handbook Multimedia Learn. **3**(2), 19–30 (2005)
31. Worthen, J.B., Hunt, R.R.: Mnemonology: Mnemonics for the 21st century. Psychology Press (2011)
32. Yang, F., Qian, J., Novotny, J., Badre, D., Jackson, C.D., Laidlaw, D.H.: A virtual reality memory palace variant aids knowledge retrieval from scholarly articles. IEEE Trans. Visual Comput. Graphics **27**(12), 4359–4373 (2020)
33. Yates, F.A.: Art of memory. Routledge (2013)

Analysis of Outcomes from the Gamification of a Collaboration Intensive Course on Computer Networking Basics

Sławomir Zieliński and Marek Konieczny(✉)

AGH University, Kraków, Poland
{slawek,marekko}@agh.edu.pl

Abstract. The paper contains a comparison of the results of conducting a computer networking basics course in a gamified and non-gamified way. In the analysis we focus rather on students' learning outcomes, measured by exam scores, rather than on their satisfaction, and other subjective metrics. The experience presented in the article was gathered during last four years (2019-2022). The COVID-19 pandemic, which affected the way the classes were conducted in 2020 and 2021, gave us an opportunity to compare the same means of gamification used in the same course in different learning environments. The evaluation results show that gamification yielded better results when applied in an in-person environment.

Keywords: unplugged gamification · collaborative education · university education · computer networking

1 Introduction

Games – played by people for a few thousand years – strongly motivate the players to devote their time, money, and skills to achieve wins, accumulate points, reach levels, collect artifacts, etc. Computer games, which build upon that phenomenon, are one of the most popular ways of spending free time.

According to the Limelight Networks report [3], in 2020, online gamers spent 5-7 h a week playing (average: 6.33). Apparently – according to market reports, the breakout of the COVID-19 pandemic did not result in significant growth in the number of gamers. That can be explained by the huge popularity of gaming before the pandemic – in 2020, 68% of adult U.S. population were gamers, while in the EU the ratio reached 50%. The numbers are expected to grow steadily in the future [7,12]. The Entertainment Software Association (ESA) in its 2020 report argues that most of the gamers declare a positive impact of

The research presented in this paper has been partially supported by the funds of Polish Ministry of Science and Higher Education assigned to AGH University of Science and Technology.

J. Mikyška et al. (Eds.): ICCS 2023, LNCS 14077, pp. 766–779, 2023.
https://doi.org/10.1007/978-3-031-36030-5_59

gaming on their lives, including mental stimulation (80-87% responses), increase in teamwork skills (81%), and help in problem solving (63%) [2,11].

The percentage of gamers in generation Z (18-34 years of age) reaches 81% [11]. In the group of 18-25 (which includes university students), weekly averages of time spent on gaming were 7.08, 7.78, and 7.48 h in 2018, 2019, and 2020, respectively [1,3]. While the numbers can vary from one country to another, it is safe to assume that majority of IT students play games. That motivated us to experiment with introducing elements of gaming into an IT course.

The goal of this paper is to share our experience obtained from 4 years of leading the courses in a gamified way. The orientation on gamification in courses led by us started just before COVID-19 outbreak (i.e., in the academic year of 2019/20). Therefore, after initial experiences gathered by leading only in-person classes, we had to adjust the teaching method to remote, or blended learning modes. On one hand, we obviously perceived such changes as obstacles, but on the other they provided for making a few interesting observations.

A list of specific questions answered in this paper is as follows:

1. is there any relevant positive overall effect of introducing gamification into IT teaching practice?
2. if yes, can we characterize the group of students, in which the effect is especially visible?
3. does the effect depend on the teaching environment (in person/remote)?

In order to answer the questions we compared the effects of teaching between groups of students which were taught in a gamified way (we call them 'gamified groups') and groups which classes were lead traditionally ('non-gamified' groups). The material of the course was identical and the number of lectures and laboratory classes was equal to all groups. Moreover, all the students were given the same exam at the end of their courses, consisting of five parts, prepared by five different professors, one of which was teaching the gamified groups. The results of the exam formed the basis for answering the questions.

The structure of the paper is as follows. Section 2 surveys the important developments in the subject area. Section 3 shows how a selected course was augmented with gamification means. Section 4 discusses the results of conducting the course in a gamified way. Section 5 concludes the paper and points out the directions of future work.

2 Background and Related Work

Gamification is commonly understood as a "process by which elements of game play are incorporated into alternative contexts" [6]. The alternative contexts are not limited to education. Gamified approaches are used, for example, in construction industry [14], marketing [16], and human resource management [4], to name just a few. However, literature surveys show that education is the most studied area of applying gamification. Luckily, the researchers point out not only advantages, but also disadvantages of using gamified approaches in various contexts. For example, authors of [10] demonstrate a case study of a question answering website, in which introducing gamification leads not only to increased quantity of answers, but also to their decreased quality. As correctly stated by the authors, most of research efforts stop on quantitative analysis, while lacking in-depth insight into operation quality. Therefore, we took *quality as one of guidelines for the presented research.*

The psychological and sociological backgrounds of gamification mechanisms are an active area of research. Many researchers devote their work to understand the mechanisms that make gamified activities popular and effective in multiple areas of daily life (consider, for example [9]). Arguably, gamification is commonly associated with technologies, especially with dedicated software. The author of [8] argues that the association is not obligatory. The paper introduces, or at least announces research on a systematic framework for so called 'unplugged gamification', not supported by IT. From our perspective, as university teachers, *"unplugged gaming" is a very interesting area for experiments.*

A number of survey articles referring to the gamification area exist. One of them [13] shows that the list of game elements used in empirical scenarios typically includes points, challenges, badges, and leaderboards. However, the actual list of elements should be selected based on the target audience, and expected outcomes. The target audience can be analyzed in terms of players' personalities, and the set of gamification means can be adjusted accordingly. An interesting discussion of the reasonability of such adjustment is presented in [15].

Typically, gamified environments incorporate elements of competition. However, we agree with authors of [17], who indicate the needs for incorporating mechanisms of cooperation or collaboration into such environments. Intuitively, we also decided for *promoting collaboration in our experimental story line.* Moreover, the design of the story line and the assignment of roles within teams are meant to promote group integration rather than in-group competition.

3 Selection of the Experimental Course

The computer networks basics course is obligatory for all IT students at our department. The course was designed to contain 15 lectures (each of which lasts 90 min), and 15 labs (also 90 min each). While the lecture is led for all the students at once, the labs – which require convenient access to hardware – are conducted for groups of 12 people each, in the way described in Sect. 3.1. At the end of the semester all the students were given a 90-minutes written exam, covering all the course material.

Because the course is obligatory, it is not necessarily of top interest for all the students. Therefore, their attitudes towards the classes differ. While the course curriculum is well-defined, we experiment with teaching methods during the laboratory classes. Introducing gamification into the course has been such an experiment conducted for three to five groups each year. The course was augmented with a story (see Sect. 3.2), inspired by a popular comic book, which encompassed both in-class, and out-of-class gamification means. The term 'in-class gamification' refers to compulsory activities, performed synchronously by groups of students, while 'out-of-class gamification' refers to activities which are voluntary, and may be performed asynchronously.

The main reason for which we selected the particular course was that during the lab classes students are split into teams of two to four people, who are collaborating on common tasks. Therefore, as opposed to courses in which the students work on their own, this course offered more possibilities of introducing gaming elements, as well as of drawing interesting conclusions.

3.1 The Laboratory Classes Sequence

Traditionally the laboratory classes were split into three stints, each of which lasted up to five weeks (see Fig. 1). The first stint, focused on OSI model layers 1 and 2, consisted of four labs (one per week) and one written test (in a lab-free week). The second stint was focused on layer 3, especially on IP addressing and routing. It also lasted five weeks, and also consisted of four labs and a written test. The third stint encompassed L3/L4 boundary (i.e., PAT and firewall basics) and case study lab exercises, built upon the previous labs. Just as in the case of the first two, it ended with a test, but that test consisted of hands-on and oral parts. The final stint lasted three to five weeks, depending on the year and even the day of week the particular group had their classes on, because of holidays, etc. In general, the organization of the course was perceived as adequate, so we decided that the gamified course should *in general follow the pattern.*

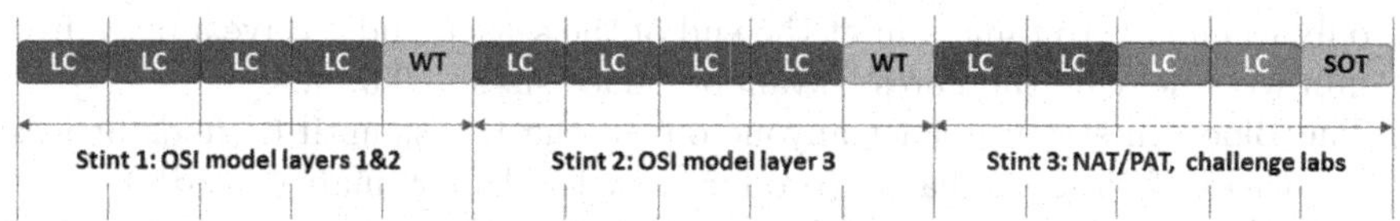

Fig. 1. The flow of computer networking basics course. Legend: LC = lab class; WT = written test; SOT = skills-based exam and oral test

Occasionally, during the lab classes, the students were requested to fill in a short test, so that their preparedness to the labs was verified. The short tests were not announced in advance. Typically, during a semester, each group of students was given three to four such tests. The progress of the course, as well as the grades received by the students, were reflected in the learning management system (LMS) operated by the university. The LMS was deployed as a customized instance of Moodle, and provided sufficient support for conducting the classes.

From the lab teacher's point of view, sustaining students' focus on learning the course contents, has always been the key for their success at the end of the semester. Week-long breaks between laboratory classes were perceived as too long, and – as the students did not have any real incentive – trusting that all of them will learn systematically, proved to be only a wishful thinking. Therefore the second of the guidelines of development of a gamified version of the course was to *add some activities in between regular classes.*

3.2 The Story Line of the Course

In this section, the story line of the course is presented. For the sake of brevity, some elements of low importance were omitted. The story was originally written in Polish, therefore the translation lacks some word-play. Nonetheless, we did our best to save its parodic style. Therefore, the professor is called a Bloodthirster, the students are Unfortunates, a grade is a mace, everything happens in a dark forest, and so forth.

Unfortunates in the Bloodthirster Forest. As every year, a bunch of Unfortunates were sent to the Dark Valley. Four different roads lead to it, including the sinister one – leading through the Bloodthirster Forest. Legend has it that the Bloodthirster strives every year to transform the Unfortunates into Utmost Thugs through elaborate tortures. Therefore, tempted by the mirage of future glory, some of the Unfortunates entered the Forest. They would spend four ghastly months there, hoping that what didn't kill them would make them stronger enough to walk away with a sneer.

General Rules

1. The forest is ruled by the Bloodthirster.
2. At the entrance to the Forest, each Unfortunate gets a barrel – empty and divided into four casks. They can fill it according to the Bloodthirster rules only to be able to tame him at the end of the season and receive a mace from him when leaving the Forest (available mace sizes: 3.0-5.0).
3. The Bloodthirster won't let anyone out of the Forest until he finds at least two casks of oil in the barrel (produced by the Unfortunate's head[1]).
4. Any Unfortunate can appeal against the Bloodthirster's judgments at any time – the Bloodthirster likes listening to the echoes of voices...

Filling the Casks. The casks that form an Unfortunate's barrel can be filled in four distinct ways.

Cask 1: Wandering in the Forest, Surprises and Oil Pressing. The Bloodthirster regularly – once a week – lets the Unfortunates go for walks in the Forest. During the walk, the Unfortunates may (but do not have to) reach the destination set by the Bloodthirster, where they can pour the oil into the first cask. Typically Unfortunates wander in groups of three, although the Bloodthirster allows smaller and larger groups. The walks may be enriched with Surprises prepared by the Bloodthirster – every Unfortunate struggles with the Surprises alone.

The First – Lonely – Walk. On their first walk, the Unfortunates meet the Bloodthirster for the first time. He introduces them with the rules of the forest.

[1] That comes from a Polish idiom 'mieć olej w głowie', which is used for describing a person as a smart one – it would be literally translated as 'to have oil in the head'.

As a result of the meeting, each Unfortunate braids a rope...
Reward: 0 (words zero, null, nada, nicht).

Subsequent Walks (At Least Nine) – Filling the Pots. Starting with the second walk, the Unfortunates huddle in teams of three. Each team tries to complete the tasks set by the Bloodthirster, using ropes (a.k.a. cables) and scrap metal found in the Forest (a.k.a. hardware). As a reward for the effort, the Bloodthirster pours a certain amount of oil into the first cask of each Unfortunate. Prizes are awarded individually to each Unfortunate and also to a group - in the event of a rare case of completing the task in the allotted time. The group reward is shared equally among the group members.

At the beginning of a walk, roles of Overseer (OS), Cablemaster (CM), and Scribe (SC) are assigned to the members of each team – by drawing lots and ordering (see Table 1). If a team is created only by two Unfortunates, there is no Overseer role. The possible fourth member takes the role of Drunkard, who helps in both topology building and documentation, but lowers the other team members' rewards (see Table 2).

Table 1. Roles assigned to team members

Role	Assigned by	Responsibilities
Overseer	draw	Supervision over the construction, configuration and documentation of the group's achievements
Cablemaster	Overseer's choice	Building topology and preparation of materials
Scribe	Overseer's choice	Lab documentation
Drunkard	Overseer's choice	Up to the group members

Table 2. Rewards assigned to team members (as of the first edition of the course)

Member count	Group reward	Overseer reward	Cablemaster reward	Scribe reward	Drunkard reward
2	6 quarts	not applicable	2 quarts	2 quarts	
3	9 quarts	sum of CM and SC rewards	2 quarts	2 quarts	
4	8 quarts	sum of CM and SC rewards	2 quarts	2 quarts	2 quarts

At the end of the season, each Unfortunate can use up to three pots of oil earned during the walks. Any excess oil can be used for gambling purposes (see: Gambling) or poured into the Sand of Indulgences.

Filling Cans (Surprises). On a given signal (and a signal can be given at any time) the Unfortunate tries to pour the oil (as ordered by the Bloodthirster) to a can. Whoever misses the signal does not fill the can. Everyone keeps three best-filled cans at the end of the season, the others can be poured into the Sands of Indulgences or used for gambling purposes (see: Gambling).

Cask 2: November Hunt
November nights make the Bloodthirster especially oppress the Unfortunates. After four walks, the Unfortunates take part in the November Hunt. During the hunt, the Unfortunates are plagued by a series of Bloodthirster's Riddles. Solving each riddle is awarded with oil, which fills the second cask. The amount of the oil collected by each Unfortunate is known about ten days after the hunt.

Cask 3: Mighty Frost
Long and cold nights lead the Bloodthirster to seek entertainment - at the expense of the Unfortunates, of course. After eight walks, it's time for the second hunt! As a side effect of the hunt, the third cask is filled in.

Cask 4: Finches' Songs
At the end of the season, the Bloodthirster organizes the Festival of Forest Songs. In the beginning, the Unfortunates, who already took at least ten walks in this forest, in groups of three appointed by the Bloodsucker, implement his bizarre ideas. After that each of them – alone – sing sad songs about the means of transport used in the forest and goods which are transported.

Additional Rules
Measurement Units. Simply put, each barrel contains four casks, each cask contains 18 pots, and each pot contains four quarts. Each can contains five pots, i.e., 20 quarts. The maximum amount of oil one can collect is 288 quarts.

Sand of Indulgences grants an Indulgence in exchange for pouring one can (20 quarts) of oil into it. The oil poured in the Sand of Indulgences must come from the first cask. Note: it is perfectly fine to borrow oil (see: Usury).

An Indulgence allows skipping one Surprise or annihilating one Bloodthirster riddle.

Riddle Annihilation. Indulgence can be used to annihilate a riddle during a hunt. As a result, only the riddles that have not been annihilated will be taken into account when evaluating the effects of hunting (for) the Unfortunate who had the Indulgence. This rule also applies to singing (Cask 4) - an Indulgence can be used to ignore one of the melodies.

Usury. In case an Unfortunate lacks oil (e.g., to pour in the Sands of Indulgence), they can borrow oil from the Bloodthirster. Repayment must be made before the end of the season. As long as the loan is repaid with oil from the first cask, the Bloodthirster will not charge interest. Repaying the loan with oil from the second or third cask results in the need to pay an additional quart for each can of oil. The Unfortunate cannot be indebted to more than one cask of oil.

(Un)fortunate Initiative. The Unfortunates can propose a hunting date common to all groups. This will allow them to take an extra walk.

Gambling - the way the Unfortunates pass the time between walks. It is about bidding and solving other people's problems. The Unfortunates organize auctions for the privilege of solving the problems of others. The winner risks a declared number of quarts of oil from the first cask. If he succeeds, he is rewarded with three quarts of oil. His bliss also generates oil for the Cablemaster and Scribe of the group that reported the problem (they get their share of the group reward) as well as the Overseer (he gets half of his reward). The Bloodthirster cares about the integrity of the Unfortunates' veins, so he allows anyone to win the bid no more than three times during the season.

Ranks. As the course progresses, each Unfortunate is assigned a rank – starting with Hamster, ending with Bonebreaker. The ranks reflect progress in the course. Moreover, in case there is only one Bonebreaker in the group, he automatically takes the Overseer role.[2]

Badges. A number of badges can be earned by an Unfortunate. Any Scribe who prepares three perfect documentations, becomes a Chronicler. Any Cablemaster who builds three perfect topologies, becomes a Tangler. Anyone who was nominated three times for helping other group during the walk, becomes an E.U.Genius. The one who won three problem auctions, becomes an A.B.Normal.

3.3 Mapping the Story to the Course Rules

While designing the story line we did not follow any specific gamification framework rules. Rather, we strictly observed the guidelines, listed in Sect. 2, and paid attention not to overwhelm the students with the story and game mechanics, and preserve the structure of the course. The course was the first one in our department to be led in a gamified way, and therefore we did not have any previous experience with gamification. As a consequence we decided to develop a story around the leveraged sequence of the course rather than to develop a new course sequence in adherence to a story. Our goal was also to gain experience from subsequent courses and adjust the usage of gamification means accordingly.

Introduction of the story did affect the course grading rules especially with regard to the lab classes and short tests (Cask 1). Following the guidelines for gamified courses, the number of points (quarts) that could possibly be collected, exceeded the number taken into account when grading. In the described case, the number of points possible to collect during ten labs and six short tests was $10*7+6*20 = 190$, more than twice as much than 72, which was set as the maximum.[3] Therefore, each of the students was able to collect up to five 'Indulgences', i.e., trade the points for eliminating questions during the stint tests. Three of the six short tests were organized online, outside the classroom, three days after last class, in order to make the learning more systematic. Note that the participation in the tests was optional – anyone who skipped the online tests still was able to collect 116 points on average (max 130) – more than 72.

Casks 2 and 3 still represented the results of written tests (after fourth and eighth lab class). Because they counted for 50% of the total grade, they still were

[2] The names of consecutive ranks follow characters of a popular comic book – they cannot be translated into English directly.

[3] On average, due to the inequality of rewards for people having different roles, the number was 176 – still two times more than the maximum in Cask 1.

the main components of grading. Each test was graded using a 72-point scale. The tests consisted of 14 questions each. The average number of points that could be earned for answering the question was just above 5. The annihilation of a question resulted in taking it out from grade calculation. For example, if a student decided to eliminate a question 'worth' 6 points, the test result was calculated as sum of the points collected from other answers multiplied by 72/66.

Similarly, Cask 4 represented skills-based and oral tests conducted during the last lab class. The maximum grade was also 72 points. Two thirds (48 points) could have been earned during the skills-based part, one during short oral exam. The oral exam consisted of three questions, which could have been annihilated – just as in the case of Casks 2 and 3. The differences between gamified and non-gamified courses with regard to point counts are summarized in Table 3.

Table 3. Comparison of maximum numbers of points that could be collected in gamified and non-gamified courses.

Course module	Labs and short tests	Stint test 1	Stint test 2	Stint test 3	Total
gamified	190	72	72	72	406
non-gamified	72	72	72	72	288

At a first glance, it seems that the gamified course was much easier to pass, because the passing threshold was not changed, and stayed set to 144. Therefore the possibility of collecting up to about 280% of the threshold value seems to make passing the course much easier. However, that's not exactly true, because simple sum of points serves only as a criteria for earning respective ranks, which are given for surpassing the thresholds of 20, 60, 110, 160, 200, and 240 points. The final mark was calculated without the excess from Cask 1, which could only be traded for Indulgences. During the four conducted courses, the real average 'worth' of an Indulgence was 3 points[4], and each of the students was able to gain two or three Indulgences. Therefore, in reality, the students could earn 6-9 points by trading.

4 Reflection on Conducting the Course in Varying Environments

As presented in Sect. 3.3, the discussed course was gamified in a conservative way. Taking out the story, there were only five elements of gaming introduced, namely: ranks and badges, assignment of roles by a draw, a pool of points that could be traded for elimination of questions during the tests, and bidding for problems to be solved. In this section we share our observations from four subsequent years

[4] A stint test contained 14 questions, each graded 72/14=5.14 point on average; the average score on such a test was 60%.

(2019-2022) the course was conducted in the gamified way. Noteworthy, our experience covers courses led both in person, and remotely (due to the COVID-19 pandemic).

4.1 Teachers' Subjective Reflection

Introduction of gaming changed the students' attitude towards the classes' organization. First of all – there was absolutely no trouble in establishing dates for stint tests, common to three or four groups of students. The reward of additional lab (which could have been 'converted' into a few points) was sufficiently motivating. Second – the engagement in conducting lab exercises was visibly increased. Taking into account that the computer networking course is obligatory for all the students, that – in our opinion – was the main positive change. Third – making the additional tests an opportunity rather than an obligation resulted in participation of more than 80% of the students.

Introduction of the Drunkard role proved to be a good idea too. Before introduction of gamified rules, the students were much more willing to work in a single team of four, than in two teams of two. The small reduction of the number of points awarded to the members of larger groups changed the preferences. During the four years of experimenting with the gamified course there were no teams of four students formed.

On the other hand, the initial design of the course proved to be too optimistic in two aspects. First, the bidding for problems was never organized. That resulted both from the limit of lab time (everybody tried to resolve the problems until the very end of the lab class), and the organization of computer network classroom – the equipment used by the students is detached from the Internet, so it was quite hard to obtain a snapshot of configuration. Moreover, the current LMS did not support such kind of activity properly. Second, the badges were assigned too late (they should have been assigned instantiately). Although the idea was welcomed by the students, who liked collecting badges 'just for fun', the lack of automation of evaluating badge assignment criteria resulted in the need to export grades from university LMS, and evaluate them in an external spreadsheet. That process simply took too much time.

Introduction of the ranks did not affect the course much. The only effect that was observed, was the positive students' reaction to first aggregate report sent to them, entitled "There are no more Hamsters in the forest", which notified them of their good progress at the start of the semester. The subsequent reports were almost unnoticed.

4.2 Course Statistics and Educational Outcomes

In the academic year of 2019/2020, 35 (out of 145) students took part in the gamified course. On average, each of the students received 28 grades, most of them evaluating their work during the laboratory classes. The average score at the end of the semester was almost 177 points (61%). That was a little greater than the baseline from the previous year, which equaled 167 point (58%). The difference was reflected in the exam score – the exam results of the students who participated in the gamified course were on average 5% better, than their counterparts whose classes were led in the traditional way. Similar statistics from the following years are presented in Table 4.

Table 4. Comparison of the average grades received by the students in gamified and non-gamified courses.

Academic year	2018/2019	2019/2020	2020/2021	2021/2022	2022/2023
Laboratory classes score	167(58%)	177(61%)	215(75%)	214(74%)	206(72%)
– gamified?	no	yes	yes	yes	yes
Average Exam score					
– in gamified groups	n/a	62%	67%	76%	50%
– in non-gamified groups	57%	57%	67%	71%	46%
Standard deviation of scores					
– in gamified groups	n/a	16%	7%	9%	13%
– in non-gamified groups	15%	20%	11%	19%	15%
Coefficient of variation (CV)					
– in gamified groups	n/a	26%	11%	11%	26%
– in non-gamified groups	27%	37%	16%	28%	34%
Exam pass ratio					
– in gamified groups	n/a	80%	98%	98%	64%
– in non-gamified groups	72%	61%	96%	89%	43%
Number of observations					
– in gamified groups	0	35	60	54	33
– in non-gamified groups	167	110	125	147	129

The numbers presented in Table 4 cannot be interpreted directly, because of the changes in courses' environments – in the years of 2019/2020 and 2022/2023 the classes, as well as exams, were led in the 100% in-person mode, as opposed to 2020/2021 and 2021/2022, when classes were organized in remote or blended modes, and exams were conducted remotely. However, a few conclusions can be drawn from our experience. First and foremost, introduction of gamification resulted in an improvement of the exam results. Although the average score increased a bit, the most important difference was visible in the passing ratios. The students who participated in the groups which classes were gamified, were much more likely to pass the exam on their first attempt, both in remote and in-person cases. On the other hand, the number of students who received the highest marks, stayed almost unchanged. Therefore – based on our four years' experience – we concluded that introduction of gamification increased motivation mainly in the group which could not be characterized as enthusiasts of the course subject. The observation is confirmed by the values of standard deviation and coefficient of variation – lower values mean that the scores were closer to the average in the gamified groups. That resulted in more students passing the 50% threshold on the exam and getting positive grades. The distributions of grades are depicted in Fig. 2.

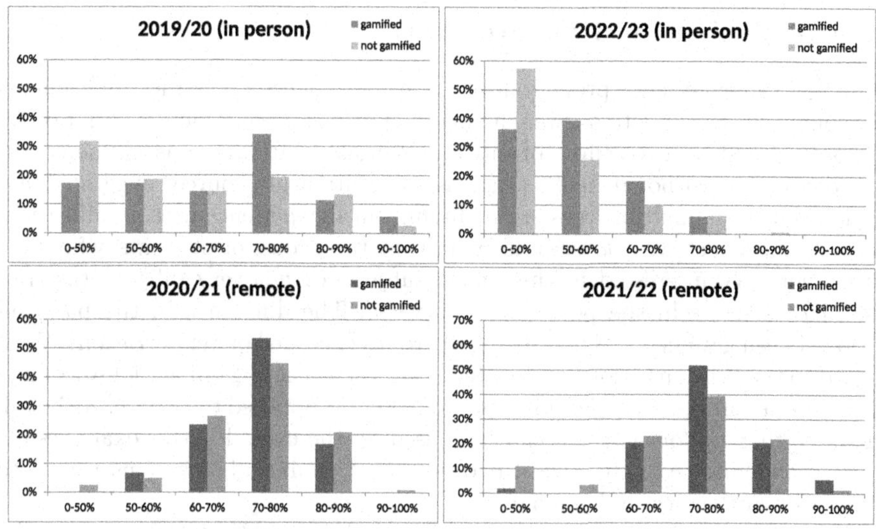

Fig. 2. Distribution of exam grades in gamified and non-gamified groups.

It should be noted that the effects of introducing gamification are more visible in cases of in-person laboratory classes and exams (2019/2020 and 2022/23). From our perspective, this could be a result of a lack of proper tools to be used cooperatively during the labs. During remote classes, the students cooperated only by sharing a screen with a network simulator run and configured by one of them, therefore making the laboratory experience not very engaging.

4.3 Adjustments to Mapping and Roles

Over the years, the initial story line, and its mapping to course rules underwent a few adjustments. First of all, the threshold set for the Bonebreaker rank proved to be too high. The first Bonebreaker ranks were awarded just before the end of the semester – too late to use the ability of grabbing the Overseer role more than two times. Therefore, the threshold was lowered to 220, and later to 200. Other thresholds were adjusted accordingly – in 2022/2023 their values were 20, 50, 80, 120, 160, and 200. The change did not result in a dramatic change during the conduct of the course, but resulted in 76% of the students reaching the highest rank at the end of the semester, thus bolstering their satisfaction. In the previous years, the ratios were 50%, 49%, and 33%, respectively.

A more important change, which was suggested by the students in the surveys conducted after the classes, applied to the rule of calculating Overseers' rewards. Starting from 2022/2023 the Overseer gets an average (not sum) number of points earned by the Scribe and the Cablemaster. This change eliminated the tensions that were sometimes observed during the role draws.

5 Conclusions and Future Work

Based on the experience presented in this paper, collected from four editions of the course, we feel able to answer the main questions listed in Sect. 1. First of all, the anticipated positive effect of introducing gamification means in the course was observable, although not as big, as some literature sources suggest. That could result from multiple reasons, including our inexperience with gamification. Second, the effect was observable mainly in the group of students who were either not very interested in the course subject, or not very diligent (as they stated themselves in the post-course surveys). The difference in the pass/fail ratio between gamified and non-gamified groups exceeded our expectations. With regard to the third question, we conclude that the effect of gamification was more visible in case of in-person learning environment, as opposed to remote. Although we agree that the observation can be biased by the overall better exam results in the pandemic years, it is worth to note that similar observations were also reported by other researchers [5].

From the perspective, we observed two main obstacles in introducing elements of gamification into daily practice. First of them was our (teachers') reluctance to exit our comfort zones, and evaluate an unusual way of teaching. Second – which can be attributed to the "unplugged" approach, which we opted for – is the amount of time needed to conduct the gamified course. Even simple calculation of the number of grades that need to be given to the students shows the overhead, which is further increased by game mechanics, such as point trading, bidding, etc. Therefore, in the future we'll opt for using IT support for out-of-class activities. Note that the support provided by a typical LMS was not sufficient.

Our future work plan is guided by the outcomes of (voluntary) post-course surveys, in which just above 50% of students took part. 94% of responses to a question 'would you choose a gamified group again?' was positive, which is very motivating for gamifying the course in the future. Asked for the reasons for the acceptance, students typically answered that the classes did not lose on merit, and gained on the atmosphere, mainly by the introduction of some humorous elements. As a drawback, some of the students pointed out the complexity of grading rules, and postulated creation of a 'tactical textbook' containing practical hints. The scopes of duties associated with the roles were in general accepted (89%), but the draws were unwelcome (37% acceptance before the change in the Overseer role award calculation – at the time of writing we did not collect the surveys from 2022/23 edition of the course). The number of points that can be traded was in general perceived as adequate (58%). Some of the students (21%) postulated organizing more online quizzes in order to increase the pool of points.

From other changes suggested by students, we selected two as the guidelines for further development of the course. First – we are going to associate special abilities with the ranks (for example, trading will be possible only after reaching a certain threshold). Second, we are going to introduce the possibility of collecting "knowledge pills", i.e., short pieces of information on the course topics.

References

1. The state of online gaming - 2019. Tech. rep., Limelight Networks (2019)
2. Essential Facts about the Video Game Industry. Tech. rep., Entertainment Software Association (2020)
3. The state of online gaming - 2020. Tech. rep., Limelight Networks (2020)
4. Abuladze, L.: The synergic effect of gamification and artificial intelligence in the process of recruiting and selecting candidates (Feb 2023)
5. Alsaadoun, A.: The impact of gamification on students' achievement in online learning environments. Int. J. Learn. Developm. **12**, 71 (2022). https://doi.org/10.5296/ijld.v12i3.20232
6. Evans, E.: Gamification in a year 10 latin classroom: Ineffective 'edutainment' or a valid pedagogical tool? J. Classics Teach. **17**(34), 1–13 (2016). https://doi.org/10.1017/S2058631016000192
7. Gilbert, N.: Number of Gamers Worldwide 2022/2023: Demographics, Statistics, and Predictions (2023). https://financesonline.com/number-of-gamers-worldwide/ (Accessed 1 Feb 2023)
8. González González, C.: Unplugged gamification: towards a definition (Oct 2022)
9. González González, C., Navarro-Adelantado, V.: The limits of gamification. Convergence **27** (2020). https://doi.org/10.1177/1354856520984743
10. Hadi Mogavi, R., Haq, E.U., Gujar, S., Hui, P., Ma, X.: More gamification is not always better: A case study of promotional gamification in a question answering website. In: Proceedings of the ACM on Human-Computer Interaction, vol. 6 (Aug 2022). https://doi.org/10.1145/3555553
11. Hadji-Vasilev, A.: 23 Video Game and Online Gaming Statistics, Facts & Trends for 2023 (2022). https://www.cloudwards.net/online-gaming-statistics/ (Accessed 1 Feb 2023)
12. Jovanovic, B.: Gamer Demographics: Facts and Stats About the Most Popular Hobby in the World (2023). https://dataprot.net/statistics/gamer-demographics/ (Accessed 1 Feb 2023)
13. Majuri, J., Koivisto, J., Hamari, J.: Gamification of education and learning: A review of empirical literature. In: The 2nd International GamiFIN conference, Pori, Finland, May 21–23 2018 (May 2018)
14. Oke, A., Aliu, J., Tunji-Olayeni, P., Abayomi, T.: Application of gamification for sustainable construction: an evaluation of the challenges. Construct. Innovat. (Jan 2023). https://doi.org/10.1108/CI-09-2022-0247
15. Subirats, L., et al.: Gamification based on user types: When and where it is worth applying. Appli. Sci. **13**, 2269 (2023). https://doi.org/10.3390/app13042269
16. Vashisht, D.: Engaging and Entertaining Customers: Gamification in Interactive Marketing, pp. 807–835 (Jan 2023). https://doi.org/10.1007/978-3-031-14961-0_35
17. Zhang, J., Jiang, Q., Zhang, W., Kang, L., Lowry, P., Zhang, X.: Explaining the outcomes of social gamification: A longitudinal field experiment. J. Manag. Inform. Syst. **2023**, 22 (2023). https://doi.org/10.2139/ssrn.4355616

Towards an Earned Value Management Didactic Simulator to Engineering Management Teaching

Manuel Castañón–Puga[1(✉)], Alfredo Tirado–Ramos[2,3], Camilo Khatchikian[3], Eugenio D. Suarez[4], Luis Enrique Palafox–Maestre[1], and Carelia Guadalupe Gaxiola–Pacheco[1]

[1] Chemistry and Engineering School, Universidad Autónoma de Baja California, Tijuana, BC 22424, Mexico
{puga,lepalafox,cgaxiola}@uabc.edu.mx

[2] Department of Epidemiology, Geisel School of Medicine at Dartmouth, Dartmouth College, Lebanon, NH 03756, USA
Alfredo.Tirado-Ramos@dartmouth.edu

[3] Department of Biomedical Data Science, Geisel School of Medicine at Dartmouth, Dartmouth College, Lebanon, NH 03756, USA
Camilo.Khatchikian@dartmouth.edu

[4] Trinity University, One Trinity Place, Northrup Hall 210Q, San Antonio, TX 78212, USA
esuarez@trinity.edu

http://www.uabc.edu.mx , https://geiselmed.dartmouth.edu/epidemiology/ , https://geiselmed.dartmouth.edu/bmds/ , https://www.trinity.edu/

Abstract. Agile development (AD) is a methodology that many small businesses have adopted for production convenience, and educators have taken notice of the trend. A need to implement some form of agile development in undergraduate programs at universities is now clear, particularly for undergraduate engineering students who should understand their role in a project focused on AD. This paper presents our preliminary evaluation of user experience (UX) using an Earned Value Management (EVM) simulator, which helps the student understand the team member's role in an agile development process. The simulator uses a Task-board interface to display task status changes, a burn-down chart to depict the remaining work, and EVM metrics to assess the efficiency of the teamwork. Using the Task-board and EVM models, the simulator offers students different agile project management experimental experiences.

Keywords: Agile Development · Earned Value Management · Agent-based modelling and simulation · Undergraduate Engineering Students

Supported by Universidad Autónoma de Baja California, Mexico.

J. Mikyška et al. (Eds.): ICCS 2023, LNCS 14077, pp. 780–792, 2023.
https://doi.org/10.1007/978-3-031-36030-5_60

1 Introduction

Agile development (AD) is a methodology that many small businesses have adopted for production convenience. That is why academics have turned to implementing some form of agile development in undergraduate programs. From a Project Management (PM) perspective, we can approach AD as a particular type of project where the team members have a developer role, and together they attend to the tasks planned in the Work Breakdown Structure (WBS).

Any team member can approach project management from different perspectives: a) Visualize the WBS using a Gantt Chart or a Task-board. Project managers widely use the Task-board in small and agile projects due to its ease of implementation. b) Earned Value Management (EVM) is the standard method to assess the performance of a project. It provides a set of metrics that fundamentally help estimate a project's cost and schedule efficiency.

This paper presents our preliminary evaluation of user experience (UX) using an Earned Value Management (EVM) simulator, helping the student understand the team member's role in an agile development process [4]. The simulator uses a Task-board interface to display task status changes, a burn-down chart to depict the remaining work, and EVM metrics to assess the teamwork efficiency [5]. Using the Task-board and EVM models, the simulator offers students different agile project management experimental experiences.

1.1 Agile Development

Agile project management is a lean production approach based on Lean Manufacturing (LM) principles [17], applied to managing projects that require extraordinary speed and flexibility in their processes. In particular, the software industry has adopted agile software development as a viable approach to managing the development of software products, inspired by the "Manifesto for Agile Software Development.[1]"

The Agile approach has gained substantial momentum and has spread to many other areas of implementation, such as in manufacturing processes, education, healthcare, and other industries that are becoming agile [9]. Complementary to the sixth edition of "The Guide to the Fundamentals of Project Management (PMBOK Guide)," published in 2017 [14] by the Project Management Institute, the special edition "The PMBOK Agile Practice Guide" was published as a companion [13]. This publication is intended exclusively for software development project managers who have adopted the agile approach. According to this guide, agile software development is a type of lean production, and it was published not only to address the use of the agile approach in the software development industry but to go beyond its original home, finding applications in environments other than software development. The application of the agile approach in manufacturing, education, healthcare, and other sectors is within the scope of this practical guide [13].

1.2 Agile Practices in Education

Education is an excellent fertile ground for extending agile practices beyond software development. Middle school and high school teachers, as well as college professors worldwide, have begun to employ the agile approach to create a learning culture that aligns with state-of-the-art in the business and engineering spheres. Stakeholders have used agile techniques to focus on prioritizing competing priorities. Face-to-face interaction, meaningful learning, self-organizing teams, and incremental and iterative learning that stimulate imagination are agile principles that can change the mindset in the classroom and advance educational goals [3].

In the case of engineering schools, particularly in software engineering programs, the application of agile learning is not only convenient as an educational strategy but is considered almost a necessity. There are experiences in applying agile learning with the aim that software engineering students learn to improve their competence in agile production processes such as Scrum [12]. This requirement is due, in general, to the fact that a large part of the software development industry has adopted some agile production methodology. This situation has motivated university programs oriented to software development to develop skills in future engineers related to these production methods [6].

However, agile learning in engineering schools, particularly software engineering, is still far from standard practice. Oftentimes, the internal processes of educational institutions do not necessarily facilitate agile learning management. In addition, not all have the human capital to implement this practice. Finally, local industry support to implement this practice must be linked appropriately [11]. This situation leads us to seek alternative solutions to the problem of agile development, especially the one that refers to teaching agile software development through an educational strategy based on agile learning.

1.3 Task-Board and Earned Value Management

The Earned Value Management (EVM) [15] is considered a fundamental part of the Project Management Body of Knowledge (PMBOK) [16] to establish practical measures. Over the last four decades, project management professionals have used this method to measure performance and assess the status of a project [8]. This notwithstanding, managing an Agile project can be a challenging endeavor, forcing managers to use a task board to visually represent the work on a project and the path to completion. The route includes pending, in-progress and completed tasks performed by teams. For example, the "Kanban" methodology uses a task board to distribute assignments and activities as a fundamental part of a production process [11].

2 Methodology

We designed a small experiment to evaluate students' experiences in using the Web Earned Value Management simulation didactic tool to learn the concepts of project management, which we summarize in the three actions below:

- Earned Value Management model adaptation.
- Class experiment design and implementation.
- Assessment questionnaire application.

2.1 The Earned Value Management Model Adaptation

The tool that we evaluated is a Web-based version of the NetLogo Earned Value Management Model 1.0. A complete model description is available in [5], with a downloadable original NetLogo model in [4].

2.2 The Class Experiment Design and Implementation

20 students in the "Software development tools" class of the Software Engineering undergraduate program at the Universidad Autónoma de Baja California in Mexico [10] participated in a learning experiment as described below:

1. First, they received a lesson about Project Management, WSB, Task-board, EVS metrics, and other basic concepts.
2. Second, they answered a short quiz to evaluate what they had learned.
3. Third, they conducted a laboratory practice using the Web Earned Value Management simulation didactic tool.
4. Fourth, they retook the same quiz to re-evaluate their learning.
5. Finally, they answered a questionnaire to evaluate their user experience when interacting with the Web tool.

2.3 The Assessment Questionnaire Application

We applied a short User Experience Questionnaire (UEQ) to evaluate the User Experience (UX) of the 20 students after interacting with the tool [19]. The questionnaire evaluates the pragmatic quality (Efficiency, perspicuity, and dependability) and the hedonic quality (Stimulation and novelty) of the tool. We can find complete information about this questionnaire in [22]. Additionally, the students gave feedback on their experience in short comments to complement the assessment.

3 Assessment Results

We have three outcomes as a result of this first approach:

- Web Earned Value Management simulation didactic tool.
- Student's UX assessment results.
- Student's quiz results and feedback.

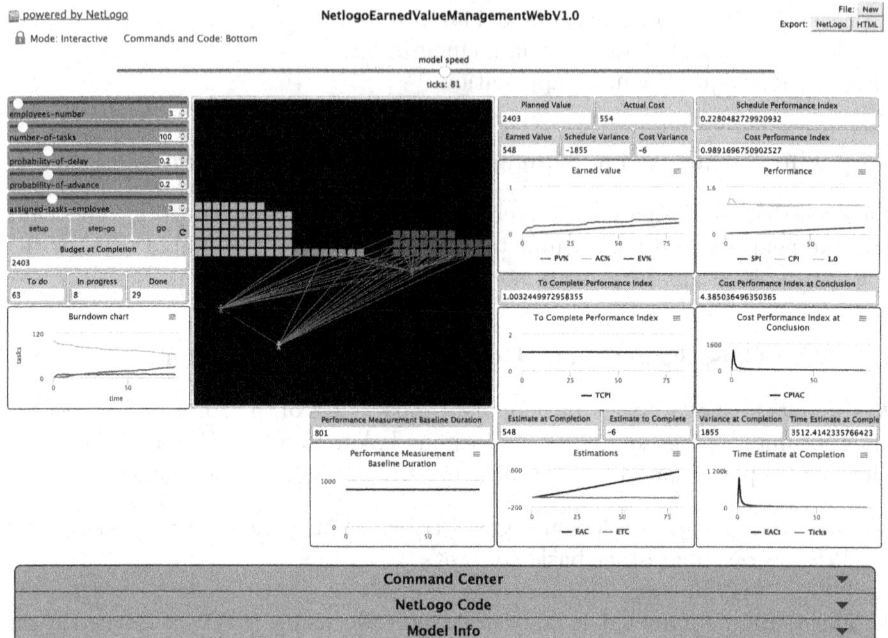

Fig. 1. Web Earned Value Management simulation didactic tool screenshot.

3.1 The Web Earned Value Management Simulation Didactic Tool

We produced a Web version from the original model [5] for better student access to the tool during the evaluation. Figure 1 shows the Web tool interface with which the participants interacted.

The tool interface has three parts to interact with: a) components where the initial conditions of the scenario are configured and the simulation is run (step-by-step or continuous execution), b) an interface that visually draws a dashboard of tasks and employees, c) components that show the Earned Value calculation at run-time.

3.2 The Student's UX Assessment Results

In this section, we present the results of applying a short questionnaire to measure the students' experience after interacting with the simulation Web tool.

Pragmatic and Hedonic Quality. We used the "Short UEQ Data Analysis Tool" [21] to analyze the questionnaire responses according to [19]. In table 1, we show the results by questionnaire item [20]. We can observe that the means results are relatively high in all assessment cases.

Table 1. Results

Item	Mean	Variance	Std. Dev.	No.	Negative	Positive	Scale
1	(High) 2.0	1.5	1.2	20	obstructive	supportive	Pragmatic Quality
2	(High) 1.4	1.2	1.1	20	complicated	easy	Pragmatic Quality
3	(High) 2.3	0.4	0.6	20	inefficient	efficient	Pragmatic Quality
4	(High) 1.4	1.3	1.1	20	confusing	clear	Pragmatic Quality
5	(High) 1.9	0.5	0.7	20	boring	exciting	Hedonic Quality
6	(High) 2.5	0.3	0.5	20	not interesting	interesting	Hedonic Quality
7	(High) 1.7	2.1	1.5	20	conventional	inventive	Hedonic Quality
8	(High) 1.3	1.8	1.3	20	usual	leading edge	Hedonic Quality

The Table 2 shows the results by quality area. We can observe that the pragmatic and hedonic qualities resulted in high assessment.

Table 2. Short User Experience Questionnaire Scales.

Short UEQ Scales.	
Pragmatic Quality	(High) 1.738
Hedonic Quality	(High) 1.825
Overall	(High) 1.781

Figure 2 depicts the mean value per questionnaire item. We can note that in all cases, the mean value results are positive, and Fig. 3 depicts the mean value per questionnaire scale. We can note that in all cases, the scale and overall value results are positive.

The "Short UEQ Data Analysis Tool" [21] provides details on how we can interpret the means of the scales as pragmatic quality and hedonic quality according to [19]. Table 3 shows the 5% confidence intervals for the means of the single items, and Table 4 shows the 5% confidence intervals for the scale means.

In table 5, we show the pragmatic and hedonic quality correlations of the items per scale and Cronbach Alpha-Coefficient [7].

Benchmark Comparison. According to [18], the benchmark data set contains data from 21175 persons from 468 studies concerning different products (business software, web pages, web shops, social networks). Currently, these benchmark data sets are based on the full UEQ, but the scale values of the short version are a reasonably good approximation of the corresponding values of the full version; thus, it is a rough approximation possible to use the data from the complete UEQ benchmark as a target for the short UEQ [19]. However, these

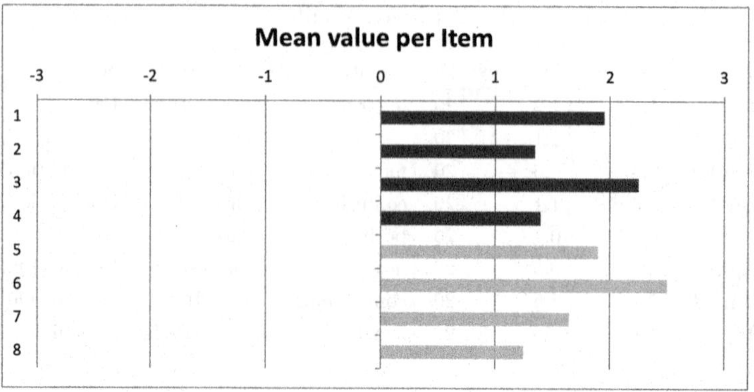

Fig. 2. Mean value per Item.

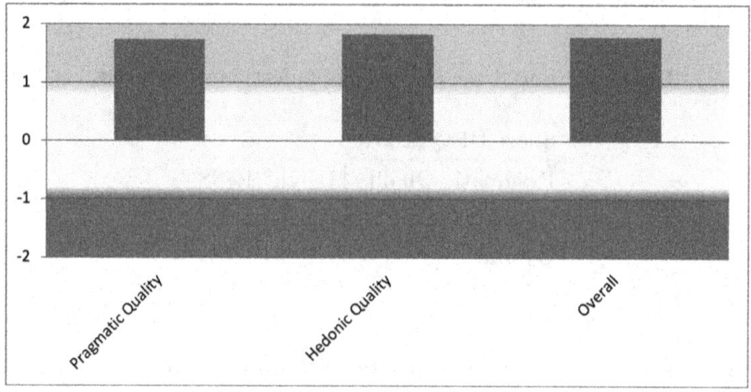

Fig. 3. Results.

are still different questionnaires, forcing us to treat the data carefully. Benchmark authors intend to replace this proposed measure with a special benchmark for the short version when enough data with the short UEQ become available. Table 6 compares to benchmark interpretation. The measured scale means are set in relation to existing values from a benchmark data set [18].

The comparison of the results for the evaluated product with the data in the benchmark allows conclusions about the relative quality of the evaluated product compared to other products. Two versions of the benchmark chart are shown. Figure 4 shows only the scale scores in relation to the benchmark categories, whereas Fig. 5 reflects the confidence intervals of the scale scores.

Table 3. Confidence interval per item.

Confidence interval (p = 0.05) per item						
Item	Mean	Std. Dev	N	Confidence	Confidence interval	
1	1.950	1.234	20	0.541	1.409	2.491
2	1.350	1.089	20	0.477	0.873	1.827
3	2.250	0.639	20	0.280	1.970	2.530
4	1.400	1.142	20	0.501	0.899	1.901
5	1.900	0.718	20	0.315	1.585	2.215
6	2.500	0.513	20	0.225	2.275	2.725
7	1.650	1.461	20	0.640	1.010	2.290
8	1.250	1.333	20	0.584	0.666	1.834

Table 4. Confidence intervals per scale.

Confidence intervals (p = 0.05) per scale						
Scale	Mean	Std. Dev	N	Confidence	Confidence interval	
Pragmatic Quality	1.738	0.719	20	0.315	1.423	2.052
Hedonic Quality	1.825	0.783	20	0.343	1.482	2.168
Overall	1.781	0.588	20	0.257	1.524	2.039

Table 5. Correlations of the items per scale and Cronbach Alpha-Coefficient. Pragmatic and hedonic quality.

Pragmatic Quality.	
Items	Correlation
1.2	0.25
1.3	0.08
1.4	0.16
2.3	0.25
2.4	0.69
3.4	0.36
Average	0.30
Alpha	0.63

Hedonic Quality.	
Items	Correlation
5.6	0.43
5.7	0.32
5.8	0.30
6.7	0.04
6.8	0.12
7.8	0.86
Average	0.34
Alpha	0.68

3.3 The Student's Quiz Results and Feedback

When applying the earned value management quiz before and after experimenting with the tool, 45% of the students increased their scores. However, 40% of the students kept their scores unchanged, and 15% of the students decreased their scores. Unfortunately, 5% did not answer the quiz. Table 7 shows the distribution of the percentages of the results obtained, and Fig. 6 depicts these proportions.

Table 6. Comparison to benchmark interpretation.

Scale	Mean	Comparison to benchmark	Interpretation
Pragmatic Quality	1.7375	Good	10% of results better, 75% of results worse
Hedonic Quality	1.825	Excellent	In the range of the 10% best results
Overall	1.78	Excellent	In the range of the 10% best results

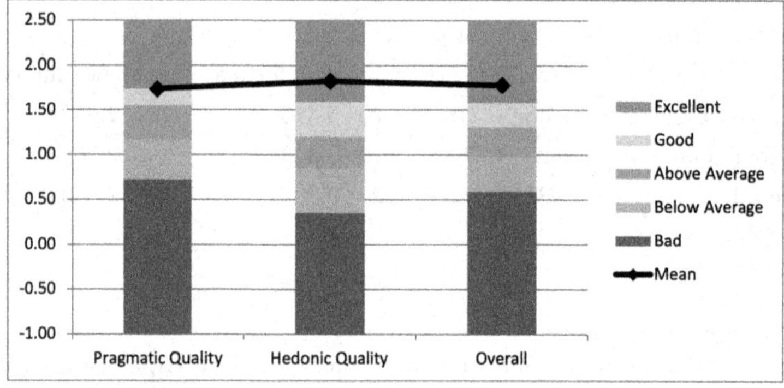

Fig. 4. Comparison to benchmark. The scale scores in relation to the benchmark categories

Although we would have liked to have had a more substantial recorded impact on learning, the written statements of students described their experience with very positive comments. After doing a sentiment analysis of the students' feedback [2], we obtained that 86% sentiments were positive, 7% neutral, and 7% negative. Many of them found the exercise interesting and enjoyed the interaction. Moreover, the exercise was considered intuitive for a significant portion of the students, who, after some interaction, understood the mechanism and found it rewarding. Figure 7 shows a word-cloud representative of students' comments.

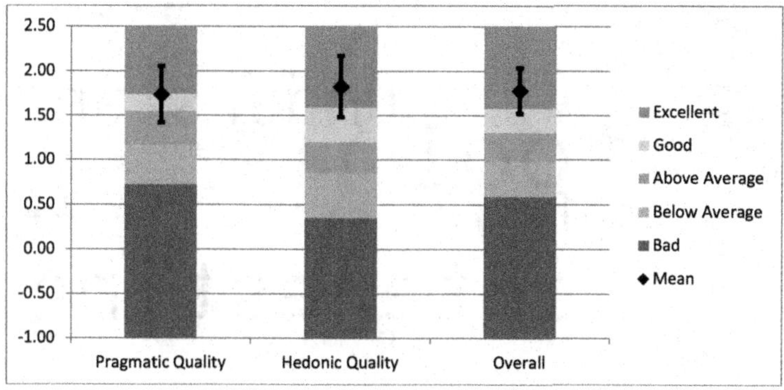

Fig. 5. Comparison to benchmark. The confidence intervals of the scale scores

Table 7. Percentage distribution of 'Score impact'

Score impact	Percentage distribution
Increase	45.00%
No change	35.00%
Decrease	15.00%
No answer	5.00%
Grand Total	100.00%

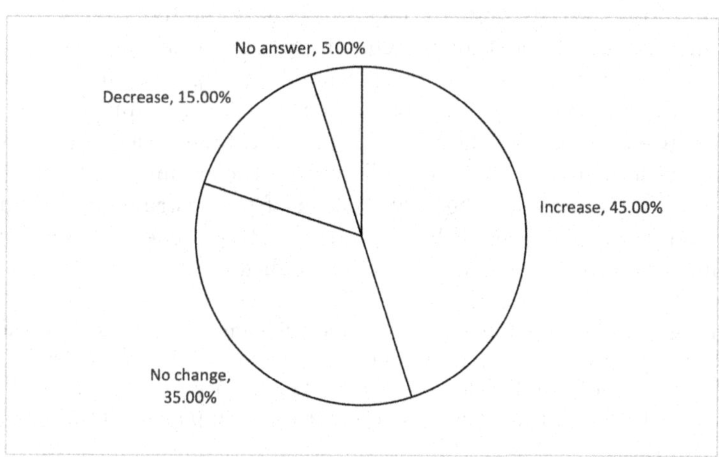

Fig. 6. Percentage distribution of 'Score impact.'

Fig. 7. The student's feedback word-cloud.

4 Conclusion and Future Work

In this paper, we have presented a preliminary study with undergraduate students in engineering and their experience using a web-based task-board simulation tool in NetLogo. The students used the tool to learn the earned value management approach. In classroom practice, they experimented with various scenarios to observe an agile project's behavior and performance. We proceeded to evaluate their experience by applying a short questionnaire and assessed their learning using a quiz before and after the interaction. The preliminary result shows an excellent practical and hedonic experience and a trim learning outcome. This result bids us to continue with the evaluation of the user experience in a full way but will be necessary to re-design the experiment to improve the learning evaluation method.

In future work, we will first continue to evaluate the tool with a more extensive and diverse sample of students. Our interest is to include different courses where we consider it essential to teach students the project management process. Secondly, we will apply the comprehensive questionnaire to evaluate the user experience and identify the most significant characteristics of the student's interaction with the tool. Finally, we will improve the learning evaluation method to more accurately measure the contribution of the instrument in comparison to other digital learning media. We will also analyze student feedback through their comments to discover new interactive qualities.

Acknowledgements. We thank the Complex Systems Laboratory at the School of Chemistry and Engineering, Universidad Autónoma de Baja California, and the Biomedical Data Science Research Software Laboratory at the Geisel School of Medicine at Dartmouth, Dartmouth College, for their support of this project.

References

1. Beedle, M., et al.: Manifesto for agile software development (2001)
2. Bird, S., Klein, E., Loper, E.: Natural Language Processing with Python. O'Reilly Media, Inc. (2009)
3. Briggs, S.: Agile based learning: What is it and how can it change education? (2014). http://www.opencolleges.edu.au/informed/features/agile-based-learning-what-is-it-and-how-can-it-change-education/
4. Castañón-Puga, M., Rosales-Cisneros, R.F., Acosta-Prado, J.C., Tirado-Ramos, A., Khatchikian, C., Aburto-Camacllanqui, E.: Netlogo earned value management model (Dec 2022). https://www.comses.net/codebases/4effb542-f69f-464d-9f96-69f19d36ae8a/
5. Castañón-Puga, M., Rosales-Cisneros, R.F., Acosta-Prado, J.C., Tirado-Ramos, A., Khatchikian, C., Aburto-Camacllanqui, E.: Earned value management agent-based simulation model. Systems **11**(2) (2023). https://www.mdpi.com/2079-8954/11/2/86
6. Chatley, R., Field, T.: Lean learning - applying lean techniques to improve software engineering education. In: Proceedings - 2017 IEEE/ACM 39th International Conference on Software Engineering: Software Engineering and Education Track, ICSE-SEET 2017, pp. 117–126 (2017). https://doi.org/10.1109/ICSE-SEET.2017.5
7. Cronbach, L.J.: Coefficient alpha and the internal structure of tests. Psychometrika **16**, 297–334 (1951). https://doi.org/10.1007/BF02310555
8. Fleming, Q.W., Koppelman, J.M.: Earned value project management a powerful tool for software projects. CROSSTALK J. Defense Softw. Eng., 19–23 (1998)
9. Henao, R., Sarache, W., Gómez, I.: Lean manufacturing and sustainable performance: Trends and future challenges. J. Clean. Prod. **208**, 99–116 (2019). https://doi.org/10.1016/j.jclepro.2018.10.116
10. López-Chico, L.E., Lara-Melgoza, R.: Herramientas de Desarrollo de Software. Programa de unidad de aprendizaje (Mar 2023). https://doi.org/10.5281/zenodo.7697335
11. Papatheocharous, E., Andreou, A.S.: Empirical evidence and state of practice of software agile teams. J. Softw. Evolut. Proc. **26**(9), 855–866 (2014). https://doi.org/10.1002/smr.1664
12. Pham, A., Pham, P.V.: Scrum in action : Agile software project management and development (2012)
13. PMI: Agile Practice Guide PMBOK. Project Management Institute Inc, 1 edn. (2017). www.pmi.org
14. PMI: A guide to the project management body of knowledge. Project Management Institute Inc, 6 edn. (2017)
15. PMI: The Standard for Earned Value Management. Project Management Institute, Inc. (2019). www.pmi.org
16. PMI: The standard for project management and a guide to the project management body of knowledge (PMBOK guide). Project Management Institute Inc., 7 edn. (2021). www.pmi.org
17. Rouse, M.: Lean manufacturing (lean production) (2018). https://searcherp.techtarget.com/definition/lean-production
18. Schrepp, M., Hinderks, A., Thomaschewski, J.: Construction of a benchmark for the user experience questionnaire (ueq). Int. J. Interact. Multimedia Artifi. Intell. **4**, 40 (2017). https://doi.org/10.9781/ijimai.2017.445

19. Schrepp, M., Hinderks, A., Thomaschewski, J.: Design and evaluation of a short version of the user experience questionnaire (ueq-s). Int. J. Interact. Multimedia Artifi. Intell. **4**, 103 (2017). https://doi.org/10.9781/ijimai.2017.09.001
20. Schrepp, M., Hinderks, A., Thomaschewski, J.: Items of the short version of the user experience questionnaire (ueq) (2018). https://www.ueq-online.org/Material/UEQS_Items.pdf
21. Schrepp, M., Hinderks, A., Thomaschewski, J.: Short ueq data analysis tool (2018). https://www.ueq-online.org/Material/Short_UEQ_Data_Analysis_Tool.xlsx
22. Schrepp, M., Hinderks, A., Thomaschewski, J.: Ueq user experience questionnaire (2018). https://www.ueq-online.org

Author Index

J. Mikyška et al. (Eds.): ICCS 2023, LNCS 14077, pp. 793–795, 2023.
https://doi.org/10.1007/978-3-031-36030-5